Rafael Bachiller • José Cernicharo

Editors

Astrophysics and Space Science

Science with the Atacama Large Millimeter Array:
A New Era for Astrophysics

Proceedings of the Conference held in Madrid (Spain), 13–17 November 2006

Reprinted from *Astrophysics and Space Science*
Volume 313, Nos. 1–3, 2008

 Springer

Rafael Bachiller
Observatorio Astronomico Nacional
Alfonso XII, 3
E-28014 Madrid
Spain

José Cernicharo
CSIC Madrid
Inst. Estructura de la Materia
Serrano, 199
28006 Madrid
Spain

* Background image: Artists' impression of ALMA (Atacama Large Millimeter/submillimeter Array), Chajnantor plateau, Northern Chile. Credit: European Organisation for Astronomical Research in the Southern Hemisphere (ESO), produced for ALMA.

* The photos of the Editors.

* The 5 astronomical images:

*** Images of dusty circumstellar disks:
Up: The large dust disk surrounding Beta Pictoris observed with a coronograph at 0.8 micron with the University of Hawaii 2.2 m telescope on Mauna Kea. Paul Kalas, UC Berkeley.

Middle: Epsilon Eridani observed at 850 µm wavelength with the Submillimetre Common-User Bolometer Array at the James Clerk Maxwell Telescope. Greaves et al. 1998, The Astrophysical Journal 506, L133.

Down: The disk around HD141569 observed with the Hubble Space Telescope. Mouillet et al. 2001, Astronomy & Astrophysics 372, L61.

*** The spiral galaxy M51.
Up: Radio image of M51 showing the location of Carbon Monoxide gas observed by Schinenerer, Weiss, Scoville, and Aalto. (Credit: Institute for Millimeter-wave Radioastronomy, IRAM, Owens Valley Radio Observatory OVRO, and National Radio Astronomy Observatory, NRAO.)
Down: M51 as seen with the Hubble Space Telescope. (Credit: Space Telescope Science Institute.)

Astronomy Subjects Classification (2007): SCP22006 Astronomy, Astrophysics and Cosmology; SCP22014 Astronomy; SCP22022 Astrophysics

Library of Congress Control Number: 2007943079

ISBN: 978-1-4020-6934-5   e-ISBN: 978-1-4020-6935-2

Printed on acid-free paper.

1

springer.com

Contents

Part 8. High redshift galaxies. Cosmology

Preface

Currently under construction in the Andean Altiplano, Northern Chile, the Atacama Large Millimeter/submillimeter Array (ALMA) is an international astronomy facility, a radio interferometer composed of 54 antennas of 12 m diameter, and twelve 7 m antennas with about 6600 m^2 of total collecting area. Initially covering the most interesting spectral wavelength ranges from 3 to 0.3 mm, ALMA will be a revolutionary telescope providing astronomy with the first detailed view of the dark and youngest objects of the Universe.

ALMA is a partnership of Europe, Japan and North America in cooperation with the Republic of Chile. ALMA is funded in Europe by the European Organisation for Astronomical Research in the Southern Hemisphere, in Japan by the National Institutes of Natural Sciences (NINS) in cooperation with the Academia Sinica in Taiwan and in North America by the U.S. National Science Foundation (NSF) in cooperation with the National Research Council of Canada (NRC). ALMA construction and operations are led on behalf of Europe by ESO, on behalf of Japan by the National Astronomical Observatory of Japan (NAOJ) and on behalf of North America by the National Radio Astronomy Observatory (NRAO), which is managed by Associated Universities, Inc. (AUI).

The scientific preparations for ALMA are being extremely active since the birth of the project. The various science committees, groups of astronomers working for ALMA, and regional communities interested in the project meet regularly to exchange ideas about the scientific capabilities and first observations to be carried out with the interferometer. A first world-wide conference on "Science with the Atacama Large Millimeter Array" took place in Washington, B.C. (USA), on 6-8 October 1999.

The conference "Science with ALMA: a new era for Astrophysics" was held in Madrid (Spain), on 13-17 November 2006. This international ALMA conference was the second world-wide meeting on ALMA science and it was envisioned as a way for the astronomers interested on ALMA, not necessarily radioastronomers, to exchange views, to plan preparatory observations in view of the scientific exploitation of the interferometer, and to obtain the information needed to orient their scientific work to the best possible use of ALMA.

More than 320 scientists from nearly 20 countries took part in the symposium. The conference covered a wide range of topics, which indeed included the main scientific drivers of ALMA: the formation and evolution of galaxies, the physics and chemistry of the interstellar medium, and the processes of star and planet formation.

Rafael Bachiller & Jose Cernicharo

Madrid, 1 April 2007

Acknowledgements

The organizers gratefully acknowledge financial help from the Spanish Ministerio de Educatión y Ciencia, Consejo Superiores de Investigaciones Científicas (CSIC), Ministerio de Fomento (Institute Geográfico Nacional, IGN) and Comunidad de Madrid (Astrocam network, PRICYT).

Thanks are also due to the European Southern Observatory (ESO), to the US National Radio Astronomy Observatory (NRAO), to the National Astronomical Observatory of Japan (NAOJ) and to the European Community's Sixth Framework Programme under project RadioNet (R113CT 2003 5058187).

We are especially grateful to the staffs of Observatorio Astronómico Nacional (OAN, IGN) and Departamento de Astrofísica Molecular e Infrarroja (DAMIR, IEM, CSIC), very particularly to Maite Alonso Gallego and Elena López Sánchez for their efforts to have a smooth, well-presented meeting and to Alicia Fernandez Clavero and Maria Rosa de Armas for devoting much time and effort to have an excellent conference and to the production of this book.

Organizing Committees

Scientific Organizing Committee

J. Cernicharo (Chairman)	(DAMIR, IEM, CSIC, Spain)
R. Bachiller (Vicechairman)	(OAN, IGN, Spain)
C. Carilli	(NRAO, USA)
P. Cox	(IRAM, France)
M. Gurwell	(CfA, USA)
R. Kawabe	(NRO, Japan)
D. Mardones	(Univ. Chile, Chile)
T. Onishi	(Nagoya Univ., Japan)
M. Tarenghi	(ALMA, Chile)
L. Testi	(Arcetri Obs., INA, Italy)
C. Wilson	(McMaster Univ., Canada)
T. Wilson	(ALMA, Germany)
A. Wootten	(NRAO, USA)
S. Yamamoto	(Tokyo Univ., Japan)

Local Organizing Committee

J. Martín-Pintado (co-Chairman)	(DAMIR, IEM, CSIC, Spain)
P. Planesas (co-Chairman)	(OAN, IGN, Spain)
R. Bachiller	(OAN, IGN, Spain)
J. Cernicharo	(DAMIR, IEM, CSIC, Spain)
F. Colomer	(OAN, IGN, Spain)
J.R. Pardo	(DAMIR, IEM, CSIC, Spain)
P. de Vicente	(OAN, IGN, Spain)
M. Agúndez	(DAMIR, IEM, CSIC, Spain)
J. Gracía	(OAN, IGN, Spain)
I. Jiménez-Serra	(DAMIR, IEM, CSIC, Spain)

Aalto, Susanne
Onsala Space Observatory
Chalmers University of Technology
S-439 92 Onsala, Sweden
susanne@oso.Chalmers.se

Adams, Mark
NRAO
520 Edgemont Road, None
22903 Virginia, U.S.A.
mtadams@nrao.edu

Agúndez, Marcelino
DAMIR, IEM, CSIC
C/ Serrano 121
28006 Madrid, Spain
marce@damir.iem.csic.es

Aikawa, Yuri
Department of Earth and Planetary Sciences
Kobe University
1-1 Rokko-dai-cho, Nada-ku,
657-8501 Kobe, Japan
aikawa@kobe-u.ac.jp

Alcolea, Javier
Observatorio Astronómico Nacional
C/ Alfonso XII N° 3 y 5,
E-28014 Madrid, Spain
j.alcolea@oan.es

Alonso, Tomás
Observatorio Astronómico Nacional
Alfonso XII, 3
E-28014 Madrid, Spain
t.alonso@oan.es

Amo-Baladrón, Mª Aránzazu
DAMIR, IEM, CSIC
C/ Serrano 121
28006 Madrid, Spain
arancha@damir.iem.csic.es

Andre, Philippe
CEA Saclay
Service d'Astrophysique
Orme des Merisiers - Bat. 709,
F-91191 Gif-sur-Yvette, France
pandre@cea.fr

Andreani, Paola
European Southern Observatory
Karl-Schwarzschild strasse 2
34131 Garching bei Muenchen, Germany
andreani@oats.inaf.it

Andrews, Sean
University of Hawaii Institute for Astronomy
2680 Woodlawn Drive
96822 Honolulu, Hawaii, U.S.A.
andrews@ifa.hawaii.edu

Anglada, Guillem
IAA-CSIC
Instituto de Astrofísica de Andalucía
18198 Granada, Spain
guillem@iaa.es

Aravena, Manuel
Max Planck Institute for Radioastronomy
Auf dem Huegel 69
53121 Bonn, Germany
maraven@astro.uni-bonn.de

Arce, Hector
American Museum of Natural History
Department of Astrophysics
Central Park West at 79th Street
10024 New York, U.S.A.
harce@amnh.org

Audard, Marc
ISDC Geneva Observatory
Ch. d'Ecogia 16
1290 Geneva, Switzerland
Marc.Audard@obs.unige.ch

Baan, Willem
ASTRON
PO Box 4
7990 AA Dwingeloo, Netherlands
baan@astron.nl

Bachiller, Rafael
Observatorio Astronómico Nacional
c/ Alfonso XII, 3
28014 Madrid, Spain
r.bachiller@oan.es

Bacmann, Aurore
Observatoire de Bordeaux
BP 89
F-33270 Floirac, France
bacmann@obs.u-bordeauxl.fr

Basu, Kaustuv
Max-Planck-Institute for Radio Astronomy
Auf dem Huegel 69
53121 Bonn, Germany
kbasu@astro.uni-bonn.de

Baudry, Alain
University of Bordeaux 1, OASU
Observatoire Aquitain des Sciences de l'Univers,
L3AB
2 rue de l'Observatoire
BP 89
33270 Floirac, France
baudry@obs.u-bordeauxl.fr

Beasley, Tony
ALMA Joint ALMA Office
El Golf 40, Piso 18
Las Condes, Santiago, Chile
tbeasley@alma.cl

Beelen, Alexandra
MPIfR/AIfA
Auf dem Hügel 71
53121 Bonn, Germany
abeelen@mpifr-bonn.mpg.de

Belloche, Arnaud
Max-Planck Institut fuer Radioastronomie
Auf dem Huegel 69
53111 Bonn, Germany
belloche@mpifr-bonn.mpg.de

Beltran, Maite
Universitat de Barcelona
Departament d'Astronomia i Meteorologia
Av. Diagonal 647
08028 Barcelona, Spain
mbeltran@am.ub.es

Benedettini, Milena
INAF - Istituto di Fisica dello Spazio
Interplanetario
Via Fosso del Cavaliere 100
00133 Roma, Italy
milena@ifsi-roma.inaf.it

Bertoldi, Frank
Argelander-Institute for Astronomy
University of Bonn
Auf dem Hügel 71
D-53121 Bonn, Germany
bertoldi@astro.uni-bonn.de

Bettoni, Daniela
INAF - Osservatorio Astronomico di Padova
Vicolo Osservatorio 5
35122 Padova, Italy
bettoni@pd.astro.it

Blundell, Katherine
Oxford University Astrophysics
Keble Road
OX1 3RH Oxford, U.K.
kmb@astro.ox.ac.uk

Bockelee-Morvan, Dominique
Observatoire de Paris
5 place Jules Janssen
92195 Meudon, France
dominique.bockelee@obspm.fr

Boffin, Henri
ESO
Karl-Schwarzschild-str. 2
85748 Garching bei Muenchen, Germany
hboffin@eso.org

Boissier, Jeremie
Observatoire de Paris
LESIA - Bat. 2
5 pi. jules Janssen
92195 Meudon, France
jeremie.boissier@obspm.fr

Bolatto, Alberto
U.C. Berkeley
601 Campbell Hall
94720 Berkeley
California, U.S.A.
bolatto@berkeley.edu

Bontemps, Sylvain
Observatoire de Bordeaux
OASU/L3AB
2, rue de l'Observatoire
BP89
33270 Floirac, France
bontemps@obs.u-bordeauxl.fr

Boone, Frederic
Observatoire de Paris, LERMA
61, avenue de l'Observatoire
75014 Paris, France
frederic.boone@obspm.fr

Brand, Jan
INAF - Istituto di Radioastronomia
Via P. Gobetti 101
40133 Bologna, Italy
brand@ira.inaf.it

Brinch, Christian
Leiden Observatory
P.O. Box 9513
2300 RA Leiden, Netherlands
brinch@strw.leidenuniv.nl

Brinks, Elias
University of Hertfordshire
Centre for Astrophysics Research
College Lane
AL10 9AB Hatfield Hertfordshire, U.K.
ebrinks@star.herts.ac.uk

Brogan, Crystal
NRAO National Radio Astronomy Observatory
520 Edgemont Rd
22903 Charlottesville, VA, U.S.A.
cbrogan@nrao.edu

Bronfman, Leonardo
Universidad de Chile
Camino del Observatorio 1515
Las Condes, Santiago, Chile
leo@das.uchile.cl

Bujarrabal, Valentín
Observatorio Astronómico Nacional
Apartado 112, Alcalá de Henares
E-28803 Madrid, Spain
v.bujarrabal@oan.es

Butler, Bryan
National Radio Astronomy Observatory
1003 Lopezville Road
87801 Socorro, NM, U.S.A.
bbutler@nrao.edu

Carilli, Chris
NRAO
1003 Lopezville Road
87801 Socorro, NM, U.S.A.
ccarilli@nrao.edu

Carpenter, John
CALTECH California Institute of Technology
Department of Astronomy
MC 105-24
91125 Pasadena, CA, U.S.A.
jmc@astro.caltech.edu

Cernicharo, José
DAMIR IEM-CSIC
Serrano 121
28006 Madrid, Spain
cerni@damir.iem.csic.es

Cerrigone, Luciano
Universita' di Catania, Catania, Italy
Harvard-Smithsonian Center for Astrophysics
60 Garden St
02138 Cambridge, MA, U.S.A.
lcerrigone@cfa.harvard.edu

Cesaroni, Riccardo
INAF - Osservatorio Astrofisico di Arcetri
Largo Fermi, 5
50125 Firenze, Italy
cesa@arcetri.astro.it

Cesarsky, Catherine
ESO European Southern Observatory
Karl-Schwarzschild-Strasse 2
85748 Garching bei Muenchen, Germany
ccesarsk@eso.org

Ciliegi, Paolo
INAF – Observatorio Astronomico di Bologna
Via Ranzai 1
40127 Bologna, Italy
paolo.ciliegi@oabo.inaf.it

Codella, Claudio
INAF Istituto di Radioastronomia
Largo E. Fermi, 5
50125 Firenze, Italy
codella@arcetri.astro.it

Colomer, Francisco
Observatorio Astronómico Nacional
Apartado 112
28803 Alcalá de Henares, Madrid, Spain
f.colomer@oan.es

Combes, Francoise
Observatoire de Paris, LERMA
61 Av. de l'Observatoire
F-75014 Paris, France
francoise.combes@obspm.fr

Comito, Claudia
Max-Planck-Institut für Radioastronomie
Auf dem Hügel 69
53121 Bonn, Germany
ccomito@mpifr-bonn.mpg.de

Corder, Stuartt
CARMA California Institute of Technology
1200 E. California Blvd.
MC105-24
91125 Pasadena, CA, U.S.A.
sac@astro.caltech.edu

Cox, Pierre
IRAM
300, rue de la piscine
Domaine Universitaire
F-38406 Saint-Martin-d'Heres, France
cox@iram.fr

Crapsi, Antonio
Leiden Observatory
P.O. Box 9513
NL-2300 RA Leiden, Netherlands
crapsi@strw.leidenuniv.nl

Crutcher, Richard
University of Illinois
1002 W Green St.
61801 Urbana, Illinois, U.S.A.
crutcher@uiuc.edu

Cunningham, Maria
University of New South Wales
School of Physics
UNSW
2052 Sydney NSW, Australia
maria.cunningham@unsw.edu.au

de Graauw, Thijs
SRON-Leiden Observatory
PO Box 800
9700AV Groningen, Netherlands
thijsdg@sron.rug.nl

de Gregorio-Monsalvo, Itziar
ESO/ALMA-LAEFF
Apartado 50.727
28080 Madrid, Spain
itziar@laeff.inta.es

De Luca, Massimo
Università di Roma "Tor Vergata"
v. Frascati, 33, Monte Porzio Catone
00040 Roma, Italy
deluca@mporzio.astro.it

de Vicente, Pablo
Observatorio Astronómico Nacional
Cerro de La Palera S/N
19080 Yebes Guadalajara, Spain
p.devicente@oan.es

Desmurs, Jean-François
Observatorio Astronómico Nacional
C/Alfonso XII, 3
28014 Madrid, Spain
desmurs@oan.es

Dessauges-Zavadsky, Miroslava
Geneva Observatory
51, Ch. des Maillettes
1290 Sauverny, Switzerland
miroslava.dessauges@obs.unige.ch

Di Francesco, James
National Research Council of Canada
5071 West Saanich Road
V9E 2E7 BC Victoria, Canada
james.difrancesco@nrc-cnrc.gc.ca

Diego Rodríguez, José
IFCA
Avda. Los Castros s/n
39005 Santander, Spain
jdiego@ifca.unican.es

Domínguez, Rosa
Universidad Autónoma de Madrid
28049 Madrid, Spain
rosa.dominguez@uam.es

Dutrey, Anne
Observatoire de Bordeaux
L3AB, 2 rue de l'observatoire
33 270 Floirac, France
Anne.Dutrey@obs.u-bordeauxl.fr

Eiroa, Carlos
Universidad Autónoma de Madrid
Dpto. Física Teórica, Facultad de Ciencias
28049 Cantoblanco, Madrid, Spain
carlos.eiroa@uam.es

Espada Fernández, Daniel
Academia Sinica Institute of Astronomy and
Astrophysics
7F of Condensed Matter Sciences and Physics
Department Building
National Taiwan University.
No.1, Roosevelt Rd., Sec. 4, Taipei, Taiwan
despada@asiaa.sinica.edu.tw

Estalella, Robert
Universitat de Barcelona
Dep. Astronomia i Meteorologia
Av. Diagonal 647
E-08028 Barcelona, Spain
robert.estalella@am.ub.es

Fazio, Giovanni
Harvard Smithsonian Center for Astrophysics
MS/65 60 Garden St.
02138 Cambridge, Massachusetts, U.S.A.
gfazio@cfa.harvard.edu

Fernández, José Ma.
C.S.I.C. Institute de Estructura de la Materia
Serrano 121
E-28006 Madrid, Spain
jmfernandez@iem.cfmac.csic.es

Ferrari, Chiara
Institute for Astrophysics
Innsbruck University
Technikerstrasse 25/8
6020 Innsbruck, Austria
chiara.ferrari@uibk.ac.at

Fomalont, Ed
NRAO
520 Edgemont Road
22903 Charlottesville Virginia, U.S.A.
efomalon@nrao.edu

Fonfría Expósito, José
Dept. Molecular and Infrared Astrophysics
(DAMIR)
IEM, CSIC
C/Serrano, 121
28006 Madrid, Spain
jpablo.fonfria@damir.iem.csic.es

Fontani, Francesco
INAF - Istituto di Radioastronomia
via P. Gobetti, 101
1-40129 Bologna, Italy
ffontani@ira.inaf.it

Frail, Dale
National Radio Astronomy Observatory
1003 Lopezville Road
87801 Socorro, NM, U.S.A.
dfrail@nrao.edu

Fuente, Asunción
Observatorio Astronómico Nacional
Apdo 112
E-28803 Alcalá de Henares, Madrid, Spain
a.fuente@oan.es

Fuller, Gary
University of Manchester
School of Physics and Astronomy
Sackville Street Building
PO Box 88
M60 1QD Manchester, U.K.
G.Fuller@manchester.ac.uk

García-Burillo, Santiago
Observatorio Astronómico Nacional
Alfonso XII, 3
28014 Madrid, Spain
s.gburillo@oan.es

Gerin, Maryvonne
LERMA - CNRS and ENS
24 Rue Lhomond
75005 Paris, France
gerin@lra.ens.fr

Giannini, Teresa
INAF - Osservatorio Astronomico di Roma
Via Frascati, 33, Monte Porzio
00040 Roma, Italy
giannini@mporzio.astro.it

Gil de Paz, Armando
Universidad Complutense de Madrid
Departamento de Astrofísica
Facultad de CC. Físicas
Avda. de la Complutense, s/n
28040 Madrid, Spain
agpaz@astrax.fis.ucm.es

Giovannini, Gabriele
Bologna University
Dipartimento di Astronomia
via Ranzani 1
40127 Bologna, Italy
ggiovann@ira.inaf.it

Girart, Josep Miquel
Institut de Ciències de l'Espai (CSIC-IEEC)
Campus UAB - Facultat de Ciències
Torre C5-parell 2ª
08193 Bellaterra, Spain
girart@ieec.uab.es

Giroletti, Marcello
INAF Istituto di Radioastronomia
via Gobetti 101
40129 Bologna, Italy
giroletti@ira.inaf.it

Goicoechea, Javier
Observatoire de Paris
24 Rue Lhomond
75231 Paris, France
javier@lra.ens.fr

Gracía Carpio, Javier
Observatorio Astronómico Nacional
Alfonso XII, 3
28014 Madrid, Spain
j.gracia@oan.es

Gregorini, Loretta
Dept. of Physics
University of Bologna
via Irnerio 46
40121 Bologna, Italy
gregorini@ira.inaf.it

Gruppioni, Carlotta
INAF - Osservatorio Astronomico di Bologna
via Ranzani 1
1-40127 Bologna, Italy
carlotta.gruppioni@bo.astro.it

Guedel, Manuel
Paul Scherrer Institut
Wuerenlingen and Villigen
CH-5232 Villigen PSI, Switzerland
guedel@astro.phys.ethz.ch

Guelin, Michel
IRAM
300 rue de la Piscine
38240 St. Martin d'Heres Isere, France
guelin@iram.fr

Gueth, Frederic
IRAM Grenoble
300 rue de la Piscine
38400 St. Martin d'Heres, France
gueth@iram.fr

Guilloteau, Stephane
L3AB, OASU
2 rue de l'Observatoire
BP 89
33270 Floirac, France
guilloteau@obs.u-bordeauxl.fr

Gupta, Neeraj
NCRA-TIFR
National Centre for Radio Astrophysics
Pune University Campus
Pune Maharashtra, India
neeraj@ncra.tifr.res.in

Hailey-Dunsheath, Steven
Cornell University
208 Space Sciences Building
14853 NY, U.S.A.
steve@astro.Cornell.edu

Hales, Antonio
National Radio Astronomy Observatory
520 Edgemont Road,
VA 22903-2, Charlottesville, U.S.A.
ahales@star.ucl.ac.uk

Hardy, Eduardo
NRAO
Apoquindo 3650, piso 18
Las Condes, Santiago, Chile
ehardy@nrao.cl

Harvey, Paul
University of Texas
Astronomy Dept., C1400
78712 Austin, TX, U.S.A.
pmh@astro.as.utexas.edu

Hasegawa, Tetsuo
ALMA-J Office
National Astronomical Observatory of Japan
2-21-1 Osawa, Mitaka
181-0015 Tokyo, Japan
tetsuo.hasegawa@nao.ac.jp

Hatchell, Jennifer
University of Exeter
School of Physics,
Stocker Road,
EX4 4QL Exeter, U.K.
hatchell@astro.ex.ac.uk

Helmich, Frank
SRON - Netherlands Institute for Space Research
Landleven 12
9747 AD Groningen, Netherlands
f.p.helmich@sron.rug.nl

Herbst, Eric
Ohio State University
Department of Physics
191 W. Woodruff Ave.
43210 Columbus Ohio, U.S.A.
herbst@mps.ohio-state.edu

Herpin, Fabrice
L3AB, Observatoire de Bordeaux
B.P.89, 33270 Floirac, France
herpin@obs.u-bordeaux1.fr

Hesser, James (Jim)
National Research Council of Canada
Herzberg Institute of Astronomy
5071 West Saanich Road,
V9E 2E7 Victoria, BC, Canada
jim.hesser@nrc-cnrc.gc.ca

Hibbard, John
NRAO/NAASC
520 Edgemont Road
22903 Charlottesville
Virginia, U.S.A.
jhibbard@nrao.edu

Hieret, Carolin
Max-Planck-Institut fuer Radioastronomie
Auf dem Huegel 69
53121 Bonn, Germany
chieret@mpifr-bonn.mpg.de

Hill, Tracey
Leiden Observatory
PO BOX 9513
2300 RA Leiden, Netherlands
thill@strw.leidenuniv.nl

Hirano, Naomi
Institute of Astronomy & Astrophysics
Academia Sinica
P.O. Box 23-141
10617 Taipei, Taiwan
hirano@asiaa.sinica.edu.tw

Hofstadter, Mark
Jet Propulsion Laboratory
Mail Stop 183-301
4800 Oak Grove Drive
91109 Pasadena, CA, U.S.A.
mark.hofstadter@jpl.nasa.gov

Hota, Ananda
NCRA-TIFR
Pune University Campus
411007 Pune, India
hota@ncra.tifr.res.in

Huggins, Patrick
New York University
Physics Department
4 Washington Place
10003 New York, NY, U.S.A.
patrick.huggins@nyu.edu

Hughes, Meredith
Harvard-Smithsonian Center for Astrophysics
60 Garden St., MS-10
02138 Cambridge, MA, U.S.A.
mhughes@cfa.harvard.edu

Hughes, David
Institute Nacional de Astrofísica, Óptica y
Electrónica
Luis Enrique Erro, 1
Tonantzintla
72000 Puebla, México
dhughes@inaoep.mx

Hunter, Todd
Harvard-Smithsonian Center for Astrophysics
60 Garden St., MS-78,
02138 Cambridge, MA, U.S.A.
thunter@cfa.harvard.edu

Iono, Daisuke
NAOJ
2-21-2 Osawa, Mitaka
181-8588 Tokyo, Japan
d.iono@nao.ac.jp

Jablonka, Pascale
EPFL - Observatoire de Genève
51 Chemin des Maillettes
1290 Sauverny, Switzerland
pascale.jablonka@obs.unige.ch

Jethava, Nikhil
Max Planck Institute for Radio Astronomy
Auf dem Hugel-69,
53121 Bonn, Germany
njethava@mpifr-bonn.mpg.de

Jiménez-Esteban, Francisco
Observatorio Astronómico Nacional
FRACTAL SLNE
Campus Universitario, Ctra. NII km 33,600
Apartado 112
E-28803 Alcalá de Henares Madrid, Spain
f.jimenez-esteban@oan.es

Jiménez-Serra, Izaskun
Dpto. de Astrofísica Molecular e Infrarroja
Instituto de Estructura de la Materia
C/ Serrano 121
E-28006 Madrid, Spain
izaskun@damir.iem.csic.es

Johnstone, Doug
NRC-HIA Univ. Victoria
5071 West Saanich Rd.
V9E 2E7 Victoria, BC, Canada
doug.johnstone@nrc-cnrc.gc.ca

Jones, Paul
University of New South Wales
740 Elizabeth Street
2017 Waterloo, NSW, Australia
Paul.Jones@csiro.au

Jorgensen, Jes
Harvard-Smithsonian Center for Astrophysics
60 Garden Street, MS42
02138 Cambridge, MA, U.S.A.
jjorgensen@cfa.harvard.edu

Josselin, Eric
GRAAL - Universite Montpellier II
cc 072, (cedex 05)
34095 Montpellier, France
josselin@graal.univ-montp2.fr

Kamaya, Hideyuki
Kyoto University
Department of Astronomy
Kitashirakawa-Oiwake-Cho
Sakyo-Ku, Kyoto
606-8502 Kyoto-Fu, Japan
kamaya@kusastro.kyoto-u.ac.jp

Kawabe, Ryohei
ALMA-J Project Office
National Observatory of Japan
2-21-1 Osawa, Mitaka
181-8588 Tokyo, Japan
ryo.kawabe@nao.ac.jp

Kawamura, Akiko
Nagoya University
Department of Astrophysics
Furo-cho, Chikusa, Nagoya
464-8602 Aichi, Japan
kawamura@a.phys.nagoya-u.ac.jp

Klein, Uli
Argelander-Institut für Astronomic
Auf dem Hügel 71
D-53121 Bonn, Germany
uklein@astro.uni-bonn.de

Knee, Lewis
National Research Council of Canada
5071 West Saanich Road,
V9E 2E7 Victoria, BC, Canada
lewis.knee@nrc-cnrc.gc.ca

Knudsen, Kirsten
Max-Planck-Institut fur Astronomic
Königstuhl 17
D-69117 Heidelberg, Germany
knudsen@mpia-hd.mpg.de

Kobayashi, Kaori
University of Toyama
3190 Gofuku
930-8555 Toyama, Japan
kaori@sci.u-toyama.ac.jp

Koda, Jin
California Institute of Technology
MS105-24, Caltech
91125 Pasadena, California, U.S.A.
koda@astro.caltech.edu

Kohno, Kotaro
University of Tokyo
2-21-1, Osawa, Mitaka
181-0015 Tokyo, Japan
kkohno@ioa.s.u-tokyo.ac.jp

Krips, Melanie
Harvard-Smithsonian Center for Astrophysics
SMA - Site
645 North A'Ohoku Place,
96720 Hilo, Hawaii, U.S.A.
mkrips@cfa.harvard.edu

Kuno, Nario
Nobeyama Radio Observatory
Minamimaki-mura,
384-1305 Nagano, Japan
kuno@nro.nao.ac.jp

Kurono, Yasutaka
The Univ. of Tokyo/NRO
Nobeyama Radio Observatory
National Astronomical Observatory Japan
Minamimaki, Minamisaku
384-1305 Nagano, Japan
kurono@nro.nao.ac.jp

Laing, Robert
European Southern Observatory
Karl-Schwarzschild-Strasse 2
D-85748 Garching-bei-Muenchen, Germany
rlaing@eso.org

Lara, Luisa
Institute de Astrofísica de Andalucía
Camino Bajo de Huétor 50
18008 Granada, Spain
lara@iaa.es

Lefloch, Bertrand
LAOG
Observatoire de Grenoble
BP 53
38041 Grenoble, France
lefloch@obs.ujf-grenoble.fr

Lellouch, Emmanuel
LESIA
Observatoire de Paris
92195 Meudon, France
emmanuel.lellouch@obspm.fr

Lemaire, Jean Louis
Observatoire de Paris et Université de Cergy-
Pontoise
LERMA
Observatoire de Meudon
92195 Meudon, France
jean-louis.lemaire@obspm.fr

Leurini, Silvia
European Southern Observatory
Karl-Schwarzschild-Str. 2
85748 Garching bei München, Germany
sleurini@mpifr-bonn.mpg.de

Levrier, Francois
Ecole Normale Supérieure
LRA - Département de physique
24 rue Lhomond
75005 Paris, France
levrier@lra.ens.fr

Lim, Jeremy
Institute for Astronomy & Astrophysics
Academia Sinica
ASIAA P.O. Box 23-141
10619 Taipei, Taiwan
jlim@asiaa.sinica.edu.tw

Lintott, Chris
University of Oxford
Dept. of Physics
Denys Wilkinson Building
OX1 3RH London, U.K.
cjl@star.ucl.ac.uk

Lis, Darek
Caltech
MC 320-47
91125 Pasadena, CA, U.S.A.
dcl@caltech.edu

Liu, Sheng-Yuan
Academia Sinica
Institute of Astronomy and Astrophysics
P.O. Box 23-141, None
10619 Taipei, Taiwan
syliu@asiaa.sinica.edu.tw

Lo, Fred
National Radio Astronomy Observatory
520 Edgemont Road
22903-2475 Charlottesville, VA, U.S.A.
brodrigu@nrao.edu

Loenen, Edo
Kapteyn Astronomical Institute / ASTRON
Landleven 12
9747 AD Groningen, Netherlands
loenen@astro.rug.nl

Loiseau, Nora
XMM-Newton Science Operations Centre
European Space Astronomy Centre (ESAC)
Apartado 50727
E-28080 Madrid, Spain
nora.loiseau@sciops.esa.int

Lorenzetti, Dario
INAF - Osservatorio Astronomico di Roma
Via Frascati, 33, Monte Porzio
00040 Roma, Italy
dloren@mporzio.astro.it

Loughnane, Robert
National University of Ireland
9 Woodhaven, Merlin Park, Dublin Road
Galway, Ireland
loughnane.robert@gmail.com

Loukitcheva, Maria
Astronomical Institute, St. Petersburg University
Universitetskii pr. 28, Peterhof,
198504 St. Petersburg, Russia
marija@peterlink.ru

Lovell, Amy
Agnes Scott College
357 S. Candler St.
30030 Decatur, GA, U.S.A.
alovell@agnesscott.edu

Lucas, Robert
IRAM Grenoble
300 rue de la Piscine
38406 Saint Martin d'Heres, France
lucas@iram.fr

Mack, Karl-Heinz
Istituto di Radioastronomia - INAF
Via P. Gobetti 101
40129 Bologna, Italy
mack@ira.inaf.it

Manzitto, Patrizia
Università di Catania
Via S. Sofia 78
95123 Catania, Italy
pmanzitto@oact.inaf.it

Marcaide, Jon
Universitat de Valencia
Dpto. Astronomia i Astrofísica
C/ Dr. Moliner, 50
46100 Burjassot, Spain
J.M.Marcaide@uv.es

Marcelino, Nuria
Institute de Radioastronomía Milimétrica
Avenida Divina Pastora 7, Local 20
E 18012 Granada, Spain
marcelino@iram.es

Mardones, Diego
Universidad de Chile
Casilla 36-D, Santiago, Chile
mardones@das.uchile.cl

Márquez, Isabel
CSIC - Institute de Astrofísica de Andalucía
Apdo. 3004
18008 Granada, Spain
isabel@iaa.es

Marrone, Daniel
University of Chicago, KICP
5640 S. Ellis AveLASR 136,
60637 Chicago, IL, U.S.A.
dmarrone@uchicago.edu

Martín, Sergio
Harvard-Smithsonian Center for Astrophysics
Avenida Divina Pastora 7, Local 20
E-18012 Granada, Spain
martin@iram.es

Martín-Pintado, Jesús
DAMIR (IEM-CSIC)
C/ Serrano 121
28006 Madrid, Spain
jmartin.pintado@iem.cfmac.csic.es

Masegosa, Josefa
Institute de Astrofísica de Andalucía, CSIC
C/Camino Bajo de Huetor, 50
18008 Granada, Spain
pepa@iaa.es

Massi, Fabrizio
INAF - Osservatorio Astrofisico di Arcetri
Largo E. Fermi, 5
1-50125 Firenze, Italy
fmassi@arcetri.astro.it

Matthews, Brenda
Herzberg Institute of Astrophysics
5071 West Saanich Road
V9E 2E7 Victoria British Columbia, Canada
brenda.matthews@nrc-cnrc.gc.ca

Mauersberger, Rainer
IRAM
Avda Divina Pastora 7, local 20
18012 Granada, Spai
mauers@iram.es

Michalowski, Michal
DARK Cosmology Centre
Copenhagen University
H.C. Orstedsvej 11 A, 2tv,
DK-1879 Frederiksberg C, Denmark
michal@astro.ku.dk

Millar, Tom
Queen's University Belfast
School of Mathematics and Physics
Queen's University Belfast
BT7 INN Belfast, Northern Ireland
Tom.Millar@qub.ac.uk

Minier, Vincent
Service d'Astrophysique/DAPNIA/DSM, CEA
Saclay
Orme des Merisiers
91191 Gif-sur-Yvette, France
vincent.minier@cea.fr

Mirabel, Felix
ESO
Alonso de Cordova 3107
7630472 Vitacura RM, Chile
fmirabel@eso.org

Miura, Rie
University of Tokyo NAOJ
Osawa 2-21-1
181-8588 Mitaka Tokyo, Japan
rie.miura@nao.ac.jp

Miyama, Shoken
National Astronomical Observatory of Japan,
NAOJ
Osawa, 2-21-1
181-8588 Mitaka, Tokyo, Japan
director-general@nao.ac.jp

Mollá, Mercedes
CIEMAT, Investigación Bàsica
Avda. Complutense 22
28040 Madrid, Spain
mercedes.molla@ciemat.es

Momose, Munetake
Institute of Astronomy & Planetary Sciences
Ibaraki University
Bunkyo 2-1-1
310-8512 Mito Ibaraki, Japan
momose@mx.ibaraki.ac.jp

Monje, Raquel
Onsala Space Observatory
Chalmers University of Technology
S-439 92 Onsala, Sweden
raquel@oso.chalmers.se

Morata, Oscar
LAEFF/INTA
Apdo. 50727
28080 Villafranca del Castillo (Madrid), Spain
omorata@laeff.inta.es

Moreno, Raphael
LESIA (LAM - bat. 18)
Observatoire de Paris-Meudon
5 place Jules Janssen
92195 Meudon, France
Raphael.Moreno@obspm.fr

Mori, Masao
Senshu University
2-1-1 Higashimita, Tama
214-8580 Kawasaki, Japan
mmori@isc.senshu-u.ac.jp

Moro-Martin, Amaya
Princeton University
Department of Astrophysical Sciences
Peyton Hall - Ivy Lane
08540 Princeton, NJ, U.S.A.
amaya@astro.princeton.edu

Motte, Frederique
Astrophysique des Interactions Multi-echelles
(UMR7158)
Service d'Astrophysique, CEA/Saclay
91191 Gif-sur-Yvette, France
motte@cea.fr

Moullet, Arielle
LESIA-Observatoire de Paris
5 place J. Janssen
92195 Meudon, France
arielle.moullet@obspm.fr

Mundy, Lee
University of Maryland
Astronomy Department
20742 Maryland, U.S.A.
lgm@astro.umd.edu

Najarro, Francisco
DAMIR, IEM, CSIC
Institute de Estructura de la Materia
Serrano 121
28006 Madrid, Spain
najarro@damir.iem.csic.es

Narayanan, Desika
Steward Observatory
University of Arizona
933 N Cherry Ave
85721 Tucson, AZ, U.S.A.
dnarayanan@as.arizona.edu

Natta, Antonella
INAF Osservatorio di Arcetri
Largo Fermi 5
50125 Firenze, Italy
natta@arcetri.astro.it

Nisini, Brunella
INAF Osservatorio Astronomico di Roma
Via di Frascati 33, Monte Porzio
00040 Roma, Italy
nisini@oa-roma.inaf.it

Nuernberger, Dieter
European Southern Observatory
Casilla 19001
19001 Santiago, Chile
dnuernbe@eso.org

Nyman, Lars-Ake
ESO
Casilla 19001
19001 Santiago, Chile
lnyman@eso.org

Ohashi, Nagayoshi
Academia Sinica Institute of Astronomy &
Astrophysics
P.O. Box 23-141
10617 Taipei, Taiwan
ohashi@asiaa.sinica.edu.tw

Olofsson, Hans
Onsala Space Observatory
Stockholm Observatory
SE-43992 Onsala, Sweden
hans@astro.su.se

Onishi, Toshikazu
Nagoya University
Department of Astrophysics
Furo-cho, Chikusa-ku
464-8602 Nagoya Aichi, Japan
ohnishi@a.phys.nagoya-u.ac.jp

Ossenkopf, Volker
I. Physikalisches Institut
Universitaet zu Koeln
Zuelpicher Str. 77
50937 Koeln NRW, Germany
ossk@ph1.uni-koeln.de

Pagani, Laurent
LERMA, Observatoire de Paris
61, Av. de l'Observatoire
75014 Paris, France
laurent.pagani@obspm.fr

Pandian, Jagadheep
Cornell University
518 Space Science
14853 Ithaca, NY, U.S.A.
jagadheep@astro.cornell.edu

Panic', Olja
Leiden Observatory
PO Box 9513
2300 RA Leiden, Netherlands
olja@strw.leidenuniv.nl

Papadopoulos, Padelis
ETH, Zurich Institute for Astronomy
ETH, Hoenggerberg
HPF D8, 8093
8093 Zurich, Switzerland
papadop@phys.ethz.ch

Pardo, Juan
CSIC
Serrano 121
28006 Madrid, Spain
pardo@damir.iem.csic.es

Parise, Bérengère
Max Planck Institut fur Radioastronomie
Auf dem Hügel 69
53121 Bonn, Germany
bparise@mpifr-bonn.mpg.de

Parma, Paola
INAF- Istituto di Radioastronomía Bologna
via Gobetti, 101
40129 Bologna, Italy
parma@ira.inaf.it

Patience, Jenny
Caltech
Department of Astronomy
1200 E. California Blvd
MS 105-24
91125 Pasadena, CA, U.S.A.
patience@astro.caltech.edu

Pearson, John
Jet Propulsion Laboratory
California Institute of Technology
4800 Oak Grove Dr.
Mail Stop 301-429
91109 Pasadena, CA, U.S.A.
John.C.Pearson@jpi.nasa.gov

Peck, Alison
Harvard Smithsonian Center for Astrophysics
645 N. Aohoku Place
96720 HI, U.S.A.
apeck@cfa.harvard.edu

Peretto, Nicolas
University of Manchester
Department of Physics & Astronomy
Sackville street
PO Box 88
M60 1QD Manchester, U.K.
Nicolas.Peretto@manchester.ac.uk

Pérez-Fournon, Ismael
Instituto de Astrofísica de Canarias
Via Láctea, s/n
E-38200 S/C de Tenerife, Spain
ipf@iac.es

Pety, Jérôme
IRAM Grenoble
300 rue de la Piscine
F-38406 St. Martin d'Heres, France
pety@iram.fr

Pineda, Jorge
Argelander-Institut fuer Astronomie
Auf dem Huegel 71
D-53121 Bonn, Germany
jopineda@astro.uni-bonn.de

Planesas, Pere
Observatorio Astronómico Nacional
Apartado 112
28803 Alcalá de Henares (Madrid, Spain)
p.planesas@oan.es

Plume, Rene
University of Calgary
Dept. of Physics & Astronomy
2500 University Dr. NW
T2N1N4 Calgary, AB, Canada
plume@ism.ucalgary.ca

Prandoni, Isabella
INAF - Istituto di Radioastronomia
Via P. Gobetti 101
40129 Bologna, Italy
prandoni@ira.inaf.it

Pratap, Preethi
MIT Haystack Observatory
Off Route 40
01886 Westford, MA, U.S.A.
ppratap@haystack.mit.edu

Puxley, Phil
National Science Foundation
MPS/AST Room 1045
4201 Wilson Blvd.
22230 Arlington, VA, U.S.A.
ppuxley@nsf.gov

Qi, Chunhua
Harvard-Smithsonian Center for Astrophysics
60 Garden Street, Mailstop 42
02138 Cambridge MA, U.S.A.
cqi@cfa.harvard.edu

Quintana-Lacaci, Guillermo
Observatorio Astronómico Nacional
Alfonso XII, 3
28014 Madrid, Spain
g.quintana@oan.es

Recillas, Elsa
INAOE
Luis Enrique Erro # 1 Sta. María Tonantzintla
72840 San Andrés Cholula Puebla, México
elsare@inaoep.mx

Redman, Matt
National University of Ireland Galway
Department of Experimental Physics
Galway, Ireland
matt.redman@nuigalway.ie

Regan, Michael
Space Telescope Science Institute
3700 San Martin Drive
21218 Baltimore, MD, U.S.A.
mregan@stsci.edu

Reid, Michael
Harvard-Smithsonian Submillimeter Array
645 North A'Ohoku Pl.
96720 Hilo Hawaii, U.S.A.
mareid@sma.hawaii.edu

Requena-Torres, Miguel
Departamento de Astrofísica Molecular e
Infrarroja
Instituto de Estructura de la Materia
C/ Serrano 121
28006 Madrid, Spain
requena@damir.iem.csic.es

Riechers, Dominik
Max-Planck-Institut fuer Astronomie (MPIA)
Koenigstuhl 17
69117 Heidelberg, Germany
riechers@mpia.de

Rigopoulou, Dimitra
University of Oxford
Denys Wilkinson Building
Keble Road
OX2 8AJ Oxford, U.K.
d.rigopouloul@physics.ox.ac.uk

Rodríguez-Franco, Arturo
Departamento de Astrofísica Molecular e
Infrarroja (DAMIR)
Calle Serrano, 121
28006 Madrid, Spain
arturo@damir.iem.csic.es

Rykaczewski, Hans
E.S.O. - European Organisation for Astronomical
Research
Karl-Schwarzschildstr. 2,
85748 Garching near Munich, Germany
hrykacze@eso.org

Sahai, Raghvendra
Jet Propulsion Laboratory - Caltech
JPL, MS 183-900
4800 Oak Grove Drive
91109 Pasadena CA, U.S.A.
raghvendra.sahai@jpl.nasa.gov

Sakai, Nami
Department of Physics
University of Tokyo
7-3-1, Hongou, Bunkyo-ku
113-0033 Tokyo, Japan
nami@taurus.phys.s.u-tokyo.ac.jp

Sakamoto, Kazushi
National Astronomical Observatory of Japan
2-21-1, Osawa
181-8588 Mitaka Tokyo, Japan
sakamoto.kazushi@nao.ac.jp

Sakamoto, Seiichi
National Astronomical Observatory of Japan
2-21-1 Osawa
181-8588 Mitaka, Tokyo, Japan
seiichi@nro.nao.ac.jp

Salome, Philippe
IRAM
300 rue de la Piscine
38400 St. Martin d'Heres, France
salome@iram.fr

Salvati, Marco
INAF - Osservatorio di Arcetri
Largo E. Fermi 5
50125 Firenze, Italy
salvati@arcetri.astro.it

Sánchez Contreras, Carmen
Departamento de Astrofísica Molecular e
Infrarroja (DAMIR)
C/ Serrano 121
28006 Madrid, Spain
carmen@damir.iem.csic.es

Santiago-García, Joaquín
Observatorio Astronómico Nacional
C/ Alfonso XII, 3
28014 Madrid, Spain
j.santiago@oan.es

Sargent, Anneila
California Institute of Technology
1200 E. California Blvd.
91107 Pasadena, CA, U.S.A.
afs@astro.caltech.edu

Schilke, Peter
Max-Planck-Institut für Radioastronomie
Auf dem Hügel 69
53121 Bonn, Germany
schilke@mpifr-bonn.mpg.de

Schinnerer, Eva
Max Planck Institute of Astronomy
Königstuhl 17
69117 Heidelberg, Germany
schinner@mpia.de

Schneider-Bontemps, Nicola
SAp/DAPNIA CEA Saclay
Orme des Merisiers
91191 Gif-sur-Yvette, France
nschneid@cea.fr

Schreier, Ethan
AUI
1400 16th Street, NW Suite 730
20036 Washington, DC, U.S.A.
ejs@aui.edu

Scoville, Nick
CALTECH
astronomy 105-24
91125 Pasadena, CA, U.S.A.
nzs@astro.caltech.edu

Sekiguchi, Tomohiko
ALMA-J Project Office
National Astronomical Observatory of Japan
2-21-1 Osawa
181-8588 Mitaka, Tokyo, Japan
t.sekiguchi@nao.ac.jp

Semenov, Dmitry
Max Planck Institute for Astronomy
Koenigstuhl 17
69117 Heidelberg, Germany
semenov@mpia.de

Shepherd, Debra
National Radio Astronomy Observatory
P.O. Box O,
87801 Socorro, NM, U.S.A.
dshepher@aoc.nrao.edu

Sorai, Kazuo
Graduate School of Science
Hokkaido University
Kita 10, Nishi 8
060-0810 Sapporo, Japan
sorai@astrol.sci.hokudai.ac.jp

Soria Ruiz, Rebeca
Joint Institute for VLBI in Europe (JIVE)
Oude Hoogeveensedijk 4
7991 PD Dwingeloo Drenthe, Netherlands
soria@jive.nl

Sugiyama, Naoshi
Department of Physics and Astrophysics
Nagoya University
Furo-cho, Chikusa-ku
464-8602 Nagoya, Japan
naoshi@a.phys.nagoya-u.ac.jp

Sunada, Kazuyoshi
Nobeyama Radio Observatory
Nobeyama 462-2, Minamimaki, Minamisaku
384-1305 Nagano, Japan
sunada@nro.nao.ac.jp

Tacconi, Linda
Max-Planck-Institut fuer extraterrestrische Physik
Giessenbachstrasse 1
85748 Garching bei Muenchen, Germany
linda@mpe.mpg.de

Tafalla, Mario
Observatorio Astronómico Nacional
Alfonso XII, 3
28014 Madrid, Spain
m.tafalla@oan.es

Takahashi, Satoko
Graduate University for Advanced Studies
National Astronomical Observatory of Japan
Osawa 2-21-1
181-8588 Tokyo,Japan
satoko.takahashi@nao.ac.jp

Takakuwa, Shigehisa
National Astronomical Observatory of Japan
ALMA Project Of
Osawa 2-21-1
181-8588 Mitaka Tokyo, Japan
s.takakuwa@nao.ac.jp

Taniguchi, Yoshiaki
Physics Department, Graduate School of Science
& Engineering
2-5 Bunkyo-cho
790-8577 Matsuyama, Japan
tani@sgr.phys.sci.ehime-u.ac.jp

Tarenghi, Massimo
ALMA
El Golf 40 Piso 18
Las Condes Santiago, Chile
mtarengh@alma.cl

Tercero Martínez, Belén
DAMIR, IEM, CSIC
C/ Serrano 121
28006 Madrid, Spain
belen@damir.iem.csic.es

Testi, Leonardo
INAF-Osservatorio Astrofísico di Arcetri
Largo E. Fermi 5
1-50125 Firenze, Italy
lt@arcetri.astro.it

Thum, Clemens
IRAM
300 rue de la picine
Domaine Universitaire de Grenoble
38406 St. Martin d'Heres, France
thum@iram.fr

Torii, Kazufumi
Department of Astrophysics
Nagoya University
Furou-cho, Chikusa-ku
464-8602 Aichi-ken, Japan
torii@a.phys.nagoya-u.ac.jp

Torrelles, José María
ICE(CSIC)-IEEC
C/ Gran Capita 2-4
08034, Barcelona, Spain
torrelles@ieec.fcr.es

Toscano, Simona
Università di Catania
via S. Sofia 78
95123 Catania, Italy
sto@oact.inaf.it

Trigilio, Corrado
INAF-Osservatorio Astrofísico di Catania
Via Santa Sofía 78
1-95123 Catania, Italy
ctrigilio@oact.inaf.it

Tsamis, Yiannis
University College London
Physics and Astronomy
Gower Street
WC 1E 6BT London, U.K.
ygt@star.ucl.ac.uk

Turner, Jean
UCLA
430 Portola Plaza UCLA Box 951547
90095-1547 Los Angeles, CA, U.S.A.
turner@astro.ucla.edu

Umana, Grazia
INAF-Osservatorio Astrofísico di Catania
Via S. Sofia 78
95123 Catania, Italy
gumana@oact.inaf.it

Usero, Antonio
Observatorio Astronómico Nacional (OAN)
C/ Alfonso XII, 3
28014 Madrid, Spain
a.usero@oan.es

van der Tak, Floris
National Institute for space Research (SRON)
Landleven 12
9747 AD Groningen, Netherlands
vdtak@sron.rug.nl

van Dishoeck, Ewine
Leiden Observatory
P.O. Box 9513
2300 RA Leiden, Netherlands
ewine@strw.leidenuniv.nl

van Kempen, Tim
Leiden Observatory
Niels Bohrweg 2,
NL-2333 CA Leiden, Netherlands
kempen@strw.leidenuniv.nl

van Langevelde, Huib
JIVE Dwingeloo/Sterrewacht Leiden
Postbus 2
7990 AA Dwingeloo, Netherlands
langevelde@jive.nl

Vanden Bout, Paul
NRAO
520 Edgemont Road
22903-2475 Charlottesville, VA, U.S.A.
pvandenb@nrao.edu

Velusamy, Thangasamy
Jet Propulsion Laboratory
MS 169-506
4800 Oak Grove Dr
CA 91109 Pasadena, CA, U.S.A.
velusamy@jpi.nasa.gov

Verdes-Montenegro, Lourdes
Institute de Astrofísica de Andalucía
Camino Bajo de Huetor 50
18008 Granada, Spain
lourdes@iaa.es

Viallefond, Francois
LERMA, Observatoire de Paris
61 av. de l'Observatoire
75014 Paris, France
fviallef@maat.obspm.fr

Vidal, Iván
Universitat de Valencia
Dpto. Astronomia i Astrofísica
Office 4.12
C/Dr. Moliner, 50
46100 Burjassot, Spain
i.marti-vidal@uv.es

Vig, Sarita
INAF-Osservatorio Astrofísico di Arcetri
Largo E. Fermi - 5
50125 Florence, Italy
sarita@arcetri.astro.it

Vila Vilaro, Baltasar
ALMA-J Project Office
National Observatory of Japan
2-21-1 Osawa
181-8588 Mitaka, Tokyo, Japan
vila.vilaro@nao.ac.jp

Viti, Serena
University College London
Physics and Astronomy
Gower Street
WC1EBT London, U.K.
sv@star.ucl.ac.uk

Vlemmings, Wouter
Jodrell Bank Observatory
University of Manchester
SK11 9DL Macclesfield Cheshire, U.K.
wouter@jb.man.ac.uk

Wagg, Jeff
NRAO
Array Operations Center
87801 Socorro, NM, U.S.A.
jwagg@cfa.harvard.edu

Walmsley, Malcolm
Osservatorio di Arcetri
Largo E Fermi 5
50125 Firenze, Italy
walmsley@arcetri.astro.it

Walter, Fabian
Max Planck Institut fur Astronomie
Königstuhl 17
69117 Heidelberg, Germany
walter@mpia.de

Wang, Wei-Hao
NRAO
1003 Lopezville Rd.
87801 Honolulu, NM, U.S.A.
wang@ifa.hawaii.edu

Warmels, Rein
ESO
Karl-Schwarzschild-Strasse 2
D-85748 Garching bei Muenchen, Germany
rwarmels@eso.org

West, Andrew
University of California
Astronomy Department
601 Campbell Hall
94720-3411 Berkeley, CA, U.S.A.
awest@astro.berkeley.edu

Wiedner, Martina
1. Physikalisches Institut
Universitaet zu Koeln
Zuelpicher Str. 77
50937 Koeln NRW, Germany
wiedner@phl.uni-koeln.de

Wild, Wolfgang
SRON Netherlands Institute for Space Research
Landleven 12
9747AD Groningen, Netherlands
W.Wild@sron.rug.nl

Williams, Jonathan
University of Hawaii
2680 Woodlawn Dr.
96816 Honolulu, HI, U.S.A.
jpw@ifa.hawaii.edu

Wilner, David
Harvard-Smithsonian Center for Astrophysics
60 Garden Street
02138 Cambridge, MA, U.S.A.
dwilner@cfa.harvard.edu

Wilson, Thomas
European Southern Observatory
K-Schwarzschild-Str. 2
85748 Garching bei Muenchen, Germany
twilson@eso.org

Wilson, Christine
McMaster University
Department of Physics & Astronomy
1280 Main St. W.
L8S 4M1 Hamilton Ontario, Canada
wilson@physics.mcmaster.ca

Wolf, Sebastian
Max Planck Institute for Astronomy
Koenigstuhl 17
69117 Heidelberg, Germany
swolf@mpia.de

Wootten, Al
NRAO/ALMA
520 Edgemont Road
22903 Charlottesville, Virginia, U.S.A.
awootten@nrao.edu

Wyrowski, Friedrich
Max Planck Institute for Radioastronomy
Auf dem Huegel 69
53121 Bonn, Germany
wyrowski@mpifr-bonn.mpg.de

Yamada, Toru
Subaru Telescope
650 N A'ohoku Place
Hi 96720 Hilo, U.S.A.
yamada@subaru.naoj.org

Yamamoto, Satoshi
Department of Physics
University of Tokyo
7-3-1 Kongo
113-0033 Bunkyo-ku Tokyo, Japan
yamamoto@phys.s.u-tokyo.ac.jp

Yun, Min
University of Massachusetts
Department of Astronomy
01003 Amherst, Massachusetts, U.S.A.
myun@astro.umass.edu

Zanichelli, Alessandra
Istituto di Radioastronomia - INAF
via Gobetti 101
40129 Bologna, Italy
a.zanichelli@ira.inaf.it

Zapata, Luis
Max Planck Institute für Radioastronomie
Auf dem Hügel 69
D-53121 Bonn, Germany
lzapata@astrosmo.unam.mx

Zhang, Qizhou
Harvard-Smithsonian Center for Astrophysics
60 Garden Street, MS 42
02138 Cambridge, U.S.A.
qzhang@cfa.harvard.edu

Zwaan, Martin
ESO
Karl-Schwarzschildstrasse 2
85748 Garching bei Muenchen, Germany
mzwaan@eso.org

Articles included in the CD
(corresponding to posters presented at the Symposium)

P. Andreani and C. Baccigalupi: Dark-energy constraints with ALMA polarization measurements. Synergies swith CMB experiments.

G. Anglada, M. Osorio, S. Lizano, and P. D'Alessio: Molecular emission as a test for the early stages of massive star formation.

M. Aravena, F. Bertoldi, C. Carilli, E. Schinnerer, H. Voss, V. Smolcic, N. Scoville, H. Aussel, A. Blain, K. M. Menten, D. Lutz, P. Capak, Y. Taniguchi, M. Brusa, B. Mobasher, S. Lilly, D. Thompson, E. Kreysa, G. Hasinger, J. Aguirre, J. Schlaerth, A. Koekemoer, and Cosmos collaboration: MAMBO 1.2 mm survey of the COSMOS field.

M. Audard, S. Skinner, M. Güdel, T. Lanz, E. Paerels, and H. Arce: A mid-infrared spitzer study of the Herbig be star R Mon and the associated HH 39 Herbig-Haro object.

A. Belloche, B. Parise, P. Schilke, and F. van der Tak: The evolutionary state of the dense core Chamaeleon-MMS1: a combined APEX and Spitzer view.

M. T. Beltrán, R. Cesaroni, C. Codella, L. Testi, R. S. Furuya, and L. Olmi: Infall and the formation of a massive star.

M. Benedettini, S. Viti, R. Bachiller, F. Gueth, and C. Codella: The blue lobe of the L1157 outflow at high spatial resolution.

M. Benedettini, J. A. Yates, S. Viti, and C. Codella: A detailed modelling of the chemically rich clumps along the CB3 outflow.

J. Boissier, D. Bockelée-Morvan, J. F. Crifo, and A.V. Rodionov: Molecular spatial distributions observed in comet Hale-Bopp with IRAM Plateau de Bure interferometer.

F. Boone, D. Schaerer, R. Pello, F. Combes, A. Hempel, and E. Egami: Millimeter observations of a gravitationally lensed Lyman-alpha emitting galaxy at $z = 6.56$.

J. Brand, C. Codella, L. DiFabrizio, F. Massi, and J.G.A. Wouterloot: Jet-driven outflows in star-forming globules.

V. Casasola, F. Combes, D. Bettoni, M. Pohlen, and G. Galletta: Are truncated stellar disks linked to the molecular gas density?

L. Cerrigone, J. L. Hora, G. Umana, and C. Trigilio: IC 4406: a radio-infrared comparison.

C. Codella, S. Cabrit, F. Gueth, F. Bacciotti, R. Cesaroni, B. Lefloch, D. Panoglou, P. García, and M.J. McCaughrean: The molecular jet driving the HH212 protostellar outflow.

C. Codella, S. Viti, D.A. Williams, and R. Bachiller: Evidence for molecular outflow-ambient interfaces in CepA-East?

J. F. Desmurs, C. Codella, J. Santiago-García, M. Tafalla, and R. Bachiller: AU-scale observations of protostellar outflows from H_2O masers.

D. Espada, A. Peck, S. Matsushita, G. Petitpas, Y. Pihlstroem, C. Henkel, D. Iono, F. Israel, K. Sakamoto, and G. Taylor: CO(2-1) SMA imaging of the molecular gas in Centaurus A.

C. Ferrari, R. W. Hunstead, L. Feretti, S. Maurogordato, C. Benoist, A. Cappi, S. Schindler, and E. Slezak: Star formation in the merging galaxy cluster Abell 3921: optical and radio observations.

J. P. Fonfría, M. Agúndez, B. Tercero, J. R. Pardo, and J. Cernicharo: High-J $v = 0$ SiS Maser emission in IRC+10216: a new case of infrared overlaps.

F. Fontani, P. Caselli, A. Crapsi, R. Cesaroni, and J. Brand: Searching for massive pre-stellar cores through observations of N_2H+ and N_2D^+.

T. Giannini, M. de Luca, D. Lorenzetti, G. Fazio, M. Marengo, F. Massi, B. Nisini, and H. A. Smith: Exploring southern star formation regions: H_2 jets driven by very young protostars.

M. Giroletti, G. Giovannini, and M. A. Pérez Torres: Millimeter VLBI detection of the TeV blazar Markarian 501.

I. de Gregorio-Monsalvo, J. F. Gómez, O. Suárez, T. B. H. Kuiper, G. Anglada, N. A. Patel, and J. M. Torrelles: Water maser emission in Bok globules.

N. Gupta, C. J. Salter, D. J. Saikia, T. Ghosh, and S. Jeyakumar: Probing radio source environments using 21-cm absorption.

C. Hieret, S. Leurini, K. M. Menten, P. Schilke, S. Thorwirth, and F. Wyrowski: APEX survey of southern high mass star forming regions.

T. Hill, M. G. Burton, M. R. Cunningham, and V. Minier: Profiling young massive stars.

N. Hirano, C. F. Lee, S. Y. Liu, and H. Schang: High velocity SiO emission in the protostellar jets imaged with the Submillimeter Array.

A. Hota, D. J. Saikia, and J. A. Irwin: Radio continuum and HI study of gas-loss processes in nearby galaxies.

M. Imanishi, K. Kohno, and K. Nakanishi: Buried AGNs in luminous infrared galaxies.

D. Iono, C. D. Wilson, S. Takakuwa, M. S. Yun, G. R. Petitpas, A. B. Peck, P. T. P. Ho, S. Matsushita, Y. M. Pihlstrom, and Z. Wang: High resolution CO(3-2) and HCO^+(4-3) imaging of the luminous infrared galaxy NGC6240.

N. Jethava, C. Henkel, J. Braatz, and K. M. Menten: Ammonia toward gravitational lens B0218+357 and PKS 1830-211.

P. A. Jones, M. R. Cunningham, and S. Stanimirovic: The CO outflow from IRAS 18316-0602 in G25.65+1.05.

H. Kamaya, R. Suzaki, K. Shibata: Challenges to detect the Magnetic activity of YSO outflow by ALMA.

M. Krips, R. Neri, S. García-Burillo, F. Combes, S. Martín, G. Petitpas, A. Peck, A. Eckart, R. Davies, E. Schinnerer, A. J. Baker, A. Usero, S. Matsushita: A multi-transition study of HCN and HCO$^+$ in active galaxies: Starburst versus AGN environments.

F. Levrier, E. Falgarone, and F. Viallefond: ALMA: Fourier phase analysis made possible.

R. M. Loughnane, M. P. Redman, and E. R. Keto: Physical and chemical diagnostics from HCN: Accounting for its hyperfine anomalies.

M. de Luca, T. Giannini, D. Lorenzetti, D. Elia, G. Fazio, M. Marengo, F. Massi, B. Nisini, and H. A. Smith: Resolving the dust emission in the Vela Molecular Ridge.

K.-H. Mack, L. Saripalli, I. A. G. Snellen, and R. T. Schilizzi: Molecular gas in radio galaxies with new and restarted activity.

N. Marcelino, R. Mauersberger, J. Martín-Pintado, J. Cernicharo, C. Thum, E. Roueff, and M. Gerin: A 3 mm line survey of prototype stellar cores.

R. Miura, T. Tosaki, S. K. Okumura, Y. Tamura, R. Kawabe, N. Kuno, T. Sawada, K. Nakanishi, K. Kohno, S. Sakamoto, and T. Hasegawa: Dense clouds and star formation on spiral arm in M33. Deep CO and HCN observation in NGC604.

M. Mollá, E. Hardy, and A. I. Díaz: The H_2 density within spiral and irregular galaxies at high redshift: estimating CO detection limits.

M. Momose, Y. Kitamura, T. Sekiguchi, R. Kawabe, M. Saito, S. Sakamoto, and T. Tsukagoshi: ASTE observations of circumstellar material around nearby pre-main sequence stars.

O. Morata, J. M. Girart, and R. Estalella: Looking for more evidence for transient clumps and gas chemical evolution in dense cores of molecular clouds. Multitransitional observations in HH 43.

R. Moreno and A. Marten: Measurements of zonal winds on Titan with mm interferometry: From IRAM PdBI to ALMA.

M. Mori and M. Umemura: Chemical and dynamical evolution of Lyman alpha emitters.

A. Moullet, E. Lellouch, R. Moreno, and M. A. Gurwell: Io's SO_2 atmosphere: first disk-resolved millimeter observations.

P. P. Papadopoulos, T. R. Greve, P. van der Werf, S. Müehle, K. Isaak, and Y. Gao: A large CO and HCN line survey of Luminous Infrared Galaxies.

G. Petitpas, C. Wilson, A. Peck, D. Iono, K. Sakamoto, A. Baker, M. Krips, S. Matsushita, and M. Yun: Observations of nearby galaxies with the Submillimeter Array.

J. Pety, J. R. Goicoechea, M. Gerin, P. Hily-Blant, D. Teyssier, E. Roueff, E. Habart, and A. Abergel: The Horsehead edge: A reference for PDR models.

I. Prandoni, R. A. Laing, P. Parma, H. R. de Ruiter, F. M. Montenegro-Montes, and T. L. Wilson: CO and dust properties of low luminosity radio galaxies.

G. Quintana-Lacaci, A. Castro-Carrizo, V. Bujarrabal, J. Alcolea, and R. Neri: Structure, kinematics and chemistry of the circumstellar envelopes around the yellow hypergiants AFGL2343 and IRC+10420.

M. A. Reid, A. B. Peck, R. E. Hills, P. G. Anathasubramian, K. G. Isaak, M. Owen, J. S. Richer, H. Smith, A. J. Stirling, R. Williamson, V. Y. Belitsky, R. Booth, M. Hagstrom, L. Helldner, M. Pantaleev, L. E. Pettersson, T. R. Hunter, S. Paine, A. E. T. Schinckel, and K. Young: Testing the ALMA water vapour radiometers at the SMA.

D. A. Riechers, F. Walter, C. L. Carilli, F. Bertoldi, and P. Cox: High-resolution imaging of molecular gas in high-redshift quasar host galaxies.

R. Rolffs, P. Schilke, C. Comito, C. Hieret, and F. Wyrowski: Hot cores in the submm—obscured by dust?

J. Santiago-García, M. Tafalla, and R. Bachiller: Chemical evolution of young bipolar molecular outflows in low-mass protostars.

J. Santiago-García, M. Tafalla, R. Bachiller, and D. Johnstone: Shells and jets in the IRAS04166+2706 outflow.

K. Sorai, N. Kuno, N. Nakai, Y. Watanabe, H. Matsui, and A. Habe: Molecular clouds and star formation in barred spiral galaxies.

R. Soria-Ruiz, J. Alcolea, F. Colomer, V. Bujarrabal, and J. F. Desmurs: SiO maser emisión in oxygen AGB Stars.

S. Toscano, C. Trigilio, G. Umana, C. Buemi, and P. Leto: Flares in binary systems as seen by ALMA.

C. Trigilio, M. E. Palumbo, C. Siringo, and P. Leto: Search for CCO and C_3O in star forming regions.

G. Umana, C. Trigilio, C. Buemi, P. Leto, L. Cerrigone, and P. Manzitto: Studying stellar ejecta with ALMA.

A. Usero, S. García-Burillo, J. Martín-Pintado, A. Fuente, and R. Neri: Large-scale molecular shocks in galaxies.

F. van der Tak, and M. Hogerheijde: Molecular data and radiative transfer needs for ALMA.

T. Velusamy, K. A. Marsh, and K. Grogan: Multiple-sized dust grains in planetary resonances in debris disks: Predictions for ALMA.

L. Verdes-Montenegro, D. Espada, U. Lisenfeld, S. Leon, M. Yun, E. García, K. Nario, and N. Sato: The molecular gas in the most isolated galaxies.

D. Vir Lal, S. Matsushita, J. Lim: Using fast-switching data to characterize atmospheric phase fluctuations at the Submillimeter Array.

The Atacama Large Millimeter/Submillimeter Array: overview & status

Massimo Tarenghi

Originally published in the journal Astrophysics and Space Science, Volume 313, Nos 1–3.
DOI: 10.1007/s10509-007-9602-9 © Springer Science+Business Media B.V. 2007

Abstract The Atacama Large Millimeter/Submillimeter Array (ALMA) is an international millimeter-wavelength radio telescope under construction in the Atacama Desert of northern Chile. ALMA will be situated on a high-altitude site at 5000 m elevation which provides excellent atmospheric transmission over the instrument wavelength range of 0.3 to 3 mm. ALMA will be comprised of two key observing components—a main array of up to sixty-four 12-m diameter antennas arranged in a multiple configurations ranging in size from 0.15 to $\sim$18 km, and a set of four 12-m and twelve 7-m antennas operating in a compact array $\sim$50 m in diameter (known as the Atacama Compact Array, or ACA), providing both interferometric and total-power astronomical information. High-sensitivity dual-polarization 8 GHz-bandwidth spectral-line and continuum measurements between all antennas will be available from two flexible digital correlators.

At the shortest planned wavelength and largest configuration, the angular resolution of ALMA will be $0.005''$. The instrument will use superconducting (SIS) mixers to provide the lowest possible receiver noise contribution, and special-purpose water vapor radiometers to assist in calibration of atmospheric phase distortions. A complex optical fiber network will transmit the digitized astronomical signals from the antennas to the correlators in the Array Operations Site Technical Building, and post-correlation to the lower-altitude Operations Support Facility where the array will be controlled, and initial construction and maintenance of the instrument will occur. ALMA Regional Centers in the US, Europe, Japan and Chile will provide the scientific portals for the use of ALMA; early science observations are expected in 2010, with full operations in 2012.

Keywords Radioastronomy · Array · Millimeter · Submillimeter

1 Introduction

ALMA has been designed to provide sensitive spectra and images in the wavelength range from 0.3 to 3 mm of atomic & molecular gas, nonthermal electrons and thermal dust in our Solar System, the Galaxy, nearby galaxies and high-redshift universe. These data will provide new and unique insights into the formation of galaxies, stars, planets and the chemical precursors necessary for life itself. ALMA will complement 8–10 meter optical/near-IR telescopes such as the Very Large Telescope, Gemini, Subaru and to the Hubble Space Telescope and its successor, the James Webb Space Telescope, with its ability to image dust enshrouded or cold molecular material.

Three exciting new observing capabilities have been used to define the primary technical specifications of ALMA: (1) the ability to detect spectral line emission from rotational spectral lines of the carbon monoxide molecule, atomic and ionized carbon in a galaxy with the properties of the Milky Way at a redshift of $z = 3$ in less than 24 hours of measurement; (2) to image the kinematics of gas in protostars and protoplanetary disks around young solar type stars out to a distance of 500 light years (this represents the distance to the nearby well-known clouds in Ophiuchus, Taurus or Corona Australis); and (3) to provide high-fidelity precise images at an angular resolution better than $0.1''$.

M. Tarenghi (✉)
Atacama Large Millimeter/Submillimeter Array, Joint ALMA Office, Santiago, Chile
e-mail: mtarengh@eso.org

ALMA's flexible design will support:

- Imaging the broadband emission from dust in evolving galaxies at epochs of formation as early as $z = 10$.
- Tracing the chemical composition of star-forming gas in galaxies throughout the history of the universe through measurements of molecular and atomic spectral lines.
- Measuring the motions of obscured galactic nuclei and Quasi-Stellar Objects on spatial scales finer than 300 light years.
- Imaging gas-rich heavily obscured regions that are collapsing to form protostars, protoplanets and pre-planetary disks.
- Measuring the crucial isotopic and chemical gradients within circumstellar shells that reflect the chronology of stellar nuclear processing.
- Producing sub-arcsecond images of cometary nuclei, hundreds of asteroids, Centaur and Kuiper belt objects together with images of planets and their moons.
- Observations of active solar regions to investigate particle acceleration on the suns surface.

2 Overview

ALMA is a partnership between Europe, Japan and North America in cooperation with the Republic of Chile. In Europe it is funded by the European Southern Observatory (ESO), in Japan by the National Institutes of Natural Sciences (NINS) in cooperation with the Academia Sinica in Taiwan, and in North America by the US National Science Foundation (NSF), in cooperation with the National Research Council of Canada (NRC). ALMA construction and operations are carried out on behalf of Europe by ESO, on behalf of Japan by the National Astronomical Observatory of Japan (NAOJ) and on behalf of North America by the National Radio Astronomy Observatory (NRAO), which is managed by Associated Universities, Inc. (AUI). ALMA project development is coordinated by the Joint ALMA Office (JAO), based in Santiago Chile.

ALMA will be built on the Chajnantor altiplano in the Atacama Desert of northern Chile (see Figs. 1–2) at an elevation of slightly over 5000 m. The site is administered by the Chilean Ministry of National Assets and set aside by Presidential decree as a protected region for science. Measurements made since 1995 of the atmospheric trans-

Fig. 1 Location of the ALMA array site near San Pedro de Atacama, northern Chile

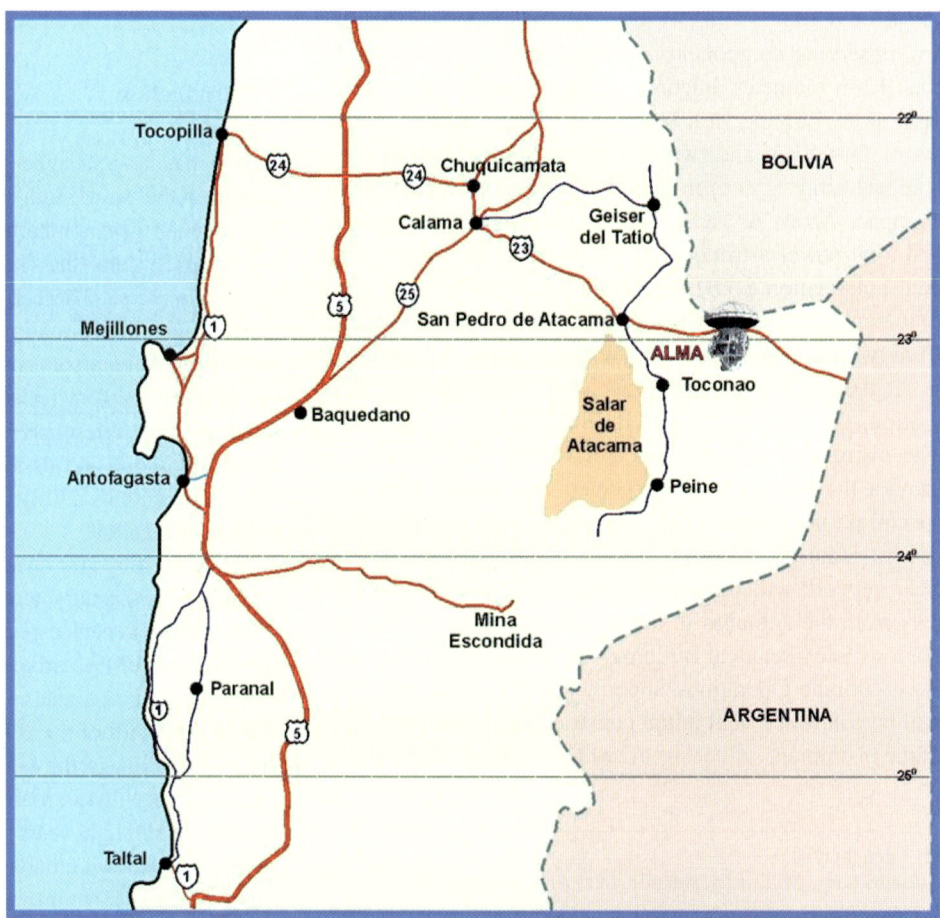

Fig. 2 The ALMA site at 5000 m (*photo: S. Radford*)

Fig. 3 Simulation of ALMA main array plus ACA on the Chajnantor site

parency and stability confirm that the site has superior conditions for millimeter and submillimeter-wavelength astronomy.

The ALMA antennas each have a primary reflecting surface 12 meters in diameter with a parabolic cross-section. The materials used in their construction have been selected to allow the antennas to maintain their performance when fully exposed to the thermal variations and wind gusts imposed by the site environment. Each antenna is fully steerable, and more than 85 percent of the celestial sphere is above the horizon at the Chajnantor site (Fig. 3). The antennas can be moved (reconfigured) among 186 prepared antenna locations (see Fig. 4) to provide a range of spatial

resolutions in the final astronomical images. Each station has a concrete foundation to support the antenna and provision for electrical power and fiber-optic based data communications. The antennas are moved by a pair of specially-designed rubber-tired antenna transporters currently under construction. ALMA will be delivered with a range of antenna configurations forming arrays as small as 150 meters in diameter (for the study of large or low surface brightness objects) and as large as 18.5 km in diameter (for the study of small, high surface brightness objects). The ACA's four 12 m and twelve 7 m antennas will be located in a more compact configuration ~50-m in diameter to allow sensitive wide-field imaging and total power measurements. Three

Astrophys Space Sci (2008) 313: 1–7

Fig. 4 The antenna locations in the center of the main array

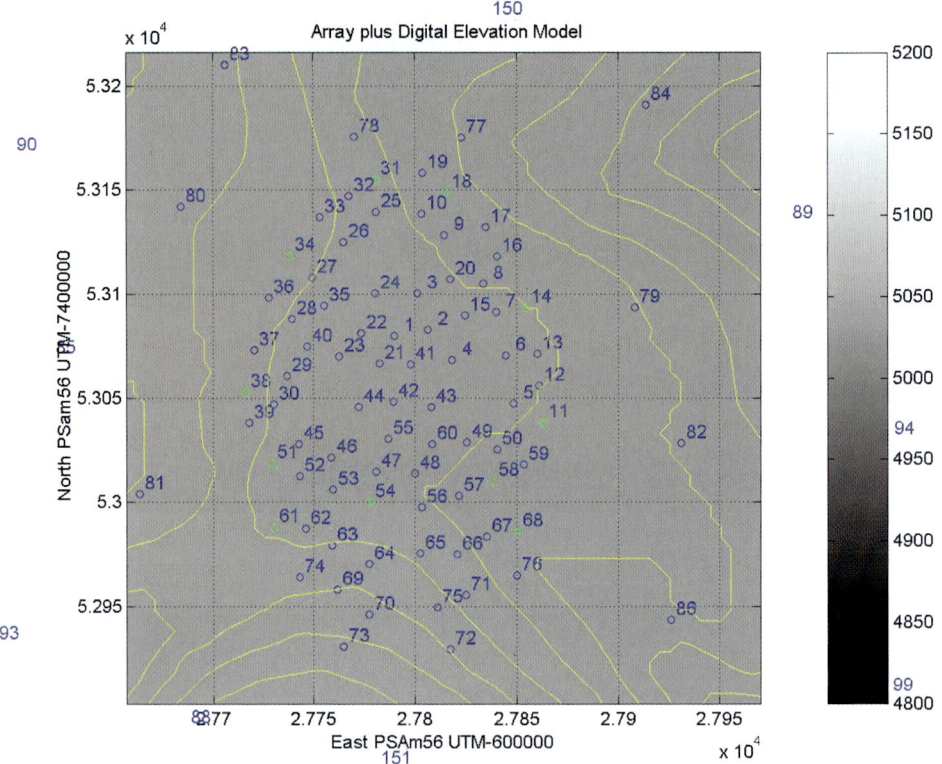

Fig. 5 The Alcatel (*left*) and Vertex (*right*) antenna prototypes, built at the VLA site in New Mexico

different manufacturers are involved in producing antennas for ALMA: in Europe, XXXXX; in the US, Vertex RSI, and Mitsubishi Electric Company in Japan. During 2002–2005 these companies built prototype antennas at the NRAO VLA site in New Mexico (see Fig. 5) to explore the designs and engineering required to meet the demanding ALMA antenna technical specifications.

Each antenna will be equipped with a receiving system (Front End) capable of detecting astronomical signals in six wavelength bands. Two of the receiver bands are to be provided by NAOJ; the design and infrastructure of ALMA will allow the installation of up to ten receiver bands, eventually covering all the millimeter/submillimeter atmospheric transmission windows from 9 mm to 0.3 mm. The ALMA

Fig. 6 The Front End cryostat holding the antenna receivers (*left*), and a Band 9 cartridge (*right*)

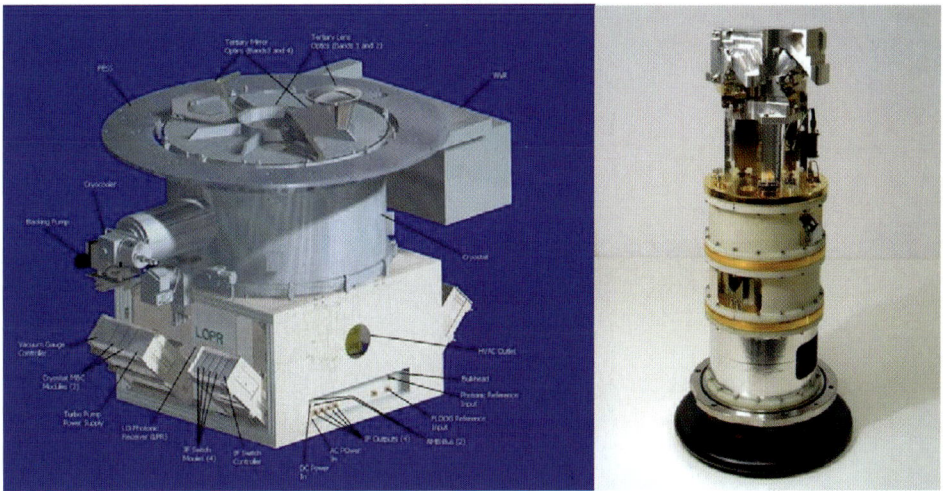

receivers are coherent detectors, meaning they strictly preserve phase across the elements of the array. To achieve this, a common local oscillator signal is distributed to all antennas to convert the received astronomical signal to a much lower intermediate frequency that is transmitted to the high-site technical building, where it is correlated with the signals from all other antennas. Each frequency band receiver cartridge includes two receivers which operate in orthogonal linear polarizations, allowing the complete polarization state of received radiation signals to be measured. All receivers utilize superconducting mixers that operate at temperatures below 4 K. The receivers for each antenna are housed in a common cryogenic dewar located at the Cassegrain focus of each antenna (see Fig. 6).

Also mounted at the Cassegrain focus, but removed from the optical axis of the telescope, is a water vapor radiometer tuned to the 183 GHz line of terrestrial water emission. These devices will be used to correct for the atmospheric phase distortions caused by fluctuations in the amount of water vapor over the site, which would otherwise seriously limit the performance of the array over long baselines.

The received signals are amplified, digitized at the antenna and returned to the control building via fiber optics connections. In order to process the 16 GHz bandwidth IF, digital electronics subdivides that signal into eight 2 GHz sub-bands for transmission to the correlator. Timing signals and reference oscillators synchronize the operation of the antennas and the data collection. Buried power and fiber optic connections link each station to the Technical Building at the Array Operations Site (AOS-TB), which houses the array correlator and support electronics (including the LO system, fiber patch panel and computers) and contains an interim control room for array operations and hardware testing.

The astronomical signals are processed in the AOS-TB (Fig. 7) by a correlator: a special-purpose digital signal processor (see Fig. 8). It combines the digitized IF signals from all the antennas pair-wise and produces a set of complex correlation coefficients (fringe amplitude and phase) as a function of baseline and frequency. Images of the angular distribution of the radio emission from the astronomical source on the sky are created by Fourier inversion of these complex (phase and amplitude) data. For these recycling correlators the product of total bandwidth with number of channels is a constant. For a 2 GHz bandwidth, two polarizations, the correlator provides 128 spectral channels for each 2016 baseline correlation. The finest frequency resolution will be 31 kHz, or 0.1 km s^{-1} at 100 GHz.

To support the construction, maintenance and operation of ALMA, an Operations Support Facility (OSF) is under construction at 3000 m. The OSF provides a pleasant working environment for staff involved in a broad range of activities. Scientific operation of the array will be from a control room at the OSF via a high speed digital link to the AOS-TB. Infrastructure at the OSF will consist of the antenna service building, array control building, electronic laboratories, and office, administrative and residential facilities. The OSF is connected to the AOS by a road constructed to transport the antennas and the operations/maintenance staff.

The ALMA computing system has the task of scheduling observations on the array, controlling all the array instruments, including pointing the antennas, monitoring instrument performance, monitoring environmental parameters, managing the data flow through the electronics and presentation of these data to the correlator. The correlator output must be processed through an image pipeline, where it is calibrated and first-look images produced. Finally, the science data and all associated calibration data, monitor data, and derived data products are archived and made available for network transfer. In full operation, the standard output from ALMA will be calibrated images that have been processed in a standard set of reduction programs linked in

Fig. 7 The center of the area planned for antenna stations; the AOS-TB is visible to the right

Fig. 8 University of Bordeaux high-speed digitizer developed for ALMA

a pipeline. The user will receive these images, together with the correlated data (*uv*-data files), calibration files, and monitor information files. The average data rate is expected to be 6 MB/s, with a peak projected rate an order of magnitude higher. These results will also be stored in a data archive and delivered to the astronomers in a timely manner.

An office in Santiago will house ALMA administrative and local scientific staff. Additional support facilities in North America and Europe (the ALMA Regional Centers, or ARCs) will provide interfaces and user support between the instrument and the regional astronomical communities. Further image processing of the astronomical data will be carried out at the ARCs. A diverse community will use and benefit from ALMA's powerful scientific capabilities, and producing an easy-to-use system for both novices and experts is a system design goal.

Major upcoming milestones for the project include:

First production antenna accepted for start of assembly, integration & verification tasks: 2007

Completion of the AOS-TB and OSF buildings: 2007
Two-antenna interferometry at the OSF: 2008; three-antenna interferometry at the AOS: 2009
Call for Early Science proposals from the community: 2009
Start of full operations: Q3 2012.

ALMA capabilities for observations of spectral line emission

Alwyn Wootten

Originally published in the journal Astrophysics and Space Science, Volume 313, Nos 1–3.
DOI: 10.1007/s10509-007-9520-x © Springer Science+Business Media B.V. 2007

Abstract The Atacama Large Millimeter/submillimeter Array (ALMA) (The Enhanced Atacama Large Millimeter/submillimeter Array (known as ALMA) is an international astronomy facility. ALMA is a partnership between North America, Europe, and Japan/Taiwan, in cooperation with the Republic of Chile, and is funded in Europe by the European Southern Observatory (ESO) and Spain, in North America by the U.S. National Science Foundation (NSF) in cooperation with the National Research Council of Canada (NRC), and in Japan by the National Institutes of Natural Sciences (NINS) in cooperation with the Academia Sinica in Taiwan. ALMA construction and operations are led on behalf of Japan/Taiwan by the National Astronomical Observatory of Japan (NAOJ), on behalf of North America by the National Radio Astronomy Observatory (NRAO), which is managed by Associated Universities, Inc. (AUI), and on behalf of Europe by ESO) combines large collecting area and location on a high dry site to provide it with unparalleled potential for sensitive millimeter/submillimeter spectral line observations. Its wide frequency coverage, superb receivers and flexible spectrometer will ensure that its potential is met. Since the 1999 meeting on ALMA Science (Wootten, ASP Conf. Ser. 235, 2001), the ALMA team has substantially enhanced its capability for line observations. ALMA's sensitivity increased when Japan joined the project, bringing the 16 antennas of the Atacama Compcat Array (ACA), equivalent to eight additional 12 m telescopes. The first four receiver cartridges for the baseline ALMA (Japan's entry has brought two additional bands to ALMA's receiver retinue) have been accepted, with performance above the already-challenging specifications. ALMA's flexibility has increased with the enhancement of the baseline correlator with additional channels and flexibility, and with the addition of a separate correlator for the ACA. As an example of the increased flexibility, ALMA is now capable of multi-spectral-region and multi-resolution modes. With the former, one might observe e.g. four separate transitions anywhere within a 2 GHz band with a high resolution bandwidth. With the latter, one might simultaneously observe with low spectral resolution over a wide bandwidth and with high spectral resolution over a narrow bandwidth; this mode could be useful for observations of pressure-broadened lines with narrow cores, for example. Several science examples illustrate ALMA's potential for transforming millimeter and submillimeter astronomy.

Keywords ALMA · Spectroscopy

1 Introduction: a boom time for IS spectroscopy

Correlator technology has benefitted from huge increases in data processing ability in recent years. The first of the new generation of correlators is already producing a flood of new spectral line data.

At NRAO, the Green Bank Telescope (GBT) combines a collecting an effective area roughly commensurate with ALMA's with powerful correlator capacity. Furthermore, it provides nearly complete frequency coverage below 50 GHz, with initial bolometric array tests at 90 GHz producing promising results. Nine new molecules have been identified in the past two years, including most recently the first negatively charged interstellar molecule, C_6H^- (McCarthy et al. 2006). Enhancement of the Very Large Array

A. Wootten (✉)
National Radio Astronomy Observatory (NRAO),
520 Edgemont Rd., Charlottesville, VA 22903, USA
e-mail: awootten@nrao.edu

Table 1 Summary of ALMA specifications

Parameter	Specification
Number of antennas	>66
Antenna diameter	12 m & 7 m
Antenna surface precision	<25 μm rss
Antenna pointing accuracy	<0$''$.6 rss
Total collecting area	>6900 m^2
Angular resolution	0$''$.015 λ (mm)
Configuration extent	150 m to ~14 km
Correlator bandwidth	16 GHz per baseline
Spectral channels	4096 per IF
Number of IFs	8

(VLA) will provide it with vastly increased correlator capacity within the next few years, as well as nearly complete frequency coverage below 50 GHz. Correlator capacity at IRAM and at CARMA has also recently increased and in the submillimeter regime, the Submillimeter Array is pioneering interferometric imaging in the last accessible atmospheric windows.

ALMA combines total power and interferometric modes of radio imaging. It will provide complete frequency coverage, initially from 84–950 GHz, a pair of flexible and powerful correlators, a high (5000 m) dry southern hemisphere location with a large collecting area (initially 6900 m^2). It will image line emission from celestial objects with about two orders of magnitude more sensitivity than has been available, and with resolution up to two orders of magnitude better than has been provided before.

2 Elements of ALMA

2.1 Science requirements

Annex B of the ALMA Bilateral Agreement set ALMA's highest level science requirements. The highest level science requirements that have determined the ALMA parameters are the ability to: (1) detect spectral line emission from rotational spectral lines of the carbon monoxide molecule, atomic and ionized carbon in a galaxy with the properties of the Milky Way at a redshift of $z = 3$ (de Breuck 2005) in less than 24 hours of measurement, (2) image the kinematics of gas in protostars and protoplanetary disks around young solar type stars out to a distance of 500 light years (Richer 2005; Wootten et al. 2004). This represents the distance to the nearby well-known clouds in Ophiuchus, Taurus or Corona Australis, and (3) provide precise images at an angular resolution better than 0.1$''$. Here "precise" means that the ratio of the most intense to weakest feature in the image can reach 1000. This applies to sources that transit at more than 20° elevation at the ALMA site (Richer 2005).

The key features of ALMA which will allow it to achieve these key science goals are routine milliJansky sensitivity (as a result of the superb site, the receivers, which define the state of the art, and the large collecting area of ALMA) and high resolution (afforded by the long baselines on the extensive site).

2.2 Progress at the superb site

ALMA construction has rapidly progressed. In ALMA labs worldwide prototypes of nearly all elements of ALMA have been tested. These elements are now being brought together at the ALMA Test Facility in New Mexico. There prototype integration of ALMA components into a functioning whole is ongoing. In February 2007, for example, fringes were detected from a transmitted signal external to the antennas; fringes from astronomical sources will soon be assessed.

The site shows excellent submillimeter transparency-atmospheric characterization shows that $\tau(490$ GHz$) \leq 1$ for 70% of the time during the six months encompassing winter. Construction of the infrastructure necessary to support ALMA has reached an advanced state. The 51 foot wide, 43 km ALMA road, passable already at the ALMA groundbreaking on 2003 November 6, has been finished. The 2900 m altitude ALMA Camp sleeps and feeds ALMA personnel in its 32 bed facility while the Contractor Camps bed and feed supervisors and workers (currently numbering more than 350) with offices and recreational facilities. ALMA personnel will move to the future Operations Support Facility, now in construction. The Technical Building at the 5000 m altitude Array Operations Site is complete. John Conway, Mark Holdaway and collaborators have produced a new 186-station design for the ALMA configurations, optimized for staged deployment of up to 64 antennas in addition to the 16 antennas of the Atacama Compact Array (ACA). The construction of the first antenna pads has been finished at the 2900 m facility. Early in 2007 the first production antenna will arrive at the Contractor's camp for assembly before it moves to the project testing area in mid-2007. As characterization of ALMA equipment is completed and that equipment moves to the site for scientific deployment, thoughts turn toward the scientific output of ALMA.

2.3 Receivers

The four first cartridges from partners in Europe and North American have been assembled and tested in the dewar in the Front End Integration Facility at the NRAO Technology Center in Charlottesville. Tests show that all of the pre-production cartridges are exceeding the specifications given in Table 2. For example, for Band 6 (1.3 mm), a receiver temperature of ~40 K has been measured in the lab. This measurement has been verified on the sky. Mixer/preamps

Table 2 Summary of ALMA receivers

Band no.	Frequency range (GHz)	Receiver noise temperature[a] (K)	Mixing scheme	IF Bandwidth
3	84–116	37	2SB	4 GHz
4	125–169	51	2SB	4 GHz
5[b]	163–211	65	2SB	4 GHz
6	211–275	83	2SB	8 GHz
7	275–373	147	2SB	4 GHz
8	385–500	98	2SB	4 GHz
9	602–720	175	DSB	8 GHz
10[b]	787–950	230	DSB	8 GHz

[a]Over 80% of the band, specification. Preproduction units tested to date have been outperforming their specifications

[b]At first light, these bands will be available on fewer than all of the antennas in the array

Table 3 Summary of ALMA line sensitivity

Frequency (GHz)	$B_{max} = 0.2$ km[a] beamsize "	ΔT (K) 1 km s^{-1}	$B_{max} = 14.7$ km beamsize "	ΔT (K) 1 km s^{-1}	ΔT (K) 25 km s^{-1}
110	2.8	0.10	0.038	532	106
140	2.2	0.10	0.030	543	109
230	1.3	0.15	0.018	780	156
345	0.9	0.23	0.012	1240	248
409	0.7	0.34	0.010	1722	344
675	0.4	0.85	0.006	4200	840

[a]For an integration time of 60 seconds, a spectral resolution of 1 km s^{-1} or 25 km s^{-1}, the rms brightness temperature sensitivity ΔT for an array combining all 54 12 m and 12 7 m antennas and a maximum baseline B_{max} is given

[b]The assumed precipitable water vapor (pwv) content varies as a function of frequency. Highest frequency observations are assumed to be carried out during 'best weather' (e.g., lowest pwv) and lower frequency observations during 'worst weather'. This implies higher noise temperatures at mm wavelengths. The assumed pwv values are: pwv = 2.3 mm for $f < 300$ GHz; pwv = 1.2 mm for $300 < f < 500$ GHz; pwv = 0.5 mm for $f > 500$ GHz. Note that pwv = 0.5 mm corresponds approximately to the 25-th percentile of the pwv distribution over time

for Band 6 have been sky-tested at the Sub-millimeter Telescope of the Arizona Radio Observatories on Mt. Graham, Arizona. The results of these tests were quite impressive, with record-breaking, single-sideband system temperatures and exceptional baseline stability over wide IF bandwidths. Typical system temperatures at elevations of about 45 degrees were around 120–140 K SSB with consistent performance across the whole frequency range of the receiver. Image rejection was also excellent. Although the ALMA image rejection specification only required 10 dB, the actual values were typically greater than 20 dB in the LSB and greater than 15 dB in the USB. The first dewar, equipped with the four first cartridges, will be tested at the ALMA Test Facility during late Spring 2007 before being mounted on the first production antenna at the OSF later in the year.

2.4 Correlators

ALMA will have two correlators, one (ALMA Correlator) serving 64 elements comprised of either antennas of the main array of 12-m antennas or an array combined of these antennas with elements of the ACA antenna complement, or another correlator (ACA Correlator) which serves the 16 elements of the ACA. To the observer, the two correlators offer nearly identical functionality and operate in a parallel fashion. The ALMA Correlator offers a great deal of flexibility (Escoffier et al. 2007); the seventy-one supported modes of the full correlator are described in an ALMA Memo (Escoffier et al. 2006). The observer interacts with the correlator through the Observing Tool (OT), software which generates a set of commands which execute the observation. In general, the observer may specify a set of disjoint or overlapping spectral regions, each characterized by bandwidth (31.25 MHz to 2 GHz); each of eight 2 GHz 'baseband' inputs drives 32 tunable digital filters. For each spectral window, the observer also specifies the central or starting frequency, the number of channels (determining spectral resolution; typically 8192 channels are available for a maximum resolution of 3.8 kHz), and the number of polarization prod-

Table 4 Extragalactic gas and dust setup

Line	CO	^{13}CO	C^{18}O	HNCO	Continuum
Transition	$J = 2$–1	$J = 2$–1	$J = 2$–1	$J = 10$–9	
Frequency (GHz)	230.5	220.4	219.6	219.8	4 GHz
Sideband	USB	LSB	LSB	LSB	USB&LSB
Resolution (km s^{-1})	0.64	0.64	0.64	0.64	21.
Quadrant	Q1	Q2	Q2	Q2	Q3&Q4
Window bandwidth	500 MHz	500 MHz	500 MHz	500 MHz	2×2 GHz
Channels	1024	1024	1024	1024	2×128
Spatial resolution	1″, B$_{max}$ = 0.3 km				

ucts. In the ALMA system, the baseband analog outputs of the antennas are digitized in a standard fashion with 3-bits at 4 Gigasamples per second; this is resampled at the correlator input with 2-bit resolution; improved sensitivity options for 4×4 bit correlation or double Nyquist modes is also available. The temporal resolution depends on the mode chosen and may range from 16 msec to 512 msec. Autocorrelation is also available; a 1 ms time resolution can be achieved for autocorrelation data.

The tremendous sensitivity of ALMA combined with its flexible correlators will enable a wide range of science. In the Solar System, for example, venting on small bodies can be studied in detail. Consider Saturn's 500 km diameter moon Enceladus orbiting at a distance of just under 4 Saturn radii with a period of 1.37 days. Among the suprising features of Enceladus discovered by Cassini is the existence of fountains, probably eruptions of subsurface water. The fountains show the spectral signature of ice particles, expected at the -201 C temperature of the surface of the satellite, but hint at the presence of liquid water deep inside the moon. The fountains project about 50 km (about the ALMA beamsize at submillimeter frequencies) from the surface in backlit images of the limb near its south pole. ALMA should be capable of imaging the water in these plumes, as it can easily resolve the moon's disk, providing data on these events long after Cassini ceases its observations. ALMA will also provide time and velocity resolved images of the SO and SO$_2$ molecules emitted from the volcanoes on Io. The ability to simultaneously image the pressure-broadened lines in planetary atmospheres in low resolution while imaging the narrow cores of the lines in high resolution mode will produce good profiles of gas distributions in planetary atmospheres.

Consider an example ALMA project, proposed by David Meier, to examine the gas and dust structure of the nearby star forming galaxy IC342 at 1.3 mm wavelength. A Hubble Wide Field Planetary Camera image, for example, shows the dusty center of the galaxy in an image 2′.7 across. The ALMA primary bean at this wavelength subtends 27″; Nyquist sampling of the complementary ALMA image would require 324 pointings to construct an image cube

of similar extent to the HST image. The ALMA correlator could image $J = 2$–1 lines of ^{12}CO, ^{13}CO and C^{18}O along with, for example, the $J = 10$–9 line of HNCO simultaneously. To attain a sensitivity of 10σ for the lines, one would aim for an rms of 0.06 K, achieved in two minutes per pointing. The entire experiment would take about eleven hours, consisting of multiple mosaics of the source. Simultaneously, continuum observations (through binning channels free of line emission) would reach a sensitivity of 65 μJy. The images would provide arcsecond resolution on the galaxy's dust and gas.

A prototype suite of high-priority ALMA projects that could be carried out in about three years of full ALMA operations has been compiled in the form of the ALMA Design Reference Science Plan (DRSP). The DRSP in comprised of more than ten dozen submissions received from a nearly equal number of astronomers. The current version is publicly available through links at any of the ALMA websites; it is currently being expanded to include projects which exploit the enhancements made possible through Japan joining ALMA.

References

de Breuck, C.: In: Wilson, A. (ed.) Proceedings of the Dusty and Molecular Universe: A Prelude to Herschel and ALMA, Paris, France, 27–29 October 2004, ESA SP-577, p. 27. ISBN 92-9092-855-7. ESA Publications, Noordwijk (2005)

Escoffier, R.P., et al.: Observational modes supported by the ALMA correlator. ALMA Memo No. 566 (2006). http://www.alma.nrao.edu/memos/html-memos/alma556/memo556.pdf

Escoffier, R.P., et al.: Astron. Astrophys. **462**, 801 (2007)

McCarthy, M.C., Gottlieb, C.A., Gupta, H., Thaddeus, P.: Astrophys. J. Lett. **652**, L141 (2006)

Richer, J.: In: Wilson, A. (ed.) Proceedings of the Dusty and Molecular Universe: A Prelude to Herschel and ALMA, Paris, France, 27–29 October 2004, ESA SP-577, p. 33. ISBN 92-9092-855-7. ESA Publications, Noordwijk (2005)

Wootten, A. (ed.): Science with the Atacama Large Millimeter Array. ASP Conf. Ser., vol. 235 (2001)

Wootten, A., Mangum, J.G., Holdaway, M.: In: Debris Disks and the Formation of Planets. ASP Conf. Ser., vol. 324, p. 277 (2004)

ALMA capabilities for observations of continuum emission

T.L. Wilson

Originally published in the journal Astrophysics and Space Science, Volume 313, Nos 1–3.
DOI: 10.1007/s10509-007-9524-6 © Springer Science+Business Media B.V. 2007

Abstract The Atacama Large Millimeter/submillimeter Array, ALMA, combines a large collecting area, very sensitive receivers and a location on a high dry site. ALMA's sensitivity for continuum measurements is increased with the added feature of an 8 GHz instantaneous bandwidth. Taken together, these four factors provide unparalleled sensitivity in the millimeter/submillimeter wavelength range. With its great sensitivity and angular resolution, ALMA will transform our view of mm/sub-mm astronomy.

Keywords ALMA · Continuum

1 Introduction

The Atacama Large Millimeter/submillimeter Array, ALMA, is an equal partnership between Europe and North America, in cooperation with the Republic of Chile. The project is funded in North America by the U.S. National Science Foundation (NSF) in cooperation with the National Research Council of Canada (NRC). In Europe, it is funded by the European Southern Observatory (ESO) and Spain. In the bilateral project, ALMA construction and operations are led on behalf of North America by the National Radio Astronomy Observatory (NRAO) which is managed by Associated Universities, Inc. (AUI), and on behalf of Europe by ESO.

A rebaselining process was carried out in 2005–2006. As a result, the North America-Europe Atacama Large Millime-ter/submillimeter Array, ALMA, the bilateral array is specified to have fifty 12 meter antennas, with a goal of sixty-four such antennas. Japan has also entered the ALMA construction project. Among other contributions, Japan is bringing: (1) twelve 7 meter antennas and four 12 meter antennas for measuring total power continuum in the Atacama Compact Array (ACA) and (2) two additional receiver bands, with a third band at 0.3 mm, to be built after R&D work is completed. ALMA combines a large collecting area, very sensitive receivers and a location on a high dry site.

2 High level science goals related to continuum measurements

One of the high level science goals of ALMA is the production of high quality images. Another high level science goal is the detection of a circumstellar dust disk around a solar mass star at the distance to the nearest molecular cloud, 140 parsecs (Richer 2005; Wootten et al. 2004).

3 Response to extended emission

For any interferometer, there is a 'short spacings' problem. That is, for ALMA (and all multiplying interferometers) the largest structure recorded is proportional to the closest spacing of two antennas. This value is given in the last 2 columns of Table 1. For ALMA there are also four 12 meter antennas which are used to record the total flux density of a source. With the 7 meter antenna array and the four 12 meter antennas of the Atacama Compact Array, ACA, extended structures can be measured. For continuum measurements, the four 12 meter antennas of the ACA are equipped with wobbling secondaries. Use of the ACA increases the sensitivity

T.L. Wilson (✉)
European Southern Observatory, K-Schwarzschild-Str. 2,
85748 Garching, Germany
e-mail: twilson@eso.org

Table 1 Sampling of spatial scales

Wavelength (mm)	FWHP primary beam		Minimum λ/D	
	(arcmin)	(arcmin)[a]	(arcmin)	(arcmin)[a]
2.72	56	99	37	64
1.30	27	46	18	31
0.87	18	31	12	21

[a]This is the largest structure to which the ACA 7 meter antennas are sensitive

Table 2 ALMA continuum sensitivity[a]

Frequency (GHz)	Wavelength (mm)	ΔS_ν (mJy)	ΔT_B (K)	θ (″)
110	2.72[c]	0.06	0.02	0.60
230	1.30[c]	0.13	0.04	0.16
345	0.87[c]	0.32	0.10	0.15
409	0.73[c]	0.42	0.13	0.09
675	0.44[c]	3.0	0.98	0.09
675	0.44[d]	0.5	0.16	0.09
850	0.35[c]	6.2	2.0	0.07
850	0.35[d]	0.9	0.3	0.07

[a]RMS noise for 2 polarizations each with 8 GHz bandwidth, elevation 50°, with a precipital water vapor column given in footnotes c and d. The brightness temperature (T_B) uncertainties are for a baseline of 1 km, with fifty 12 meter antennas

[b]For a few frequencies in selected receiver bands

[c]For PWV = 1.5 mm which is median for the ALMA site

[d]For PWV = 0.2 mm which is the best 5% value for the ALMA site

of ALMA, but the main use of the ACA is to provide information about extended emission.

3.1 Continuum sensitivity

In Table 2, we give the sensitivities for a selection of ALMA bands. These have been chosen to illustrate the response of ALMA at a number of different wavelengths. As noted, this sensitivity is provided by the state of the art receivers, which have the best receiver noise temperatures reached to date. These receivers also have 8 GHz wide bandwidths, which is crucial for the sensitivity.

3.2 Science and sensitivity

The flux densities of dust grains are rather small at 100 GHz but increase rapidly with frequency. Thus ALMA will have a unique capability to detect and image a wide range of objects. Except for black body emission from planets, meter and centimeter wavelength range emission is dominated by the synchrotron or free-free emission. In the millimeter, sub-mm and far infrared range, broadband continuum emission from interstellar sources is caused by radiation from dust particles. The flux densities increase with frequency ν as $\nu^{2+\beta}$, where β is in the range 1 to 2. For sources with well determined flux densities, synchrotron radiation can be separated from dust continuum emission. For a given dust to gas ratio, the continuum intensity from dust is proportional to the product of T_D and $N(H)$, where $N(H)$ is the column density of protons and T_D the dust temperature.

A simulation by another first priority science goal of ALMA is measurements of the Sun. However these observations require additional equipment to attenuate the high level of solar emission. In addition to the solar system, the photospheres of O and B stars emit with spectral indices that are close to ν^2. For later type stars with warm dusty envelopes, one can image the dust continuum as a function of distance from the star. A few examples are high redshift sources, galaxies, molecular clouds, planets (normal or minor) or asteroids. Because of the *inverse k correction*, ALMA can detect active star forming galaxies such as M82 even at redshifts of $z = 12$ (de Breuck 2005).

References

de Breuck, C.: In: Wilson, A. (ed.) Proceedings of the Dusty and Molecular Universe: a Prelude to Herschel and ALMA, Paris, France, 27–29 October 2004, p. 27, ISBN 92-9092-855-7. ESA Publications, Noordwijk (2005), ESA SP-577

Richer, J.: In: Wilson, A. (ed.) Proceedings of the Dusty and Molecular Universe: a Prelude to Herschel and ALMA, Paris, France, 27–29 October 2004, p. 33, ISBN 92-9092-855-7. ESA Publications, Noordwijk (2005), ESA SP-577

Wootten, A., Mangum, J.G., Holdaway, M.: Debris disks and the formation of planets. ASP Conf. Ser. **324**, 277 (2004)

Star and planet-formation with ALMA: an overview

Ewine F. van Dishoeck · Jes K. Jørgensen

Originally published in the journal Astrophysics and Space Science, Volume 313, Nos 1–3.
DOI: 10.1007/s10509-007-9600-y © Springer Science+Business Media B.V. 2007

Abstract Submillimeter observations with ALMA will be the essential next step in our understanding of how stars and planets form. Key projects range from detailed imaging of the collapse of pre-stellar cores and measuring the accretion rate of matter onto deeply embedded protostars, to unravelling the chemistry and dynamics of high-mass star-forming clusters and high-spatial resolution studies of protoplanetary disks down to the 1 AU scale.

Keywords Star formation · Protoplanetary disks

1 Introduction

The formation of stars and planets occurs deep inside clouds and disks of gas and dust with hundreds of magnitudes of extinction, and can therefore only be studied at long wavelengths. In the standard scenario for the formation of an isolated low-mass star, a cold core contracts as magnetic and turbulent support are lost and subsequently collapses from the inside out to form a protostar with a surrounding disk. Soon after formation, a stellar wind breaks out along the rotational axis of the system and drives a bipolar outflow entraining surrounding cloud material. The outflow gradually disperses the protostellar envelope, revealing an optically visible pre-main sequence star with a disk. Inside this disk,

grains collide and stick owing to the high densities, leading to pebbles, rocks and eventually planetesimals which interact to form planets. The original interstellar gas and dust is gradually lost from the disk through a combination of processes, including accretion onto the new star, formation of gas-rich planets, photoevaporation and stellar winds.

These different evolutionary stages in star- and planet formation are traditionally linked to their Spectral Energy Distributions (SEDs) (Lada 1999), which illustrate how the bulk of the luminosity shifts from far- to near-infrared wavelengths as matter moves from envelope to disk to star. So far, most tests of this scenario have been done using spatially unresolved data which encompass the entire star-disk-envelope system in a single beam. ALMA will be the first telescope capable of spatially and spectrally resolving the individual components and tracing the key physical and chemical processes on all scales.

The strengths of ALMA are (i) its high angular resolution, combined with enough sensitivity to image continuum and lines down to $0.01''$ (~ 1 AU = terrestrial planet-forming zone at 150 pc, 30 AU = disk of high-mass YSO at 3 kpc); (ii) its high spectral resolution down to 0.01 km s^{-1} so that the details of the dynamics and kinematics can be probed; (iii) access to thousands of lines from hundreds of species allowing a wide variety of physical and chemical regimes to be probed; and (iv) its ability to detect optically thin dust emission and thus directly derive dust masses.

ALMA probes the wavelength range of 0.3–9 millimeter, which is on the Rayleigh–Jeans tail of the SEDs of young stellar objects. For a complete picture of these sources complementary space and ground-based observations at shorter wavelengths are necessary. Compared with other missions, ALMA is less well suited for large area surveys because of its small field of view. Mid-infrared observatories such as the *Spitzer Space Telescope* at 3–70 µm probe the peak

E.F. van Dishoeck (✉)
Leiden Observatory, P.O. Box 9513, 2300 RA Leiden,
The Netherlands
e-mail: ewine@strw.leidenuniv.nl

J.K. Jørgensen
Harvard-Smithsonian Center for Astrophysics, 60 Garden Street
MS 42, Cambridge, MA 02138, USA

of the SED for low-mass YSOs and can scan large areas much more rapidly, albeit at lower spatial and spectral resolutions. Near-infrared imaging is a powerful tool to characterize the stellar component, whereas the *Herschel Space Observatory* and ground-based single-dish submillimeter telescopes equipped with large format bolometers can rapidly search large areas for cold dust emission. These missions will provide complete unbiased catalogs with thousands of sources covering all the nearby molecular clouds and star-forming regions within a few hundred pc (contained within Gould's Belt), many of the young clusters and associations within 1 kpc, and most of the prominent high-mass star-forming clouds to the outer edge of the Galaxy. Thus, the primary source lists for ALMA will come from these missions. They will also provide the main statistical results from which, for example, timescales for the different phases can be derived.

This paper outlines a number of key questions for each evolutionary state where ALMA can make a major contribution. It focusses mostly on low-mass star formation and protoplanetary disks, but many of the same arguments are also valid for high-mass star formation. Inspiration for this review was provided by the many beautiful paintings by Juan Miró displayed in Madrid and elsewhere around the world. Most appropriate for this topic are 'Birth of the world', 'Chiffres and constellations', 'Red disk' and 'Serpent looking at comet'. A challenge for the reader is to find the relations between these paintings and the topics described here.

2 Low-mass star formation

2.1 Pre-stellar cores

Question 1 What are the initial conditions for low-mass star formation, in particular the physical structure and kinematics of the densest part of the core?

In recent years, a number of cold, highly extincted clouds have been identified which have a clear central density condensation. These so-called pre-stellar cores are believed to be on the verge of collapse and thus represent the earliest stage in the star-formation process (e.g., Tafalla et al. 1998). The physical and chemical state of these clouds is now well established on scales of few thousand AU by single dish millimeter observations combined with extinction maps. The cores are cold, with temperatures varying from 10–15 K at the edge to as low as 7–8 K at the center, and have density profiles that are well described by Bonnor–Ebert profiles. It is now widely accepted that most molecules are highly depleted in the inner denser parts of these cores (Caselli et al. 1999; Bergin et al. 2002): images of clouds such as B68 show only a ring of $C^{18}O$ emission, with more than 90% frozen out toward its center.

ALMA will be particularly powerful in probing the central part of the core on scales of 100 AU and search for signs of collapse in the very earliest stages. Important probes are the lines of N_2H^+ and H_2D^+ at 372 GHz, with the latter line a unique probe of the kinematics in regions where all heavy molecules are depleted (van der Tak et al. 2005).

2.2 Very low luminosity objects and formation of brown dwarfs

Question 2 What prevents some clouds from collapsing? Why do some low-luminosity sources have such a low accretion rate in spite of the much larger reservoir of gas and dust?

About 75% of so-called 'starless' cores (i.e., dark cores with no IRAS source) remain starless down to $0.01L_\odot$ or less even after deep surveys with *Spitzer* (Kirk et al. 2007). However, *Spitzer* has revealed a small set of cores with so-called Very Low Luminosity Objects (VeLLOs). Examples include L1014 ($\sim 0.1L_\odot$, Young et al. 2004), L1521F ($\sim 0.05L_\odot$, Crapsi et al. 2005; Bourke et al. 2006), and IRAM 04191 ($\sim 0.08L_\odot$, Dunham et al. 2006). These VeLLOs are embedded in cores with typical masses of $1M_\odot$, but their low luminosities suggest that their central stellar masses are low and that they (currently) have low accretion rates. They also show very different outflow properties ranging from a large well developed outflow in IRAM 04191 to a miniscule outflow in L1014 only detectable through high angular resolution millimeter observations (Fig. 1). This suggests that accretion in these cores may be episodic. It remains an interesting question whether these VeLLOs constitute a separate stage in the evolution of low-mass protostars or are precursors of substellar objects, but without a better handle on their dynamical structure it is difficult to predict the "end result" of the ongoing star formation in these cores. ALMA will be able to zoom in on these sources, image their disks (whose presence is inferred from the SEDs) and small scale outflows and furthermore constrain the kinematics of their envelopes.

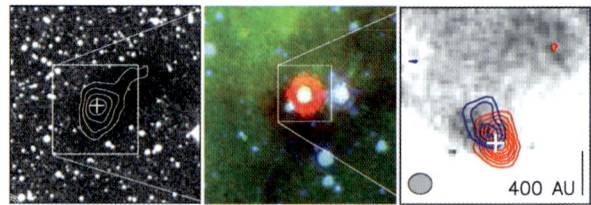

Fig. 1 The VeLLO L1014-IRS. *Left*: optical image with 1.2 mm dust continuum emission overlayed. *Middle*: Spitzer mid-infrared image with 4.5 μm (*blue*), 8.0 μm (*green*) and 24 μm (*red*). *Right*: CO 2–1 map of the innermost region of the core from the SubMillimeter Array (SMA). Images from Young et al. (2004) (*left, middle*) and Bourke et al. (2005) (*right*)

Question 3 How do brown dwarfs form? Like stars or like planets?

The VeLLOs described above could be the precursors of substellar objects like brown dwarfs. If so, the presence of outflows and accretion disks imply a formation process similar to that of stars. Formation in, or fragmentation of, a disk around a more massive primary star is not consistent with these data. Many other lines of evidence based on infrared imaging and spectroscopy point in the same direction (e.g., Testi et al. 2002; Mohanty et al. 2004). Large samples of young brown dwarfs now exist, several of them in wide binaries which are easily disrupted, providing further clues on their origin (see for a review Luhman et al. 2007). ALMA will be able to determine whether brown dwarfs in the earliest stages have similar statistics in terms of binary fraction and separation.

2.3 Formation of stellar clusters and origin of the IMF

Question 4 What is the relation between cloud or clump structure and the IMF?

The advent of large bolometer arrays on submillimeter telescopes has revived detailed studies of the structure of molecular clouds and cores just prior to and during the star formation process. On large pc-size scales, molecular clouds have a highly inhomogeneous or 'clumpy' structure, from which the mass distribution $\Delta N/\Delta M$ can be measured (see Williams et al. 2000, for a review). Interestingly, the mass spectrum of cores follows a law $\propto M^{-1.5}$ below $0.5 M_\odot$ and $\propto M^{-2.5}$ above $0.5 M_\odot$ (Alves et al. 2007), similar to the stellar initial mass function (IMF). This suggests that the IMF may already be determined at the pre-stellar stage during the fragmentation of a (turbulent) molecular cloud, a result with wide-ranging implications for studying the evolution of molecular clouds in our Galaxy and galaxies as a whole. The sensitivity and spatial resolution of ALMA are needed, however, to separate the lower density cloud material from the dense cores and to link this work with the optical and infrared determinations of the low-mass end of the IMF in young clusters. For example, the mass spectrum is still uncertain for clump masses below $0.1 M_\odot$, whereas ALMA can probe the mass function down to planetary masses and study the origin and distribution of brown dwarfs and free-floating Jupiter-mass exo-planets using also kinematic information.

Question 5 What fraction of stars forms as binaries or multiples? Are all cluster members co-eval? Is there evidence for dynamical interactions?

Related to Question 4 is the question whether cloud fragmentation leads to a young cluster or to distributed star formation. In contrast with previous claims, *Spitzer* finds evidence for both processes, including a significant fraction of young stars distributed throughout the clouds (e.g., Allen et al. 2007). The formation of the more massive members of a cluster through competitive accretion is also heavily debated (e.g., Bate and Bonnell 2005; Krumholz et al. 2005). ALMA can address this question by determining masses of YSOs in the earliest, deeply embedded stages when accretion is still taking place. The fraction of binaries and multiples can be compared with those of optically visible T Tauri stars and field main-sequence stars, providing clues on binary evolution and dynamical interactions (e.g., ejection).

Spitzer data show a wide variety of SEDs of clusters of YSOs on 0.1 pc scales (e.g., Rebull et al. 2007; Jørgensen et al. 2006). What causes this diversity? The normal assumption is that the different SEDs (rising, falling) reflect a real age spread from <0.1 to >1 Myr as part of a more-or-less uniform star formation process. However, on scales as small as 0.1 pc it is also reasonable to assume that all objects form quasi-simultaneously from a fragmenting core. In such a coeval scenario, the diversity in SEDs would imply that objects go through the evolutionary states at different rates. ALMA and mid-infrared data will be needed to settle this.

2.4 Embedded YSOs: infall vs. outflow

Question 6 What are the accretion rates during the earliest stages of star formation and how do they vary with time?

Deeply embedded young stellar objects (the so-called 'Class 0' objects) have a complex physical and kinematical structure, with envelopes, disks and outflows all blurred together in current single-dish observations. High spatial and spectral resolution ALMA data will be essential to disentangle the infall, outflow and rotation components of these systems and study their evolution. Of particular importance will be to measure directly the accretion rates onto the disk and star. Redshifted absorption against the continuum is thought to be the most direct tracer (Fig. 2) but this absorption is completely overwhelmed by large-scale emission in single-dish data. Interferometer data reveal such red-shifted absorption in a few (but not all) Class 0 sources but only in low-density tracers perhaps indicative of large-scale infall rather than small-scale accretion (e.g., Jørgensen et al. 2007; Di Francesco et al. 2001). ALMA will have orders of magnitude higher sensitivity to infall tracers due to the combination of larger collecting area and smaller beam.

Question 7 What drives outflows and how does the outflow structure change with time?

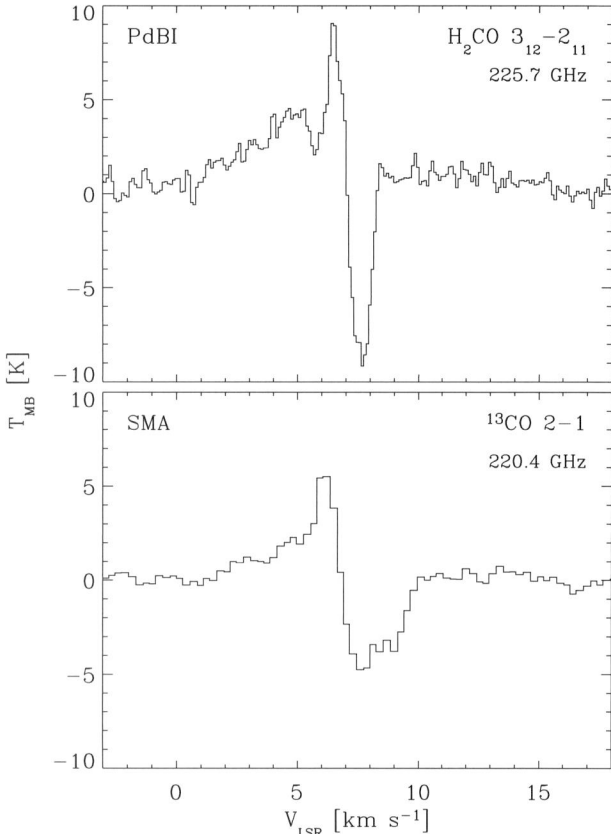

Fig. 2 Inverse P Cygni profiles toward the Class 0 YSO NGC1333-IRAS4A in lines of H_2CO from IRAM Plateau de Bure observations (Di Francesco et al. 2001) and ^{13}CO from SMA observations (Jørgensen et al. 2007)

Violent outflows are a key characteristic of star formation (Richer et al. 2000). Although they have been studied with single-dish telescopes for more than 25 years, the mechanism of their formation remains poorly understood (Shang et al. 2007). In the most deeply embedded objects, highly collimated jet-like molecular outflows with extreme velocities up to 200 km s^{-1} are observed, but when the protostar evolves, both the mechanical power and the collimation seem to decrease, suggesting that the former is due to a decline in the overall mass accretion rate (e.g., Bachiller and Tafalla 1999; Arce and Sargent 2006) (Fig. 3). ALMA will permit studies of many different kinds of YSOs, and, combined with independent estimates of the mass accretion rate (see above), test magnetohydrodynamical models of their evolution. The actual location and mechanism by which outflows are launched, and whether they are episodic in nature, is still a subject of intense debate. ALMA will provide detailed images of the disk/outflow interface down to a few AU scales where the outflow is accelerated and where the most intense interactions between the outflow and its surroundings take place.

Question 8 What is the role of outflows in determining the final mass of a star and in dispersing the core?

Comparisons between cloud core and stellar mass functions show great similarity in shape (see Sect. 2.3) but with a shift in mass of typically a factor of $\sim$3 (Alves et al. 2007). Is this largely due to the action of outflows dispersing the cores? Also, what is the role of outflows in carrying off angular momentum? Attempts to measure the mass dispersion rate in embedded YSOs have traditionally suffered from poor spatial resolution. Interferometer data are needed to measure the local specific angular momentum as a function of radius in the protostellar envelope, from the large-scale core to the rotationally supported disk-like structures (e.g., Hogerheijde 2001; Takahashi et al. 2006).

3 High-mass star formation

Question 9 Is high-mass star formation a scaled-up version of low-mass star formation? What triggers high-mass star formation?

High-mass stars ($\gtrsim 10M_\odot$, $\gtrsim 10^4 L_\odot$) play a major role in the interstellar energy budget and the shaping of the Galactic environment (e.g., Cesaroni 2005). Phenomena associated with massive stars such as photoionization, powerful winds, shocks, expanding H II regions and supernovae drastically modify the interstellar medium. Due to large distances, short time scales, and heavy extinction, the formation of high mass stars is still poorly understood compared to that of their lower mass counterparts. The earliest stages of massive star formation have been revealed as dark clouds seen in absorption against mid-infrared emission (the so-called 'Infrared Dark Clouds') (e.g., Simon et al. 2006) and systematic surveys are attempting to put the various observational signposts in an evolutionary sequence starting with centrally condensed clouds with masers (so-called High-Mass Protostellar Objects), followed by hot cores with high temperature (>100 K) and high abundances complex organic molecules (Hot Molecular Cores), and subsequently ultra-compact H II regions showing significant amounts of ionized gas.

The similarity of some of these stages with those of their lower-mass counterparts, coupled with the detection of outflows and signs of rotating disks (e.g., Shepherd and Churchwell 1996; Shepherd and Kurtz 1999), suggest a common formation mechanism for all stars. However, while these similarities may hold for young B-type stars, the situation is much less clear for O-type stars and the debate between the turbulent cloud fragmentation scenario and that of competitive accretion or mergers is still far from settled (McKee and Tan 2003; Krumholz et al. 2005; Bonnell and Bate 2006). A related question is whether the formation of massive stars

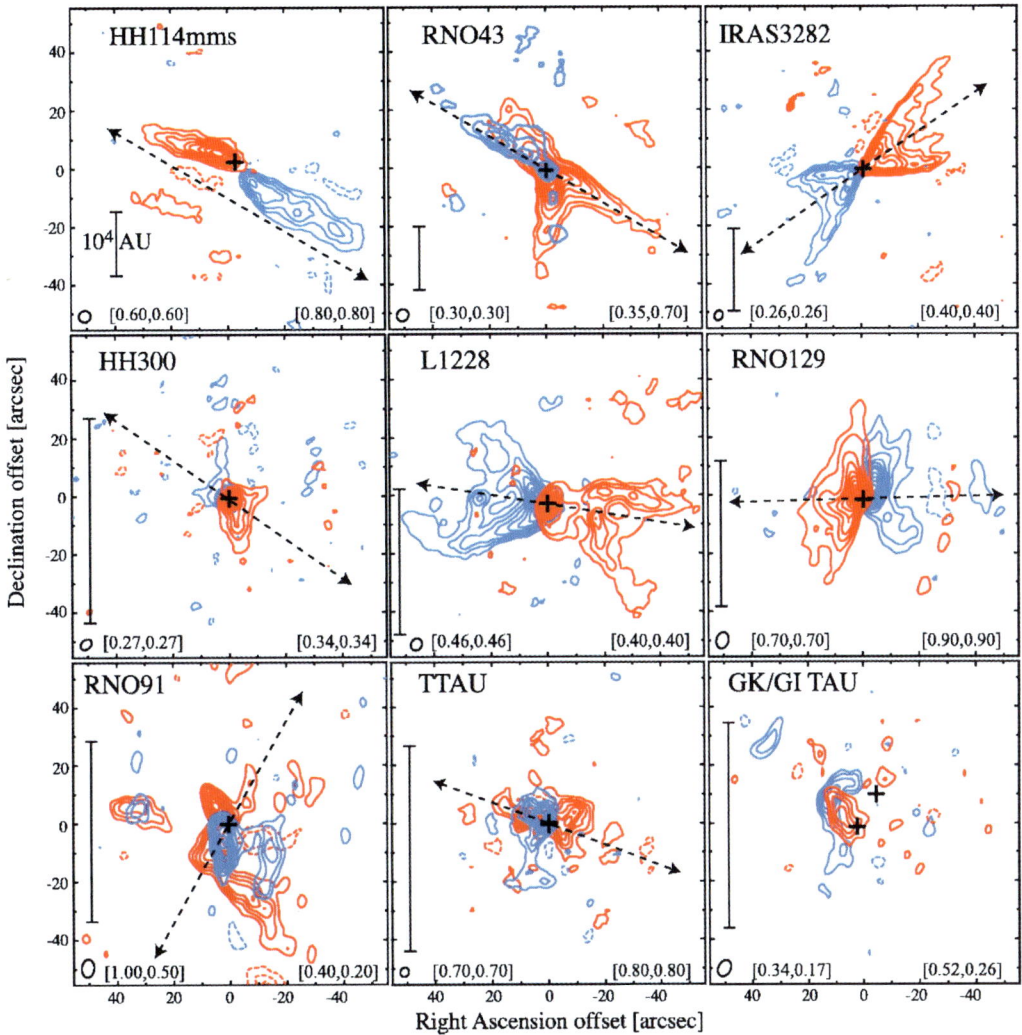

Fig. 3 Gallery of ^{12}CO 1–0 outflows from the OVRO survey of Arce and Sargent (2006) with Class 0 objects in the *top*, Class I in the *middle* and Class II in the *bottom* panels. In each panel the cross mark the position of the protostar given by millimeter continuum observations

is largely triggered, either by shocks compressing the cloud material ('collect and collapse') or by 'radiation-driven implosion'. Clear observational evidence for triggered star formation is still lacking. Examination of cluster properties and separating the more diffuse clump and denser core material will be particularly relevant to determine which process dominates in which environment. Contrasting star formation in the outer Galaxy, where metallicity, densities, radiation field and gravitational potential well are lower compared with the inner Galaxy may also be revealing.

4 Astrochemistry

Question 10 Which molecule is best suited to trace which physical component of the YSO environment? Can we use chemistry as a 'clock' of the evolutionary state of the object?

Systematic studies of molecules in YSOs are starting to reveal the different chemical characteristics associated with the various stages star formation (e.g., Jørgensen et al. 2004; Ceccarelli et al. 2007; van Dishoeck 2006). The coldest pre-stellar cores show heavy freeze-out of virtually all gas-phase molecules onto the cold grains (see Sect. 2.1), where grain-surface chemistry can lead to more complex species. Once the protostar starts to heat the envelope, the ices will evaporate in a sequence according to their sublimation temperatures, with the most volatile species like CO coming off at temperatures as low as 20 K and the most strongly bound species like H_2O around 100 K. The impact of outflows on the inner envelope can also liberate molecules from the ices and sputter grain cores. The evaporated ices subsequently drive a rapid gas-phase chemistry for a period of 10^4–10^5 yr. For example, reactions with evaporated CH_3OH are thought to lead to high abundances of CH_3OCH_3 and $HCOOCH_3$, although a grain surface origin of these molecules is also possible (e.g., Bisschop et al. 2007; Bottinelli et al. 2007). These so-called 'hot cores' are signposts of the earliest stages of high-mass star formation, and are now also found around some low-mass YSOs. After $\sim 10^5$ yr, the abundances are reset by ion-molecule chemistry to their normal cloud values, leading to the potential of molecules to act as chemical clocks.

So far, this scenario is almost entirely based on spatially unresolved single-dish data, with smaller-scale structure extracted from observations of multiple lines of the same molecule with different excitation conditions. Limited interferometer data confirm the different chemical zones (e.g., Jørgensen 2004), but only ALMA will have the combined sensitivity, spatial resolution and (u, v) coverage to make chemical 'images' of YSOs in large sets of lines necessary to directly test chemical models and explore and develop the use of molecules as clocks of star formation.

Question 11 How far does chemical complexity go? Can we find (the building blocks of) pre-biotic molecules?

Of the >130 different molecules detected in interstellar clouds, the majority ($\sim 75\%$) are organics, including species as complex as ethyl-cyanide (C_2H_5CN), acetamide (CH_3CONH_2, the largest interstellar molecule with a peptide bond) and glycol-aldehyde (CH_2OHCHO, the first interstellar sugar) (e.g., Hollis et al. 2000, 2006). However, in spite of literature claims, the simplest amino-acid glycine (NH_2CH_2COOH) has not yet been convincingly detected. Current instrumentation prevents deep searches for more complex molecules for several reasons: (a) the regions of high chemical complexity are often very small ($<1''$), the typical sizes of hot cores; (b) the crowding of lines is usually so high that the confusion limit is reached; and (c) the largest molecules have many close-lying energy levels so that the intensity is spread over many different lines, each of them too weak to detect. ALMA will be able to push the searches for prebiotic molecules two orders of magnitude deeper to abundances of $<10^{-13}$ with respect to H_2, because it will have a much higher sensitivity to compact emission and will resolve the sources so that spatial information can be used to aid identifications of lines.

Question 12 Which fraction of complex molecules will end up unaltered in the protoplanetary disk? How are they modified before incorporation into planetary systems?

The dynamics of gas in the inner few hundred AU of protostellar envelopes are not yet well understood, but are important to determine whether some of the observed (complex) molecules end up in the rotating disk. Also, the disk entry point is highly relevant, since gas falling in too close to the star will experience such a strong accretion shock onto the disk that all molecules will dissociate (Neufeld and Hollenbach 1994) and only molecules entering at much larger distances survive. ALMA's high spatial and kinematic resolution will obviously be needed to address this issue.

Once in the disk, the chemistry is governed by similar gas-phase and gas-grain interactions as in envelopes, but at higher densities. Also, UV and X-rays from the young star dissociate molecules and modify the chemistry in the optically thin surface layers. This results in a layered chemical structure, with a top layer consisting mostly of atoms, a mid-plane layer where most molecules are frozen out, and an intermediate layer where the dust grains are warm enough to prevent complete freeze-out and where molecules are sufficiently shielded from radiation to survive (for a review, see Bergin et al. 2007). So far, chemical images have been limited to just a few pixels across a handful of disks in a few lines (e.g., Qi et al. 2003; Piétu et al. 2007). Obviously, ALMA will throw this field wide open. A particularly exciting topic is the chemistry in the inner disk, i.e., inside the 'snow-line' where all molecules evaporate and the chemistry approaches that at LTE. For example, *Spitzer* data have revealed highly abundant and hot HCN in the inner disk (Lahuis et al. 2006). The brightness temperatures of the submillimeter lines are predicted to be several hundred K, sufficient for ALMA to image the inner few AU in the nearest disks.

5 Protoplanetary disks

5.1 Young disks in the embedded phase

Question 13 How do disks form and grow with time? How hot or cold is the disk? Can we find evidence for gravitational instabilities?

The study of disk formation in the earliest stages will be a central scientific goal of ALMA. Key questions include whether most of the disk mass is already assembled in the earliest Class 0 phase (e.g., Fig. 4), or whether disk growth—and thus stellar growth—continues in the later stages. Also, the dynamics of disk formation and how this depends on initial core parameters (e.g., core rotation) are unclear. Through high spatial resolution kinematic data, ALMA can image the Keplerian motions of the gas and thus obtain direct estimates of stellar mass. Tracing M_{env}/M_{disk} and M_{disk}/M_* for a wide variety of stellar types as a function of evolution and testing scenarios such as that shown in Fig. 5 will be a major legacy of ALMA. Multi-line observations can determine the gas temperature and the amount of heating through accretion shocks vs. UV radiation. Finally, images of young disks can reveal asymmetries or spiral arms (e.g., Lin et al. 2006), indicative of gravitational instabilities which may lead to giant planet formation (Boss 2003).

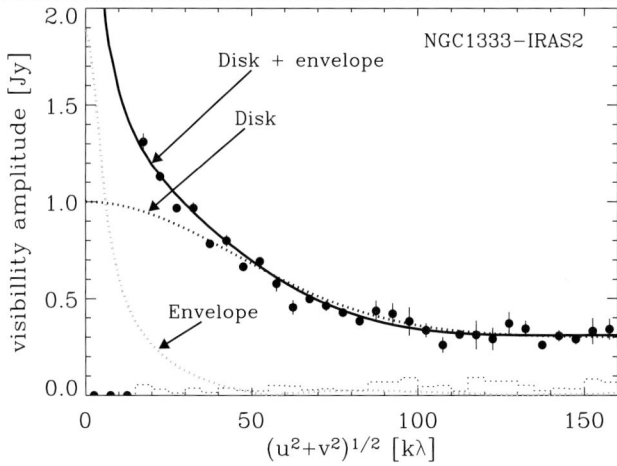

Fig. 4 Interferometric 850 μm continuum observations of the NGC 1333-IRAS2A Class 0 YSO from the SMA (Jørgensen et al. 2005). The data show the presence of a compact source of emission not accounted for by the larger scale envelope models. Rather, the data can be well-fit by a combination of the extended envelope and a 300 AU circumstellar disk

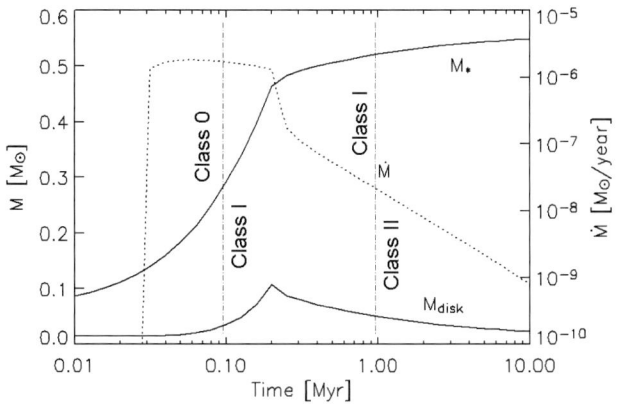

Fig. 5 Model for the evolution of disk and stellar masses (*solid lines*) and accretion rate from the disk onto the star from the Class 0 through Class II stages of young stellar objects. Figure from Dullemond et al. (2007) after Hueso and Guillot (2005)

5.2 Disks around pre-main sequence stars

Question 14 What is the physical and chemical structure of the inner planet-forming zones of disks?

Large gas disks with masses of $\sim 10^{-2} M_\odot$ have been revealed around classical T Tauri and Herbig Ae stars with ages of a few Myr by millimeter continuum and CO line emission, providing constraints on masses and sizes of disks, their velocity patterns and even the level of turbulence (e.g., Koerner and Sargent 1995; Dutrey et al. 1996; Andrews and Williams 2007; Qi et al. 2003). However, these data only probe the outer (>50 AU) disk region. ALMA will have the sensitivity to detect and image all the dust in the disk down to the 1 AU scale. Through multifrequency ob-

servations, ALMA will also be able to measure the change of dust properties within disks, perhaps showing direct evidence for radius-dependent grain growth in the midplane, up to sizes of several mm. ALMA will have the sensitivity to map optically thick lines at a few AU resolution, providing information about the gas content and its chemistry and kinematics down to the planet-forming zones. The stellar masses inferred from kinematics form important direct tests of pre-main-sequence stellar evolution models. By imaging lines with different excitation conditions, maps of the H_2 density distribution will become possible. Together, such ALMA data can provide unbiased surveys of disks in different star-forming regions, down to an equivalent sensitivity of a few Earth masses of dust and gas, and probe the distribution of disk parameters with stellar mass, luminosity, age and environment.

5.3 Disk evolution and gap formation

Question 15 When and how do gas and dust disappear from the disks? Do they disappear at the same time? Are there multiple paths from gas-rich disks to the debris disk stage?

Near-infrared surveys have shown that inner dust disks (<few AU) disappear on timescales of a few Myr (Haisch et al. 2001). Surveys with *Spitzer* at mid-infrared wavelengths are starting to reveal a similar trend for the planet-forming zones out to ~ 10 AU, with a significantly lower disk fraction for weak-line T Tauri stars than for their classical counterparts (Cieza et al. 2007). Indeed, examination of hundreds of SEDs of stars with disks in *Spitzer* surveys show that there may be multiple evolutionary paths from the massive gas-rich disks to the tenuous gas-poor debris disks, involving both grain growth and gap opening. ALMA will be critical to study these transitional objects by imaging the holes or gaps in their dust disks down to a few AU and measure the remaining gas mass through tracers like CO and [C I], some of which may be left inside the holes. Giant planet formation, grain growth and photoevaporation are the three major contending theories for explaining holes in the dust disks but they have different predictions for the gas vs. large dust distribution. Overall, surveys for gaps in disks can provide statistics on the frequency and timescale for planet formation.

6 Conclusions

ALMA will be vital and unique to answer key questions in star- and planet formation, by resolving the physical processes taking place during the collapse of molecular clouds, imaging the structure of protostars and of proto-planetary disks, and determining the chemical composition of the material from which future solar systems are made. Many of the ALMA source lists will come from unbiased surveys being carried out now, in particular *Spitzer, Herschel*, near-infrared and single-dish submillimeter surveys. To extract information from ALMA data, however, sophisticated analysis and modeling tools are needed. The community needs to invest now in those tools to ensure that they are ready by the time that ALMA is fully commissioned.

Other major facilities in the timeframe of ALMA operations include the *James Webb Space Telescope* and ground-based extremely large optical telescopes (ELTs). These facilities will be highly complementary to ALMA, each addressing a different part of the star- and planet formation puzzle. There is no doubt, however, that ALMA will be *the* key instrument for much of the physics and chemistry associated with star- and planet formation.

References

Allen, L., Megeath, S.T., Gutermuth, R., et al.: In: Reipurth, B., Jewitt, D., Keil, K. (eds.) Protostars and Planets V, pp. 361–376. University of Arizona Press, Tucson (2007)

Alves, J., Lombardi, M., Lada, C.J.: Astron. Astrophys. **462**, L17 (2007)

Andrews, S.M., Williams, J.P.: Astrophys. J. **659**, 705 (2007)

Arce, H.G., Sargent, A.I.: Astrophys. J. **646**, 1070 (2006)

Bachiller, R., Tafalla, M.: In: Lada, C.J., Kylafis, N.D. (eds.) The Origin of Stars and Planetary Systems, p. 227. Kluwer Academic, Dordrecht (1999)

Bate, M.R., Bonnell, I.A.: Mon. Not. Roy. Astron. Soc. **356**, 1201 (2005)

Bergin, E.A., Alves, J., Huard, T., Lada, C.J.: Astrophys. J. **570**, L101 (2002)

Bergin, E.A., Aikawa, Y., Blake, G.A., van Dishoeck, E.F.: In: Reipurth, B., Jewitt, D., Keil, K. (eds.) Protostars and Planets V, pp. 751–766. University of Arizona Press, Tucson (2007)

Bisschop, S.E., Jørgensen, J.K., van Dishoeck, E.F., de Wachter, E.B.M.: Astron. Astrophys. **465**, 913 (2007)

Bonnell, I.A., Bate, M.R.: Mon. Not. Roy. Astron. Soc. **370**, 488 (2006)

Boss, A.P.: Astrophys. J. **599**, 577 (2003)

Bottinelli, S., Ceccarelli, C., Williams, J.P., Lefloch, B.: Astron. Astrophys. **463**, 601 (2007)

Bourke, T.L., Crapsi, A., Myers, P.C., et al.: Astrophys. J. **633**, L129 (2005)

Bourke, T.L., Myers, P.C., Evans, N.J., et al.: Astrophys. J. (2006, in press)

Caselli, P., Walmsley, C.M., Tafalla, M., Dore, L., Myers, P.C.: Astrophys. J. **523**, L165 (1999)

Ceccarelli, C., Caselli, P., Herbst, E., Tielens, A.G.G.M., Caux, E.: In: Reipurth, B., Jewitt, D., Keil, K. (eds.) Protostars and Planets V, pp. 47–62. University of Arizona Press, Tucson (2007)

Cesaroni, R.: Astrophys. Space Sci. **295**, 5 (2005)

Cieza, L., Padgett, D.L., Stapelfeldt, K.R., et al.: Astrophys. J. (2007, in press)

Crapsi, A., Caselli, P., Walmsley, C.M., et al.: Astrophys. J. **619**, 379 (2005)

Di Francesco, J., Myers, P.C., Wilner, D.J., Ohashi, N., Mardones, D.: Astrophys. J. **562**, 770 (2001)

Dullemond, C.P., Hollenbach, D., Kamp, I., D'Alessio, P.: In: Reipurth, B., Jewitt, D., Keil, K. (eds.) Protostars and Planets V, pp. 555–572. University of Arizona Press, Tucson (2007)

Dunham, M.M., Evans, N.J. II, Bourke, T.L., et al.: Astrophys. J. **651**, 945 (2006)

Dutrey, A., Guilloteau, S., Duvert, G., et al.: Astron. Astrophys. **309**, 493 (1996)

Haisch, Jr. K.E., Lada, E.A., Lada, C.J.: Astrophys. J. **553**, L153 (2001)

Hogerheijde, M.R.: Astrophys. J. **553**, 618 (2001)

Hollis, J.M., Lovas, F.J., Jewell, P.R.: Astrophys. J. **540**, L107 (2000)

Hollis, J.M., Lovas, F.J., Remijan, A.J., et al.: Astrophys. J. **643**, L25 (2006)

Hueso, R., Guillot, T.: Astron. Astrophys. **442**, 703 (2005)

Jørgensen, J.K.: Astron. Astrophys. **424**, 589 (2004)

Jørgensen, J.K., Schöier, F.L., van Dishoeck, E.F.: Astron. Astrophys. **416**, 603 (2004)

Jørgensen, J.K., Bourke, T.L., Myers, P.C., et al.: Astrophys. J. **632**, 973 (2005)

Jørgensen, J.K., Harvey, P.M., Evans, N.J. II, et al.: Astrophys. J. **645**, 1246 (2006)

Jørgensen, J.K., Bourke, T.L., Myers, P.C., et al.: Astrophys. J. **659**, 479 (2007)

Kirk, J.M., Ward-Thompson, D., André, P.: Mon. Not. Roy. Astron. Soc. **375**, 843 (2007)

Koerner, D.W., Sargent, A.I.: Astron. J. **109**, 2138 (1995)

Krumholz, M.R., McKee, C.F., Klein, R.I.: Nature **438**, 332 (2005)

Lada, C.J.: In: Lada, C.J., Kylafis, N.D. (eds.) The Origin of Stars and Planetary Systems, p. 143. Kluwer Academic, Dordrecht (1999)

Lahuis, F., van Dishoeck, E.F., Boogert, A.C.A., et al.: Astrophys. J. **636**, L145 (2006)

Lin, S.-Y., Ohashi, N., Lim, J., et al.: Astrophys. J. **645**, 1297 (2006)

Luhman, K.L., Joergens, V., Lada, C., et al.: In: Reipurth, B., Jewitt, D., Keil, K. (eds.) Protostars and Planets V, pp. 443–457. University of Arizona Press, Tucson (2007)

McKee, C.F., Tan, J.C.: Astrophys. J. **585**, 850 (2003)

Mohanty, S., Jayawardhana, R., Natta, A., et al.: Astrophys. J. **609**, L33 (2004)

Neufeld, D.A., Hollenbach, D.J.: Astrophys. J. **428**, 170 (1994)

Piétu, V., Dutrey, A., Guilloteau, S.: Astron. Astrophys. **467**, 163 (2007)

Qi, C., Kessler, J.E., Koerner, D.W., Sargent, A.I., Blake, G.A.: Astrophys. J. **597**, 986 (2003)

Rebull, L.M., Stapelfeldt, K.R., Evans, N.J. II, et al.: Astrophys. J. Suppl. **171**, 447 (2007)

Richer, J.S., Shepherd, D.S., Cabrit, S., Bachiller, R., Churchwell, E.: In: Mannings, V., Boss, A.P., Russell, S.S. (eds.) Protostars and Planets IV, p. 867. University of Arizona Press, Tucson (2000)

Shang, H., Li, Z.-Y., Hirano, N.: In: Reipurth, B., Jewitt, D., Keil, K. (eds.) Protostars and Planets V, pp. 261–276. University of Arizona Press, Tucson (2007)

Shepherd, D.S., Churchwell, E.: Astrophys. J. **457**, 267 (1996)

Shepherd, D.S., Kurtz, S.E.: Astrophys. J. **523**, 690 (1999)

Simon, R., Jackson, J.M., Rathborne, J.M., Chambers, E.T.: Astrophys. J. **639**, 227 (2006)

Tafalla, M., Mardones, D., Myers, P.C., et al.: Astrophys. J. **504**, 900 (1998)

Takahashi, S., Saito, M., Takakuwa, S., Kawabe, R.: Astrophys. J. **651**, 933 (2006)

Testi, L., Natta, A., Oliva, E., et al.: Astrophys. J. **571**, L155 (2002)

van der Tak, F.F.S., Caselli, P., Ceccarelli, C.: Astron. Astrophys. **439**, 195 (2005)

van Dishoeck, E.F.: Proc. National Acad. Sci. **103**, 12249 (2006)

Williams, J.P., Blitz, L., McKee, C.F.: In: Mannings, V., Boss, A.P., Russell, S.S. (eds.) Protostars and Planets IV, p. 97. University of Arizona Press, Tucson (2000)

Young, C.H., Jørgensen, J.K., Shirley, Y.L., et al.: Astrophys. J. Suppl. Ser. **154**, 396 (2004)

High-mass star forming regions: An ALMA view

R. Cesaroni

Originally published in the journal Astrophysics and Space Science, Volume 313, Nos 1–3.
DOI: 10.1007/s10509-007-9596-3 © Springer Science+Business Media B.V. 2007

Abstract The advent of ALMA is bound to improve our knowledge of OB star formation dramatically. Here, we present an overview of this topic outlining how high angular resolution and sensitivity may contribute to shed light on the structure of high-mass star forming regions and hence on the process itself of massive star formation. The impact of this new generation instrument will range from establishing the mass function of pre-stellar cores inside IR-dark clouds, to investigating the kinematics of the gas from which OB stars are built up, to assessing or ruling out the existence of circumstellar accretion disks in these objects.

Keywords Stars: formation · HII regions · Circumstellar matter · ISM: molecules · ISM: jets and outflows · Submillimeter

1 Introduction

Massive stars are commonly defined as those in excess of a few $10^3 L_\odot$ or equivalently with masses above $\sim 8 M_\odot$, corresponding, on the zero-age main sequence (ZAMS), to spectral types earlier than B3. Such a definition is intimately related to the formation process of these stars. According to theory (Shu et al. 1987), star formation proceeds through mass accretion onto a protostellar nucleus, which slowly contracts until the density and inner temperature are large enough to ignite hydrogen burning. The problem with stars above $\sim 8 M_\odot$ (Palla and Stahler 1993) is that their ZAMS luminosity exerts a strong radiation pressure on the infalling

gas, thus halting the accretion process. Without discussing the many proposed solutions to solve this problem, here we note that the existence itself of such a problem demonstrates that there are likely fundamental differences between the formation of high-mass stars and their low-mass analogues. Understanding this process requires knowledge of the environment where OB stars are born, from the scale of molecular clouds (a few parsecs) to that of the (proto)star. As shown in the following, ALMA will be the perfect instrument for this purpose.

In our discussion, we will describe what are thought to be the main steps in the process of OB star formation. A reasonable hypothesis is that the process starts with the fragmentation of cold (IR-dark) clouds and that subsequent infall inside each fragment (core) eventually leads to the formation of multiple stars. Although this scenario is commonly accepted as a reasonable zero-th order approximation, the details are still a matter of debate. In particular, the relative importance of gravitation, magnetic field, and turbulence is unclear.

The scope of the present contribution is to describe the physical characteristics of the objects involved in the evolutionary sequence going from pc-scale clouds to newly born OB stars and discuss how ALMA will improve our knowledge of these objects and, as a consequence, of the formation mechanism of massive stars.

2 IR-dark clouds

In recent years, observations in the mid-IR with improved sensitivity and resolution with respect to the old IRAS survey have revealed absorption against the bright background of small grains and PAHs (Perault et al. 1996; Egan et al. 1998). This is an indication of the existence of clouds with

R. Cesaroni (✉)
INAF—Osservatorio Astrofisico di Arcetri, Firenze, Italy
e-mail: cesa@arcetri.inaf.it

low gas temperature and large column density, as indeed confirmed by subsequent observations of the dust continuum and line emission (Carey et al. 1998; Pillai et al. 2006). These "IR-dark" clouds have typical masses of 10^3–$10^4 M_\odot$, sizes of 1–10 pc, densities of $\sim 10^5$ cm^{-3}, and temperatures of 10–20 K, which are significantly less than those measured towards clouds containing high-mass (proto)stellar objects (Sridharan et al. 2005). The presence of sub-structure and (in some cases) of IR sources—possibly deeply embedded protostars (Rathborne et al. 2005)—demonstrates that these clouds are not homogeneous and quiescent but instead may be undergoing fragmentation. This suggests that IR-dark clouds could be the high-mass analogues of the pre-stellar clumps detected in low-mass star forming regions (Ward-Thompson et al. 2007).

If this is the case, it is important to obtain a detailed picture of the cloud structure and kinematics. In fact, one question is whether the mass spectrum of the cores into which clouds appear to fragment is similar to the stellar initial mass function (IMF). This will allow us to assess whether the IMF is established before or after fragmentation. The former would support theories which consider OB star formation a scaled-up version of low-mass star formation, where one or few stars form inside each core (Beuther et al. 2007). In the latter case, the stellar mass spectrum would instead be determined in each core by other physical processes (turbulence, merging of lower-mass stars, competitive accretion Ballesteros-Paredes et al. 2006; Bonnell et al. 2007). Knowledge of the kinematics is also

important to complement the spatial information and hence discriminate between different star formation theories: the velocity field expected both in the gas and the newly formed stars should carry the imprint of the formation process, and hence, e.g., allow us to decide between "turbulent" models and collapse models.

With this in mind, it is clear that the challenge for ALMA will be twofold: to detect all relevant cores and thus reconstruct their mass spectrum, as done by Beuther and Schilke (2004) in their pivotal study of a massive star forming clump; and to find out their distribution both in space and velocity. A similar experiment has been done by Rodríguez et al. (2005) on (proto)stars in Orion, by measuring the proper motions of the BN object and radio source I. ALMA will easily resolve regions as small as 100 AU at a distance of 10 kpc, comparable to the typical structures predicted by numerical simulations of star forming clumps (Bate et al. 2003). Such a resolution will permit to repeat the experiment performed in Orion up to a distance of 10 kpc and not only on the stellar component but also on the compact, dense cores where massive stars form. These might be used as test particles to obtain a direct measurement of the velocity field of the gas in the associated molecular cloud. The question is whether ALMA will be sensitive enough to detect such cores. The minimum mass of interest for an accurate estimate of the core mass function is the Jeans mass, which is $\sim 0.5 M_\odot$ for temperatures and densities typical of IR-dark clouds. Figure 1 demonstrates that dust continuum emission from cold cores of 10 K, with masses in excess of $0.5 M_\odot$ can be detected by ALMA all over the Galaxy, whereas the

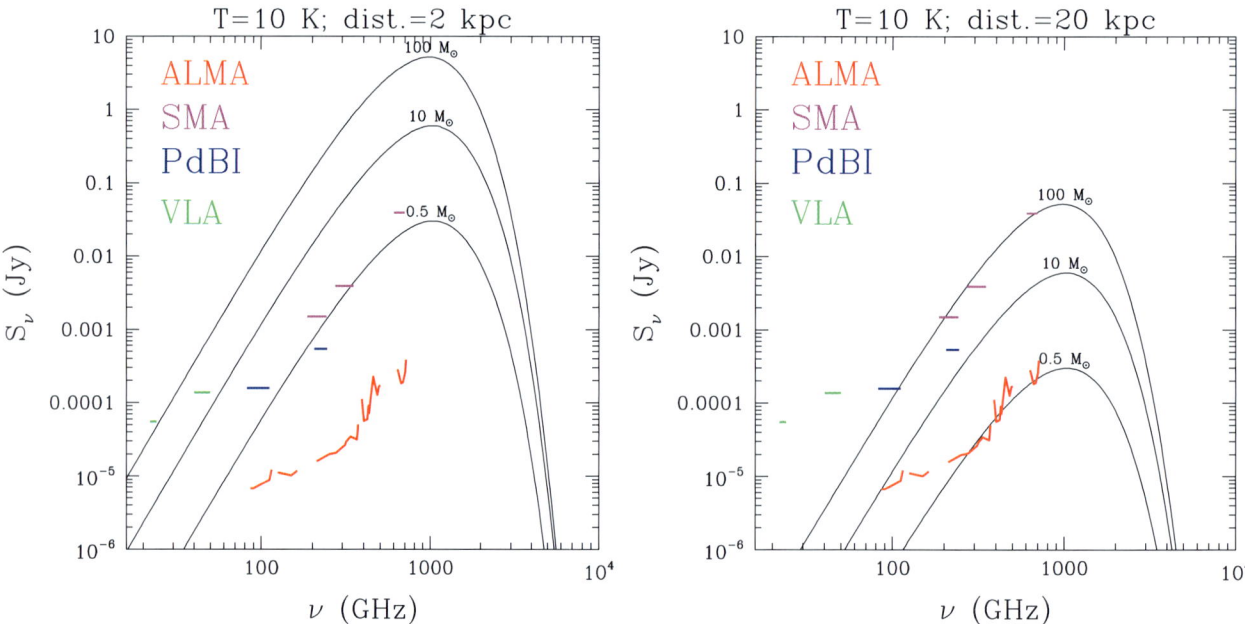

Fig. 1 Spectra of the dust continuum emission from molecular cores with temperature of 10 K and masses as indicated beside each curve. The *left* and *right panels* correspond respectively to distances of 2 and 20 kpc. The 3σ sensitivities of current interferometers and ALMA are indicated by the *coloured lines*, assuming an integration time of 5 hr

same result can be obtained with currently available interferometers only up to a few kpc.

In conclusion, ALMA will be the ideal instrument to determine the mass function of prestellar cores and investigate their velocity distribution.

3 Hot molecular cores

Newly formed OB (proto)stars heat up their molecular surroundings and possibly evaporate dust grain mantles enriching the gas phase with rare molecular species. These hot, chemically rich molecular regions are named "hot molecular cores" and are believed to be the cradle of high-mass stars (Kurtz et al. 2000; Cesaroni 2005b). Consistent with this idea, one might expect to find infall, outflow, and rotational motions in such cores, as these three phenomena are naturally associated with the process of star formation (Cesaroni 2005a). Indeed, these phenomena have been detected in massive star forming regions and ALMA will help dramatically improving our knowledge of them, as illustrated in the following.

3.1 Molecular outflows

Given the typical size of outflow lobes (>0.5 pc or $10''$ at 10 kpc), high angular resolution is not necessary to image the structure of a single outflow, but is needed to resolve multiple outflows present in young protoclusters. It has been found that what appears as a single bipolar outflow when observed with single-dish telescopes, often turns out to be the result of the overlap of multiple flows in the plane of the sky (Beuther et al. 2002; Beuther et al. 2003; Beuther et al. 2004). ALMA's resolution will be more than sufficient for this purpose, since the mean separation in a cluster between OB stars powering the outflows is $\sim$0.05 pc, i.e. 0.5–10 arcsec. Spectral resolution and sensitivity are not an issue in this case, as the molecular lines tracing outflows are typically as broad as several km s^{-1} and as strong as a few K, compared to an ALMA sensitivity of <0.1 K for 1 km s^{-1} resolution, 1 hour integration, and $1''$ beam. ALMA can study any outflow all over the Galaxy.

Our knowledge of outflows in massive stars will enormously benefit from ALMA observations. Measurements of lobe proper motions will be feasible, as an expansion speed of 100 km s^{-1} corresponds to 20 mas yr^{-1} at 1 kpc, namely 1/3 of the ALMA beam at 3 mm. This in turn will allow us to derive the inclination of the outflow (from the ratio between velocity along the line of sight and that on the plane of the sky) and thus to deproject the outflow parameters. Finally, knowledge of the shape of the outflow lobes from a few 100 AU to 1 pc will help revealing the presence of precession, as already found in some cases (Shepherd et al. 2000; Cesaroni et al. 2005).

3.2 Infall

Despite the fact that infall plays a crucial role in star formation, this process is very difficult to detect, mostly because, while infall causes line broadening, this can also occur due to other effects (outflow, large optical depths, turbulence). A very effective method of identifying the presence of infalling gas is to observe red-shifted line absorption towards bright embedded sources. In the case of OB stars, such sources may be the associated H II regions, provided they are still very compact. Hypercompact H II regions have been indeed used to sample the velocity field along the lines of sight across their surfaces (Keto et al. 1988; Sollins and Ho 2005; Sollins et al. 2005; Beltrán et al. 2006), thus establishing the presence of infall towards the H II region itself. Note that the free-fall velocity at a distance <0.1 pc (the typical size of a hot molecular core) from a $\sim$10$M_\odot$ star is >1 km s^{-1}: the spectral resolution of current interferometers (and even more so of ALMA) is more than sufficient to detect velocity shifts like this.

Up to now, the previous technique has been used only at centimeter wavelengths, whereas ALMA will operate shortward of 7 mm: will a similar "absorption experiment" be feasible also in the (sub)mm regime? A positive answer to this question requires three conditions to be satisfied: (i) the instrumental beam must be less than the H II region diameter; (ii) the free-free emission from the H II region must be optically thick; (iii) the dusty core enshrouding the H II region must be optically thin. From these, one obtains a relationship between the spectral type of the star ionising the H II region and the maximum distance up to which the absorption experiment can be made with that H II region. Similarly, a relationship between the star spectral type and the maximum frequency usable for that experiment is also obtained. These two relationships are shown in Fig. 2. In the calculation, for the molecular core we have adopted a radius of 0.05 pc, a mean H$_2$ density of 10^7 cm^{-3}, a density profile $n \propto R^{-3/2}$, and a dust absorption coefficient equal to 0.005 cm^2 g$^{-1} \times (\nu/230$ GHz$)^2$. One can see that infall onto B-type stars can be detected only up to 1 kpc, while for O stars this region extends at least up to the Galactic centre. The right panel shows, as expected, that low frequencies are to be preferred for the experiment, since dust absorption is minimum. Typical targets are hence hypercompact H II regions as small as 1000 AU or less. Objects like these have been detected (De Pree et al. 1998; De Pree et al. 2000) and even more numerous and smaller H II regions should be seen with ALMA.

According to recent observations, it appears that not only the molecular surroundings of newly formed OB stars are undergoing infall, but this should continue through the ionised gas inside the H II region (Keto 2002). Recombination line studies with ALMA will improve on our knowledge of this topic thus establishing whether large accretion

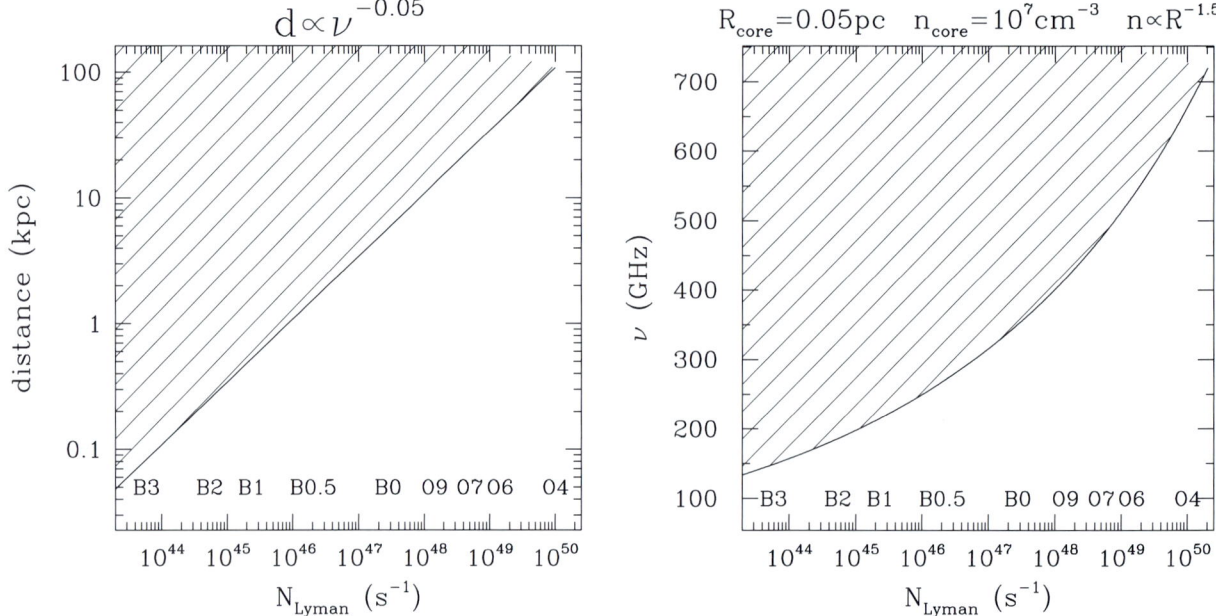

Fig. 2 *Left panel.* Maximum distance (d) at which an HII region with a given Lyman continuum can be used to detect red-shifted line absorption from the molecular gas infalling towards the star at the center of the HII region. The relationship is only weakly dependent on the observing frequency ($d \propto \nu^{-0.05}$). *The shaded area* denotes the region of the plot that cannot be used for this absorption experiment. *Right panel.* Same as *left panel* for the maximum frequency to be used for the absorption experiment

rates may not only overcome the radiation pressure from the star, but also significantly increase the lifetime of HII regions in their most compact phases. It is worth noting that the (sub)millimeter range is better suited than the centimeter one for observations of recombination lines. This is not only because at shorter wavelengths the free-free continuum is optically thinner, but also because pressure broadening becomes less important and this permits a better understanding of the kinematics of the ionised gas.

3.3 Rotation

Circumstellar disks are a natural outcome of the star formation process, due to angular momentum conservation of the material infalling onto the protostar. However, so far only limited evidence for rotating disks around high-mass stars has been found. In most cases, this is limited to B stars, while in association with (presumed) O stars only massive, rotating toroids (likely transient circum-cluster structures) are seen. A detailed discussion of disks and toroids (see Beltrán et al. 2004, 2005) in OB stars has been made in recent reviews (Cesaroni et al. 2007; Cesaroni et al. 2006) and will not be repeated here. We stress that establishing the existence of true, stable, Keplerian disks around O stars would lend support to accretion models. So, the question is whether the lack of evidence for such disks in O stars is real or an observational bias due to the limited resolution and sensitivity of current interferometers.

To investigate this issue, in Fig. 3 we plot the maximum distance up to which a circumstellar disk can be detected in line emission, as a function of the mass of the corresponding star. This is done for both the IRAM-Plateau de Bure interferometer and ALMA, under a number of reasonable assumptions, namely: the disk is Keplerian, the instrumental beam is equal to 1/4th of the disk radius, the line width is equal to the rotation velocity at the outer radius of the disk, the disk mass is proportional to the stellar mass, the mean surface density is the same for all disks, and the line brightness temperature is >20 K. For the estimate of the sensitivities we require a signal-to-noise ratio to be ≥ 20 and assume an observing frequency of 230 GHz, a spectral resolution of 0.2 km s^{-1}, and an integration time on-source of 5 hours.

Both an edge-on disk (solid lines) and one with an inclination angle of 35 deg[1] (dashed lines) are considered. Comparison with disks known to date (both in low- and high-mass stars) shows that, as expected, all of them lie below the PdBI curve. Noticeably, Fig. 3 also shows the existence of a "region of avoidance" above $\sim 20 M_\odot$, where no disks have been found. This result has a natural explanation in that the more massive is a star, the larger will be the distance inside which such a star can be found. There will thus be a region in the plot where it is very unlikely to find any young embedded star (no matter whether with or without a

---

[1] 35 deg is the mean inclination angle for a distribution of disks randomly oriented.

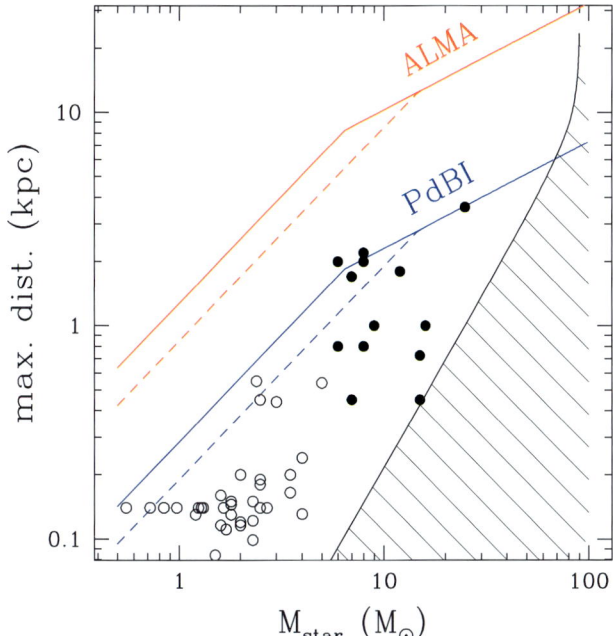

Fig. 3 Maximum distance at which a Keplerian circumstellar disk can be detected with a given instrument (ALMA or PdBI) as a function of the mass of the star at the center of the disk. The *solid lines* correspond to edge-on disks, the *dashed ones* to disks with an inclination of 35 deg (the mean inclination angle assuming random orientation of the disk axes). *Solid* and *empty circles* correspond respectively to known circumstellar disks in high-mass (Cesaroni et al. 2007; Fuente et al. 2006) and low-mass (Simon et al. 2000; Garcia Lopez et al. 2006) (proto)stars. The *shaded area* denotes the region where the probability to find a star is low

circumstellar disk) and this corresponds to the shaded area in Fig. 3. The conclusion is that it will be necessary to span the whole Galaxy to find a circumstellar disk associated with a star more massive than a few $10 M_\odot$, namely an early O star: ALMA is the instrument needed for this type of search, as it will be sensitive enough to detect such a disk up to distances as large as 20 kpc.

4 Conclusions

The usage of ALMA for the study of high-mass star forming regions is bound to shed light on a number of critical issues which are still matter of debate among observers and theorists involved in the study of OB star formation. Among these, the determination of the mass spectrum of pre-stellar cores, the detection of infall in young, embedded OB (proto)stars, and the detailed investigation of the associated bipolar outflows on scales ranging from a few 100 AU to a few pc. Equally important will be the possibility to establish the existence/absence of Keplerian circumstellar disks around newly formed OB (proto)stars, which might be a crucial test capable of discriminating between differ-

ent star formation theories. With ALMA one will be able to study these questions all over the Galaxy.

Notwithstanding the dramatic improvement in angular resolution, sensitivity, and frequency coverage provided by ALMA, and the plethora of important outcomes expected, it is worthwhile to conclude this review with a word of caution. Albeit powerful, ALMA is not flawless: in particular, it would be very important to obtain an accurate estimate of the luminosity of the deeply embedded stars, which in turn will allow us to determine their spectral type and evolutionary stage. The spectral coverage of ALMA will not suffice for this purpose. The spectral energy distribution of a hot molecular core hosting deeply embedded OB (proto)stars peaks at a few 10 µm, while the shortest wavelength covered by ALMA will be an order of magnitude greater. Sub-arcsec resolution in the far-IR will be needed to achieve an accurate determination of the luminosity of embedded OB (proto)stars, but this has to await for the advent of FIRI, a space far-IR interferometer.

Acknowledgements   It is a pleasure to thank Malcolm Walmsley for critically reading the manuscript.

References

Ballesteros-Paredes, J., Klessen, R.S., Mac Low, M.-M., Vazquez-Semadeni, E.: Molecular cloud turbulence and star formation. In: Protostars and Planets V, p. 63 (2006) (astro-ph/0603357)

Bate, M.R., Bonnell, I.A., Bromm, V.: The formation of a star cluster: predicting the properties of stars and brown dwarfs. Mon. Not. R. Astron. Soc. **339**, 577 (2003)

Beltrán, M.T., Cesaroni, R., Neri, R., Codella, C., Furuya, R.S., Testi, L., Olmi, L.: Rotating disks in high-mass young stellar objects. Astrophys. J. **601**, L187 (2004)

Beltrán, M.T., Cesaroni, R., Neri, R., Codella, C., Furuya, R.S., Testi, L., Olmi, L.: A detailed study of the rotating toroids in G31.41+0.31 and G24.78+0.08. Astron. Astrophys. **435**, 901 (2005)

Beltrán, M.T., Cesaroni, R., Codella, C., Testi, L., Furuya, R.S., Olmi, L.: Infall of gas as the formation mechanism of stars up to 20 times more massive than the Sun. Nature **443**, 427 (2006)

Beuther, H., Schilke, P.: Fragmentation in massive star formation. Science **303**, 1167 (2004)

Beuther, H., Schilke, P., Gueth, F., McCaughrean, M., Andersen, M., Sridharan, T.K., Menten, K.M.: IRAS 05358+3543: Multiple outflows at the earliest stages of massive star formation. Astron. Astrophys. **387**, 931 (2002)

Beuther, H., Schilke, P., Stanke, T.: Multiple outflows in IRAS 19410+2336. Astron. Astrophys. **408**, 601 (2003)

Beuther, H., Schilke, P., Gueth, F.: Massive molecular outflows at high spatial resolution. Astrophys. J. **608**, 330 (2004)

Beuther, H., Churchwell, E.B., McKee, C.F., Tan, J.C.: The formation of massive stars. In: Protostars and Planets V, p. 165 (2007) (astro-ph/0602012)

Bonnell, I.A., Larson, R.B., Zinnecker, H.: The origin of the initial mass function. In: Protostars and Planets V, p. 149 (2007) (astro-ph/0603447)

Carey, S.J., Clark, F.O., Egan, M.P., Price, S.D., Shipman, R.F., Kuchar, T.A.: The physical properties of the midcourse space experiment galactic infrared-dark clouds. Astrophys. J. **508**, 721 (1998)

Cesaroni, R.: Outflow, infall, and rotation in high-mass star forming regions. Astrophys. Space Sci. **295**, 5 (2005a)

Cesaroni, R.: Hot molecular cores. IAU Symposium 227, p. 59 (2005b)

Cesaroni, R., Neri, R., Olmi, L., Testi, L., Walmsley, C.M., Hofner, P.: A study of the Keplerian accretion disk and precessing outflow in the massive protostar IRAS 20126+4104. Astron. Astrophys. **434**, 1039 (2005)

Cesaroni, R., Galli, D., Lodato, G., Walmsley, C.M., Zhang, Q.: The critical role of disks in the formation of high-mass stars. Nature **444**, 703 (2006)

Cesaroni, R., Galli, D., Lodato, G., Walmsley, C.M., Zhang, Q.: Disks around young O-B (proto)stars: Observations and theory. In: Protostars and Planets V, p. 197 (2007)

De Pree, C.G., Goss, W.M., Gaume, R.A.: Ionized gas in Sagittarius B2 Main on scales of 0.065 arcsecond (600 AU). Astrophys. J. **500**, 847 (1998)

De Pree, C.G., Wilner, D.J., Goss, W.M., Welch, W.J., McGrath, E.: Ultracompact H II regions in W49N at 500 AU scales: Shells, winds, and the water maser source. Astrophys. J. **540**, 308 (2000)

Egan, M.P., Shipman, R.F., Price, S.D., Carey, S.J., Clark, F.O., Cohen, M.: A population of cold cores in the galactic plane. Astrophys. J. **494**, L199 (1998)

Fuente, A., Alonso-Albi, T., Bachiller, R., Natta, A., Testi, L., Neri, R., Planesas, P.: A Keplerian gaseous disk around the B0 star R Monocerotis. Astrophys. J. **649**, L119 (2006)

Garcia Lopez, R., Natta, A., Testi, L., Habart, E.: Accretion rates in Herbig Ae stars. Astron. Astrophys. **459**, 837 (2006)

Keto, E.: An ionized accretion flow in the ultracompact H II region G10.6-0.4. Astrophys. J. **568**, 754 (2002)

Keto, E.R., Ho, P.T.P., Haschick, A.D.: The observed structure of the accretion flow around G10.6-0.4. Astrophys. J. **324**, 920 (1988)

Kurtz, S., Cesaroni, R., Churchwell, E., Hofner, P., Walmsley, C.M.: Hot molecular cores and the earliest phases of high-mass star formation. In: Protostars and Planets IV, p. 299 (2000)

Palla, F., Stahler, S.W.: The pre-main-sequence evolution of intermediate-mass stars. Astrophys. J. **418**, 414 (1993)

Perault, M., Omont, A., Simon, G., Seguin, P., Ojha, D., Blommaert, J., Felli, M., Gilmore, G., Guglielmo, F., Habing, H., Price, S., Robin, A., de Batz, B., Cesarsky, C., Elbaz, D., Epchtein, N., Fouque, P., Guest, S., Levine, D., Pollock, A., Prusti, T., Siebenmorgen, R., Testi, L., Tiphene, D.: First ISOCAM images of the Milky Way. Astron. Astrophys. **315**, L165 (1996)

Pillai, T., Wyrowski, F., Carey, S.J., Menten, K.M.: Ammonia in infrared dark clouds. Astron. Astrophys. **450**, 569 (2006)

Rathborne, J.M., Jackson, J.M., Chambers, E.T., Simon, R., Shipman, R., Frieswijk, W.: Massive protostars in the infrared dark cloud MSXDC G034.43+00.24. Astrophys. J. **630**, L181 (2005)

Rodríguez, L.F., Poveda, A., Lizano, S., Allen, C.: Proper motions of the BN object and the radio source I in Orion: Where and when did the BN object become a runaway star? Astrophys. J. **627**, L65 (2005)

Shepherd, D.S., Yu, K.C., Bally, J., Testi, L.: The molecular outflow and possible precessing jet from the massive young stellar object IRAS 20126+4104. Astrophys. J. **535**, 833 (2000)

Shu, F.H., Adams, F.C., Lizano, S.: Star formation in molecular clouds—Observation and theory. Annu. Rev. Astron. Astrophys. **25**, 23 (1987)

Simon, M., Dutrey, A., Guilloteau, S.: Dynamical masses of T Tauri stars and calibration of pre-main-sequence evolution. Astrophys. J. **545**, 1034 (2000)

Sollins, P.K., Ho, P.T.P.: The molecular accretion flow in G10.6-0.4. Astrophys. J. **630**, 987 (2005)

Sollins, P.K., Zhang, Q., Keto, E., Ho, P.T.P.: Spherical infall in G10.6-0.4: Accretion through an ultracompact H II region. Astrophys. J. **624**, L49 (2005)

Sridharan, T.K., Beuther, H., Saito, M., Wyrowski, F., Schilke, P.: High-mass starless cores. Astrophys. J. **634**, L57 (2005)

Ward-Thompson, D., André, P., Crutcher, R., Johnstone, D., Onishi, T., Wilson, C.: An observational perspective of low-mass dense cores II: Evolution toward the initial mass function. In: Protostars and Planets V, p. 33 (2007)

Early stages of star formation

The ALMA promise

Philippe André · Patrick Hennebelle · Nicolas Peretto

Originally published in the journal Astrophysics and Space Science, Volume 313, Nos 1–3.
DOI: 10.1007/s10509-007-9601-x © Springer Science+Business Media B.V. 2007

Abstract The study of the earliest stages of star formation in molecular clouds is one of the fields that should benefit most from ALMA. Improving our understanding of these deeply embedded stages is crucial to gain insight into the origin of stellar masses and binary systems. While the use of large single-dish (sub)millimeter radiotelescopes and existing interferometers has led to good progress on the overall density structure of isolated prestellar cores and young protostars, many questions remain open concerning, e.g., their fragmentation properties and detailed kinematics. Furthermore, the classical paradigm for the formation of single low-mass stars in well-separated, magnetized prestellar cores has been challenged on the grounds that most young stars actually belong to multiple systems and/or coherent clusters. A new paradigm based on supersonic turbulence has emerged which emphasizes the role of dynamical interactions between individual (proto)stars in cluster-forming clumps. The debate is far from settled and ALMA will greatly help to discriminate between these two paradigms.

Keywords Stars: formation · Stars: low-mass · Brown dwarfs · Stars: binaries · ISM: structure

P. André (✉)
CEA Saclay—Service d'Astrophysique, 91191 Gif-sur-Yvette, France
e-mail: pandre@cea.fr

P. Hennebelle
ENS/LERMA, Paris, France

N. Peretto
Department of Physics & Astronomy, University of Manchester, Manchester, UK

1 Introduction: formation and fragmentation of prestellar cores

Two major limitations in our present understanding of the star formation process are that we do not know well (a) how prestellar dense cores are generated in molecular clouds, and (b) how such prestellar cores subsequently collapse into protostars or, more likely, protostellar systems. The problem of the formation of cloud cores is currently the subject of a lively debate between two schools of thought: The classical picture of slow, quasi-static core formation by ambipolar diffusion in magnetically-supported clouds (e.g. Mouschovias and Ciolek 1999; Shu et al. 1987, 2004) has been seriously challenged by a new, more dynamic picture, which emphasizes the role of supersonic turbulence in supporting clouds on large scales and generating density fluctuations on small scales (e.g. Padoan and Nordlund 2002; Mac Low and Klessen 2004). In the first picture, prestellar cores are quasi-equilibrium structures which evolve over several dynamical timescales toward higher degrees of central condensation before collapsing dynamically. By contrast, in the second picture, cloud cores are always far from hydrostatic equilibrium and form by collision of large-scale, converging supersonic flows in molecular clouds (e.g. Hartmann et al. 2001; Klessen et al. 2005).

The detailed manner in which individual prestellar cores fragment (or not) during collapse to form multiple (or single) stars is another major open problem (see Goodwin et al. 2007 for a recent review). In particular, the typical outcome of cloud core collapse (e.g., single star, binary, or higher-order multiple system) remains unclear (cf. Lada 2006), and the influence of magnetic fields on core fragmentation is still poorly understood (see Sect. 3 below).

Improving our understanding of the formation and fragmentation of prestellar cores is of fundamental importance

Astrophys Space Sci (2008) 313: 29–34

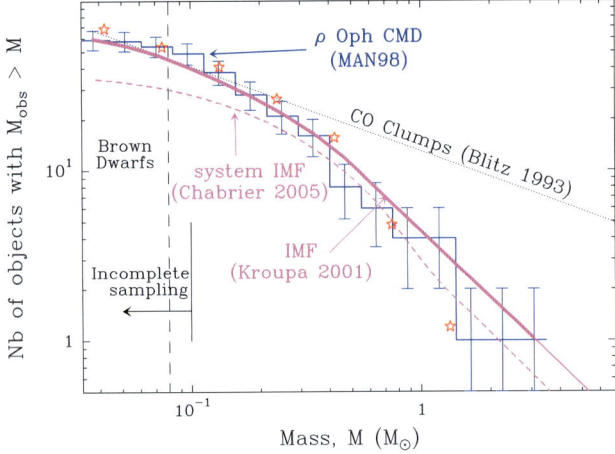

Fig. 1 Cumulative mass distribution of a sample of 57 prestellar condensations, complete down to ~$0.1 M_\odot$, in the ρ Oph protocluster (histogram with *error bars*—from Motte et al. 1998 MAN98). For comparison, the *solid curve* shows the shape of the field star IMF (e.g. Kroupa 2001), while the *dashed curve* corresponds to the IMF of multiple systems (e.g. Chabrier 2005). The *star markers* represent the mass function of ρ Oph (primary) pre-main sequence objects as derived from a mid-IR survey with ISOCAM (Bontemps et al. 2001). The *thin dotted line* shows a $N(>M) \propto M^{-0.6}$ power-law distribution corresponding to the typical mass spectrum found for CO clumps (see Blitz 1993)

for at least two reasons: (a) there is good evidence that dense cores represent the basic units of star formation on galactic scales (e.g. Gao and Solomon 2004), and (b) the stellar Initial Mass Function (IMF) appears to be largely determined at the prestellar core stage (e.g. Motte et al. 1998). Several (sub)-millimeter dust continuum surveys of nearby, compact cluster-forming clouds such as ρ Ophiuchi, Serpens, and Orion B have indeed uncovered 'complete' (but small) samples of self-gravitating prestellar condensations whose associated mass distributions resemble the stellar IMF from ~$0.1 M_\odot$ to ~$5 M_\odot$ (e.g. Motte et al. 1998 and Fig. 1; see also Testi and Sargent 1998; Motte et al. 2001; Stanke et al. 2006). These findings strongly support scenarios according to which there is a one-to-one (or one-to-two) correspondence between prestellar core mass and stellar mass, and the bulk of the IMF is at least partly determined by pre-collapse, gravo-turbulent cloud fragmentation (cf. Larson 1985; Klessen and Burkert 2000; Padoan and Nordlund 2002). Subsequent, additional processes such as rotationally-driven core fragmentation *during collapse* (e.g. Goodwin et al. 2007) are however required to account for the formation of binary systems and fully explain the low-mass end of the IMF.

Observationally, the best way to make progress on these issues is through detailed submillimeter line and continuum studies of prestellar cores and Class 0 protostars (see André et al. 2000; Di Francesco et al. 2007; Ward-Thompson et al. 2007 for reviews). *Herschel*, the Far InfraRed and Submillimeter Telescope to be launched by ESA in 2008, will make

possible complete surveys for prestellar cores and Class 0 objects down to the proto-brown dwarf regime in the nearby cloud complexes of the Gould Belt (cf. André and Saraceno 2005). These *Herschel* surveys will provide rich databases, including in the Southern hemisphere, for follow-up high-resolution studies of statistically representative samples of prestellar cores and protostars with ALMA. ALMA will be a powerful tool to carry out detailed studies of the structure and kinematics of the sources found with *Herschel*, which will greatly help to discriminate between competing models of core formation and evolution. Here, we discuss the unique potential of ALMA for identifying proto-brown dwarfs (Sect. 2), constraining the formation process of binary (proto)stars (Sect. 3), and investigating the structure and dynamics of protoclusters (Sect. 4).

2 Searching for proto-brown dwarfs

Accounting for the ultra-low-mass end of the IMF ($5 M_{\mathrm{Jup}} \lesssim m < 0.075 M_\odot$) is a particularly severe test for any detailed theory of star/core formation. In the turbulent fragmentation scenario (Padoan and Nordlund 2002), brown dwarfs form essentially in the same way as normal stars, from self-gravitating starless gas condensations of substellar mass or (pre)proto-brown dwarfs. In this theoretical scenario, gas condensations with a wide range of masses are formed from material compressed by shocks resulting from supersonic turbulence. At low masses, this includes both gravitationally bound and unbound condensations. A fraction of the low-mass condensations produced by turbulence are dense enough to be gravitationally unstable to collapse: their masses exceed the local Jeans or Bonnor–Ebert mass. Turbulence can indeed generate large enough pressure or density enhancements ($\delta n > n_{\mathrm{BD}}$) for the corresponding critical Bonnor–Ebert mass, $m_{\mathrm{BE}} = 0.1 M_\odot\,(T/10\,\mathrm{K})^{3/2}(\delta n/10^6\,\mathrm{cm}^{-3})^{-1/2}$ to become smaller than the maximum brown-dwarf mass $m_{\mathrm{BD}} = 0.075 M_\odot$. Padoan and Nordlund (2004) have shown that the abundance of proto-brown dwarfs produced by turbulent fragmentation is consistent with the observed abundance of brown dwarfs, assuming initial cloud conditions typical of dense cluster-forming regions such as the ρ Oph protocluster. To summarize, the Padoan and Nordlund scenario generates a large number of proto-brown dwarfs, which can be approximately described as critical Bonnor–Ebert isothermal spheres of mass $m_{\mathrm{BE}} < m_{\mathrm{BD}}$. As illustrated in Fig. 2, these proto-brown dwarfs are expected to be dense ($n \gg n_{\mathrm{BD}} \approx 2 \times 10^6\,\mathrm{cm}^{-3}$), compact ($R \sim R_{\mathrm{BE}} \ll R_{\mathrm{BD}} \approx 800\,\mathrm{AU}$), to exhibit small, thermally-dominated velocity dispersions, and to be associated with large velocity jumps tracing strong shock compressions.

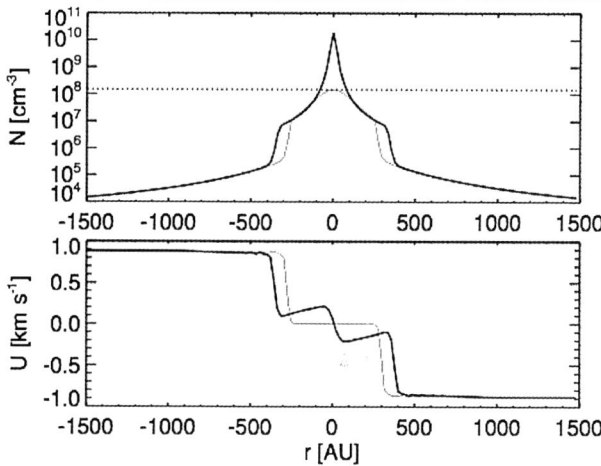

Fig. 2 Density and velocity profiles through a proto-brown dwarf dense core forming from a convergent flow in a model of turbulent cloud fragmentation (from Whitworth et al. 2007). Note the large density and velocity jumps at the location of the shock (∼300 AU) near the core boundary. ALMA will have the resolution and sensitivity to observe these proto-brown dwarf signatures up to a few kpc

A major alternative theoretical picture has been proposed, however, in which there is no direct correspondence between prestellar core mass and stellar mass, and the IMF results entirely from dynamical interactions and competitive accretion at the protostellar/pre-main sequence stages (e.g. Bate et al. 2002, 2003). Here, each prestellar core typically fragments into an unstable small-N body system (with $N \gtrsim 3$–5) during protostellar collapse, which then disintegrates into binaries and single stars in less than $\sim 10^5$ yr. In this alternative picture, *proto-brown dwarfs do not exist* and brown dwarfs are formed as aborted stellar embryos, dynamically ejected from such unstable multiple systems before they can accrete enough mass to become true stars (Reipurth and Clarke 2001; Bate et al. 2002; Goodwin et al. 2004).

Because proto-brown dwarfs are expected to be very compact and to be characterized by large velocity contrasts over their parent clouds (cf. Fig. 2), they will be easy dust continuum targets for ALMA in nearby cluster-forming clouds, provided of course that they exist. Future single-dish submillimeter continuum surveys with *Herschel* and SCUBA2 will have the sensitivity to identify candidate proto-brown dwarfs (see Greaves et al. 2003), but only a sensitive millimeter interferometer can unambiguously confirm the nature of such compact objects. Alternatively, if brown dwarfs are formed as aborted stellar embryos ejected during the collapse phase, a large number of substellar companions should be detectable with ALMA in the outer envelopes of solar-type Class 0 protostars (cf. Reipurth and Clarke 2001).

3 Testing binary fragmentation models

It is generally believed that multiple systems form by dynamical fragmentation at the end of the first collapse phase of prestellar cores when the central H_2 density reaches values close to $n_{crit} \sim 3 \times 10^{10}$ cm^{-3} and the equation of state of the gas switches from isothermality to adiabacity (cf. Goodwin et al. 2007). Observationally, this corresponds to the very beginning of the Class 0 phase (cf. André et al. 2000), suggesting that the fragmentation physics may still be imprinted in the structure of young Class 0 objects.

Purely hydrodynamic SPH simulations of rotating cloud core collapse show that a low level of initial core turbulence (e.g. $E_{turb}/E_{grav} \sim 5\%$) is sufficient to lead to the formation of a multiple system (Goodwin et al. 2004; Hennebelle et al. 2004—see Fig. 3a). In such SPH simulations, fragmentation occurs in large ($\gtrsim 100$ AU) disk-like structures or "circumstellar accretion regions" (CARs—cf. Goodwin et al. 2007). These CARs are not rotationally supported and are highly susceptible to spiral instabilities which fragment them into small-N systems with typical values of $N \sim 3$–5 (Goodwin et al. 2004; Delgado-Donate et al. 2003—see Fig. 3a).

On the other hand, a very different outcome is found in simulations of *magnetized* cloud collapse, as shown by the results recently obtained by (Fromang et al. 2006) with an MHD version of the RAMSES code using the AMR technique (cf. Fig. 3b). The simulations illustrated in Fig. 3a and Fig. 3b were both obtained using this code and started from the *same initial conditions* (rotating core of radius 0.015 pc and uniform density $n_{H_2} = 1.2 \times 10^6$ cm^{-3}), *except for the initial value of the magnetic field* ($B = 0$ in Fig. 3a; moderate field $\sim 1/2$ of the value required to hold the core against gravitational collapse in Fig. 3b). The presence of a non-zero magnetic field strongly modifies angular momentum transport during collapse and at least partly suppresses core fragmentation, leading to the formation of a single (as opposed to quadruple) object in the example shown in Fig. 3.

High-resolution imaging with ALMA will provide critical tests of such binary fragmentation models. The $0.2''$-resolution continuum images shown in Fig. 3 were obtained with the ALMA simulator available in the IRAM GILDAS software (cf. Pety et al. 2001) and used the above-described RAMSES simulations as synthetic input models assuming a distance of 140 pc. ALMA will clearly have the capability to discriminate between these two models ($B = 0$ vs. $B = B_{crit}/2$). This example illustrates the potential of ALMA for gaining insight into the physics of core fragmentation and multiple star formation.

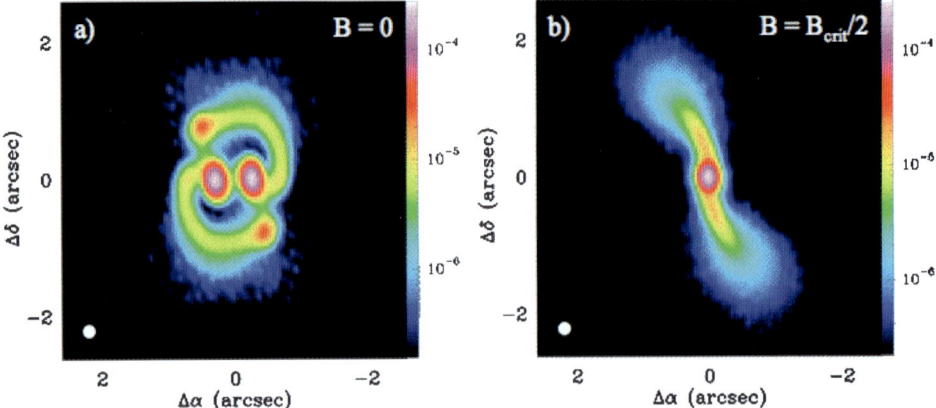

Fig. 3 Synthetic 0.2″-resolution images representing snapshot views of the column density distribution during the collapse and fragmentation of a model prestellar core observed at $d = 140$ pc. Both maps were obtained by processing model images resulting from MHD simulations of protostellar collapse (Fromang et al. 2006) with the ALMA simulator included in the GILDAS software (Pety et al. 2001). The magnetic field was set to 0 in the (purely hydrodynamic) simulation shown on the *left*, while its initial strength was half the value required to prevent collapse in the simulation shown on the *right*. In both cases, the snapshot was taken $\sim 10^4$ yr after the start of core collapse, corresponding to the early Class 0 phase of protostellar evolution. Note the presence of a quadruple system when $B = 0$, contrasting with the single object and disk-like structure obtained when $B = B_{crit}/2$

4 Structure and dynamics of protoclusters

Another exciting prospect with ALMA is that it can shed light on the demography and dynamical structure of embedded protoclusters throughout the Galaxy, while present studies are limited to the nearest cluster-forming clouds (cf. Motte et al. 1998; Testi and Sargent 1998; Peretto et al. 2006, 2007; André et al. 2007).

As illustrated by our recent study (Peretto et al. 2006, 2007) of the NGC 2264-C clump in the Mon OB1 giant molecular cloud complex ($d \sim 800$ pc), combining comprehensive, high-resolution millimeter continuum/line observations of embedded protoclusters with numerical simulations (cf. Fig. 4) can be used to set strong constraints on cluster formation models. Dust continuum mapping of NGC 2264-C at 1.2 mm with MAMBO on the IRAM 30 m telescope (e.g. Fig. 4a) resolved the column density structure of the region, uncovering a total of 12 compact prestellar/protostellar cores, at least 8 of which are Class 0-like objects with associated near-IR H_2 jets (Wang et al. 2002). The three most massive cores of the complex (C-MM2, C-MM3, C-MM4, with masses of $\sim 8 M_\odot$, $30 M_\odot$, and $25 M_\odot$, respectively) are located near the center of NGC 2264-C. Mapping in the optically thick $HCO^+(3–2)$ and $CS(3–2)$ line tracers combined with radiative transfer modelling (Peretto et al. 2006) established the presence of large-scale collapse motions (see also Williams and Garland 2002), converging onto the central, most massive core, C-MM3. A total mass inflow rate $\sim 10^{-3} M_\odot$ yr^{-1} was inferred onto C-MM3, which is comparable to the accretion rate of the McKee and Tan (McKee and Tan 2003) model for high-mass protostars. Mapping of the same region in the low-optical-depth trac-

ers $N_2H^+(1–0)$ and $H^{13}CO^+(1–0)$ revealed a sharp velocity discontinuity (~ 2 km s^{-1} in amplitude) at the location of C-MM3, tracing of a strong dynamical interaction in the central part of NGC 2264-C. This led us (Peretto et al. 2006) to propose a simple model, according to which the Class 0-like protostellar cores C-MM2 and C-MM4 (cf. Fig. 4a) were moving toward each other and dynamically interacting with C-MM3. We speculated that a massive, ultra-dense protostellar core was in the making in the central part of the NGC 2264-C clump as a result of the gravitational merger of two or more lower-mass Class 0 objects. High-resolution observations of the inner part of NGC 2264-C with the IRAM Plateau de Bure interferometer (PdBI) allowed us to resolve the central velocity discontinuity (cf. Fig. 4b). Detailed comparison of the 30 m/PdBI observations with numerical SPH simulations of the evolution of a $1000 M_\odot$ Jeans-unstable, elongated clump (Peretto et al. 2007—see Figs. 4c, d) then confirmed the view that NGC2264-C is an elongated clump collapsing and fragmenting along its long axis. Furthermore, our SPH simulations (Peretto et al. 2007) indicate that NGC 2264-C is observed at a very early stage of global clump collapse, typically $\lesssim 10^5$ yr after the start of dynamical contraction. Surprisingly, a very low level of initial turbulence (i.e., an initial turbulent to gravitational energy ratio $\alpha_{turb} \sim 5\%$) was required in the simulations to get a good match to the observations (cf. Fig. 4). This suggests that the NGC 2264-C cluster-forming clump is *structured primarily by gravity rather than by interstellar turbulence*, which, taken at face value, is at variance with the purely turbulent fragmentation scenario of (Padoan and Nordlund 2002). A significant shortcoming of our SPH simulations, however, is that they only produced the observed level of clump fragmentation when the total mass of dense ($>10^4$ cm^{-3}) gas

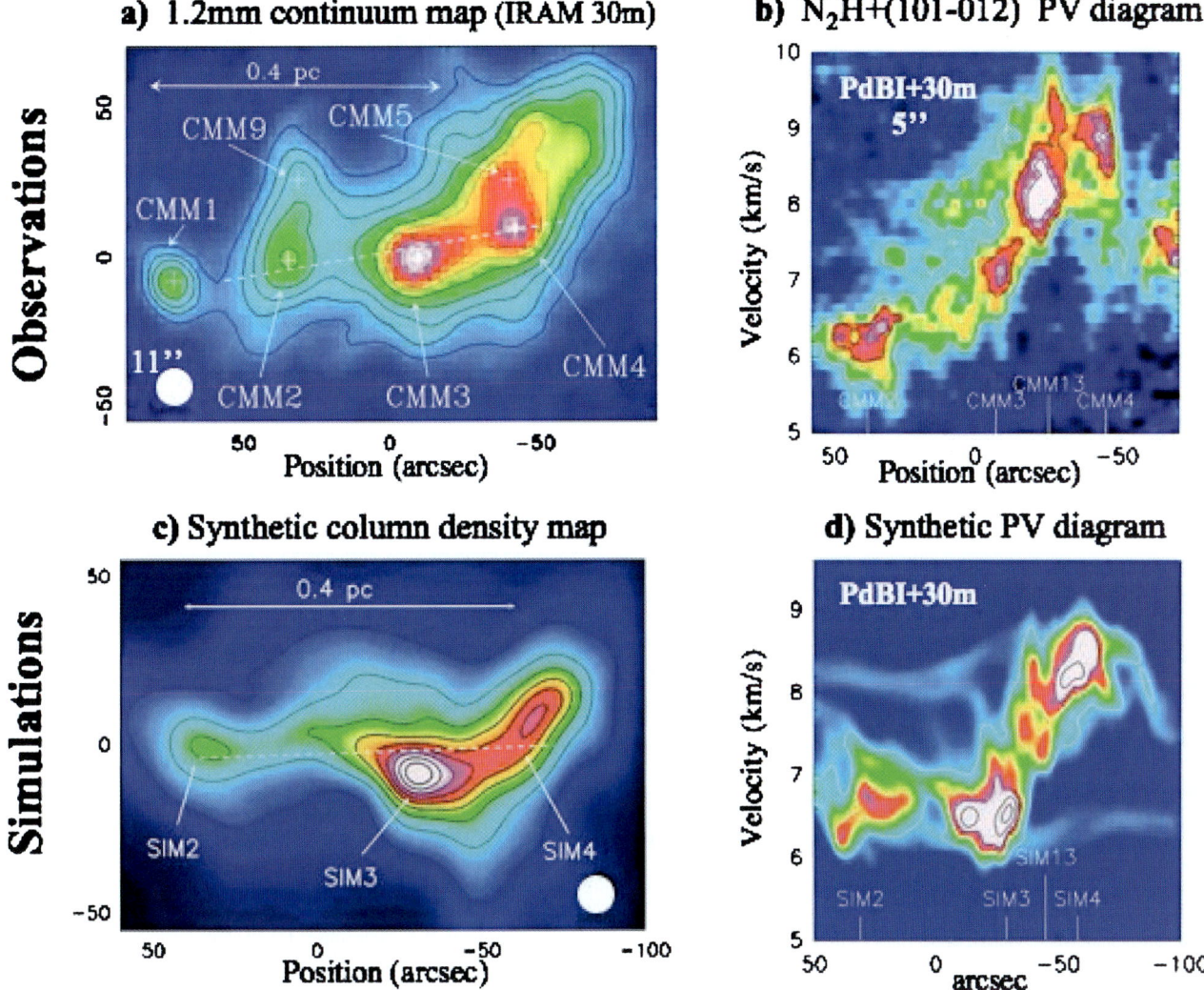

Fig. 4 Comparison between millimeter observations of the NGC 2264-C cluster-forming clump (*top row*) and numerical SPH simulations of the collapse and fragmentation of a Jeans-unstable, elongated clump (*bottom row*). **a** 1.2 mm dust continuum map obtained with MAMBO at the IRAM 30 m telescope, to be compared with the synthetic column density map shown in (**c**). **b** $N_2H^+(101-012)$ position-velocity diagram taken along the long axis of the NGC 2264-C clump, resulting from the combination of Plateau de Bure and 30 m data (effective synthesized beam $\sim 5''$). **d** Synthetic position-velocity diagram for an optically thin line tracer, convolved to the same resolution as in (**b**). The SPH simulations shown in (**c**) and (**d**) have an initial turbulent to gravitational energy ratio $\alpha_{turb} \sim 5\%$ and best fit the observations at a time step $\sim 10^5$ yr after the start of large-scale, dynamical clump contraction. (From Peretto et al. 2007.)

in the model was a factor of ~ 10 lower than in the actual NGC 2264-C clump. This points to the need for extra support against gravity, not included in the current simulations, such as support provided by magnetic fields or feedback from protostellar outflows.

Overall, our results in NGC 2264-C suggest a picture of clustered star formation intermediate between the two extreme scenarios described in Sect. 2 for brown dwarfs (Padoan and Nordlund 2002 vs. Bate et al. 2003), whereby self-gravity, (protostellar) turbulence, magnetic fields, and dynamical interactions all play some role in shaping the structure and emergent distribution of stellar masses in a protocluster.

The great advantage of ALMA in this area is that it will allow systematic studies of more distant, more massive protoclusters, including in the extreme environment of the Galactic center region which is possibly reminiscent of the star-forming conditions found in starburst galaxies. In nearby protoclusters such as NGC 2264-C, ALMA will also make it possible to probe (ultra-)low-mass star formation and intermediate- to high-mass star formation *simultaneously*. Since clustered star formation appears to be the dominant star formation mode in the Milky Way and other galaxies (e.g. Lada and Lada 2003), such studies are of fundamental importance.

5 Conclusions

ALMA will undoubtedly revolutionize our understanding of the early stages of star formation, which likely hold the key to unlocking the crucial issue of the origin and possible universality of the stellar IMF. In this short paper, we discussed three specific key problems (proto-brown dwarfs, binary fragmentation, and protocluster dynamics) for which the potential of ALMA seems particularly promising. Of course, these were only examples and ALMA will be used to make progress in many other areas of star formation research (see, e.g., papers by van Dishoeck, Cesaroni, Shepherd, Aikawa, Crutcher, and Tafalla in this volume).

References

André, P., Saraceno, P.: In: The Dusty and Molecular Universe: A Prelude to *Herschel* and ALMA, ESA SP-577, p. 179 (2005)
André, P., Ward-Thompson, D., Barsony, M.: In: Protostars and Planets IV, p. 59. University of Arizona Press, Tucson (2000)
André, P., Belloche, A., Motte, F., Peretto, N.: Astron. Astrophys. (2007, in press). ArXiv:0706.1535
Bate, M.R., Bonnell, I.A., Bromm, V.: Mon. Not. R. Astron. Soc. 332, L65 (2002)
Bate, M., Bonnell, I., Bromm, V.: Mon. Not. R. Astron. Soc. 339, 577 (2003)
Blitz, L.: In: Protostars and Planets III, p. 125. University of Arizona Press, Tucson (1993)
Bontemps, S., André, P., Kaas, A.A., Nordh, L., et al.: Astron. Astrophys. 372, 173 (2001)
Chabrier, G.: In: The Initial Mass Function 50 Years Later. ASSL vol. 327, p. 41 (2005)
Delgado-Donate, E.J., Clarke, C.J., Bate, M.R.: Mon. Not. R. Astron. Soc. 342, 926 (2003)
Di Francesco, J., Evans, N.J., Caselli, P., Myers, P.C., Shirley, Y., Aikawa, Y., Tafalla, M.: In: Protostars and Planets V, p. 17. University of Arizona Press, Tucson (2007)
Fromang, S., Hennebelle, P., Teyssier, R.: Astron. Astrophys. 457, 371 (2006)
Gao, Y., Solomon, P.: Astrophys. J. 606, 271 (2004)
Goodwin, S., Whitworth, A., Ward-Thompson, D.: Astron. Astrophys. 414, 633 (2004)
Goodwin, S.P., Kroupa, P., Goodman, A., Burkert, A.: In: Protostars and Planets V, p 133. University of Arizona Press, Tucson (2007)
Greaves, J.S., Holland, W.S., Pound, M.W.: Mon. Not. R. Astron. Soc. 346, 441 (2003)
Hartmann, L., Ballesteros-Paredes, J., Bergin, E.: Astrophys. J. 562, 852 (2001)
Hennebelle, P., Whitworth, A., Cha, S.-H., Goodwin, S.: Mon. Not. R. Astron. Soc. 348, 687 (2004)
Klessen, R., Burkert, A.: Astrophys. J. Suppl. Ser. 128, 287 (2000)
Klessen, R.S., Ballesteros-Paredes, J., Vázquez-Semadeni, E., Durán-Rojas, C.: Astrophys. J. 620, 786 (2005)
Kroupa, P.: Mon. Not. R. Astron. Soc. 322, 231 (2001)
Lada, C.J.: Astrophys. J. 640, L63 (2006)
Lada, C., Lada, E.: Annu. Rev. Astron. Astrophys. 41, 57 (2003)
Larson, R.B.: Mon. Not. R. Astron. Soc., 214, 379 (1985)
Mac Low, M.-M., Klessen, R.S.: Rev. Mod. Phys., 76, 125 (2004)
McKee, C., Tan, J.: Astrophys. J. 585, 850 (2003)
Motte, F., André, P., Neri, R.: Astron. Astrophys. 365, 440 (1998)
Motte, F., André, P., Ward-Thompson, D., Bontemps, S.: Astron. Astrophys. 372, L41 (2001)
Mouschovias, T.C., Ciolek, G.E.: In: Lada, C.J., Kylafis, N.D. (eds.) The Origin of Stars and Planetary Systems, p. 305. Kluwer, Dordrecht (1999)
Padoan, P., Nordlund, A.: Astrophys. J. 576, 870 (2002)
Padoan, P., Nordlund, A.: Astrophys. J. 617, 559 (2004)
Peretto, N., André, P., Belloche, A.: Astron. Astrophys. 445, 979 (2006)
Peretto, N., Hennebelle, P., André, P.: Astron. Astrophys. 464, 983 (2007)
Pety, J., Gueth, F., Guilloteau, S.: ALMA Memo No. 386 (2001)
Reipurth, B., Clarke, C.: Astron. J. 122, 423 (2001)
Shu, F.H., Adams, F.C., Lizano, S.: Annu. Rev. Astron. Astrophys. 25, 23 (1987)
Shu, F.H., Li, Z.-Y., Allen, A.: Astrophys. J. 601, 930 (2004)
Stanke, T., Smith, M.D., Gredel, R., Khanzadyan, T.: Astron. Astrophys. 447, 609 (2006)
Testi, L., Sargent, A.I.: Astrophys. J. Lett. 508, L91 (1998)
Wang, H., Yang, J., Wang, M., Yan, J.: Astron. Astrophys. 389, 1015 (2002)
Ward-Thompson, D., André, P., Crutcher, R., Johnstone, D., Onishi, T., Wilson, C.: In: Protostars and Planets V, p. 33. University of Arizona Press, Tucson (2007)
Whitworth, A., Bate, M.R., Nordlund, A., Reipurth, B., Zinnecker, H.: In: Protostars and Planets V, p. 459. University of Arizona Press, Tucson (2007)
Williams, J., Garland, C.: Astrophys. J. 568, 259 (2002)

Chemistry in low-mass star forming regions

ALMA's contribution

Yuri Aikawa

Originally published in the journal Astrophysics and Space Science, Volume 313, Nos 1–3.
DOI: 10.1007/s10509-007-9593-6 © Springer Science+Business Media B.V. 2007

Abstract We review molecular evolution in low-mass star-forming regions and discuss what we can observe with ALMA. Recent observations have revealed chemical fractionation, i.e. spatial variation of molecular abundances, in dense prestellar cores. In the central regions of cold prestellar cores, CO is heavily depleted, while the depletion of N-bearing species are rare. Models show that CO is frozen onto grains, while N-bearing species survive because of the CO depletion and slow formation of N_2 in the gas phase. CO depletion also enhances the molecular D/H ratio. Chemical fractionation and its variation among cores can be an indicator of evolutionary stage and/or accumulation process of cores.

As the core contracts, central region of the core is eventually heated by compressional heating and a new-born protostar. CO is sublimated back to the gas phase, if the temperature reaches 20 K. Warm temperature enhances the endothermic reactions which were negligible in the prestellar core stage, and also enhances grain-surface reactions among heavy-element species to form large organic molecules, which sublimate when the temperature reaches ~ 100 K. Warm regions with high abundances of the gaseous organic species are called hot corinos or low-mass hot cores. Adopting a theoretical model of core contraction, we present the temporal variation of the radius inside which CO and large organic species are sublimated. We also investigate the molecular evolution in infalling shells to derive molecular distribution in a protostellar core.

Y. Aikawa (✉)
Department of Earth and Planetary Sciences, Kobe University,
657-8501, Kobe, Japan
e-mail: aikawa@kobe-u.ac.jp

Keywords Stars: formation · ISM: molecules · ISM: clouds

1 Introduction

Stars are formed by contraction of molecular cloud cores. Since the kinetics of cores are probed by molecular lines, chemistry is important in deriving the star-formation processes from the observational data. Prestellar cores are dense cores without infrared sources. Some prestellar cores show the asymmetric line profile which indicates gravitational contraction. Protostellar cores are dense cores with infrared sources (i.e. protostars). They are still accreting towards the protostar or interacting with the outflow. Recent observations show that the molecular abundances in cores vary spatially and temporally.

In Sect. 2, we review the observations and models of chemistry in prestellar cores, namely the chemical fractionation and deuterium enrichment. Chemistry in protostellar stage, i.e. formation of large organic species and sublimation of ice, is described in Sect. 3.

2 Chemistry in prestellar cores

2.1 Chemical fractionation

Chemical fractionation in prestellar cores has been intensively studied in last several years (e.g. Caselli et al. 1999; Tafalla et al. 2004, 2006). Figure 1a is the integrated intensity map of CCS and N_2H^+ towards L1544 (Ohashi et al. 1999; Aikawa et al. 2001). Cross indicates the peak position of the dust continuum. While the dust continuum and N_2H^+ intensities have a peak at the core center, CCS emission is

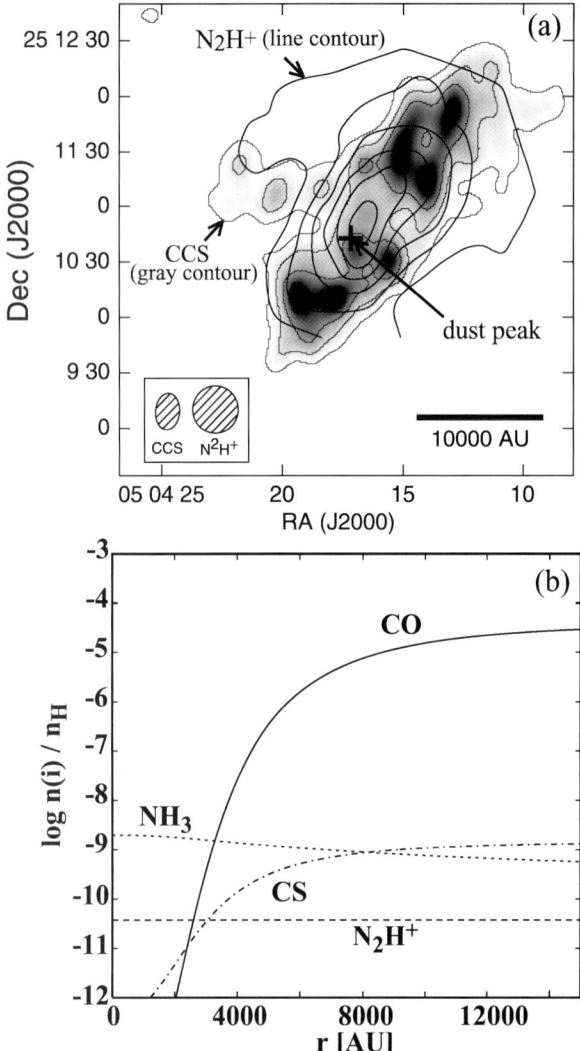

Fig. 1 (**a**) Intensity map of L1544 (from Aikawa et al. 2001). (**b**) Radial distribution of molecular abundance in L1544 (Tafalla et al. 2002; Aikawa et al. 2005)

stronger in the outer radius. Comparison between dust continuum and molecular lines reveals the radial distribution of molecular abundances (Fig. 1b). Carbon-bearing species are depleted in the central region, while the N-bearing species keep almost constant abundances (Tafalla et al. 2002).

Theoretical models show that early-phase species such as CS and CCS are destroyed by gas-phase reactions, and CO is frozen onto grains (Bergin and Langer 1997; Aikawa et al. 2001, 2005). Under the typical gas density in molecular clouds ($n_H \sim 10^3$–10^4 cm^{-3}), the freeze-out timescale is about Myr, which is comparable to the gravitational contraction timescale. Hence it is natural to find CO depletion right before the star formation. Theoretical models also show that non-depletion of N_2H^+ and NH_3 are caused by depletion of CO (Aikawa et al. 2001, 2005; Maret et al. 2006). When CO is abundant N_2H^+ reacts mainly with CO to produce N_2. Although N_2 is also subject to freeze-out, the destruc-

tion rate of N_2H^+ temporarily decreases as CO depletes. After the CO depletion, a fraction of N_2H^+ recombines with an electron to produce NH, which is transformed to NH_3 by gas-phase reactions. Slow-formation of N_2 also helps to reproduce the observed distribution of N-bearing species. Non-depletion of N_2H^+ and NH_3 is, however, temporal. Eventually they freeze-out onto grains in the form of N_2 and NH_3. In fact, depletion of N_2H^+ is observed in some objects such as B68 and L183 (Bergin et al. 2002; Pagani et al. 2005).

With ALMA we can get deeper look at the freeze-out region, which gives more severe constraint on the depletion factor and on the chemical interactions between freeze-out and gas-phase reactions.

2.2 Deuterium enrichment

Prestellar cores also show deuterium enrichment. For example, Bacmann et al. (2003) found D_2CO/H_2CO column density ratio of 0.01–0.14 in several prestellar cores. See Lis et al. in this volume for more examples.

Deuterium enrichment originates in some exothermic D-H exchange reactions such as

$$H_3^+ + HD \rightarrow H_2D^+ + H_2. \tag{1}$$

The back reactions are endothermic, and thus are negligible at 10 K. Multiply deuterated H_3^+ (i.e. HD_2^+ and D_3^+) are similarly produced subsequently (Roberts et al. 2003).

CO depletion further enhances the D/H ratio. H_3^+ and its isotopomers are mainly destroyed by the reaction with CO and recombination

$$H_2D^+ + CO \rightarrow HCO^+ + HD \tag{2}$$

$$H_2D^+ + e \rightarrow D + H_2$$

$$\text{or } H + HD. \tag{3}$$

The abundance ratio of H_2D^+/H_3^+ is given as a function of the CO and electron abundances, and reaction rate coefficients k of the reactions (1)–(3),

$$\frac{n(H_2D^+)}{n(H_3^+)} = \frac{k_1 n(HD)}{k_2 n(CO) + k_3 n(e)}. \tag{4}$$

As CO depletes, deuterated H_3^+ (not only H_2D^+ but also HD_2^+ and D_3^+) become more abundant relative to H_3^+.

Deuterium enrichment in H_3^+ propagates to other molecules, because H_3^+ is the key reactant in the ion-molecule reaction network. Deuteration also occurs on grain surfaces. Because of the dissociative recombination of deuterated H_3^+, atomic D/H ratio is enhanced in the gas phase. These atoms are adsorbed onto grains and hydrogenate other atoms and molecules. Recently Nagaoka et al. (2005) found that

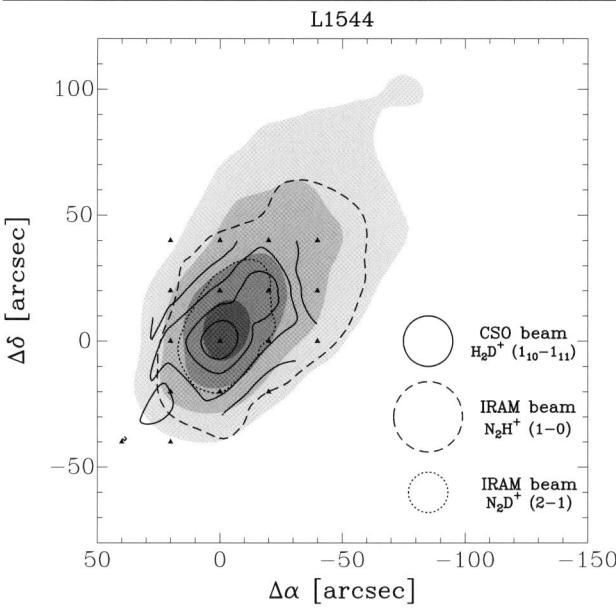

Fig. 2 Integrated intensity maps of H_2D^+ ($1_{1,0}$–$1_{1,1}$), N_2H^+ (1–0) and N_2D^+(2–1) superposed on the 1.3 mm continuum emission map of the prestellar core L1544 (Vastel et al. 2006)

H-D exchange reactions on grain surfaces further enhance the D/H ratio in methanol in the laboratory experiment.

It should be noted that H_2D^+ and HD_2^+ have emission lines at 372 GHz and 692 GHz, respectively, and are very good tracers of the central region of the cold prestellar cores where heavy element species are depleted. Figure 2 is the integrated intensity map of H_2D^+ and other tracers in L1544 (Vastel et al. 2006). ALMA will provide such maps with higher spatial resolutions.

It should also be noted that as soon as the core is heated by the star-formation, H_2D^+ decreases rapidly. At temperature of 50 K and density of $n_H \sim 10^6$ cm^{-3}, for example, the timescale of the back reaction of (1) is only 10^4 sec. Other deuterated species without such direct exchange will survive to be observed in protostellar stage (see Lis et al. in this volume).

2.3 Variation among cores

Degree of the deuterium enrichment and molecular depletion varies among cores. L1544 is the best-studied core with high D/H ratios and heavy CO depletion. On the other hand, L1521B shows relatively low D/H ratios and centrally peaked CCS distribution (Hirota et al. 2004). NH_3 line is weak and N_2H^+ is not detected towards this object. L492 is an intermediate type, with low D/H ratios, centrally peaked CCS, but with N_2H^+ and NH_3 lines are stronger than in L1521B (Hirota and Yamamoto 2006). In comparison with chemistry models, L1521B is considered to be chemically young, while L1544 is chemically old. Interestingly, such classification coincides with physical properties;

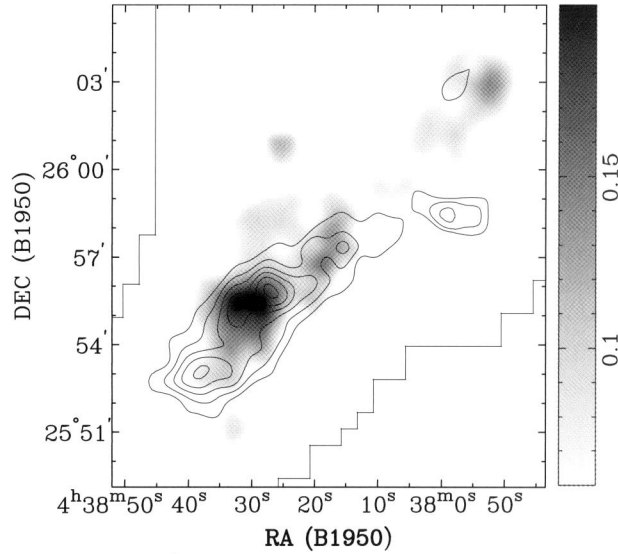

Fig. 3 Intensity map of HCO^+ (*line contour*) and CH_3OH (*gray contour*) towards TMC-1C (by courtesy of S. Takakuwa)

L1544 shows an infall signature, while L1521B does not. It indicates that chemistry can be a probe of evolutionary state of cores (Hirota et al. 2004).

However, some cores have similar central densities in spite of the different chemical status. Then another interpretation of the chemical variation among cores is these cores have different contraction and/or accumulation timescale, because molecular distribution is determined by a balance between chemical timescale and dynamical time scale (Aikawa et al. 2001, 2005; Lee et al. 2003).

ALMA will provide more statistics on correlation between chemical and physical properties of cores.

2.4 Chemical fractionation in lower density regions

Although chemical fractionation in the dense prestellar cores is relatively well understood, lower density regions are more mysterious. Figure 3 shows the intensity map of HCO^+ and CH_3OH towards TMC-1C (Takakuwa et al. 1998). Intensity distribution varies significantly with species. Furthermore, a high-resolution observation of the methanol core revealed very small clumps, which are estimated to be gravitationally unbound (Takakuwa et al. 2003). Correlation between molecular abundances and physical parameter is not yet found. Such small clumps and chemical fractionation are of importance not only from a chemical point of view, but also in relation to the formation of cores and clouds (see also Takakuwa et al. in this volume).

3 Chemistry in protostellar cores

When the gravity overwhelms the thermal pressure, turbulent pressure, and/or magnetic pressure, the core contracts to form a protostar (or protostars). Figure 4 shows the temporal variation of (a) density, (b) temperature, and (c) infall velocity in a spherical core (Masunaga and Inutsuka 2000). Labels in panel (b) depict the time; the protostar is born at $t_{core} = 0$. The gray lines represent the prestellar phase (i.e. $t_{core} < 0$), while the black lines represent the protostellar phase ($t_{core} \geq 0$). A core stays almost isothermal as long as the radiation cooling is efficient. When the compressional heating overwhelms the cooling, the core center gets warmer. Eventually the protostar is born, which further heats the cores. As the core temperature rises, ice sublimates. CO sublimates at 20 K, water at 160 K, and large organic species such as CH_3OH sublimates at around 100 K. It should be noted that ice is formed not only by direct adsorption of gaseous species during the prestellar stage but also modified by grain-surface reactions. Sublimates experience further reactions in the gas phase.

3.1 Sublimation of CO

CO sublimation is an important event for radio observers, not only because CO becomes observable again, but also because it significantly changes the abundances of other species. For example, N_2H^+ is destroyed by the reaction with CO (Lee et al. 2004).

Sublimation radius, inside which a certain molecular species is thermally desorbed to the gas phase, changes with time. Figure 5 depicts the sublimation radius of CO as a function of the total luminosity of the core, which increases with time (Aikawa et al. 2007). The plot starts about 800 years before the birth of a protostar, when the first core is about to be formed. The first core collapses due to the dissociation of hydrogen molecule. When the dissociation completes, the second core (i.e. protostar) is formed. Then CO is sublimated within 100 AU. When the protostar is about 10^5 years old, the total luminosity of the core reaches $25\ L_\odot$, and CO is sublimated within several 1000 AU. In the nearest star-forming regions, ALMA can spatially resolve the CO sublimation region at the moment of protostar formation.

3.2 Low-mass hot cores

Figure 5 also shows the radius of 100 K, where large organic species such as CH_3OH are sublimated. These species have been known to be abundant in hot cores in high-mass star-forming regions, but are found to be also abundant around low-mass protostars in recent years. Spatial distribution of large organic species varies with species in low-mass hot cores as well as in high-mass hot cores for a reason yet to be investigated (Kuan et al. 2004; Remijan and Hollis 2006).

Formation of these species was previously investigated in the context of hot-cores in high-mass star forming regions.

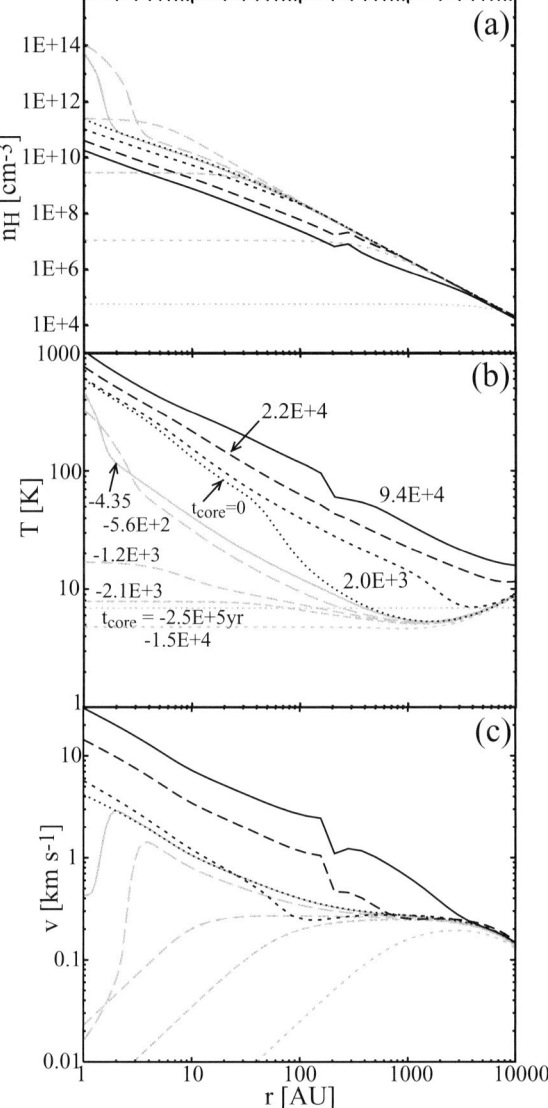

Fig. 4 Temporal variation of (**a**) temperature, (**b**) density, and (**c**) infall velocity in a collapsing spherical core. The labels in (**b**) depict the age of the core t_{core} in yr. A protostar is born at $t_{core} = 0$

A classical scenario goes as follows. In prestellar cores before star-formation, molecules freeze-out onto dust grains. Grain-surface reactions hydrogenate, for example, CO to methanol. When the core gets warm, the ice sublimates and reacts with each other in the gas phase to form large organic species. However, recent works suggest the gas-phase reactions may not produce these organic species so efficiently. Theoretical calculation shows formation of methyl formate ($HCOOCH_3$) in the gas phase is inefficient (Horn et al. 2004). Laboratory experiments revealed that large ion species are broken to pieces when they recombine (Geppert et al. 2006).

Garrod and Herbst (2006) suggested that grain-surface reactions during warm-up phase can be an alternative or additional mechanism to form large organic species. Grain-

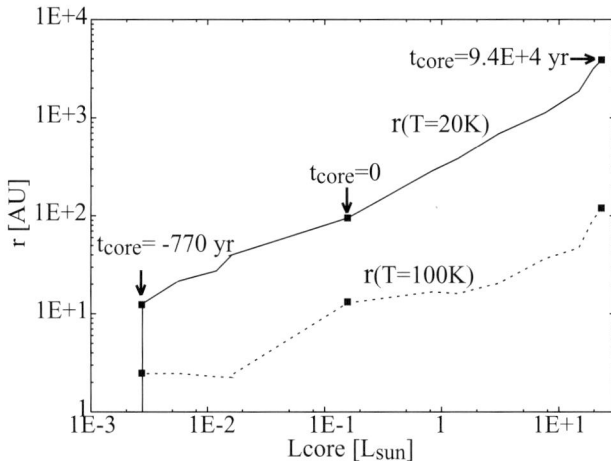

Fig. 5 Sublimation radius of CO ($r(T = 20K)$) and large organic species ($r(T = 100$ K)) as a function of the total core luminosity

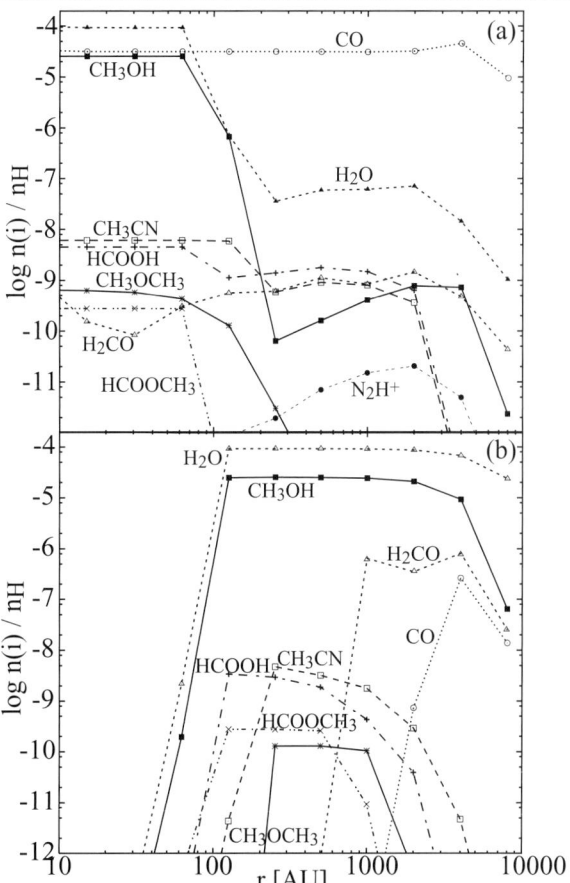

Fig. 6 Radial distribution of molecular abundances (**a**) in the gas-phase and (**b**) in the ice mantle in a protostellar core after 9×10^4 yrs after the birth of a protostar

surface reactions are mostly hydrogenation under low temperature ($\sim$10 K). In the warm-up phase of several tens K, on the other hand, hydrogen atom sublimates so quickly, that hydrogenation is not efficient. Instead, heavy-element species can now migrate and react with each other to form large organic species.

In order to evaluate the efficiency of these mechanisms in low-mass cores, we adopt the chemical model of Garrod and Herbst (2006) to calculate the molecular evolution in infalling shells in the model core of Masunaga and Inutsuka (2000) (Fig. 4) (Aikawa et al. 2007). It should be noted that the warm (or high) temperature region is much smaller in low-mass cores, and thus the fluid parcels pass through the warm temperature region in much shorter time scale than in high-mass hot cores. For example, at the latest stage in the model (i.e. $t_{\text{core}} = 9 \times 10^4$ yr), the temperature is higher than 100 K inside $\sim$100 AU (Fig. 4). Since the infall velocity of shells are $\gtrsim$1 km s^{-1}, we have only 100 yrs to cook material before it falls onto the central star, if large organic species are formed in the gas-phase after the sublimation of CH$_3$OH at 100 K.

Figure 6 shows the radial distribution of (a) gas and (b) ice in a protostellar core at $t_{\text{core}} = 9 \times 10^4$ yr (Aikawa et al. 2007). Gaseous organic species are abundant in the central regions. CH$_3$CN and HCOOH extend to $\sim$1000 AU, while CH$_3$OH and CH$_3$OCH$_3$ abundance sharply rises inwards at 100 AU. These large organic species are mostly formed via grain-surface reactions at 20–40 K, and then desorbed to the gas phase at the sublimation temperature of each species.

ALMA will greatly contribute to the chemical studies in low-mass protostellar cores. The high sensitivity enables detections of weak lines of complex species, and tells us how complex the interstellar species can be, although a line contamination would impede the detection of very weak lines (see Herbst, in this volume). For example, if we do not pur-

sue spatial resolution, 4 min integration at ALMA corresponds to 18 hr integration at Nobeyama 45 m (N. Sakai, private communication). Since emission regions of large organic species are mostly compact, spatially resolved observations are important in order to evaluate the molecular abundances without suffering from the beam dilution. Spatial distribution of molecules and statistical observation of cores with various evolutionary stage are key to understand the formation mechanism of large organic species (e.g. Sakai et al., in this volume).

In the model described above, we assumed spherical symmetry. But in reality magnetic field and rotation will break the spherical symmetry; they not only elongate the core (envelope) structure, but also forms a circumstellar disk, which is the birth place of a planetary system. With ALMA's high spatial resolution we can discriminate the forming disk from the envelope component. Such observation will naturally connect the chemical studies in cores to the chemistry in disks and planetary systems.

Acknowledgements I would like to thank Charlotte Vastel and Shigehisa Takakuwa for providing figures from their works. I also thank Valentine Wakelam, Eric Herbst and Robin T. Garrod for useful discussions on chemistry in protostellar cores. This work is supported by Grants-in-Aid for Scientific Research (17039008, 18026006) and "The 21st Century COE Program of Origin and Evolution of Planetary Systems" of the Ministry of Education, Culture, Sports, Science, and Technology of Japan (MEXT).

References

Aikawa, Y., Ohashi, N., Inutsuka, S., Herbst, E., Takakuwa, S.: Molecular evolution in collapsing prestellar cores. Astrophys. J. **552**, 639 (2001)

Aikawa, Y., Herbst, E., Roberts, H., Caselli, P.: Molecular evolution in collapsing prestellar cores. III. Contraction of a Bonnor-Ebert sphere. Astrophys. J. **620**, 330 (2005)

Aikawa, Y., Wakelam, V., Herbst, E., Garrod, R.T.: Astrophys. J. (2007, submitted)

Bacmann, A., Lefloch, B., Ceccarelli, C., Steinacker, J., Castets, A., Loinard, L.: CO depletion and deuterium fractionation in prestellar cores. Astrophys. J. **585**, L55 (2003)

Bergin, E.A., Langer, W.D.: Chemical evolution in preprotostellar and protostellar cores. Astrophys. J. **486**, 316 (1997)

Bergin, E.A., Alves, J., Huard, T., Lada, C.J.: N_2H^+ and $C^{18}O$ depletion in a cold dark cloud. Astrophys. J. **570**, L101 (2002)

Caselli, P., Walmsley, C.M., Tafalla, M., Dore, L., Myers, P.C.: CO depletion in the starless core L1544. Astrophys. J. **523**, L165 (1999)

Garrod, R.T., Herbst, E.: Formation of methyl formate and other organic species in the warm-up phase of hot molecular cores. Astron. Astrophys. **457**, 927 (2006)

Geppert, W.D., Thomas, R.D., Ehlerding, A. et al.: Faraday Discuss. 133, paper 13 (2006)

Hirota, T., Yamamoto, S.: Molecular line observations of carbon-chain-rich core L492. Astrophys. J. **646**, 258 (2006)

Hirota, T., Maezawa, H., Yamamoto, S.: Molecular line observations of carbon-chain-producing regions L1495B and L1521B. Astrophys. J. **617**, 399 (2004)

Horn, A., et al.: The gas-phase formation of methyl formate in hot molecular cores. Astrophys. J. **611**, 605 (2004)

Kuan, Y.J., et al.: Organic molecules in low-mass protostar hot cores: submillimeter imaging of IRAS 16293-2422. Astrophys. J. **616**, L27 (2004)

Lee, E.-L., Evans, N.J.II., Shirley, Y.L., Tatematsu, K.: Chemistry and dynamics in pre-protostellar cores. Astrophys. J. **583**, 789 (2003)

Lee, J.E., Bergin, E.A., Evans, N.J.II: Evolution of chemistry and molecular line profiles during protostellar collapse. Astrophys. J. **617**, 360 (2004)

Maret, A., Bergin, E.A., Lada, C.J.: A low fraction of nitrogen in molecular form in a dark cloud. Nature **442**, 425 (2006)

Masunaga, H., Inutsuka, S.: A radiation hydrodynamics model for protostellar collapse. II. The second collapse and the birth of a protostar. Astrophys. J. **531**, 350 (2000)

Nagaoka, A., Watanabe, N., Kouchi, A.: H-D substitution in interstellar solid methanol: A key route for D enrichment. Astrophys. J. **624**, L29 (2005)

Ohashi, N., Lee, S.W., Wilner, D.J., Hayashi, M.: CSS imaging of the starless core L1544: an envelope with infall and rotation. Astrophys. J. **518**, L41 (1999)

Pagani, L., Pardo, J.-R., Apponi, A.J., Bacmann, A., Cabrit, S.: L183 (L134N) revisited III. The gas depletion. Astron. Astrophys. **429**, 181 (2005)

Remijan, A.J., Hollis, J.M.: IRAS, 16293-2422: Evidence for infall onto a counterrotating protostar accretion disk. Astrophys. J. **640**, 842 (2006)

Roberts, H., Herbst, E., Millar, T.J.: Enhanced deuterium fractionation in dense interstellar cores resulting from multiply deuterated H_3^+. Astrophys. J. **591**, L41 (2003)

Tafalla, M., Myers, P.C., Caselli, P., Walmsley, C.M., Comito, C.: Systematic molecular differentiation in starless cores. Astrophys. J. **569**, 815 (2002)

Tafalla, M., Myers, P.C., Caselli, P., Walmsley, C.M.: On the internal structure of starless cores. I. Physical conditions and the distribution of CO, CS, N_2H^+ and NH_3 in L1498 and L1517B. Astron. Astrophys. **416**, 191 (2004)

Tafalla, M., Santiago-Garcia, J., Myers, P.C., Caselli, P., Walmsley, C.M., Crapsi, A.: Astron. Astrophys. **455**, 577 (2006)

Takakuwa, S., Mikami, H., Saito, M.: $H^{13}CO^+$ and CH_3OH line observations of prestellar dense cores in the TMC-1C region. Astrophys. J. **501**, 723 (1998)

Takakuwa, S., Kamazaki, T., Saito, M., Hirano, N.: $H^{13}CO^+$ and CH_3OH line observations of prestellar dense cores in the TMC-1C region. II. internal structure. Astrophys. J. **584**, 818 (2003)

Vastel, C., et al.: Distribution of ortho-$H_2D^+(1_{1,0}-1_{1,1})$ in L1544: Tracing the deuteration factory in prestellar cores. Astrophys. J. **645**, 1198 (2006)

Molecular outflows observed with ALMA

Debra S. Shepherd

Originally published in the journal Astrophysics and Space Science, Volume 313, Nos 1–3.
DOI: 10.1007/s10509-007-9594-5 © Springer Science+Business Media B.V. 2007

Abstract Optical, infrared and radio (single dish and inter-ferometric) observations of jets and outflows from newly formed stars have helped to improve our understanding of molecular outflows and the outflow/accretion connection. However, once the Atacama Large Millimeter Array (ALMA) is completed, it will provide a significant increase in sensitivity and resolution at millimeter and sub-millimeter wavelengths that will allow astronomers to address critical issues that cannot be explored with established observatories. Of particular importance is that ALMA will recover both extended and compact emission from the large scale molecular cloud and outflows to compact cores and disks. Thus, we will be able to study the detailed kinematics of the outflows, entrainment properties, momentum transfer and feedback, and collimation. I will review our current observational limitations and provide examples of how ALMA will contribute to the study of molecular outflows in star forming regions.

Keywords Stars: formation · ISM: clouds · H(II) regions · Jets · Outflows

1 Introduction

Molecular outflows from young stellar objects (YSOs) provide critical angular momentum transport from the accretion disk into the surrounding environment allowing accretion to proceed beyond $\sim 0.1 M_\odot$. Indeed, outflows affect more than just the central stars. Outflow and infall dynamics affect the energy input and turbulence in molecular clouds, the chemical composition of the host cloud, dissipation of molecular gas, and, finally, disk (and planet) evolution. Molecular outflows also provide a valuable fossil record of the mass-loss history of a protostar or protostellar cluster.

The Atacama Large Millimeter Array (ALMA) will provide a significant increase in sensitivity and resolution at millimeter and sub-millimeter wavelengths that will allow astronomers to address critical issues associated with molecular outflows and the link between accretion and outflow. In this short overview, I will briefly discuss some observational issues that ALMA will be able to contribute to and provide a few examples of how one might go about setting up a project to explore outflow kinematics.

2 Outflow properties

There are a wide range of observational issues that ALMA will be able to contribute to including outflow collimation, precession & entrainment, chemistry and cluster formation and interaction.

Outflow collimation: Molecular outflows from low-mass YSOs tend to have relatively well-collimated geometries early in their development and then evolve to have both collimated and a wider-opening angle flow component in later stages. Recent examples of low-mass YSOs that exhibit both collimated jet-like components and wide-angle molecular flows include L1228 (e.g. Arce and Sargent 2004, 2005) and HH 30 (Pety et al. 2006).

The National Radio Astronomy Observatory is a facility of the National Science Foundation operated under cooperative agreement by Associated Universities, Inc.

D.S. Shepherd (✉)
Scientific Services, National Radio Astronomy Observatory, 1003 Lopezville Rd., P.O. Box 0, Socorro, NM 87801, USA
e-mail: dshepher@aoc.nrao.edu

The general trend of collimated to wide-angle flow evolution appears to hold for more massive stars (up to about late O stars) however the high-velocity jet-component in older flows ($\sim 10^5$ years) is most often either missing or undetectable with current telescopes. A few early B stars show a well-collimated jet (HH 80-81) or a moderately collimated jet (IRAS 20126+4104) in conjunction with a wide-angle molecular flow. The evolutionary time scale for massive stars is much shorter and the underlying physics that affects the flow geometry, energetics and chemistry will likely have a different balance due to the intense radiation field of the star as it first reaches the main sequence while it is still accreting and driving an energetic outflow (see, e.g., recent reviews by Shepherd 2005; Cesaroni 2005 and Arce et al. 2007).

At early times (less than 10^3–10^4 years) collimated flows are relatively compact and tend to be marginally resolved in the width. Older, wide-angle flows are larger on the sky, and, while they can have significant diffuse emission, they can also have compact structure such as shells, ridges and knots within the flow. Thus, it can be difficult to obtain high fidelity images of older outflows with either interferometers or single dish telescopes. To make this problem more tractable, a standard ALMA observing mode will be to collect both interferometric and single dish observations of a field and perform a joint-deconvolution on the combined image.

Outflow precession & entrainment: The underlying wind or jet that creates the molecular flow is known to precess in many sources. In lower mass YSOs this precession is typically only a few degrees. Flows with tens of degrees of precession are occasionally seen from more luminous objects. Precession tends to widen the outflow opening angle with age and increases the interaction with the core and molecular cloud. Thus, precession increases the entrainment of molecular material into the outflowing gas and helps to shorten the lifetime of the accretion process by removing gas from the immediate surroundings. Examples of sources that show significant precession include the low luminosity YSO L1157 (e.g. Beltrán et al. 2004; Stojimirović et al. 2006), and the early-B (proto)star IRAS 20126+4104 (e.g. Lebrón et al. 2006).

Mass entrainment does not appear to be constant nor is it a simple function of the YSO luminosity or the mass of the cloud core. There is clear evidence that both the momentum rate and energy rate in a molecular flow scales as a function of the bolometric luminosity of the source $\dot{P}_f$, $\dot{E}_f \propto L_{bol}^{0.6}$ (Cabrit and Bertout 1992; Shepherd and Churchwell 1996; Henning et al. 2000). However, the flow mass per unit time, $\dot{M}_f$, does not have a clear correlation. Rather, it has an upper limit defined by $\dot{M}_f \propto L_{bol}^{0.6}$ but it can be lower than this by nearly 2 orders of magnitude (Beuther et al. 2002). Thus,

how much molecular material that can be entrained into an outflow appears to vary by roughly 2 orders of magnitude. Our current understanding of the entrainment process is qualitative due mostly to our inability to trace the entrainment process on a micro-level. With the increased sensitivity and spatial resolution of ALMA, we can uncover the full complexity of how molecular gas is entrained in the jet or wind.

Chemistry in molecular flows: Outflows are comprised of a mix of warm (20–30 K) and hot (>1000 K) gas with velocities that can range from a few km/s to more than 1000 km/s. In general, higher J transitions allow a better determination of the temperature, density and energetics of the bulk gas and the shock-enhanced gas in the flows (e.g. Bachiller et al. 2001; Leurini et al. 2006; Parise et al. 2006; Arce et al. 2007). Further, the chemical species that may be enhanced in a molecular flow varies with position, age and energy in the flow. For example, in L1157, Bachiller et al. (2001) show that the morphology of chemical species such as CO, H_2H^+, HCO^+, SiO, CH_3OH, CS and SO_2 can vary significantly and they appear to trace different physical processes. The flexibility of the ALMA correlator will allow many spectral lines to be observed at a given time and should improve our ability to understand the detailed chemistry in the shocked outflowing gas as a function of time.

Clusters of young stellar objects: YSOs tend to form in clusters and the more massive the central star(s), the more dense the cluster tends to be. Recent examples of clusters of interacting outflows include: OMC-1 South (Zapata et al. 2005, 2006), IRAS 05358+3543 (Beuther et al. 2002, 2004; Sridharan et al. 2002) and IRAS 18507+0121 (G34.4, Shepherd et al. 2007). In all of these clusters, the analysis is fundamentally limited by resolution, sensitivity and the uv-coverage of the existing observations. Although the flows cannot be disentangled for independent study, ALMA will allow astronomers to study how outflow clusters interact with their environment and examine the details of each outflow close to the individual sources.

Comparison with theory: A recent, comprehensive review of molecular outflow models is presented in Arce et al. (2007). Outflow models can be separated into four general classes: (1) wind-driven shells; (2) jet-driven bow shocks; (3) jet-driven turbulent flows; and (4) circulation flows. In the last decade, computational power has increased sufficiently that it is now possible to run simulations of multi-dimensional hydrodynamical or magneto-hydrodynamical protostellar outflows that include a reduced chemical network. Although high-resolution observations of molecular flows can be compared directly with theoretical predictions, the data are often inconclusive due to lack of resolution and/or incomplete coverage of the uv-plane. ALMA will have the resolution and sensitivity to examine shock-enhanced chemistry along the jet axis to determine detailed jet physics and it will allow a more detailed comparison between observations, models and simulations beyond what can currently be done.

3 ALMA: A mosaicing machine

The key to observing a high-fidelity mosaic of an outflow with both bright, extended emission and compact, possibly faint structures is to combine observations from the 12 m array, 7 m compact array, and the total power dishes. The ability to recover all size scales will allow astronomers to relate outflows and embedded sources to larger scale structure and kinematics observed at the same resolution. Further, ALMA will have matched resolution with the expanded Very Large Array (EVLA) which will allow a detailed comparison of the ionized gas observed at centimeter wavelengths with the molecular gas imaged with ALMA.

3.1 Example project 1: A $5' \times 5'$ or $2' \times 12.5'$ outflow mosaic

Say you want to make a relatively small, $5' \times 5'$ mosaic of the inner region of the HH 80-81 outflow or a $2' \times 12.5'$ mosaic of the collimated BHR 71 flow as mapped by Parise et al. (2006). Choosing Band 6 (230 GHz) to make the observations will provide a $27''$ primary beam and roughly $1''$ spatial resolution in a compact configuration of the 12 m array. To obtain uniform sensitivity across the overlapping mosaic region, the fields should be Nyquist spaced with $11''$ spacing giving rise to a total of 730 fields.

As any good astronomer will do, you want to use all available resources of the correlator to get as many spectral lines as possible to observe outflow gas as well as neutral cloud and dense core gas. Such a configuration might be to observed selected spectral lines at 164 km/s bandwidth with 512 channels (0.32 km/s resolution) and use all remaining resources to obtain thermal continuum:

- Correlator Quadrant 1: lines CO($J = 2$–1), CH_3OH & SO_2
- Correlator Quadrant 2: lines $^{13}CO(J = 2$–1), $C^{18}O$, SO, CH_3OH & CH_3CN
- Correlator Quadrants 3 & 4: continuum 2 GHz bandwidth, 64 channels in the upper and lower sidebands

To obtain 15 mJy/beam RMS in the spectral lines you will need roughly 42 seconds of on-source integration time per field (this includes a root 2 increase in sensitivity due to overlapping fields). For 730 fields, you will need 8.6 hours of on-source integration time plus additional time for calibration and slew.

To obtain the best image fidelity in post-processing, it will be necessary to try to minimized differences in the uv-coverage for each field. To minimize differences it is best to observe the entire mosaic in a 'round-robin' fashion, getting through the mosaic as fast as possible and then repeating the mosaic observations multiple times to increase sensitivity. The final image fidelity will have to be balanced against the time required for extra slew and calibrations over multiple days of observation. Assuming 6 s integration time, the data rate (for 64 antennas in the 12 m array) will be about 11.3 MB/s on source leading to a total data product size of 230 GB.

To obtain complementary 7 m compact array observations of the same region covered by the 12 m array, it may take up to 3 or 4 times longer because it will take longer to reach a similar sensitivity on data with over-lapping uv-coverage because there will be fewer antennas. The actual time needed will most likely depend on the source structure. Compact structures will likely require less time to reach adequate sensitivity, very extended structure may require more time.

It will take even longer to obtain complementary observations of the mosaic using the total power dishes because one must observe a "guard band" of extra pointings around the source to improve the joint-deconvolution during post-processing imaging. If significant emission exists outside of the immediate outflow field (e.g. you are observing an outflow in the galactic plane—very common!) then you may have to observe an even larger region to get outside of the emission.

There is no doubt about it, mosaics will be complicated. And the larger the mosaic, the more complicated it will be. Eventually, the observing, processing and imaging heuristics will be such that mosaics will become straight forward for the observer. But ALMA won't start out that way. As one of the first PIs, the best thing to do will be to start out small when ALMA has relatively few antennas, and increase the complexity and size of the mosaic as antennas and experience with the system are gained.

3.2 Example project 2: A single field outflow study

Not all outflow studies will require mosaicing. For example, consider a project in which you want to study the outflow-disk interaction in band 7 (345 GHz) line emission. The primary beam will be about $18''$ in diameter and resolution between about $1''$ and $0.015''$ can be obtained. A 100 AU accretion disk at a distance of 300 pc will be about $0.3''$—well within the primary beam so no mosaicing is required. A possible correlator setup to observe both outflow and disk tracers might look like:

- Correlator Quadrant 1, line observations in the upper sideband:

– CO($J = 3$–2) & SO$_2$ with 220 km/s bandwidth, 512 channels at 0.43 km/s spectral resolution.
– HCOOCH$_3$, H^{13}CN(4–3) & CS(7–6) with 107 km/s bandwidth, 512 channels at 0.21 km/s spectral resolution.

- Correlator Quadrant 2, line observations in the lower sideband:
  – ^{13}CO and two SO$_2$ lines with 220 km/s bandwidth, 512 channels at 0.43 km/s spectral resolution.
  – CH$_3$CN(18–17) multiple lines with 220 km/s bandwidth, 1024 channels at 0.21 km/s spectral resolution.
  – HCOOCH$_3$ with 54 km/s bandwidth, 512 channels at 0.1 km/s resolution.
- Correlator Quadrants 3 & 4: continuum 2 GHz bandwidth, 64 channels in the upper and lower sidebands

Note that it is not possible to obtain C^{18}O($J = 2$–1) simultaneously in this observation because the difference in frequency is too great to fit in the same sideband as ^{13}CO.

Assuming 1 hour on-source integration time per configuration, the estimated RMS noise levels will be: 4.2, 5.9 & 8.5 mJy/beam for 0.43, 0.21 & 0.10 km/s spectral resolution, respectively. The continuum RMS noise level is expected to be 0.065 mJy/beam. And finally, the expected on-source data rate for 64 antennas (assuming 5 s integrations) will be 18.6 MB/s bringing the total data size to 6.7 GB for 1 hour.

4 Summary

In summary, ALMA will have the resolution, sensitivity and bandwidth to address major issues associated with the outflow driving mechanism, entrainment, feedback, shock chemistry and outflow collimation. By combining data from the 12 m array, the 7 m array and the total power dishes, ALMA will recover all emission on a wide range of size scales making it ideal for mosaicing large-scale molecular outflows. However, the primary beam sizes are sufficiently small that large mosaics will require hundreds to thousands of fields and it will be challenging to optimize the observational strategy to obtain uniform uv-coverage for each field which is needed high-fidelity final images.

Large mosaics with ALMA will be difficult at first and it will be a good idea to start small and increase the size and complexity of the field as ALMA capabilities increase. Eventually, the ability to observe interferometric mosaics while the antennas are slewing in an "on-the-fly" raster will significantly decrease the slew time required for large mosaics. However, this is a subject of state-of-the-art research and will not likely be available during the first several years of ALMA operations.

References

Arce, H.G., Sargent, A.I.: Outflow-infall interactions in early star formation and their impact on the mass-assembling process in L1228. Astrophys. J. **612**, 342–356 (2004)

Arce, H.G., Sargent, A.I.: Pushing the envelope: The impact of an outflow at the earliest stages of star formation. Astrophys. J. **624**, 232–245 (2005)

Arce, H.G., Shepherd, D., Gueth, F., Lee, C.-F., Bachiller, R., Rosen, A., Beuther, H.: Molecular outflows in low- and high-mass star-forming regions. In: Protostars & Planets, vol. V, pp. 245–260. University of Arizona Press, Tucson (2007)

Bachiller, R., Pérez, G.M., Kumar, M.S.N., Tafalla, M.: Chemically active outflow L 1157. Astron. Astrophys. **372**, 899 (2001)

Beltrán, M.T., Gueth, F., Guilloteau, S., Dutrey, A.: L1157: Interaction of the molecular outflow with the Class 0 environment. Astron. Astrophys. **416**, 631–640 (2004)

Beuther, H., Schilke, P., Sridharan, T.K., Menten, K.M., Walmsley, C.M., Wyrowski, F.: Massive molecular outflows. Astron. Astrophys. **383**, 892–904 (2002)

Beuther, H., Schilke, P., Gueth, F.: Massive molecular outflows at high spatial resolution. Astrophys. J. **608**, 330–340 (2004)

Cabrit, S., Bertout, C.: CO line formation in bipolar flows. III—The energetics of molecular flows and ionized winds. Astron. Astrophys. **311**, 858–872 (1992)

Cesaroni, R.: Outflow, infall, and rotation in high-mass star forming regions. Astrophys. Space Sci. **295**, 5–17 (2005)

Henning, Th., Schreyer, K., Launhardt, R., Burkert, A.: Massive young stellar objects with molecular outflows. Astron. Astrophys. **353**, 211–226 (2000)

Lebrón, M., Beuther, H., Schilke, P., Stanke, Th.: The extremely high-velocity molecular outflow in IRAS 20126+4104. Astron. Astrophys. **448**, 1037–1042 (2006)

Leurini, S., Schilke, P., Parise, B., Wyrowski, F., Güsten, R., Philipp, S.: The high velocity outflow in NGC 6334 I. Astron. Astrophys. **454**, L83–L86 (2006)

Parise, B., Belloche, A., Leurini, S., Schilke, P., Wyrowski, F., Güsten, R.: CO and CH3OH observations of the BHR71 outflows with APEX. Astron. Astrophys. **454**, L79–L82 (2006)

Pety, J., Gueth, F., Guilloteau, S., Dutrey, A.: Plateau de Bure interferometer observations of the disk and outflow of HH 30. Astron. Astrophys. **458**, 841–854 (2006)

Shepherd, D.S.: Massive molecular outflows. In: IAU Symp. 227: Massive Star Birth: A Crossroads of Astrophysics, pp. 237–246. Cambridge University Press, Cambridge (2005)

Shepherd, D.S., Churchwell, E.: Bipolar molecular outflows in massive star formation regions. Astrophys. J. **472**, 225–239 (1996)

Shepherd, D.S., Povich, M.S., Whitney, B.A., Robitaille, T.P., Nürnberger, D.E.A., Bronfman, L., Stark, D.P., Indebetouw, R., Meade, M., Babler, B.: Molecular outflows and a mid-infrared census of the massive star formation region associated with IRAS 18507+0121, Astrophys. J. (2007, in press)

Sridharan, T.K., Beuther, H., Schilke, P., Menten, K.M., Wyrowski, F.: High-mass protostellar candidates. I. The sample and initial results. Astrophys. J. **566**, 931–944 (2002)

Stojimirović, I., Narayanan, G., Snell, R.L., Bally, J.: Entrainment mechanisms for outflows in the L1551 star-forming region. Astrophys. J. **649**, 280–298 (2006)

Zapata, L.A., Rodríguez, L.F., Ho, P.T.P., Zhang, Q., Qi, C., Kurtz, S.E.: A highly collimated, young, and fast CO outflow in OMC-1 South. Astrophys. J. **630**, L85–L88 (2005)

Zapata, L.A., Ho, P.T.P., Rodríguez, L.F., O'Dell, C.R., Zhang, Q., Muench, A.: Silicon monoxide observations reveal a cluster of hidden compact outflows in the OMC 1 South region. Astrophys. J. **653**, 398–408 (2006)

Unveiling the chemistry of hot protostellar cores with ALMA

M. Guélin · N. Brouillet · J. Cernicharo · F. Combes ·
A. Wooten

Originally published in the journal Astrophysics and Space Science, Volume 313, Nos 1–3.
DOI: 10.1007/s10509-007-9684-4 © Springer Science+Business Media B.V. 2007

Abstract High angular resolution mm-wave observations
of the Orion-KL region, made with the IRAM Plateau de
Bure interferometer (PdBI), reveal the presence of several
cores of size 10^3 AU, which have distinct spectral signatures. Complex molecules such as ethanol, vinyl cyanide and
dimethyl ether show different distributions and their relative abundance varies from core to core by orders of magnitude. The molecular column densities derived in the cores
also differ widely from the beam-averaged column densities
observed with large single-dish telescopes. Obviously, the
predictions of hot core chemistry models must be checked
against high resolution observations. ALMA, which allies
sensitivity and high angular resolution, will be a key instrument for this type of studies.

The PdBI observations were part of a search for interstellar glycine, also carried out with the IRAM 30-m telescope and the Green Bank Telescope. We derive a 3σ upper
limit on the column density of glycine of 1×10^{15} cm^{-2} per
$2'' \times 3''$ beam in the Orion *Hot Core* and *Compact Ridge*.

Keywords Astrochemistry · Astrobiology · Line:
identification · Molecular processes · Techniques: high
angular resolution · ISM: abundances

Based on observations made with the IRAM PdB Interferometer, the
IRAM 30-m telescope and the NRAO Green-Bank telescope. IRAM
is supported by CNRS, MPG and IGN.

M. Guélin (✉)
IRAM, 300 rue de la piscine, 38406 St. Martin d'Hères, France
e-mail: guelin@iram.fr

N. Brouillet
L3AB, CNRS UMR5804, OASU, 2 rue de l'Observatoire, BP 89,
33270 Floirac, France
e-mail: nathalie.brouillet@obs.u-bordeaux1.fr

J. Cernicharo
Dept. Molecular and Infrared Astrophysics, Instituto de Estructura
de la Materia, CSIC, Serrano 121, 28006, Madrid, Spain
e-mail: cerni@damir.iem.csic.es

F. Combes
Observatoire de Paris, LERMA, 61 av. de l'Observatoire,
75014 Paris, France
e-mail: combes@obspm.fr

A. Wooten
National Radio Astronomy Observatory, 520 Edgemont Road,
Charlottesville, VA 22903-2475, USA
e-mail: awootten@nrao.edu

1 Introduction

Considering the extremely low densities and temperatures
(by terrestrial standards) prevailing in interstellar space, the
observation of molecules with up to 12 atoms is a puz-
zling result. Both gas-phase and grain-surface chemistry
have been evoked, but they have difficulties in reproducing
the abundance of "complex" species such as methyl formate
($HCOOCH_3$) and dimethyl ether (CH_3OCH_3) in "hot" protostellar cores. The main reasons for this failure are the tendency of dissociative recombination, the last step of the gas-
phase formation channel, to break up into small pieces the
backbones of large saturated ions (Geppert et al. 2005) and
the narrowness of the range of temperatures favorable to the
formation of heavy molecules on grain surfaces.

Laboratory measurements and quantum mechanical calculations of reaction rates, on the one hand, and better models of protostellar core evolution, on the other hand, now put
on firmer grounds the hot core chemistry (Horn et al. 2004).

A recent model developed by Garrod and Herbst (2006) couples gas-phase and grain-surface processes all along the initial collapse of the core and its subsequent warm-up. During the cold collapse phase, small molecules, such as H_2O, H_2CO and CH_3OH form by hydrogenation of CO deposited on the grains. During the warm up phase, larger molecules (e.g. $HCOOCH_3$ and CH_3CH_2OH) are formed from H_2CO through gas-phase and grain-surface reactions. A subtle interplay between condensation of molecules, when the density becomes large, and evaporation, when the temperature rises, maximizes the production of complex molecules for the different types of cores. Methyl formate, for example, is produced in abundance in low mass cores via grain-surface reactions, but is mainly formed through gas-phase reactions in massive cores with a short warm-up phase. Similarly, dimethyl ether forms preferentially through gas-phase or grain-surface reactions, depending on the actual products of the dissociative recombination of $C_2H_6OH^+$ (Peeters et al. 2006).

The current hot core models predict the abundance of large molecules as a function of the core age and radius, and of the duration of the warm-up phase. Unfortunately, except in a few cases, the observational data are not at the level of the predictions. Those are mostly compared with low resolution single-dish observations that mix-up the signals from nearby cores and/or pick-up the emission from foreground and background gas. For example, Garrod and Herbst (2006) compare their model predictions to the Orion Hot Core abundances derived by Sutton et al. (1995) from JCMT 0.8-mm observations. At this wavelength, the JCMT has a HPBW of $13.7''$, definitely too broad to resolve the multiple dense cores that constitute the Orion "Hot-Core" and "Compact Ridge". As will be seen below, these hot and dense cores have a quite different molecular content, so that the JCMT beam-averaged column densities and abundances are irrelevant. No wonder that the model predictions disagree for some species by two orders of magnitude, as is the case for formic acid. The situation is even worse for more distant Hot Cores, such as those of Sgr B2, for which the spatial resolution is even poorer.

Besides mixing up the signals from unrelated sources, low spatial resolution increases the spectral line confusion, a severe limitation to the detection of weak lines in the overcrowded spectra of the richest hot core regions. Those regions not only contain a large variety of molecules, but are hot and dense enough to populate excited bending states, multiplying the number of lines in the mm domain, hence the chances of line confusion. Examples of this will be given below.

Obviously, model predictions have to be compared with high spatial resolution data, which, at millimeter wavelengths, can only be obtained with interferometers. The following illustrates the drastic changes in our representation

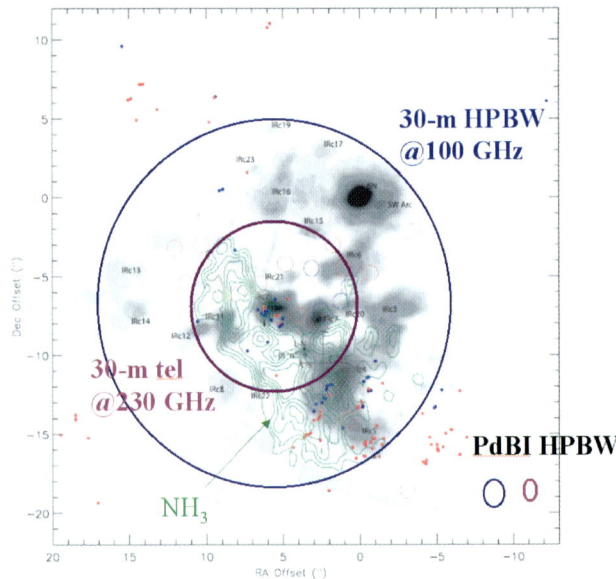

Fig. 1 The Orion KL region as observed in the NH_3 (4,4) inversion line emission (*contours*) and in the 12.5 µm continuum (Shuping et al. 2004). The *circles* represent the 30-m telescope HPBW at 101 GHz and 223 GHz and the ellipses the PdBI synthetized beam

of a complex region, the Orion Kleinman-Low "Hot Core" region, when one increases tenfold the linear spatial resolution. The observations (at 43 GHz, 101 GHz and 223 GHz) were made in the course of a search for interstellar glycine in Orion KL with the NRAO Green Bank Telescope (GBT), the IRAM Pico Veleta 30-m telescope and the IRAM Plateau de Bure interferometer (PdBI).

2 Single-dish line surveys of the Orion KL hot core region

2.1 The Orion KL region

The target of our search was the Orion KL star formation region, which hosts a number of infrared sources and bright molecular cores. Figure 1 shows the emission from the (4,4) inversion line of ammonia, observed with the VLA by Wilson et al. (2000) at a resolution of $\simeq 1''$, superimposed on the 12.5 µm continuum emission (in grey—Shuping et al. 2004). The thick circles show the 30-m telescope HPBW at 101 GHz ($24''$) and 223 GHz ($11''$), while the ellipses show the PdBI synthetized beams. The JCMT half-power beam at 330 GHz ($14''$) and the GBT beam at 43 GHz ($17''$) are comprised between the blue and the purple circles. The ammonia emission reveals the presence of a dozen of cores of sizes $\simeq 2''$. These cores are detected in several high excitation transitions of ammonia and must be hot (≥ 150 K) and dense ($> 10^6$ cm^{-3} see e.g. Wilson et al. 2000). Their velocities differ by 1–2 km s^{-1}, i.e. by less than the individual

Fig. 2 Portion of the 7-mm spectral survey of Orion-IRC2 carried out with the GBT telescope

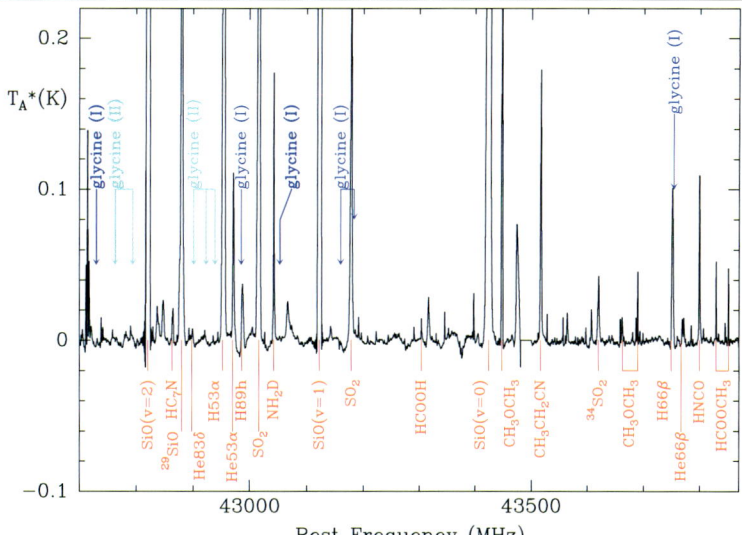

core line widths. Obviously, only the interferometer is capable of resolving individual cores, whilst the JCMT, the GBT and the 30-m telescope beams are essentially covering the entire region. The distance to Orion KL being 450 pc, the KL region size is about 10^4 AU and the typical core size ($2''$) about 1000 AU.

2.2 The GBT 43–45 GHz spectral survey

We have observed the Orion KL region with the GBT equipped with its dual-beam Q-band receiver. The observations were made in April 2004 in the double-beam switching mode, with one beam pointing on the SiO maser (IRC2) and the other beam on a reference position. The surveyed frequency band was 42.7–45.6 GHz and the spectral resolution 0.39 MHz ($2.7\ \mathrm{km\,s^{-1}}$). The radio frequency being relatively high for the GBT, we checked frequently the telescope pointing and focusing and monitored the telescope gain by measuring at every scan the SiO maser intensity, T_{SiO}, which we assumed to be equal to 240 K.

Figure 2 shows a portion of the spectrum obtained after rescaling the observations in intensity and averaging them together with a weight proportional to $(T_{\mathrm{SiO}}/\sigma)^2$, where σ is the individual scan r.m.s. noise. Some 100 lines with an intensity >10 mK are observed in the 2.9 GHz-wide band. 70 of them are identified as molecular rotational transitions and recombination lines of H, He, and C. These include lines of CH_3CH_2CN, CH_3OCH_3 and $HCOOCH_3$. Twenty clear lines (>20 mK) remain unidentified. Eighty strong (transition strength $S > 1$, level energy $E_u/k < 500\ \mathrm{cm^{-1}}$) transitions from glycine (2/3 from conformer I and 1/3 from conformer II) lie in the observed band. Although many of these transitions do coincide in frequency with observed spectral lines, many others have no counterpart. From our data, we estimate a 3σ upper limit on the beam-averaged column

density of glycine (conformer I) of $1.5 \times 10^{14}\ \mathrm{cm^{-2}}$, assuming a velocity width of $4\ \mathrm{km\,s^{-1}}$ and $T_{\mathrm{rot}} = 100$ K (Combes et al. 1996, hereafter CRW). Because of the larger dipole moment, the limit on conformer II, which lies higher in energy than conformer I, is an order of magnitude lower. We note that these limits result in part from baseline ripples.

2.3 The 30-m telescope and PdBI 101 GHz and 223 GHz spectral surveys

The results of this search for glycine with the IRAM 30-m telescope have already been reported by CRW. Figure 3 shows the 3-mm spectrum. The number of lines per GHz increases with increasing frequency (compare with Fig. 2) and line confusion becomes all the more a problem that the line widths also increase with frequency. Although, here again, several glycine transitions coincide with observed spectral lines, strong glycine transitions from both conformer I and II are missing in the spectrum or are buried in the noise and/or the pseudo-continuum. CRW derived an upper limit of $5 \times 10^{13}\ \mathrm{cm^{-2}}$.

Obviously, a molecule as complex as glycine is unlikely to be widespread and its mm-wave emission must be mainly restricted to the hot cores detected in ammonia. This is observed, for example, in the case of methyl formate and formic acid (see Hollis et al. 2003 and below) the emissions of which fill 1/20–1/10 of the GBT and 30-m beams. The beam-averaged limits derived from the GBT and the 30-m telescope should then be multiplied by large factors and are not as significant as they seem at first glance. In order to decrease confusion and improve the detection limit, we have re-observed with the PdBI the same frequency intervals as CRW. The synthetized beam (FWHP) was $3.3'' \times 1.7''$ at 101 GHz and $1.6'' \times 0.9''$ at 223 GHz, some 100 times smaller in area than the 30-m telescope beam, and the primary beam area was twice larger.

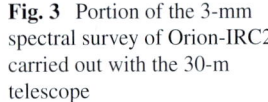

Fig. 3 Portion of the 3-mm spectral survey of Orion-IRC2 carried out with the 30-m telescope

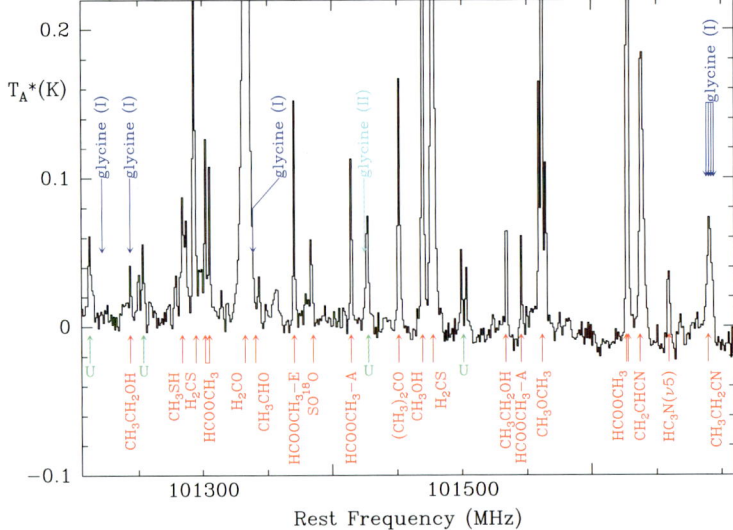

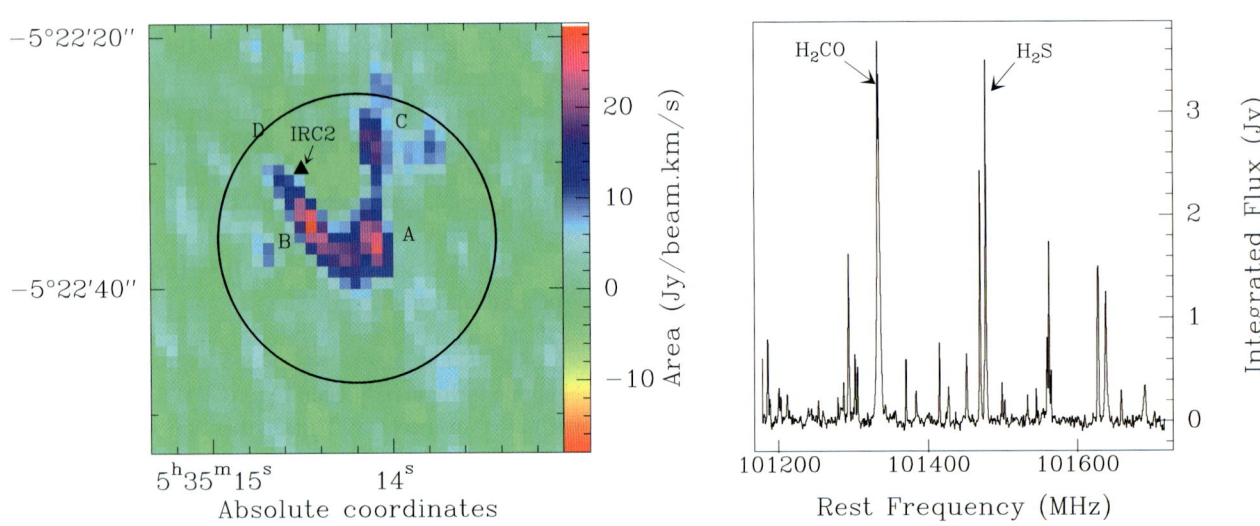

Fig. 4 *Left*: map of the frequency-integrated emission (after subtraction of the continuum). *Right*: spectrum integrated over the circular area defined by the 30-m telescope HPB (*circle* on the left map)

Figure 4 (left) shows the global line emission map integrated over all the molecular lines of Fig. 3 (more precisely the signal detected by the PdB interferometer, integrated from 101200 MHz to 101700 MHz, after subtraction of the continuum), whereas Fig. 4 (right) shows the interferometer spectrum integrated over the 24″ 30-m HPBW.

The lines of H_2CO and H_2CS, whose emissions are known to extend over the entire field of view, appear much weaker on the interferometer spectrum (Fig. 4) than on the 30-m telescope spectrum (Fig. 3), due to the lack of short spacings. For the other lines, both spectra look much alike. Within the calibration uncertainties, 100% of the 30-m flux is recovered by the interferometer, showing that the emission of most molecules arises from compact sources.

We identify the most conspicuous sources by the letters A–D. A, the brightest source, is located some 10″ to the SW of IRC2 and of the SiO maser; it is usually referred to at the *Compact Ridge*. B, the second brightest source, corresponds to the so-called *Hot Core*; C is located 10″ N of the compact ridge A, whereas D, which is strong in NH_3, is barely visible on the integrated map, is located a few arcsec NE of IRC2. A, B, D consist each of several bright NH_3 cores and are known to be hot and dense (Wilson et al. 2000). C, although weaker, is detected in NH_3 (4,4); it is conspicuous in the $12_{2,10}–11_{3,9}$ transition of $(CH_3)_2O$ ($E_u/k = 73$ K) and must also be hot and dense.

The right frames of Fig. 5 show the spectra observed toward A, B, C, and D, while the maps in the left frames show the emission in the rotational lines of $HCOOCH_3$, CH_3CH_2OH, CH_3OCH_3 and CH_3CH_2CN. These lines arise

Fig. 5 *Right*: PdB interferometer spectra observed toward sources A, B, C and D. *Left*: maps of the emission in the 101 GHz lines of HCOOCH$_3$, CH$_3$CH$_2$OH, CH$_3$OCH$_3$ and CH$_3$CH$_2$CN (marked with thin vertical lines). The continuum emission was subtracted

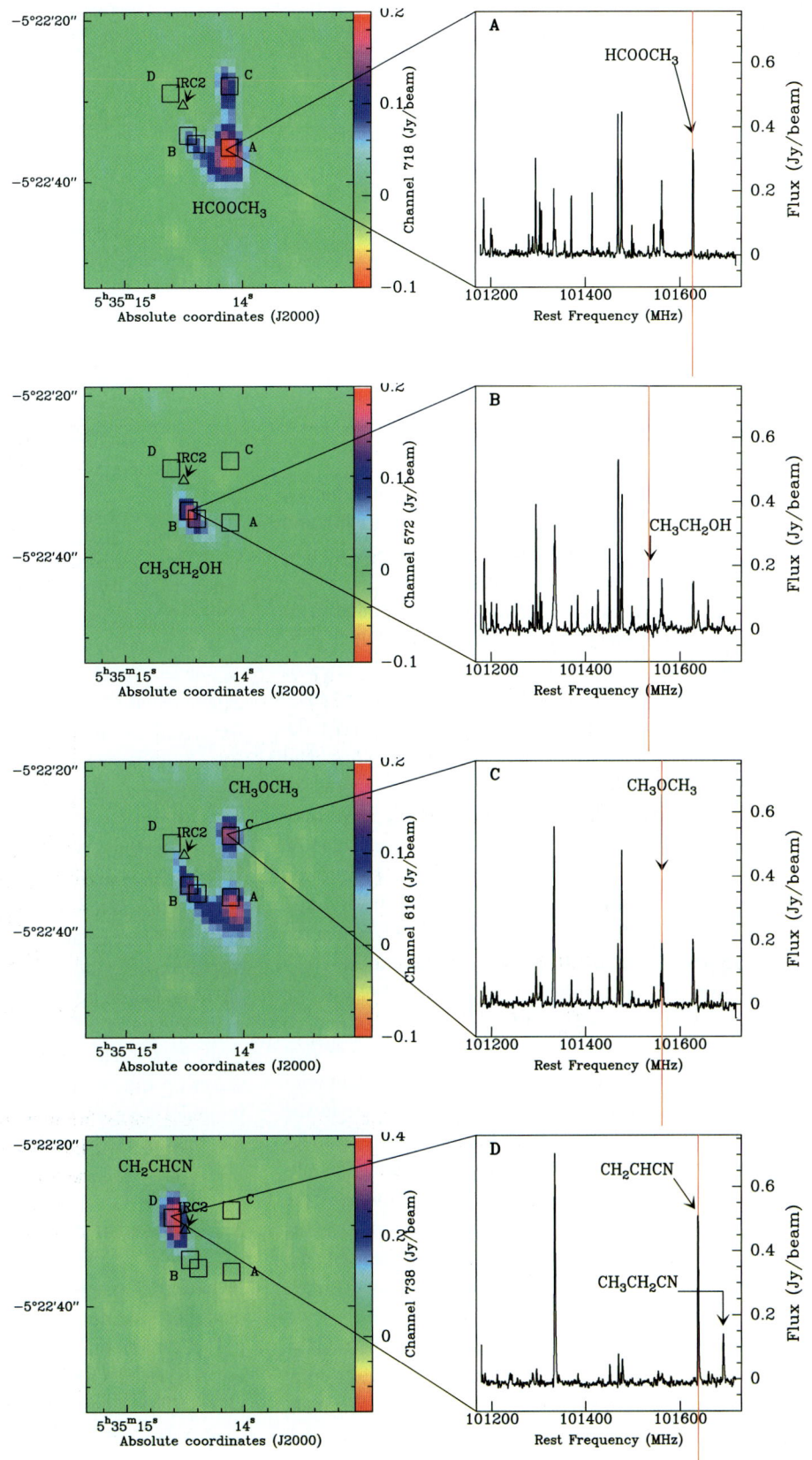

Table 1 Parameters of relevant molecular lines

Molecule	Transition	Frequency (MHz)	E_l/k (K)
NH_2CH_2COOH	$16_{2,15}-15_{2,14}$	101221.5	44
CH_3OH	$8_{2,7}-8_{2,8}$	101469.7	109
CH_3CH_2OH	$12_{1,12}-12_{0,12}$	101533.5	125
$(CH_3)_2O$	$9_{1,9}-8_{1,8}$	101562.2	78
$HCOOCH_3$	$9_{1,9}-8_{1,8}$	101628.2	25
CH_2CHCN	$6_{3,3}-5_{3,2}$	101637.2	32
NH_2CH_2COOH	$17_{0,17}-16_{1,16}$	101688.7	45
CH_3CH_2CN	$27_{2,25}-27_{1,26}$	101690.0	170
NH_2CH_2COOH	$17_{1,17}-16_{1,16}$	101690.6	45
NH_2CH_2COOH	$17_{0,17}-16_{0,16}$	101692.4	45
NH_2CH_2COOH	$17_{1,17}-16_{0,16}$	101694.4	45

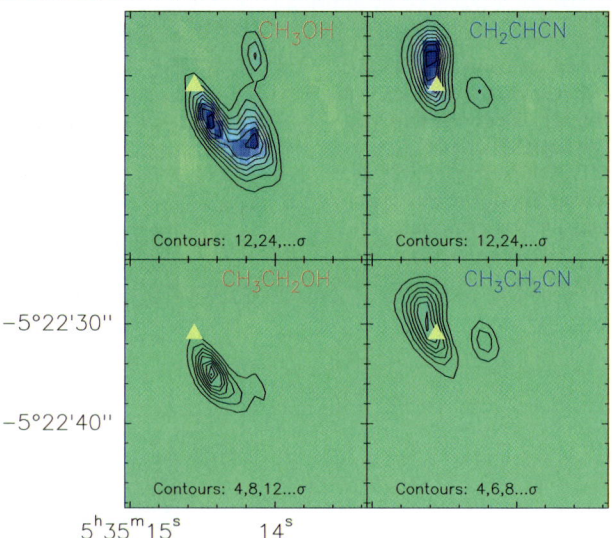

Fig. 6 Maps of the emission in the 101 GHz lines of CH_3OH, CH_3CH_2OH, CH_2CHCN and CH_3CH_2CN. The continuum emission has been subtracted. *First contour* and *contour steps* are 4 or 12σ, as indicated. The *triangle* shows the position of IRC2. Axis are J2000 coordinates

all from levels with energies $E_u/k \leq 170$ K (see Table 1), low enough in energy to be excited throughout these hot and dense sources, so that the maps reflect more the spatial distribution than the excitation conditions.

Yet the cores A–D have very different spectra and must have different chemical compositions. As seen on Fig. 5, the spectrum toward core D is dominated by the lines of vinyl cyanide, ethyl cyanide and formaldehyde, but shows no emission of methyl formate or ethanol; conversely, the spectrum towards A has only weak H_2CO emission and about no vinyl- or ethyl-cyanide emission. The contrast in the emission of those molecules (and presumably the ratio of their abundances) exceeds a factor 20–30 between D and B. The tendency of CN-bearing molecules to peak more to the north than methyl formate and acetic acid has been noted previously, but with an angular resolution and a S/N insufficient to see how exclusive theses molecules could be.

As can be seen on the maps of Fig. 6, the bulk of the emission from ethyl cyanide, vinyl cyanide, ethanol and, to a lesser degree, methanol arise from compact sources that cover only a small fraction (1/10 to 1/30) of the 30-m telescope beam. The column densities of these species in the 10^3 AU diameter sources are more than one order of magnitude larger than the beam-averaged column densities observed with the 30-m telescope and the JCMT. Moreover, the abundance ratios between e.g. ethyl- or vinyl cyanide, on the one hand, and ethanol or methyl formate, on the other hand, also differ by similar factors from those derived with single dish telescopes.

The BIMA map of ethyl cyanide emission at 80.6 GHz, reported by Wright et al. (1996), seems to give a different picture, as it extends over a $12'' \times 10''$ region. We note, however, that the signal-to-noise ratio in this map is much smaller than in ours and that the 80.6 GHz lines observed at BIMA (a set of 8 lines distributed over 4 MHz) are blended with a line of CH_3CH_2OH. Although, the 101.7 GHz ethyl cyanide line has a higher energy than the 80.6 GHz lines

($E_u/k = 170$ K vs. 40–60 K), we do not believe that the extent of ethyl cyanide emission in the hot cores A–C is limited in our map by weak excitation, as the vinyl cyanide $6_{3,3}-5_{3,2}$ line, whose upper level energy is only 32 K, shows a very similar distribution (Fig. 6). The physical conditions in A and B ($T_K > 160$ K, $n(H_2) \simeq 10^7$ cm^{-3}) should insure that all the lines from Table 1 are properly excited.

The concentration of vinyl- and ethyl cyanide in a single core, D, NE of the "*Hot Core*" B, where ethanol is confined, has yet to be explained. It does not result from a lack of nitrogen in B, or of oxygen in D, as HCN, NH_3 and H_2CO are abundant in both cores (as a matter of fact, the NH_3 column density is a factor of 8 larger in B than in D, Wilson et al. 2000). The chemical differences may come from the warm-up time of the cores, a critical parameter according to models (e.g. Garrod and Herbst 2006), which depends on the mass of the protostar.

3 Limits on the column density of glycine

Proposed channels for the formation of glycine in the interstellar medium are the reaction of NH_2 with acetic acid (CH_3COOH) on grain surfaces, and the reaction of the ion $NH_2HCH_2OH_2^+$ with formic acid (HCOOH) in the gas phase. The latter reaction is endothermic and may proceed only in hot gas. Similarly, the evaporation of glycine and formic acid from grain surfaces requires the that dust is heated. In other words, glycine, like methyl formate, dimethyl ether, or ethanol, must form (and most probably remains) in hot dense cores. We therefore assume that glycine, like those molecules, is concentrated in compact (2–3″ diameter) sources.

The most conspicuous spectral feature that could be assigned to glycine in our 101 GHz spectrum is the pile-up of 4 lines connecting the $17_{0,17}$ and $17_{1,17}$ levels of conformer I to the $16_{0,16}$ and $16_{1,16}$ levels (see Fig. 3 and Table 1). These lines, whose level energies are $E_u/k = 33$ K and which have similar transition strengths ($S = 15$–17, see CRW), have frequencies comprised between 101688 MHz and 101694 MHz. Unfortunately, they are blended with the $27_{2,25}$–$27_{1,26}$ transition of CH_3CH_2CN (101690 MHz). The 80 mK line observed at that frequency in the 30-m telescope spectrum (Fig. 3) is readily assigned to ethyl cyanide, since many lines from ethyl cyanide are observed with this telescope toward Orion-IRC2 and since there is no trace in Fig. 3 of another strong line of glycine, at 101221 MHz (the glycine conformer I $16_{2,15}$–$15_{2,14}$ line is similar in energy and strength to the 4 lines mentioned above).

Since ethyl cyanide is essentially confined into source D, the new PdBI observations allow to remove the line confusion. They make it possible to derive low lower limits on the strength of the main glycine feature in A and B, where the complex carboxyl molecules are observed and where glycine is the most likely to be found. There, we derive a 3σ upper limit of 1×10^{15} cm^{-2} for the column density of glycine in a $2″ \times 3″$ beam—the lowest to date at this scale.

The detection of glycine and other pre-biotic molecules, and the mapping at high spatial and spectral resolutions of hot cores and distant hot cores will require higher spatial resolution and much higher sensitivity than those of the present observations. ALMA, will combine these assets. No doubt, it will become the key instrument for studying the chemical evolution of protostars.

References

Combes, F., Nguyen-Q-Rieu, Wlodarczak, G.: Search for interstellar glycine. Astron. Astrophys. **308**, 618 (1996)

Garrod, R.T., Herbst, E.: Grain-surface formation of methyl formate. Astron. Astrophys. **457**, 927 (2006)

Geppert, W.D., Hellberg, F., Osterdahl, F., et al.: Dissociative recombination of $CD_3OD_2^+$. In: IAU Symposium 231: Astrochemistry: Recent Successes and Current Challenges, pp. 117–124. Cambridge University Press, Cambridge (2005)

Hollis, J.M., Pedelty, J.A., Snyder, L.E., et al.: A sensitive VLA search for small-scale glycine emission toward OMC-1. Astrophys. J. **588**, 353 (2003)

Horn, A., Moellendal, H., Sekiguchi, O., et al.: The gas-phase formation of methyl formate in hot molecular cores. Astrophys. J. **611**, 605 (2004)

Peeters, Z., Rodgers, S.D., Charnley, S.B., et al.: Astron. Astrophys. **445**, 197 (2006)

Shuping, R.Y., Morris, M., Bally, J.: A new mid-infrared map of the BN/KL region using the Keck telescope. Astron. J. **128**, 363 (2004)

Sutton, E.D.C., Peng, R., Danchi, W.C., et al.: Astrophys. J. Suppl. Ser. **97**, 455 (1995)

Wilson, T.L., Gaume, R.A., Gensheimer, P., Johnston, K.J.: Kinematics, kinetic temperatures and column densities of NH_3 in the Orion Hot Core. Astrophys. J. **538**, 665 (2000)

Wright, M.C.H., Plambeck, R.L., Wilner, D.J.: A multiline aperture synthesis study of Orion-KL. Astrophys. J. **469**, 216 (1996)

High resolution submillimeter observations of massive protostars

C.L. Brogan · T.R. Hunter · R. Indebetouw ·
C.J. Chandler · Y.L. Shirley · R. Rao · A.P. Sarma

Originally published in the journal Astrophysics and Space Science, Volume 313, Nos 1–3.
DOI: 10.1007/s10509-007-9598-1 © Springer Science+Business Media B.V. 2007

Abstract We describe results from recent Submillimeter
Array observations of massive protostellar objects (CepheusA-East, NGC7538 IRS1, and G5.89-0.39) with resolutions ranging from $0.8''$ to $2''$. A wide range of spectral
and continuum properties are observed, with one unifying
theme: at these resolutions all of the studied sources reveal multiple submillimeter cores. Some are observed to
have cm-wavelength counterparts, and others not, suggesting a range of evolutionary stages coexisting in close proximity. In the presence of such complexity and multiplicity of
sources, these data suggest that the interpretation of diagnostics such as kinematic velocity gradients and temperatures
that are strongly dependent on spatial resolution should be
approached cautiously.

C.L. Brogan (✉) · T.R. Hunter
NRAO, 520 Edgemont Rd, Charlottesville, VA 22903, USA
e-mail: cbrogan@nrao.edu

R. Indebetouw
University of Virginia, P.O. Box 3818, Charlottesville,
VA 22903-0818, USA

C.J. Chandler
NRAO, P.O. Box 0, Socorro, NM 87801, USA

Y.L. Shirley
University of Arizona, 933 N. Cherry Ave., Tucson, AZ 85721,
USA

R. Rao
Institute of Astronomy and Astrophysics, Academia Sinica,
Taipei, Taiwan

A.P. Sarma
Depaul University, 2219 North Kenmore Avenue, Byrne Hall 211,
Chicago, IL 60614, USA

Keywords Astrochemistry · Stars: formation

1 Introduction

There are still many unanswered questions regarding the
formation of high mass stars, including: (1) Is the process of
high mass star formation merely a scaled up version of low
mass star formation, complete with the disks and outflows
that are ubiquitous toward low mass sources? (2) What is the
accretion mechanism (e.g., accretion dominated processes
(McKee and Tan 2003), versus collisionally dominated
processes (Bonnell and Bate 2005))? (3) What is the density
distribution and multiplicity in high mass cores? One way
to answer the first question is to look for circumstellar disks
around massive YSOs. One problem with directly detecting
disks, jets, and outflows toward *individual* high mass YSOs
is that they tend to be located in more confused clustered
environments compared to the typically more isolated low
mass YSOs. In addition, most high mass YSOs are at much
greater distances than many of the well studied low mass
YSOs. Indirect evidence for such disks comes from bipolar outflows and the observation of H_2O masers which are
thought to trace them (see for example Beuther et al. 2002).
The only way to solve these problems is to observe such objects in the submillimeter at high spatial resolution in both
the continuum and spectral lines—which is only now becoming possible with submillimeter interferometers like the
Submillimeter Array[1] (SMA) and in the near future ALMA.

We are engaged in a long term project to image a wide
range of massive protostars using the SMA at ~ 345 GHz

---

[1]The Submillimeter Array is a joint project between the Smithsonian
Astrophysical Observatory and the Academia Sinica Institute of Astronomy and Astrophysics.

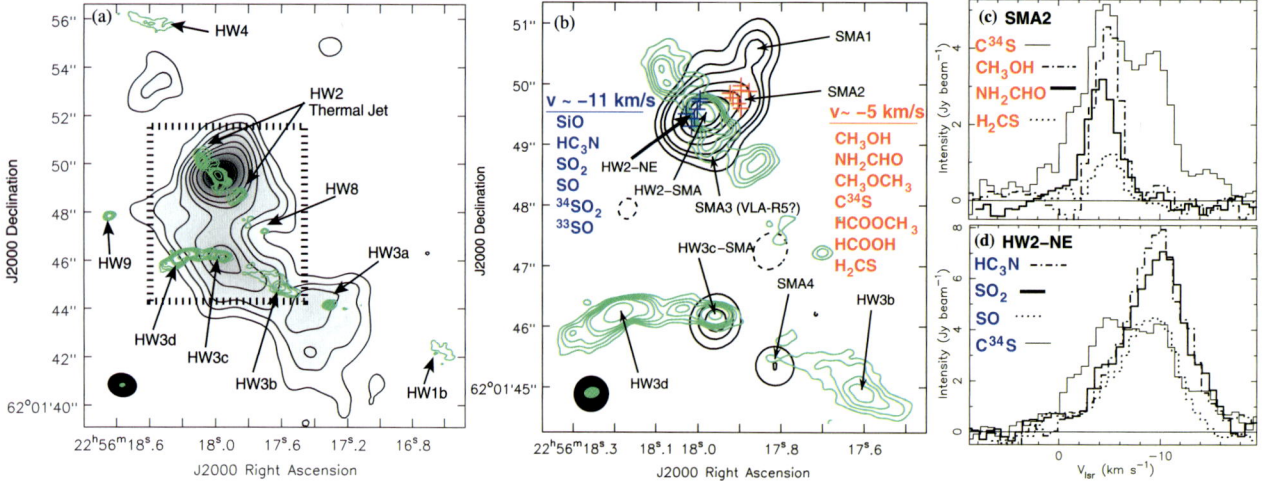

Fig. 1 a Combined CepA SMA 875 μm continuum image (greyscale and black contours) with a resolution of $1\rlap{.}''3 \times 1\rlap{.}''0$ (P.A. = 79°) and contour levels of −40, 40 (3σ), 80, 120, 200, 300, 400, 600, 1000, and 1600 mJy beam^{-1}. Green VLA 3.6 cm contours at 0.06 (3σ), 0.12, 0.2, 0.3, 0.4, 0.5, and 1.5 mJy beam^{-1} and $0\rlap{.}''23$ resolution are superposed. The region shown in (**b**) is indicated by the dashed box. **b** Superuniform weighted 875 μm contour map (*black*) created using baselines longer than 40 kλ and restored with a $0\rlap{.}''6$ beam; the contour levels are −50, 50 (4σ), 100, 150, 300, 500, 700, and 900 mJy beam^{-1}.

The *green contours* are the same as in (**a**). *Colored crosses* mark the peak positions of the molecular species listed. Prominent cm-λ and submm sources are also labeled. Sample spectral line profiles from the positions indicated on (**b**) for (**c**) SMA2 and (**d**) HW2-NE. The displayed transitions are C^{34}S (7–6), CH$_3$OH ($14_{7,8}A^{\pm}$–$15_{6,9}A^{\pm}$), NH$_2$CHO ($16_{2,15}$–$15_{2,14}$), H$_2$CS ($10_{0,10}$–$9_{0,9}$), HC$_3$N (38–37), SO$_2$ ($16_{7,9}$–$17_{6,12}$), and SO (11_{10}–10_{10}). Note that at the spectral line angular resolution of these transitions (∼2″) the two positions are not completely independent. This figure is from Brogan et al. (2007)

in order to investigate their formation mechanism(s). In this proceeding we briefly describe some of our results for three of the observed sources: CepheusA-East, NGC7538 IRS1, and G05.89-0.39.

2 CepheusA-East

CepheusA-East (hereafter CepA) is one of the closest massive star forming regions with a luminosity of 2.2×10^4 L$_\odot$ at a distance of 0.73 kpc (Mueller et al. 2002). CepA contains many tracers of ongoing massive star formation including very energetic molecular outflows and copious H$_2$O, CH$_3$OH, and OH maser emission (Codella et al. 2005; Vlemmings et al. 2006). At cm-wavelengths, CepA consists of several compact sources called HW1, HW2, ..., HW9, which lie along a roughly inverted Y-like structure (Hughes and Wouterloot 1984). It is currently unclear how many of these ionized structures correspond to individual protostars, since they could also be due to ionized jets emanating from as yet unidentified deeply embedded protostars. For example much of the cm-wavelength emission in the vicinity of the HW2 object is due to a well studied ionized bipolar jet (Curiel et al. 2006). Indeed, using SMA 345 GHz continuum and methyl cyanide data, Patel et al. (2005) recently reported the detection of a massive molecular gas and dust disk toward the HW2 source.

We have investigated the submm continuum and spectral line characteristics of CepA using three separate ∼345 GHz

SMA tunings (including the archival data of Patel et al., this work is described more fully in Brogan et al. (2007). The combined 875 μm continuum image from this work with $1\rlap{.}''3 \times 1\rlap{.}''1$ resolution is shown in Fig. 1a, and shows several previously unknown submm cores. To further delineate the morphology of the submm continuum emission in the vicinity of HW2, a superuniform weighted image restored with a $0\rlap{.}''6$ beam (equivalent to 1.8 times the longest baseline sampled) which emphasizes the locations of the clean components is presented in Fig. 1b. This image shows that there are at least two distinct submm sources in the vicinity of HW2, HW2-SMA and SMA1, in addition to an extension NW of HW2-SMA which we denote SMA2 (see Sect. 3). A weak extension south of HW2-SMA (denoted SMA3) is also visible, which may be a submm counterpart to the low mass protostar VLA-R5 (Curiel et al. 2002). The combined morphology of HW2-SMA, SMA2, and SMA3 is the structure reported by Patel et al. (2005) as a dust disk. The positions of the HW2-SMA 875 μm continuum peak and the proposed location of the powering source of the HW2 thermal jet (Curiel et al. 2006) agree to within our absolution position uncertainty of $0\rlap{.}''15$. Two additional submm cores are detected to the south of HW2, one coincident with the cm-wavelength source HW3c (called HW3c-SMA) and another (called SMA4) located at the NE tip of the cm-wavelength source HW3b. No distinct compact 875 μm counterparts to HW8, HW9, HW3a, HW3b or HW3d are detected in the SMA data.

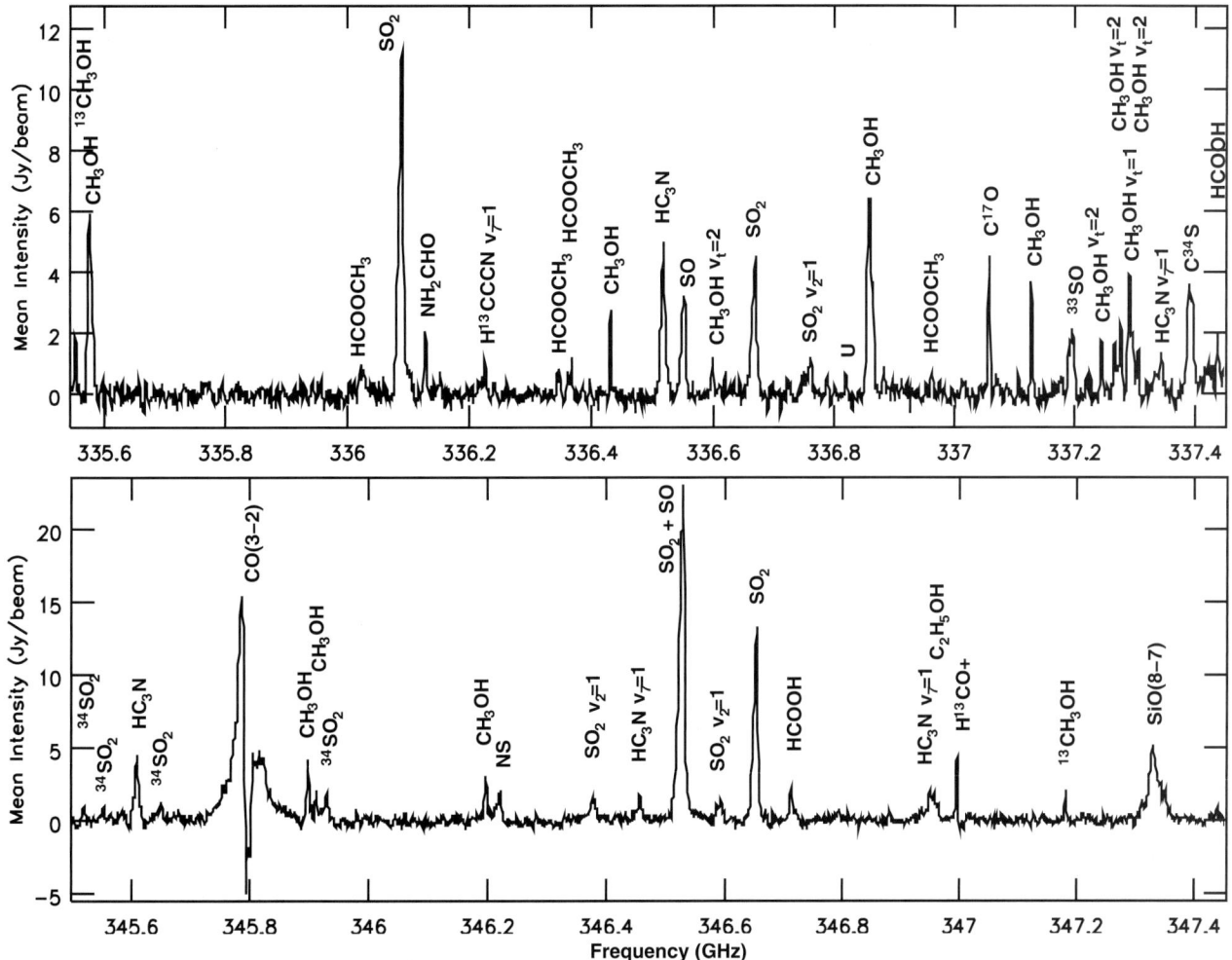

Fig. 2 Average profile over an area encompassing HW2-SMA and SMA2 of two of the four high spectral resolution frequency setups. The spectra, rich in organic molecules, are typical of other hot cores. Identified lines are indicated by species, while the strongest unidentified line is marked by a "U"

More than 20 distinct species and a number of their isotopologues have been detected in these SMA data (Fig. 2 shows two of our four 2 GHz wide sidebands). The aggregate CepA line emission resembles that of Orion A observed with $20''$ resolution, which encompasses both the energetic shock environment of BN/KL, as well as more quiescent hot cores (Schilke et al. 1997). A few transitions with low excitation energies (^{12}CO(3–2), CS(7–6), H^{13}CO$^+$(4–3)) show clumps scattered about the primary beam ($\sim36''$) and may trace the outflow emission (Codella et al. 2005). However, most species originate only from compact cores ($<2''$) which coincide with the submm continuum peaks (Fig. 1b). At $1''-2''$ resolution, the compact spectral line species in the vicinity of HW2 are strongest at one of two distinct velocities: -5.0 ± 0.5 or -10.5 ± 0.5 km s^{-1}. We find these two kinematic features to be spatially distinct: molecules that are strongest at ~-10.5 km s^{-1} exhibit their peak $\sim0''.25$ E/NE of HW2-SMA (a position we

denote as HW2-NE), while molecules that are strongest at ~-5 km s^{-1} exhibit their peak on SMA2 (see Fig. 1b–d). Although this dichotomy is present in all observed species, a few abundant high density tracers (e.g. CH$_3$CN and C^{34}S) show emission of nearly equal strength, indicating the presence of dense gas at both positions. For CH$_3$OH towards SMA2 (at ~-5 km s^{-1}), and SO$_2$ toward HW2-NE (at ~-10.5 km s^{-1}), we find a large (>100 K) difference in rotation temperature: 125 ± 6 K for SMA2 at ~-5 km s^{-1} and 237 ± 27 K for HW2-NE at ~-10.5 km s^{-1}.

Two interpretations for the distribution of the submm continuum emission in the vicinity of HW2 ($\pm0.5''$) are possible: a single elongated structure or two (or more) marginally-resolved individual sources. Although the kinematic dichotomy between the positions labeled HW2-NE and SMA2 (Fig. 1b) could be interpreted as a velocity gradient across a continuous structure (e.g. Patel et al. 2005), the dramatic chemical and thermal differentiation demonstrated by our multi-species analysis is difficult to explain with a single source scenario. We also note that beyond a velocity gradient, the kinematic evidence for a Keplerian rotating disk is quite weak. For example, the weakness of the emission at the central HW2 position in position-velocity (P-V) diagrams such as those in Patel et al. and in our own data, are inconsistent with theoretical expectations unless a central hole is invoked (see e.g. Richer and Padman 1991). Even with this modification, it remains difficult to explain the unequal position offsets of the two velocity peaks from the HW2 stellar position. In contrast, the presence of multiple sources at different velocities naturally explains the observed behavior. We therefore favor the multiple source hypothesis (also see Martín-Pintado et al. 2005), with at least three sources in the vicinity of HW2 (HW2-SMA, HW2-NE, and SMA2).

3 NGC7538 IRS1

The NGC7538 star forming complex is located at a kinematic distance of 2.8 kpc and is comprised of several distinct regions of star formation in various evolutionary stages (Ojha et al. 2004). This study focuses on the young, luminous (1.5×10^5 L$_\odot$) IRS 1 region. As shown in Fig. 2, at $2''$ resolution, IRS1 is composed of two submm continuum sources. The strongest is coincident with bipolar cm-wavelength emission and an infrared point source (see for example (Kraus et al. 2006)), while the other located about $4''$ to the NW has no cm-wavelength or near-infrared counterpart (Kraus et al. 2006).

Copious, strong, compact hot core line emission is detected toward the IRS1 submm continuum peak (see Fig. 3). In contrast to the other massive star forming regions that we have studied with the same spectral coverage, a wide range of C_2H_5OH transitions are detected toward NGC7538 IRS1. At $2''$ resolution, several of the detected molecular line species show a NE/SW velocity gradient similar to that seen on larger scales, tracing a massive outflow (Kraus et al. 2006). Additionally, high column density tracers like $C^{17}O$ and $C^{34}S$ are also detected toward the "cold core" to the NW, though no hot core tracers are detected toward this source.

4 G05.89-0.39

G05.89-0.39 (hereafter G5.89) is located at a distance of 2 kpc and harbors a well known, bright nearly circular shell-like ultracompact H II region. The luminosity of G5.89, 3×10^5 L$_\odot$ suggests that it harbors at least one early O star (Watson et al. 2007). G5.89 also harbors a very energetic outflow with CO linewidths in excess of 150 km s^{-1} (Watson et al. 2007). Puga et al. (2006) recently attempted

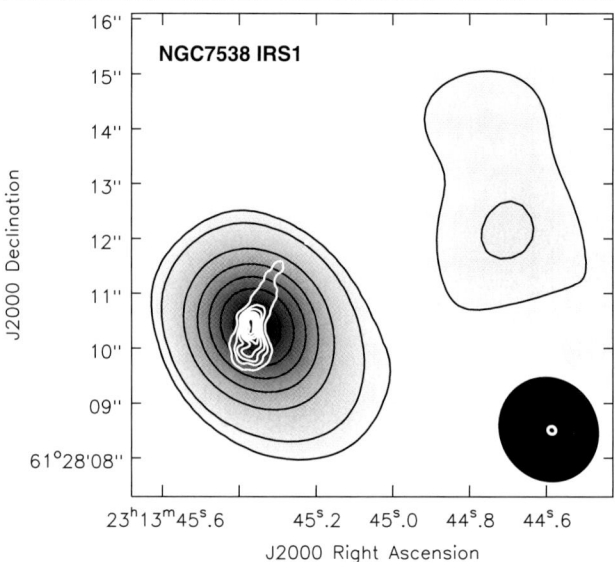

Fig. 3 Greyscale and black 875 μm SMA continuum contours of NGC7538 IRS1 at $0.25 \times (1, 2, 6, 10, 14, 18, 22)$ Jy beam^{-1}, superposed with white 3.6 cm VLA contours $0.5 \times (1, 5, 10, 20, 30, 40, 50, 60)$ mJy beam^{-1}. The beams are shown in the *lower right corner*

to identify the ionizing source of the ultracompact H II region using deep near-IR observations. The surprising result of this study is that while these authors do find strong evidence for an early O7 star, it is located on the NE rim of the ionized shell, not near the center as might be expected from the cm-wavelength morphology. One explanation for this unexpected result is that the identified O7 star is not the powering source, which remains undetected near the center of the shell due to a very high column of obscuring dust.

In an effort to better understand this enigmatic source, we have obtained extended configuration ~345 GHz SMA data toward G5.89. The resulting 875 μm continuum image with ~$0.''8$ resolution is shown in Fig. 4 superposed with comparable resolution 3.6 cm VLA continuum contours. To zeroth order, the 875 μm and 3.6 cm continuum morphologies are very similar, indicating that much of the 875 μm emission arises from free-free emission. There is no submm emission from the central region as would be expected from a region of high column density unless that emission is distributed in such a smooth, extended, way as to be resolved out by the interferometer.

As with the other sources described here, an additional "isolated" submm core is also detected SW of the ultracompact H II region. A notable aspect of G5.89 is that while it does show strong molecular emission in shock tracing molecules like SO_2, SiO, and HC_3N, it shows almost no organic molecular emission. The lack of organic emission is unusual compared to other young massive protostars, suggesting that G5.89 may be at a significantly later stage in its evolution. Similar to the continuum, no molecular emission is detected toward the central region, and is instead con-

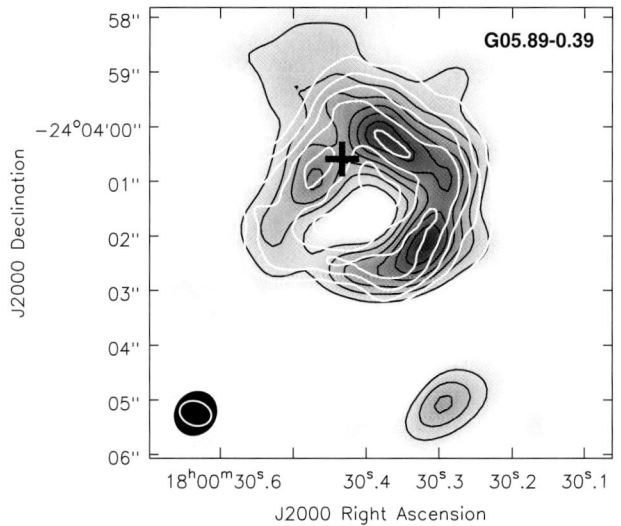

Fig. 4 Greyscale and black 875 μm SMA continuum contours of G5.89 at $0.1 \times (1, 2, 3, 4, 5, 6)$ Jy beam^{-1}, superposed with white 3.6 cm VLA continuum contours at $0.023 \times (1, 2, 3, 4, 5)$ Jy beam^{-1}. The beams are shown in the *lower left corner*. The + symbol shows the location of the O7 star detected by Puga et al. (2006)

centrated in a roughly circular morphology either coincident with the 875 μm emission or in some cases just outside it.

References

Beuther, H., Walsh, A., Schilke, P., Sridharan, T.K., Menten, K.M., Wyrowski, F.: Astron. Astrophys. **390**, 289 (2002)

Bonnell, I.A., Bate, M.R.: Mon. Not. R. Astron. Soc. **362**, 915 (2005)

Brogan, C.L., Chandler, C.J., Hunter, T.R., Shirley, Y.L., Sarma, A.P.: Astrophys. J. Lett. **660**, 133 (2007), astro-ph/0703626

Codella, C., Bachiller, R., Benedettini, M., Caselli, P., Viti, S., Wakelam, V.: Mon. Not. R. Astron. Soc. **361**, 244 (2005)

Curiel, S., et al.: Astrophys. J. **564**, L35 (2002)

Curiel, S., et al.: Astrophys. J. **638**, 878 (2006)

Hughes, V.A., Wouterloot, J.G.A.: Astrophys. J. **276**, 204 (1984)

Kraus, S., et al.: Astron. Astrophys. **455**, 521 (2006)

Martín-Pintado, J., Jiménez-Serra, I., Rodríguez-Franco, A., Martín, S., Thum, C.: Astrophys. J. **628**, L61 (2005)

McKee, C.F., Tan, J.C.: Astrophys. J. **585**, 850 (2003)

Mueller, K.E., Shirley, Y.L., Evans, N.J., Jacobson, H.R.: Astrophys. J. Suppl. Ser. **143**, 469 (2002)

Ojha, D.K., et al.: Astrophys. J. **616**, 1042 (2004)

Patel, N.A., et al.: Nature **437**, 109 (2005)

Puga, E., Feldt, M., Alvarez, C., Henning, T., Apai, D., Le Coarer, E., Chalabaev, A., Stecklum, B.: Astrophys. J. **641**, 373 (2006)

Richer, J.S., Padman, R.: Mon. Not. R. Astron. Soc. **251**, 707 (1991)

Schilke, P., Groesbeck, T.D., Blake, G.A., Phillips, T.G.: Astrophys. J. Suppl. Ser. **108**, 301 (1997)

Vlemmings, W.H.T., Diamond, P.J., van Langevelde, H.J., Torrelles, J.M.: Astron. Astrophys. **448**, 597 (2006)

Watson, C., Churchwell, E., Zweibel, E.G., Crutcher, R.M.: Astrophys. J. **657**, 318 (2007)

High-resolution mm interferometry and the search for massive protostellar disks: the case of Cep-A HW2

Claudia Comito · Peter Schilke · Ulrike Endesfelder ·
Izaskun Jiménez-Serra · Jesus Martín-Pintado

Originally published in the journal Astrophysics and Space Science, Volume 313, Nos 1–3.
DOI: 10.1007/s10509-007-9589-2 © Springer Science+Business Media B.V. 2007

Abstract The direct detection of accretion onto massive
protostars through rotating disks constitutes an important
tile in the massive-star-formation-theory mosaic. This task
is however observationally very challenging. A very interesting example is Cepheus A HW2. The properties of the
molecular emission around this YSO seems to suggest the
presence of a massive rotating disk (cf. Patel et al. in Nature
437:109, 2005). We have carried out sub-arcsec-resolution
PdBI observations of high-density and shock tracers such as
SO_2, SiO, CH_3CN, and CH_3OH towards the center of the
outflow. A detailed analysis of the spatial distribution and
of the velocity field traced by all observed species leads us
to conclude that, on a ~ 700 AU scale, the Cep-A "disk"
is actually the result of the superposition of multiple hot-
core-type objects, at least one of them ejecting an outflow
at a small angle with respect to the line of sight. Together
with the well-known large-scale outflow ejected by HW2,
this setup makes for a very complex spatial and kinematic
picture.

Keywords High-mass star formation ·
Millimeter-wavelength radioastronomy · Millimeter
interferometry · High-mass disks

Based on observations carried out with the IRAM Plateau de Bure
Interferometer. IRAM is supported by INSU/CNRS (France), MPG
(Germany) and IGN (Spain).

C. Comito (✉) · P. Schilke · U. Endesfelder
Max-Planck-Institut für Radioastronomie, Auf dem Hügel 69,
53121 Bonn, Germany
e-mail: ccomito@mpifr-bonn.mpg.de

I. Jiménez-Serra · J. Martín-Pintado
Instituto de Estructura de la Materia, Consejo Superior de
Investigaciones Científicas, Departamento de Astrofísica
Molecular e Infrarroja, C/Serrano 121, 28006 Madrid, Spain

1 Introduction

Several theories are being considered to explain the formation of massive ($M \geq 8\ M_\odot$) stars, which can be roughly
grouped into accretion-driven and coalescence-driven models (cf. Stahler et al. 2000). In the latter case, high-mass
stars would form by merging of two or more lower-mass objects, making the presence of stable massive accretion disks
around the protostar very unlikely. However, only models
based on disk-protostar interactions are capable of explaining the existence of jets and outflows: hence, the high in-
cidence, in large samples of massive YSOs, of long, collimated outflows (cf. Beuther et al. 2002) has been interpreted
as indirect evidence for the existence of high-mass disks.

It is undoubted that the direct detection of accretion
onto massive protostars through rotating disks constitutes
an important tile in the massive-star-formation-theory mosaic. From an observational point of view, this task is mainly
made difficult by two factors: (i) massive star-forming regions typically are far away, a few kpc on average, making
the direct observation of small-scale structure such as disks
virtually impossible with current instruments; and (ii), massive stars form in clusters, making the surrounding region
extremely complex, both spatially and kinematically.

Although claims of the existence of rotating disks around
massive protostars have become popular in the literature, to
our knowledge only the case for IRAS 20126+4104 has been
convincingly made (cf. Cesaroni et al. 1997, 2005; Zhang et
al. 1998; Edris et al. 2005; Sridharan et al. 2005). Cepheus A
is also considered a very promising candidate for the detection of a massive disk. Its well-studied bipolar outflow (cf.
Gómez et al. 1999, hereafter G99, and references therein)
is thought to be powered by the radio-continuum source
HW2 ($\sim 10^4\ L_\odot$, Rodríguez et al. 1994). The distribution of
H_2O masers (Torrelles et al. 1996, hereafter T96) and of the

SiO emission (G99) around HW2, both oriented perpendicularly with respect to the direction of the flow, have been interpreted as strongly supporting the existence of accretion shocks onto a rotating and contracting molecular disk of ~700 AU diameter, centered on HW2, with the outflow being triggered by the interaction between such disk and HW2 itself. The relative small distance of this region from the Sun (~725 pc, Johnson 1957) could allow such rotating object to be resolved with interferometric techniques. However, the G99 data did not have enough spatial and spectral resolution to establish a kinematical proof of a disk, and the water masers in T96 share the ambiguity of most similar studies about what the masers actually trace. Based on SMA observations of CH_3CN and dust emission, Patel et al. (2005) have recently claimed the presence of an 8 $M_\odot$ rotating disk, accreting onto HW2, and extending for about 330 AU around the protostar. Our PdBI observations do not support this interpretation: our conclusion is that, on the $1''$ scale, the Cep-A "disk" is actually the superposition of at least three different hot-core-type sources, at least one of them being the exciting source for a second molecular outflow.

2 Observations

In 2003 and 2004, with the Plateau de Bure Interferometer in AB (extended) configuration, we have carried out observations of high-density and shock tracers, such as SO_2, SiO, CH_3CN, CH_3OH, $H_2^{18}O$ and HDO towards the HW2 position ($\alpha_{J2000} = 22$ h 56 m 17.9 s, $\delta_{J2000} = +62°01'49.6''$). High-spectral-resolution correlator units were employed to achieve a channel width of up to ~0.3 km s⁻¹. A list of some of the observed transitions and beam sizes can be found in Table 1.

3 Results

Figure 1 (upper panel) shows the inner ~2800 AU of the Cep-A HW2 star-forming region. The peak of the 241 GHz dust emission (grey scale) coincides with the HW2 position and with the center of the large-scale outflow. The integrated CH_3CN emission is also centered on HW2 (contours), and somewhat elongated almost perpendicularly to the direction of the large-scale outflow.

Like other molecular tracers (cf. Comito et al. 2007; Schilke et al. 2007), CH_3CN displays two different velocity components, centered at ~−5 and ~−10 km s⁻¹ respectively. The solid contours in Fig. 1, lower panel, show the emission of the $CH_3CN(12_3-11_3)$ transition, integrated between −7 and −3 km s⁻¹, whereas the emission in the range between −11.5 and −7.5 km s⁻¹ is represented by the dashed contours (see Sect. 3.2).

Table 1 Summary of observed transitions

Transition	Frequency (GHz)	HPBW
SiO(2-1)	87	$2'' \times 1''\!.6$
HCN(1-0)	89	$1''\!.9 \times 1''\!.8$
$H_2^{18}O(3_{1,3}-2_{2,0})$	203	$1''\!.0 \times 0''\!.8$
$SO_2(12-11)$	203	$1''\!.0 \times 0''\!.8$
$^{13}CO(2-1)$	220	$0''\!.9 \times 0''\!.7$
$CH_3CN(12-11)$	220	$0''\!.9 \times 0''\!.7$
$CH_3OH(5-4)$	241	$0''\!.7 \times 0''\!.6$
$HDO(2_{1,1}-2_{1,2})$	241	$0''\!.7 \times 0''\!.6$

SiO peaks about $0''\!.3$ eastwards of HW2, close to the peak of the −10 km s⁻¹ CH_3CN component. A more detailed discussion on this transition can be found in Sect. 3.1.

In what follows, we will discuss in more detail the analysis of the CH_3CN and SiO transitions. Analysis and discussion on the other observed lines will be published in Comito et al. (2007) and Schilke et al. (2007).

3.1 SiO

In spite of the relatively low spatial resolution achieved in the imaging of the 89 GHz SiO line, this transition is the key to understand the dataset. Our data confirm that the spatial distribution of this shock tracer is mainly concentrated in the HW2 region (its presence in the large-scale outflow is limited to a few bullets at large distances from the center), although not centered on the HW2 position. This does indeed suggest that shock processes are taking place in the (projected) immediate vicinities of HW2. However, if the SiO emission were arising from accretion shocks onto a rotating disk (as proposed by G99), we would expect to observe a similar velocity structure to that observed for the other molecular tracers peaking around HW2. Instead, SiO seems to be tracing a completely different kinematic picture: unlike any other line in our dataset, the (2-1) line has a velocity spread of ~30 km s⁻¹ at the zero-flux level (~15 km s⁻¹ FWHM). A mass of about 90 $M_\odot$ would be required to produce such large line width in a gravitationally bound environment (assuming virial equilibrium, and that the emission arises in a region of ~350 AU radius). This value is about one order of magnitude larger than the estimated mass of HW2, which is expected to be a B0.5 star once in ZAMS (Rodriguez et al. 1994).

We carried out a two-dimensional Gaussian fit of the SiO(2-1) spatial distribution for every spectral channel, thus deriving a distribution of the centroids of SiO emission as a function of velocity. As shown in Fig. 1, lower panel, the centroid positions occupy a well-defined two-lobed area, centered about $0''\!.3$ eastwards of HW2 and of the dust continuum emission peak. Although the error on every single

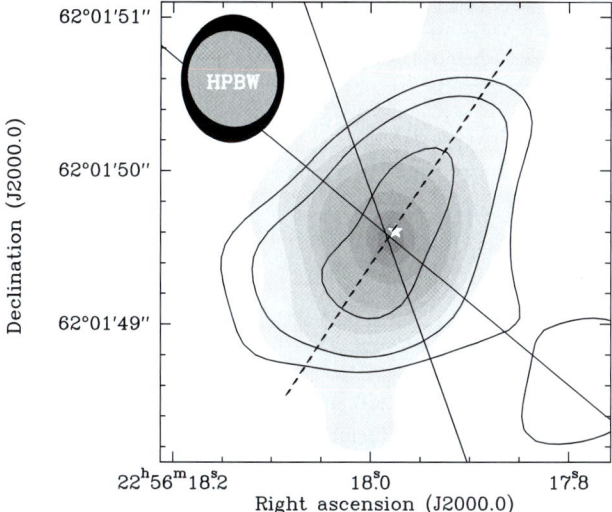

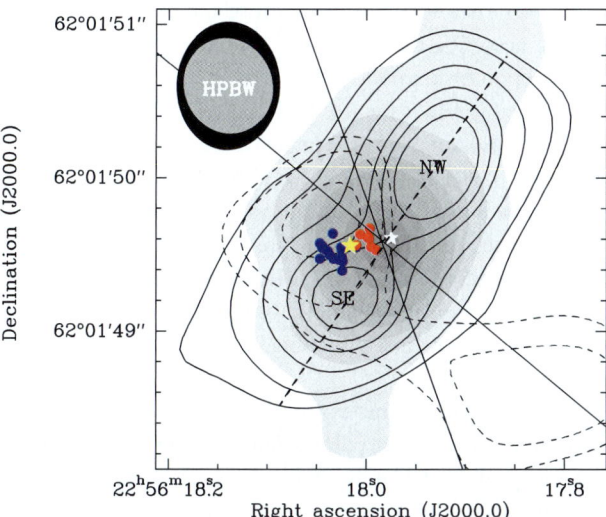

Fig. 1 *Upper and lower panel:* the levels of *grey* represent the dust emission at 241 GHz. Lowest level is 3.3 mJy/beam or 2σ, highest is 22σ. The HW2 position is indicated by the white star. The *solid, crossing lines* show the direction of the large-scale outflow, inferred from our PdBI HCN and ^{13}CO data. The *contours* trace the CH$_3$CN emission at 220 GHz, and in the top left corner, the HPBW for the dust (*grey foreground*) and CH$_3$CN (*black background ellipse*) are shown. *Upper panel:* the integrated emission of the CH$_3$CN(12_3-11_3) is shown by *solid contours*. *Lower panel:* here the ~ -5 and ~ -10 km s^{-1} velocity components of CH$_3$CN(12_3-11_3) are plotted separately (*solid and dashed contours* respectively). The *dots* show the centroid positions for SiO(2-1), *blue-shifted* on the left, *red-shifted* on the right lobe. The yellow star between the SiO lobes points to the position of the Martín-Pintado hot-core

centroid position is large (up to 30%), as a whole their distribution describes a very clear velocity trend, with all the emission at $v_{lsr} < -10$ km s^{-1} clustering in the left lobe, and all the emission at $v_{lsr} > -10$ km s^{-1} clustering in the right lobe. This result suggests that *a second molecular outflow*

is being ejected in the HW2 region, at a small angle with respect to the line of sight. Our interpretation is supported by the recent discovery of an intermediate-mass protostar, surrounded by a hot molecular core (Martin-Pintado et al. 2005), at a position which matches perfectly the inferred center position of the SiO outflow (Fig. 1, lower panel), hence a very likely candidate to be its powering engine.

3.2 CH$_3$CN

The upper panel of Fig. 1 shows the distribution of integrated intensity for the CH$_3$CN(12_3-11_3) line. Indeed, the dense molecular gas appears to be distributed around the HW2 position, and elongated in a direction roughly perpendicular to the projected direction of the large-scale outflow on the plane of the sky. From a morphological point of view, therefore, the data are very suggestive of the presence of a $\sim$300 AU-radius disk-like structure around HW2.

The kinematical picture is more complex. A position-velocity cut along the major axis of the elongated structure (indicated in Fig. 1 with a dashed line) reveals a velocity spread of about 6 km s^{-1} (see Fig. 2), also observed by Patel et al. (2005). However, the two intensity peaks along the axis share the same systemic velocity (~ -5 km s^{-1}). The weaker, blue-shifted component of emission (~ -10 km s^{-1}), appears to trace rather the outskirt of a physically separated component than a rotation-induced velocity gradient along the axis of the alleged "disk". The peak of the -10 km s^{-1} CH$_3$CN emission is spatially and kinematically close to the center of the small-scale SiO outflow (see Fig. 1, lower panel), it may therefore be associated to it and/or to its exciting source.

The CH$_3$CN integrated intensity is dominated by the two -5 km s^{-1} peaks, which lie respectively about $0\farcs6$ to the northwest, and $0\farcs5$ to the southeast of the HW2 position. In

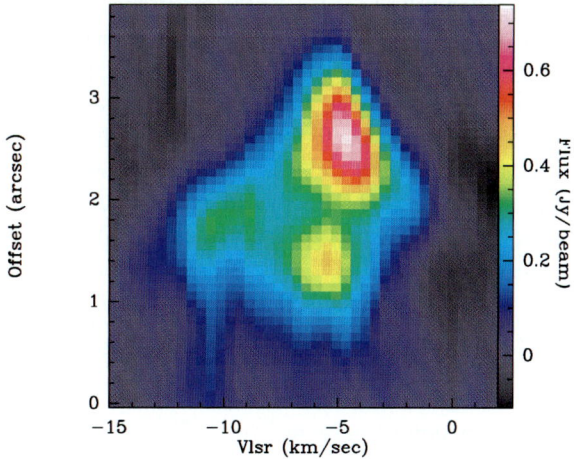

Fig. 2 Position-velocity plot for CH$_3$CN(12_3-11_3), along the major axis of the elongated structure (*dashed line* in Fig. 1)

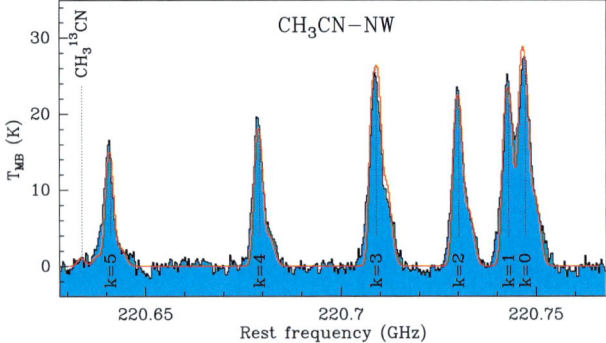

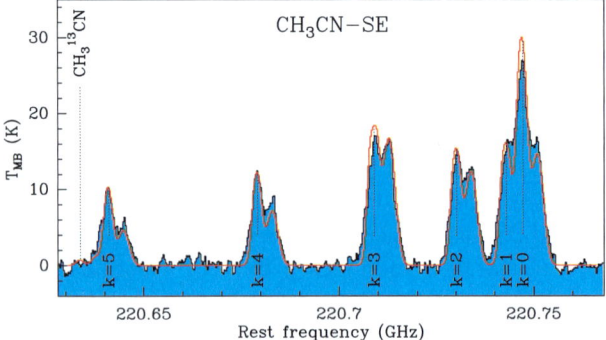

Fig. 3 High-spectral-resolution spectra of the $CH_3CN(12\text{-}11)$ emission at 200 GHz, towards the CH_3CN-NW and -SE positions (see Fig. 1, lower panel). Overlayed in red is the model spectrum, resulting from the parameters listed in Table 2

Table 2 LTE model results for the CH_3CN emission towards the NW and SE cores (Fig. 3). For a discussion on the error estimate for these numbers, see Comito et al. (2007)

	v_{lsr} (km s^{-1})	Source size	N(CH_3CN) (cm^{-2})	T_{rot} (K)	Δv (km s^{-1})
CH_3CN-NW	−4.5	0″.3	~3×10^{16}	250	2.9
	−8.9	1″	~8×10^{14}	150	4.0
CH_3CN-SE	−4.2	0″.25	~3×10^{16}	250	3.2
	−10.0	0″.45	~5×10^{15}	150	4.5

what follows, we will refer to them respectively as CH_3CN-NW and CH_3CN-SE. Figure 3 compares the spectra observed towards the two positions. It is clear that, in both cases, both the −5 and −10 km s^{-1} components are present along the line of sight, although the contribution from the latter is more intense in the CH_3CN-SE region, i.e., close to the peak of the −10 km s^{-1} SiO emission.

We have assumed LTE approximation to fit the physical parameters associated with the two different velocity components. All transitions in the spectrum are fitted simultaneously, in order to take line blending and optical depth effects properly into account (a detailed description of the method can be found in Comito et al. 2005). At −5 km s^{-1}, the line intensity ratios of the $k = 0$ through $k = 4$ transitions clearly indicate that these lines are optically thick towards both positions. The data at this velocity can only be reproduced by including a very compact, hot, dense object in the model. The emission centered at −10 km s^{-1} can be modeled by a cooler, more extended component. The results of the fit, for the two positions, are summarized in Table 2.

4 Discussion and open issues

The observed elongation of the molecular gas distribution around HW2, over a radius of ~0″.5 (~360 AU), appears

to be due to the projected superposition, on the plane of the sky, of at least two protostellar objects, of which at least one is triggering a molecular outflow at a small angle with respect to the line of sight (Sects. 3.1, 3.2). All lines in our dataset are consistent with this interpretation (cf. Comito et al. 2007). The observed chemical differentiation between clumps (cf. Brogan et al. 2007; Schilke et al. 2007; Comito et al. 2007) is a further indication of the unlikelihood of a rotating accretion disk over a 1″ spatial extension. A rotating accretion disk around HW2 is likely to exist, but it must be searched for on a smaller scale.

A few questions are still unanswered: first of all, the nature of the CH_3CN-NW and CH_3CN-SE condensations remains to be understood. The analysis of the CH_3CN spectra (Sect. 3.2) seems to suggest the presence of internally heated compact hot-core-type objects. However, the dust emission at 1mm, which peaks unambiguously on the HW2 position and does not show any clumping, does not seem consistent with this picture.

Another open issue is the physical location of the −10 km s^{-1} molecular component. Though it seems likely that the peak of CH_3CN emission is associated with the powering source of the small-scale SiO outflow, its connection to the somewhat more extended molecular emission at this systemic velocity (cf. Brogan et al. 2007) remains to be confirmed.

Although no follow-up studies will be possible from the Atacama site towards the Cep-A East cloud, over a few-hundred AU scale this source provides a very good template for the upcoming ALMA observations of high-mass star-forming regions: in fact, ALMA's high spatial resolution will open a window on a new degree of complexity for massive star-forming regions, in which what currently is known as giant massive hot-core-type sources is likely to break up in a cluster of dozens of smaller sources.

References

Beuther, H., et al.: Astron. Astrophys. **387**, 931 (2002)

Brogan, C., et al.: Astrophys. Space Sci. (2007). doi: 10.1007/s10509-007-9598-1

Cesaroni, R., et al.: Astron. Astrophys. **325**, 525 (1997)
Cesaroni, R., et al.: Astron. Astrophys. **434** (2005)
Comito, C., et al.: Astrophys. J. Suppl. Ser. **156**, 127 (2005)
Comito, C., et al.: Astron. Astrophys. **469**, 207 (2007)
Edris, K.A., et al.: Astron. Astrophys. **434**, 213 (2005)
Gómez, J.F., et al.: Astrophys. J. **514**, 287 (1999)
Johnson, H.L.: Astrophys. J. **126**, 121 (1957)
Martín-Pintado, J., et al.: Astrophys. J. Lett. **628**, L61 (2005)
Patel, N.A., et al.: Nature **437**, 109 (2005)
Rodríguez, L.F., et al.: Astrophys. J. **430**, L65 (1994)

Schilke, P., et al.: In preparation
Sridharan, T.K., Williams, S.J., Fuller, G.A.: Astrophys. J. Lett. **631**, L73 (2005)
Stahler, S.W., et al.: In: Mannings, V., Boss, A.P., Russel, S.S. (eds.) Protostars and Planets IV, p. 327. University of Arizona Press, Tucson (2000)
Torrelles, J.M., et al.: Astrophys. J. **457**, L107 (1996)
Zhang, Q., et al.: Astrophys. J. **505**, L151 (1998)

The Class 0 source Barnard 1c

Most recent results

**Brenda Matthews · Edwin Bergin · Antonio Crapsi ·
Michiel Hogerheijde · Jes Jørgensen · Dan Marrone ·
Ramprasand Rao**

Originally published in the journal Astrophysics and Space Science, Volume 313, Nos 1–3.
DOI: 10.1007/s10509-007-9591-8 © Springer Science+Business Media B.V. 2007

Abstract We present our most recent results from an ongoing study of the Class 0 source Barnard 1c in Perseus. This source is of particular interest because it exhibits evidence of strong alignment of grains all the way to the core's centre, which is contrary to all other low-mass protostellar cores observed to date. Our goal is to clarify the source of poor alignment in other sources by identifying the source of strong alignment in B1c. A central cavity has been identified in N_2H^+ emission; its anticorrelation with $C^{18}O$ emission suggests that heating in the centre has released CO from grain mantles, in turn destroying N_2H^+. We present sensitivity-limited, high spatial resolution polarimetry data from the SubMillimeter Array and discuss the potential implications of these data.

Keywords Polarization · Stars: formation · ISM:
individual (Barnard 1c, B1c) · ISM: magnetic fields

1 Introduction

Barnard 1c (B1c) was identified as a potential star-forming core in dust emission polarimetry measurements of the main molecular core of Barnard 1 by Matthews and Wilson (2002). The core shows strong polarization, with evidence that the polarization percentage is constant to the core centre. Typically, polarization percentage diminishes toward peaks in intensity, resulting in what are called 'polarization holes'. The apparent absence of a polarization hole in B1c at 850 μm led to a follow-up spectroscopic study using the Berkeley-Illinois-Maryland Association (BIMA) array and total power measurements from the Five Colleges Radio Astronomical Observatory (FCRAO). By examining the internal structure of this object, we hope to explain the strong single dish polarization from this source and in turn help explain the diminished polarization at high densities in other protostellar cores.

2 The outflow and inner envelope of B1c

Figure 1 reveals the morphology of the young stellar object B1c, which drives an impressive molecular jet and outflow. The former has been detected in 4.5 μm emission from the Spitzer Space Telescope (Jørgensen et al. 2006). A comparison of the 4.5 μm (arising from a pure rotational line of H_2 (Noriego-Crespo et al. 2004) emission to the CO

B. Matthews (✉)
Herzberg Institute of Astrophysics, 5071 West Saanich Road,
Victoria, BC, Canada V9E 2E7
e-mail: brenda.matthews@nrc-cnrc.gc.ca

E. Bergin
Dept. of Astronomy, University of Michigan, 825 Dennison
Building, 500 Church Street, Ann Arbor, MI 48109, USA

A. Crapsi · M. Hogerheijde
Leiden Observatory, P.O. Box 9513, 2300 RA, Leiden,
The Netherlands

J. Jørgensen
Harvard-Smithsonian Center for Astrophysics, 60 Garden Street
MS42, Cambridge, 02128, USA

D. Marrone
Jansky Fellow, University of Chicago/KICP, 5640 South Ellis
Avenue, Chicago, IL 60637, USA

R. Rao
Academia Sinica, Institute of Astronomy and Astrophysics,
645 North Aohoku Place, Hilo, HI 96720, USA

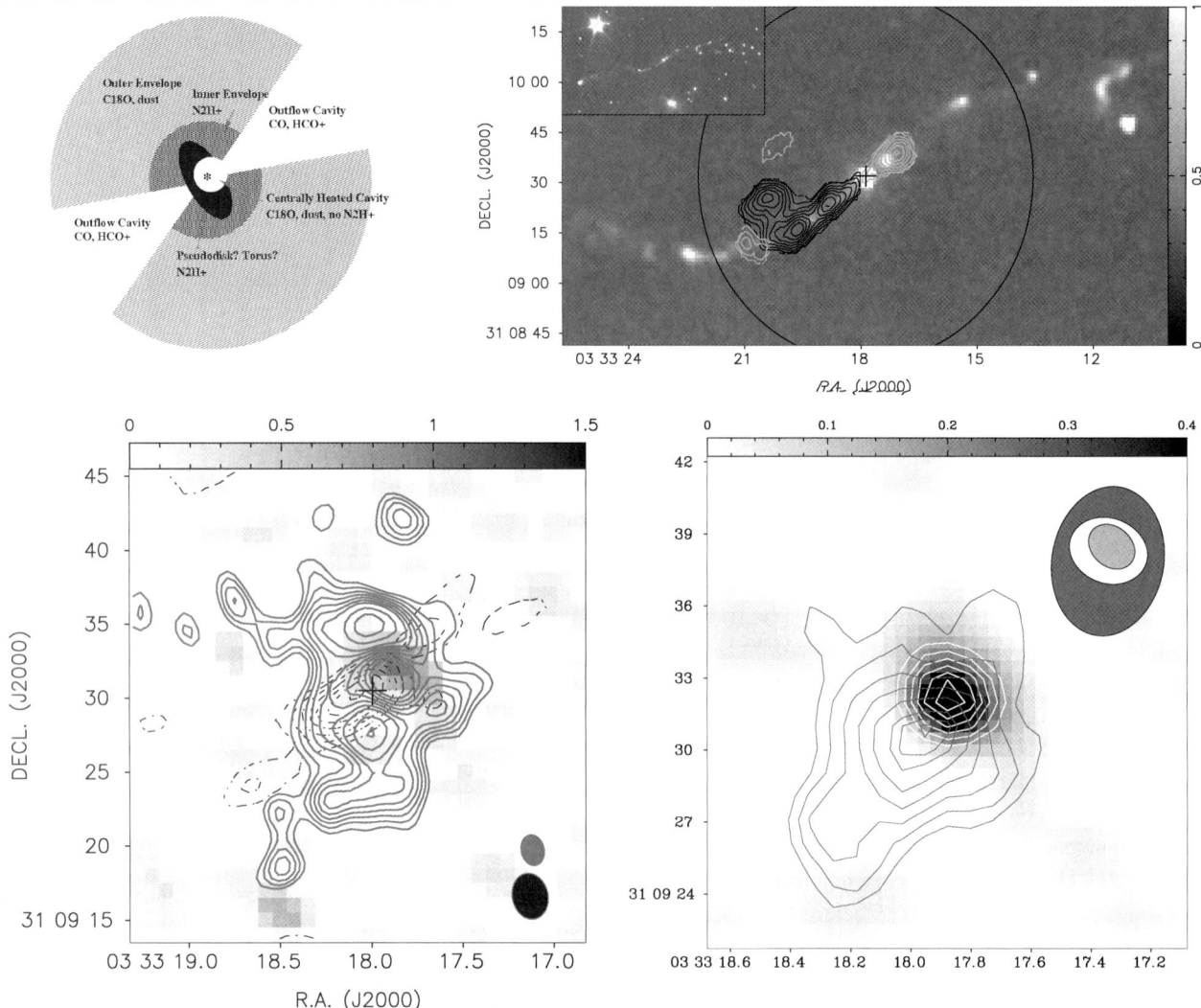

Fig. 1 *Top Left:* Schematic of the proposed morphology of B1c, including the centrally heated cavity and outflow cavity, both of which are indicated by absence of N_2H^+ emission. *Top Right:* CO 1–0 emission, showing blue-shifted (*black contours*) and red-shifted (*grey contours*) over a 4.5 μm Spitzer image (*greyscale*) from Jørgensen et al. (2006). The cross marks the position of the 3 mm continuum peak as measured with the BIMA array by Matthews et al. (2006). *Bottom Left:* N_2H^+ emission (*contours*) from the BIMA array at uniform

weighting is summed over all hyperfine transitions to give the total column. The cavity in the cold dense gas is evident. $C^{18}O$ $J = 2 - 1$ emission from the SMA in greyscale shows the anticorrelation of this species with N_2H^+. *Bottom Right:* Continuum emission at 3 mm (*grey contours*) from the BIMA array, 1 mm (*white contours*) and 850 μm (*greyscale*) from the SMA show that the shorter wavelength data peak within the cavity, while the 3 mm peak lies to the southeast

molecular outflow emission is shown in Fig. 1 (Matthews et al. 2006). The most strongly red-shifted and blue-shifted emission lies close to the protostar. The material at the leading edge of the blue-shifted lobe appears red-shifted relative to the velocity of the source. Matthews et al. note the same behavior in HCO^+ data from BIMA array and FCRAO (Matthews et al. 2006).

Figure 1 also shows that the central cavity in N_2H^+ 1–0 emission is resolved in uniformly weighted BIMA array plus FCRAO data (summed over all hyperfine components) as shown by Matthews et al. (2006). The anticorrelation of this central cavity in the dense cold gas with $J = 2–1$ emis-

sion from $C^{18}O$, as measured with the SubMillimeter Array (SMA), is clearly demonstrated. Because $C^{18}O$ is depleted onto grains with other isotopomers of CO in cold gas, its presence indicates that the grains have been heated and the CO species released from grain surfaces. CO and its isotopes are a strong destroyer of N_2H^+, which accounts for its absence at the very centre of B1c.

3 Polarimetry at high resolution

Whittet et al. (2001) observed that grains without ice mantles are more effective at polarizing radiation in the diffuse interstellar medium. The presence of a heated cavity in B1c is evidence that ice mantles may be minimized within the very centre of the inner envelope. To determine whether or not the degree of polarization is related to the heated region, it is essential that the temperature within the cavity be established and the polarization both inside and outside the cavity be measured. To measure the polarization at high resolution requires the SMA. The declining emissivity of grains and the lower fluxes of dust emission at longer wavelengths makes 850 μm measurements far superior to 1.3 mm observations, despite the more stringent weather constraints this imposes.

Figure 2 illustrates the polarization signature measured in a single track on the SMA in superior weather in September 2006. Five vectors are detected above 2 σ. If these, only three lie within the cavity defined by the N_2H^+ emission and two lie on the boundary. The vectors within the cavity have the lowest polarization percentage (but are also associated with the highest intensity positions as is typically observed). There is evidence for relative little change in polarization percentage within the cavity even as intensity continues to increase. Since the key to understanding the polarization percentage's relation to the heating is the variation of the polarization percentage, our current dataset is not sufficient to establish whether the variation in polarization inside and outside the cavity is uniquely related to the degree of heating those grains have experienced. Naively, if our hypothesis about a correlation between the destruction of ice mantles and higher polarization were correct, we would have expected the grains within the cavity to have the highest degree of polarization. In fact, the grains at the centre have low polarizations compared to the two vectors detected outside the cavity, a distribution which is completely consistent with the pervasive picture of polarization holes in other star-forming cores. All we can determine thus far is that the polarization within the cavity appears roughly constant, a conclusion consistent with the lower threshold in polarization percentage found in single dish polarization measurements at 850 μm by Matthews and Wilson (2002).

As well as a polarization measurement, we simultaneously obtain an excellent measurement of the continuum flux at 850 μm at ~2″ resolution from our polarization data. The contours of Fig. 2 show the Stokes I component of the polarized signal. The S/N of the continuum detection is 20σ on the peak. We measure an integrated flux of 1.5 Jy in an aperture of 5″, which is significantly less than the single-dish flux of 2.4 Jy (Kirk et al. 2006) due to spatial filtering.

4 Implications

A comparison of the polarization percentages measured outside and inside the cavity of B1c may reflect two different behaviors. Outside the cavity, the polarization percentage should decline with increasing density (intensity) since

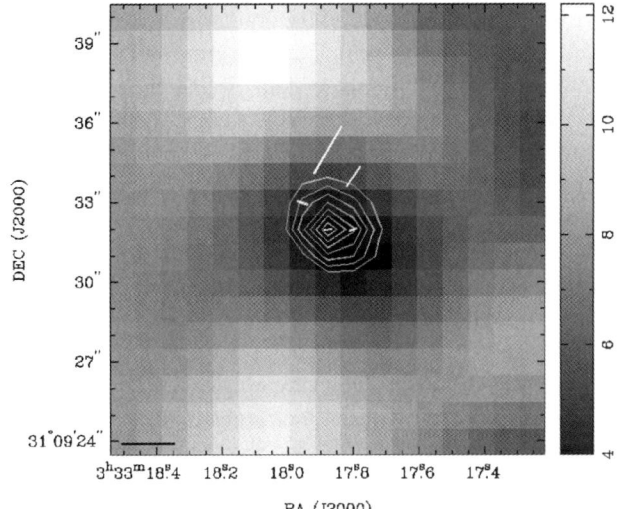

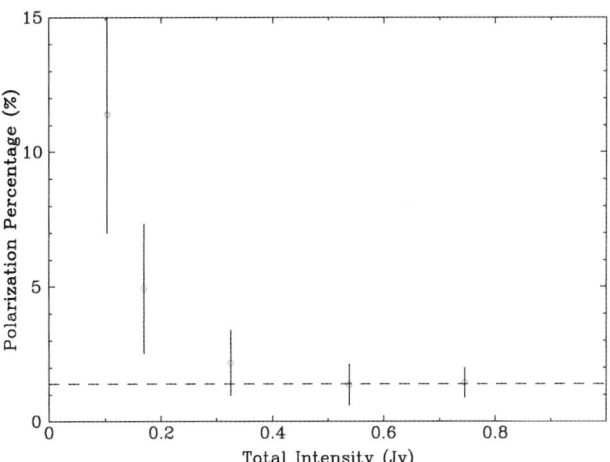

Fig. 2 *Top:* Polarization within the heated cavity of B1c. N_2H^+ emission is shown in greyscale and defines the location of dense, cold N_2H^+ gas (BIMA array + FCRAO). *White* indicates high emission, while *black* indicates diminished N_2H^+ emission. The 850 μm continuum data (2.25″ × 1.73″ beam) from the SMA (*grey contours*) clearly show that the dust emission peaks inside the cavity in N_2H^+. The polarization vectors have S/N better than 2 σ and are sampled at 1″ intervals. The length of the vectors indicates polarization percentage and the direction is the position angle of the polarization vectors +90°. The scale vector at bottom left is the value of the maximum vector, 11.3%. *Bottom:* The polarization percentage as a function of measured intensity for the five detected vectors. Those vectors within the cavity have virtually constant polarization percentage. The *dashed line* indicates 1.4% polarization

grains are colder at high densities prior to the formation of the protostar. This can be attributed to the depletion of molecules onto grains and the formation of icy mantles. However, we do see a suggestion of constant polarization in the cavity in Fig. 2). This could be a result of heating gradually increasing the polarizing efficiency of the grains (Whittet et al. 2001). The uniform value of polarization percentage, and the anticorrelation of the N_2H^+ and $C^{18}O$ emission in the cavity would indicate that all the grains are heated to a temperature sufficient to desorb CO from grain surfaces.

What is clear from these data is that more polarization data are required to effectively sample the polarization across the cavity and improve the signal-to-noise ratio of the current weak detections. It is also evident that much of our interpretation hinges on the gas and dust within the cavity having reached sufficient temperature to desorb ices from the dust grains and further on the establishment of a correlation between the temperature and the degree of polarization emitted.

In order to measure temperature, we are conducting a joint project on the Green Bank Telescope and the Very Large Array to measure the kinetic temperature at resolutions of $\sim 2''$ across the cavity using the inversion lines of NH_3. This is comparable to the resolution achievable in the compact configuration of the SubMillimeter Array at 345 GHz (850 μm). Therefore, with both datasets, we will be able to compare the variation in temperature across the cavity with the polarization percentage.

References

Jørgensen, J., et al.: The Spitzer c2d survey of large, nearby, interstellar clouds, III: Perseus observed with IRAC. Astrophys. J. **645**, 1246 (2006)

Kirk, H., Johnstone, D., Di Francesco, J.: The large- and small-scale structures of dust in the star-forming Perseus molecular cloud. Astrophys. J. **646**, 1009 (2006)

Matthews, B., Hogerheijde, M., Jørgensen, J., Bergin, E.: The rotating molecular cloud core and precessing outflow of the young stellar object Barnard 1c. Astrophys. J. **652**, 1374 (2006)

Matthews, B., Wilson, C.: Magnetic fields in star-forming molecular clouds, V: submillimeter polarization of the Barnard 1 dark cloud. Astrophys. J. **574**, 822 (2002)

Noriego-Crespo, A., et al.: A new look at stellar outflows: Spitzer observations of the HH 46/47 system. Astrophys. J. Suppl. Ser. **154**, 352 (2004)

Whittet, D., Gerakines, P., Hough, J., Shenoy, S.: Interstellar extinction and polarization in the Taurus dark clouds: the optical properties of dust near the diffuse/dense cloud interface. Astrophys. J. **547**, 317 (2001)

APEX and ATCA observations of the southern hot core G327.3-0.6 and its environs

Friedrich Wyrowski · Per Bergman · Karl Menten ·
Jürgen Ott · Peter Schilke · Sven Thorwirth

Originally published in the journal Astrophysics and Space Science, Volume 313, Nos 1–3.
DOI: 10.1007/s10509-007-9590-9 © Springer Science+Business Media B.V. 2007

Abstract There is no generally accepted evolutionary scheme for high mass star formation yet. A simple approach to address this problem is to cover several of the known stages during the formation of massive stars in the same cloud and then investigate their properties trying to construct an evolutionary sequence. Here we present such a project conducted with complementary APEX and ATCA observations. These observations show a compact and bright single hot core in the G327.3-0.6 region on a 0.03 pc scale with a mass of 500 $M_\odot$ and 0.5–1.5 10^5 $L_\odot$. Additionally a clumpy filament is seen in N_2H^+. Together with cm continuum observations, the data reveal like pearls on a string several stages of massive star formation, with likely the youngest stages hiding in the cold N_2H^+ cores analysed with a multi-level study of the APEX and ATCA observations.

Keywords G327.3-0.6 · Massive star formation

F. Wyrowski (✉) · K. Menten · P. Schilke · S. Thorwirth
Max-Planck-Institut für Radioastronomie, Bonn, Germany
e-mail: wyrowski@mpifr-bonn.mpg.de

P. Bergman
European Southern Observatory, Santiago, Chile

P. Bergman
Onsala Space Observatory, 43992 Onsala, Sweden

J. Ott
National Radio Astronomy Observatory, Charlottesville, VA, USA

1 High mass star formation: the quest for an evolutionary scheme

The quest for an evolutionary scheme, comparable to the CLASSes framework that exists for the early evolution of low-mass protostars, is currently one of the main topics of research in the field of high mass star formation. In analogy to low mass prestellar cores, precluster cores are expected in the high mass case which are cold (≤ 20 K) and massive (~ 100–1000 $M_\odot$). The observational evidence for these cores is still scarce. Massive young stellar objects will eventually start to build up mass within these cores and develope from a still relatively cold and infrared dark phase into hot molecular cores. These are internally heated to temperatures high enough to evaporate grain mantles ($T > 100$ K) and are dense ($n \sim 10^7$ cm^{-3}) with infall still ongoing, so that HII regions can be quenched for some time (Walmsley 1995) before observable hyper- and ultra-compact HII region form around the young massive cluster.

This evolutionary scheme can be studied by either surveys of a large sample of young massive star forming regions selected to cover a wide range in evolutionary phases (see for example Hieret et al. this volume) or by studying in detail template regions, which harbor several of these phases simultaneously. An example of such a region is the giant molecular cloud associated with the bright southern hot core G327.3-0.6, which has the potential of becoming a southern hemisphere hot core template for upcoming observatories like ALMA and was therefore studied by us with APEX (Wyrowski et al. 2006) and ATCA (Wyrowski et al., in preparation).

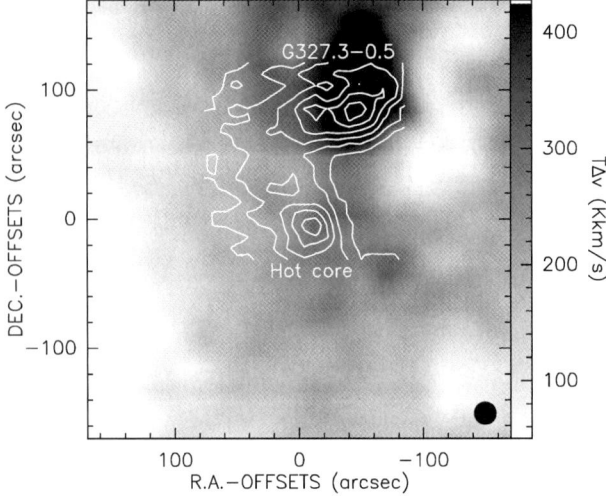

Fig. 1 APEX ^{12}CO (*greyscale*) and C^{18}O (3–2) (*contours*) images. The contour levels start at 23.6 with steps of 15.6 K km/s

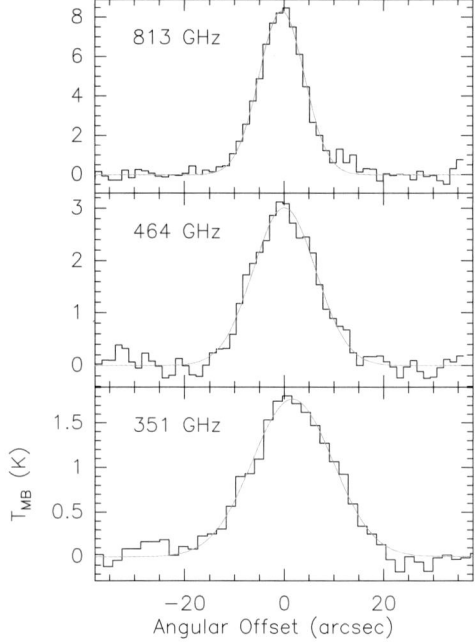

Fig. 2 APEX Continuum scans of the hot molecular core

2 The massive star forming region G327.3-0.6

The hot molecular core associated with the HII region G327.3-0.6 is situated at a kinematical distance of 2.9 kpc (Bergman 1992) about 2 arcmin south of a bright complex of infrared sources. It was discovered by its association with prominent H_2O, OH, and CH_3OH masers, and its chemistry has been studied in two papers, one reporting ethylene oxide and acetaldehyde observations (Nummelin et al. 1998), while the other investigates the chemical inventory of this source (Gibb et al. 2000). The source is remarkable for its exceptionally rich molecular line spectra with relatively narrow, well-behaved (Gaussian) line profiles (Schilke et al. 2006). Compared to hot molecular cores accessible from the northern hemisphere almost nothing is known about its environs. Only Bergman (1992) reports some SEST maps that reveal two adjacent dense cores in this molecular cloud: one relatively cold ($T_{kin} \sim 30$ K) molecular clump and one hot ($T = 100$–200 K) core. Hence, this region offers the possibility to study cores that have formed from the same parental cloud, but that are in different stages of evolution.

3 Observations

Single dish observations of G327.3-0.6 were done with the Atacama Pathfinder Experiment (APEX, Güsten et al. 2006). To probe the large scale environs of the sources, arcmin^2-sized On-The-Fly maps were observed in ^{12}CO and C^{18}O (3–2) (see Fig. 1). A smaller part of the cloud was observed with raster mapping in N_2H^+ (3–2). Higher N_2H^+ transitions were observed towards the peak of the N_2H^+ (3–2) emission.

Interferometric observations of the source at 3 mm were performed with the Australia Telescope Compact Array

(ATCA) in the H75 and H214 configurations, resulting in an angular resolution down to $2''$. A 4 point mosaic was observed to cover the cold and the hot core. N_2H^+ (1–0) was observed with a spectral resolution of 0.2 km/s and the 3 mm continuum and the (5–4) K ladder of CH_3CN were observed with 32 channels and a total bandwidth of 128 MHz.

4 The hot molecular core

The large scale CO emission as observed with APEX is shown in Fig. 1. The hot molecular core, while relatively inconspicuous in ^{12}CO, shows up as a strong column density peak in C^{18}O. In the northern part of the map strong CO emission traces the hot surface of a bright photon dominated region associated with the strong HII region G327.3-0.5 and the complex of infrared sources around IRS3 (Goss and Shaver 1970; Epchtein and Lepine 1981). The mass of the core derived from C^{18}O is about 500 $M_\odot$.

Simultaneously with C^{18}O, CH_3OH 7_1-$6_1 A^-$ was observed, which is expected to originate from hot molecular gas. Its peak is aligned with the C^{18}O column density peak. Towards this position, a typical line-rich hot core spectrum is observed (Fig. 3, lower panel). Continuum cross scans with APEX (Fig. 2) peak also at the CH_3OH and C^{18}O peaks and can be used to derive a dust mass of the core. Depending on the assumed dust properties, the dust mass range from 400 to 1000 $M_\odot$, consistent with the C^{18}O derived value. The mm/submm continuum fluxes can be used together with infrared flux densities to construct the spectral energy distribution of the sources. With a 20 and 100 micron fluxes from

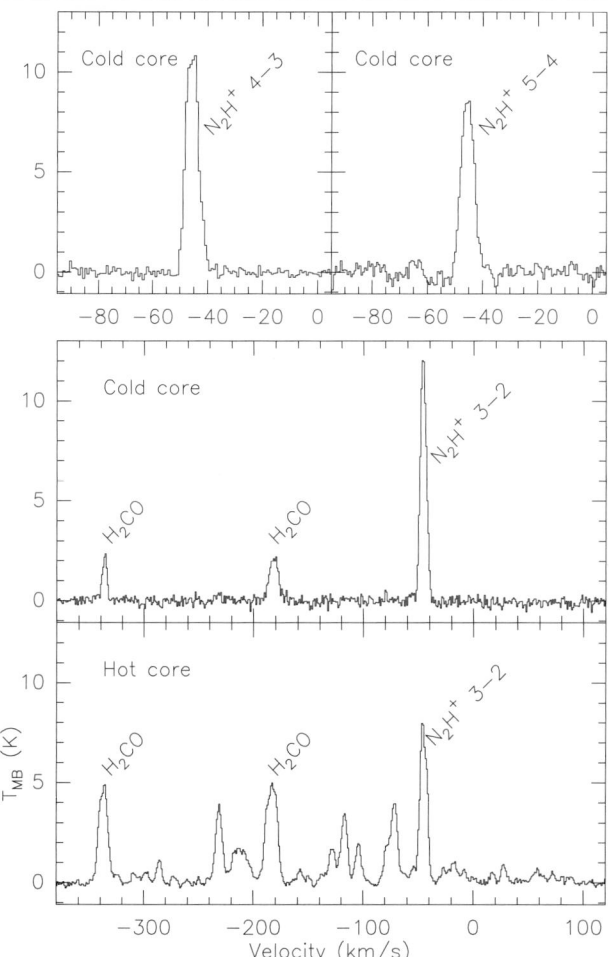

Fig. 3 APEX spectra observed toward the cold core ($(12'', 6'')$, *upper panels*) and the hot core position ($(-6'', -6'')$, *lower panel*)

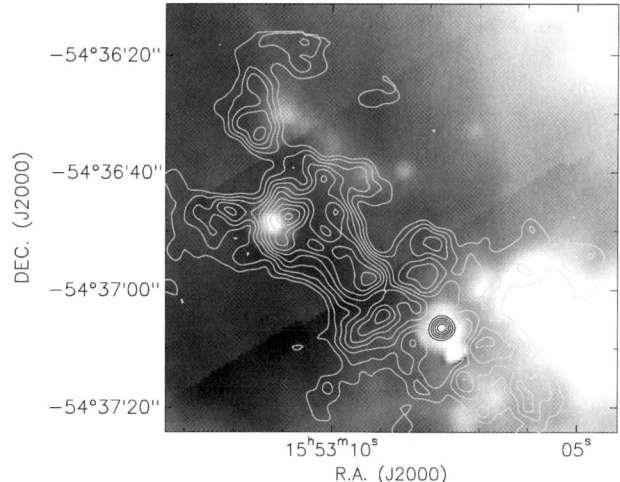

Fig. 4 ATCA $N_2H^+(1–0)$ (*grey contours*) and 3mm continuum (*black*) with GLIMPSE 8 micron as *greyscale background*

optical depth and the excitation temperature and, in turn, the N_2H^+ column density of the cores can be determined. These parameters can be used together with Non-LTE molecular radiative transfer modelling (using RADEX on-line)[1] of the higher excited lines observed with APEX to constrain the density of the cores. To reproduce the strong emission of the higher-J N_2H^+ lines, densities of at least 5×10^6 cm^{-3} are needed. Together with the observed sizes this points to masses of order 500 $M_\odot$, hence similar to the hot molecular core. The observed line widths of N_2H^+ can be used to estimate the virial masses of the cold cores, which are only several 10 solar masses, hence these cores are likely gravitationally unstable and therefore represent a promising and rare example for massive pre-protocluster cores.

IRAS HIRES and MSX images of the region, a luminosity between $5–15 \times 10^4$ L$_\odot$ is estimated for the hot molecular core.

5 The cold molecular clump

Spectra of the cold molecular clump to the north-east of the hot core are shown in Fig. 3. Extremely bright N_2H^+ was detected and even emission from H_2CO, probing an embedded "hot corino" within the cold clump. This finding prompted us to observe the fundamental N_2H^+ rotational transition with the ATCA at high angular resolution. The $N_2H^+(1–0)$ integrated intensity distribution is shown in Fig. 4. N_2H^+ probes a mid-infrared dark clump with a size of about 1 parsec. With the high angular resolution of ATCA, fragments within this clump with core size of 0.1 to 0.2 parsec can be detected. The N_2H^+ emission avoids the hot core itself. The ^{14}N hyperfine structure of the (1–0) lines is partly resolved in the observations and therefore the

6 Summary

The combined APEX/ATCA study of the environs of the hot molecular core G327.3-0.6 shows the power of combining the higher angular resolution of millimeter interferometers with observations of high excitation lines accessible to sub-millimeter telescopes. The observations resolve the region into several different stages of massive star formation:

- A cold clump harboring promising pre-protocluster cores.
- A luminous hot molecular core offset from a UC HII region, visible in Fig. 4 west of the core.
- A large HII region with bright photon dominated region to the north.

G327.3-0.6 will therefore be an ideal laboratory for ALMA to study and test evolutionary schemes of massive star formation.

[1] http://www.sron.rug.nl/~vdtak/radex/radex.php.

References

Bergman, P.: PhD thesis, Göteborg (1992)

Epchtein, N., Lepine, J.R.D.: Astron. Astrophys. **99**, 210 (1981)

Gibb, E., Nummelin, A., Irvine, W.M., Whittet, D.C.B., Bergman, P.: Astrophys. J. **545**, 309 (2000)

Goss, W.M., Shaver, P.A.: Austr. J. Phys. Astrophys. Suppl. **14**, 1 (1970)

Güsten, R., Nyman, L.Å., Schilke, P., Menten, K., Cesarsky, C., Booth, R.: Astron. Astrophys. **454**, L13 (2006)

Nummelin, A., Dickens, J.E., Bergman, P., Hjalmarson, A., Irvine, W.M., Ikeda, M., Ohishi, M.: Astron. Astrophys. **337**, 27 (1998)

Schilke, P., Comito, C., Thorwirth, S., Wyrowski, F., Menten, K.M., Güsten, R., Bergman, P., Nyman, L.-Å: Astron. Astrophys. **454**, L41 (2006)

Walmsley, M.: In: Revista Mexicana de Astronomia y Astrofisica Conference Series 1, p. 137 (1995)

Wyrowski, F., Menten, K.M., Schilke, P., Thorwirth, S., Güsten, R., Bergman, P.: Astron. Astrophys. **454**, L91–L94 (2006)

The physical conditions in the BHR71 outflows

Bérengère Parise · Arnaud Belloche · Silvia Leurini · Peter Schilke

Originally published in the journal Astrophysics and Space Science, Volume 313, Nos 1–3.
DOI: 10.1007/s10509-007-9588-3 © Springer Science+Business Media B.V. 2007

Abstract Highly-collimated outflows are believed to be the earliest stage in outflow evolution, so their study is essential for understanding the processes driving outflows. The BHR71 Bok globule is known to harbour such a highly-collimated outflow, which is powered by a protostar belonging to a protobinary system. Using the APEX telescope on Chajnantor, we mapped the BHR71 highly-collimated outflow in CO(3-2), and observed several bright points of the outflow in the molecular transitions CO(4-3), CO(7-6), ^{13}CO(3-2), C^{18}O(3-2), CH$_3$OH(7-6) and H$_2$CO(4-3). We use an LVG code to characterise the temperature enhancements in these regions. These observations are particularly interesting for investigating the interaction of collimated outflows with the ambient molecular cloud. In our CO(3-2) map, the second outflow driven by IRS2, which is the second source of the binary system, is completely revealed and shown to be bipolar. We also measure temperature enhancements in the lobes. The CO and methanol LVG modelling points to temperatures between 30 and 50 K in the two lobes. The methanol emission in the southern lobe bright knot is barely resolved with the APEX single-dish. ALMA will thus be a central tool to study the shock chemistry in these regions.

Keywords Astrochemistry · Molecular data · Stars: formation · ISM: jets and outflows · Submillimeter

B. Parise (✉) · A. Belloche · P. Schilke
MPIfR, Auf dem Hügel 69, 53121 Bonn, Germany
e-mail: bparise@mpifr-bonn.mpg.de

S. Leurini
ESO, Karl Schwarzschild Str. 2, 85748 Garching bei München, Germany

1 Introduction

Low-mass stars form in molecular clouds from the collapse of dense cores. In the earliest stage, known as Class 0, the newly-born protostar is deeply embedded in a thick envelope of gas and dust from which it will accrete most of its mass, and it drives powerful outflows. However, the mechanisms driving bipolar outflows, as well as the interaction of the outflow with the surrounding material, are still not understood well. It is commonly believed that highly-collimated outflows represent the earliest stage in outflow evolution, making their observation central to getting new insights into the mechanisms driving outflows.

The BHR71 isolated Bok globule, located at ~200 pc, harbours one of the best examples of such a highly-collimated outflow. This outflow, which is almost lying in the plane of the sky, was first discovered and mapped in ^{12}CO(1-0) and ^{13}CO(1-0) by Bourke et al. (1997). It appears to be driven by the very young stellar object IRAS11590−6452, classified as a Class 0/I protostar by Froebrich (2005), with a total luminosity of ~9 L$_\odot$. The CO emission was observed to peak towards the outflow lobes, implying that the ambient gas is heated by interaction with the outflow.

Subsequent observations by Garay et al. (1998) aimed at constraining the magnitude of the temperature enhancement in the lobes, by mapping several transitions of CS, SiO, CH$_3$OH, and HCO$^+$ with the SEST telescope. Assuming n = 10^5 cm^{-3}, the CH$_3$OH (2-1) and (3-2) observations were found to be consistent with kinetic temperatures in the 50–100 K range.

ISOCAM observations of BHR71 showed that the IRAS source is in fact a multiple system (Myers and Mardones 1998). Two protostellar objects IRS1 and IRS2 are present in BHR71, and the large collimated outflow is driven by IRS1

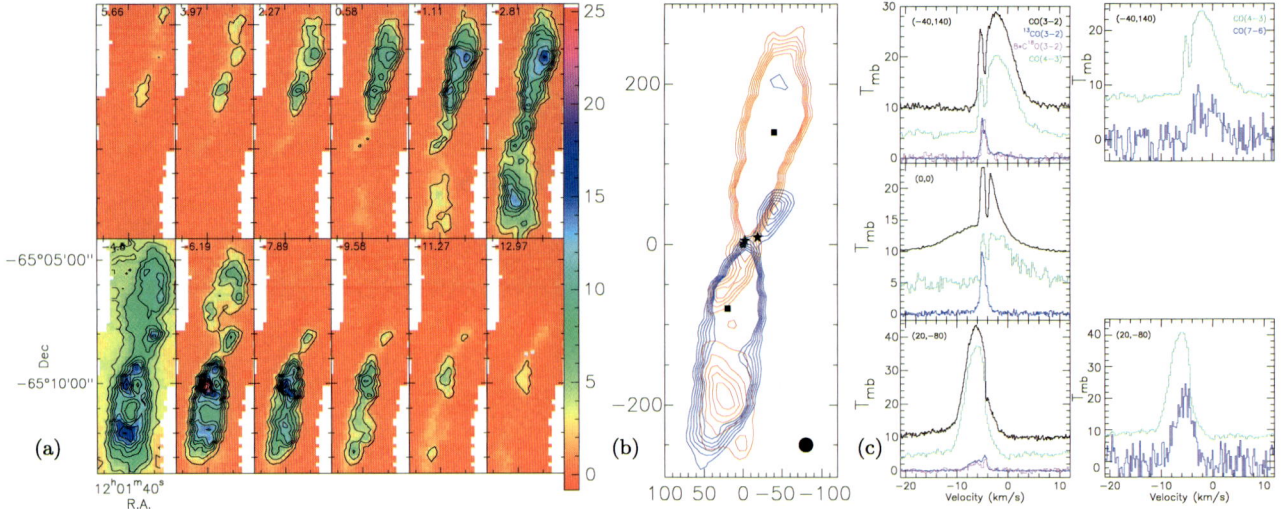

Fig. 1 (**a**) CO(3-2) channel maps (centred on $\alpha(2000) = $ 12 h 01 m 37 s, $\delta(2000) = -65°08'53.5''$) on a $\int T_A^*$ dv scale (the v_{lsr} of the cloud is -4.5 km s^{-1}). (**b**) Map of the integrated CO(3-2) emission in the $-14 < v < -6$ (*blue*) and $-3 < v < 5$ km s^{-1} (*red*) intervals, enlightening the two bipolar outflows. Contours are from 12 to 32 in steps of 4 K km s^{-1}. Contours for higher flux (in the north–south outflow) have not been plotted in order to show the fainter outflow better. The two black stars show the position of the two protostellar sources. The *black circle* in the lower right corner shows the beam size. (**c**) CO(3-2), ^{13}CO(3-2), C^{18}O(3-2), CO(4-3) and CO(7-6) observations on the three positions marked as *black squares* in (**b**)

(Bourke 2001). Combination of CO(2-1), NIR, and cm observations shows evidence of the presence of the blue lobe of a second, fainter outflow driven by IRS2 (Bourke 2001).

In this contribution, we present CO(3-2) observations that fully unveil the presence of this second outflow, and show precession in the bigger outflow. We characterise the temperature enhancement towards bright knots of the larger outflow by observations of CO(4-3), CO(7-6), ^{13}CO(3-2), ^{18}CO(3-2), and CH$_3$OH(7-6). Small maps obtained in CH$_3$OH(7-6) and H$_2$CO(4-3) show that the shock regions are barely resolved by APEX.

2 Observations

Observations towards BHR71 were carried out with the Atacama Pathfinder Experiment (APEX[1], Güsten et al. 2006), using the APEX2a receiver (Risacher et al. 2006) and the MPIfR FLASH receiver (Heyminck et al. 2006), along with the MPIfR FFTS backend (Klein et al. 2006). The beam of the telescope is 17.3'' at 345 GHz and 13.3'' at 460 GHz (Güsten et al. 2006). More details about the observations can be found in Parise et al. (2006). In addition to the observations published in Parise et al. (2006), we present here recent observations of CO(4-3) towards two bright spots of the

outflow, obtained under better weather conditions, as well as detection of CO(7-6) on the same positions. We also mapped the CH$_3$OH(7-6) and H$_2$CO(5-4) emission on the bright CO bullet of the southern lobe.

3 Outflow morphology

The channel map of the CO(3-2) transition (Fig. 1a) displays the beautifully collimated outflow discovered by Bourke et al. (1997). The cloud emission arises between -6 and -3 km s^{-1}. Note that the two lobes are well separated spatially, and that there is a small amount of red emission in the blue lobe and vice versa. These two facts confirm that the inclination of the outflow with respect to the plane of the sky is close to its semi-opening angle ($\sim 15°$), as already noted by Bourke et al. (1997). Some hints of precession of the outflow are visible on the CO(3-2) channel map, but a detailed study would strongly benefit from better spatial resolution observations, as for example with the forthcoming CHAMP$^+$ receiver allowing CO(6-5) observations at APEX, and ultimately ALMA.

The channel map also shows hints of a second fainter outflow that is a bit tilted relative to the bigger one, with the blue and red lobes inverted. This is clearer in Fig. 1b, where contours from emission for velocities in the $[-14, -6]$ and $[-3, 5]$ km s^{-1} ranges have been plotted. The second outflow is centred on the second source of the binary, IRS2. Its PA is between $-35°$ and $-30°$ and its projected semi-opening angle $\sim 25°$. Its projected collimation factor is >2.

[1]This publication is based on data acquired with the Atacama Pathfinder Experiment (APEX). APEX is a collaboration between the Max-Planck-Institut für Radioastronomie, the European Southern Observatory, and the Onsala Space Observatory.

No red emission seems associated with the blue lobe, and the blue and red lobes are well separated spatially, so the inclination of this outflow is between 25° and 65° (Cabrit and Bertout 1986).

4 Temperature enhancement in the lobes

4.1 CO excitation

Using a standard spherically symmetric large-velocity-gradient (LVG) code, we can derive the kinetic temperature in the outflow lobes from our CO(3-2), ^{13}CO(3-2), CO(4-3) and CO(7-6) observations. The inputs of the code are the CO column density, the kinetic temperature, and the density.

Modelling of the CO emission shows temperature enhancements from 30 to 40 K towards the northern position of the IRS1 outflow at the $(-40'', 140'')$ offset position, and of $\sim$50 K towards the southern lobe at the $(20'', -80'')$ offset position, which is consistent with the temperature derived from methanol analysis (see Sect. 4.2).

4.2 Methanol statistical equilibrium calculations

The CH$_3$OH $7_k \rightarrow 6_k$ band at 338 GHz was observed towards the core position and the two bright positions of the main outflow with the APEX-2a receiver tuned in the lower side band (Fig. 2). This frequency setup allows for simultaneous detection in the upper sideband of the 4_{04}-3_{-13} line at 350.688 GHz and of the 1_{11}-0_{00} transition at 350.905 GHz, which have low excitation energies ($T_{low} < 20$ K) compared to the $7_k \rightarrow 6_k$ band ($T_{low} > 44$ K). The three positions differ significantly in the methanol emission. Towards the central position only weak emission is detected from the 4_{04}-3_{-13}, the 1_{11}-0_{00}, and maybe the 7_{07}-6_{06} transitions. In contrast, lines from the $7_k \rightarrow 6_k$ are also observed towards the two lobe positions, the lower energy transitions toward the northern lobe, and moderately higher energy transitions towards the south position. In the $(20'', -80'')$ offset position, the methanol emission is blueshifted, and thus associated with the IRS1 outflow. In the $(-40'', 140'')$ offset position, the lines are double-peaked, with a redshifted peak and a contribution at the cloud velocity (see inset in Fig. 2), which might be associated with emission from the preshock material, where methanol is released from the grains, but not accelerated yet.

Leurini et al. (2004) concluded that low-excitation CH$_3$OH transitions often trace the density of the gas, while highly excited lines are also sensitive to the kinetic temperature. Methanol can therefore be used as a tracer of both temperature and density. To determine the physical parameters we used a spherically symmetric large velocity gradient statistical equilibrium code, with the cosmic background as

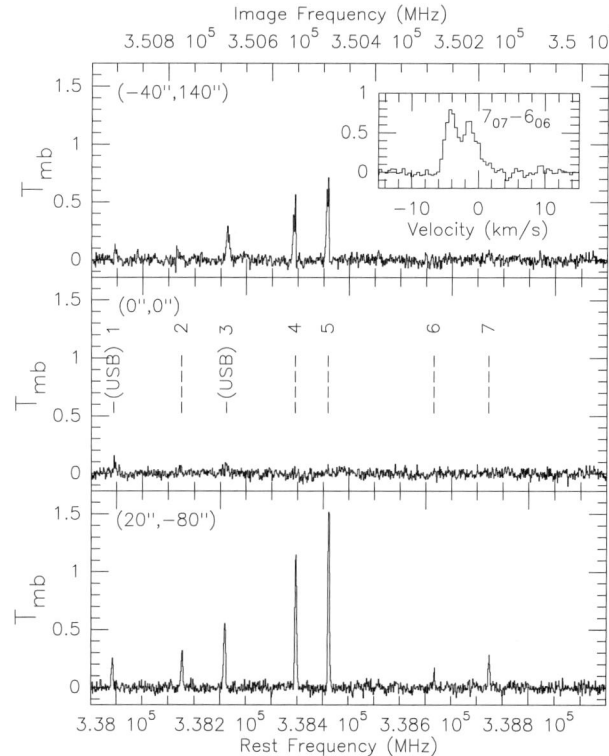

Fig. 2 Methanol spectra observed towards the 3 positions shown in Fig. 1b, c. The vertical dashed numbered lines in the middle panel refer to the following transitions: (1) $1_{1,1}$-$0_{0,0}$ [USB], (2) $7_{0,7}$-$6_{0,6}$, (3) $4_{0,4}$-$3_{-1,3}$ [USB], (4) $7_{-1,7}$-$6_{-1,6}$, (5) $7_{0,7}$-$6_{0,6}$, (6) $7_{+1,6}$-$6_{+1,5}$, (7) $7_{2,5}$-$6_{2,4}$, & $7_{-2,6}$-$6_{-2,5}$. The inset shows the double-peak structure of the $7_{0,7}$-$6_{0,6}$ line towards the northern position

the only radiation field. Line ratios between different transitions can be used to constrain the temperature and density. We assumed that all lines are emitted in the same gas; i.e. we suppose that all lines have the same beam-filling factor. This assumption is likely to be correct, as all the detected transitions have similar lower energy (<75 K, and even between 44 and 75 K, if we only focus on the $7_k \rightarrow 6_k$ band).

From our modelling, we conclude that, towards the $(20'', -80'')$ position, CH$_3$OH comes from a relatively warm ($30 < T < 50$ K) and dense ($10^5 < n(H_2) < 3 \times 10^5$ cm^{-3}) gas. The temperature enhancement that we derive is a bit lower than the range suggested by Garay et al. (1998, between 50 and 100 K, as deduced from analysis of lower energy lines). The transitions they used span a narrower range of excitation conditions than the presently observed ones, so they are less sensitive to temperature, as shown by Leurini et al. (2004). The detection of only the 7_{-17}-6_{-16} and 7_{07}-6_{06} lines towards the red lobe implies that the gas must be around 20 to 30 K and not denser than a few 10^5 cm^{-3}. Finally, the non detection of the $7_k \rightarrow 6_k$ lines towards the

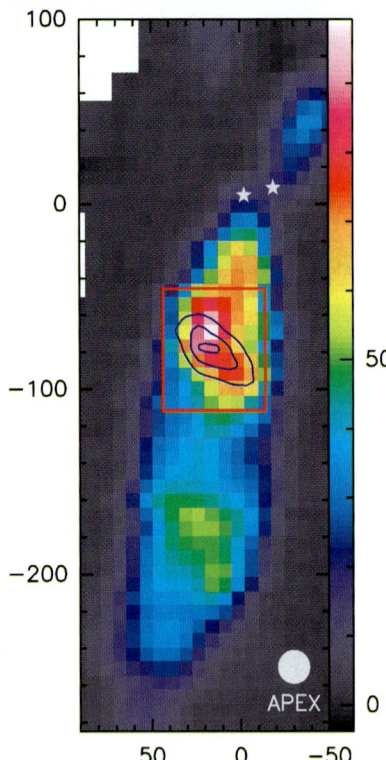

Fig. 3 CH$_3$OH contour map (*blue*) towards the CO(3-2) bullet in the southern lobe (*background*). Contours are 1.5, 2.6 and 3.7 K km/s. The *red square* delimitates the region mapped in methanol

core position infers a temperature $\sim$10 K (computation for $N_{CH_3OH} = 10^{16}$ cm^{-2}).[2]

5 Mapping of the CO bullet in methanol and formaldehyde

We mapped the CH$_3$OH(7-6) and the H$_2$CO(4-3) emission around the position of the CO bullet in the southern lobe of the IRS1 outflow. The distribution of the brightest CH$_3$OH(7-6) line is shown in Fig. 3, overlaid on the CO(3-2) map. The methanol emission follows roughly the

[2]For lower column densities, more appropriate for the core position, the detection of only the 1_{11}-0_{00} transition points even more clearly to low temperatures.

CO emission. It is barely resolved in the transverse direction at the resolution of the APEX single-dish. Study of the shocks will thus greatly benefit from the spatial resolution that will be provided by ALMA.

6 Conclusion

From CO and methanol observations, we have constrained the density and temperature enhancement in the lobes of the highly-collimated outflow of BHR71. Excitation analysis points to temperatures between 30 and 50 K, while the density is around 10^5 cm^{-3}. Our observations have allowed us to isolate and determine the structure of the second outflow in BHR71 driven by IRS2. Studies of shocks in this outflow will greatly benefit from the high spatial resolution that ALMA will provide.

Acknowledgements BP is grateful to the *Alexander von Humboldt Foundation* for a Humboldt Research Fellowship.

References

Bourke, T.L.: Astrophys. J. **554**, L91 (2001)
Bourke, T.L., Garay, G., Lehtinen, K.K., Koehnenkamp, I., Launhardt, R., Nyman, L., May, J., Robinson, G., Hyland, A.R.: Astrophys. J. **476**, 781 (1997)
Cabrit, S., Bertout, C.: Astrophys. J. **307**, 313 (1986)
Froebrich, D.: Astrophys. J. Suppl. Ser. **156**, 169 (2005)
Garay, G., Köhnenkamp, I., Bourke, T.L., Rodríguez, L.F., Lehtinen, K.K.: Astrophys. J. **509**, 768 (1998)
Güsten, R., Nyman, L.Å., Schilke, P., Menten, K., Cesarsky, C., Booth, R.: Astron. Astrophys. **454**, L13 (2006)
Heyminck, S., Kasemann, C., Güsten, R., de Lange, G., Graf, U.U.: Astron. Astrophys. **454**, L21 (2006)
Klein, B., Philipp, S.D., Krämer, I., Kasemann, C., Güsten, R., Menten, K.M.: Astron. Astrophys. **454**, L29 (2006)
Leurini, S., Schilke, P., Menten, K.M., Flower, D.R., Pottage, J.T., Xu, L.-H.: Astron. Astrophys. **422**, 573 (2004)
Myers, P.C., Mardones, D.: Star formation with the infrared space observatory. In: ASP Conf. Ser., vol. 132, pp. 173–182 (1998)
Parise, B., Belloche, A., Leurini, S., Schilke, P., Wyrowski, F., Güsten, R.: Astron. Astrophys. **454**, L79 (2006)
Risacher, C., Vassilev, V., Monje, R., Lapkin, I., Belitsky, V., Pavolotsky, A., Pantaleev, M., Bergman, P., Ferm, S.-E., Sundin, E., Svensson, M., Fredrixon, M., Meledin, D., Gunnarsson, L.-G., Hagström, M., Johansson, L.-Å., Olberg, M., Booth, R., Olofsson, H., Nyman, L.-Å.: Astron. Astrophys. **454**, L17 (2006)

Interstellar deuteroammonia

Tracing physical conditions in dense, cold interstellar medium

D.C. Lis · M. Gerin · E. Roueff · T.G. Phillips ·
D.R. Poelman

Originally published in the journal Astrophysics and Space Science, Volume 313, Nos 1–3.
DOI: 10.1007/s10509-007-9595-4 © Springer Science+Business Media B.V. 2007

Abstract Close to 30 deuterated molecules have now been detected in the ISM, including doubly-deuterated species D_2H^+, ND_2H, D_2CO, CHD_2OH, D_2S, and D_2CS, as well as *triply*-deuterated ammonia and methanol. We review the current understanding of depletion and deuteration processes in cold, dense interstellar medium (ISM) and discuss the utility of deuteroammonia as a tracer of the physical conditions and kinematics of cold, dense gas.

Keywords ISM: abundances · ISM: molecules · Stars: formation

D.C. Lis (✉) · T.G. Phillips
California Institute of Technology, MC 320-47, Pasadena,
CA 91125, USA
e-mail: dcl@caltech.edu

T.G. Phillips
e-mail: tgp@submm.caltech.edu

M. Gerin
LERMA, UMR 8112 du CNRS, Observatoire de Paris and Ecole
Normale Supérieure, 24 Rue Lhomond, 75231 Paris cedex 05,
France
e-mail: maryvonne.gerin@lra.ens.fr

E. Roueff
LUTH and UMR 8102 du CNRS, Observatoire de Paris, Section
de Meudon, Place J. Janssen, 92195 Meudon, France
e-mail: evelyne.roueff@obspm.fr

D.R. Poelman
Kapteyn Astronomical Institute, P.O. Box 800, 9700 AV,
Groningen, The Netherlands
e-mail: dieter@astro.rug.nl

1 Introduction

Under low-temperature, high-density conditions, characteristic of pre-stellar cores and mid-planes of protoplanetary disks, a complex network of non-equilibrium chemical reactions take place that efficiently transfers deuterium atoms into heavy molecules. These reactions, referred to as fractionation, are caused by differences in molecular binding energies (Solomon and Woolf 1973; Gerlich and Schlemmer 2002; Gerlich et al. 2002). Although the initial detection of deuteration in heavy molecules was carried out over third of a century ago (DCN in Orion Nebula, Jefferts et al. 1973), it is only in the past few years that we started fully comprehending the complexity of these processes. The discovery of multiply-deuterated molecules (Ceccarelli 2002; Millar 2005), in particular triply-deuterated ammonia (Lis et al. 2002; van der Tak et al. 2002), has lead to significant revisions of low-temperature ion-molecule chemistry. ND_3 is present in some regions at levels of order 10^{-3} with respect to regular ammonia. This is truly amazing given the cosmic D/H ratio of order 10^{-5}—an enhancement of more than 11 orders of magnitude above the LTE ratio. The key revision to the ion-molecule chemistry, necessary to explain the high abundances of multiply-deuterated species, was the realization that the deuteration of H_3^+ has to be extended beyond H_2D^+, to D_2H^+ and D_3^+ (Phillips and Vastel 2003; Roberts et al. 2003). Shortly after, D_2H^+ was detected in the ISM using the Caltech Submillimeter Observatory (Vastel et al. 2004).

2 Deuteration and depletion

The high deuteration levels observed in cold gas are intimately related to the process of depletion of the abundant gas-phase species onto grain mantles. Molecular differentiation in starless cores is now firmly established (e.g., Bergin et al. 2002). Classical tracers of molecular gas, such as CO, CS and their optically thin isotopologues have been shown to deplete onto grain mantles at densities above a few times 10^4 cm^{-3}. Nitrogen-bearing species, such as N$_2$H$^+$, are unaffected by depletion at densities up to a few times 10^5–10^6 cm^{-3}, and ammonia abundance has been actually shown to be enhanced in central regions of starless cores (Tafalla et al. 2002, 2004).

Recent chemical calculations (Flower et al. 2004; Walmsley et al. 2004), suggest that at densities in excess of 10^6 cm^{-3} even the N-bearing species should eventually condense onto grain mantles and disappear from the gas phase. Under such "complete freeze-out" conditions, the H$_3^+$ ion and its deuterated isotopologues become the only tracers of molecular hydrogen (Fig. 1). However, as the density threshold for complete freeze-out is time and model dependent, good *observational* constraints are needed.

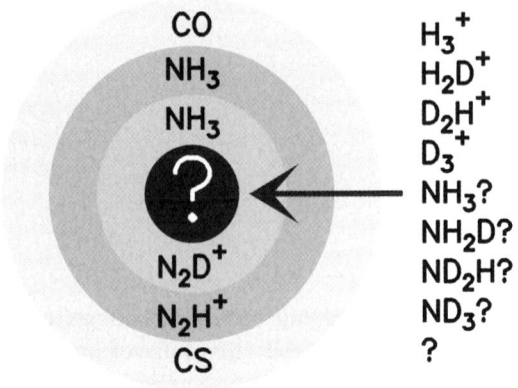

Fig. 1 Chemical differentiation of starless core as predicted by "complete freeze-out" models. The "classical" molecular tracers, such as CO and CS, are only abundant in the outer layers of pre-stellar cores

3 Interstellar deuteroammonia

Ammonia has been shown to be unaffected by depletion and to remain in the gas phase at densities up to a few times 10^5–10^6 cm^{-3}. The ground state submillimeter lines of its isotopologues have critical densities well matched to the physical conditions in the central regions of prestellar cores, making them excellent tracers of dense, cold regions. The 572 GHz line of NH$_3$ (Keene et al. 1983; Liseau et al. 2003) is not observable from the ground. The 309 GHz line of ND$_3$ has now been detected in a number of sources (Roueff et al. 2005), but the line is relatively weak and mapping observations are challenging. However, the two ground state rotational lines of ND$_2$H at 336 and 389 GHz have been shown to offer a good compromise be-

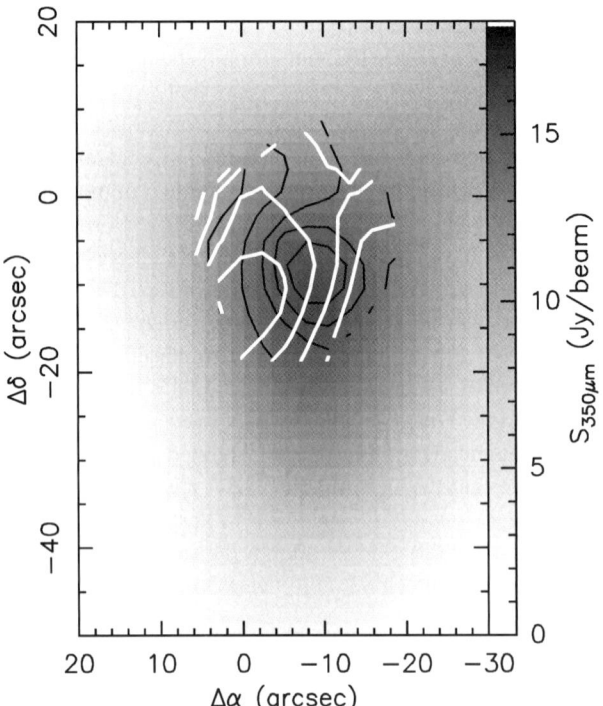

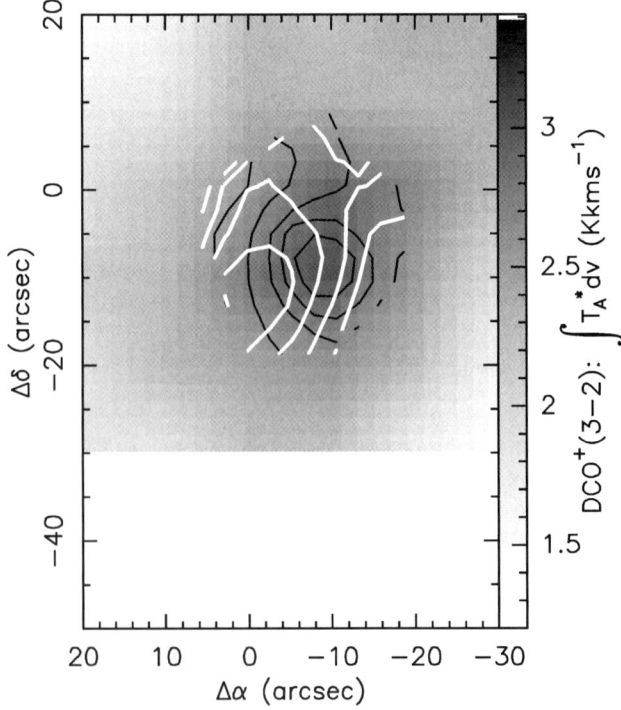

Fig. 2 *Left:* Color image of the 350 μm dust continuum emission in LDN 1689N obtained using the SHARC II bolometer camera at the CSO, convolved to 20″ angular resolution. White and black contours show the distribution of the 336 GHz ND$_2$H and H^{13}CO$^+$ (4-3) emission, respectively. *Right:* Same contours plotted on a DCO$^+$ (3-2) image. Figure from Gerin et al. (2006)

tween the line strength and atmospheric transmission (Lis et al. 2006).

The submillimeter lines of ND_2H have now been detected in LDN 1689N, Barnard 1, and LDN 1544. It is generally accepted that LDN 1689N is a shock-interaction region between an outflow emanating from the solar type protostar IRAS 16293-2422 and a nearby pre-stellar core (Lis et al. 2002; Stark et al. 2004). The observed stratification between deuterated ions and neutrals (Fig. 2) is consistent with this explanation (Gerin et al. 2006), although the comparison is qualitative, as multiply-deuterated species are not included in the current C-shock models. Barnard 1 shows a complicated pattern of high-velocity CO emission, indicating multiple outflows. The dense core B1-b, where abundances of deuterated molecules are strongly enhanced, does not appear to be driving an outflow. Nevertheless, it is possible that shocks or UV/X-ray radiation from nearby embedded sources may play a role in removing the ice mantle material and affecting gas-phase abundances. However, LDN 1544 is a quiescent pre-stellar core that does not harbor embedded luminous sources.

Strong ND_2H emission is detected in both submillimeter transitions in LDN 1544 (Fig. 3). The high critical density and narrow line width suggest that the emission originates from dense, quiescent gas rather than low-density turbulent envelope. According to the latest chemical models (Flower et al. 2006), ammonia starts depleting at densities of $\sim 10^6$ cm^{-3}. In the case of LDN 1544, this corresponds to a radius of ~ 1600 AU or $\sim 11''$ at a distance of 140 pc (Doty et al. 2005) (an angular size well matched to the size of the CSO beam; $\sim 21''$ at 345 GHz). The centrally depleted region in LDN 1544 may thus be detectable with the current submillimeter telescopes and mapping observations are critically needed. ALMA will be able to map the distribution of ND_2H and detect the presence of centrally depleted holes, if present, in more distant pre-stellar cores.

4 Radiative transfer modeling

The two submillimeter ND_2H lines have very different Einstein A-coefficients (1.26×10^{-5} and 4.77×10^{-4} s^{-1} for the 336 and 389 GHz lines, respectively (Lis et al. 2006)). The two lines thus have different critical densities and excitation requirements. In addition, each line has three hyperfine components (Fig. 3), with opacities differing by a factor of 5. Simultaneous observations of the two transitions can thus provide excellent constraints on the physical conditions (temperature and density) and chemistry (ND_2H abundance) in cold clouds. We are in the process of setting up a radiative transfer code for ND_2H, based on the escape probability method (Poelman and Spaans 2006)—the main obstacle being the availability of the collisional

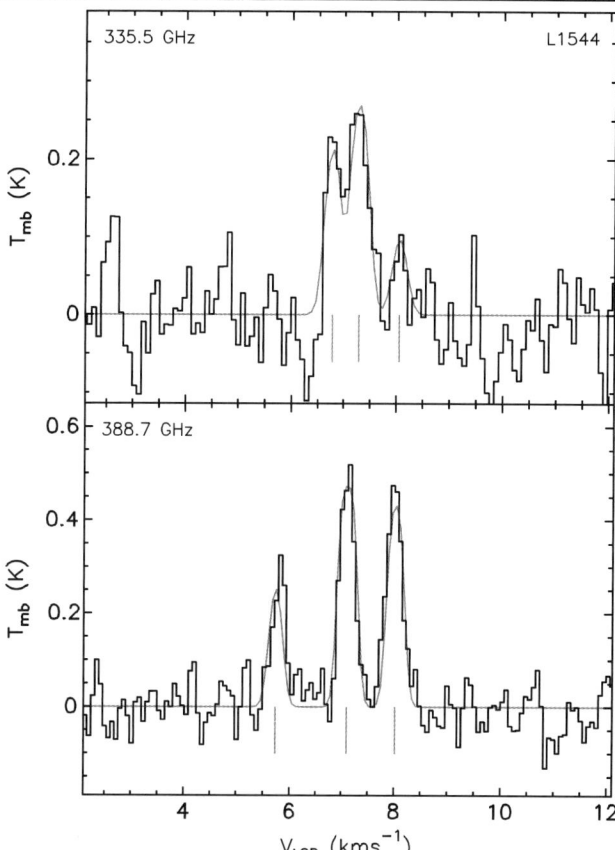

Fig. 3 Submillimeter lines of ND_2H in LDN 1544. The hyperfine splitting is clearly detected providing a measure of the optical depth and excitation temperature, and consequently ND_2H column density. Figure from Lis et al. (2006)

cross-sections. Collisional cross-sections of ND_2H with He have now been computed and are very low (Machin and Roueff 2007). Our preliminary computations suggest that, with these cross-sections, sufficiently high level populations cannot be obtained in the excited ND_2H states to explain the observed line intensities of the submillimeter ND_2H transitions in LDN 1544, given the density and temperature profiles derived for this source. More detailed modeling will have to await the availability of collisional cross-sections with H_2, work currently in progress.

Acknowledgements This research has been supported by NSF grant AST-0540882 to the Caltech Submillimeter Observatory. MG and ER acknowledge travel support from the CNRS/INSU research program PCMI.

References

Bergin, E.A., Alves, J., Huard, T., Lada, C.J.: Astrophys. J. **570**, L101 (2002)
Ceccarelli, C.: Astrophys. Space Sci. **50**, 1267 (2002)
Doty, S.D., Everett, S.E., Shirley, Y.L., Evans II, N.J., Palotti, M.L.: Mon. Not. Roy. Astron. Soc. **359**, 228 (2005)

Flower, D.R., Walmsley, C.M., Pineau de Forêts, G.: Astron. Astrophys. **427**, 887 (2004)

Flower, D.R., Walmsley, C.M., Pineau de Forêts, G.: Astron. Astrophys. **456**, 215 (2006)

Gerin, M., Lis, D.C., Philipp, S., Güsten, R., Roueff, E., Reveret, V.: Astron. Astrophys. **454**, 63 (2006)

Gerlich, D., Schlemmer, S.: Astrophys. Space Sci. **50**, 1287 (2002)

Gerlich, D., Herbst, E., Roueff, E.: Astrophys. Space Sci. **50**, 1277 (2002)

Jefferts, K.B., Penzias, A.A., Wilson, R.W.: Astrophys. J. **179**, L57 (1973)

Keene, J.B., Blake, G.A, Phillips, T.G.: Astrophys. J. **271**, L27 (1983)

Lis, D.C., Gerin, M., Phillips, T.G., Motte, F.: Astrophys. J. **569**, 322 (2002)

Lis, D.C., Roueff, E., Gerin, M., Phillips, T.G., Coudert, L.H., van der Tak, F.F.S., Schilke, P.: Astrophys. J. **571**, L55 (2002)

Lis, D.C., Gerin, M., Roueff, E., Vastel, C., Phillips, T.G.: Astrophys. J. **636**, 916 (2006)

Liseau, R., Larsson, B., Brandeker, A., Bergman, P., et al.: Astron. Astrophys. **402**, L73 (2003)

Machin, L., Roueff, E.: Astron. Astrophys. **465**, 647 (2007)

Millar, T.J.: Astron. Geophys. **46**, 29 (2005)

Phillips, T.G., Vastel, C.: In: Curry C.L., Fich M. (eds.) Chemistry as a Diagnostic of Star Formation, p. 3. NRC, Ottawa (2003)

Poelman, D., Spaans, M.: Astron. Astrophys. **453**, 615 (2006)

Roberts, H., Herbst, E., Millar, T.J.: Astrophys. J. **591**, L41 (2003)

Roueff, E., Lis, D.C., van der Tak, F.F.S., Gerin, M., Goldsmith, P.F.: Astron. Astrophys. **438**, 585 (2005)

Solomon, P.M., Woolf, N.J.: Astrophys. J. **180**, L89 (1973)

Stark, R., Sandell, G., Beck, S., Hogerheijde, M., et al.: Astrophys. J. **608**, 341 (2004)

Tafalla, M., Myers, P.C., Caselli, P., Walmsley, C.M., Comito, C.: Astrophys. J. **569**, 815 (2002)

Tafalla, M., Myers, P.C., Caselli, P., Walmsley, C.M.: Astron. Astrophys. **416**, 191 (2004)

van der Tak, F.F.S., Schilke, P., Müller, H.S.P., Lis, D.C., Phillips, T.G., Gerin, M., Roueff, E.: Astron. Astrophys. **388**, L53 (2002)

Vastel, C., Phillips, T.G., Yoshida, H.: Astrophys. J. **606**, L127 (2004)

Walmsley, C.M., Flower, D.R., Pineau de Forêts, G.: Astron. Astrophys. **418**, 1035 (2004)

Massive star formation in the southern Milky Way

From large scale surveys to high resolution observations

Leonardo Bronfman

Originally published in the journal Astrophysics and Space Science, Volume 313, Nos 1–3.
DOI: 10.1007/s10509-007-9597-2 © Springer Science+Business Media B.V. 2007

Abstract During the past decade we have compiled a large molecular line data base of massive star forming regions in the southern Milky Way. These regions are confined into giant molecular clouds that trace the galactic spiral arms. Their radial distribution has a pronounced peak midway between the Sun and the galactic center, which in the IV quadrant corresponds to the location of the Norma Spiral arm. We study in some detail one of the foremost regions of massive star formation in the Norma arm, using millimeter continuum and line emission maps obtained with the SEST, APEX, and ASTE telescopes. It is a multiple system evolving along a complete GMC core, candidate for future ALMA observations.

Keywords Galaxy: structure · ISM: molecules · Stars: formation

1 Introduction

Massive stars ($M \geq 8M_o$) form in the dense cores of giant molecular clouds ($M \geq 5 * 10^5 M_o$). Their ultraviolet photons heat the surrounding dust, which reradiates the energy in the infrared, peaking at a wavelength of about 100 um. This emission is clearly seen in IRAS maps of the galactic disk, and has a large-scale correlation with the velocity integrated emission of CO, a tracer of molecular gas. After completion of the IRAS survey it became clear that identification of the individual IRAS galactic sources via molecular line emission would yield the distribution of massive star formation in the Galaxy. Because of the ubiquity of the CO molecule, however, it was rather difficult to use the velocity information yielded by the available CO surveys of the Galaxy to deconvolve the IRAS survey, determine the distances and luminosities of the sources, and produce a face-on map of massive star formation in the Milky Way. The CO(1-0) profiles, for example, are normally very complex toward the inner Galaxy, and it is very difficult to assign a single velocity component to a given infrared feature. A better tracer of high density gas would be needed for such purpose, and at an angular resolution comparable with that of the IRAS maps. However, to produce such a map of the Milky Way would take way too long; for a single pixel detector, for instance, observing one square degree of the sky at a resolution of 1 arcmin, at 6 minutes of time per position, would take 30 days. A way out of the dilemma is to carry on pointed observations of a large number of suitable representative sources. Even allowing a 100% overhead for changing the telescope position, it is possible to perform 1800 single-point observations in the same time it would take to map 1 square degree of the sky at a resolution of 1 arcmin.

2 A CS(2-1) survey of UC H II regions

Ultraviolet photons from massive stars, as well as heating the nearby dust, ionize the surrounding gas forming ultra-compact (UC) H II regions. Most of these regions are detected as IRAS point-like sources. The SED (Spectral Energy Density) of these sources has a characteristic shape, that can be used to identify them from the vast number of point-like sources in the IRAS catalog. Using such criterion, Wood and Churchwell (1989) identified 1646 UC H

L. Bronfman (✉)
Astronomy Department, Universidad de Chile, Casilla 36-D, Santiago, Chile
e-mail: leo@das.uchile.cl

II region candidates which fall mostly in the galactic disk. To determine the distances of these sources, and hence their luminosities, it is necessary to carry on line observations of molecular species associated with the dense molecular gas cores surrounding the UC H II regions.

A very good tracer of high molecular gas density is the CS(2-1) transition line. It becomes excited at densities above 10^4–10^5 cm^{-3}, characteristic of star forming regions. The line is fairly insensitive to high column density resulting from large regions, along the line of sight, of low and intermediate gas volume density. In other words, it probes the small dense cores but not the large molecular cloud envelopes. A CS(2-1) survey of IRAS point-like sources with FIR colors characteristic of UC H II regions, along the whole galactic plane, was carried out with the SEST and the OSO telescopes in the early nineties (Bronfman et al. 1996), yielding 843 detections out of a total of 1427 sources observed. More recently the undetected sources have been re-observed using new detectors with better sensitivity; the latitude extent of the survey has been increased too, and a database of 1200 regions of massive star formation in the Galaxy has been compiled. This is the largest database in its kind so far.

3 Mean radial distribution of massive star formation in the Milky Way

A simple axisymmetric model fit to the sample of CS-IRAS sources from the survey yields the mean radial distribution of massive star formation in the galactic disk. The main feature in the mean radial distribution is a well defined annulus, peaking at $R = 0.55R_o$, with a FWHM of $0.28R_o$ (Bronfman et al. 2000). Such peak coincides with the so called molecular ring, which is somewhat broader ($0.51R_o$ FWHM). A lower limit for the total FIR luminosity originated by embedded massive stars in the galactic disk is of $1.5 * 10^8 L_o$, for a total H_2 mass of $2 * 10^9 M_o$ (Fig. 1). The face-on FIR luminosity at the peak of the massive star formation annulus is higher ($1.6 L_o$ pc^{-2}) in the IV galactic quadrant (southern Galaxy) than in the I galactic quadrant ($1.2 L_o$ pc^{-2}). Massive star formation extends, in the outer Milky Way, to galactocentric radii of about 17 kpc.

4 Molecular gas and massive star formation in the southern spiral arms

Spiral arms can be traced, in a longitude velocity diagram of the IV galactic quadrant, by identifying the largest molecular cloud complexes and estimating their distances using

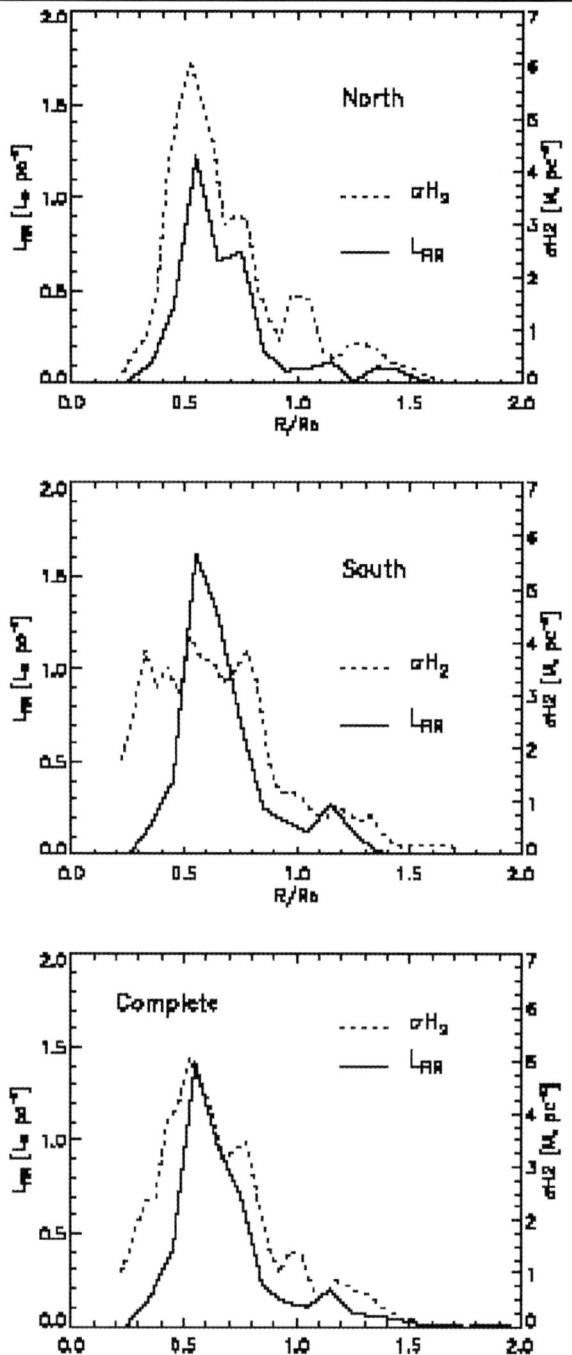

Fig. 1 The *solid line* shows the face-on FIR surface luminosity originated by dust heated by embedded young massive stars, as a function of galactocentric radius. The *dotted line* indicates the face-on surface density of molecular hydrogen (Bronfman et al. 2000, Fig. 10)

a rotation curve. A two-fold distance ambiguity must be resolved, using various methods, like latitude effect (high latitude clouds are most probably at the near distance), and line absorption against associated H II regions. The tangent regions of such spiral arms are found *a priori* from bumps in the rotation curve, originated from systematic de-

Fig. 2 Spatial maps of CO(1-0) emission integrated in velocity, within ranges associated with spiral arms. The *crosses* indicate the positions of IRAS point-like sources, detected in CS(2-1), within the same velocity ranges

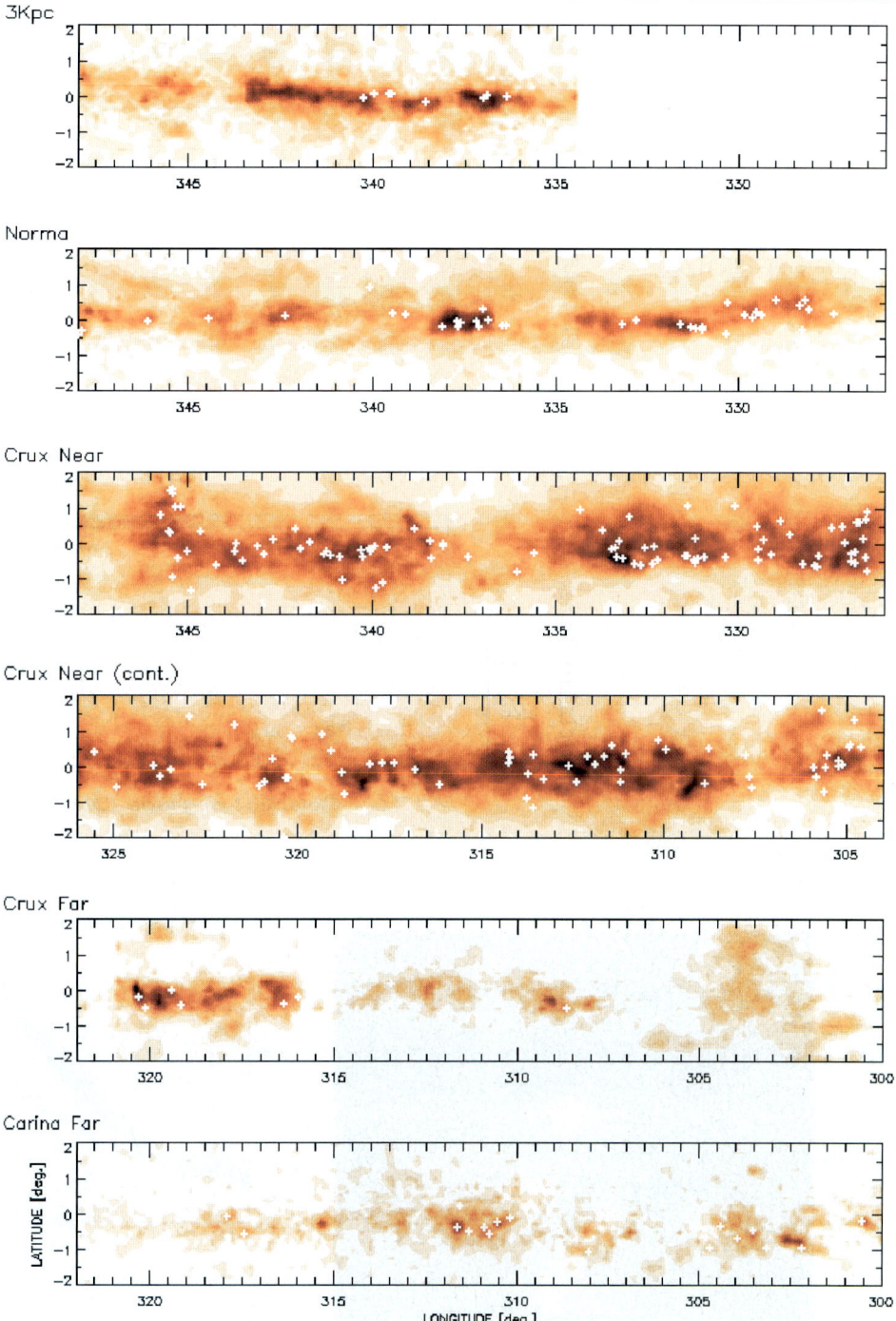

viations from pure circular motion at the spiral arms (Luna et al. 2006). These tangent regions are, in the inner Galaxy, $l \sim 308$ deg for the Crux arm; $l \sim 328$ deg for the Norma arm; and $l \sim 336$ deg for the 3-kpc arm. Figure 2 shows a panel of spatial maps of CO(1-0) emission, integrated over velocity ranges associated with spiral arms from the longitude velocity diagram. The crosses indicate the position of IRAS point-like sources with FIR colors of UC H II regions, with their velocities obtained from the CS(2-1) profiles.

5 Statistical properties of dense cores harboring massive star formation

The CS(2-1) survey of IRAS point-like sources yields the distances an FIR luminosities of massive star forming regions in the Galaxy. To study the extension and morphology of the dense gas and dust cores, a set of 146 maps was obtained with the SIMBA bolometer at SEST at a wavelength of 1.2 mm (Faúndez et al. 2004). The sources selected, in the

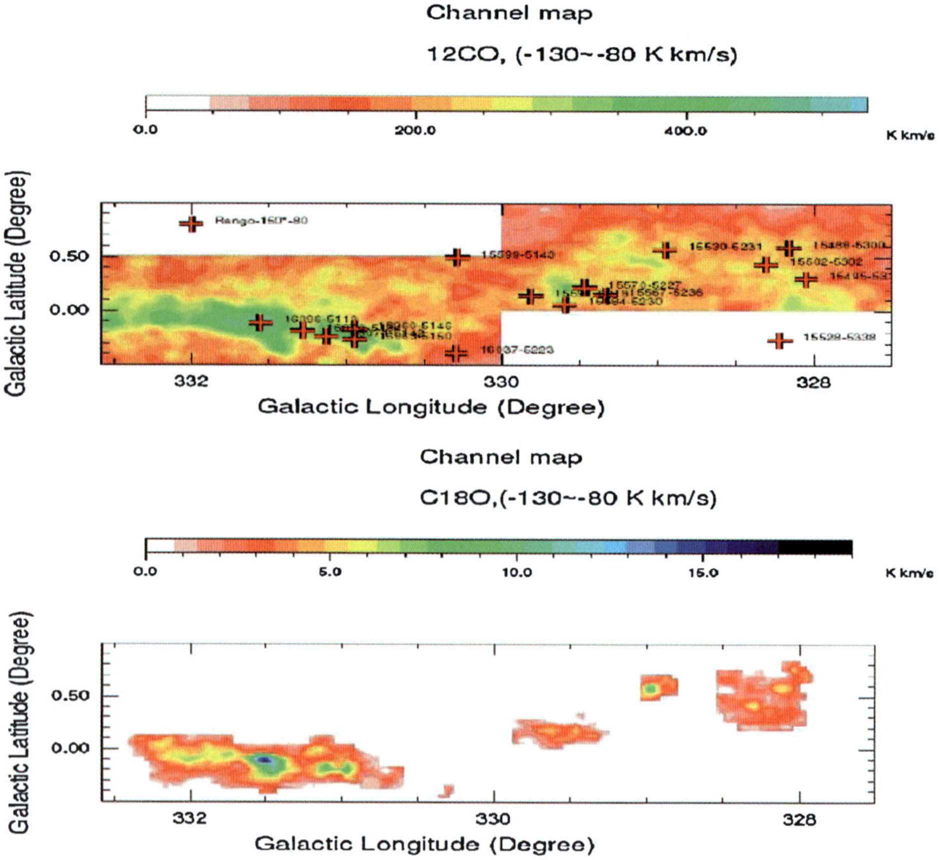

Fig. 3 Spatial maps obtained with the NANTEN Telescope, at a resolution of 2.5 arcmin, of CO(1-0) and C^{18}O(1-0) emission integrated in velocity, within the velocity range associated with the Norma Spiral Arm. The *crosses* indicate the positions of IRAS point-like sources, detected in CS(2-1), within the same velocity range

Channel map

12CO, (-130~-80 K km/s)

Channel map

C18O,(-130~-80 K km/s)

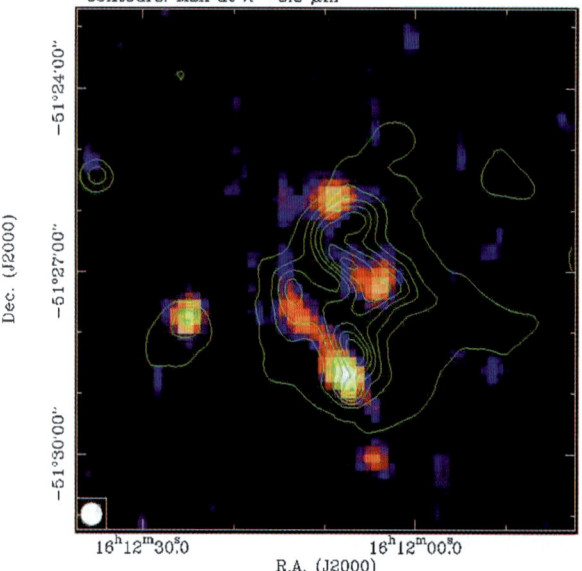

Fig. 4 The G331.5 massive star forming region in Norma. In color, 1.2 mm continuum emission observed with the SIMBA bolometer at the SEST Telescope. *Contours* show the mid infrared emission observed with MSX at 8.3 um

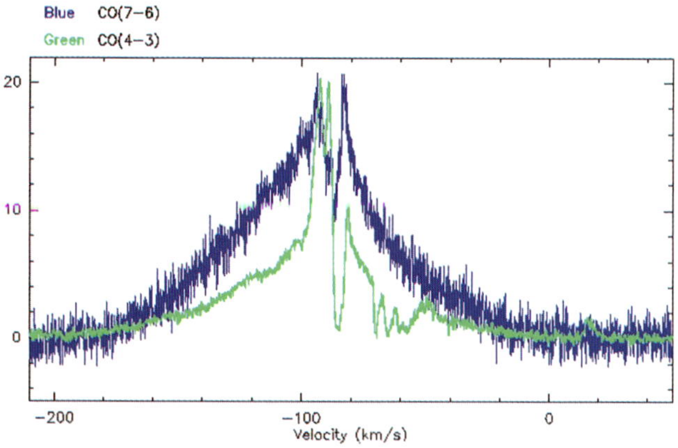

Fig. 5 Massive molecular outflow in G331.5, as observed with APEX telescope in CO(7-6) and CO(4-3) emission

I and IV galactic quadrants, had the most intense CS(2-1) emission in the Bronfman et al. (1996) catalog. The mean characteristics of these continuum sources were: a diameter of 0.4 pc; a mass of 5000 M_o; a luminosity of 230000 L_o; and a temperature of 32 K.

6 The G331.5 massive star forming region in the Norma Spiral Arm

The Norma Spiral arm contains the most massive giant molecular clouds and the most luminous regions of massive star formation in the Milky Way. A map of the tangent region of the arm, in CO(1-0), has been obtained with the NAN-TEN telescope from Nagoya University (Fig. 3). The high opacity of the CO line does not allow to discern the densest gas clumps; however, the $C^{18}O$ transition, more transparent, allows to detect an extended dense condensation at $l = 331.5$ deg. When observed with the SIMBA bolometer at SEST, at a wavelength of 1.2 mm, a region of multiple massive star formation can be readily seen (Fig. 4). The spatial extent of such region, at a non ambiguous distance of 7.4 kpc, is of 20 pc. The total gas mass, derived from the dust observations, is of 24000 M_o, while the mass of the brightest component is of 10000 M_o. It is suspected that the cluster components are at different evolutionary stages, a proposition that has to be tested via higher resolution observations of molecular transition lines. In any case, one of the 1.2 mm components is associated with a high mass molecular outflow (Fig. 5), with a very high velocity width (160 km s^{-1}).

Analysis of these recent observations, obtained with the ASTE and APEX telescopes, is presently underway and will be published elsewhere.

Acknowledgements LB gratefully acknowledge support by Chilean Center of Astrophysics FONDAP 15010003.

References

Bronfman, L., Nyman, L.-A., May, J.: A CS(2-1) survey of IRAS point sources with color characteristics of ultra-compact HII regions. Astron. Astrophys. Suppl. Ser. **115**, 81–95 (1996)

Bronfman, L., Casassus, S., May, J., Nyman, L.-A.: The radial distribution of OB star formation in the Galaxy. Astron. Astrophys. **358**, 521–534 (2000)

Faúndez, S., Bronfman, L., Garay, G., Chini, R., Nyman, L.-A., May, J.: SIMBA survey of southern high-mass star forming regions. I. Physical parameters of the 1.2 mm IRAS sources. Astron. Astrophys. **426**, 97–103 (2004)

Luna, A., Bronfman, L., Carrasco, L., May, J.: Molecular gas, kinematics, and OB star formation in the spiral arms of the Southern Milky Way. Astrophys. J. **641**, 938–948 (2006)

Wood, D.O.S., Churchwell, E.: Massive stars embedded in molecular clouds—their population and distribution in the Galaxy. Astrophys. J. **340**, 265–272 (1989)

SMA observations of the magnetic fields around a low-mass protostellar system

J.M. Girart · R. Rao · D.P. Marrone

Originally published in the journal Astrophysics and Space Science, Volume 313, Nos 1–3.
DOI: 10.1007/s10509-007-9592-7 © Springer Science+Business Media B.V. 2007

Abstract Observations of the submillimeter polarized dust emission is an important tool to study the role of the magnetic fields in the evolutions of molecular clouds and in the star formation processes. The Submillimeter Array (SMA) is the first imaging submillimeter interferometer. The installation of quarter wave plates in front of the 345 GHz receivers has allowed to carry out polarimetric observations. We present high angular resolution 345 GHz SMA observations of polarized dust emission towards the low-mass protostellar system NGC 1333 IRAS 4A. We show that in this system the observed magnetic field morphology is in agreement with the standard theoretical models of formation of low-mass stars in magnetized molecular clouds at scales of a few hundred AU; gravity has overcome magnetic support and the magnetic field traces a clear hourglass shape. The magnetic field is substantially more important than turbulence in the evolution of the system and the initial misalignment of the magnetic and spin axes may have been important in the formation of the binary system.

Keywords ISM: magnetic fields · ISM: clouds · ISM: polarization · Stars: formation

J.M. Girart (✉)
Institut de Ciències de l'Espai (CSIC-IEEC), Campus UAB,
Facultat de Ciències; C-5 parell, 2ª, 08193 Bellaterra, Catalonia,
Spain
e-mail: girart@ieec.cat

R. Rao
Academica Sinica, Institute of Astronomy and Astrophysics, 645
North Aohoku Place, Hilo, HI 96720, USA

D.P. Marrone
Harvard-Smithsonian Center for Astrophysics, 60 Garden St,
Cambridge, MA 02138, USA

1 Linear polarization of the dust emission

Magnetic fields are believed to play a very important role in the star formation process (Heiles et al. 1993). Observations of the polarization of radiation is the main tool to study the interstellar magnetic fields. There are three main techniques to study the magnetic field in molecular dense environments. Linear polarization (the Goldreich–Kylafis effect: Goldreich and Kylafis, 1981, 1982) and circular polarization (Zeeman effect: e.g. Heiles and Crutcher 2005) of molecular emission allows to trace the magnetic field morphology in the plane of the sky and the magnetic field strength along the line of sight, respectively. The third technique is to observe the linear polarization of the dust emission. The polarized emission at far-IR and (sub)mm wavelengths is caused by dust grains that become aligned perpendicular to the direction of the magnetic field. Even though the exact details of the alignment mechanism are uncertain (Lazarian 2003), the most widely accepted one is based on the work by Davis and Greenstein (1951). The basic idea is that large enough grains (≥ 0.1 μm) have suprathermal spin velocities due to torques that can be caused by the formation of H_2 (Purcell 1979) or by the radiation field (Drain and Weingartner 1996). Under these circumstances, the paramagnetic relaxation of spinning grains is an efficient mechanism in removing the components of rotation perpendicular to the magnetic field.

Due to the low temperature of the molecular clouds, the dust emission emits mainly in the far-IR and submillimeter wavelengths. The first detections of the polarized emission were carried out with airborne telescopes in the far-IR (Cudip et al. 1982; Hildebrand et al. 1984). The development of state-of-the-art submillimeter bolometric cameras in the previous decade, in particular at the JCMT and CSO, allowed a burst of polarized dust emission observations towards star forming molecular clouds at angular scales of ∼15″ (Vallée and Bastien 1995; Dotson et al. 2000; Matthews et al. 2001; Houde et al. 2004). The use of quarter wave plates in front of the receivers in the millimeter BIMA array also allowed to observe the dust polarization at higher angular resolution (3–5″) (Rao et al. 1998; Lai et al. 2001, 2002, 2003).

2 NGC 1333 IRAS 4A

NGC 1333 is an active low and intermediate mass star forming region, located in the Perseus molecular cloud complex. One of its members, NGC 1333 IRAS 4A (IRAS 4A), is a very embedded (Class 0) low-mass protostellar system with a bolometric luminosity of ∼12 L⊙ (Sandell et al. 1991). It has associated a strong and highly collimated molecular outflow in the north-south direction (Blake et al. 1995). High angular resolution interferometric continuum observations indicate that IRAS 4A is a binary system with a separation ∼1.8″ (Lay et al. 1995; Looney et al. 2000). High angular resolution observations of H_2CO show inverse P Cygni profiles towards IRAS 4A, indicating infall motions (Di Francesco et al. 2001). The molecular line analysis done by Di Francesco et al. (2001) suggests a high accretion rate. Recent work by Belloche et al. (2006) suggests that the onset of the gravitational collapse has being induced, or at least helped, by a fast external compression possibly produced by the outflow activity of the near cluster of young stellar objects.

The dust emission associated with this source is among the strongest ones associated with low mass protostars. This makes this source a suitable candidate for polarization observations. This first polarization observations towards IRAS 4A were carried out with single-dish submm telescopes at angular resolutions of ∼15″ (Minchin et al. 1995; Tamura et al. 1995). Millimeter interferometric observations resolved partially the polarized emission (Akeson et al. 1996; Girart et al. 1999). In particular, observations carried with the BIMA array showed that the polarization map of the dust continuum is roughly consistent with an hourglass magnetic field morphology (Girart et al. 1999). These observations also detected and obtained the first mapping of linearly polarized spectral line emission due to the Goldreich-Kylafis effect: the polarization of the CO 2–1 line is detected long the molecular outflow. The CO polarization detection and mapping make it possible to extend the study of the magnetic field morphology beyond the region of strong dust continuum emission to the bipolar outflow.

3 Observations

The observations were taken on December 5 and 6, 2004 at the SMA in Mauna Kea (Hawaii). Six of the eight antennas

were used for these observations. The array was tuned such that the CO 3-2 line at 345.796 GHz was located toward the middle of the upper sideband. The continuum bandwidth was approximately 2 GHz and the LSB and USB center frequencies were 336.804 and 346.804 GHz respectively. The gain calibrator that was used was QSO 3C84 and the absolute flux scale was determined from observations of Ganymede. The bandpass was obtained from observations of QSO 3C279.

Linear polarization measurements with interferometers are best done with circularly polarized feeds. This minimizes the effect of gain errors, as the cross-correlation of opposite circular polarizations does not involve the Stokes total intensity (I) parameter. Quarter wave plates are used in front of the linearly polarized SMA feeds to obtain left (L) or right (R) circular polarizations. Since only a single polarization is received, L and R are time-multiplexed on each antenna using a fast Walsh function switching pattern in order to sample all possible cross correlations (LL, LR, RL, RR) on every baseline. The data are averaged over the Walsh cycle to produce quasi-simultaneous dual polarization measurements. The instrumental polarization response, or "leakage," for each antenna was calibrated by observing a strong point source over a wide hour angle range to provide good parallactic angle coverage. For antennas with Nasmyth (modified alt-az) mounts and orthogonal circular feeds, the fringe phases for a linearly polarized source vary with parallactic angle while the instrumental leakages remain constant, so one can solve simultaneously for the leakages and the source polarization. The average leakage amplitudes were less than 0.5% in the upper sideband and less than 3% in the lower sideband and their response with frequency is consistent with theoretical calculations.

The data were reduced with the Caltech millimeter array software package MIR that is modified for SMA and with the MIRIAD software package. The averaged spectral data were corrected for the leakages, and continuum bands in both the lower and upper sidebands without obvious spectral line contamination were used to produce maps of the I, Q, and U Stokes parameters. These maps were deconvolved independently and then combined to produce maps of the linear polarization intensity, the fractional polarization and the position angle. The synthesized beam size was 1.5″ × 1.0″ with a position angle of 83°.

4 Results

In Fig. 1, we present the SMA map of the emission for the three Stokes parameters, I, Q and U, towards IRAS 4A (Girart et al. 2006). The total flux density that we measure is 6.4 Jy, which is about 50% of the measured with the JCMT

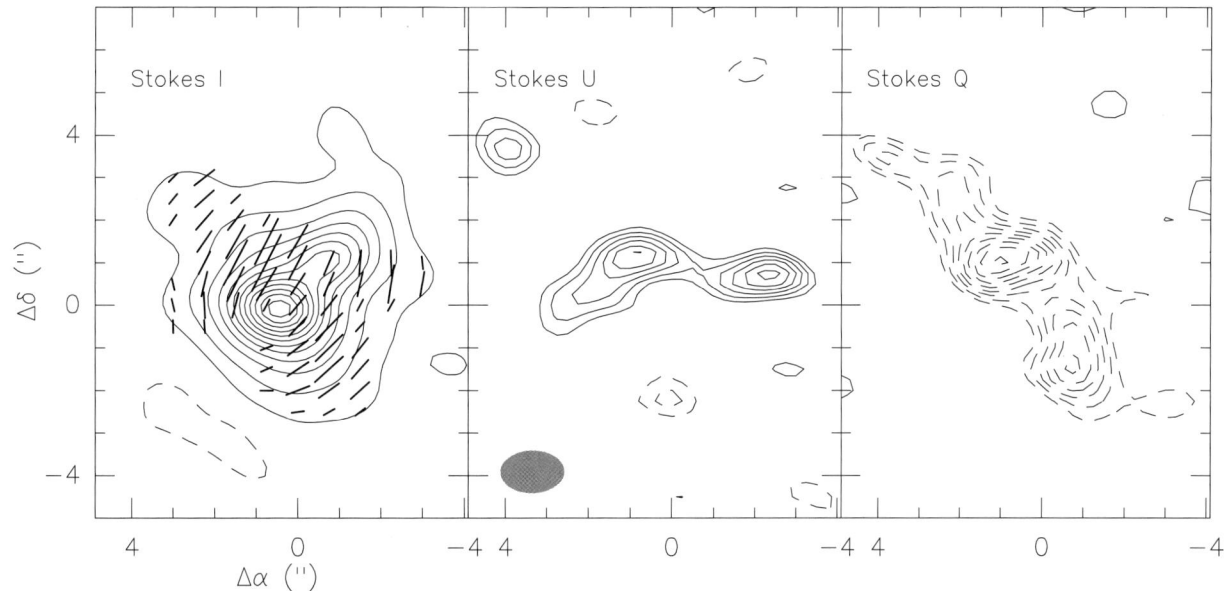

Fig. 1 345 GHz SMA contour maps of the Stokes parameters I (total intensity, *left panel*), U (*central panel*) and Q (*right panel*) for the dust emission towards the low-mass protostellar system NGC 1333 IRAS4A. *Thick bars* on the *left panel* show the polarization vectors, with their length proportional to the fractional polarization.

Contour levels for Stokes Q and U are ± 3, ± 4, ± 5, ... ± 12 times 3.2 mJy beam^{-1} (the *rms* noise of the maps). Contour levels for the Stokes I is -1, 1, 3, 6, 9, ... 33 times 55 mJy beam^{-1}. The synthesized beam is shown in the bottom left corner of the *central panel*

(Sandell et al. 1991). The missing flux comes from the large scale structure in the IRAS 4A envelope. The dust continuum emission (Stokes I) is resolved in two components that have a separation of 1.8″ with a position angle of PA of 130°. This is in agreement with previous interferometric observations (Lay et al. 1995; Looney et al. 2000). At the scales traced by the SMA (few hundreds of AU), the dust emission arises at a volume density of about $n(H_2) = 4 \times 10^7$ cm^{-3}. IRAS 4A has also associated strong linear polarization from the dust emission, although the Stokes Q and U emission have a different distribution than Stokes I emission (see Fig. 1). The total polarized flux is ~ 0.3 Jy and the integrated polarization fraction is 4.7% with a position angle of 145°. These values are in excellent agreement with the previous observations (Minchin et al. 1995; Tamura et al. 1995; Girart et al. 1999). The fractional polarization is lowest at the position of the peak intensity. This can be caused by a decrease of the alignment efficiency at high densities ($n(H_2) \geq 10^8$ cm^{-3}) or by an unresolved complex geometrical distribution of the magnetic fields (Lazarian et al. 1997; Gonçalves et al. 2005).

In Fig. 2 we present a comparison of the dust emission and the magnetic field distribution obtained with the BIMA array at 1.3 mm (Girart et al. 1999) and with the SMA array at 0.87 mm (Girart et al. 2006). It is clear that the marginal hour-glass magnetic field morphology that can be seen in the BIMA image is well resolved with the SMA. This is the morphology predicted in standard core collapse models for magnetized clouds (Galli and Shu 1993;

Fiedler 1993). The hour-glass magnetic field morphology is achieved during the supercritical regime of the core collapse, when the initially uniform magnetic field is warped and strengthened by the accreting material. At the derived volume density for the SMA dust emission, $n(H2) = 4 \times 10^7$ cm^{-3}, we estimate that the magnetic field strength for the plane-of-sky component is ~ 5 mG. Comparing the magnetic field properties with other properties of the collapsing core we found that the magnetic field is substantially more important than turbulence in the evolution of this system.

The axis normal to the dusty envelope (44°) lies between the magnetic field axis (61°) and the main outflow axis (19°) (Girart et al. 2006). This suggests that when the collapse initiated, the spin and magnetic axes were not aligned. Interestingly, it is possible that this misalignment could cause the fragmentation of the core, leading to the formation of a binary stellar system (Machida et al. 2006).

5 Conclusion and future perspectives with ALMA

The observations carried out with the SMA toward NGC 1333 IRAS 4A show that high angular resolution polarization observations with high sensitivity in the submm wavelengths provides important information on the role of the magnetic field during the earliest phases of the star formation processes at scales of only few hundreds AU. The ALMA array is going to provide an improvement so large in sensitivity with respect to the past and present arrays that the

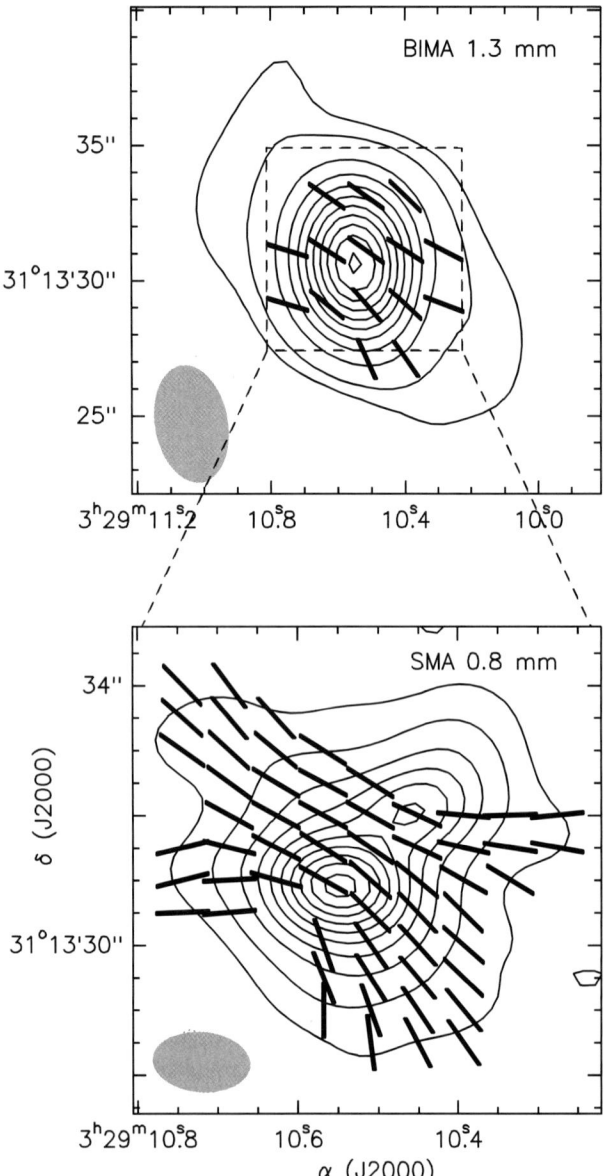

Fig. 2 Maps of the dust total intensity, overlaid with the magnetic field vectors obtained at 1.3 mm with BIMA (Girart et al. 1999) and at 0.8 mm with the SMA (Girart et al. 2006)

polarization observations of the dust emission (nowadays still restricted to few sources) will be done successfully in many sources at a very good spatial resolution (≤ 100 AU), e.g. probably in most of the known low mass protostars (Class 0 and I).

Acknowledgements The SMA is a joint project between the Smithsonian Astrophysical Observatory and the Academica Sinica Institute of Astronomy and Astrophysics. J.M.G. acknowledges support from AGAUR (Generalitat de Catalunya) and SEUI (Ministerio de Educación y Ciencia, Spain) for support through grants 2004BE00370 and AYA2005-08523-C03-02, respectively.

References

Akeson, R.L., Carlstrom, J.E., Phillips, J.A., Woody, D.P.: Astrophys. J. **456**, l45 (1996)
Belloche, A., Hennebelle, P., André, P.: Astron. Astrophys. **453**, 145 (2006)
Blake, G.A., et al.: Astrophys. J. **441**, 689 (1995)
Cudip, W., Furniss, I., King, K.J., Jenings, R.E.: Mon. Not. Roy. Astron. Soc. **200**, 1169 (1982)
Davis, L.J., Greenstein, J.L.: Astrophys. J. **114**, 206 (1951)
Di Francesco, J., et al.: Astrophys. J. **562**, 770 (2001)
Dotson, J.L., et al.: Astrophys. J. Suppl. Ser. **128**, 335 (2000)
Drain, B.T., Weingartner, J.C.: Astrophys. J. **470**, 551 (1996)
Fiedler, R.A., Mouschovias, T.Ch.: Astrophys. J. **415**, 680 (1993)
Galli, D., Shu, F.H.: Astrophys. J. **544**, 243 (1993)
Girart, J.M., Crutcher, R.M., Rao, R.: Astrophys. J. **525**, L109 (1999)
Girart, J.M., Rao, R., Marrone, D.P.: Science **313**, 812 (2006)
Goldreich, P., Kylafis, N.D.: Astrophys. J. **243**, L75 (1981)
Goldreich, P., Kylafis, N.D.: Astrophys. J. **253**, 606 (1982)
Gonçalves, J., Galli, D., Walmsley, A.: Astron. Astrophys. **430**, 979 (2005)
Heiles, C., Crutcher, R.M.: Cosmic Magnetic Fields. Lecture Notes in Physics, vol. 664, p. 137 (2005)
Heiles, C., Goodman, A.A., McKee, C.F., Zweibel, E.G.: In: Protostars and Planets III, p. 279 (1993)
Hildebrand, R.H., Dragovan, M., Novak, G.: Astrophys. J. **284**, L51 (1984)
Houde, M., et al.: Astrophys. J. **604**, 717 (2004)
Lai, S.-P., Crutcher, R.M., Girart, J.M., Rao, R.: Astrophys. J. **561**, 864 (2001)
Lai, S.-P., Crutcher, R.M., Girart, J.M., Rao, R.: Astrophys. J. **566**, 925 (2002)
Lai, S.-P., Girart, J.M., Crutcher, R.M.: Astrophys. J. **598**, 392 (2003)
Lay, O.P., Carlstrom, J.E., Hills, R.E.: Astrophys. J. **452**, L73 (1995)
Lazarian, A.: J. Quant. Spectrosc. Radiat. Transf. **79**, 881 (2003)
Lazarian, A., Goodman, A.A., Myers, P.C.: Astrophys. J. **490**, 273 (1997)
Looney, L.W., Mundy, L.G., Welch, W.J.: Astrophys. J. **529**, 477 (2000)
Machida, M.N., Matsumoto, T., Hanawa, T., Tomisaka, K.: Astrophys. J. **645**, 1227 (2006)
Matthews, B.C., Wilson, C.D., Fiege, J.D.: Astrophys. J. **562**, 400 (2001)
Minchin, N.R., Sandell, G., Murray, A.G.: Astron. Astrophys. **293**, L61 (1995)
Purcell, E.M.: Astrophys. J. **231**, 404 (1979)
Rao, R., et al.: Astrophys. J. **502**, L75 (1998)
Sandell, G., et al.: Astrophys. J. **376**, L17 (1991)
Tamura, M., Hough, J.H., Hayashi, S.S.: Astrophys. J. **448**, 346 (1995)
Vallée, J.P., Bastien, P.: Astron. Astrophys. **294**, 831 (1995)

ASTE observations of the massive-star forming region Sgr B2: a giant impact scenario

**Tetsuo Hasegawa · Takaaki Arai ·
Nobuyuki Yamaguchi · Fumio Sato · the ASTE team**

Originally published in the journal Astrophysics and Space Science, Volume 313, Nos 1–3.
DOI: 10.1007/s10509-007-9523-7 © Springer Science+Business Media B.V. 2007

Abstract We report mapping observations of a 35 pc × 35 pc region covering the Sgr B2 molecular cloud complex in the ^{13}CO (3-2) and the CS (7-6) lines using the ASTE 10 m telescope with high angular resolution. The central region was mapped also in the $C^{18}O$ (3-2) line. The images not only reproduce the characteristic structures noted in the preceding millimeter observations, but also highlight the interface of the molecular clouds with a large velocity jump of a few tens of km s^{-1}. These new results further support the scenario that a cloud–cloud collision has triggered the formation of massive cloud cores, which form massive stars of Sgr B2. Prospects of exciting science enabled by ALMA are discussed in relation to these observations.

Keywords ISM : clouds · ISM : individual (Sgr B2) · Stars : formation

1 Introduction

Sagittarius B2, embedded in the bar-shaped complex of molecular clouds in the Galactic center (Sawada et al. 2004), is one of the most active regions of star formation in the Galaxy. It contains many compact and ultracompact HII regions and molecular masers clustered to the three centers, Sgr B2 (N), (M), and (S). They are associated with massive cores of dense molecular gas embedded in a giant molecular cloud complex. It has long been noted that the velocity structure around these cores is very complex with a steep velocity shift across a north–south line that runs through the three star-forming centers (e.g., Rogstad et al. 1974; Martín-Pintado et al. 1990). Based on the wide-area ^{13}CO (1-0) image taken at high resolution with the Nobeyama 45-m telescope, Hasegawa et al. (1994) identified the characteristic kinematic structures, Shell ($V_{LSR} = $ 20–40 km s^{-1}), Hole ($V_{LSR} = $ 40–50 km s^{-1}), and Clump ($V_{LSR} = $ 70–80 km s^{-1}), which they interpreted as results of a cloud–cloud collision; a relatively compact giant cloud colliding with and punching through another less dense giant cloud (Hasegawa et al. 1994; Sato et al. 2000). The star forming centers Sgr B2 (N), (M), and (S) are located along the interface of the two colliding clouds, suggesting that the massive star formation there has been triggered by the cloud–cloud collision.

In this paper, we report a new set of molecular line images taken with the Atacama Submillimeter Telescope Experiment (ASTE) in ^{13}CO (3-2), CS (7-6) and $C^{18}O$ (3-2). A full account of the results will be published in a separate paper (Arai et al. 2007).

2 ASTE observations

ASTE is a 10-m submillimeter telescope located at Pampa la Bola at an elevation of 4,800 m next to the ALMA site (Ezawa et al. 2004). It is jointly operated by the National Astronomical Observatory of Japan and researchers in Japanese universities. The observations reported here were

T. Hasegawa (✉) · N. Yamaguchi
National Astronomical Observatory of Japan, Osawa, Mitaka,
Tokyo 181-8588, Japan
e-mail: tetsuo.hasegawa@nao.ac.jp

T. Arai
Department of Astronomy, The University of Tokyo, Hongo,
Bunkyo-ku, Tokyo 113-0033, Japan

F. Sato
Department of Astronomy and Earth Sciences, Tokyo Gakugei
University, Koganei, Tokyo 184-8501, Japan

made in 2005 August in the on-the-fly mapping mode using a double-sideband SIS mixer receiver. A $14' \times 14'$ (35 pc $\times$ 35 pc at the adopted distance to the Galactic center of 8.5 kpc) area was covered in ^{13}CO (3-2) and CS (7-6), and a $5' \times 8'$ (12.5 pc $\times$ 20 pc) area was covered in C^{18}O (3-2), with a HPBW of $22''$ (0.9 pc). The beam efficiency was measured to be 59% during observations and the pointing accuracy was better than $5''$. The ^{13}CO (3-2) emission is widespread and detected in most positions in the mapping area, while the CS (7-6) and C^{18}O (3-2) emissions are much more concentrated toward Sgr B2 (N) and (M) with faint extended emission surrounding them.

3 Results and discussion

3.1 Overall kinematics

Figure 1 is the velocity channel maps of the ^{13}CO (3-2) emission observed with ASTE. It shows the large-scale velocity field in the Sgr B2 giant molecular cloud complex, and reproduces all the characteristic structures identified in

the preceding millimeter-wave observations (Hasegawa et al. 1994; Sato et al. 2000); the Shell in the frames of 22.5–37.5 km s^{-1}, the Hole at 37.5–47.5 km s^{-1} near the map center, and the Clump at 62.5–77.5 km s^{-1}. The morphological similarity between the Hole and the Clump is also beautifully reproduced (e.g., compare the panels of 42.5 and 72.5 km s^{-1}).

3.2 Brightened *rim* with *lumps*

In addition to the known features noted above, the ASTE results reveal a new feature, i.e., the brightening of the rim surrounding the Hole. It is recognized in the panels at 42.5 to 52.5 km s^{-1} in Fig. 1. Figure 2 displays the ^{13}CO (3-2) emission integrated over these velocity ranges to show the brightened "Rim" clearly. The Rim is oval-shaped, and lumpy particularly in the eastern part. We could identify several "Lumps" as indicated by arrows in Fig. 2. The ^{13}CO (1-0) data do not show the brightened Rim with Lumps, and velocity channel maps of the ^{13}CO 3-2/1-0 line intensity ratio (with the 1-0 data from (Oka et al. 1998)) show a marked enhancement of the ratio on the eastern Rim at corresponding velocities. A similar pattern is found also in the maps of

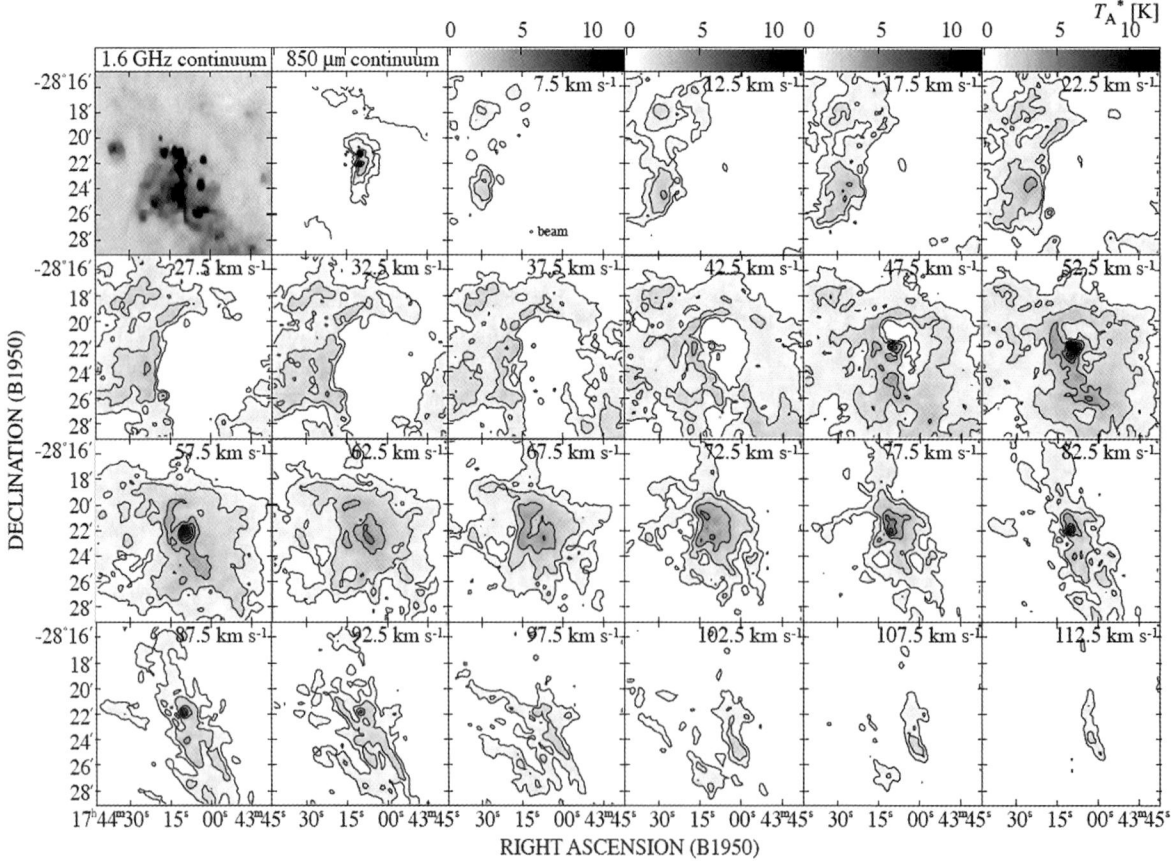

Fig. 1 Velocity channel maps of ^{13}CO (3-2) observed with ASTE. *Each panel* shows the map of T_A^* averaged over a 5 km s^{-1} interval centered at the velocity indicated in the panel. The *top left panel* shows the VLA image of the 1.616 GHz radio continuum emission (Liszt 1992) and the *next panel* shows the SCUBA image at 850 μm (Pierce-Price et al. 2000) for comparison

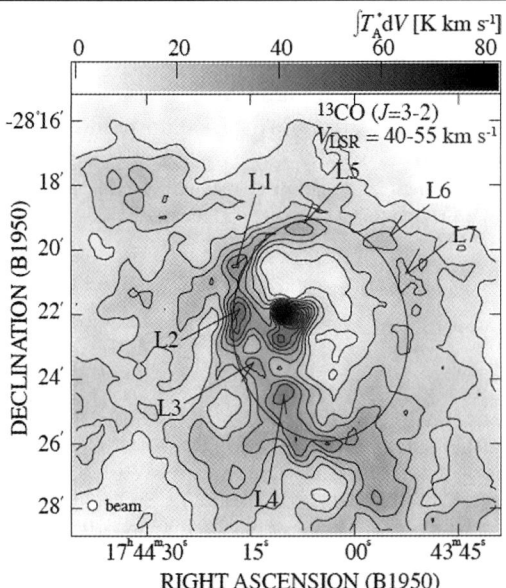

Fig. 2 The ^{13}CO (3-2) emission integrated over V_{LSR} = 40–55 km s^{-1}. Around the hole, we can see the brightened Rim (indicated by the *oval*) with the Lumps (*arrows*)

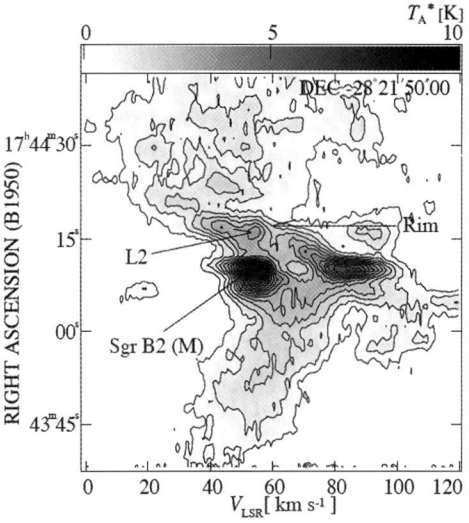

Fig. 3 The position–velocity diagram of the ^{13}CO (3-2) emission along an E-W line at δ_{1950} = $-28°21'50.0''$. The positions of the eastern Rim and a Lump (L2) are shown by *arrows*

the C^{18}O 3-2/1-0 line intensity ratio (with the 1-0 data from (Sato et al. 2007)). An LVG analysis of the ratios indicates that the product of the density and temperature, $n_{H_2}T_k$, on the Rim is enhanced by a factor of >5 compared with the region outside the Rim. Indeed, some of the Lumps are dense enough to be detected also in CS (7-6) emission.

Figure 3 shows a position-velocity diagram along an east–west line at δ_{1950} = $-28°21'50.0''$. We immediately note that the Rim has a large velocity shear at the interface of clouds at two different velocities; V_{LSR} = 20–60 km s^{-1} on the eastern (upper in Fig. 3) side (i.e., Shell and Hole), and V_{LSR} = 50–100 km s^{-1} on the western side of the Rim (i.e., Clump). Figure 3 also shows the intense and broad emission of Sgr B2 (M), split into two by the foreground absorption at $V_{LSR} \approx$ 65 km s^{-1}. Sgr B2 (M) is located adjacent to the Rim with a separation of ~1′ (2.5 pc).

3.3 The *hot ring*

From a detailed analysis of their CH$_3$CN mapping observations, de Vicente et al. (1997) have revealed a ring-like region of higher gas temperature ("Hot Ring") that surrounds Sgr B2 (N) and (M) (de Vicente et al. 1997). A comparison between their temperature maps and the ^{13}CO (3-2) intensity maps at exactly the same velocity ranges of integration indicates that the higher temperature is found toward the Rim, which delineates the interface of the two clouds at different velocities. The Hot Ring fits in the overall picture naturally as a result of dissipation of the kinetic energy of the colliding clouds.

3.4 The giant impact scenario

In the preceding subsections, we have seen that the new ASTE images of the Sgr B2 giant molecular cloud complex in ^{13}CO (3-2) not only reproduce the Shell, Hole, and Clump identified in millimeter-wave images, but also newly reveal the brightened Rim surrounding the Hole (Figs. 1 and 2). The Rim coincides with the interface between the two clouds, i.e., one with the Shell/Hole and another as the Clump, which are ~30 km s^{-1} apart in radial velocity. Indeed, a steep velocity shear is observed in the eastern and northern part of the Rim (Fig. 3). The higher 3-2/1-0 line intensity ratios in ^{13}CO and C^{18}O on the Rim, the detection of the CS (7-6) emission in some part of it, and the coincidence of the Rim with the region of high temperature (the Hot Ring (de Vicente et al. 1997)) all indicate that the density and temperature are higher in the Rim. These new observations lend another support to the scenario that a large-scale collision of giant molecular clouds has occurred in this region.

The dense molecular cloud cores associated with Sgr B2 (N), (M) and (S) are located along the interface of the colliding clouds, tempting us to speculate that they may have been formed at the compressed interface layer between the colliding clouds, with the high density and temperature and the large shear velocity field. Although the current position of the Rim is displaced from the dense cores by a few parsecs, it is probable that the expanding Rim was at the position of Sgr B2 (N), (M) and (S) when the formation of the dense cores was triggered ~10^5 years ago. The cores were formed in high energy density environment (gravitational, turbulent and magnetic), and once they become gravitationally unstable, they support large mass accretion rates that may naturally result in a vigorous formation of massive stars.

We do not have definite observational information yet as to how the massive cores have formed in the cloud collision interface. In this context, the Lumps found in the Rim are quite interesting, because they might tell us how density structure builds up in the compressed layer. Follow-up observations with high density tracers are needed to explore this possibility.

4 Prospects for ALMA

Formation of massive stars is an important process that exerts strong and sometimes essential influence to the evolution of galaxies and their interstellar media. It is much less understood compared with the formation of solar-type stars, mainly because of the lack of observations with sufficient resolution (see, e.g., Evans II et al. 2002; Beuther et al. 2007).

Formation of massive dense cores in collisions of molecular clouds is probably one of the viable scenarios of massive star formation. However, as illustrated in the case of present ASTE observations of Sgr B2, the spatial resolution of the current generation of instruments can only barely resolve the structure that may have critical information to understand the formation mechanism of massive cores needed to form massive stars. ALMA will not only break this limit of spatial resolution by two orders of magnitude, but also highlight hot and dense regions under strong dynamical influence by enabling imaging observations at submillimeter wavelengths. With ALMA, we will get necessary information to fully understand what happens in the regions of massive star formation. In the interface of colliding clouds in Sgr B2, for example, the shock structure and dense cores in formation would be studied with much less ambiguity, contributing to a real understanding of the massive star formation process.

The scenario of massive core formation in a cloud–cloud collision provides an attractive possibility for explaining the mechanism of a starburst, which is often observed in nuclear regions of galaxies (e.g., Scoville 2000; Koda et al. 2005). With ALMA, nearby starburst galaxies can be observed with a linear resolution (in pc) comparable with that in the current ASTE observations of Sgr B2. This will provide us with great opportunities to see the detailed kinematics of molecular clouds there. For example,

we could reveal how common the cloud–cloud collision events are in the central regions of galaxies and how such events are related to the star formation.

We can hardly wait until ALMA starts producing images!

Acknowledgements   A part of this study was financially supported by the MEXT Grant-in-Aid for Scientific Research on Priority Areas No. 15071202.

References

Arai, T., Yamaguchi, N., Hasegawa, T., Sato, F.: ASTE observations of massive star forming region Sgr B2. (2007, in preparation)

Beuther, H., Churchwell, E.B., McKee, C.F., Tan, J.C.: The formation of massive stars. In: Protostars and Planets V, p. 165 (2007)

de Vicente, P., Martín-Pintado, J., Wilson, T.L.: A hot ring in the Sagittarius B2 molecular cloud. Astron. Astrophys. **320**, 957 (1997)

Evans II, N.J., Shirley, Y.L., Mueller, K.E., Knez, C.: Early phases and initial conditions for massive star formation. In: Hot Star Workshop III: the Earliest Stages of Massive Star Birth. ASP Conf. Ser., vol. 267, p. 17 (2002)

Ezawa, H., Kawabe, R., Kohno, K., Yamamoto, S.: The Atacama submillimeter telescope experiment (ASTE). Proc. SPIE **5489**, 763 (2004)

Hasegawa, T., Sato, F., Whiteoak, J.B., Miyawaki, R.: A large-scale cloud collision in the Galactic center molecular cloud near Sagittarius B2. Astrophys. J. **429**, L77 (1994)

Koda, J., Okuda, T., Nakanishi, K., Kohno, K., Ishizuki, S., Kuno, N., Okumura, S.K.: Starbursting nuclear CO disks of early-type spiral galaxies. Astron. Astrophys. **431**, 887 (2005)

Liszt, H.S.: Radio images of Sagittarius. I. Overview, Sagittarius D and Sagittarius E. Astrophys. J. Suppl. Ser. **82**, 495 (1992)

Martín-Pintado, J., de Vicente, P., Wilson, T.L., Johnston, K.J.: Dust and gas in the cores and the envelope in Sagittarius B2. Astron. Astrophys. **236**, 193 (1990)

Oka, T., Hasegawa, T., Sato, F., Tsuboi, M., Miyazaki, A.: A large-scale CO survey of the Galactic center. Astrophys. J. Suppl. Ser. **118**, 455 (1998)

Pierce-Price, D., et al.: A deep submillimeter survey of the Galactic center. Astrophys. J. **545**, L121 (2000)

Rogstad, D.H., Lockhart, I.A., Whiteoak, J.B.: Aperture synthesis of formaldehyde absorption in Sgr B2. Astron. Astrophys. **36**, 253 (1974)

Sato, F., Hasegawa, T., Whiteoak, J.B., Miyawaki, R.: Cloud collision-induced star formation in sagittarius B2. I. Large-scale kinematics. Astrophys. J. **535**, 857 (2000)

Sato, F., et al.: (2007, in preparation)

Sawada, T., Hasegawa, T., Handa, T., Cohen, R.J.: A molecular face-on view of the Galactic centre region. Mon. Not. Roy. Astron. Soc. **349**, 1167 (2004)

Scoville, N.Z.: Ultra-luminous IR galaxies at low and high redshift, dynamics of galaxies: from the early universe to the present. In: ASP Conf. Ser., vol. 197, p. 301 (2000)

A new view of proto-planetary disks with ALMA

Stéphane Guilloteau · Anne Dutrey

Originally published in the journal Astrophysics and Space Science, Volume 313, Nos 1–3.
DOI: 10.1007/s10509-007-9666-6 © Springer Science+Business Media B.V. 2007

Abstract The dynamical, physical and chemical processes which lead to planet formation constitute an astrophysical domain which will strongly benefit from ALMA in terms of frequency coverage, sensitivity and angular resolution. Recent results from current mm/submm interferometers obtained on molecules and dust in proto-planetary disks are presented. The observational coupling between gas and dust is discussed and it is shown that dust disks must be analyzed with the knowledge provided by gas disks, and respectively, both from the chemical and physical points. For these purposes, the methods of analysis of mm/submm interferometric data specific to disks are summarized. Emphasis is given on recent, unexpected, findings obtained in the highest sensitivity and resolution observations obtained so far, as they provide a hint of what ALMA could discover. A comparison with the expected sensitivities for ALMA illustrates how ALMA can enhance our knowledge of the disk physics, either by providing statistics or by allowing much more detailed studies of representative objects.

Keywords Star formation · Planetary system formation · Circumstellar matter

1 Introduction

The existence of disks around young, pre-main-sequence stars, was initially inferred from the IR SED, until the early 1990s, where the advent of the HST, the progress of adaptive optics and of millimeter interferometers provided the first images, either in scattered light at near-IR or optical wavelengths (Roddier et al. 1996; Burrows et al. 1996), or in the thermal dust emission and CO rotational lines (e.g. Koerner et al. 1993; Dutrey et al. 1994). These disks are understood in the framework of viscous accretion disks. The viscosity allows quasi-Keplerian rotation of the gas (with a tiny residual due to the gas pressure), and mass accretion through the disk by re-distributing the angular momentum through outward disk diffusion (Hartmann 2001). Beyond about 10 AU from the star, the energy dissipated by the accretion process becomes small compared the energy flux from by the central star: passive irradiation dominates the overall thermal structure (D'Alessio et al. 2001). These disks are often called "proto-planetary" disks, because the ones studied so far contain sufficient amount of dust and gas to form Jupiter-like giant planets.

These disks are small, 100 to 900 AU, and cold, with typical temperatures of order 20–30 K at 100 AU from the star. Keplerian rotation give projected velocities of order 2 km s^{-1} at 100 AU, but the intrinsic linewidth due to turbulence and thermal motion is much smaller (typically 0.1 km s^{-1}). Given the distance of the nearest star forming regions, 120 to 140 pc (except for the TW Hya association which is about 60 pc), angular sizes are at most a few arcseconds. These 3 properties (low temperature, small size, narrow linewidth) make mm interferometers essential tools to study the properties of these objects.

2 Deriving disk properties

A fundamental property to understand disk observations is the Keplerian rotation. Because of the rotation, the maxi-

S. Guilloteau (✉) · A. Dutrey
LAB, OASU, CNRS, Université de Bordeaux, BP 89, 2 rue de l'Observatoire, 33670 Floirac, France
e-mail: guilloteau@obs.u-bordeaux1.fr

A. Dutrey
e-mail: dutrey@obs.u-bordeaux1.fr

mum fraction of the disk area which can emit at any given projected velocity is of order

$$\delta V / (2 V (r_{\text{out}})) \sin i$$

i.e. only 10–20% for typical values of the local linewidth δV (Guilloteau et al. 2006). This simple geometric effect has two consequences. For optically thick lines which are significantly excited by collisions and appear in emission, such as the CO $J = 1$–0 or 2–1 transitions, it implies a total line flux proportional to the local line width

$$S_\nu \propto R_{\text{out}}^2 \delta V \cos i \, T_{\text{ex}}.$$

For lines which have very low excitation temperatures, and would appear in absorption against the dust continuum, the geometric factor implies that the equivalent width is less than δV. Optically thin lines will of course have a smaller flux (emission) or equivalent width (absorption).

Geometry and kinematics can be derived from the spatial distribution of line emission as a function of velocity. Disk masses are in general derived from dust emission, as the surface brightness

$$T_b(r) \propto T_{\text{dust}}(r) \kappa_\nu(r) \Sigma_{\text{dust}}(r). \tag{1}$$

However, because the dust properties are expected to evolve in proto-planetary disks, the dust emissivity is ill constrained and the disk masses are often uncertain by at least an order of magnitude. The gas temperature and molecular abundances are derived from spectral line observations. For example, for the CO $J = 1$–0 transition, which is thermalized,

$$T_b(r) \propto \Sigma_{\text{mol}}(r) / T_k(r) \tag{2}$$

(because of the partition function which goes as $1/T_k^2$ at high enough temperatures), until the line becomes optically thick, in which case

$$T_b(r) = T_k(r). \tag{3}$$

However, with current mm arrays, sensitivity and angular resolution are limiting factors to separate the effects of surface density and line excitation. The derivation is in general made under the assumption that both T and Σ are power laws of the radius r

$$T_k(r) = T_0(r/r_0)^{-q} \quad \text{and} \quad \Sigma(r) = \Sigma_0(r/r_0)^{-p}$$

(Piétu et al. 2007), with sharp inner and outer edges. This is in general a good approximation. Moreover, the surface density law is in general much steeper than the temperature law. Using (2–3), it can be seen that provided $p > q/2$, an optically thick core always exists for the fundamental rotation transitions. If this core is resolved, it is then possible to derive the temperature (in the optically thick part) and the surface density (from the optically thin part). For other transitions than the fundamental, the requirements to have an optically thick core are actually less stringent.

3 What have we learned from mm interferometers?

After the pioneering studies of GM Aur (Koerner et al. 1993), and of the circum-binary disk of GG Tau (Dutrey et al. 1994), the first quantitative study using the optically thick CO $J = 1$–0 revealed the radial temperature gradient in DM Tau (Guilloteau and Dutrey 1998). Simon et al. (2000) made a small survey of a dozen disks in CO $J = 2$–1, showing that they were in Keplerian rotation, and constraining the stellar mass. By comparing the CO isotopologues, Dartois et al. (2003) proved the existence of a vertical temperature gradient in DM Tau. This study has recently been extended to other sources (LkCa 15, MWC 480, and AB Aur) by Piétu et al. (2005, 2007), showing that the radial and vertical temperature gradients agree qualitatively with the expectation of models of passively heated disks. Another important result from the CO isotopologue observations is the smaller (true, not just apparent) outer radius in ^{13}CO and C^{18}O than in ^{12}CO, which is highly suggestive of photodissociation effects at the outer disk edge. The variation of the CO to dust ratio from source to source, and its relation with the disk temperature, clearly suggests that condensation on dust grains plays a significant role in the CO abundance. Sub-mm observations are still rare, but have provided important results. In particular, CO $J = 3$–2 and $J = 6$–5 observations of TW Hya with the SMA (Qi et al. 2006) have shown that the gas disk is warmer than expected in simple models, perhaps as a result of X-ray heating in the upper layers.

Besides CO, other molecules have been detected in disks: HCO$^+$, CN, HCN, C$_2$H, H$_2$CO, and CS were reported by Dutrey et al. (1997). However, with the exception of HCO$^+$, whose emission is relatively strong (e.g. in GG Tau Guilloteau et al. 1999), the lines are weak, and very few interferometric observations have been published so far. H$_2$CO was observed by Aikawa et al. (2003). A recent result illustrating the difficulties is the detection of N$_2$H$^+$ by Dutrey et al. (2007): because of the hyperfine structure of the $J = 1$–0 transition, the line is hardly visible, and only a sophisticated analysis reveals the detection and allows to derive the N$_2$H$^+$ content. The derived column densities are almost 100 times smaller than in an earlier claim from Qi et al. (2003), in part due to more appropriate assumptions about the molecule distribution, and in part due to confusion with the weak underlying continuum. The new measurements bring the [N$_2$H$^+$]/[HCO$^+$] ratio down to 0.02, like in dark clouds, and no longer requires a different behaviour of N$_2$ and CO regarding condensation on dust grains, in agreement with recent laboratory measurements from Bisschop et al. (2006).

Deuterated molecules have also been discovered, but not with interferometers: DCO$^+$ has been detected in two sources: in TW Hya by van Dishoeck et al. (2003) and in DM Tau by Guilloteau et al. (2006). H$_2$D$^+$ was reported by Ceccarelli et al. (2004), but the detection is only 2σ (Guilloteau et al. 2006) and requires confirmation.

4 Not all disks are the same

The disks studied so far are *not* representative of all disks: because of limited angular resolution and sensitivity, we picked the largest and most massive disks first. Indeed, the smallest is the closest one: TW Hya disk is 250 AU in radius, and the other ones are at least a factor 2 larger. Recent observations have started revealing disks which differ from the general picture that one could be tempted to derive from the best studied cases, DM Tau, MWC 480 and LkCa 15.

Small disks exist: Dutrey et al. (2003) showed that BP Tau is surrounded by a small (120 AU), warm (50 K at 100 AU) disk in which CO is heavily depleted (under-abundance of order 100). Because of the high temperature, this depletion cannot be due to sticking on dust grains, and other mechanisms must be invoked.

A highly unanticipated discovery concerns AB Aur. The disk displays a large inner cavity, of about 100 AU in radius, and of a "spiral-like" structure around it, both in dust and gas emission (Piétu et al. 2005; Lin et al. 2006). Furthermore, the disk displays clear evidence of non-Keplerian rotation, although the non-circularity of the structure affects the interpretation of the projected velocities. The cavity is not totally devoid of gas: the apparent inner radius increases from the most optically thick tracer, CO $J = 2$–1, to the most optically thin one, dust emission. The disk is not massive enough to be unstable, so cannot be the cause of the spiral structure. It has been suggested that the cavity is due to a low mass companion orbiting around 40 AU, companion whose trace may have been found since (Baines et al. 2006; Rodriguez et al. 2006). However, it is unclear if such a companion can sustain the large spiral structure. An alternative (not necessarily exclusive, as stars are seldom born single) may be that the disk is much younger than anticipated, so that the Keplerian rotation is not yet fully established. The existence of a massive envelope is an argument in this direction.

Surprises also appear in the "standard" disks. Using the new 800-m baselines of the IRAM array, Piétu et al. (2006) have shown that a hole of 50 AU radius also exists in the LkCa 15 disk (see Fig. 1). The LkCa 15 disk is also unusual in that no significant vertical temperature gradient was observed.

In the same experiment, the optically thick dust core of MWC 480 was partially resolved at 1.3 mm, offering, in combination with the 2.8 mm data obtained at the same time, the first direct measurement of the dust temperature in a proto-planetary disk. The dust was found to be much colder than extrapolated from the Spectral Energy Distribution, with $T_{\mathrm{dust}} \simeq 20$ K at 20 AU. In fact, the near-IR

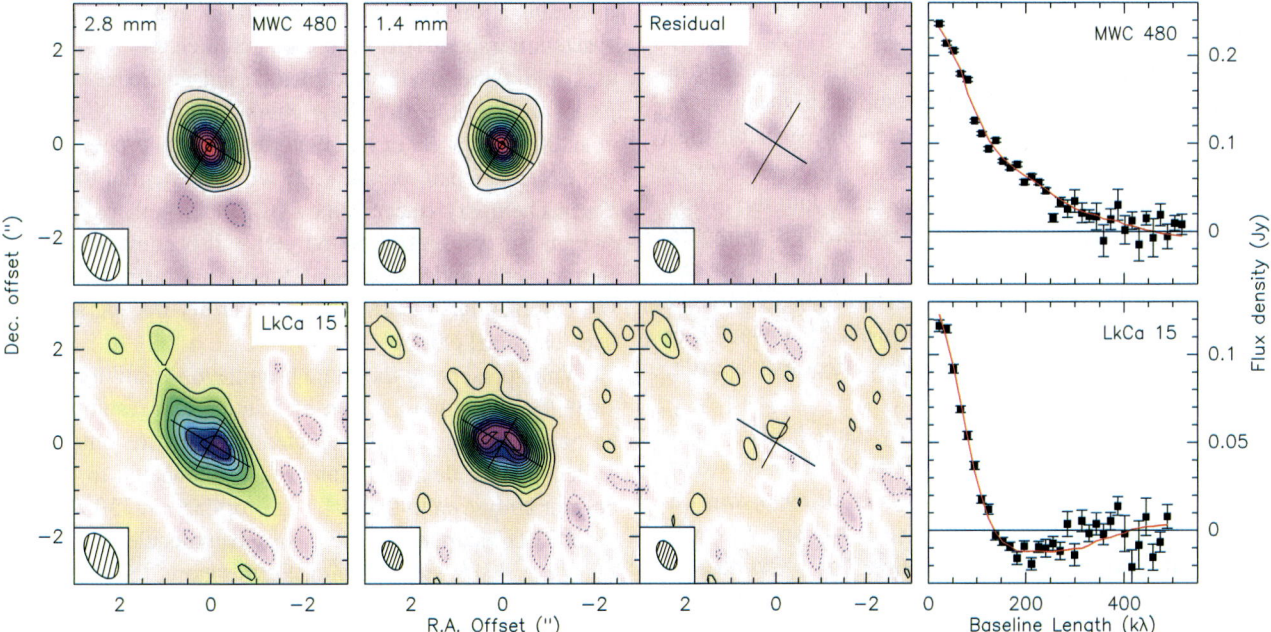

Fig. 1 Sub-arcsec resolution observations of the dust disks of LkCa 15 and MWC 480. The LkCa 15 disk exhibits a central hole of radius about 45 AU. From left to right: 2.8 mm emission, 1.4 mm emission, residual emission after best model subtraction, and (deprojected) radial average of the visibilities, with best fit model superimposed

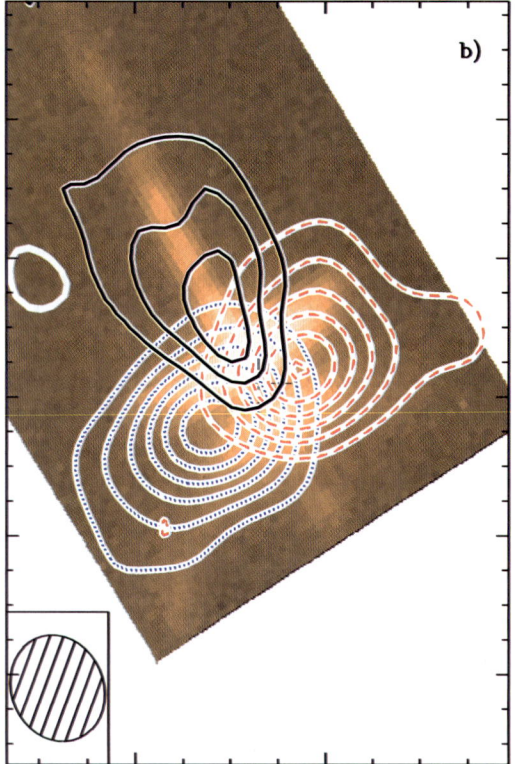

Fig. 2 Arcsec resolution observations of the HH 30 system. The color background is the HST image from Burrows et al. (1996). *Red and blue contours* indicate *red*-shifted (resp. *blue*-shifted) ^{13}CO $J = 2$–1 emission, showing the disk rotation. The *black contours* are the ^{12}CO $J = 2$–1 emission from the one-sided molecular outflow, delineating a cone of 30° angle. Adapted from Pety et al. (2006)

SED is often used to obtain an estimate of the dust temperature. However, the SED only probes the upper layers of the disk, and in the presence of a vertical temperature gradient, it can lead to significant over-estimate of the temperature of the dust emitting at mm wavelengths, since this dust is largely located in the disk mid-plane. As a consequence, since the dust mass required to explain the mm flux densities is inversely proportional to the assumed dust temperature (see (1)), the new low temperature implies larger disk masses (by a factor of order 2–3) than originally thought. Although MWC 480 is so far the only object in which such a determination has been possible, this effect may apply to all disks.

An archetypical object was recently studied: HH 30. This beautiful example of the flared disk with an on-axis optical jet revealed by the HST (Burrows et al. 1996) was observed in CO $J = 2$–1 and ^{13}CO $J = 2$–1 and 1–0 with the IRAM array by Pety et al. (2006). The ^{13}CO data reveal that the disk is orbiting a 0.45 $M_\odot$ star, and indicates a temperature of 12 K, much lower than the 35 K derived for the scale height of the scattering dust, again indicating significant vertical temperature gradient. However, the disk is unusual because a prominent CO outflow originates from

HH 30. The outflow delineates an empty (as seen through the CO $J = 2$–1 line) cone of 30° semi-opening angle, with a constant 12 km s^{-1} flow speed. The most surprising fact is that this outflow is purely one-sided, while the optical jet is two-sided (see Fig. 2).

5 A new view with ALMA

The above results can serve to guide us for what studies can be performed and what results should be searched for with ALMA. ALMA provides two major breakthroughs compared to current arrays. In the mm domain, ALMA will be about 25 times more sensitive than the limit reached in images from current mm arrays. Second, ALMA will truly open the sub-mm domain (the SMA being limited to studies of the CO transitions because of its small collecting area). There are several major ways to use this large gain

- search for weaker lines
- improve the angular resolution
- perform surveys
- look for more distant objects.

These aspects can be applied to several major aspects of the study of disks.

5.1 Disk chemistry

Current mm arrays only observe the most abundant molecule(s), often only CO. Molecules which are essential in the disk characterization produce much weaker lines. DCO$^+$, in combination with HCO$^+$ and N$_2$H$^+$ can be used as an ionization tracer. H$_2$D$^+$ could be a unique tool to study the disk mid-plane, if all molecules do indeed freeze out onto grains as predicted by some chemical models (Ceccarelli et al. 2004). Tracers of the photodissociation processes, like the CN/HCN pair, or like C$_2$H will be essential tools too. Grain chemistry tracers, like H$_2$CO and perhaps CH$_3$OH (which has not been detected yet) will become accessible. Sulfur compounds, of which CS only has been detected so far, can play an important role as density tracers, and perhaps chemical clocks.

While preserving the same brightness sensitivity than in current observations, a gain of 5 in angular resolution can be obtained. This gain is sufficient to allow to compare the molecular and dust emission at 0.2″, compared to 1″ with current arrays, up to the disk edge for disks at 150 pc. Such detailed comparison of the dust and molecule distribution will be a major step forward. Furthermore, the power law approximation, or other model dependent a priori distributions which are used nowadays, can be replaced by direct measurements of the radial distribution.

Spatially resolved, multi-molecule, multi-transition studies of disks will determine the excitation conditions of the molecules, and combined with disk models (physics and chemistry) will reveal where above the disk plane the molecules are located, and what is the (3-D) density structure of the disks.

5.2 Measuring the disk masses and disk shape

The LkCa 15 and MWC 480 studies reveal that the dust distribution (surface density law) is ill constrained. ALMA will improve our knowledge in 3 directions: by allowing higher angular resolution, by allowing multi-wavelength studies at similar (high) angular resolution, and by tracing the dust at larger distances from the star. Besides measurement of the dust temperature in the optically thick core, ALMA will provide the derivation of the dust emissivity index β as a function of radius, thereby directly tracing grain growth in the mm-size domain. The H_2 density can be constrained from excitation conditions of (especially) sub-mm lines. Hence, it is expected that ALMA will provide measurement of the dust to gas ratio, as a function of radius.

The linear resolution offered by ALMA for dust emission is high enough (<5–10 AU) to provide a direct measurement of the vertical stratification of dust for edge-on disks. Piétu et al. (2007) showed that information on the disk thickness can also be obtained for inclined disks from the spectral line studies. Finally, disk masses can perhaps be measured directly from the departure of the rotation velocity from the Keplerian value, because of the disk self-gravity. This will be a difficult experiment, because the disk geometry has to be adequately known, but it would be the only direct measurement of the disk mass.

5.3 Probe the disk evolution

Disk studies have been limited so far to the brightest disks in the Taurus or TW Hya regions. The BP Tau example shows that they are probably not representative of the whole disk population. ALMA will allow to go much further (a factor 5) in distance, even beyond the Orion cluster distance, and thus allow to study disk properties in different stellar environments. The capabilities of studying the disk properties in dense clusters will be a major breakthrough. Much fainter disks will be within reach, e.g. disks around very low mass stars, or perhaps even young "debris" disks.

A major step forward will be the capability of performing surveys, and observing a statistically significant sample of sources, not just half a dozen objects like today. Such samples can be used to probe the disk properties as function of stellar ages and stellar masses. However, to do so, ALMA will have to spend only "short" integration time on each object, and will thus be limited in object distance and/or in number of transitions studied for each object. The choice of the tracers and of the analysis methods will be essential.

These will have to be developed and validated by detailed studies on a few sample objects, using much higher angular resolution for such cross-validation.

One of the key questions needing such a survey approach is why, when and how disks disappear. Is the disappearance triggered by planet formation? Or does photo-evaporation (Alexander et al. 2006) play the critical role? The time evolution of outflows is a similarly important question.

6 Conclusion

ALMA has the potential to truly revolutionize our views of "proto-planetary" disks. However, to do so will require very significant amounts of observing time. To illustrate this point, consider that current mm arrays used about 12 to 24 hours per observed transition, at angular resolution of order 1–2″, reasonably appropriate for the Taurus region. ALMA can do that in less than 1 hour, but observing 8 molecules, 3 transitions each (to study excitation effects) will still require a day of observation per object at that same 1″ resolution. So even a sample of 30 sources at the Taurus distance will consume a month of ALMA time. If higher angular resolution is desired, e.g. 0.2″ or 30 AU, a single multi-molecule, multi-transition study will require several days of integration.

Because of these significant integration times, using appropriate line combinations offered by the wide instantaneous bandwidth of the ALMA receivers and by the flexibility of its correlator will be absolutely essential to maximize the scientific return of ALMA.

References

Aikawa, Y., Momose, M., Thi, W.-F., et al.: Publ. Astron. Soc. Jpn. **55**, 11 (2003)

Alexander, R.D., Clarke, C.J., Pringle, J.E.: Mon. Not. R. Astron. Soc. **369**, 229 (2006)

Baines, D., Oudmaijer, R.D., Porter, J.M., Pozzo, M.: Mon. Not. R. Astron. Soc. **367**, 737 (2006)

Bisschop, S.E., Fraser, H.J., Öberg, K.I., van Dishoeck, E.F., Schlemmer, S.: Astron. Astrophys. **449**, 1297 (2006)

Burrows, C.J., Stapelfeldt, K.R., Watson, A.M., et al.: Astrophys. J. **473**, 437 (1996)

Ceccarelli, C., Dominik, C., Lefloch, B., Caselli, P., Caux, E.: Astrophys. J. **607**, L51 (2004)

D'Alessio, P., Calvet, N., Hartmann, L.: Astrophys. J. **553**, 321 (2001)

Dartois, E., Dutrey, A., Guilloteau, S.: Astron. Astrophys. **399**, 773 (2003)

Dutrey, A., Guilloteau, S., Simon, M.: Astron. Astrophys. **286**, 149 (1994)

Dutrey, A., Guilloteau, S., Guélin, M.: Astron. Astrophys. **317**, L55 (1997)

Dutrey, A., Guilloteau, S., Simon, M.: Astron. Astrophys. **402**, 1003 (2003)

Dutrey, A., Henning, T., Guilloteau, S., et al.: Astron. Astrophys. **464**, 615 (2007)

Astrophys Space Sci (2008) 313: 95–100

Guilloteau, S., Dutrey, A.: Astron. Astrophys. **339**, 467 (1998)

Guilloteau, S., Dutrey, A., Simon, M.: Astron. Astrophys. **348**, 570 (1999)

Guilloteau, S., Piétu, V., Dutrey, A., Guélin, M.: Astron. Astrophys. **448**, L5 (2006)

Hartmann, L.: Accretion Processes in Star Formation. Cambridge University Press, Cambridge (2001) 237 pp., ISBN 0521785200

Koerner, D.W., Sargent, A.I., Beckwith, S.V.W.: Icarus **106**, 2 (1993)

Lin, S.-Y., Ohashi, N., Lim, J., et al.: Astrophys. J. **645**, 1297 (2006)

Pety, J., Gueth, F., Guilloteau, S., Dutrey, A.: Astron. Astrophys. **458**, 841 (2006)

Piétu, V., Guilloteau, S., Dutrey, A.: Astron. Astrophys. **443**, 945 (2005)

Piétu, V., Dutrey, A., Guilloteau, S., Chapillon, E., Pety, J.: Astron. Astrophys. **460**, L43 (2006)

Piétu, V., Dutrey, A., Guilloteau, S.: Astron. Astrophys. **467**, 163 (2007)

Qi, C., Kessler, J.E., Koerner, D.W., Sargent, A.I., Blake, G.A.: Astrophys. J. **597**, 986 (2003)

Qi, C., Wilner, D.J., Calvet, N., et al.: Astrophys. J. **636**, L157 (2006)

Roddier, C., Roddier, F., Northcott, M.J., Graves, J.E., Jim, K.: Astrophys. J. **463**, 326 (1996)

Rodriguez, L.F., Zapata, L., Ho, P.T.P.: ArXiv Astrophysics e-prints (2006)

Simon, M., Dutrey, A., Guilloteau, S.: Astrophys. J. **545**, 1034 (2000)

van Dishoeck, E.F., Thi, W.-F., van Zadelhoff, G.-J.: Astron. Astrophys. **400**, L1 (2003)

Observational signature of planet formation: The ALMA view

Nagayoshi Ohashi

Originally published in the journal Astrophysics and Space Science, Volume 313, Nos 1–3.
DOI: 10.1007/s10509-007-9667-5 © Springer Science+Business Media B.V. 2007

Abstract Protoplanetary disks are the most probable sites where planet formation takes place. According to theory, planet formation in protoplanetary disks should show remarkable signatures, such as a gap/hole or a spiral structure. In fact, recent high-angular and high-sensitivity observations in millimeter and submillimeter wavelengths, as well as optical/near-IR wavelengths, have shown such structures in protoplanetary disks. Two particular examples of such disks around AB Aurigae and HD 142527 are discussed here, with an emphasis on results obtained using the Submillimeter Array. These disks—and their probable planet formation—will be very important future targets for ALMA to study the physical process of planet formation in detail.

Keywords Planet formation · Circumstellar disk

1 Introduction

How are planets like earth formed? This is one of the most outstanding questions in modern astrophysics. In particular, as more extrasolar planets have been discovered since the first—around 51 Pegasi in 1995 (Mayor and Queloz 1995)—people have started to recognize that planet formation is a very fundamental process in this universe.

Although studies of planet formation were first led mainly by theoreticians to explain how our own solar system formed, recent observations made possible by dramatically developed technology have opened up a new vista of observational studies of planet formation. The discovery of protoplanetary disks around young stellar objects has had a particularly significant impact on the observational study of planet formation because these disks are considered to be the most probable sites for planet formation. Some main-sequence-stars like Vega were also found to be associated with disks. Although these disks have been considered to be debris left over after planet formation, recent observations of debris disks have revealed structures that are probably formed due to existing planets with Jupiter-like masses (e.g., Wilner et al. 2002). In order to observe ongoing planet formation, it is essential for us to observe protoplanetary disks around young stellar objects.

According to theoretical simulations, planet formation in protoplanetary disks should show remarkable signatures, such as a gap/hole or a spiral structure (e.g., Bryden et al. 1999; Wolf et al. 2002). In fact, recent optical and near-infrared observations have shown that some protoplanetary disks have such structures. Examples are AB Aurigae (Fukagawa et al. 2004) and HD142527 (Fukagawa et al. 2006). Disks with these signatures are considered to be associated with ongoing planet formation, although some of them have stellar companions which could be responsible for these signatures (e.g., HD 141569A Augereau and Papaloizou 2004). However, optical and near-infrared images trace scattered light from dust disks, and provide no information beyond the surface of the optically thick dust disk. The apparent spiral structures may just be surface features, and the existence of density spiral arms needs to be verified. It would also be intriguing to measure the kinematics of the arms and to look for streaming motion, if any, along or across the spirals to examine the possibility of dynamical perturbations from underlying planetary bodies. In order to investigate real structures of these interesting disks, it is essential for us to ob-

N. Ohashi (✉)
Academia Sinica Institute of Astronomy and Astrophysics,
P. O. Box 23-14, Taipei 106, Taiwan
e-mail: ohashi@asiaa.sinica.edu.tw

serve emissions from molecules and dust grains in disks at millimeter and submillimeter wavelengths.

The Atacama Large Millimeter/Submillimeter Array (ALMA) will provide extremely high sensitivity and very high-angular resolution at millimeter and submillimeter wavelengths. ALMA will be an ideal instrument to observe circumstellar disks, including those with possible ongoing planet formation, and readers can find articles discussing ALMA's ability to study circumstellar disks elsewhere in the proceedings. In this article, we focus on how we can be ready to use ALMA to study ongoing planet formation in circumstellar disks. One obvious way to prepare for our work with ALMA is to use existing mm and submm arrays to observe the most promising disks.

The Submillimeter Array (SMA),[1] which consists of eight 6-m antennas built at the top of Mauna Kea, HI, is currently the only interferometer working at submillimeter wavelengths in the world (Ho et al. 2004). SMA provides a unique opportunity to observe circumstellar disks at submillimeter wavelengths with higher angular resolutions. There are a couple of advantages of SMA:

- Dust emission from disks becomes stronger at the 3–4th power of the observation frequency, which enables us to observe them with a higher S/N and, as a result, with a higher angular resolution.
- Molecular line emission in submillimeter wavelengths traces denser and warmer gas, which enables us to observe inner regions of disks.
- Multitransitions of molecular lines, such as CO (4–3), (3–2), and (2–1), are accessible in submillimeter wavelengths. Different transitions trace different areas of the disk, and multitransition observations provide better understanding of disk structures, such as vertical structures.

We have been using SMA to observe circumstellar disks, which are probably associated with ongoing planet formation. In the following sections, results obtained using SMA are described.

2 AB Aurigae

AB Aur ($V = 7.06 \pm 0.06$, spectral type = A0 Ve + sh, $d = 144$ pc; van den Ancker et al. 1997) is one of the nearest and best-studied Herbig Ae stars. With an age of 2–5 Myr (DeWarf et al. 2003), it not only has a bright dust disk, but is also rich in gas. Previous millimeter wavelength observations with an angular resolution of $5''$ made by Mannings and Sargent (1997) with the OVRO interferometer

in CO (1–0) revealed a velocity gradient along the major axis that might be interpreted as a Keplerian disk. Optical imaging with the HST (Grady et al. 1999) showed a north–south asymmetry in the dust disk. Near-IR imaging with the Coronagraphic Imager with Adaptive Optics (CIAO) on the Subaru Telescope shows an even more complicated structure (Fukagawa et al. 2004; hereafter F04). F04 identified at least two apparent spiral-like structures in the dust disk: a prominent inner arm with a radius of 230 AU from the east to northeast, and an outer arm with a radius of 330 AU from the south to northeast. We will use "the inner arm" and "the outer arm" to refer to these two arms, respectively. We observed AB Aur in CO (3–2) and 345 GHz continuum using SMA (Lin et al. 2006). Piétu, Guilloteau, and Dutrey (hereafter P05) (Piétu et al. 2005) also observed AB Aur in CO (2–1) and (1–0) as well as continuum at 110 GHz and 230 GHz.

2.1 345 GHz continuum emission

We clearly detected 345 GHz continuum emission arising from the dust disk of AB Aur. The dust continuum map, made with natural weighting, is superimposed on the Subaru near-IR image in Fig. 1a. The continuum emission has an overall size of 450 AU × 270 AU (P.A. ∼ 66°), derived by a Gaussian fit to the data of the most compact array configuration, which only marginally resolved the disk. The emission does not peak at the center, nor does it exhibit an intensity distribution decreasing monotonically in the radial direction, as is generally found for circumstellar disks associated with pre-main-sequence stars. Instead, the disk shows two distinct peaks—one in the northeast direction (NE) and the other southwest (SW)—and an extension towards the northeast (labeled ET in Fig. 1). The two peaks form a ring-like structure with a radius of 150 AU (measured at the maximum of the two peaks) centered at the stellar position. The peaks also coincide with the inner region of the prominent inner arm inferred by F04 from the Subaru image. P05 also observed AB Aur in 230 GHz continuum with PdBI, finding a similar structure of the dust disk. Compared with the 230 GHz continuum image, we find that the SW peak is in agreement with the local column density enhancement ∼1″ west from the center in their 230 GHz continuum map, while there is no apparent peak (no local maxima) in that map at the NE position.

We measured a total flux density of 235 ± 42 mJy for the dust continuum emission within a radius of ∼350 AU from the central star. This is about 65% of the flux density observed by the JCMT. It is not surprising since the JCMT is sensitive to the extended envelope as well as the circumstellar disk, while the extended envelope was resolved out by the SMA. Note that the JCMT primary beam is even smaller than the SMA primary beam, which implies that the missing flux may be a lower limit.

---

[1] SMA is a joint project between the Smithsonian Astrophysical Observatory and the Academia Sinica Institute of Astronomy and Astrophysics, and is funded by the Smithsonian Institution and the Academia Sinica.

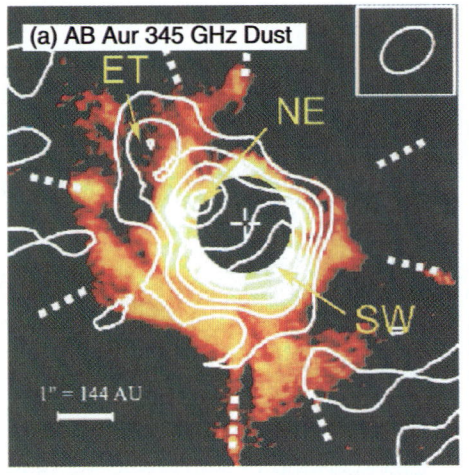
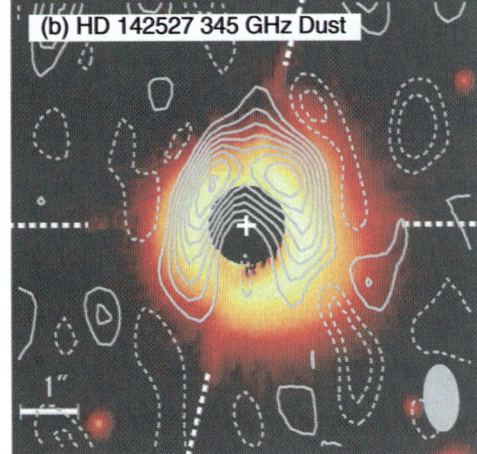

Fig. 1 SMA 345 GHz dust continuum maps of two Hebig Ae stars. **a**: Contour map of AB Aur is superimposed on the Subaru coronagraphic image. *Contours* start from 2σ with a spacing of 2σ. The angular resolution, $1.0'' \times 0.7''$, is shown at the *top right* corner, while the *white cross* shows the position of AB Aur. **b**: Contour map of HD 142527 is superimposed on the Subaru coronagraphic image. Contours start from 2σ with a spacing of 2σ. The angular resolution, $1.2'' \times 0.6''$, is shown at the *bottom right* corner, while the *white cross* shows the position of HD 142527

2.2 ^{12}CO (3–2) emission

We significantly detected ^{12}CO (3–2) emission ($>3\sigma$) in velocity channels between 4.1 km s^{-1} to 8.0 km s^{-1} with a peak brightness temperature of 34.3 ± 4.0 K. We detected only 20% of the ^{12}CO (3–2) flux measured by the JCMT (Thi et al. 2001), suggesting that the single-dish observation contains considerable emission contributed from the large-scale envelope. The ^{12}CO line emission did not suffer from the large-scale ripple effect as did the dust emission, perhaps because in a given velocity channel the ^{12}CO emission is not as extended as the dust. Therefore the inner 30kλ data were retained for maximum sensitivity. Figure 2a shows the integrated intensity (moment 0) map of ^{12}CO over the velocity channels from 4.1 km s^{-1} to 8.0 km s^{-1}, superimposed on the Subaru coronagraphic image. The gas disk with a FWHM size of 530 AU $\times$ 330 AU (P.A. $\sim 66.7°$), derived from Gaussian fitting of the data of the lowest resolution data set, appears to be larger than the dust disk. In contrast with the dust disk, the gas disk exhibits a peak at the stellar position. This is mostly because ^{12}CO (3–2) is optically thick and the dust continuum at 345 GHz is optically thin. Indeed, the gas disk also shows a central depression in optically thinner lines, such as ^{13}CO and C^{18}O (P05), consistent with this interpretation.

In addition to the central peak, there is a secondary peak in the ^{12}CO (3–2) intensity map coincident with the NE peak in the dust disk, which possibly traces the inner spiral arm. The ^{12}CO (2–1) emission observed by P05 does not peak at this position. This is most probably due to insufficient angular resolution of their ^{12}CO (2–1) map. When we convolve our ^{12}CO (3–2) map with the same beam size

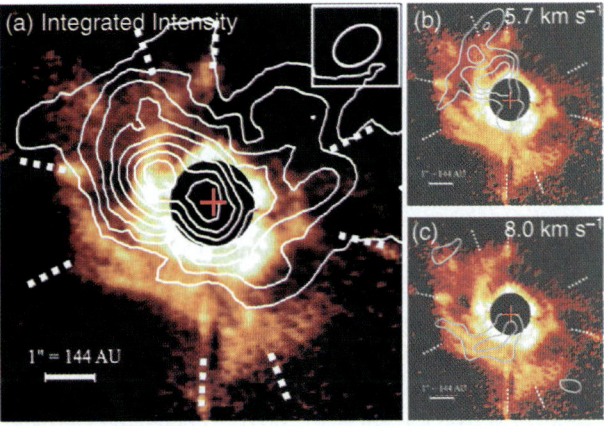

Fig. 2 AB Aur ^{12}CO (3–2) contour maps obtained with SMA are superimposed on the Subaru coronagraphic image. The angular resolution of the maps, $1.0'' \times 0.7''$, is shown at the *top right* corner in the panel **a**. The *red crosses* show the position of AB Aur. **a**: ^{12}CO (3–2) integrated intensity map. **b**: ^{12}CO (3–2) channel map at LSR velocity of 5.7 km s^{-1}. **c**: ^{12}CO (3–2) channel map at LSR velocity of 8.0 km s^{-1}

of the ^{12}CO (2–1) map ($2.0'' \times 1.6''$), the secondary peak disappears and the resultant map looks similar to the PdBI map. The ^{12}CO (3–2) emission, which probably traces the spiral arms, are even more obvious in some of ^{12}CO (3–2) channel maps. Figures 2b and c show two examples, one at $V_{lsr} = 5.7$ km s^{-1} and the other at $V_{lsr} = 8.0$ km s^{-1}. Our results show not only dust emission but also ^{12}CO (3–2) emission, which most likely traces the spiral structures in the disk. This strongly suggests that the spiral structures seen at near-IR are not just surface features of the disk, but are due to real density structures.

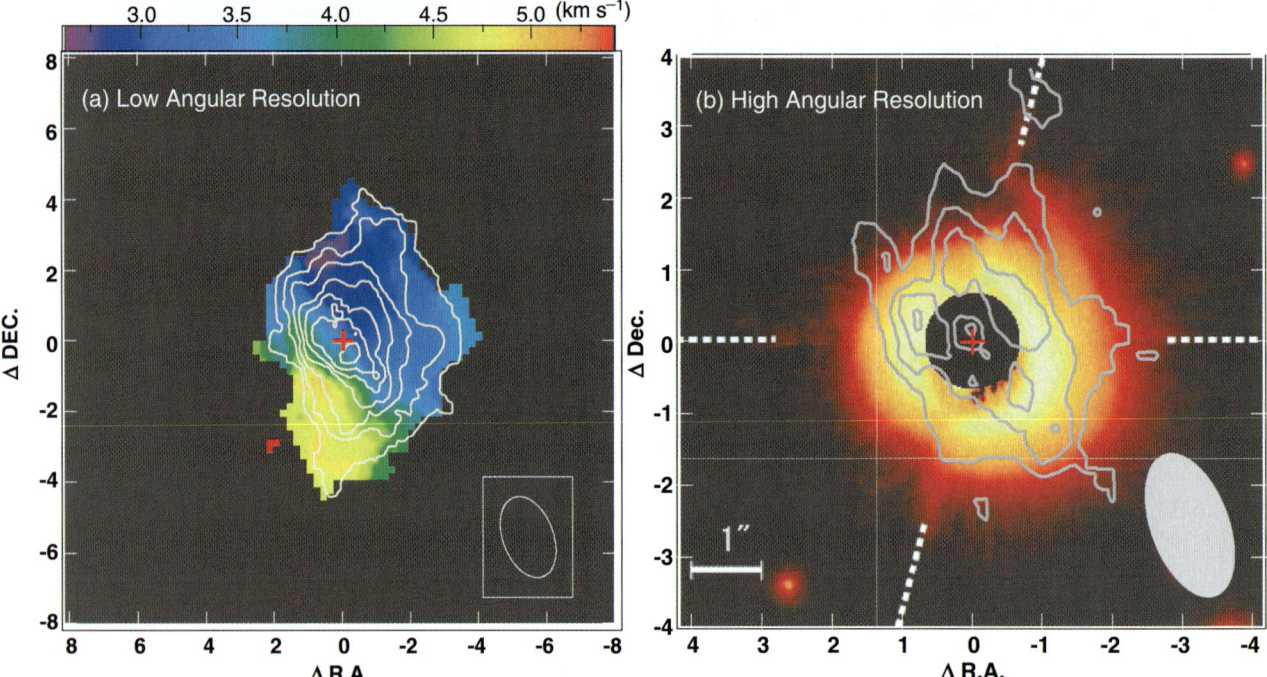

Fig. 3 SMA ^{12}CO3-2 integrated intensity maps of HD 142527 at different angular resolutions. **a**: ^{12}CO contour maps at a lower angular resolution of 2.4″ × 1.5″ with contours showing the ^{12}CO integrated intensity and false color showing ^{12}CO mean velocity. Contours start from 2σ with a spacing of 2σ. **b**: ^{12}CO integrated intensity map (contour) at a higher angular resolution of 2.1″ × 1.1″ is superimposed on the Subaru coronagraphic image. Contours start from 2σ with a spacing of 2σ. The *red cross* shows the position of HD 142527

In the ^{12}CO mean velocity map (see Fig. 2 in Lin et al. 2006), the largest velocity gradient of ^{12}CO was found along the major axis of the disk, consistent with the pattern of a rotating disk. Therefore the molecular gas emission is interpreted as a large rotating disk as previous millimeter observations have shown (Mannings and Sargent 1997; Corder et al. 2005; Piétu et al. 2005). Although the bulk motion of the gas disk looks agreeable with rotation, there are also velocity gradients along the minor axis. In particular, at the south end of the minor axis there is an abrupt change in velocity, suggestive of noncircular motion.

The channel maps of the ^{12}CO (3–2) emission (see Fig. 3 in Lin et al. 2006) show a general velocity gradient along the disk major axis as seen in the mean velocity map. The channel maps, however, do not show a simple "butterfly" diagram that Keplerian disks show (e.g., Beckwith and Sargent 1993), suggestive of deviation from Kepler rotation. When we compare the ^{12}CO channel maps with a simple disk model having Kepler motion, it is found that some part of the ^{12}CO emission does not follow simple Kepler rotation. Interestingly, ^{12}CO emission tracing spiral arms show such deviation more clearly, suggesting that spiral arms are the main source of the non-Kepler motion. Careful inspection of the kinematics of the ^{12}CO emission tracing the inner spiral arm shows that its velocity is redshifted with respect to the expected Keplerian motion. This redshifted motion may

be explained by possible streaming motion along or across the spiral arm (see more details in Lin et al. 2006).

3 HD 142527

HD 142527 (F6 IIIe) is classified as a Herbig Ae star. Its stellar mass and age was estimated to be $1.9 \pm 0.3 M_\odot$ and 2 Myr, respectively (Fukagawa et al. 2006). The distance to the source was estimated to be 200^{+60}_{-40} pc (van den Ancker et al. 1998). Fukagawa et al. (2006) observed HD 142527 using the coronagraphic camera with AO on the Subaru telescope, and discovered a banana-split structure (two arcs facing each other) as well as a spiral arm in the disk. The disk appears to be almost face-on, whereas it shows slight ellipticity with PA $\sim 60°$. Fukagawa et al. (2006) suggested that unseen eccentric binary and a recent stellar counter would be responsible for these unique disk structures. Similar disk structures are also observed in middle-infrared wavelengths (Fujiwara et al. 2006).

3.1 345 GHz continuum emission

Continuum emission arising from the disk around HD 142527 was clearly detected at 345 GHz. Figure 1b shows a map of the 340 GHz continuum emission at an angular resolution of $1.2'' \times 0.6''$ (PA $\sim 4°$), revealing an arc-like structure enclosing the central star. Although the continuum emission is clearly weak at the stellar position, two bright peaks—one at $\sim 1''$ northeast and the other at $\sim 1''$ northwest of the star—are detected in the arc-like structure. Even though the morphology of the continuum emission seen in Fig. 1b is similar to the banana-split structure the Subaru image showed at first sight, careful comparison between the SMA and Subaru images indicates a couple of remarkable differences between the two maps. First of all, the 345 GHz continuum emission is weaker at the southern side of the disk, where near-infrared emission is clearly detected. Second of all, positions of peaks are different between the two maps: the two bright peaks in the Subaru map are shifted slightly to the south as compared to the two peaks in the SMA map. As a result, the morphology of the 345 GHz continuum emission appears as a single arc rather than a banana-split shown by near-infrared emission. There is no clear continuum emission corresponding to the spiral structure found in the Subaru image.

3.2 ^{12}CO (3–2) emission

Figure 3a shows an integrated intensity (moment 0) map of ^{12}CO (3–2) at a lower angular resolution of $2.4'' \times 1.5''$ (PA $\sim 22°$), superimposed on its mean velocity map (moment 1). ^{12}CO channel maps within LSR velocity ranging from 2.2 km s^{-1} to 5.4 km s^{-1}, in which ^{12}CO was detected at more than four sigma level, were combined to make this integrated intensity map. The ^{12}CO emission, having a single peak close to the central stellar position, represents an elongated structure from the northeast to the southwest at higher contour levels, with weaker extensions to the north and the south at lower contour levels. The elongation at higher contour levels looks similar to ellipticity of the scattered light seen in the Subaru image, although the position angles are different between them. We note that the synthesized beam shown at the corner of the map is also elongated from northeast to southwest, which makes the interpretation of the elongation of ^{12}CO from northeast to southwest difficult. The appearance of the ^{12}CO emission does not clearly show the disk direction.

It is quite obvious that the ^{12}CO map shown in Fig. 3a looks very different from the high-angular resolution dust continuum map shown in Fig. 1b. The difference between the continuum map and the ^{12}CO map may be attributed to the lower angular resolution of the ^{12}CO map. We therefore made another ^{12}CO integrated intensity map using the same data set but different data weighting to emphasize small-scale structures. The resulting map at an angular resolution of $2.1'' \times 1.1''$ (PA $\sim 21°$) is shown in Fig. 3b. The same velocity channels we integrated for the low-resolution integrated intensity map were integrated although less significant ^{12}CO emission was detected at each channel because

of limited sensitivity. The new map at a higher angular resolution demonstrates that ^{12}CO is also in decline at the stellar position, with peaks located at $1''$ east, $1''$ north, and $1''$ west of the central star. These peaks make a structure corresponding to the arc-like structure seen in the high-angular resolution dust map. In addition, there is weak emission elongated from the northeast to the southwest. On the other hand, the structure elongated from north to south seen in the lower contour levels in the low-resolution map does not appear in the high-angular resolution map. The similar distributions between ^{12}CO and dust suggests that not only the dust emission but also ^{12}CO emission is most probably optically thin in the circumstellar disk of HD 142527.

A clear velocity gradient from northwest to southeast was shown in the ^{12}CO mean velocity map (see Fig. 3a). The direction of this velocity gradient is similar to the weaker extension of the ^{12}CO emission seen in the lower angular resolution map. If this velocity gradient is due to disk rotation, then the direction of the disk major axis is northwest–southeast, and the weak extension from north to south of the ^{12}CO emission is probably due to disk elongation. This disk direction is consistent with the brightness asymmetry seen in infrared emission (Fukagawa et al. 2006).

In addition to the major velocity gradient from northwest to southeast, the northwestern side of the ^{12}CO emission seems to show a weak velocity gradient along the northeast–southwest direction. Similar to AB Aur, this weak velocity gradient might be due to noncircular motion. A comparison between the observed data and a model disk having Kepler motion shows that there is deviation from Kepler motion. Because of limited angular resolution and sensitivity, it is difficult for us to investigate relationship between the deviation and the spiral arm.

3.3 Disk mass and molecular abundance

The mass of the disk can be estimated from the total dust flux density, which was measured to be 1.2 Jy. For this purpose, it is necessary for us to derive the dust opacity and temperature. The dust opacity at the observed frequency, κ_ν, is often derived by extrapolating the dust opacity measured at 250 μm (0.1 cm^2 g^{-1}) using an appropriate power index, the so called β-index. When dust emission is optically thin, its β-index and spectral index, α, have the relationship $\alpha = 3 + \beta$ (e.g., Beckwith et al. 1990). We estimated a spectral index of the dust emission to be 3.7 using our SMA measurements at 345 GHz and those at 2.9 mm using the ATCA (van den Ancker et al. 1998). This implies that the β-index is 0.7, giving $\kappa_\nu = 4.8$ cm^2 g^{-1} at 345 GHz. On the other hand, deriving the dust temperature is not so straightforward. In our case, we simply assume the dust temperature to be ~ 50 K. Using the mass opacity and the dust temperature, the total disk mass (gas + dust) is estimated to be $\sim 1.5 \times 10^{-2} M_\odot$. Note that we assumed the gas-to-dust mass ratio to be 100, which is generally accepted.

The mass of the disk is also derived from the total integrated intensity of the ^{12}CO emission, ~ 6.9 Jy km s^{-1}, under the LTE assumption, if the ^{12}CO emission is optically thin. In order for us to do this, we are required to estimate the excitation temperature of ^{12}CO. Similar to the dust temperature, there is no straightforward way to estimate the excitation temperature of ^{12}CO. We simply assume that the dust temperature and the excitation temperature would be approximately the same provided that the dust grains and the ^{12}CO molecules are mixed well. If this is the case, the ^{12}CO excitation temperature is ~ 50 K. With this ^{12}CO excitation temperature, the total disk mass is estimated to be $\sim 3.0 \times 10^{-6} M_\odot$. Note that we assumed the ^{12}CO abundance relative to the H$_2$ molecule to be 10^{-4}.

As you can clearly see here, the total gas mass derived from dust is four orders of magnitude larger than that derived form ^{12}CO. The biggest uncertainty in the mass estimation described above would be the dust temperature and the ^{12}CO excitation temperature. It is impossible, however, to explain the difference in the estimated disk mass by the uncertainty of the dust temperature and the ^{12}CO excitation temperature; for example, even if we assume that both the dust temperature and the ^{12}CO excitation temperature are 100 K, the difference in the estimated total disk mass is still three orders of magnitude.

Another uncertainty in the disk mass estimation is the ^{12}CO fractional abundance. It is considered that ^{12}CO molecules freeze out on dust grains in circumstellar disks if ^{12}CO molecular gas is colder than 20 K, making the ^{12}CO fractional abundance lower than its terrestrial value (e.g., Aikawa et al. 1999). In such a case, a disk mass derived from ^{12}CO becomes smaller than that derived from dust if we use the terrestrial ^{12}CO fractional abundance. In fact, it is often found that masses of circumstellar disks derived from ^{12}CO are smaller than those derived from dust. Disks around Herbig Ae stars, however, might be warmer than those around T Tauri stars, which may make CO freezing more difficult in circumstellar disks around Herbig Ae stars.

We should note that the difference in the derived disk mass can also be explained by depletion of molecular gas including H$_2$ in the disk. In this case, mass derived from dust using the terrestrial gas-to-dust mass ratio overestimates the disk mass. Since it might be difficult for CO to be depleted in the disk around HD 142527, as mentioned earlier, we may have to consider gas (H$_2$) depletion as well as CO depletion to explain the difference in mass.

We also note that even though it is often found that masses of circumstellar disks derived from ^{12}CO are smaller than those derived from dust, the differences are factors of 10–200 (Thi et al. 2001). In the case of HD 142527, the difference is 3–4 orders of magnitude, which seems to be an extreme case and would be similar to the cases of debris disks, where there is almost no gas (e.g., Thi et al. 2001). It is interesting that a disk around a relatively young pre-main-sequence star and debris disks shows high gas (CO and/or H$_2$) depletion factors. There might be some mechanism that quickly depletes gas in circumstellar disks around Herbig Ae stars, such as photoevaporation (Takeuchi et al. 2005). C(I) (atomic carbon) observations of circumstellar disks with ALMA might provide us with a hint of photoevaporation in disks and open the door for study of gas dissipation.

Acknowledgements This work was partially supported by 95NSC-2112-M-001-037-MY2.

References

Aikawa, Y., Umebayashi, T., Nakano, T., Miyama, S.M.: Evolution of molecular abundances in proto-planetary disks with accretion flow. Astrophys. J. **519**, 705 (1999)

Augereau, J.C., Papaloizou, J.C.B.: Structuring the HD 141569A circumstellar dust disk. Astron. Astrophys. **414**, 1153 (2004)

Beckwith, S.V.W., Sargent, A.I.: Molecular line emission from circumstellar disks. Astrophys. J. **402**, 280 (1993)

Beckwith, S.V.W., Sargent, A.I., Chini, R.S., Güesten, R.: A survey for circumstellar disks around young stellar objects. Astron. J. **99**, 924 (1990)

Bryden, G., Chen, X., Lin, D.N.C., Nelson, R.P., Papaloizou, J.C. B.: Tidally induced gap formation in protoplanetary disks: gap clearing and suppression of protoplanetary growth. Astrophys. J. **514**, 344 (1999)

Corder, S., Eisner, J., Sargent, A.: AB Aurigae resolved: evidence for spiral structure. Astrophys. J. **622**, L113 (2005)

DeWarf, L.F., Sepinsky, J.F., Guinan, E.F., Ribas, I., Nadalin, I.: Intrinsic properties of the young stellar object SU Aurigae. Astrophys. J. **590**, 357 (2003)

Fukagawa, M., et al.: Spiral structure in the circumstellar disk around AB Aurigae. Astrophys. J. **605**, L53 (2004)

Fukagawa, M., Tamura, M., Itoh, Y., Kudo, T., Imaeda, Y., Oasa, Y., Hayashi, S.S., Hayashi, M.: Near-infrared images of protoplanetary disk surrounding HD 142527. Astrophys. J. **636**, L153 (2006)

Fujiwara, H., Honda, M., Kataza, H., Yamashita, Y., Onaka, T., Fukagawa, M., Okamoto, Y.K., Miyata, T., Sako, S.: The asymmetric thermal emission of the protoplanetary disk surrounding HD 142527 seen by Subaru/COMICS. Astrophys. J. **644**, L133 (2006)

Grady, C.A., Woodgate, B., Bruhweiler, F.C., Boggess, A., Plait, P., Lindler, D., Clampin, M., Kalas, P.: *Hubble Space Telescope* space telescope imaging spectrograph coronagraphic imaging of the Herbig AE star AB Aurigae. Astrophys. J. **523**, L151 (1999)

Ho, P.T.P., Moran, J.M., Lo, K.Y.: The submillimeter array. Astrophys. J. **616**, L1 (2004)

Lin, S.-Y., Ohashi, N., Lim, J., Ho, P.T.P., Fukagawa, M., Tamura, M.: Possible molecular spiral arms in the protoplanetary disk of AB Aurigae. Astrophys. J. **645**, 1297 (2006)

Mannings, V., Sargent, A.I.: Astrophys. J. **490**, 792 (1997)

Mayor, M., Queloz, D.: A Jupiter-mass companion to a solar-type star. Nature **378**, 355 (1995)

Piétu, V., Guilloteau, S., Dutrey, A.: Sub-arcsec imaging of AB Aur molecular disk and envelope at millimeter wavelengths: a non Keplerian disk. Astron. Astrophys. **443**, 945 (2005)

Takeuchi, T., Clarke, C.J., Lin, D.N.C.: The differential lifetime of protostellar gas and dust disks. Astrophys. J. **627**, 286 (2005)

Thi, W.F., et al.: H_2 and CO emission from disks around T Tauri and Herbig Ae pre-main-sequence stars and from debris disks around young stars: warm and cold circumstellar gas. Astrophys. J. **561**, 1074–1094 (2001)

van den Ancker, M.E., Thé, P.S., Djie, H.T.A., Catala, C., de Wilner, D., Blondel, P.F.C., Waters, L.B.: HIPPARCOS data on Herbig Ae/Be stars: an evolutionary scenario. Astron. Astrophys. **324**, L33 (1997)

van den Ancker, M.E., de Wilner, D., Tjin A Djie, H.R.E.: HIPPARCOS photometry of Herbig Ae/Be stars. Astron. Astrophys. **330**, 145 (1998)

Wilner, D.J., Holman, M.J., Kuchner, M.J., Ho, P.T.P.: Structure in the dusty debris around Vega. Astrophys. J. **569**, L115 (2002)

Wolf, S., Gueth, F., Henning, T., Kley, W.: Detecting planets in protoplanetary disks: a prospective study. Astrophys. J. **566**, L97 (2002)

Detecting protoplanets with ALMA

Sebastian Wolf

Originally published in the journal Astrophysics and Space Science, Volume 313, Nos 1–3.
DOI: 10.1007/s10509-007-9660-z © Springer Science+Business Media B.V. 2007

Abstract Theoretical investigations show that planet-disk interactions cause structures in circumstellar disks, which are usually much larger in size than the planet itself and thus more easily detectable. The specific result of planet-disk interactions depends on the evolutionary stage of the disk. Exemplary signatures of planets embedded in disks are gaps and spiral density waves in the case of young, gas-rich protoplanetary disks and characteristic asymmetric density patterns in debris disks. Numerical simulations convincingly demonstrate that high-resolution imaging performed with observational facilities which are already available or will become available in the near future will allow to trace these "fingerprints" of planets in protoplanetary and debris disks. These observations will provide a deep insight into specific phases of the formation and early evolution of planets in circumstellar disks.

In this context, the Atacama Large Millimeter Array (ALMA) will play a crucial role by allowing to trace features in disks which are indicative for various stages of the formation and early evolution of planets in circumstellar disks.

Keywords Infrared excess · Debris disks · Protoplanetary disks · Exo-zodiacal dust · Extrasolar planetary systems

1 Introduction

The detection of extrasolar planets and planetary systems has enormously stimulated and invigorated the stud-

ies of planet formation during the last decade. In particular, a detailed picture of the evolution of circumstellar/protoplanetary disks, which provide the material and environment from and in which planets are expected to form, has been developed. However, the planet formation process itself is in major parts still under discussion. In order to improve our understanding of planet formation and to refine existing hypotheses for the various phases of this process, adequate observational constraints are required.

2 Young planets in protoplanetary disks

Both, the initial conditions of the dust phase in the circumstellar environment, i.e., the distribution of submicron-sized dust grains, and the early stages of the planet formation process (particle growth via coagulation to ∼ mm size) can be observed directly in the optical to millimeter wavelength range. In contrast to this, even larger bodies can hardly be traced directly. However, bodies in this size regime are predicted to undergo collision events (Weidenschilling 1997), creating fragments which may allow to trace the location, abundance, and chemical composition of the centimeter to ∼ few kilometer-sized parent bodies. Beyond this, numerical simulations have shown that sufficiently massive planets may cause characteristic large-scale signatures in the disk density distribution. In young circumstellar disks, with a structure dominated by gas dynamics, the most important of these signatures are gaps and spiral density waves (Bryden et al. 1999). The importance of investigating these signatures lies in the possibility to use them in the search for embedded young planets. Therefore, these disk features can provide constraints on the processes and timescales of planet formation.

S. Wolf (✉)
Max Planck Institute for Astronomy, Königstuhl 17,
69117 Heidelberg, Germany
e-mail: swolf@mpia.de

Disk structures caused by the planet-disk interaction may be traced by high spatial resolution mapping of the thermal dust reemission. We performed simulations with the goal to investigate whether the planet itself and/or its surrounding environment, heated by the planet and through accretion onto it, could be detected. The detection of a gap would already represent a strong indication of the existence of a planet, thus providing information about the planetary mass, viscosity, and pressure scale-height of the disk. However, the detection/non-detection of warm dust close to the planet would additionally give valuable constraints on the temperature and luminosity of the planet, the accretion process onto the planet, and the density structure of the surrounding medium. In order to achieve these goals, we tested different environments of a planet located in a circumstellar disk for the resulting temperature structure, which, in combination with the density distribution, mainly determines the likelihood of detecting any of the features characterizing the embedded planet. A detailed description of all considered model configurations is given in Wolf and D'Angelo (2005). Based on these simulations we make the following predictions about the observability of giant protoplanets in young circumstellar disks: (1) The resolution of the images to be obtained with ALMA will allow detection of the warm dust in the vicinity of the planet only if the object is at a distance of not more than ∼50–100 pc (see Fig. 1). For larger distances, the contrast between the planetary region and the adjacent disk in any of the considered planet/star/disk configurations will be too low to be detectable. (2) Even at a distance of 50 pc a resolution being high enough to allow a

study of the circumplanetary region can be obtained only for those configurations with the planet on a Jupiter-like orbit but not when it is as close as 1 AU to the central star. (3) The observation of the emission from the dust in the vicinity of the planet will be possible only in the case of the most massive and thus young circumstellar disks. (4) The planetary radiation significantly affects the dust reemission spectral energy distribution (SED) only in the near- to mid-infrared wavelength range. Since this spectral region is strongly influenced also by the warm upper layers of the disk and the inner disk structure, the planetary contribution and thus the temperature/luminosity of the planet cannot be derived from the SED alone.

For completeness, it is important to emphasize that further studies have demonstrated that larger-scale features induced by the planet's interaction with the disk are expected to be observable much easier. One example is the gap, which can be observed also for objects at larger distances, such as in nearby rich star-forming region; e.g., in Taurus (Wolf et al. 2002). Another example are apparent inner cavities in disks—inner holes with radii which are much larger than the sublimation radius of interstellar medium-like submicron-sized grains. Examples among T Tauri disks are GM Aurigae (Rice et al. 2003) and TW Hydrae (Calvet et al. 2002). These cavities might be due to the influence of a giant planet on the circumstellar disk (e.g., Kley 1999). Furthermore, viscous accretion and photoevaporation by stellar radiation are assumed to clear the inner region of circumstellar disks (e.g., Goto et al. 2006). However, an alternative explanation could be the consequence of the dust evo-

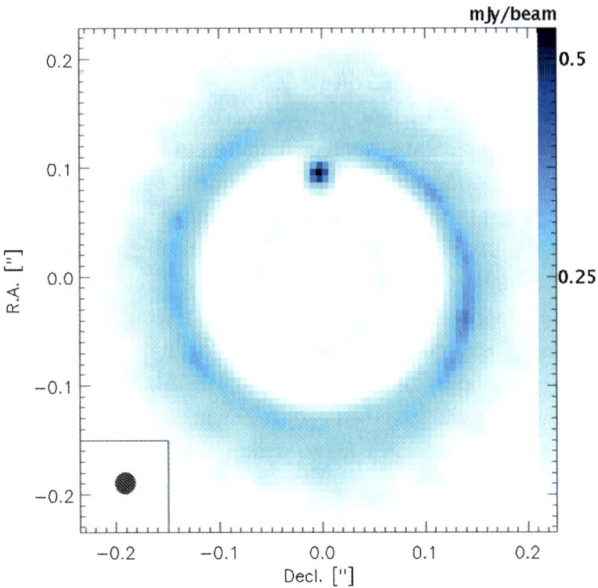
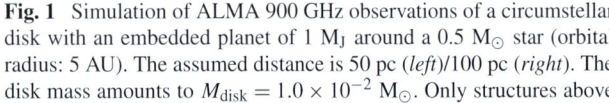

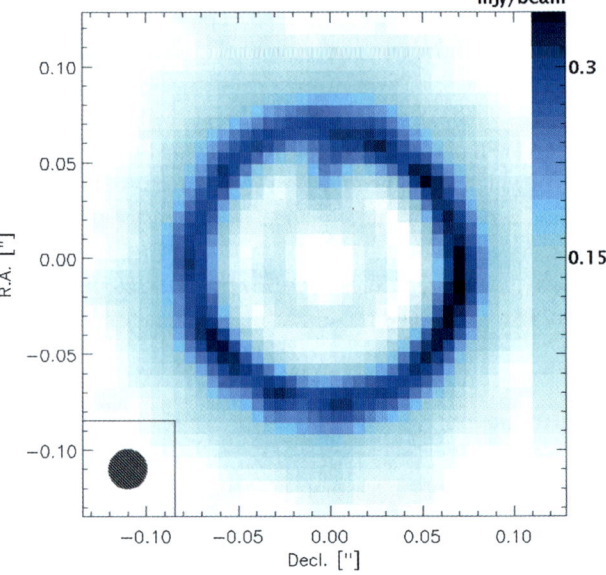

Fig. 1 Simulation of ALMA 900 GHz observations of a circumstellar disk with an embedded planet of 1 M_J around a 0.5 $M_\odot$ star (orbital radius: 5 AU). The assumed distance is 50 pc (*left*)/100 pc (*right*). The disk mass amounts to $M_{disk} = 1.0 \times 10^{-2}$ $M_\odot$. Only structures above the 2σ-level are shown. The size of the synthesized beam is symbolized in the *lower left edge* of each image. Note the reproduced shape of the spiral wave near the planet and the slightly shadowed region behind the planet in the *left* image (from Wolf and D'Angelo 2005)

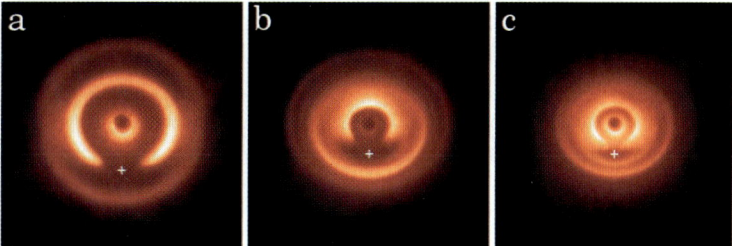

Fig. 2 Simulated scattered light images of debris disks with an embedded planet at a wavelength of 1.1 μm. The position of the planet is marked by the *white cross*. In order to emphasize faint structures, the intensity has been scaled as the cubic root of the flux density. Furthermore, the images were convolved with the corresponding HST/NICMOS F110W filter point spread function. (Selected simulation parameters: Size of the displayed region: 400 AU × 400 AU. Disk: Density distribution based on a dynamical model taking into account the gravitational perturbations by the planet, photon radiation pressure, and dissipative drag force due to the Poynting–Robertson effect and stellar wind. Disk inclination = 31.7°; with the upper part facing towards the observer. Planet: Mass = 3.0 M_J (**a, b**)/0.3 M_J (**c**); Orbit radius: 86.3 AU (**a**)/54.4 AU (**b, c**). Dust: radius = 14 μm, composition: astronomical silicate (Weingartner and Draine 2001)) (from Wolf 2007; see also Rodmann 2006)

lution, resulting in a depletion of small grains in the inner region. Complementary high-resolution interferometric observations at mid-infrared to millimeter wavelengths, tracing larger grains, are best-suited to confirm or rule out the second versus the first scenario.

3 Planets in debris disks

Planetary debris disks are assumed to represent the almost final stage of the circumstellar disk evolution, i.e., they are the evolutionary products of ongoing or completed planet formation. More specifically, debris disks are solar system-sized dust disks produced as by-products of collisions between asteroid-like bodies and the activity of comets left over from the planet formation process. In contrast to optically thick young circumstellar disks around Herbig Ae/Be and T Tauri stars, the much lower optical depth and lower gas-to-dust mass ratio in debris disks let the stellar radiation—in addition to gravity—be responsible for the disk structure (e.g., Zuckerman et al. 1995). Besides, fragmentation becomes a typical outcome of particle collisions because relative velocities of grains are no longer damped by gas. Thus, the Poynting–Robertson effect, radiation pressure, collisions, and gravitational stirring by embedded planets are all important in determining the dust population and disk structure (e.g., Liou and Zook 1999).

High-resolution images of debris disks in scattered light in the optical/near-infrared and in thermal emission at mid-infrared to millimeter wavelengths show complex structures, such as rings, gaps, arcs, warps, offset asymmetries and clumps of dust (e.g., Greaves et al. 1998; Holland et al. 1998, 2003; Koerner et al. 2001; Schneider et al. 1999; Wilner et al. 2002). In evolved, optically thin debris disks some of these features are likely to be the result of gravitational perturbations by one or more massive planets on the dust disk, i.e. characteristic density patterns are expected to provide the strongest indirect hints on the existence of planets embedded in these disks. The dominance of any of these structures mainly depends on the mass of the planet and the eccentricity of its orbit (see Fig. 2 for illustration).

Although debris disks represent a rich source of information about the formation and evolution of planetary systems, they also impose problems on the observations of exoplanetary systems. The exozodiacal dust disk around a target star, even at solar level, will likely be the dominant signal originating from the extrasolar system. In the case of a solar system twin, its overall flux over the first 5 AU is about 400 times larger than the emission of the Earth at 10 μm. Besides, one has to make sure that the exozodiacal signature will not mimic planetary signals such as would be the case if the disk is significantly clumpy. If the origin of this clumpiness is in the outlined perturbations of planets, then detecting clumps can help to pinpoint those planets. However, one has to be aware that collisionally regenerated debris disks are also intrinsically clumpy because dust created by collisions between large planetesimals starts out in a clumpy dust distribution (Wyatt and Dent 2002).

Beside resonant structures, inner cavities have been found in several prominent debris disks: β Pic (inner radius: 20 AU), HR 4796A (30–50 AU), ε Eri (50 AU), Vega (80 AU), and Fomalhaut (125 AU)—(e.g., Dent et al. 2000; Greaves et al. 2000; Holland et al. 2003; Wilner et al. 2002). The analysis of the mid-infrared SED of further debris disks discovered recently with the Spitzer Space Telescope shows that the occurrence of inner cavities, i.e. inner regions with strong dust depletion, is a frequent phenomenon in these systems (e.g., Hines et al. 2006; Kim et al. 2005; Quillen et al. 2004; Silverstone et al. 2006). These cavities may be created by gravitational scattering with an inner planet: Dust grains drifting inwards due to the Poynting–Robertson effect are likely to be scattered into larger orbits resulting in a lower dust number density within the planet's orbit.

Another mechanism for the possible influence of a planet on the disk structure has been discussed in the case of the β Pictoris disk. The Northeast and Southwest extensions of the dust disk have been found to be asymmetric in scattered light as well as in thermal emission. This warp is assumed to be caused by a giant planet on an inclined orbit that gravitationally perturbs the dust disk (Augereau et al. 2001).

A detailed study of the influence of planets on the SED of debris disks has been performed by Wolf and Hillenbrand (2003) and Moro-Martín et al. (2005). The authors find that there exist degeneracies that can complicate the interpretation of the SED in terms of determining the location of embedded planets. For example, the mid-infrared SED of a dust disk dominated by weakly absorbing grains (e.g., Fe-poor silicates) has its minimum at wavelengths longer than those of a disk dominated by strongly absorbing grains (e.g., carbonaceous and Fe-rich silicate). Because the minimum of the mid-infrared SED also shifts to longer wavelengths when the gap radius increases, there might be a degeneracy between the chemical composition of the dust and the semi-major axis of the planet clearing the gap. The degeneracy in the SED analysis illustrates the importance of obtaining high-resolution images, allowing to spatially resolve the debris disk structure, in addition to spectroscopic observations constraining the chemical composition of the dust. High resolution imaging with ALMA will therefore be essential to resolve these degeneracies.

Acknowledgements S.W. was supported by the German Research Foundation (DFG) through the Emmy Noether grant WO 857/2.

References

Augereau, J.C., Nelson, R.P., Lagrange, A.M., Papaloizou, J.C.B., Mouillet, D.: Dynamical modeling of large scale asymmetries in the beta Pictoris dust disk. Astron. Astrophys. **370**, 447–455 (2001)

Bryden, G., Chen, X., Lin, D.N.C., Nelson, R.P., Papaloizou, J.C.B.: Tidally induced gap formation in protostellar disks: gap clearing and suppression of protoplanetary growth. Astrophys. J. **514**, 344–367 (1999)

Calvet, N., D'Alessio, P., Hartmann, L., Wilner, D., Walsh, A., Sitko, M.: Evidence for a developing gap in a 10 Myr old protoplanetary disk. Astrophys. J. **568**, 1008–1016 (2002)

Dent, W.R.F., Walker, H.J., Holland, W.S., Greaves, J.S.: Models of the dust structures around Vega-excess stars. Mon. Not. R. Astron. Soc. **314**, 702–712 (2000)

Goto, M., Usuda, T., Dullemond, C.P., Henning, T., Linz, H., et al.: Inner rim of a molecular disk spatially resolved in infrared CO emission lines. Astrophys. J. **652**, 758–762 (2006)

Greaves, J.S., Holland, W.S., Moriarty-Schieven, G., Jenness, T., Dent, W.R.F., et al.: A dust ring around epsilon Eridani: analog to the young solar system. Astrophys. J. **506**, L133–L137 (1998)

Greaves, J.S., Mannings, V., Holland, W.S.: The dust and gas content of a disk around the young star HR 4796A. Icarus **143**, 155–158 (2000)

Hines, D.C., Backman, D.E., Bouwman, J., Hillenbrand, L.A., Carpenter, J.M., et al.: The formation and evolution of planetary systems (FEPS): discovery of an unusual debris system associated with HD 12039. Astrophys. J. **638**, 1070–1079 (2006)

Holland, W.S., Greaves, J.S., Zuckerman, B., Webb, R.A., McCarthy, C., et al.: Submillimetre images of dusty debris around nearby stars. Nature **392**, 788–790 (1998)

Holland, W.S., Greaves, J.S., Dent, W.R.F., Wyatt, M.C., Zuckerman, B., et al.: Submillimeter observations of an asymmetric dust disk around Fomalhaut. Astrophys. J. **582**, 1141–1146 (2003)

Kim, J.S., Hines, D.C., Backman, D.E., Hillenbrand, L.A., Meyer, M.R., et al.: Formation and evolution of planetary systems: cold outer disks associated with Sun-like stars. Astrophys. J. **632**, 659–669 (2005)

Kley, W.: Mass flow and accretion through gaps in accretion discs. Mon. Not. R. Astron. Soc. **303**, 696–710 (1999)

Koerner, D.W., Sargent, A.I., Ostroff, N.A.: Millimeter-wave aperture synthesis imaging of Vega: evidence for a ring arc at 95 AU. Astrophys. J. **560**, L181–L184 (2001)

Liou, J.-C., Zook, H.A.: Signatures of the giant planets imprinted on the Edgeworth–Kuiper Belt dust disk. Astron. J. **118**, 580–590 (1999)

Moro-Martín, A., Wolf, S., Malhotra, R.: Signatures of planets in spatially unresolved debris disks. Astrophys. J. **621**, 1079–1097 (2005)

Quillen, A.C., Blackman, E.G., Frank, A., Varnière, P.: On the planet and the disk of COKU TAURI/4. Astrophys. J. **612**, L137–L140 (2004)

Rice, W.K.M., Wood, K., Armitage, P.J., Whitney, B.A., Bjorkman, J.E.: Constraints on a planetary origin for the gap in the protoplanetary disc of GM Aurigae. Mon. Not. R. Astron. Soc. **342**, 79–85 (2003)

Rodmann, J.: Dust in circumstellar disks. Ph.D. Thesis, University of Heidelberg (2006)

Schneider, G., Smith, B.A., Becklin, E.E., Koerner, D.W., Meier, R., et al.: NICMOS imaging of the HR 4796A circumstellar disk. Astrophys. J. **513**, L127–L130 (1999)

Silverstone, M.D., Meyer, M.R., Mamajek, E.E., Hines, D.C., Hillenbrand, L.A., et al.: Formation and evolution of planetary systems (FEPS): primordial warm dust evolution from 3 to 30 Myr around Sun-like stars. Astrophys. J. **639**, 1138–1146 (2006)

Weidenschilling, S.J.: The origin of comets in the solar nebula: a unified model. Icarus **127**, 290–306 (1997)

Weingartner, J.C., Draine, B.T.: Dust grain-size distributions and extinction in the Milky Way, Large Magellanic Cloud, and Small Magellanic Cloud. Astrophys. J. **548**, 296–309 (2001)

Wilner, D.J., Holman, M.J., Kuchner, M.J., Ho, P.T.P.: Structure in the dusty debris around Vega. Astrophys. J. **569**, L115–L119 (2002)

Wolf, S.: Signatures of planets and of their formation process in circumstellar disks. Habilitation Thesis, University of Heidelberg (2007)

Wolf, S., D'Angelo, G.: On the observability of giant protoplanets in circumstellar disks. Astrophys. J. **619**, 1114–1122 (2005)

Wolf, S., Hillenbrand, L.A.: Model spectral energy distributions of circumstellar debris disks. I. Analytic disk density distributions. Astrophys. J. **596**, 603–620 (2003)

Wolf, S., Gueth, F., Henning, T., Kley, W.: Detecting planets in protoplanetary disks: a prospective study. Astrophys. J. **566**, L97–L99 (2002)

Wyatt, M.C., Dent, W.R.F.: Collisional processes in extrasolar planetesimal discs—dust clumps in Fomalhaut's debris disc. Mon. Not. R. Astron. Soc. **334**, 589–607 (2002)

Zuckerman, B., Forveille, T., Kastner, J.H.: Inhibition of giant planet formation by rapid gas depletion around young stars. Nature **373**, 494 (1995)

The study of young substellar objects with ALMA

Antonella Natta · Leonardo Testi

Originally published in the journal Astrophysics and Space Science, Volume 313, Nos 1–3.
DOI: 10.1007/s10509-007-9635-0 © Springer Science+Business Media B.V. 2007

Abstract We discuss the potential of ALMA for studying the formation of substellar objects. We first review briefly the various formation mechanisms proposed so far and stress the unique capability of ALMA to detect and study pre-brown dwarf cores and to confirm the core-collapse scenario to the lowest possible masses. We then discuss the properties of disks around substellar objects. We show how it will be possible to detect with ALMA most disks around objects with mass as low as few Jupiter masses, and to resolve spatially their emission in the more favorable cases.

Keywords Stars: formation · Stars: brown dwarfs · Accretion disks

1 Introduction

The definition "substellar" applies to all objects with mass lower than the hydrogen-burning limit of 0.075 $M_\odot$. They include brown dwarfs, with mass in the interval 0.075–0.013 $M_\odot$, and planetary mass objects, below the limit of 0.013 $M_\odot$ of deuterium burning. These definitions are relevant only after pre-main sequence evolution, when, e.g., an object of mass >0.075 $M_\odot$ will burn hydrogen in its center, while one with mass <0.075 $M_\odot$ will not. Before that, all objects derive their energy from accretion and contraction, and precise mass boundaries have, in fact, no meaning. In the following, we will use very loosely definitions such as very low mass objects, substellar objects or brown dwarfs.

In principle, during the formation and pre-main sequence evolution phases, one should not expect stellar and substellar mass objects to behave differently. Gravitational collapse of molecular cores is possible for cores of mass as low as ∼1–5 Jupiter masses (opacity limit; e.g., Whitworth et al. (2007), provided that such cores are produced in molecular clouds and are gravitationally unstable. So far, it has been very difficult to identify and study very low mass cores, and a number of alternative scenarios for the formation of brown dwarfs have been proposed (Sect. 2). Although there is no compelling evidence that alternatives to the core collapse hypothesis are required, one needs ALMA capabilities to investigate quantitatively the properties of the very low mass cores identified in nearby molecular clouds.

A second aspect which we will discuss is disks around very low mass objects. Disks are a necessary part of the core collapse scenario, and it has been known since several years that brown dwarfs have disks, detected firstly with ISO through their mid-IR excess. However, the mere existence of disks does not constrain the formation mechanism, as most of them predict the existence of small/light disks at least for some fraction of the pre-main sequence evolution. One needs large samples of objects and, even more important, measurements of the disk masses.

Brown dwarf disks are interesting not only as tracers of the formation process, but also in their own right. The properties and evolution of these disks, less massive and cooler than those around solar type stars, may shed light on the mechanisms of angular momentum dissipation which control the disk life. *Spitzer* is providing a wealth of data on disks around brown dwarfs in a variety of star forming regions, which we will briefly discuss in Sect. 3. However, at millimeter wavelengths, brown dwarf disks are still below or very close to the detection limit of current interferometer,

A. Natta (✉) · L. Testi
INAF—Osservatorio Astrofisico di Arcetri, Firenze, Italy
e-mail: natta@arcetri.inaf.it

and we will discuss what ALMA will do in this area, and which questions we may hope to answer in the future.

2 The formation of very low mass objects

The various scenarios for the formation of very low mass objects have been recently reviewed by Whitworth et al. (2007) in *Protostars and Planets V*, and we refer to that paper for a comprehensive discussion and references. As we have already mentioned, it is possible that brown dwarfs form in similar fashion to their more massive counterparts, if fragmentation of molecular clouds produces gravitationally unstable very low mass cores. Such cores may also be the result of the fragmentation of solar-mass collapsing cores into lower-mass gravitationally unstable fragments. Alternative possibilities include the "failed" stars hypothesis, namely that brown dwarfs are the result of competitive accretion in a multi-seed core, ejected before they reach their final mass by dynamical interactions within the core itself; the formation in disks around more massive stars and subsequent ejection by dynamical effects of other planets or stars; the formation in specially harsh environment, such as an expanding HII region, which evaporates the contracting core before complete collapse.

All these mechanisms are possible, and may occur in specific cases. However, if we expect that one mechanism dominates above a certain mass M and a different one for objects below this value, we expect that the statistical properties of young stars will have a break at M. In other words, we expect a discontinuity of some sort in the properties of young stars at the mass where the dominant mechanism of formation changes.

Recently, there has been a large effort aimed at obtaining an accurate description of properties such as the initial mass function, binarity, clustering, velocity dispersion, disk fraction and accretion properties for complete samples covering a large range of masses, from solar to well into the brown dwarf range. None of these studies has revealed any "break" in the observed trends. We can only repeat the conclusion of Whitworth et al. "Brown dwarfs form as hydrogen-burning stars, i.e., on a dynamical time scale by gravitational instability". It should, however, be pointed out that very few objects with mass <10 Jupiter masses have been found so far. It is possible that a change in the main formation process occurs below ~40–50 Jupiter masses, which is the current completeness limit of most surveys.

To confirm the core-collapse scenario, one needs now to search for proto-brown dwarfs and pre-brown dwarf cores, the analogues to Class 0 and pre-stellar cores for solar-mass objects. Very low mass clumps are found in many molecular clouds (e.g., Motte et al. 1998; Walsh et al. 2007), but it is not clear if they are gravitationally bound or not. They

are good candidates for ALMA, which will be able to spatially resolve them and measure their density profile (see Ph. André chapter in these proceedings).

3 Disks

3.1 Brown dwarf disks in the mid-infrared

Disks around brown dwarfs in star-forming regions were discovered several years ago, when their excess emission at mid-IR wavelengths was clearly detected by ISO (e.g., Comerón et al. 1998). Early models (Natta and Testi 2001; Natta et al. 2002) showed that this excess was consistent with the emission of disks heated by the radiation of the central objects, in analogy to disks around T Tauri stars. Subsequent observations, from ground-based large telescopes (e.g., Mohanty et al. 2004) and, in particular, with Spitzer, have increased by large factors the number of objects and of star-forming regions studied, showing that the fraction of brown dwarfs with disks is similar to the fraction of T Tauri stars (TTS in the following) with disks (see the review by Luhman et al. 2007 and, e.g., Lada et al. 2006; Hernández et al. 2007).

The *Spitzer* observations have confirmed earlier suggestions (Apai et al. 2005; Natta et al. 2002) that a large fraction of the detected disks are not fully flared (Fig. 1). The geometry of disks, as revealed by their SED, is very likely linked to the process of grain growth and settling (Dullemond and Dominik 2005; D'Alessio et al. 2001; D'Alessio et al. 2006), as one expects that disks with larger grains and/or lower dust-to-gas mass ratio on the surface are less flared. This, and the apparent large fraction of crystalline silicates (Apai et al. 2005), suggests that brown dwarf

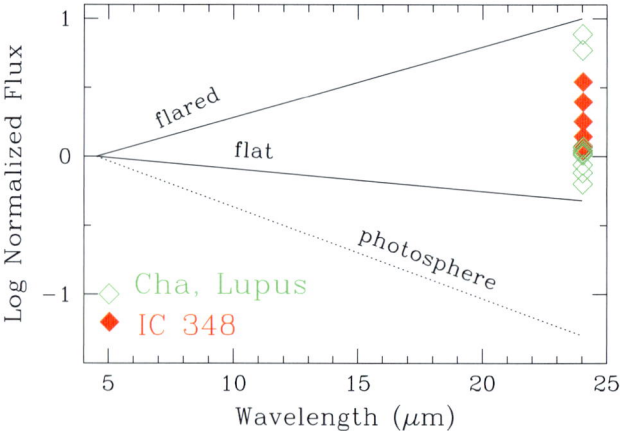

Fig. 1 Ratio of the 24 to 4.8 μm fluxes measured by Spitzer for brown dwarfs in Cha I, Lupus and IC 348 (Apai et al. 2005; Allers et al. 2006; Muzerolle et al. 2006). The *solid lines* show the ratios predicted by models of flared and flat disks, as labeled. The *dotted line* shows the flux ratio of a typical brown dwarf photosphere

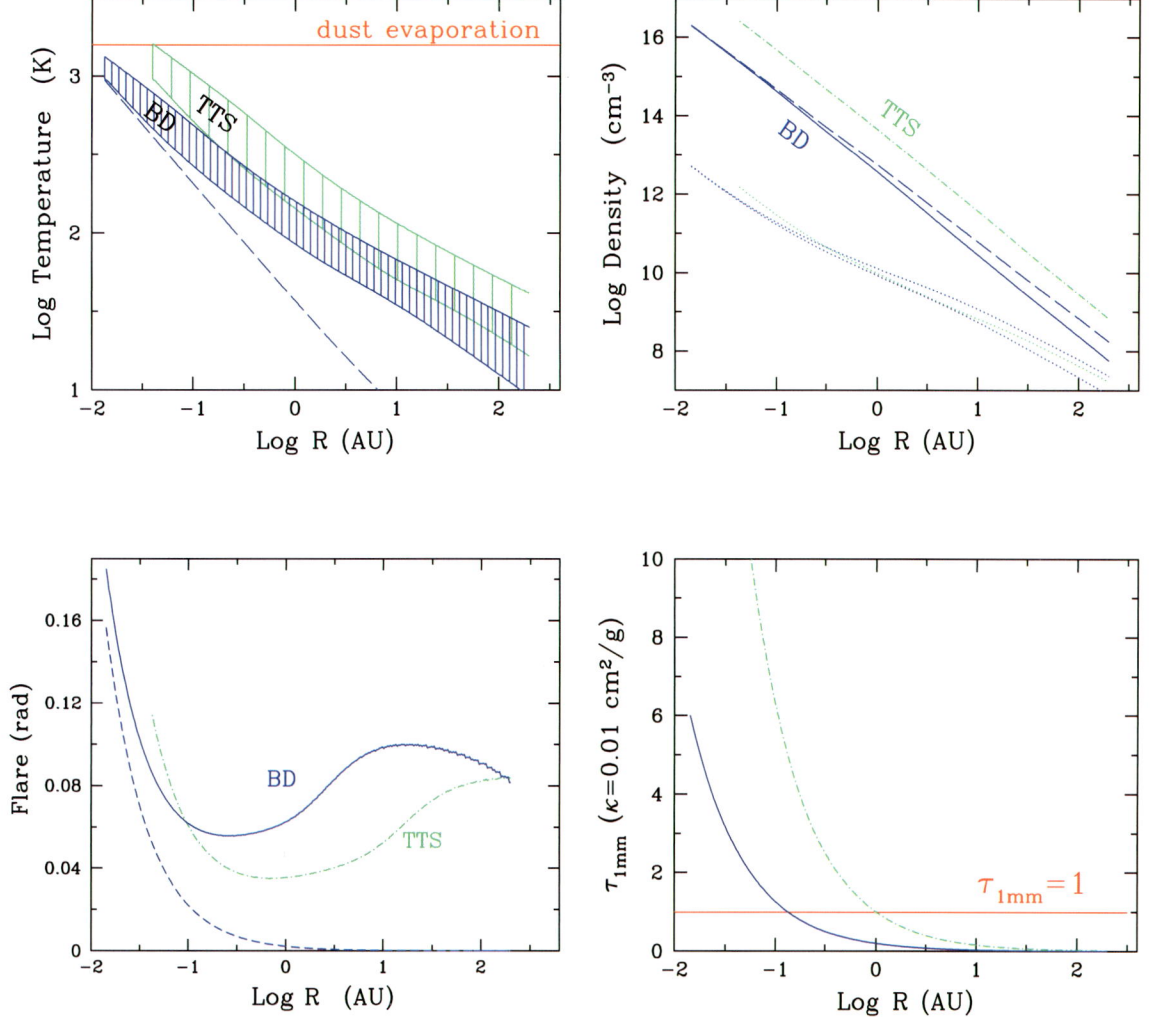

Fig. 2 Physical properties of typical brown dwarf disks, compared to those of a TTS disk. In all cases, the disk mass is about 8% of the central object mass. The *top-left panel* shows the temperature as function of radius. At each radius, the temperature ranges from a maximum value on the disk surface to a minimum value on the midplane (*shaded regions*). The *dashed line* is the midplane temperature for a flat brown dwarf disk. The surface temperature does not depend on the disk geometry. The *bottom-left panel* shows the flaring angle for the same disks. The *solid line* is a flared brown dwarf disk, the *dashed line* a flat one, the *dash-dotted* line a TTS flared disk. The *top-right panel* shows the density in the midplane for the same disks; the *solid curve* is for a brown dwarf flared model, the *dashed curve* a flat one, the dash-dotted curve a flared TTS disk. The three very similar *dotted lines* at the bottom plot the density at the disk surface, where most of the stellar radiation is absorbed. The *bottom-right panel* shows the run of the 1 mm optical depth for the brown dwarf disks (*solid line*) and the TTS disk (*dash-dotted line*) seen face-on

disks are more "evolved" than their TTS analogs, i.e., that they may be an optimal place to look for evidence of forming planets. However, if indeed brown dwarf disks are not fully flared, they should also be cooler and weaker at all wavelengths. Figures 2 and 4 illustrate the expected physical structure of brown dwarf disks and the corresponding SED for the two extreme cases of fully flared and flat geometries.

Note that a fully flared brown dwarf disk is actually more flared than an analogous T Tauri disk, and that, even if its mass is probably lower by a large factor (see following), the density in the midplane and surface are always very high.

3.2 Millimeter wavelengths

The capability to measure and study the emission of disks at millimeter wavelengths has been fundamental for understanding disks around TTS and intermediate mass stars. A large fraction of these disks have been detected with submillimeter and millimeter single dish and interferometers and a few of them have been spatially resolved. For a large number of disks, we have been able to measure (albeit with very large uncertainties, due to the uncertainties on the grain opacity) the disk mass. In several cases, there is evidence that most of the original solid mass is now in very large grains (see, e.g., Natta et al. 2007 and references therein),

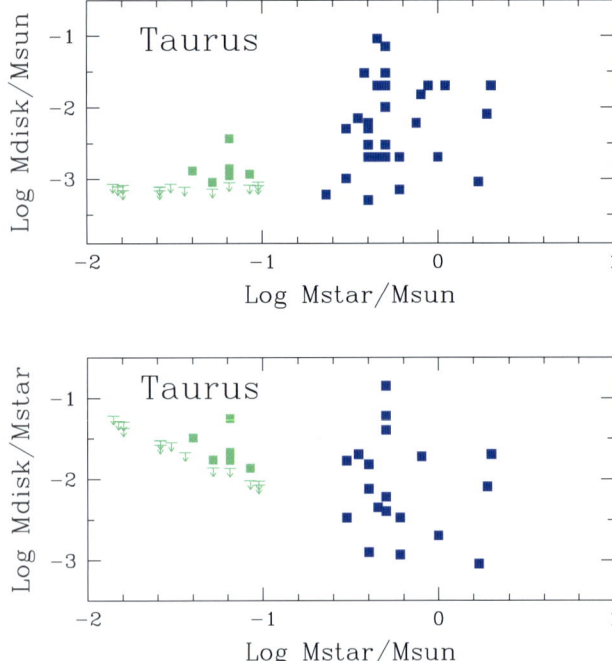

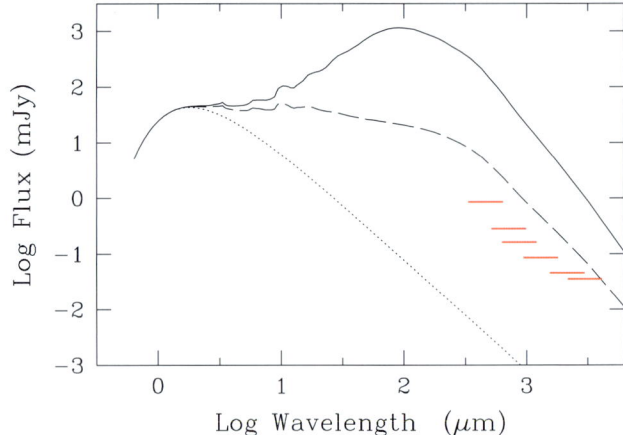

Fig. 4 Spectral energy distribution (SED) for brown dwarf disks with fully flared (*solid curve*) and flat (*dashed curve*) geometries. The central object has mass 0.05 $M_\odot$, luminosity 0.07 $L_\odot$, distance 140 pc. The disk mass is $\sim$10% of the brown dwarf mass. At wavelengths >450 μm, the opacity in the disk midplane is a power law $\kappa = 0.01$ $(\lambda/1 \text{ mm})^{-1}$ cm^2/g of gas. The *dotted line* shows the photospheric contribution, described, for simplicity, by a black body at $T = 2700$ K. The horizontal markers show the ALMA sensitivity (5σ in 1 hour integration time) in the various bands

Fig. 3 *Top panel*: disk mass as function of the mass of the central object for Taurus objects. Values for stars (M > 0.1M$_\odot$) are from Andrews and Williams (2005), for brown dwarfs from Scholz et al. (2006), normalized to the same value of the opacity ($\kappa_{300\,\mu\text{m}} = 10$ cm^2/g). Only one of the brown dwarf measurements is a 3σ detection. *Bottom panel*: same for the ratio of the disk to the stellar mass

suggesting that disks are already quite evolved. In a few selected objects (the brightest) Keplerian rotation pattern has been detected. Very recently, deviations from the Keplerian pattern and structures (spiral arms?) in the mass distribution have been detected in some disks, as summarized in these proceedings by S. Guilloteau. One can expect that ALMA will provide detailed disk images at various wavelengths for most TTS disks in nearby star forming regions.

So far, only very few disks around brown dwarfs have been detected at millimeter wavelengths. In Taurus, Scholz et al. (2006) have observed a group of 20 brown dwarfs at 1.3 mm, detecting 6 of them (only one at 3σ level). The corresponding masses are shown in Fig. 3 as a function of the mass of the central object, together with a group of well known TTS (Andrews and Williams 2005). The brown dwarf disk masses are of the order of few M_J, similar to the lower values for TTS disks. The bottom panel of Fig. 3 shows the ratio of the disk to the stellar mass for the same sample. One can clearly see that only disks relatively massive with respect to the central object can be detected with the current instruments.

3.3 ALMA potential

ALMA will provide a major step forward in our understanding of disks around very low mass objects, even if with sig-

nificant limitations. As shown in Fig. 4, it will be possible to measure the disk emission at all ALMA wavelengths for a large fraction of brown dwarf disks in nearby star forming regions. The examples in Fig. 4 have been computed for a rather massive disk ($\sim$10% of the central brown dwarf mass, which is 0.05 M$_\odot$ in this particular example), and a relatively large opacity ($\kappa = 0.01$ $(\lambda/1 \text{ mm})^{-1}$ cm^2/g of gas). For this combination of parameters, brown dwarf disks of all geometry will be easily detected at all wavelengths. Lower values of the product disk mass $\times$ opacity will reduce the flux roughly proportionally, and make disks with very large grains (reduced opacity) or less massive hard to detect. Disks around planetary-mass objects may be difficult to detect, unless they are fully flared; however, those with the highest ratio of the disk to central mass may be observable.

Spatially resolving brown dwarf disks will be possible in few, favorable cases. Figure 5 shows the intensity profile at 450, 850, 1300 and 3000 microns for the same disk models shown in Fig. 4. One can see that only a small disk region (roughly between 15 and 40 AU for a distance of 140 pc) can be resolved with a resolution of 0.1 arcsec, and only if the disk is fully flared.

Summarizing our considerations, we expect that ALMA will detect most disks around brown dwarfs and some (the most favorable ones) around planetary mass objects in nearby star forming regions. In many cases, it will be possible to measure the disk fluxes at all ALMA wavelengths and to have information on the dust properties and on the disk mass, in analogy to what is possible now for TTS disks. Spatially resolving the continuum emission of disks will be

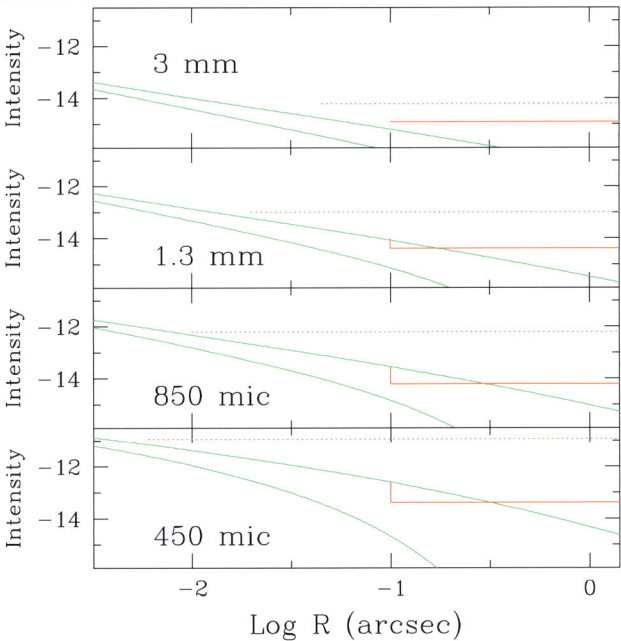

Fig. 5 Intensity profile of disks around brown dwarfs assumed to be at a distance of 140 pc (same as Fig. 4) at 3 mm, 1.3 mm, 850 μm and 450 μm, as labeled. In each panel, the top curve shows the predicted intensity of a fully flared disk, the bottom one that of a geometrically flat disk. The *dotted horizontal line* shows the ALMA sensitivity at the maximum resolution; the *solid horizontal line* that at a resolution of 1 arcsec (5σ in 1 hour). One can see that none of these disks is observable at the maximum resolution. With a resolution of 1 arcsec, it will be possible to detect and resolve fully flared disks in a region which will extend between ∼15 and 40 AU at 450 and 850 μm, and ∼15 and 23 AU at 1.3 mm, while it will be impossible to detect them at longer wavelengths

only feasible in a fraction of cases, again as is currently the case for TTS.

References

Allers, K.N., Kessler-Silacci, J.E., Cieza, L.A., Jaffe, D.T.: Young, low-mass brown dwarfs with mid-infrared excesses. Astrophys. J. **644**, 634 (2006)

Andrews, S.M., Williams, J.P.: Circumstellar dust disks in Taurus-Auriga: the submillimeter perspective. Astrophys. J. **631**, 1134 (2005)

Apai, D., Pascucci, I., Bouwman, J., Natta, A., Henning, Th., Dullemond, C.P.: The onset of planet formation in brown dwarf disks. Science **310**, 864 (2005)

Comerón, F., Rieke, G.H., Claes, P., Torra, J., Laureijs, R.J.: ISO observations of candidate young brown dwarfs. Astron. Astrophys. **335**, 522 (1998)

D'Alessio, P., Calvet, N., Hartmann, L.: Accretion disks around young objects, III: grain growth. Astrophys. J. **553**, 221 (2001)

D'Alessio, P., Calvet, N., Hartmann, L., Franco-Hernández, R., Servín, H.: Effects of dust growth and settling in T Tauri disks. Astrophys. J. **638**, 314 (2006)

Dullemond, C.P., Dominik, C.: Dust coagulation in protoplanetary disks: a rapid depletion of small grains. Astron. Astrophys. **434**, 971 (2005)

Hernández, J., Hartmann, L., Megeath, T., Gutermuth, R., Muzerolle, J. et al.: A Spitzer space telescope study of disks in the young σ Orionis cluster. Astrophys. J. **662**, 1067 (2007)

Lada, C.J., Muench, A.A., Luhman, K.L., Allen, L., Hartmann, L., et al.: Spitzer observations of IC 348: the disk population at 2–3 million years. Astron. J. **131**, 1574 (2006)

Luhman, K.L., Joergens, V., Lada, C., Muzerolle, J., Pascussi, I., White, R.: The formation of brown dwarfs: observations. In: Reipurth, B., Jewitt, D., Keil, K. (eds.) Protostars and Planets V, p. 443. The University of Arizona Press, Tucson (2007)

Mohanty, S., Jayawardhana, R., Natta, A., Fujiyoshi, T., Tamura, M., Barrado y Navascués, D.: Flared disks and silicate emission in young brown dwarfs. Astrophys. J. **609**, L33 (2004)

Motte, F., André, P., Neri, R.: The initial conditions of star formation in the rho Ophiuchi main cloud: wide-field millimeter continuum mapping. Astron. Astrophys. **336**, 150 (1998)

Muzerolle, J., Adame, L., D'Alessio, P., Calvet, N., Luhman, K.L., et al.: 24 μm detections of circum(sub)stellar disks in IC 348: grain growth and inner holes? Astrophys. J. **643**, 1003 (2006)

Natta, A., Testi, L.: Exploring brown dwarf disks. Astron. Astrophys. **376**, L22 (2001)

Natta, A., Testi, L., Comerón, F., Oliva, E., D'Antona, F., Baffa, C., Comoretto, G., Gennari, S.: Exploring brown dwarf disks in rho Ophiuchi. Astron. Astrophys. **393**, 597 (2002)

Natta, A., Testi, L., Calvet, N., Henning, Th., Waters, R., Wilner, D.: Dust in protoplanetary disks: properties and evolution. In: Reipurth, B., Jewitt, D., Keil, K. (eds.) Protostars and Planets V, p. 767. The University of Arizona Press, Tucson (2007)

Scholz, A., Jayawardhana, R., Wood, K.: Exploring brown dwarf disks: a 1.3 mm survey in Taurus. Astrophys. J. **645**, 1498 (2006)

Walsh, A.J., Myers, P.C., Di Francesco, J., Mohanty, S., Bourke, T.L., Gutermuth, R., Wilner, D.: A large-scale survey of NGC 1333. Astrophys. J. **655**, 958 (2007)

Whitworth, A., Bate, M.R., Nordlund, A., Reipurth, B., Zinnecker, H.: The formation of brown dwarfs: theory. In: Reipurth, B., Jewitt, D., Keil, K. (eds.) Protostars and Planets V, p. 459. The University of Arizona Press, Tucson (2007)

A submillimeter view of protoplanetary dust disks

Sean M. Andrews · Jonathan P. Williams

Originally published in the journal Astrophysics and Space Science, Volume 313, Nos 1–3.
DOI: 10.1007/s10509-007-9614-5 © Springer Science+Business Media B.V. 2007

Abstract We present some results from our submillimeter single-dish and aperture synthesis imaging surveys of protoplanetary disks using the JCMT, CSO, and Submillimeter Array (SMA) on Mauna Kea, Hawaii. Employing a simple disk model, we simultaneously fit the spectral energy distributions and spatially resolved submillimeter continuum emission from our SMA survey to constrain disk structure properties, including surface density profiles and sizes. The typical disk structure we infer is consistent with a fiducial accretion disk model with a viscosity parameter $\alpha \approx 0.01$. Combined with a large, multiwavelength single-dish survey of similar disks, we show how these observations provide evidence for significant grain growth and rapid evolution in the outer regions of disks, perhaps due to an internal photo-evaporation process. In addition, we discuss SMA observations of the disks in the Orion Trapezium (proplyds) in the context of disk evolution in a more extreme environment.

Keywords Stars: circumstellar matter · Planetary systems: protoplanetary disks

1 Introduction

Circumstellar disks play integral roles in early stellar evolution and the genesis of planetary systems. A key goal of star formation research has been to place some constraints on the origins of planets by studying the properties and evolution of the circumstellar disks from which they form. While direct observations of the planet formation process remain just out of reach for now, the evolutionary behavior of disk properties can be used to help infer the conditions, processes, and timescales involved in producing planetary systems. In measuring disk properties, observations at submillimeter wavelengths are critical because the continuum emission is a diagnostic of the spatial distribution of mass and the range of dust particle sizes in the disk. Since the outer regions of a disk are cool and emit primarily at these wavelengths, this emission can be spatially resolved with an interferometer.

The focus of our work has been to learn about the dominant mechanisms that govern the evolution of protoplanetary disks: viscous accretion onto the central star, photoevaporation, and the growth of solid particles. This article highlights some of our work in this area, centered around three basic categories:

1. Basic disk properties and evolution: a statistical look at the distribution of masses, the spectral behavior of opacities, and radial evolution with a large multi-wavelength photometry survey;
2. Structure of the outer disk: placing constraints on densities, temperatures, opacities, and sizes using a large interferometric survey to investigate viscous accretion properties and planet formation prospects;
3. The effects of extreme environments: studying the impact of harsh external irradiation from nearby massive stars on some basic disk properties.

2 A statistical look at young dust disks

The cool outer parts of disks, representing the vast majority of their mass and volume, emit thermal continuum emission

S.M. Andrews (✉) · J.P. Williams
University of Hawaii Institute for Astronomy, Honolulu,
HI 96822, USA
e-mail: andrews@ifa.hawaii.edu

J.P. Williams
e-mail: jpw@ifa.hawaii.edu

at submillimeter wavelengths. Since most of this emission is optically thin, the submillimeter luminosity is a diagnostic of the total disk mass (Beckwith et al. 1990) and the shape of the long-wavelength spectral energy distribution (SED) is related to the spectral behavior of the disk opacity (Beckwith and Sargent 1991). To study these properties in a statistical sense, we used the SCUBA and SHARC-II cameras on the JCMT and CSO telescopes, respectively, to conduct the largest and most sensitive multiwavelength (350, 450, and 850 μm) submillimeter photometry survey of disks to date (Andrews and Williams 2005).

With a sample of 153 disks in the Taurus-Auriga star formation region, we determined the distribution of disk masses over ∼3 orders of magnitude. The median disk is 0.5% as massive as the central star, with a typical mass of 0.005 $M_\odot$. This is a factor of 2–3 less than the expected minimum mass for the primordial disk that gave rise to the Solar System (Weidenschilling 1977). The submillimeter SED follows a power-law $F_\nu \propto \nu^n$ with $n \approx 2$, much lower than the $n \approx 4$ seen in the ISM (Hildebrand 1983). The low n values for disks can be caused by optically thick emission or the inapplicability of the Rayleigh-Jeans criteria, but may also be associated with an altered disk opacity due to the collisional growth of dust grains (Draine 2006; D'Alessio et al. 2006). Figure 1 shows significant decreases in the masses and submillimeter colors (n) of disks along

an empirical evolution sequence based on the infrared SED. These trends may be due to changes in disk structure, a change in the submillimeter emission efficiency due to particle growth, or both.

The relative detection frequencies of these disks in the infrared (warm inner disk) and submillimeter (cool outer disk) are statistically identical. We suggested that the scarcity of transitional disks that have cleared inner regions and show only submillimeter emission implies that the timescale over which disk signatures disappear *at all radii* is short, ∼10^5 yr. This rapid evolution timescale may be caused by the combined effects of viscous accretion in the inner disk and the photoevaporation of the outer disk by high-energy irradiation from the central star (Alexander et al. 2006).

ALMA will play a key role in determining the radial evolution timescales in the outer parts of circumstellar disks. High resolution, sensitive image mosaics of more distant star clusters with a range of ages are required to track the decline of submillimeter disk luminosities as a function of time. The results from such a census will provide an important complement to infrared observations (Haisch et al. 2001) and should turn up a substantial population of transition disks that would allow a more detailed study of the metamorphosis of disk material into planetary systems.

3 Resolving disk structure

To build on the results of our single-dish survey with a more detailed examination of disk properties, we used the SMA interferometer to conduct the largest ever high spatial resolution submillimeter continuum survey of young disks (Andrews and Williams 2007). Aperture synthesis images at either 880 μm or 1.3 mm for 24 disks are shown in Fig. 2. By simultaneously modeling the SMA continuum visibilities and SEDs, we determined key disk structure parameters including their sizes and radial distributions of temperatures and surface densities. The results indicate that a typical disk has a radius of ∼200 AU, a temperature distribution $T \propto r^{-q}$ with $q \approx 0.6$, a density distribution $\Sigma \propto r^{-p}$ with $p \approx 1$, and an opacity $\kappa \propto \nu^\beta$ with $\beta \approx 1$.

These results are in good agreement with accretion disk models where the viscosity varies linearly with radius (Hartmann et al. 1998). In the context of those α-disk models we showed that these disks have a typical level of turbulent viscosity consistent with $\alpha \approx 0.01$ using a wide range of observations including the median spatial density distribution, SED, and submillimeter continuum surface brightness profile, along with the variations of disk sizes, submillimeter flux densities, and accretion rates with stellar age. Some of the latter trends are shown in Fig. 3, including the expected behavior of fiducial accretion disk models with different levels of viscosity. The value of α plays a key role in setting

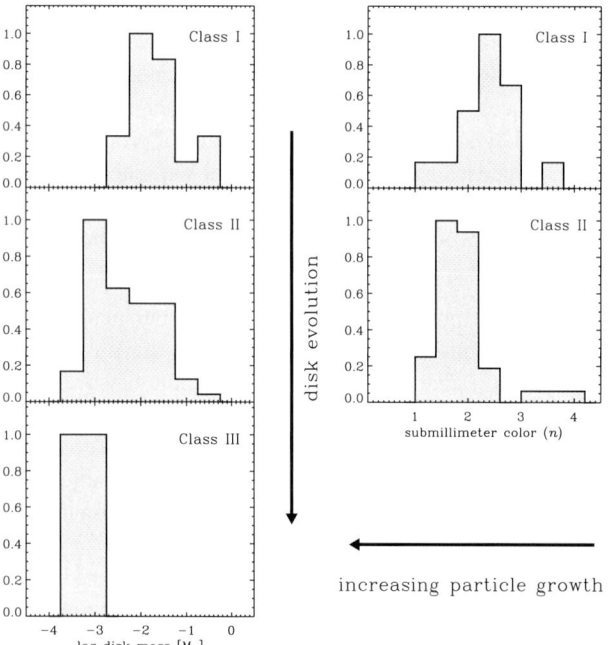

Fig. 1 Normalized histograms of disk masses (*left*) and submillimeter colors (*n*; *right*) from a multiwavelength photometry survey of Tau-Aur members (Andrews and Williams 2005). The different panels represent disk evolution classifications based on the infrared SED. There is empirical evidence for decreasing disk masses and increasing particle growth as disks evolve

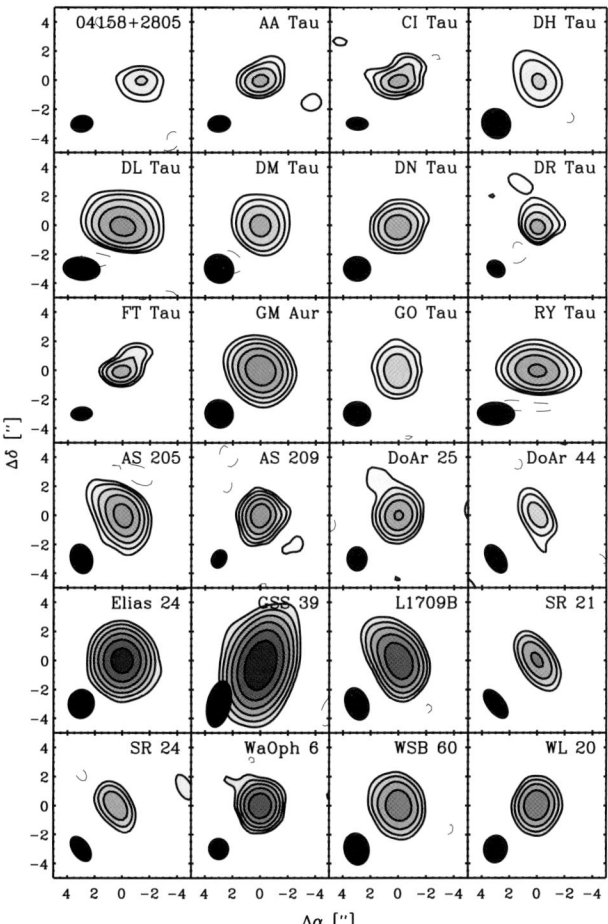

Fig. 2 Submillimeter continuous images from an SMA survey of T Tauri disks (Andrews and Williams 2007). Each panel is 10″ (~1500 AU) on a side

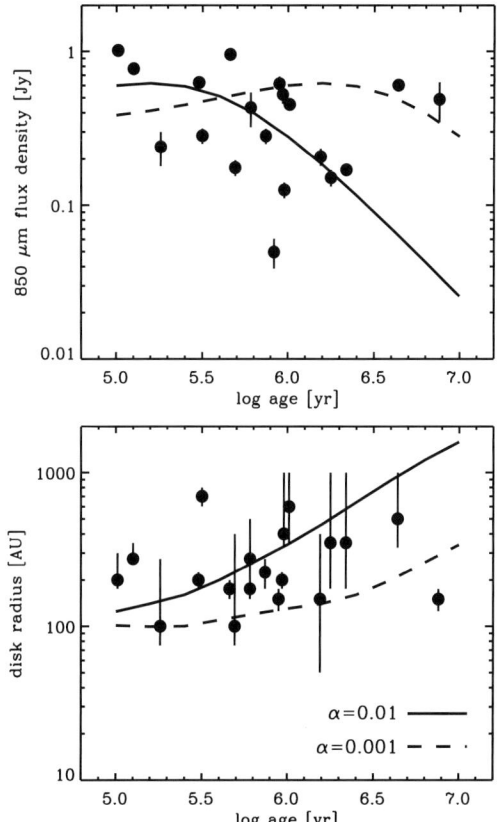

Fig. 3 The 850 μm flux density (*top*) and disk radius (*bottom*) as a function of stellar age. Overlaid are the expected behaviors for a fiducial viscous accretion disk model with two values of the viscosity parameter, $\alpha = 0.01$ (*solid*) and 0.001 (*dashed*). The constraints on disk structure from resolved submillimeter data can potentially help determine the typical level of turbulent viscosity in these disks, which sets the internal timescale for their evolution

the viscous timescale for the evolution of disk structure, and therefore the window of opportunity when disk densities are high enough to initiate the planet formation process.

In terms of the spectral behavior of the disk opacity, we noted that the derived value of β is significantly lower than in the ISM ($\beta_{ISM} \approx 1.7$; Draine 2006). Doing so required high resolution constraints on the disk structure to properly account for the contribution of optically thick emission from the inner disk. The derived β values are consistent with the effects of solid particle growth to at least millimeter size scales (Draine 2006; D'Alessio et al. 2006). Extrapolating the inferred surface density distributions into the unresolved planet formation region would lead us to believe that the values are too low to produce planets by the traditional mechanisms (Pollack et al. 1996). However, this is not necessarily the case. The assumed opacity values for these models are only relevant if the maximum particle sizes in the disk are ~1 mm: larger particles would result in lower opacities and larger densities (D'Alessio et al. 2006). The signatures of such particle growth in these disks may be manifest in the spatial distributions of their submillimeter colors. High spatial resolution multiwavelength observations of young disks with ALMA will aid in addressing this issue, and therefore in determining the potential that any given disk has for developing a planetary system.

4 Environmental impact on disk properties

In general, the disks studied in the work described above are representative of a simple picture where only *internal* processes affect their properties and evolution. However the majority of stars in the Galaxy, likely including the Sun, were created in environments that have a high potential to *externally* influence the development of their accompanying disks. One of the most relevant sources of such an environmental impact is the intense ultraviolet radiation from nearby massive stars.

The disks around low-mass stars in Orion are subject to intense UV irradiation from the OB stars in the Trapezium

Astrophys Space Sci (2008) 313: 119–122

Fig. 4 (*left*) 880 µm continuum map of the proplyds in the Orion Trapezium (contours at 3, 5, and 7 σ). Star symbols mark the OB stars in the Trapezium, and crosses denote proplyds seen at optical wavelengths. (*right*) The corresponding optical image from *HST* (Bally et al. 1998) with SMA contours overlaid

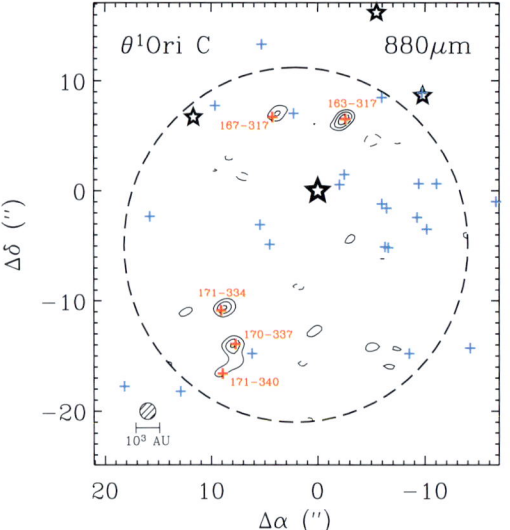 

region. Given the photoevaporation rates for these disks, it seemed unlikely that enough raw material would still be available for the planet formation process to run its natural course. We used the increased resolution and sensitivity of the SMA interferometer to observe these photoevaporating disks (see Fig. 4) and provide the first estimates of their masses, which turned out to be similar to disks in the comparatively quiescent Taurus region (Williams et al. 2005). Depending on the orbits of these disks in the Trapezium, the data suggest that planet formation may not be doomed after all in such a harsh environment.

ALMA will be a crucial tool to study disks in these more extreme environments. Its capabilities will permit us to resolve the detailed structure of disks like the proplyds, representing a significant step forward in understanding how external effects factor into the efficiency and viability of the planet formation process.

References

Alexander, R.D., Clarke, C.J., Pringle, J.E.: Photoevaporation of protoplanetary disks II: evolutionary models and observable properties. Mon. Not. R. Astron. Soc. **369**, 229 (2006)

Andrews, S.M., Williams, J.P.: Circumstellar dust disks in Taurus-Auriga: the submillimeter perspective. Astrophys. J. **631**, 1134 (2005)

Andrews, S.M., Williams, J.P.: High resolution submillimeter constraints on circumstellar disk structure. Astrophys. J. **695**, 705 (2007)

Bally, J., Sutherland, R.S., Devine, D., Johnstone, D.J.: Externally illuminated young stellar environments in the Orion Nebula: Hubble space telescope planetary camera and ultraviolet observations. Astron. J. **116**, 293 (1998)

Beckwith, S.V.W., Sargent, A.I.: Particle emissivity in circumstellar disks. Astrophys. J. **381**, 250 (1991)

Beckwith, S.V.W., Sargent, A.I., Chini, R.S., Güsten, R.: A survey for circumstellar disks around young stellar objects. Astron. J. **99**, 924 (1990)

D'Alessio, P., Calvet, N., Hartmann, L., Franco-Hernández, R., Servín, H.: Effects of dust growth and settling in T Tauri disks. Astrophys. J. **638**, 314 (2006)

Draine, B.T.: On the submillimeter opacity of protoplanetary disks. Astrophys. J. **636**, 1114 (2006)

Haisch, K.E., Lada, E.A., Lada, C.J.: Disk frequencies and lifetimes in young clusters. Astrophys. J. **553**, L153 (2001)

Hartmann, L., Calvet, N., Gullbring, E., D'Alessio, P.: Accretion and the evolution of T Tauri disks. Astrophys. J. **495**, 385 (1998)

Hildebrand, R.H.: The determination of cloud masses and dust characteristics from submillimetre thermal emission. Q. J. R. Astron. Soc. **24**, 267 (1983)

Pollack, J.B., Hubickyj, O., Bodenheimer, P., Lissauer, J.J., Podolak, M., Greenzweig, Y.: Formation of the giant planets by concurrent accretion of solids and gas. Icarus **124**, 62 (1996)

Weidenschilling, S.J.: The distribution of mass in the planetary system and solar nebula. Astrophys. Space Sci. **51**, 153 (1977)

Williams, J.P., Andrews, S.M., Wilner, D.J.: The masses of the Orion proplyds from submillimeter dust emission. Astrophys. J. **634**, 495 (2005)

Studies of dense cores with ALMA

Mario Tafalla

Originally published in the journal Astrophysics and Space Science, Volume 313, Nos 1–3.
DOI: 10.1007/s10509-007-9630-5 © Springer Science+Business Media B.V. 2007

Abstract Dense cores are the simplest star-forming sites that we know, but despite their simplicity, they still hold a number of mysteries that limit our understanding of how solar-type stars form. ALMA promises to revolutionize our knowledge of every stage in the life of a core, from the pre-stellar phase to the final disruption by the newly born star. This contribution presents a brief review of the evolution of dense cores and illustrates particular questions that will greatly benefit from the increase in resolution and sensitivity expected from ALMA.

Keywords ISM: clouds · ISM: molecules · ISM: jets and outflows · Stars: formation

1 Introduction

Nearby dark clouds like Taurus and Perseus contain dozens of dense molecular cores where stars like our Sun are currently forming or have done so in the recent past (Myers 1995). Their large number, together with their proximity and simple structure, make cores unique targets to study the complex physics involved in the formation of a star. Dense cores that have not yet formed stars, the so called starless or pre-stellar cores, inform us of the initial conditions of star formation, and their study can help us elucidate the process by which pockets of cloud material condense and become gravitationally unstable. Cores with deeply embedded young stellar objects ("protostellar cores") are unique targets to study the complex motions that occur during the period of accretion, when a combination of infall, outflow, and rotation is necessary to assemble the star and redistribute the gas angular momentum. Finally, evolved cores are primary targets to study the interaction between the newly born star and its environment. These feedback effects are responsible for the transition of the protostar from embedded to visible, and may be important determining the final mass of the star and stabilizing the nearby gas via turbulence generation.

The observational study of dense cores has advanced enormously over the last decade thanks to the increase in resolution provided by the new millimeter and submillimeter interferometers, and also due to the systematic combination of observations of dust and molecular tracers (e.g., Bergin and Tafalla 2007). This brief review summarizes some new results from dense cores studies and presents a number of current issues that will greatly benefit from ALMA observations. The limited space of this article makes any attempt to review the field necessarily incomplete, and the reader is referred for further information to the other contributions on star formation in these proceedings, in particular to those by van Dishoeck, André, Shepherd, Aikawa, Wilner, Johnstone, and Crutcher.

Despite significant recent progress, our understanding of the structure and evolution of dense cores is still incomplete due in part to limitations in the resolution and sensitivity of the available observations. Even the highest resolution data of nearby dense cores cannot discern details finer than about 100 AU, which is still insufficient to disentangle the complex kinematics of infall and outflow motions in the vicinity of a protostar. Probably more important, the low temperatures of the gas and the dust in cores (≈ 10 K) make the emission of any core tracer intrinsically weak, so any increase in the resolution needs to be accompanied by a par-

M. Tafalla (✉)
Observatorio Astronómico Nacional, Alfonso XII 3, E-28014
Madrid, Spain
e-mail: m.tafalla@oan.es

allel increase in the sensitivity, or the observations will not achieve enough S/N to provide useful information. This is particularly important when using weak, optically thin tracers to sample the innermost gas in the core. These tracers, in addition, often present extended emission, which poses a problem to the current generation of interferometers that cover sparsely the uv plane and therefore suffer systematically from missing flux. The high resolution and collecting area afforded by ALMA, combined with its great sensitivity to extended emission, promises to revolutionize the field of dense cores studies. On the one hand, ALMA will allow studying the dense cores of nearby clouds with the greatest detail, achieving subarcsecond resolution with high sensitivity. On the other hand, ALMA will permit the systematic study of dense cores in more distant clouds, enlarging the sample of available targets from the current set of the nearest clouds to cores at distances of at least 1 kpc.

2 Pre-stellar cores

The earliest phase of a core, the so-called starless or pre-stellar stage, is characterized by the lack of a point-like object at its center (e.g., Di Francesco et al. 2007). This characterization is of course dependent on the current sensitivity limits of the observations, and is therefore susceptible of misclassifying a core with an embedded source of very low luminosity (see the case of VeLLOs below). Still, the significant number of dense cores with no pointlike source detected even after deep Spitzer Space Telescope observations suggests that a population of truly starless cores exists in nearby clouds like Taurus (Werner et al. 2006).

Starless cores present systematically a close to constant density of 10^5–10^6 cm^{-3} over the central 5000–10000 AU followed by an almost power-law drop at large distances. This central flattening of the density profile has been observed in a number of cores using different observational techniques, like millimeter dust continuum emission (Ward-Thompson et al. 1999), MIR absorption (Bacmann et al. 2000), and NIR extinction (Alves et al. 2001), and therefore constitutes a robust result of recent core studies. The presence of a density flattening provides further evidence that starless cores have not yet developed a central singularity, and that they are of pre-stellar nature. The physical origin of the flattening, however, is still a matter of debate, as a number of interpretations are consistent with it. The most natural one is that the profile results from an equilibrium configuration in which the pressure of an isothermal gas balances its gravitational attraction, the so called Bonnor-Ebert profile (e.g., Alves et al. 2001). Indeed, the gas temperature in a core is typically close to constant (≈ 10 K), and the associated thermal pressure dominates the turbulent component by a factor of several (e.g., Tafalla et al. 2004). The Bonnor-Ebert interpretation, however, seems in conflict with the

non-spherical shape of most cores (typical axial ratio is 2:1, Myers et al. 1991), and with the fact that the density contrast observed in cores often exceeds the factor of 14 limit for stability of the Bonnor-Ebert analysis (Bacmann et al. 2000). Additional magnetic field support could be responsible for these deviations from the theoretical expectation, but unfortunately, the observation of this magnetic component is extremely hard to make (see contribution from Crutcher in this volume). Even the apparently "simple" structure of the cores still eludes our understanding.

When the density distribution of a core, as inferred from dust measurements, is compared with the observed emission from most molecular tracers, it is commonly found that they disagree significantly. As illustrated in Fig. 1 for L1498 in Taurus, the dust emission of a core often appears centrally concentrated (with of course a relative flattening at the center), while all molecular species but NH$_3$ and N$_2$H$^+$ present ring-like distributions around the continuum peak. Radiative transfer analysis of the molecular emission indicates that the abundance of most species drops by at least a factor of 10 towards the high density peak of the molecular core (Caselli et al. 1999; Bergin et al. 2002; Tafalla et al. 2002). Such strong abundance decrease is suffered by all the C-bearing molecules as well as other species (like SO), while it does not affect significantly NH$_3$ or N$_2$H$^+$ (see Di Francesco et al. 2007 and Bergin and Tafalla 2007 for reviews). NH$_3$ seems in fact to be enhanced toward the center of most cores (Tafalla et al. 2002), while the N$_2$H$^+$ abundance tends to have a constant value or may drop at the very center of some cores (Bergin et al. 2002; Pagani et al. 2005). Cores therefore have a differentiated (onion-like) molecular composition, with a center rich in NH$_3$ and N$_2$H$^+$ and a series of outer layers containing C-bearing species.

The inhomogeneous composition of the starless dense cores most likely results from the freeze out of the main molecular species onto the cold dust grains at the center (Bergin and Langer 1997; Aikawa et al. 2005). The high densities and low temperatures typical of dense core centers make the freeze out time ($\approx 5 \times 10^9/n_{H_2}$ yr) become much shorter than the core dynamical scale (≈ 1 Myr), and as a consequence, species like CO disappear rapidly from the gas phase. Other molecular species suffer the same fate as CO, but more importantly, the original chemical balance, characterized by a relative large CO abundance ($\sim 10^{-4}$), is changed dramatically by freeze out. A new chemical balance emerges, and it is characterized by the enhancement of certain N-bearing species, like N$_2$H$^+$, which are daughter products of N$_2$ and whose abundance is controlled by the amount of CO in the gas phase (CO is the main destroyer of N$_2$H$^+$). Even as N$_2$ freezes out on the dust grains with a similar binding energy as CO (Öberg et al. 2005), the N$_2$H$^+$ abundance can increase relatively from its value

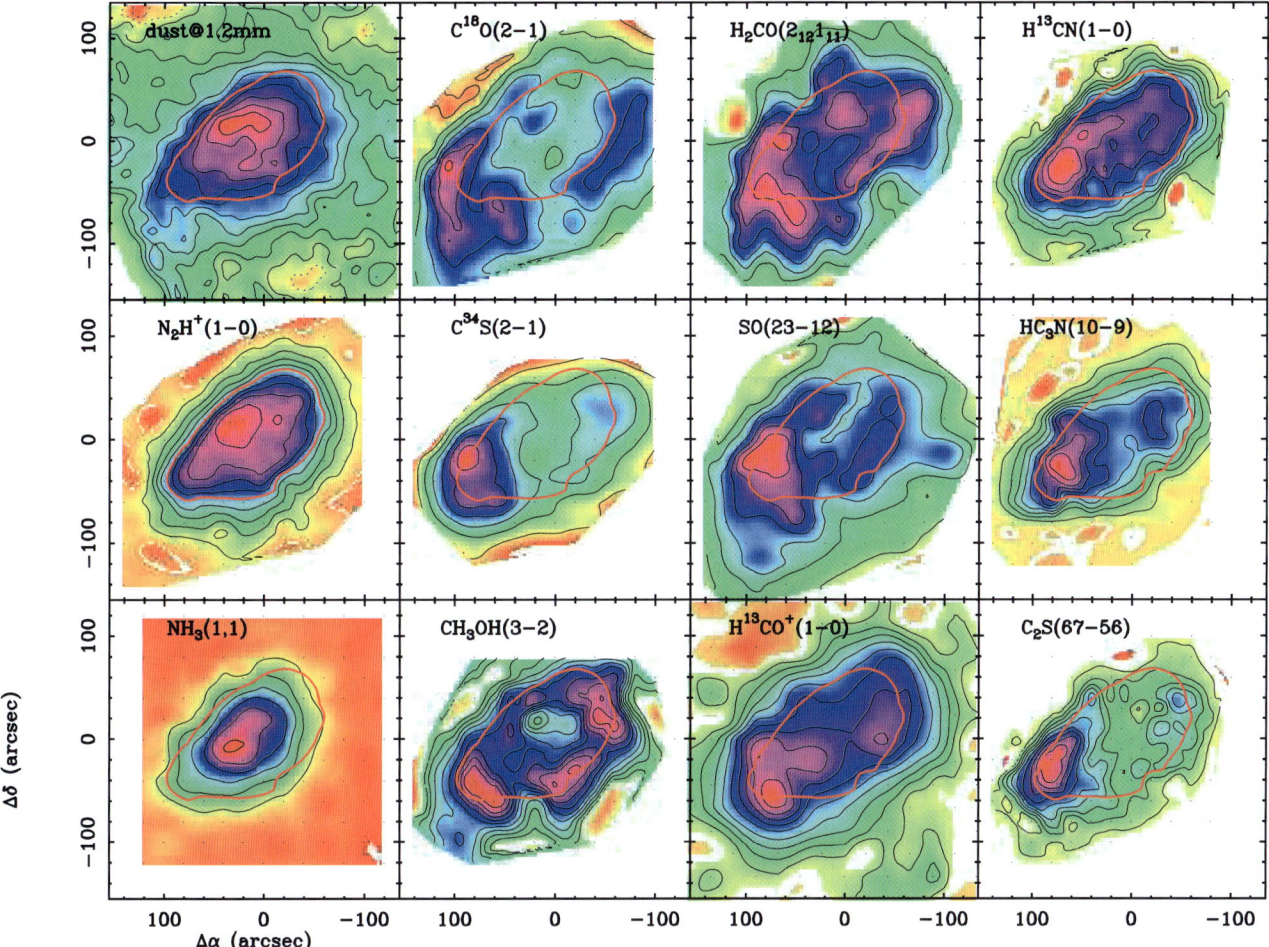

Fig. 1 Sample of maps of the L1498 dense core in Taurus illustrating its differentiated chemical composition. The *three panels* in the *left* show tracers that are sensitive to the core interior, and therefore present a centrally-concentrated emission (dust continuum map is in the *top left panel*). All *other panels* show ring-like distributions of emission that result from the depletion of the emitting molecules in the core interior (data from Tafalla et al. 2006)

in the diffuse cloud (where CO is undepleted) and give rise to the relatively "high" abundances (few 10^{-10}) typical of dense cores. NH_3 can then form from N_2H^+ via dissociative recombination (Geppert et al. 2004), giving rise to the observed central enhancement (Aikawa et al. 2005).

Another effect of the CO depletion in cores is the enhancement of deuterated species. Deuteration at the low (10 K) temperature of dense cores occurs via the enhancement of H_2D^+, which then passes the deuterium atom to other species via ion-molecule reactions (Dalgarno and Lepp 1984). As H_2D^+ is mainly destroyed by CO, the depletion of CO further enhances the H_2D^+ abundance, which in turn enriches in deuterium a number of additional species. High abundance of H_2D^+ has in fact been observed in the heavily CO-depleted dense core L1544 (Caselli et al. 2003), and a correlation of CO depletion and high deuteration has been reported by Bacmann et al. (2003) and Crapsi et al. (2005). This deuteration in the cold and dense pre-stellar phase is responsible for the extreme deuteration values of

species like H_2CO, CH_3OH, and NH_3 seen toward protostellar cores (Ceccarelli et al. 1998; Roueff et al. 2000; van der Tak et al. 2002).

3 From cores to protostars

As cores evolve, they are expected to become more and more centrally concentrated until they reach the point of gravitational instability. One of the most pressing issues in star formation studies is to understand whether this process of concentration is driven by the loss of magnetic field support via ambipolar diffusion (e.g., Shu et al. 1987; Mouschovias and Ciolek 1999) or by the dissipation of turbulence via shocks (e.g., MacLow and Klessen 2004). Observations of dense cores cannot yet distinguish between these scenarios, but do show a systematic correlation between central concentration and other indicators of evolution, like CO depletion and deuterium fractionation (Crapsi et al. 2005). Evidence for inward motions also seems correlated with central concentration, and this suggests that some cores that we see now as starless have already begun collapsing to form stars. One of the best candidates for such a collapsing system is the L1544 core in Taurus, whose pattern of inward motions has been studied in a number of molecules (Tafalla et al. 1998; Williams et al. 1999; Caselli et al. 2002). The L1544 dense core is characterized by a high central density and concentration (Ward-Thompson et al. 1999; Tafalla et al. 2002), a high degree of CO depletion and deuterium fractionation (Caselli et al. 1999, 2002), and seems starless despite deep Spitzer Space Telescope observations in the IR (Bourke, private communication). Clearly this core, an similar objects, will be prime targets for ALMA observations.

Cores more evolved than L1544 are expected to contain already a luminous object surrounded by an envelope of accreting material. The little observable difference between the pre and proto-stellar phases of a core is illustrated by the case of L1521F, a core initially thought from molecular data to be an almost twin of L1544 (Crapsi et al. 2004) and later found with Spitzer observations to have a luminous central star (Bourke et al. 2006). The central object in L1521F has a luminosity close to 0.1 $L_\odot$, and is characteristic of a new group of objects identified by the Spitzer telescope and usually referred as VeLLOs (Very Low Luminosity Objects). These VeLLOs seem associated with very weak NIR nebulosity and low velocity bipolar outflows (Bourke et al. 2005), and their status in the evolutionary sequence of protostars is still unclear. Although some VeLLOs could represent precursors of substellar objects ("proto brown dwarfs"), it seems more likely that in the case of L1521F we are witnessing the very first moments of accretion, when the central source has an extremely low mass. The proto brown dwarf alternative is unlikely in this case because the dense core has about 5 $M_\odot$ of mass (Crapsi et al. 2004), and no clear perturbation seems stopping the accretion (the outflow has too little mechanical power).

The pristine nature of VeLLOs makes them ideal candidates to study star-forming infall motions. The study of these motions has a long and rich tradition, and is plagued by difficulties as illustrated by the case of B335. This dense core harbors a very young (Class 0) object whose inward motions were first characterized by Zhou et al. (1993). These authors found that the spectral signatures from this core are in good agreement with the expectation from the inside-out collapse model of Shu (1977). High resolution observations with the Plateau de Bure Interferometer by Wilner et al. (2000), however, have shown that some of the signatures of "infall" (like the high velocity wings in the CS lines) arise in fact from outflow acceleration, and not from an increase in velocity of the infalling material as it approaches the central object. A revisit of B335 (and similar objects) making use of ALMA's high angular resolution and selecting appropriate (i.e., depletion resistant) tracers is therefore

needed to clarify the still confusing picture of star-forming infall motions. The clean appearance of some VeLLOs, together with their weaker outflow emission, offers an interesting alternative to the more evolved (and massive) objects like B335, that have fully developed outflows. Because of their lower mass, VeLLOs may present weaker signatures of infall and may be tracing the very first moments of collapse. The combined study of VeLLOs and more luminous Class 0 and Class I sources should therefore allow us to reconstruct the sequence of star-forming accretion as a function of time.

The presence of a protostar at the center of a core affects not only the gas kinematics but its chemistry. The newly born star heats up the nearby gas and dust introducing a temperature gradient in its vicinity. In the $\sim$1000 AU region where the dust temperature exceeds the CO evaporation temperature ($\approx$20–30 K), this molecule returns to the gas phase and undoes part of the chemical processing that occurred during the pre-stellar phase (Jørgensen et al. 2004; Jørgensen 2004). Closer to the protostar ($\sim$100 AU), the dust temperature reaches the 90–100 K value at which water evaporates from the grains, further enriching the chemistry. Observations of some very young protostellar objects, like IRAS 16293–2422, show that these very small regions have extreme abundance of a number of complex molecules like $HCOOH$, $HCOOCH_3$, and CH_3OCH_3 (Cazaux et al. 2003; Bottinelli et al. 2004). The chemical richness of these regions rivals that of the hot cores around massive protostars, justifying their common denomination as "hot corinos" (Ceccarelli et al. 2007). The exact origin of the complex molecules in these regions, however, is still not fully understood. One possibility is that they result from direct evaporation of species trapped in the water ice, while an alternative is that they result from the processing of simpler evaporated molecules. Even the geometry of hot corinos remains unknown, with the innermost part of the envelope or a more stable disk-like distribution as the most likely locations. Despite these temporary uncertainties, hot corinos offer a unique opportunity to study the innermost vicinity of low-mass protostars. Their distinctive chemical composition makes them highly selective tracers of the most complex and interesting region of the protostar, where inflow, outflow, and rotation motions play comparable roles, and angular momentum is transferred between different gas components. Hot corino studies with ALMA will surely constitute some of the first scientific projects of the instrument.

4 Outflow acceleration and core disruption

At the same time that protostars accrete material, they eject powerful bipolar outflows of supersonic speed. CO observations of these outflows reveal masses that are too large to originate directly from the central protostar, and indicate that most of the moving gas is core ambient material accelerated by a collimated stellar wind (Lada 1985). The lobes of bipolar outflows, in addition, commonly coincide with evacuated cavities seen via scattered light from the protostar, further illustrating how the outflow phenomenon represents a major disruption in the core internal structure (Padgett et al. 1999).

Despite more than two decades of intense outflow research, a number of outstanding problems remain, and ALMA observations represent our current best hope to solve them (see also contribution by D. Shepherd in this volume). The properties of the underlying wind, for example, are not yet understood, and several alternative models have been proposed over the years. The two main types of models that attempt to fit the observations are the jet-driven outflow and the wind-driven shell, each of them with a number of flavors (see Bachiller 1996 for a review). Despite significant successes, however, neither type of model can reproduce the rich variety of kinematic properties found by observations, so each of them is necessarily incomplete (Lee et al. 2002). In the jet driven model, a highly collimated agent shocks and sweeps cloud material along an almost straight line. This model succeeds in explaining the highly collimated CO outflows often found toward Class 0 objects, but fails to reproduce observations of less collimated flows (usu-

ally powered by Class I sources), where the CO emission arises from gas along limb-brightened shells (like L1551, see Moriarty-Schieven et al. 1987). To fit these less collimated systems, the jet models need to broaden the outflow path, and this has been done by either invoking jet precession/"wandering" (Masson and Chernin 1993) or large-scale bow shocks (Raga and Cabrit 1993). None of these elements however seems consistent with observations (see Arce et al. 2007 for more details), and this leaves the jet models limited to fitting the youngest, and admittedly more spectacular, bipolar outflows. Wind-driven models, on the other hand, naturally produce shell-like structures thanks to a wide-angle agent that sweeps ambient material (Shu et al. 1991). These models, unfortunately, do not reproduce the appearance of the highly collimated outflows or the mass-velocity distribution commonly observed even in the poorly collimated flows (Masson and Chernin 1992).

A combination of high resolution observations and new developments in outflow modeling are starting to show a possible solution to the current impasse. Interferometer mapping of the outflow powered by the very young source IRAS 04166+2706 in Taurus shows both jet and shell features simultaneously (see Fig. 2 and poster contribution by Santiago-García et al.). The jet-like feature in this outflow,

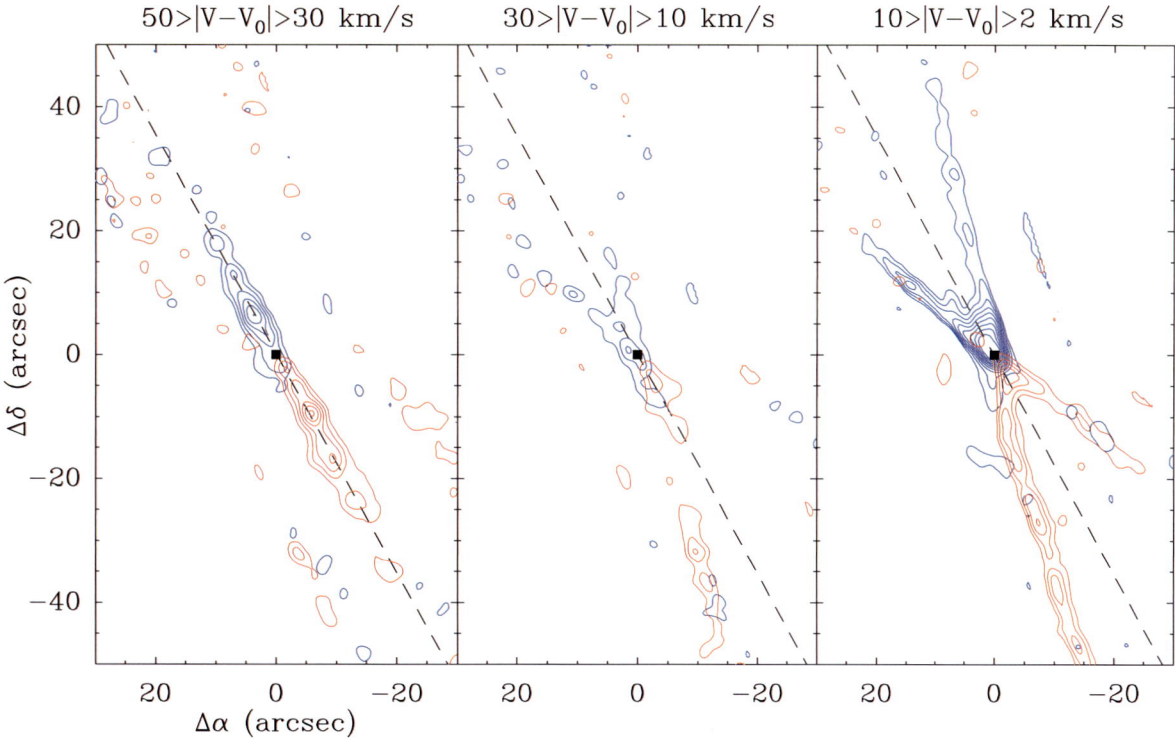

Fig. 2 CO(J = 2–1) emission from the IRAS 04166+2706 outflow (PdBI data, Santiago-García et al. in preparation). The highest velocity gas (*left panel*) forms two opposed jets that emerge from the IRAS source (*filled square*) and travel toward the north-east (blue gas) and south-west (red gas). The intermediate velocity regime (*middle panel*) is almost absent, while the low velocity gas (*right panel*) lies along the walls of two opposed evacuated cavities. The combination of highly collimated jets and limb-brightened shells requires an underlying wind with both highly-collimated and wide-angle components, and shows the limitation of single-component models. See poster contribution from Santiago-García et al. for further details

seen in both CO and SiO emission, is extremely rectilinear, appears only at the highest velocities (between 30 and 50 km s^{-1}), and shows no evidence for precession or wandering. The shell-like part appears at low velocities (2 to 10 km s^{-1}) and seems to delineate two opposed cavities with the IRAS source at their vertex. This cavity interpretation is supported by the fact that the blue outflow shell coincides with the walls of a NIR scattering nebula seen in Spitzer images, as expected from its more favorable projection. In addition, the high velocity jet runs along the axis of the two cavities showing a remarkable degree of symmetry (see poster contribution for further details). The data from IRAS 04166+2706, therefore, leads to the inevitable conclusion that, at least in some cases, both highly collimated and wide-angle components coexist in the outflow driving agent, and that a model that considers both components simultaneously is needed to explain the observations. Interestingly enough, recent realistic modeling of the interaction between the X-wind of Shu et al. (1994) and a toroidal core shows that both jet and shell components should be observed simultaneously in very young outflows (Shang et al. 2006). This so-called "unified" model of bipolar flows shows in fact a remarkable likeness with the IRAS 04166+2706 observations, both in geometry and kinematics (compare Fig. 2 and the models in Shang et al. 2006).

The unified outflow model not only unifies the jet and wide-angle aspects of the outflows, but also brings together the evolution of flows and the dense cores, two elements often treated separately. Evidence for outflow-core interaction has been reported in a number of systems (e.g., Tafalla and Myers 1997; Arce and Sargent 2006), but no unified framework of how this interaction happens or how outflows and cores evolve in parallel exists yet. The beautiful simulations of Shang et al. (2006) illustrate how the most important elements of this interaction occur inside the central 1000 AU region, which corresponds to less than 10″ even towards the most nearby clouds. High angular resolution observations with ALMA are clearly needed to sample the complex geometry and kinematics inside this critical region, and thus compare real outflows with their simulated counterparts. Producing a unified picture of the different and interacting processes occurring during the formation of a solar-type star can be one of most significant achievements of ALMA.

References

Aikawa, Y., Herbst, E., et al.: Astrophys. J. **620**, 330 (2005)
Alves, J.F., Lada, C.J., Lada, E.A.: Nature **409**, 159 (2001)
Arce, H.G., Sargent, A.I.: Astrophys. J. **646**, 1070 (2006)
Arce, H.G., Shepherd, D., et al.: In: Reipurth, B., Jewitt, D., Keils, K. (eds.) Protostars and Planets V, p. 245. University of Arizona Press, Tucson (2007)
Bachiller, R.: Annu. Rev. Astron. Astrophys. **34**, 111 (1996)
Bacmann, A., André, P., et al.: Astron. Astrophys. **361**, 555 (2000)
Bacmann, A., Lefloch, B., et al.: Astrophys. J. **585**, L55 (2003)
Bergin, E.A., Alves, J., et al.: Astrophys. J. **570**, L101 (2002)
Bergin, E.A., Langer, W.D.: Astrophys. J. **486**, 316 (1997)
Bergin, E.A., Tafalla, M.: Annu. Rev. Astron. Astrophys. **45**, 339 (2007)
Bottinelli, S., Ceccarelli, C., et al.: Astrophys. J. **617**, L69 (2004)
Bourke, T.L., Crapsi, A., et al.: Astrophys. J. **633**, L129 (2005)
Bourke, T.L., Myers, P.C., et al.: Astrophys. J. **649**, L37 (2006)
Caselli, P., Walmsley, C.M., et al.: Astrophys. J. **523**, L165 (1999)
Caselli, P., Walmsley, C.M., et al.: Astrophys. J. **565**, 331 (2002)
Caselli, P., van der Tak, F.F.S., et al.: Astron. Astrophys. **403**, L37 (2003)
Cazaux, S., Tielens, A.G.G.M., et al.: Astrophys. J. **593**, L51 (2003)
Ceccarelli, C., Castets, A., et al.: Astron. Astrophys. **338**, L43 (1998)
Ceccarelli, C., Caselli, P., et al.: In: Reipurth, B., Jewitt, D., Keils, K. (eds.) Protostars and Planets V, p. 47. University of Arizona Press, Tucson (2007)
Crapsi, A., Caselli, P., et al.: Astron. Astrophys. **420**, 957 (2004)
Crapsi, A., Caselli, P., et al.: Astrophys. J. **619**, 379 (2005)
Dalgarno, A., Lepp, S.: Astrophys. J. **287**, L47 (1984)
Di Francesco, J., Evans, N.J. II, et al.: In: Reipurth, B., Jewitt, D., Keils, K. (eds.) Protostars and Planets V, p. 17. University of Arizona Press, Tucson (2007)
Geppert, W.D., Thomas, R., et al.: Astrophys. J. **609**, 459 (2004)
Jørgensen, J.K.: Astron. Astrophys. **424**, 589 (2004)
Jørgensen, J.K., Schöier, F.L., van Dishoeck, E.F.: Astron. Astrophys. **416**, 603 (2004)
Lada, C.J.: Annu. Rev. Astron. Astrophys. **23**, 267 (1985)
Lee, C.F., Mundy, L.G., et al.: Astrophys. J. **576**, 294 (2002)
MacLow, M.M., Klessen, R.S.: Rev. Mod. Phys. **76**, 125 (2004)
Masson, C.R., Chernin, L.M.: Astrophys. J. **387**, L47 (1992)
Masson, C.R., Chernin, L.M.: Astrophys. J. **414**, 230 (1993)
Moriarty-Schieven, G.H., Snell, et al.: Astrophys. J. **319**, 742 (1987)
Mouschovias, T.C., Ciolek, G.E.: In: Lada, C.L., Kylafis, D. (eds.) The Origin of Stars and Planetary Systems, p. 305. Kluwer, Dordrecht (1999)
Myers, P.C.: In: Yuan, C., You, J. (eds.) Molecular Clouds and Star Formation, p. 47. World Scientific, Singapore (1995)
Myers, P.C., Fuller, G.A., et al.: Astrophys. J. **376**, 561 (1991)
Öberg, K.I., van Broekhuizen, F., et al.: Astrophys. J. **621**, L33 (2005)
Padgett, D.L., Brandner, W., et al.: Astron. J. **117**, 1490 (1999)
Pagani, L., Pardo, J.R., et al.: Astron. Astrophys. **429**, 181 (2005)
Raga, A., Cabrit, S.: Astron. Astrophys. **278**, 267 (1993)
Roueff, E., Tiné, S., et al.: Astron. Astrophys. **354**, L63 (2000)
Santiago-García et al.: in preparation
Shang, H., Allen, A., et al.: Astrophys. J. **649**, 845 (2006)
Shu, F.H.: Astrophys. J. **214**, 488 (1977)
Shu, F.H., Adams, F.C., Lizano, S.: Annu. Rev. Astron. Astrophys. **25**, 23 (1987)
Shu, F.H., Ruden, S.P., et al.: Astrophys. J. **370**, L31 (1991)
Shu, F., Najita, J., et al.: Astrophys. J. **429**, 781 (1994)
Tafalla, M., Myers, P.C.: Astrophys. J. **491**, 653 (1997)
Tafalla, M., Mardones, D., et al.: Astrophys. J. **504**, 900 (1998)
Tafalla, M., Myers, P.C., et al.: Astrophys. J. **569**, 815 (2002)
Tafalla, M., Myers, P.C., et al.: Astron. Astrophys. **416**, 191 (2004)
Tafalla, M., Santiago-García, J., et al.: Astron. Astrophys. **455**, 577 (2006)
van der Tak, F.F.S., Schilke, P., et al.: Astron. Astrophys. **388**, L53 (2002)
Ward-Thompson, D., Motte, F., André, P.: Mon. Not. R. Astron. Soc. **305**, 143 (1999)
Werner, M., Fazio, G., et al.: Annu. Rev. Astron. Astrophys. **44**, 269 (2006)
Williams, J.P., Myers, P.C., et al.: Astrophys. J. **513**, L61 (1999)
Wilner, D.J., Myers, P.C., et al.: Astrophys. J. **544**, L69 (2000)
Zhou, S., Evans, N.J., et al.: Astrophys. J. **404**, 232 (1993)

Chemistry in the ISM: the ALMA (r)evolution

The cloudy crystal ball of one astrochemist

Eric Herbst

Originally published in the journal Astrophysics and Space Science, Volume 313, Nos 1–3.
DOI: 10.1007/s10509-007-9639-9 © Springer Science+Business Media B.V. 2007

Abstract Increases in knowledge in various subfields of galactic astrochemistry that will ensue as ALMA becomes operational are discussed. A distinction is made between evolutionary and revolutionary changes. It is proposed that the most revolutionary enhancements will occur in our knowledge of small-scale structure and the attendant need for more complex chemical models that contain spatial inhomogeneities and dynamics as well as chemical processes.

Keywords ISM: molecules · Stars: formation · Telescopes

1 Introduction

When astronomers think of the results that will come from a massive new telescope project, they often use the word "revolutionary". More mature reflection, however, suggests that progress will be revolutionary in some aspects of astronomy and evolutionary in others. Despite its huge improvements in sensitivity and spatial resolution compared with existing single dishes and interferometers, ALMA is no exception. To understand what will be truly revolutionary about ALMA and what will be merely evolutionary in the field of interstellar astrochemistry, we need to adopt an historical analysis to determine how far we have come in this field and how much more remains to be accomplished.

The field of astrochemistry dates back at least to the detections of the gas-phase polyatomic molecules ammonia (NH_3) and formaldehyde (H_2CO) in the galactic interstellar medium (Cheung et al. 1968; Snyder et al. 1969). Soon

afterward, a detailed theory of how interstellar molecules are both formed and destroyed, the ion-molecule theory, was reported and used successfully to understand the exotic nature of the chemistry (Herbst and Klemperer 1973). This theory actually predated the confirmation of the first known polyatomic ion, HCO^+, which occurred several years later (Snyder et al. 1976). The use of ion-molecule chemistry to understand deuterium fractionation followed shortly (Guélin et al. 1977; Watson et al. 1978). By the early 1980's, more powerful computers allowed the construction of larger and more detailed gas-phase models of interstellar chemistry (Prasad and Huntress 1980). About this time, the era of looking for new molecules began to be overtaken by the use of molecules as probes of the interstellar medium. It was recognized that the larger interstellar clouds of gas and dust are quite heterogeneous, better described as assemblies of both denser and more diffuse material (Blake et al. 1986). Interest was soon focused on the substructure, much of which has to do with the evolutionary stages of stellar formation, both low- and high-mass. Models of so-called photon-dominated regions came in the mid 1980's; these regions consist of matter close to young stars and strongly affected by their radiation (Tielens and Hollenbach 1985). In the late 1980's, models of hot cores, warm regions (300 K) surrounding young high-mass stellar objects, came on the scene; the models contained a historical analysis, showing that the current gas in these objects stems partially from material frozen onto mantles of dust grains before the onset of star formation (Brown et al. 1988). The field of low-mass star formation gained prominence with the observation and study of protoplanetary disks (Aikawa and Herbst 1999) and the observation and study of pre-stellar cores, which are objects collapsing isothermally to form protostars. Pre-stellar cores show the absence of neutral molecules in their dense centers (Caselli

E. Herbst (✉)
Department of Physics, The Ohio State University, Columbus, OH 43210, USA
e-mail: herbst@mps.ohio-state.edu

et al. 1999), and a strongly enhanced deuterium fractionation related to this absence (Walmsley et al. 2004). Molecular observations and models have begun to probe protostars, which are complex collections of outflows, warm quiescent material, and disks, all surrounding the nascent star (Bottinelli et al. 2004). Even the short-lived stages of high-mass star formation have become targets of investigation (Boonman et al. 2003). Meanwhile, the study of the diffuse interstellar medium, which heretofore had been regarded as a well-understood field of study except for isolated problems such as the abundance of CH^+ (Elitzur and Watson 1980), became much more complex with detections of polyatomic molecules in the gas (Liszt et al. 2004). Along with this vast increase in our knowledge of the interstellar medium, the list of known interstellar molecules grew to over 140, excluding isotopologues, and containing molecules up to 13 atoms in size, with different evolutionary stages showing very different abundances of these species. The most revolutionary detection in recent years is that of the negative molecular ion C_6H^- (McCarthy et al. 2006).

So, much has already been learned. Given this background, we can look at various sub-fields of interstellar chemistry to try to guess what will happen when ALMA becomes operational.

2 Detection of new molecules

Will the use of ALMA result in detections of many new interstellar molecules? Will molecules much larger in size than 13 atoms be seen? Will little viruses be found to inhabit the interstellar medium? Caution is clearly in order here. The best sites for the detection of complex organic molecules of a terrestrial nature, the hot cores, are already difficult places to detect new molecules because of the dense, strong, and only partially assigned spectra of certain hydrogen-rich molecules known as "weeds". For the hot cores in Orion KL (Comito et al. 2005), it is estimated that, at current levels of sensitivity, at least half of the lines may be unassigned (Cernicharo, private communication). These molecular pests, including methanol (CH_3OH), methyl formate ($HCOOCH_3$), and dimethyl ether (CH_3OCH_3), are not rigid rotors, as are most small molecules, but also contain a large-amplitude motion known as torsion, or internal rotation, which adds more energy levels and interacts with rotation to add layers of complexity. An increase in the detected intensity by several orders of magnitude, as will happen when ALMA is fully operational, is likely to yield a continuum of lines from weeds and other larger molecules such as ethyl cyanide (C_2H_5CN), with dense rotational spectra because of their small moments of inertia, especially in the submillimeter. Much effort will have to be directed towards either erasing the spectra of weeds and other large species or accounting

for them in rather complex algorithms if ALMA is to be able to detect any but the strongest of molecular sources at these frequencies. The situation is probably better at lower frequencies, especially if searches are directed at lukewarm regions such as those in the Galactic Center.

Other types of sources are likely to be affected less strongly since their current spectral density is much smaller. For example, the starless core TMC-1, the complex molecules of which are mainly hydrogen-poor linear species, will be successfully probed for much larger species in this class. Carbon-rich circumstellar envelopes and protoplanetary nebulae occupy a middle ground, since their spectra are rich and complex, yet not as overwhelming as Orion KL. For example, the millimeter-wave spectrum of CRL 618 has been almost fully assigned and analyzed, but with great effort, and over a period of years (J. Pardo, private communication).

3 New chemical processes

Before polyatomic molecules were discovered in the interstellar medium, the refrain was that molecules could not exist under such conditions. Once molecules were indeed discovered, the refrain was that all gas-phase molecules were formed on the surfaces of dust particles, in analogy to the formation of molecular hydrogen. After the semi-quantitative ion-molecule theory gained vogue in the early 1970's, this model held sway in low temperature regions, supplemented by high temperature processes in shock models (Elitzur and Watson 1980) and the outer layers of photon-dominated regions (Tielens and Hollenbach 1985). Perhaps a decade later, surface chemistry began a comeback, aided by the increasing evidence for ices on cold interstellar dust particles. By the early 1990's, models containing both gas-phase and surface chemistry began to be published (Hasegawa et al. 1992). Major roadblocks towards the use of such models, however, include the complexity of processes occurring on surfaces, the need for computer-intensive stochastic methods to treat the chemistry exactly (Chang et al. 2005), and the uncertainty in our understanding of desorption mechanisms (Garrod et al. 2006a). Nevertheless, the role of surface chemistry has gained prominence in recent years with the failure of gas phase chemistry to produce molecules such as methanol (Garrod et al. 2006a). The detection of complex organic molecules in many sources throughout the Galactic Center appears to indicate that widespread sputtering of these species from grain surfaces occurs. Nevertheless, the recent detection of a negative molecular ion, the anion C_6H^-, in both the molecular-rich cold core TMC-1 and the circumstellar envelope IRC+10216 (McCarthy et al. 2006), reminds us that unusual gaseous processes such as radiative attachment of electrons to neutrals occur in the interstellar medium.

During this rather complex history, a wide variety of chemical processes has been considered. Will the advent of ALMA result in the study of yet new types of processes or the use of different types of models? Of course, the answer is intertwined with the possibility of detecting new classes of molecules. One possibility is that as ALMA peers more closely at warmer regions near sources of radiation, such as the inner parts of protoplanetary disks, the outer parts of dense PDR's, and small hot cores, kinetic temperatures above 1000 K will be encountered. The use of the standard databases by modelers for such warm regions is questionable at best, since the backwards endothermic reactions may be as rapid as the forward exothermic reactions. It may even be the case that thermodynamic models, such as those used for the chemistry of stellar photospheres, may be more useful, at least as a high-temperature limit. Another area of investigation, pretty much untouched by modelers, is the inclusion of PAH's and their chemistry into the models. Since there is direct evidence for PAH's in PDR's, it is unreasonable to neglect them entirely, yet to the best of our knowledge there is no model that includes them. Finally, if new and larger molecules are discovered by ALMA, new processes to create them must be considered such as efficient radiative association and recombination reactions.

4 New large-scale and small-scale structure

What will ALMA add to our current knowledge of the large-scale structure of giant molecular clouds and cloud assemblies in our galaxy? Again, an historical analysis proves useful. In the original era of astrochemistry, when the discovery of new interstellar molecules held sway, there was little emphasis on structure. But as the subject matured, structure became a more important subject. We now know that large clouds comprise cold starless cores, collapsing pre-stellar cores, protostars of assorted classes, hot cores, hot corinos, protoplanetary disks, PDR's, and violent outflows. We have a good sense of the evolutionary stages by which a single low-mass star is formed, and are beginning to learn much more about the stages leading to high-mass star formation. Of course, the detailed physics of the evolutionary pathways is still a matter of great dispute. Even the initial formation of cold dense regions and whether or not they expand or contract are topics still in their infancy (Bergin et al. 2004).

Although it is possible that ALMA will lead us, perhaps indirectly, to the detection of new large-scale structures in large heterogeneous galactic clouds, it is more likely, indeed almost certain, that ALMA will result in a revolution of our understanding of the small-scale structure of the assorted components of dense clouds. The situation is entirely different in external galaxies, where the unsurpassed sensitivity and spatial resolution of ALMA will allow us to look at the interstellar medium in far more detail than heretofore. Where there are now a few hot cores to study, there will be hundreds more, mainly in nearby galaxies!

One obvious advance in our study of the Milky Way will be in a more detailed knowledge of the clumpy structure of objects on the smallest distance scales and their dynamic history. This advance will lead to, indeed *require*, a similar advance in our chemical modeling of such objects (Garrod et al. 2006b). Consider, as an example, the current status of PDR models. Although detailed in terms of chemical and excitation processes, these models are still mainly one-dimensional in nature with various slabs representing different physical conditions. Such models may represent some PDR's well, but recent observations have shown that the best-known PDR of all, the Orion Bar, is full of clumps, some of which are cold enough to support extensive deuterium fractionation and even depletion of heavy species from the gas-phase (Leurini et al. 2006). But, it would be unlikely that large clumps are themselves homogeneous, and much more likely that they are themselves inhomogeneous although a perfect fractal nature down to very small scales is undoubtedly an oversimplification. In any event, the increase in our understanding of small-scale structure will certainly add to the complexity of new generations of chemical models.

This increasing complexity is a controversial topic. Two anecdotes will help to illustrate the disagreement. When I was a graduate student at Harvard University, I studied the molecular spectroscopy of a bizarre molecule, which mercifully has not been found in space. The analysis of this molecule required, in my view, the diagonalization of large matrices. My supervisor felt that such an analysis was inappropriate because we would lose our detailed intuitive understanding of the physics of the situation. He preferred the standard perturbative approach, but as I explained, such an approach did not converge! Here, complexity was indeed necessary to make any progress at all. But this is not always the case. More recently, I was part of a collaboration that explained the nature of the deuterium fractionation in the cores of the Orion Bar. To understand this new and unexpected observation, we used a relatively simple steady-state gas-phase model, which clarified the processes leading to the deuteration quite nicely in our view. The referee, however, brought up the unpleasant facts that we had failed to consider surface chemistry as well as the role of PAH's, both of which cannot be understood in terms of steady-state chemistry. Besides the increase in chemical complexity, moving to a time-dependent picture would then require us to understand the dynamic history of the cores. The result would indeed be to probably lose our simple but correct understanding of the basic chemical processes involved in deuteration. These anecdotes are meant to explain my view that complexity is often necessary but should not be used to obscure simple explanations of small amounts of data. Others may disagree.

But assuming that complexity is with us to stay, it may well be that the optimum environment for research will not be the research university with individual professors in distinct departments heading groups of students, but the institute in which groups of researchers combine their talents.

5 Individual objects

To give a better sense of the history of our knowledge of small-scale structure and the likelihood of what may be discovered in the future and how it should be modeled, we will now consider three specific galactic sources: the starless core TMC-1, the pre-stellar core L1544, and the hot-core region towards Orion KL.

5.1 TMC-1

Originally a small and insignificant part of the Taurus molecular assembly, TMC-1 gained fame as that cold dark, or starless, core with the largest assortment and abundances of organic molecules, including unsaturated species such as the cyanopolyyne family ($HC_{2n}CN$) and the radicals C_nH. The abundances of this source were fit by a number of so-called pseudo-time-dependent gas-phase models, in which the physical conditions are homogeneous ($T = 10$ K, $n_H = 2 \times 10^4$ cm^{-3}) and the chemistry starts from atomic abundances with the exception of molecular hydrogen (Wakelam et al. 2006). The abundances are better reproduced when so-called 'low-metal' abundances, which contain severe depletions of metals, sulfur, and silicon, are used. Best agreement occurs at so-called "early times" of 10^{5-6} yr. As opposed to other cold dark cores such as L134N, reasonable agreement with observation for the 60-odd molecules in TMC-1 requires carbon-rich abundances ($C/O > 1$). If modern gas-grain models are used (Ruffle and Herbst 2001), in which surface processes are taken account of, more normal oxygen-rich abundances work well, with up to 90% of the gas-phase molecular abundances being fit at early times (Garrod et al. 2006a).

But further observations of TMC-1 have revealed that this core contains at least six smaller units, or clumps, now labeled A, B, C, CP, D, and E, with distinct chemical properties. For example, most of the abundances of the complex molecules are at their highest in clump CP, where the letters stand for cyanopolyyne. A conservative manner of looking at the diversity of abundances is to use a pseudo-time-dependent model for each clump and obtain the best age; in this manner, ages of 10^{5-7} yr have been obtained. A more radical manner of looking at the diversity is to assume that the gradients from one clump to another are produced by sequential sputtering off of icy grain mantles caused by MHD waves from a nearby source (Markwick et al. 2000). This

explanation then requires a more ancient period in which the mantles were produced in the first place, suggesting yet another level of complexity that may include the original dynamics of clump formation. To make headway in deducing the proper scenario will require a better understanding of the vexing question of how clumpy cores form and the chemistry and elemental depletions that occur along with this formation. Presumably, observations of ALMA may indeed allow us to understand clump formation better.

5.2 L1544

Perhaps the best-known pre-stellar core, a class of objects still at a temperature of 10 K but collapsing to form a dense central condensation of density 10^{5-7} cm^{-3}, L1544 is also located in the Taurus molecular complex. Recent observations have shown that some heavy molecules are depleted from the gas in the central condensation of this and other pre-stellar cores. Among the molecules depleted towards the center are CO, CCS, and HCO^+, while the molecules NH_3 and HN_2^+ are not depleted (Caselli et al. 1999; Aikawa et al. 2005). The depletion of some heavy species enhances the process of deuterium fractionation, because the primary deuterated ions are destroyed more slowly by reactions with heavy neutral species (Walmsley et al. 2004). Detailed chemical modeling at the present time contains the assumption of spherical shells, which may or may not be collapsing (Roberts et al. 2004; Aikawa et al. 2005). These models show good agreement with observations, but one wonders what will happen when both small-scale structure and the non-spherical nature of the source are taken into account.

5.3 Orion Hot Core/Compact Ridge

These sources, observed towards Orion KL (Blake et al. 1986), are two hot cores with differing chemistries. Both contain high gas-phase abundances of hydrogen-rich (saturated) organic molecules, but these tend to be oxygen-containing (e.g. $HCOOCH_3$, CH_3OCH_3) in the Compact Ridge and nitrogen-containing (e.g. C_2H_5CN) in the Hot Core. For many years, the chemistry of hot cores was discussed in terms of two historical eras: (a) a cold era, in which gas-phase chemistry and grain-surface chemistry occur, and (b) the current warm era, which starts with star formation, at which time the grain mantles from the previous era evaporate, enriching the gas in hydrogen-rich molecules and ushering in a second era of gas-phase chemistry at temperatures from 100–300 K depending on the particular core. During the second era, gas-phase processes are supposed to create the more complex hydrogen-rich species seen in hot cores but not in cold ones. In one class of model, the chemistry is started at the time of star formation with an initial collection of molecules thought to be formed during the colder era, and the warm gas-phase chemistry is followed (Charnley et al. 1992). In the second class (Caselli et al. 1993), the actual gas-phase and surface-chemistry in the cold era are followed and, after a sufficient time, the grain mantles are allowed to evaporate. The gas-phase chemistry is then treated in the same manner as in the first class of models. The results from a model of the latter class (Caselli et al. 1993) were actually able to explain the difference between the Compact Ridge and the Hot Core by the temperature of the preceding era: above a certain temperature, CO does not stick to grains. Since CO is the precursor of methanol, little methanol is formed, and since methanol, once desorbed into the gas phase in the warm era, is the precursor of the more complex oxygen-containing organic molecules, only the nitrogen-containing molecules are produced in high abundance. Such an historical picture may be appropriate, and gains some stature from the diversity of hot cores in other regions of the ISM. Nevertheless, these early models lacked two important features: any degree of heterogeneity in the hot core and a dynamic picture of what actually happens as the material is being heated up during the era of star formation. High-spatial-resolution observations by ALMA will push modelers to include these levels of complexity. Already, recent studies have begun to consider the chemistry occurring during the era of star formation (Viti and Williams 1999; Garrod and Herbst 2006). Although space precludes a discussion of the study of hot corinos such as IRC 16293-2422, which are the analog of hot cores for low-mass protostellar regions, such objects are even more complex since it remains unclear whether the organic molecules detected are produced in warm ambient gas as in hot cores, or in outflows, or even in early disks. ALMA will clearly help to resolve this issue.

6 Revolutionary or evolutionary?

What fields will experience revolutionary changes and what fields merely evolutionary changes when ALMA is fully operational? Based on the above discussion, our cloudy crystal ball leads to the following prognostications:

– *new molecules* The growth in this field will be evolutionary at best. New molecules are likely to be detected but the problem of unwanted interference from weeds and other more abundant species may be a formidable one.
– *new chemical processes* Dependent somewhat on the detection of new classes of molecules, growth in this specialty will also likely be evolutionary.
– *structure* The study of small-scale structure will be truly revolutionary, opening up new vistas of complexity and understanding. These new vistas will be built upon prior studies by other interferometers such as the SMA.

– *chemical modeling* The need for complex models, which take account of gas-phase and grain-surface chemistry, radiative transfer, small-scale structure, and dynamics will increase in a revolutionary manner. This turn towards complexity will hasten the nascent movement of collaborations replacing individual investigators because different areas of expertness will be needed in models of the future.

Acknowledgements I would like to acknowledge the support of the National Science Foundation (US) for my research program in astrochemistry.

References

Aikawa, Y., Herbst, E.: Molecular evolution in protoplanetary disks. Two-dimensional distributions and column densities of gaseous molecules. Astron. Astrophys. **351**, 233–246 (1999)

Aikawa, Y., Herbst, E., Roberts, H., Caselli, P.: Molecular evolution in collapsing prestellar cores, III: contraction of a Bonnor–Ebert sphere. Astrophys. J. **620**, 330–346 (2005)

Bergin, E.A., Hartmann, L.W., Raymond, J.C., Ballesteros-Paredes, J.: Molecular cloud formation behind shock waves. Astrophys. J. **612**, 921–939 (2004). doi: 10.1086/422578

Blake, G.A., Masson, C.R., Phillips, T.G., Sutton, E.C.: The rotational emission-line spectrum of Orion A between 247 and 263 GHz. Astrophys. J. Suppl. Ser. **60**, 357 (1986)

Boonman, A.M.S., Doty, S.D., van Dishoeck, E.F., Bergin, E.A., Melnick, G.J., Wright, C.M., Stark, R.: Modeling gas-phase H_2O between 5 μm and 540 μm toward massive protostars. Astron. Astrophys. **406**, 937–955 (2003)

Bottinelli, S., Ceccarelli, C., Lefloch, B., Williams, J.P., Castets, A., Caux, E., Cazaux, S., Maret, S., Parise, B., Tielens, A.G.G.M.: Complex molecules in the hot core of the low-mass protostar NGC 1333 IRAS 4A. Astrophys. J. **615**, 354–358 (2004)

Brown, P.D., Charnley, S.B., Millar, T.J.: A model of the chemistry in hot molecular cores. Mon. Not. R. Astron. Soc. **231**, 409–417 (1988)

Caselli, P., Hasegawa, T.I., Herbst, E.: Chemical differentiation between star-forming regions: the Orion hot core and compact ridge. Astrophys. J. **408**, 548–558 (1993)

Caselli, P., Walmsley, C.M., Tafalla, M., Dore, L., Myers, P.C.: CO depletion in the starless cloud core L1544. Astrophys. J. **523**, L165–L169 (1999)

Chang, Q., Cuppen, H.M., Herbst, E.: Continuous-time random-walk simulation of H_2 formation on interstellar grains. Astron. Astrophys. **434**, 599–611 (2005)

Charnley, S.B., Tielens, A.G.G.M., Millar, T.J.: On the molecular complexity of the hot cores in Orion A—grain surface chemistry as "the last refuge of the scoundrel". Astrophys. J. **399**, L71–L74 (1992)

Cheung, A.C., Rank, D.M., Townes, C.H., Thornton, D.D., Welch, W.J.: Detection of NH_3 molecules in the interstellar medium by their microwave emission. Phys. Rev. Lett. **21**, 1701–1705 (1968). doi: 10.1103/PhysRevLett.21.1701

Comito, C., Schilke, P., Phillips, T.G., Lis, D.C., Motte, F., Mehringer, D.: A molecular line survey of Orion KL in the 350 micron band. Astrophys. J. Suppl. Ser. **156**, 127–167 (2005)

Elitzur, M., Watson, W.D.: Interstellar shocks and molecular CH^+ in diffuse clouds. Astrophys. J. **236**, 172–181 (1980)

Garrod, R.T., Herbst, E.: Formation of methyl formate and other organic species in the warm-up phase of hot molecular cores. Astron. Astrophys. **457**, 927–936 (2006)

Garrod, R., Park, I.H., Caselli, P., Herbst, E.: Are gas-phase models of interstellar chemistry tenable? The case of methanol. Disc. Faraday Soc. **133**, 51–62 (2006a)

Garrod, R.T., Williams, D.A., Rawlings, J.M.C.: Molecular clouds as ensembles of transient cores. Astrophys. J. **638**, 827–838 (2006b)

Guélin, M., Langer, W.D., Snell, R.L., Wootten, H.A.: Observations of DCO$^+$—the electron abundance in dark clouds. Astrophys. J. **217**, L165–L168 (1977)

Hasegawa, T.I., Herbst, E., Leung, C.M.: Models of gas-grain chemistry in dense interstellar clouds with complex organic molecules. Astrophys. J. Suppl. Ser. **82**, 167 (1992)

Herbst, E., Klemperer, W.: The formation and depletion of molecules in dense interstellar clouds. Astrophys. J. **185**, 505 (1973)

Leurini, S., Rolffs, R., Thorwirth, S., Parise, B., Schilke, P., Comito, C., Wyrowski, F., Güsten, R., Bergman, P., Menten, K.M., Nyman, L.A.: APEX 1 mm line survey of the Orion Bar. Astron. Astrophys. **454**, L47–L50 (2006)

Liszt, H., Lucas, R., Black, J.H.: The abundance of HOC$^+$ in diffuse clouds. Astron. Astrophys. **428**, 117–120 (2004)

Markwick, A.J., Millar, T.J., Charnley, S.B.: On the abundance gradients of molecules along the TMC-1 ridge. Astrophys. J. **535**, 256–265 (2000)

McCarthy, M.C., Gottlieb, C.A., Gupta, H., Thaddeus, P.: Laboratory and astronomical identification of the negative molecular ion C$_6$H$^-$. Astrophys. J. **652**, L141–L144 (2006)

Prasad, S.S., Huntress Jr., W.T.: A model for gas phase chemistry in interstellar clouds, I: the basic model, library of chemical reactions, and chemistry among C, N, and O compounds. Astrophys. J. Suppl. Ser. **43**, 1–35 (1980)

Roberts, H., Herbst, E., Millar, T.J.: The chemistry of multiply deuterated species in cold dense interstellar cores. Astron. Astrophys. **424**, 905 (2004)

Ruffle, D.P., Herbst, E.: New models of interstellar gas-grain chemistry, III: solid CO$_2$. Mon. Not. R. Astron. Soc. **354**, 1054 (2001)

Snyder, L.E., Buhl, D., Zuckerman, B., Palmer, P.: Microwave detection of interstellar formaldehyde. Phys. Rev. Lett. **22**, 679–681 (1969)

Snyder, L.E., Hollis, J.M., Lovas, F.J., Ulich, B.L.: Detection. identification, and observations of interstellar HC^{13}O$^+$. Astrophys. J. **209**, 67–74 (1976)

Tielens, A.G.G.M., Hollenbach, D.: Photodissociation regions, I: basic model; II a model for the Orion photodissociation region. Astrophys. J. **291**, 722–754 (1985)

Viti, S., Williams, D.A.: Time-dependent evaporation of icy mantles in hot cores. Mon. Not. R. Astron. Soc. **305**, 755–762 (1999)

Wakelam, V., Herbst, E., Selsis, F.: The effect of uncertainties on chemical models of dark clouds. Astron. Astrophys. **451**, 551–562 (2006)

Walmsley, C.M., Flower, D.R., Pineau des Forêts, G.: Complete depletion in prestellar cores. Astron. Astrophys. **418**, 1035–1043 (2004)

Watson, W.D., Snyder, L.E., Hollis, J.M.: The DCO$^+$-to-HCO$^+$ abundance ratio and the electron density in cool interstellar clouds. Astrophys. J. **222**, L145–L147 (1978)

High angular resolution imaging of the circumstellar material around intermediate mass (IM) stars

A. Fuente

Originally published in the journal Astrophysics and Space Science, Volume 313, Nos 1–3.
DOI: 10.1007/s10509-007-9628-z © Springer Science+Business Media B.V. 2007

Abstract In this paper we present high angular resolution imaging of 3 intermediate-mass (IM) stars using the Plateau de Bure Interferometer (PdBI). In particular we present the chemical study we have carried out towards the IM hot core NGC 7129–FIRS 2. This is the first chemical study in an IM hot core and provides important hints to understand the dependence of the hot core chemistry on the stellar luminosity. We also present our high angular resolution (0.3″) images of the borderline Class 0-Class I object IC1396 N. These images trace the warm region of this IM protostar with unprecedented detail (0.3″ ∼ 200 AU at the distance of IC1396 N) and provide the first detection of a cluster of IM hot cores. Finally, we present our interferometric continuum and spectroscopic images of the disk around the Herbig Be star R Mon. We have determined the kinematics and physical structure of the disk associated with this B0 star. The low spectral index derived from the dust emission as well as the flat geometry of the disk suggest a more rapid evolution of the disks associated with massive stars (see Alonso-Albi et al., arXiv:astro-ph/0702119, 2007). In the Discussion, we dare to propose a possible evolutionary sequence for the warm circumstellar material around IM stars.

Keywords Stars: formation · Stars: pre-main sequence: Herbig Be · Stars: circumstellar disk · Stars: individual (NGC 7129–FIRS 2, IC1396 N, R Monocerotis)

A. Fuente (✉)
Observatorio Astronómico Nacional (OAN), Apdo 112,
28800 Alcalá de Henares, Spain
e-mail: a.fuente@oan.es

1 Introduction

Luminous intermediate-mass young stellar objects (IMs) (protostars and stars with $M_* \sim 5$–10 $M_\odot$) are crucial in star formation studies. They share many characteristics with high-mass stars (clustering, PDRs) but their study presents an important advantage: there are many located closer to the Sun ($d \sim 1$ Kpc), and in regions less complex than massive star forming regions. Thus, they can be studied with high spatial resolution. The study of important problems of massive star formation like the physical and chemical structure of hot cores, clustering and the occurrence and physical properties of the disks around massive stars, requires of high spatial resolution. With the current instrumentation, these problems can only be addressed in IMs.

During the last 3 years, we have mapped a sample of IMs in different evolutionary stages using the Plateau de Bure Interferometer (PdBI). In particular we have detected and carried out a chemical study towards the IM hot core embedded in the Class 0 object NGC 7129–FIRS 2. Moving to more evolved objects, we have studied in detail the circumstellar disk associated with the Herbig Be star R Mon (see Poster by Alonso-Albi et al.). Finally, we have imaged the borderline Class 0-Class I object IC1396 N using the new A configuration of the PdBI which provides the highest angular resolution that can be achieved with the current millimeter instrumentation (0.3″ ∼ 200 AU at the distance of IC1396 N). These observations have allowed us to detect a cluster of hot core/corinos and have a first glance at the evolutionary stage of each cluster component. In this paper, we revise the results obtained from these high spatial resolution studies and discuss the implications for the understanding of the massive star formation process.

Fig. 1 Relative abundances of the complex O- and N-bearing molecules as a function of the protostellar luminosity for a sample of hot cores/corinos. Note that the $X(HCOOH)/X(HC_3OH)$ decreases with the protostellar luminosity (Fuente et al. 2005b)

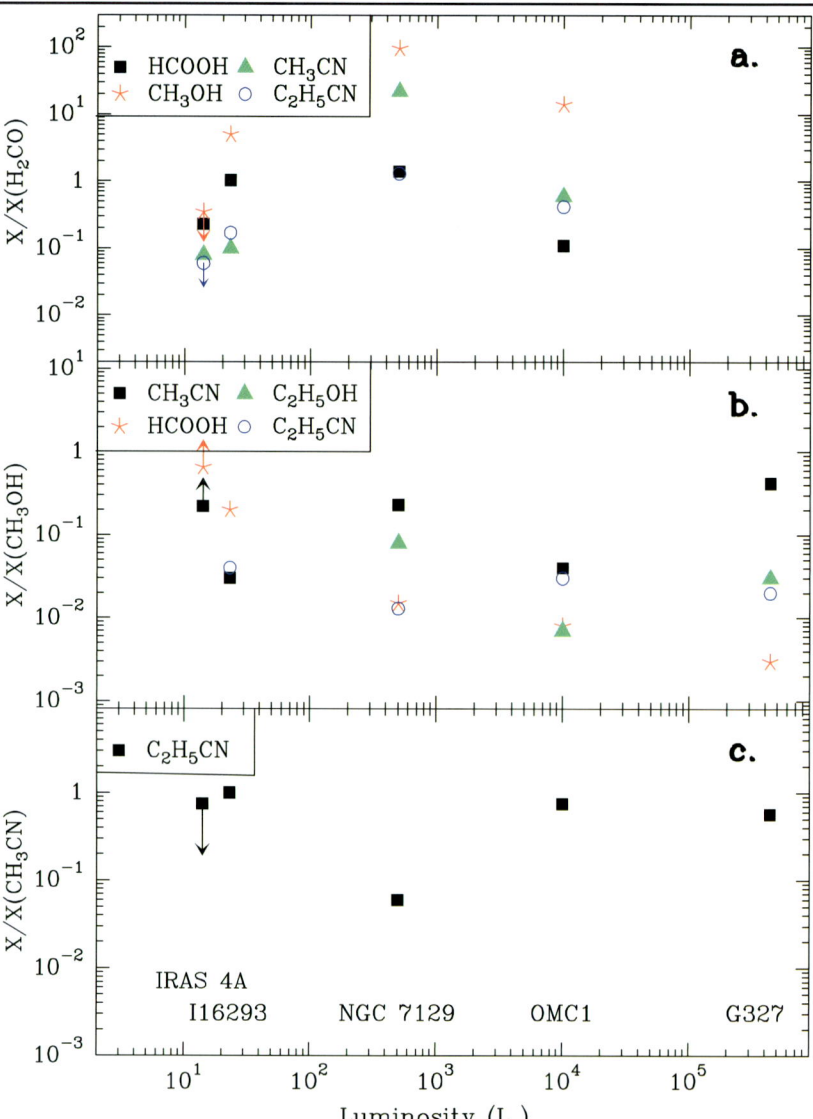

2 NGC 7129–FIRS 2

NGC 7129–FIRS 2 (hereafter, FIRS 2), with a luminosity $\sim$500 $L_\odot$ and a stellar mass $\sim$5 $M_\odot$, is very likely the youngest IM object known at present (Fuente et al. 2001, 2005a). Recent PdBI observations in the continuum and spectroscopic lines carried out by our team show the existence of an IM hot core towards this young protostar (Fuente et al. 2005b). We estimate a size of 650×900 AU and a mass of 2 $M_\odot$ for the hot core associated with this object. The dimensions and mass of this IM hot core are intermediate between those measured in hot corinos ($r \sim 150$ AU, $M < 1$ $M_\odot$) and massive stars ($r \sim 3000$ AU, $M > 10$ $M_\odot$) and consequently, a differentiated chemistry is expected. This IM hot core provides a unique opportunity to study the dependence of the hot core chemistry on the stellar luminosity.

A large number of molecular lines have been detected in our interferometric spectra towards FIRS 2. Most of these lines are identified as belonging to deuterated (D_2CO, c-C_3D and c-C_3HD), sulphuretted (^{13}CS, OCS), and complex O-/N-bearing species (HCOOH, C_2H_5OH, C_2H_5CN). One chemical difference between hot corinos and hot cores is the enhanced abundance of deuterated species in the former. Loinard et al. (2003) searched for the doubly deuterated form of formaldehyde (D_2CO) in a large sample of young stellar objects. D_2CO was detected in all low-mass protostars with $[D_2CO]/[H_2CO]$ ratios of 2–40%. On the other hand, no detection was obtained towards more massive protostars, where $[D_2CO][H_2CO] < 0.5\%$. We estimate a $[D_2CO]/[H_2CO] \sim 0.14$ towards the IM hot core FIRS 2. This value is 4 orders of magnitude larger than the cosmic D abundance and similar to those found in pre-stellar clumps and low-mass protostars.

The sulphuretted and complex compounds are characteristic of hot cores in both the low-mass and the high-mass regimes. We have compared the abundances of complex molecules in FIRS 2 with those in hot corinos and the massive hot cores OMC1 and G327.3–0.6. Contrary to model predictions, we did not detect any dependence of the O-/N-complex molecules ratio on the protostellar luminosity. However, we detected differences between the behavior of the O-bearing species with the stellar luminosity. While H_2CO and HCOOH are more abundant in low luminosity sources, CH_3OH seems to be more abundant in massive objects (see Fig. 1). Fuente et al. (2005b) proposed that this could be due to a different mantle composition in the two classes of regions, caused by different physical conditions (mainly gas density and dust temperature) during the pre-stellar and accretion phase. However, this could also be due to other factors, such as the different spatial scale of the observations or a possible contribution of the shocked gas associated with the bipolar outflow to the emission of these molecules. The detection and detailed study of other intermediate-mass and low mass hot cores are necessary to establish firm conclusions.

3 R Mon

R Mon is the most massive disk detected in dust continuum emission at mm wavelengths around a Herbig Be star (Fuente et al. 2003). Moreover, it is the only one detected in molecular lines and thus far, our unique opportunity to investigate the physical structure and kinematics of the disks associated with these stars.

The high angular resolution continuum images at 2.7 mm and 1.3 mm reported by Fuente et al. (2006) allow us to determine the position (R.A. = 06 : 39 : 09.954 Dec = +08 : 44 : 09.55 (J2000) and size ($\sim$150 AU) of the dusty disk. Moreover, by fitting the SED at cm and mm wavelengths we determine a disk mass of 0.007 $M_\odot$ and $\beta = 0.3$–0.5. Values of β between 0.5 and 1 are usually found in circumstellar disks around HAE and TTs and are thought to be evidence for grain growth in these disks (see e.g. Natta et al. 2007). The low value of β in R Mon suggests that grain growth has proceeded to very large sizes already in the short lifetime of its disk.

Alonso-Albi et al. (2007) report interferometric ^{12}CO and ^{13}CO observations towards this disk. They conclude that the disk is in Keplerian rotation around the star. Keplerian rotation has also been found in most of the TTs and Herbig Ae stars studied thus far and indicates a similar formation mechanisms for the stars in the range 1–8 $M_\odot$. However, contrary to the low mass T Tauri and Herbig Ae stars that are usually surrounded by flared disks, the observed $^{12}CO/^{13}CO$ intensity ratio in R Mon shows that the disk is

geometrically flat (see Alonso-Albi et al. 2007). This result is in line with previous near-IR and optical measurements that suggest that Herbig Be stars have geometrically flatter disks than Herbig Ae and T Tauri stars (see e.g. Meeus et al. 2001).

The flattening of the disk in Herbig Be stars can be due to the rapid grain growth. The grain growth causes the optical depth of the disk to drop and allows the UV radiation to penetrate deep into the circumstellar disk and photo-evaporate the disk external layers (Dullemond and Dominik 2004). Thus, we can propose an evolutionary sequence in which the disks associated with Herbig Be stars start with a flaring shape but become flat during the pre-main sequence and lose most of their mass ($>90\%$) before the star becomes visible ($<10^5$ year).

4 IC1396 N

The new A configuration of the PdBI allows us to study the warm interior of protostellar envelopes with unprecedented detail. In this paper, we present interferometric continuum observations of the IM protostar IC1396 N (Neri et al. 2007).

IC1396 N is a 440 $L_\odot$ protostar located at a distance of 750 pc and Classified as a Class 0/I borderline source. Figure 2 shows the continuum image at 1.3 mm. This image shows the presence of at least three bright continuum emission sources in the center of IC1396 N, all three associated with the source identified as BIMA 2 by Beltrán et al. (2002). The source BIMA 3 is also detected in our continuum image although lies outside the region shown in Fig. 2. In addition to the 3 compact cores we detect some kind of extended emission in BIMA 2. In fact, the bulk of the mm-emission is emerging from a large region ($\sim$3000 AU) centered on the triple-system. According to these results, we envision two different models for the continuum emission: (a) an envelope with sharp boundaries in which the three compact cores are embedded, (b) a region harboring a cluster of lower brightness cores from which we have detected the three most intense ones. The lack of sensitivity to large scale emission and emission distributed on a large number of cores makes it difficult to argue against one or the other model. For simplicity, we favor the model of the dusty 'cocoon' in which the three intense cores are embedded.

Neri et al. (2007) modeled the emission at 2.7 mm and 1.3 mm in the UV-plane assuming a 'cocoon+cores' structure. In Table 1 we show the fluxes and spectral indexes derived from this model. The weaker cores were not resolved by the interferometer. The primary core $41.86 + 13.2$ is resolved in the 1.3 mm emission to a size of $\sim$300 AU $\times$ 150 AU, i.e. an order of magnitude larger than the size measured in hot corinos. This is consistent with this source being the precursor of a Herbig Ae/Be star. The 'cocoon' accounts for the 80% of the 1.3 mm continuum emission and

Fig. 2 Interferometric continuum image at 1.3 mm obtained with the Plateau de Bure Interferometer (PdBI) towards the Class 0 protostar IC 1396N. A cluster of at least 3 compact cores is detected towards BIMA 2 (Neri et al. 2007)

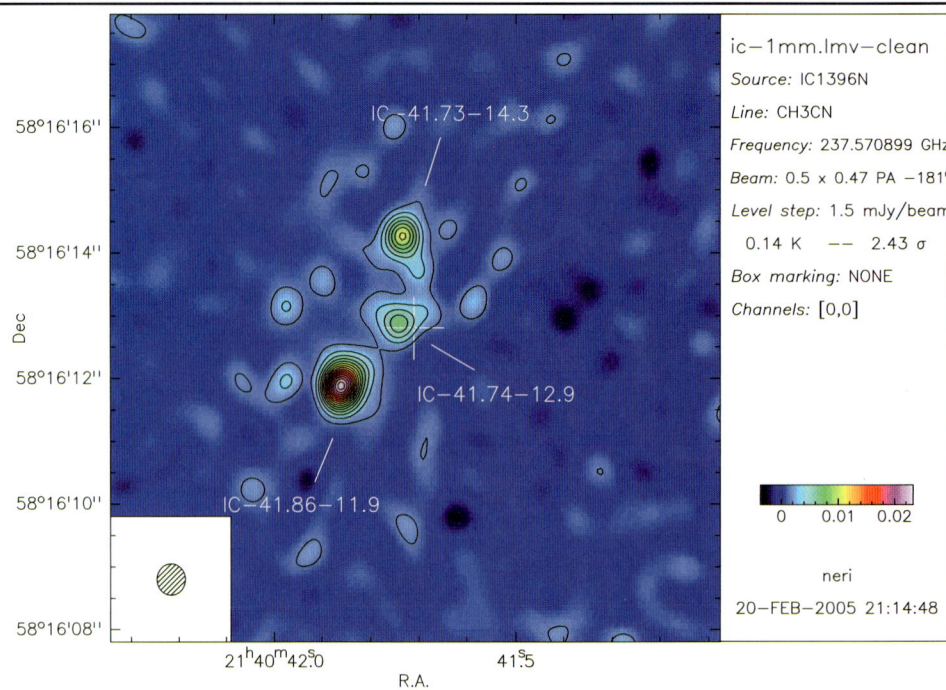

Table 1 Millimeter flux densities, sizes, spectral indexes and masses

3.3 mm (91.7 GHz)

	BIMA 3	BIMA 2			
		Cocoon	41.86 + 11.9	41.73 + 12.8	41.73 + 14.3
α (J2000)	21:40:42.84	21:40:41.86	21:40:41.85	21:40:41.73	21:40:41.72
δ (J2000)	58:16:01.4	58:16:13.2	58:16:11.9	58:16:12.8	58:16:14.3
Size ($''$)	0.8 × 0.5	4.3 × 3.1	unresolved	unresolved	unresolved
S (mJy)	8	16	5.9	1.5	2.7

1.3 mm (237.6 GHz)

	BIMA 3	BIMA 2			
		Cocoon	41.86 + 11.9	41.73 + 12.8	41.73 + 14.3
α (J2000)	21:40:42.84	21:40:41.85	21:40:41.86	21:40:41.73	21:40:41.73
δ (J2000)	58:16:01.4	58:16:13.1	58:16:11.9	58:16:12.8	58:16:14.3
Size ($''$)	0.8 × 0.4	4.5 × 3.1	0.4 × 0.2	unresolved	unresolved
S (mJy)	30	245	35	6	10
Mean Spec. Index	1.4			1.9	
Spec. Index	1.4	2.8	1.9	1.5	1.4
Mass ($M_\odot$)	0.05	0.4	0.06	0.01	0.01
β	≤0	~1.0			

has a different spectral index from the compact cores. While the spectral index in the extended emission is $\sim$2.8 as expected for dust continuum emission with a standard value of $\beta = 1$, the spectral indexes of the compact hot cores are all <2. We propose that this change in the spectral index is very likely associated with a change of the grain properties. The grains in the compact hot cores might be similar to those found in the evolved circumstellar disks. In fact, the compact hot cores could be actually disks. We cannot discard, however, other possible interpretations for these small compact regions (see Neri et al. 2007 for a more detailed discussion).

5 Summary and discussion

A major interest in the Astrophysics today is the understanding of hot cores. These warm regions in the interior of the protostellar envelopes are the prime material from which the proto-planetary disks are formed. However the detailed physical and chemical structure of these hot regions as well as their evolution to become proto-planetary disks are not known. High spatial resolution imaging of the warm regions of protostellar objects is required to have a deeper insight into this problem.

In this paper we present high angular resolution imaging of 3 IMs using the Plateau de Bure Interferometer (PdBI). These stars are thought to be in a different evolutionary stage. NGC 7129–FIRS 2 is a young Class 0 object hosting a massive ($\sim$2 M$_\odot$) and compact hot core. IC 1396 N is a borderline Class 0-Class I object. Our interferometric images reveal a massive 'cocoon' in which three compact cores are embedded. R Mon is a visible star surrounded by a circumstellar disk. We can propose a simple evolutionary scheme in which a IM star starts its life surrounded by a massive and dense hot core. The newly formed star(s) disperses part

of the hot core material producing a less dense 'cocoon' in which the circumstellar disks are immersed. At this stage, the density contrast between the 'cocoon' and the molecular cloud is still high enough for the cocoon to be detected with the interferometer. Finally the cocoon is dispersed and only the circumstellar disk is left. Of course, this is only a rough description of what the evolution of the warm circumstellar material could be and future interferometric observations are needed to confirm and extend this scheme.

Acknowledgements This work has been partially supported by the Spanish MEC and Feder funds under grant ESP2003-04957 and by SEPCT/MEC under grants AYA2003-07584.

References

Alonso-Albi, T., Fuente, A., Bachiller, R., Natta, A., Testi, L., Neri, R., Planesas, P.: Astrophys. J. (2007, submitted). arXiv:astro-ph/0702119

Beltrán, M.T., Girart, J.M., Estalella, R., Ho, P.T.P., Palau, A.: Astrophys. J. **573**, 246 (2002)

Dullemond, C.P., Dominik, C.: Astron. Astrophys. **417**, 159 (2004)

Fuente, A., Neri, R., Martín-Pintado, J., Bachiller, R., Rodríguez-Franco, A., Palla, F.: Astron. Astrophys. **366**, 873 (2001)

Fuente, A., Rodríguez-Franco, A., Testi, L., Natta, A., Bachiller, R., Neri, R.: Astrophys. J. Lett. **598**, L39 (2003)

Fuente, A., Rizzo, J.R., Caselli, P., Bachiller, R., Henkel, C.: Astron. Astrophys. **433**, 535 (2005a)

Fuente, A., Neri, R., Caselli, P.: Astron. Astrophys. **444**, 481 (2005b)

Fuente, A., Alonso-Albi, T., Bachiller, R., Natta, A., Testi, L., Neri, R., Planesas, P.: Astrophys. J. **649**, L119 (2006)

Loinard, L., et al.: In: Curry, C.L., Fich, M. (eds.) SFChem 2002: Chemistry as a Diagnostic of Star Formation, p. 351. NRC Press, Ottawa (2003)

Meeus, G., Waters, L.B.F.M., Bouwman, J., van den Ancker, M.E., Waelkens, C., Malfait, K.: Astron. Astrophys. **365**, 476 (2001)

Natta, A., Testi, L., Calvet, N., Henning, T., Waters, R., Wilner, D.: In: Reipurth, B., Jewitt, D., Keil, K. (eds.) Protostars and Planets V, p. 767. University of Arizona Press, Tucson (2007)

Neri, R., Fuente, A., Ceccarelli, C., Caselli, P., Johnstone, D., van Dishoeck, E., Wyrowski, F.: Astron. Astrophys. **468**, L33 (2007)

Polarization measurements of molecular lines

Richard M. Crutcher

Originally published in the journal Astrophysics and Space Science, Volume 313, Nos 1–3.
DOI: 10.1007/s10509-007-9613-6 © Springer Science+Business Media B.V. 2007

Abstract ALMA observations of circular (Zeeman) and linear (Goldreich–Kylafis) polarization in spectral lines will significantly enhance sensitivity and angular resolution over currently available data, and should lead to major breakthroughs in our understanding of the role of magnetic fields in the star formation process.

Keywords Magnetic fields · Polarization · Star formation · Zeeman effect · Goldreich–Kylafis effect

1 Introduction

It has become increasingly clear that cosmic magnetic fields are pervasive, ubiquitous, and likely important in the properties and evolution of almost everything in the Universe, from planets to quasars (Wielebinski and Beck 2005). One area where the role of magnetic fields is far from being understood is star formation. In spite of significant progress in recent years, there remain unanswered fundamental questions about the basic physics of star formation. In particular, what drives the star formation process? The prevailing view for the past 30 years has been that self-gravitating dense clouds are supported against collapse by magnetic fields (Mouschovias and Ciolek 1999). Since magnetic fields are frozen only into the ionized gas and dust, the neutral material (by far the majority of the mass) can contract gravitationally unaffected directly by the magnetic field. Neutrals will collide with ions in this process, so there will be support against gravity for the neutrals as well as the ions. But there will be a drift of neutrals into the core without a significant increase in the magnetic flux in the core; this is ambipolar diffusion. Eventually the core mass will become sufficiently large that the magnetic field can no longer support the core, and dynamical collapse and star formation can proceed. The other extreme from the magnetically dominated star formation scenario is that molecular clouds are intermittent phenomena in an interstellar medium dominated by turbulence (MacLow and Klessen 2004), and the problem of cloud support for long time periods is irrelevant. Clouds form and disperse by the operation of compressible supersonic turbulence, with clumps sometimes achieving sufficient mass to become self-gravitating. Even if the turbulent cascade has resulted in turbulence support, turbulence then dissipates rapidly, and the cores collapse to form stars. Hence, there are two competing models for driving the star formation process.

Observing polarization of electromagnetic radiation from the cosmos is the principal and most direct way to learn about cosmic magnetic fields. However, such observations are always sensitivity limited, since the polarized intensities are at most only a few percent of the total intensities. Observations of the polarization of the continuum emission from dust will be addressed by others at this conference. This paper will focus on polarization of molecular spectral lines. To date, although the JCMT and the IRAM 30-m telescope have achieved detections of polarization in mm-wave spectral lines, the only high angular-resolution (interferometric) results have been from BIMA. The enormous increase in sensitivity that ALMA will provide should lead to a tsunami of polarization data that, hopefully, will finally enable us to understand the role that magnetic fields play in the star formation process.

This work was partially supported by NSF grants AST 0540459 and 0606822.

R.M. Crutcher (✉)
Astronomy Department, University of Illinois, Urbana, IL 61801, USA
e-mail: crutcher@uiuc.edu

2 Observational techniques

Both the circular and linear polarization of spectral lines discussed here are due, directly or indirectly, to the Zeeman effect. If a spectral-line forming region is permeated by a field **B**, the radiation is split by the normal Zeeman effect into three separate frequencies, $\nu_{\sigma-} = \nu_0 - \nu_Z$, $\nu_\pi = \nu_0$, and $\nu_{\sigma+} = \nu_0 + \nu_Z$, where ν_0 is the line rest frequency, $\nu_Z = B \times Z$, and Z is the Zeeman sensitivity, typically given in Hz/μG. For a magnetic field in the plane of the sky, the three Zeeman components are linearly polarized with the π component parallel to and the σ components perpendicular to the magnetic field direction. For a magnetic field along the line of sight, $I_\pi = 0$ and the I_σ are oppositely circularly polarized. If the magnetic field is parallel (antiparallel) to the direction of propagation of the radiation, $I_{\sigma+}$ is right (left) circularly polarized. In the general case of arbitrary angle θ between the line of sight and the magnetic field, the σ components will be elliptically polarized. The specific expressions for the Stokes parameters in the general case are given by Crutcher et al. (1993).

Only those species with an unpaired electron will have a magnetic moment that scales with the Bohr magneton, $M_B = eh/4\pi m_H c = 1.4$ Hz/μG. Otherwise, the Zeeman effect will scale with the nuclear magneton, 1840 times smaller than M_B. Unfortunately, most molecules do not have unpaired electrons and large Zeeman splitting factors Z, so the possibilities for Zeeman observations are limited. Lines for which the Zeeman effect has been detected that have been used for measurements of interstellar magnetic fields are the 21-cm line of H I, the 18-cm, 6-cm, 5-cm, and 2-cm Λ-doublet lines of OH, the 1.3-cm H_2O maser line (a nuclear magneton case), and the 3-mm $N = 1 - 0$ lines of CN. Other possible Zeeman molecules that might become useful probes of interstellar magnetic fields include CH, CCS, and SO.

Generally $\nu_Z \ll \Delta\nu$ (the line width), and it is not possible to infer complete information about **B** from Zeeman observations. Although the spectral-line profiles in the Stokes parameters V, Q, and U provide in principle full information about magnetic field strength and direction, in practice full information on **B** cannot be obtained owing to the extreme weakness of Q and U (by a factor $\sim \nu_Z/\Delta\nu$ compared with V) (Crutcher et al. 1993). The Stokes V spectra reveal the sign (i.e., direction) and magnitude of the line-of-sight component $B_\parallel$. That is, $V(\nu) = [dI(\nu)/d\nu]\nu_Z \cos\theta$. The sign convention is that $B_\parallel$ positive means that the line-of-sight component of **B** points away from the observer. By fitting the frequency derivative of the Stokes parameter $I(\nu)$ spectrum $dI(\nu)/d\nu$ to the observed $V(\nu)$ spectrum, $B_\parallel = \nu_Z \cos\theta/Z$ may be inferred. For a large number of clouds whose magnetic fields are randomly oriented with respect to the observed line of sight, $\overline{B}_\parallel = \frac{1}{2}|\mathbf{B}|$.

A bias can occur if the angular resolution is poor. For example, if rotational twisting of a poloidal field takes place, the field may retain a significant toroidal component even after magnetic braking has reduced the rotation of a cloud to very small values. Since a toroidal field will have $B_\parallel$ with opposite signs on the two sides of the cloud, partial cancelation of the Zeeman effect will take place unless the field is spatially resolved. An example of reversal of $B_\parallel$ over a molecular cloud is W3 (Roberts et al. 1993).

Linear polarization may also arise in radio-frequency spectral lines formed in the interstellar medium, even when Zeeman splitting is negligible (Goldreich and Kylafis 1981). Frequency-shifted σ Zeeman components come from levels that only emit or absorb circularly polarized radiation parallel to the magnetic field or linearly polarized radiation perpendicular to the magnetic field, while the unshifted π Zeeman component requires linearly polarized radiation parallel to the magnetic field. Collisional excitation populates the magnetic sublevels such that no net linearly polarized line emission occurs. However, anisotropic radiative excitation can unequally populate the magnetic sublevels and result in a net linear polarization of the spectral line. The direction of the polarization can be either parallel or perpendicular to the magnetic field, depending on the relationship between the line of sight, the direction of the magnetic field, and the direction of maximum spectral-line optical depth. Although the theory makes specific predictions for whether the field is parallel or perpendicular to the line polarization, in general the observations do not provide all of the necessary information. This ambiguity is unfortunate, but if structure in a cloud causes a flip by 90° in a map of polarization position angles, it would easily be recognized and not confused with random magnetic fields. This Goldreich–Kylafis effect is therefore a valuable tool in the measurement of magnetic field direction and in the degree of randomness of the field. It of course requires radiative dominating collisional excitation, so the effect is quenched at high densities. The Zeeman splitting of the magnetic sublevels need only be larger than the natural width of the spectral line for this Goldreich–Kylafis effect to be possible; since the observed molecules need not have a large Zeeman effect, it is applicable to abundant molecules like CO. Maps of the position angles of the polarization yield the field morphology and open the prospect of exploring this morphology as a function of radial velocity. The Chandrasekhar–Fermi relationship (Chandrasekhar and Fermi 1953) may be applied to infer the mean value of $B_\perp$. Goldreich and Kylafis (1981) and Kylafis (1983) discussed spectral-line linear polarization in terms of a velocity gradient that would produce anisotropic trapping of molecular-line photons, which then would produce anisotropic radiative excitation when re-absorbed. However, a continuum excitation source that occupies only a fraction of the sky seen from a line-emitting region may also produce anisotropic molecular excitation and linearly polarized line radiation.

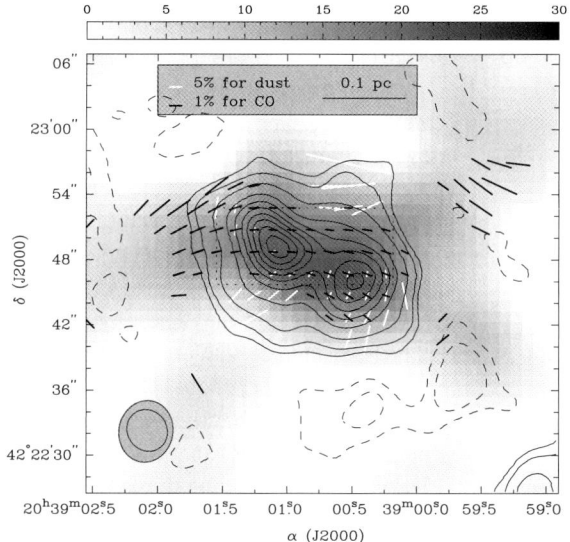
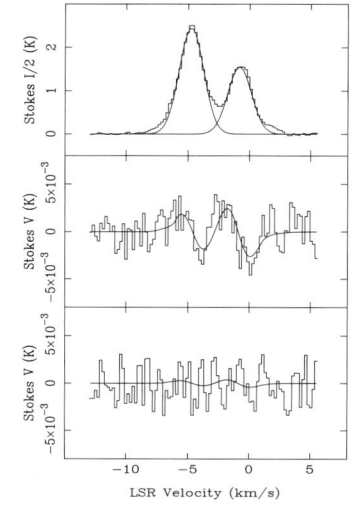

Fig. 1 *Left*: BIMA map of DR21OH (Lai et al. 2003). *Contours* show the 1.3-mm dust emission, grey scale the integrated CO 2-1 line emission, and line segments the spectral-line and dust linear polarization. *Right*: IRAM 30-m CN 1-0 line profiles toward DR21OH (Crutcher et al. 1999). Observed data are *histogram plots*, fits are *lines*. *Top panel* shows the Stokes I spectrum with two Gaussians fitted. *Middle panel* shows the mean Stokes V spectrum for the four hyperfine components that have strong Zeeman splitting coefficients Z; the *bottom panel* shows the three components with weak Z. $B_{\parallel}$ was fitted independently for the two Gaussian lines

3 Currently available line polarization results

An example (see Fig. 1) of the data currently available is provided by studies of the high-mass star formation region DR21OH. In millimeter-wave dust emission the main component of DR21OH consists of two compact cores with a total mass of $\sim 100 M_{\odot}$. The two CN velocity components (Crutcher et al. 1999) are each centered on a different one of the two compact cores. The results from the dust and CO 2-1 linear polarization maps toward the cores (Lai et al. 2003) suggest that the magnetic field direction in DR21OH is parallel to the CO polarization and therefore parallel to the major axis of DR21OH. The dust polarization data north of the cores suggest that **B** is along the minor axis to the north of the cores. The CO and dust polarization maps suggest that magnetic fields are remarkably uniform throughout the region. Both the dust emission and the CN lines sample a density $n(H_2) \approx 1 \times 10^6$ cm^{-3}. The Chandrasekhar–Fermi technique yields $B_{\perp} \approx 1$ mG, compared with $B_{\parallel} = -0.4 \pm 0.1$ mG and $B_{\parallel} = -0.7 \pm 0.1$ mG inferred from the CN Zeeman detections (Crutcher et al. 1999) shown in Fig. 1. Combining these results, $B_{\text{total}} \approx 1.2$ mG and **B** is at an angle $\theta \sim 55°$ to the line of sight. If we accept that the total field strength is 1.2 mG and make no geometrical correction, the mass-to-flux ratio is 2.0 times larger than the critical value where gravity and magnetic support are equal.

Recently, Cortes et al. (2005) mapped the linear polarization of the CO 1-0 line toward DR21OH. Remarkably, the CO 1-0 polarization is *orthogonal* to the CO 2-1 polarization. A single source of radiative excitation should produce the same sense of polarization in the two transitions. They carried out a detailed modeling of the Goldreich–Kylafis effect toward DR21OH, hypothesizing that radiation from the double dust core could be a second source of CO anisotropic radiative excitation; the first source would be a velocity gradient in the CO gas. Conclusions were the following. Fairly low H_2 densities, $n(H_2) \sim 100$ cm^{-3}, are required to produce linear polarization in the CO lines. At higher densities collisional excitation dominates, at lower densities the path lengths needed to produce the observed CO column densities are unrealistically long. With orthogonal orientations with respect to **B**, the two sources of radiative excitation overpopulate different magnetic sublevels. At very low line optical depths, photon trapping is not important, and excitation by photons from the dust cores dominates. At intermediate optical depths, photon trapping dominates the excitation. Therefore, the sense of the linear polarization switches from perpendicular to **B** at low optical depths to parallel to **B** at higher optical depths. However, the optical depths of the CO 2-1 and 1-0 transitions are not the same, so the switch occurs at different line optical depths. There is a range in τ over which the linear polarizations of the two lines are orthogonal. These results help define the geometry of the region and the density over which the observed CO linear polarization occurs. The CO polarization results apply to the envelope region of DR21OH, while the dust and CN polarization results apply to the core. Application of the Chandrasekhar–Fermi method to the CO data yields $B_{\perp} \approx 10 \, \mu$G, or two orders of magnitude smaller than in the dense core. The mass-to-flux ratio in the envelope is about 0.2 critical. This suggests that the mass-to-flux ratio has increased from the envelope to the core, as predicted by ambipolar diffusion.

4 The future with ALMA

ALMA will provide greatly improved sensitivity and resolution for imaging of linear polarization in spectral lines and greatly improved resolution for imaging of the Zeeman effect. Due to sensitivity limitations, most polarization observations will probably be in the compact array, with resolution of $\sim 4''$ at 3-mm wavelength. However, for very bright sources, considerably higher resolution will certainly be possible.

The current standard for spectral-line linear polarization mapping is BIMA. ALMA will have ≈ 30 times the collecting area, dual rather than single receivers, and much lower system noise temperatures than the BIMA system employed to produce results such as those for DR21OH. The result will be an improved sensitivity of more than two orders of magnitude at the same angular resolution, and the ability to push to higher angular resolution than was possible with BIMA. Imaging of spectral-line polarization over core regions with lines and transitions that probe a range of gas densities will make it possible to study the 3-D structure of magnetic field morphology and strength (using the Chandrasekhar–Fermi technique). Although more difficult due to the high angular resolution required, it should be possible to map linearly polarized spectral-line emission in protostellar and protoplanetary disks. Finally, following up the BIMA results for NGC1333 IRAS4 (Girart et al. 1999), the magnetic fields in CO bipolar outflows can be imaged. We will therefore be able to study the details of the magnetic fields in the cloud envelopes, protostellar cores, protoplanetary disks, and outflows from protostars.

Zeeman observations at millimeter wavelengths (of the 3-mm lines of CN) have been made with the IRAM 30-m telescope, which has $\sim 10\%$ the collecting area of ALMA and dual receivers. Hence, the sensitivity improvement at the same angular resolution would be only about one order of magnitude. However, the angular resolution of the 30-m telescope at the CN 1-0 frequency (113 GHz) is $23''$, and no mapping has been done. The major advantages of ALMA will be high angular resolution and mapping capability. This may well lead to detections of significantly higher magnetic field strengths than the single-dish results due to small-scale structure in the fields, and will complement the linear polarization results in giving direct measurements of magnetic field strengths.

References

Chandrasekhar, S., Fermi, E.: Astrophys. J. **118**, 133 (1953)
Cortes, P.C., Crutcher, R.M., Watson, W.D.: Astrophys. J. **628**, 780 (2005)
Crutcher, R.M., Troland, T.H., Goodman, A.A., et al.: Astrophys. J. **407**, 175 (1993)
Crutcher, R.M., Troland, T.H., Lazareff, B., et al.: Astrophys. J. **514**, L121 (1999)
Girart, J.M., Crutcher, R.M., Rao, R.: Astrophys. J. **525**, L109 (1999)
Goldreich, P., Kylafis, N.D.: Astrophys. J. **243**, L75 (1981)
Kylafis, N.D.: Astrophys. J. **275**, 135 (1983)
Lai, S.-P., Girart, J.M., Crutcher, R.M.: Astrophys. J. **598**, 392 (2003)
MacLow, M.-M., Klessen, R.S.: Rev. Mod. Phys. **76**, 125 (2004)
Mouschovias, T.C., Ciolek, G.E.: Magnetic fields and star formation: a theory reaching adulthood. In: Lada, C.J., Kylafis, N.D. (eds.) The Origin of Stars and Planetary Systems, pp. 305–339. Kluwer, Dordrecht (1999)
Roberts, D., Crutcher, R.M., Troland, T.H., Goss, W.M.: Astrophys. J. **412**, 675 (1993)
Wielebinski, R., Beck, R.: Cosmic Magnetic Fields. Springer, Berlin (2005)

Molecular clouds and star formation in the Magellanic Clouds and the Milky Way

Akiko Kawamura

Originally published in the journal Astrophysics and Space Science, Volume 313, Nos 1–3.
DOI: 10.1007/s10509-007-9646-x © Springer Science+Business Media B.V. 2007

Abstract Star formation is a fundamental process that dominates the life-cycle of various matters in galaxies: Stars are formed in molecular clouds, and the formed stars often affect the surrounding materials strongly via their UV photons, stellar winds, and supernova explosions. It is therefore revealing the distribution and properties of molecular gas in a galaxy is crucial to investigate the star formation history and galaxy evolution. Recent progress in developing millimeter and sub-millimeter wave receiver systems has enabled us to rapidly increase our knowledge on molecular clouds. In this proceedings, the recent results from the surveys of the molecular clouds in the Milky Way and the Magellanic Clouds as well as the Galactic center as the most active regions in the Milky Way are presented. The high sensitivity with unrivaled high resolution of ALMA will play a key role in detecting denser gas that is tightly connected to star formation.

Keywords Molecular clouds · Star formation · Magellanic Clouds · The Galaxy · The Galactic center · Sub-millimeter

1 Introduction

Stars are formed in molecular clouds and regulate evolution of galaxies in various respects. It is therefore necessary to understand how stars are formed in molecular clouds. Formation of solar type stars begins with the formation of protostellar molecular cores whose density is greater than

A. Kawamura (✉)
Department of Astrophysics, Nagoya University, Furocho, Nagoya 464-8602, Japan
e-mail: kawamura@a.phys.nagoya-u.ac.jp

10^5 cm^{-3}. The formation of the first stellar core, the first opaque object in hydrostatic equilibrium, with a size of ~ 1 AU and the mass of $0.01 M_\odot$ or less is occurred in the central part of the dense core. The first core represents the actual moment of star formation and it is crucial to find out the object at this stage and to test theories of protostar evolution. Already some candidates of the first core object are identified. Onishi et al. (1999) finds an interesting object M27, the densest core in Taurus, which has properties consistent with the first core object from a systematic search for dense cloud cores in the whole Taurus complex (Onishi et al. 2002), though the low angular resolution 30″ limits to make a detailed comparison at 10 AU scale. Optically thick spectra in millimeter/sub-millimeter exhibit self-absorption feature in the center and asymmetry consistent with contraction motion. These spectra have been analyzed by a Monte Carlo simulation to derive the radial distribution of density and infall velocity by assuming a spherical symmetry (Onishi et al. 1999). The results indicate that the outer envelope of MC27 at 1000 AU scales show the behavior predicted for an object very close to the moment of first core formation.

Formation of high mass stars is more difficult to observe than low mass stars because of the much smaller frequency of high mass star formation. It is however impossible to complete our understanding of star formation without understanding formation of high mass stars.

2 Molecular clouds and star formation in the Milky Way

High mass stars are formed exclusively in GMCs and it is of vital importance to understand star formation processes in GMCs whose mass is in a range, 10^5–$10^6 M_\odot$ in Solar vicinity (Biltz 1993). It has been widely accepted that almost

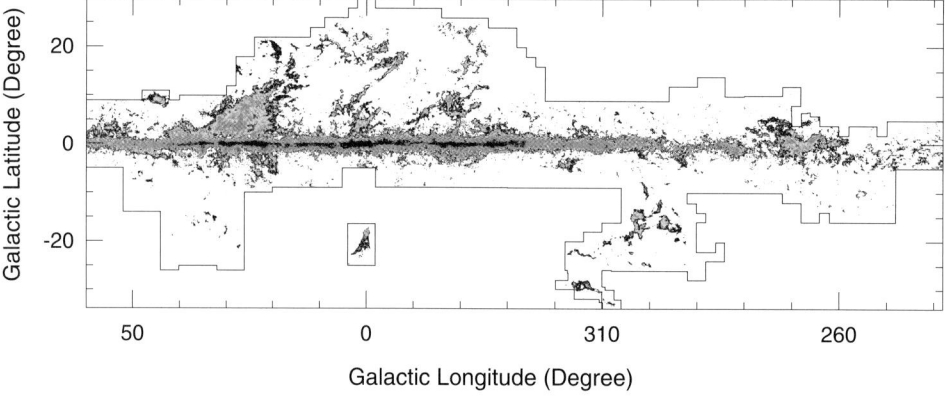

Fig. 1 A grayscale map of the ^{12}CO ($J = 1$–0) velocity integrated intensity shown in galactic coordinate for the Galactic plane survey by the NANTEN telescope. This map consists of more than 1,100,000 spectra. The observing grid spacing is $4'$ between 5 degrees from the galactic plane and $8'$ for the area above 5 degrees in the galactic latitude with a $2'.6$ beam. In the longitudinal direction almost 200 degrees, i.e., $L = 220°$ to $60°$ have been covered. The velocity resolution and coverage are 0.65 km s^{-1} and ~ 500 km s^{-1}, respectively

all GMCs are forming massive OB stars actively as indicated by accompanying HII regions and/or OB associations, in addition to a number of young solar-type low-mass stars. There have been several large scale surveys of molecular clouds toward the Galactic plane to understand the distribution and properties of the GMCs (e.g., Solomon et al. 1987; Dame et al. 2001; Matsunaga et al. 2001, see also Fig. 1) and there are many massive-star-forming GMCs such as M17, Orion A & B, W3/W4/W5, η Carinae, etc. There is only one known GMC, which lacks signs of massive star formation in the solar neighborhood, i.e., the Maddalena's cloud (Maddalena et al. 1986). The reason for this may be either that it is at a very young stage prior to massive star formation (Maddalena et al. 1986) or that it is in a late stage after active star formation (Lee et al. 1994). We should note that Maddalena's cloud is forming T Tauri stars, young low-mass stars, and its peculiarity is in its lack of OB stars. It has been discussed that some fractions of GMCs show no sign of massive star formation in the inner disk (Scoville and Good 1989). It is however to be kept in mind that the individual GMCs studied in detail so far are limited only to the solar vicinity because of heavy contamination or obscuration in the Galactic disk.

High mass stars are formed in clusters. In the Milky Way, such clusters are open clusters or OB associations, which contain about 100–1000 stars. Molecular cloud cores associated with young high mass stars are identified as "hot cores" whose mass is typically $1000 M_\odot$. Most of them are however deeply obscured at optical wavelengths and the confusion in the galactic disk is generally quite heavy, too. There is no doubt that ALMA can provide a significant progress in resolving these cloud cores to better understand the initial conditions for high mass star formation and several of recent reviews may well give a perspective on this (e.g., Mardones 2003).

We started to understand the properties of the ISM and formation processes of the stars from the observational studies to our own Milky Way with theoretical works. Its close distance enables us to carry out the highest resolution observations, however, we cannot escape from the relative uncertainties in the distance estimates as well as the extinctions to many of the objects, especially towards the Inner Galaxy. In order to better understand GMCs and star formation, extend further our survey for *spatially resolved GMCs* in galaxies.

3 Molecular clouds and star formation in the Magellanic Clouds

The Magellanic System is the nearest neighbor to our own and thus is the most suitable target to make a detailed study of ISM and star formation activities: the Large and Small Magellanic Clouds (the LMC and SMC) are located at $\sim$50 kpc from the sun, 16 times closer than the M31 group. It is also notable that the LMC is nearly face-on, making it feasible to identify associated objects with GMCs without suffering from the contamination along the line of sight. Because of this observational advantage, observations covering almost entire regions of the Magellanic Clouds have been carried out in a wide range of wavelengths and have been bringing us important knowledge of the properties of the stars and ISM.

The environments are different from those in the Galaxy: low metallicity (e.g., Dufour 1984) and a relatively strong UV field (e.g., Israel et al. 1986). Star formation activities are also different. Stellar clusters called "populous clusters", which are self-gravitating like Galactic globular clusters but with mass of $\sim 10^4$–$10^5 M_\odot$, which are by an order of magnitude smaller than those of the Galactic globular clusters but by an order of magnitude larger than those of the galactic open clusters (van den Bergh 1991), are found by photometric studies (e.g., Hodge 1961). It is notable that more than a hundred of the populous clusters are significantly young, a few to 100 Myr, and some are still being formed at present, such as R136 (e.g., Massey and Hunter 1998). This suggests that we can study the formation process of globular-like rich clusters through the studies of the molecular clouds in the LMC. To date, optical indicators of the massive star formation or cluster formation, such as HII regions and stellar clusters, have been studied in a large area of the LMC (e.g., Henize 1956; Bica et al. 1996). Recent surveys by IR satellites also started to bring us the details of the dust properties as well as young stellar object down to a few solar masses; Spitzer observations by e.g., Meixner et al. (2006) and Bolatto et al. (2007), AKARI observations (PI Onaka) by Doi et al. (2007).

A CO $J = 1$–0 survey of the Magellanic Clouds made with a 4 m mm-wave telescope NANTEN has provided an opportunity to study a rich sample of GMCs at a spatial resolution of 40 pc, well below a typical size of a GMC, ~ 100 pc (Fukui et al. 2007). This survey complement to the previous surveys covering an entire galaxy but with low resolution (e.g., Cohen et al. 1988) and those with high resolution by covering limited regions (e.g., Israel et al. 1993). The NANTEN CO survey achieved a sensitivity equivalent to the detection limit corresponding to $N(H_2) \sim 1 \times 10^{21}$ cm^{-2} (Fukui et al. 2007). It has revealed distributions of GMCs as shown in Fig. 2. The sample of GMCs has been used to

study how star formation is taking place by comparing the GMC distributions with optical signposts of star formation (both optical and radio continuum data have been employed in the comparison; Filipovic et al. 1998; Fukui et al. 2007; Kawamura et al. 2007). The galaxy is actively forming stars as populous clusters (small globular clusters) or OB associations as well as HII regions and are yet not too much contaminated with these, allowing to discern reliable assignments of young optical objects to individual GMCs. This comparison clearly indicates that the distribution of the youngest clusters and the HII regions are sharply peaked within 100 pc of a GMC, while the older clusters show much weaker correlations.

The NANTEN CO survey of the LMC has revealed that there are three Types of GMCs according to the association with HII regions and clusters or associations as young as 30 Myrs (SWB 0 by Bica et al. 1996). It shows that 24% of the GMCs are starless (Type I) in the sense that they are not associated with HII regions or young clusters. It should be noted that "starless" here indicates no associated early O star capable of ionizing HII regions, and that it does not exclude the possibility of associated young stars later than B-type not detectable through HII regions. About a half of the GMCs (Type II) are associated with small HII region(s) only but without stellar clusters and the rest (Type III) is most actively forming stars as shown by huge HII regions and young stellar clusters. A comparison of physical parameters among these three types indicates that size and mass

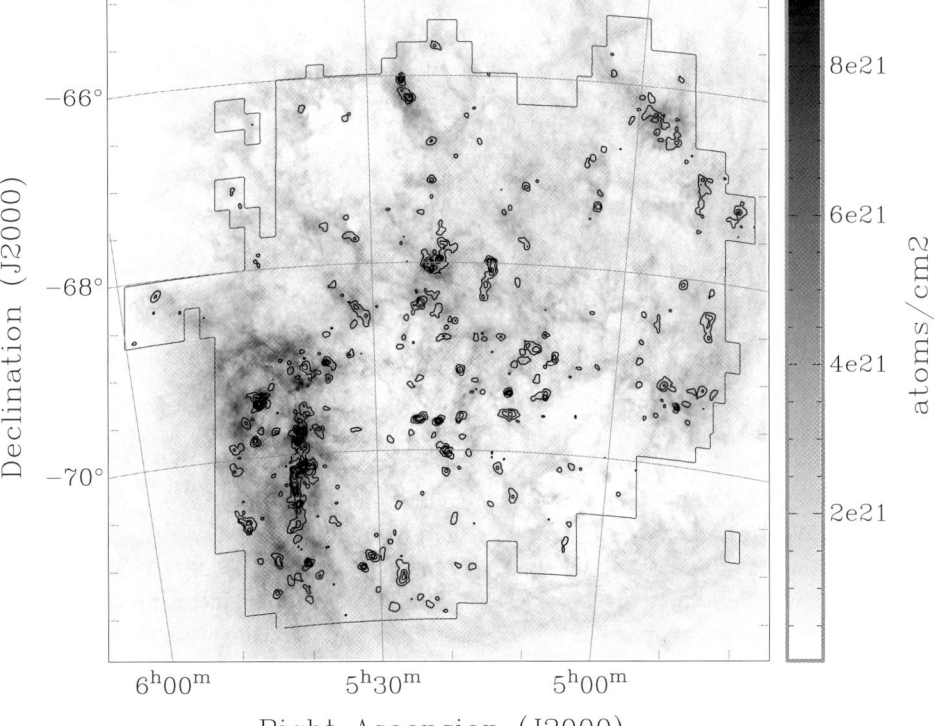

Fig. 2 HI image (Kim et al. 2003) with the CO contours (Fukui et al. 2007). The *contours* are from 1.2 K km s^{-1} ($= 3\sigma$ noise level) with 2.4 K km s^{-1} intervals. The *broken lines* indicate the observed area (Fukui et al. 2007)

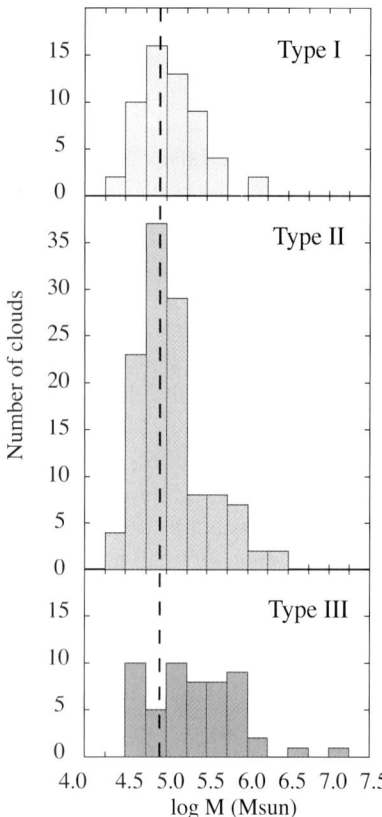

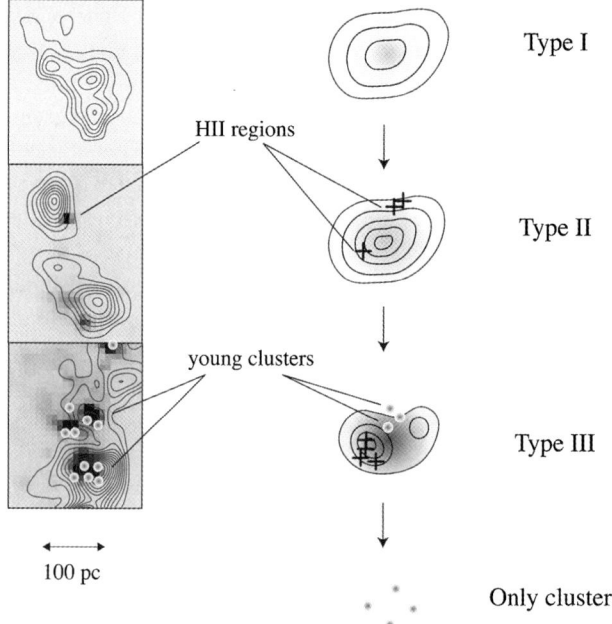

Fig. 4 Evolutionary sequence of the GMCs in the LMC. An example of the GMCs and illustration at each class are shown in the *left panels* and the *middle column*, respectively. The *images* and *contours* in the *left panels* are Hα (Kim et al. 1999) and CO integrated intensity by NANTEN (Fukui et al. 2007)

Fig. 3 Mass distribution of Type I (*top*), Type II (*middle*), and Type III (*bottom*) GMCs, respectively ($X = 7 \times 10^{20}$ cm^{-2} K km s^{-1} is assumed; Fukui et al. 2007)

tend to increase from Type I to Type III, and Type III GMC has the largest size and mass among the three, while the line width have a similar distribution among the three types (Fig. 3).

These Types can be interpreted as indication of the evolutionary sequence from I to III (Fig. 4) and the life time of a GMC is estimated to be a few × 10 Myrs in total by comparing the lifetime of stellar clusters based on a steady-state assumption (Kawamura et al. 2007). The stage after Type III is perhaps a very violent dissipation of GMCs due to UV photons and stellar winds from formed clusters as seen in the region of 30Dor most spectacularly.

In order to study star formation activities in more detail, higher resolution surveys are necessary towards the individual GMCs. SEST 15 m telescope has been used to carried out such studies and distribution and properties of dense cores have been studied. Recently, sub-mm telescopes, such as APEX and ASTE have been used to study more detail in sub-millimeter bands in the selected GMCs. Figure 5 shows CO $J = 3$–2 emission at 345 GHz detected by ASTE superposed on the NANTEN CO $J = 1$–0 map of one of the most active star and cluster forming regions, N159. The clump mass are estimated as several times $10^4 M_\odot$ to a few times $10^5 M_\odot$ by using the virial theorem,

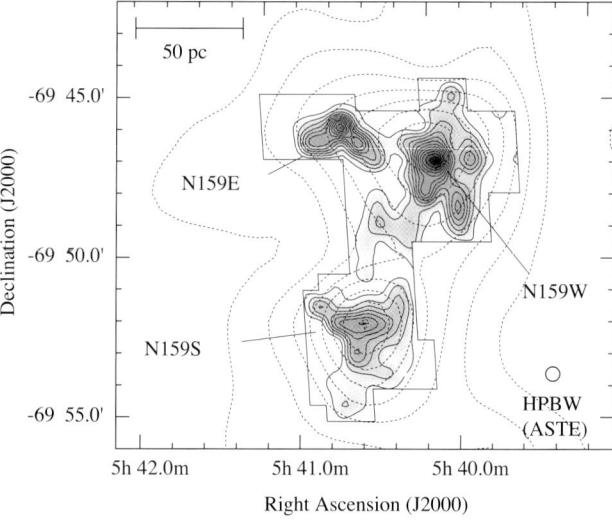

Fig. 5 Dense cores in the N159 GMC observed in ^{12}CO (3–2) by ASTE. The image with thick contours indicates the integrated intensity. The contours are from 2 K km/s with 3 K km/s intervals. Thin contours show the CO integrated intensity by NANTEN; the lowest contours and the intervals are 1.2 K km/s, respectively

$M_{\mathrm{vir}} = 200\Delta v$ [km/s]^2 R [pc] $M_\odot$, where R is the geometrical mean measured at the half intensity contour level. The beam size of ATE, 22 arcsec, corresponds to 5 pc, and the typical clump size here is several pc.

Figure 6 shows plots of the intensity ratio of CO $J = 3$–2 by ASTE to CO $J = 1$–0 by MOPRA or SEST vs. the Hα

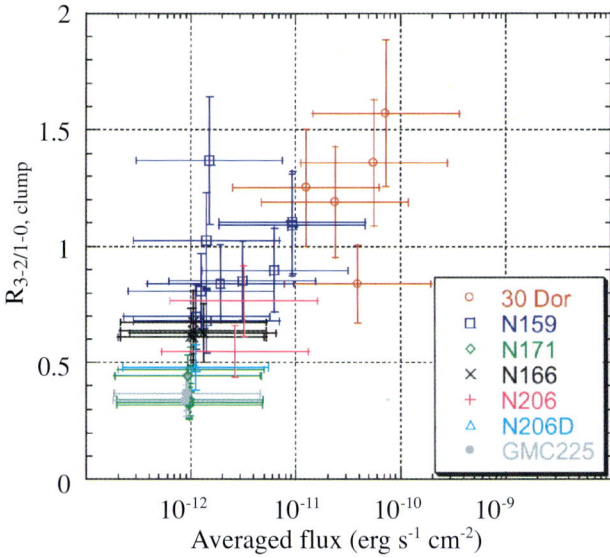

Fig. 6 Plots of $R_{3-2/1-0,clump}$ as a function of Hα flux (Kim et al. 1999) averaged over a clump

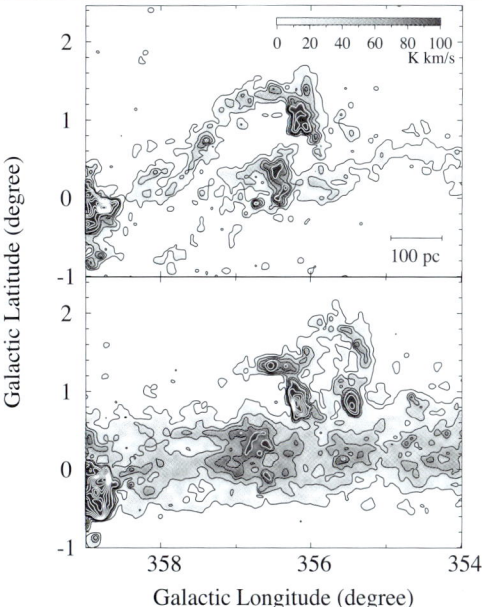

Fig. 7 Integrated intensity map of loops 1 (*top panel*) and 2 (*lower panel*) in CO ($J = 1$–0). Loop 1 is located in $l \sim 356$ deg to 358 deg, and $b \sim 0$ deg to -1.5 deg. The velocities are integrated from -180 to -90 km s^{-1}. Loop 2 is identified in $l \sim 355$ deg to 356 deg, and $b \sim 1$ deg to 2 deg. The velocities are integrated from -90 to -40 km s^{-1}. *Contours* are from 7.1 K km s^{-1} ($= 10\sigma$) with 20 K km s^{-1} intervals; *contours in white* represent those from 40 K km s^{-1} (Fukui et al. 2006)

flux averaged over each clump. This shows a clear trend that the intensity ratio becomes higher at active star forming regions. Together with the fact that the masses and sizes of the clumps are nearly similar to those of the populous stellar clusters, it is suggested that these clumps are at similar scale as the individual sites of massive stars or cluster formation, i.e. possible candidates for the precursors to populous clusters.

It will be extremely interesting to investigate the structure and properties of denser gas in these massive clumps, which are now being undertaken with APEX or NANTEN2 sub-mm telescopes at higher frequencies. It would be particularly important to find very high mass and dense cloud cores of 10^5–$10^6 M_\odot$, a promising candidate for proto-globular clusters, in order to understand the origin of difference in mass or in number of member stars among various stellar clusters. This target is perhaps to be attained with ALMA in high J submillimeter spectra that can probe dense molecular gas within GMCs.

4 The Galactic center

Center of the galaxies are most active regions in terms of both star formation and ISM dynamics (e.g. Morris 1996; Oka et al. 1998). In our Milky way, the most massive young clusters are found at the Galactic center. For decades, observations in a wide variety of wavelengths are carried out toward the Galactic center and the nice reviews of the recent studies of the center of our Galaxy as well as extra-galaxies are found in this volume. In this paper, we shall introduce interesting features revealed from the molecular clouds survey in the Galactic plane in the following.

They are two loop-like structure, loop 1 and loop 2 at $l = 355$ deg to 358 deg (Fukui et al. 2006; poster by Torii et al. in this conference) identified from the Galactic plane survey by NANTEN (Fig. 1 and Fig. 7). Their velocities are -180 to -90 km s^{-1} and -90 to -40 km s^{-1} with a large velocity gradient along the loop, respectively. The foot points of the loops show a very broad line width up to 80 km s^{-1}, which are the characteristics of the molecular gas near the Galactic center, making us possible to consider as being located in the Galactic center. Then, the projected lengths of loop 1 and loop 2 are estimated as $\sim$500 pc and $\sim$300 pc, respectively and velocity gradients corresponds to $\sim$80 km s^{-1} per 250 pc along loop 1 and $\sim$60 km s^{-1} per 150 pc along loop 2 by assuming the distance of 8.5 kpc. Furthermore, the heights of these loops are estimated as $\sim$220 to $\sim$300 pc from the Galactic plane, significantly higher than the typical scale height of the nuclear disk.

The mass of each of the loops was estimated as $0.8 \times 10^5 M_\odot$ as a lower limit and the kinetic energy 0.9×10^{51} erg for a velocity dispersion of 30 km s^{-1}; this large energy and velocity dispersion cannot be explained by supernova explosions. To explain the phenomenon, Fukui et al. (2007) offer a model incorporating MHD instability. Theoretical simulation with the gas number density of 100 cm^{-3} in and the magnetic field of 150 μm produces loop-like features consistent with the observed results (Nozawa 2005). The Alfven speed was calculated as 24 km s^{-1} with these parameters and this model offers significant heating of in the molecular gas at the foot points. The velocity dispersion of the broad CO features corresponds to kinetic temperature higher than about 10^4 K if the shock is completely converted into thermal energy at the foot points. They suggest this new model has the potential to apply to the other salient broad velocity features in the Galactic center and to the heating of the molecular gas at their foot points. Not only this idea but also many will be examined certainly by higher resolution observations toward the center of galaxies in the near future.

5 NANTEN2 projects

The "NANTEN2" is an upgrade of the 4-m mm telescope, NANTEN, which was operated at Las Campanas Observatory, Chile. The upgrade started by moving NANTEN from Las Campanas to Atacama in Northern Chile at an altitude of 4,800 m in 2004 to realize a large-scale survey at sub-mm wavelengths. The purpose of this project is to reveal the physical and chemical states of interstellar gas in various density regions with the highly excited CO and CI spectra in the millimeter to sub-millimeter wavelength (100–800 GHz) from large-scale surveys toward the Galaxy and the nearby galaxies including the Magellanic Clouds. With thorough extensive surveys, we shall make studies of evolution of ISM and star formation process in the Local Group. This NANTEN2 Project is collaboration among universities in Japan (Nagoya University and Osaka Prefecture University), Germany (University of Cologne and University of Bonn), South Korea (Seoul National University), Chile (University of Chile), Australia (University of New South Wales), and Switzerland (ETH de Zurich).

We installed a new main dish, which consists of 33 aluminum panels adjustable with actuators, and a light carbon fiber back structure to achieve the sub-mm observations. After adjustment of the main reflector using photogarmmetry and holography, the expected surface accuracy is 15 micron rms. The telescope is enclosed in a dome with a Gore-Tex membrane to prevent perturbations such as strong wind and sunlight. The installation started at the beginning of 2004 and the first wave in millimeter was received in Sept. 2005 and sub-millimeter in May 2006. The scientific observation started in Sept. 2006 with single beam receivers; the receivers for the highest observing frequencies will be upgraded by KOSMA SMART (Sub-Millimeter Array Receivers for Two frequencies) receiver, a multi-beam receiver capable of observing both at 490 GHz and 800 GHz simultaneously and effectively.

NANTEN2 is equipped with such low-noise superconducting receivers and the field of view of NANTEN2 is larger than those of ASTE, APEX, and ALMA. NANTEN2 is suitable to cover a large sky area within a short observation time while the resolution is coarser than those of other sub-millimeter telescopes. In this sense NANTEN2 and other telescopes in Atacama are complementary to one another. The database provided by NANTEN2 must be a useful guide for the further high-resolution observations by larger telescope, and surely a number of on-going project in sub-mm region as well as the surveys in other wavelength will bring us variety of targets to ALMA.

Acknowledgements The NANTEN project is based on a mutual agreement between Nagoya University and the Carnegie Institution of Washington. We greatly appreciate the hospitality of all the staff members of the Las Campanas Observatory. We, NANTEN team, are thankful to many Japanese public donors and companies who contributed to the realization of the project. Some studies are financially supported in part by a Grant-in-Aid for Scientific Research from the Ministry of Education, Culture, Sports, Science and Technology of Japan (No. 15071203) and from JSPS (No. 14102003, core-to-core program 17004 and No. 18684003).

References

Bica, E., Claria, J.J., Dottori, H., Santos, J.F.C. Jr., Piatti, A.E.: Astrophys. J. Suppl. Ser. **102**, 57 (1996)

Biltz, L.: In: Levy, E.H., Lunine, J.I. (eds.) Protostars and Planets III, p. 125. University of Arizona Press, Tucson (1993)

Bolatto, A., et al.: Astrophys. J. **655**, 212 (2007)

Cohen, R.S., Dame, T.M., Garay, G., Montani, J., Rubio, M., Thaddeus, P.: Astrophys. J. **331**, 95 (1988)

Dame, T.M., Hartmann, D., Thaddeus, P.: Astrophys. J. **547**, 792 (2001)

Doi, et al.: Publ. Astron. Soc. Jpn. (2007, submitted)

Dufour, R.J.: In: van den, S. (ed.) Structure and Evolution of the Magellanic Clouds, p. 353. Reidel, Dordrecht (1984)

Filipovic, M.D., Jones, P.A., White, G.L., Haynes, R.F.: Astron. Astrophys. Suppl. Ser. **130**, 441 (1998)

Fukui, Y., et al.: Science **314**, 5796 (2006)

Fukui, Y. et al.: Astrophys. J. Suppl. Ser. (2007, submitted)

Henize, K.G.: Astrophys. J. Suppl. Ser. **2**, 315 (1956)

Hodge, P.W.: Astrophys. J. **133**, 413 (1961)

Israel, F.P., de Graauw, T., van de Stadt, H., de Vries, C.P.: Astrophys. J. **303**, 186 (1986)

Israel, F.P., et al.: Astron. Astrophys. **276**, 25 (1993)

Kawamura, A., et al.: In: Elmegreen, B., Palous, J. (eds.) Triggered Star Formation in a Turbulent ISM, p. 101. Cambridge University Press, Cambridge (2007)

Kim, S., Dopita, M.A., Staveley-Smith, L., Bessel, M.S.: Astron. J. **118**, 2797 (1999)

Kim, S., Staveley-Smith, L., Dopita, M.A., Sault, R.J., Freeman, K.C., Lee, Y., Chu, Y.-H.: Astrophys. J. Suppl. Ser. **148**, 473 (2003)

Lee, Y., Snell, R.L., Dickman, R.D.: Astrophys. J. **432**, 167 (1994)

Maddalena, R.J., Moscowitz, J., Thaddeus, P., Morris, M.: Astrophys. J. **303**, 375 (1986)

Mardones, D.: In: Galactic Star Formation Across the Stellar Mass Spectrum. ASP Conference Series, vol. 287, p. 3 (2003)

Massey, P., Hunter, D.A.: Astrophys. J. **493**, 18 (1998)

Matsunaga, K., et al.: Publ. Astron. Soc. Jpn. **53**, 1003 (2001)

Meixner, M., et al.: Astron. J. **132**, 2268 (2006)

Morris, M.: Annu. Rev. Astron. Astrophys. **34**, 645 (1996)

Oka, T., Hasegawa, T., Hayashi, M., Handa, T., Sakamoto, Y.: Astrophys. J. **493**, 730 (1998)

Nozawa, S.: Publ. Astron. Soc. Jpn. **57**, 995 (2005)

Onishi, T., Mizuno, A., Fukui, Y.: Publ. Astron. Soc. Jpn. **51**, 257 (1999)

Onishi, T., Mizuno, A., Kawamura, A., Tachihara, K., Yasuo, F.: Astrophys. J. **575**, 950 (2002)

Scoville, N.Z., Good, C.: Astrophys. J. **339**, 149 (1989)

Solomon, P.M., Rivolo, A.R., Barrett, J., Yahil, A.: Astrophys. J. **319**, 730 (1987)

Complex organic molecules in an early stage of protostellar evolution

Nami Sakai · Takeshi Sakai · Satoshi Yamamoto

Originally published in the journal Astrophysics and Space Science, Volume 313, Nos 1–3.
DOI: 10.1007/s10509-007-9625-2 © Springer Science+Business Media B.V. 2007

Abstract We have detected the rotational lines of $HCOOCH_3$ toward a Class 0 low-mass protostar, NGC1333 IRAS4B, which is reported to be extremely young according to the dynamical age of the molecular outflow (a few 100 yr). This suggests that the complex organic molecules appear from the very early stage of protostellar evolution. On the other hand, the complex organic molecules are not detected in a more evolved protostar, L1527. We have also found a similar trend in a massive star forming region, NGC2264. The $HCOOCH_3$ emission is almost absent toward IRS1, whereas it is concentrated near MMS3, which is younger than IRS1. In addition, the $HCOOCH_3$ intensity peak is slightly shifted from the dust emission peak, as is seen in the Orion KL Compact Ridge, giving an important clue to solve its origin.

Keywords ISM: individual (NGC1333 IRAS4B) · ISM: individual (L1527) · ISM: individual (NGC2264) · ISM: molecules · Stars: formation

1 Background

Complex organic molecules such as $HCOOCH_3$, $(CH_3)_2O$ and C_2H_5CN have been recognized exclusively in hot cores ($T \geq 100$ K, $n \geq 10^6$ cm^{-3}) of massive star forming regions

N. Sakai (✉) · S. Yamamoto
Department of Physics, The University of Tokyo, 7-3-1, Hongo, Bunkyo-ku, Tokyo 113-0033, Japan
e-mail: nami@taurus.phys.s.u-tokyo.ac.jp

T. Sakai
Nobeyama Radio Observatory, Minamimaki, Minamisaku, Nagano 384-1305, Japan

like Orion KL (e.g. Blake et al. 1987). Such highly saturated molecules are scarcely produced in the gas-phase at low-temperature, and hence, the grain surface chemistry is thought to be important in their production (e.g. Millar et al. 1991). When the grains are heated up by various activities of newly formed stars, the mantle species like H_2CO and CH_3OH are sublimated, and form more complex molecules through gas phase reactions. Therefore, the complex organic molecules have been observed extensively in hot cores in massive star forming regions, whereas they have not been studied in low-mass star forming regions.

Recently, Cazaux et al. (2003) succeeded to detect the complex organic molecules toward a low-mass protostar, IRAS16293–2422 (hereafter IRAS16293). The abundances relative to H_2 are by about an order of magnitude higher than those found in the hot core of Orion KL. Furthermore, Bottinelli et al. (2004a) detected $HCOOCH_3$ toward another Class 0 protostar, NGC1333IRAS4A (hereafter IRAS4A). These detections clearly demonstrated importance of the complex organic molecules in the chemical evolution of solar-type protostars. Furthermore crucial information on the formation mechanism of these molecules could be obtained because of simplicity of low-mass star forming regions. With these in mind, we have conducted the following observations.

2 When the complex organic molecules appear?

2.1 NGC1333IRAS4B and L1527

Both IRAS16293 and IRAS4A are Class 0 protostars. This means that the complex organic molecules exist in the early stage of protostellar evolution. In order to understand when these species first appear, we observed another Class 0

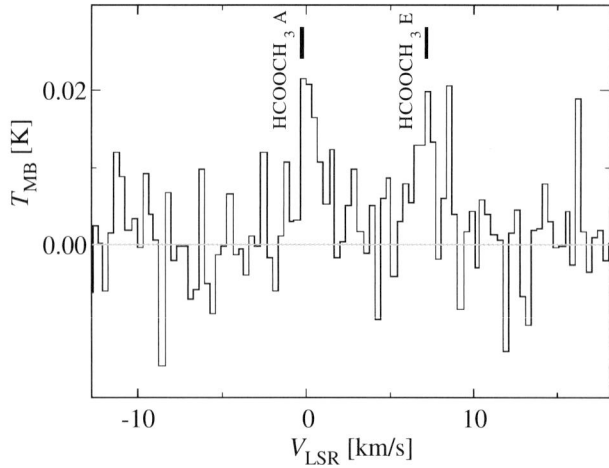

Fig. 1 Spectral line profile of HCOOCH$_3$ toward IRAS4B. V_{LSR} is calculated on the basis of the rest frequency of the E state line

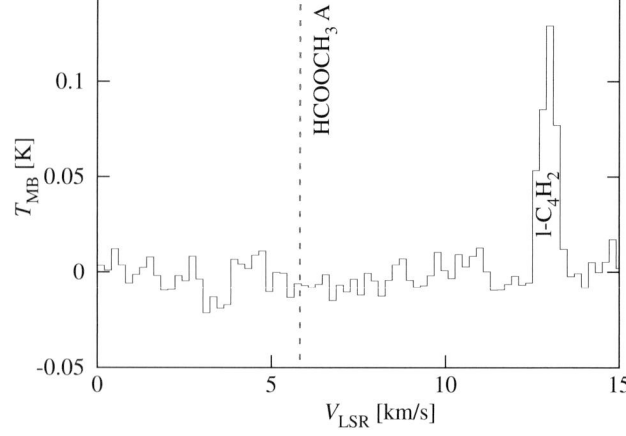

Fig. 2 Spectral line profile of HCOOCH$_3$ toward L1527 ($\alpha(2000.0) = 4^h39^m53.^s9$ and $\delta(2000.0) = 26°03'11''$). V_{LSR} is calculated on the basis of the rest frequency of the A state line

Table 1 HCOOCH$_3$ in low-mass star forming regions

Object	L_{bol} Maret et al. (2004) ($L_\odot$)	Distance Maret et al. (2004) (pc)	Cloud	Dynamical age (yr)	N(HCOOCH$_3$) 10^{15} (cm^{-2})	f(HCOOCH$_3$) 10^{-8}	Ref.
IRAS16293–2422	27	160	ρ-Oph.	(Multiple)	9.8	2.3	Cazaux et al. (2003)
NGC1333IRAS4A	6	220	Per.	6500	58	3.6	Bottinelli et al. (2004a)
NGC1333IRAS2A	16	220	Per.	–	$\leq 0.29^a$	$\leq 67^a$	Bottinelli et al. (2007), Jørgensen et al. (2005)
NGC1333IRAS4B	6	220	Per.	a few 100	≥ 3–16	≥ 1–5	Sakai et al. (2006)
L1527	2	140	Tau.	15000	≤ 0.3–1.9^b	≤ 1.2–6.6^b	This work

[a]The values averaged over 10''

[b]The source size is assumed to be 1''. T_{ex} range is from 50 K to 200K. N(H$_2$) is taken from Jørgensen et al. (2002)

protostar, NGC1333IRAS4B (hereafter IRAS4B), which is apart from IRAS4A by 0.03 pc. According to the dynamical age of the molecular outflow (a few 100 yr), IRAS4B is considered to be in a younger evolutionary stage than IRAS4A (Choi 2001).

The observations were carried out with the Nobeyama 45 m telescope (hereafter NRO 45 m). We observed HCOOCH$_3$, (CH$_3$)$_2$O, and C$_2$H$_5$CN in the 89 GHz region, and succeeded to detect two lines (A and E states) of the $8_{1,8}$–$7_{1,7}$ transition of HCOOCH$_3$ (Fig. 1) (Sakai et al. 2006). This is the third detection of HCOOCH$_3$ in low-mass star forming regions. IRAS4B is reported to be extremely young, and hence, it seems likely that the complex organic molecules appear from the very early stage of low-mass protostellar evolution.

In derivation of the column density, we assume the range for the excitation temperature to be from 50 K to 200 K according to the results for the other sources (Cazaux et al. 2003; Bottinelli et al. 2004a). The size of the emitting region

constitutes another uncertain factor. Here we assume the size to be 1'' according to the size in IRAS16293 reported by interferometric observations (e.g. Kuan et al. 2004). Under these assumptions, the range of the column density of HCOOCH$_3$ is obtained (Table 1), being comparable to that found in IRAS 16293. The column density is lower than that of IRAS4A by an order of magnitude. This is probably due to the smaller source size (0.''5) assumed for IRAS4A. When we assume the same source size for IRAS4B, the column density is similar to that found in IRAS4A. In NGC1333IRAS2A (hereafter IRAS2A), HCOOCH$_3$ is not found (Bottinelli et al. 2007), although Jørgensen et al. detected (CH$_3$)$_2$O (Jørgensen et al. 2005). All these results clearly established the existence of "hot cores" even in the Class 0 sources. Bottinelli et al. (2004a) named the hot region emitting the lines of complex organic molecules as a hot corino.

Next, we focused on a famous protostar, L1527 in Taurus. This source is known to be in a transient phase from

Class 0 to Class I, and the dynamical age of the outflow is older than 15,000 yr (Bachiller et al. 2006). Toward this source, we also made a very deep observation with NRO 45 m in the 89 GHz region. But we could not find any features of $HCOOCH_3$, $(CH_3)_2O$, and C_2H_5CN, with the rms noise level of 4.2 mK (T_{MB}) (Fig. 2). According to the gas-phase chemical model calculations (e.g. Nomura et al. 2004), the abundance of $HCOOCH_3$ has a peak at 10^4 yr after injection of the parent molecules like CH_3OH. Since $HCOOCH_3$ is destroyed by ionic species with a timescale of 10^{4-5} yr, the abundance would be lower in the evolved stage. This may be a possible reason why $HCOOCH_3$ is deficient in L1527. Such a chemical evolutionary effect may also affect the abundance of $HCOOCH_3$ in IRAS2A, which is more evolved than IRAS4A and IRAS4B. Alternatively, there might be a regional difference in chemical evolution, making the $HCOOCH_3$ abundance lower in L1527. However, the number of sources studied is very limited (Table 1), and more observations are still needed to establish the evolutionary scenario.

2.2 NGC2264IRS1 and NGC2264MMS3

The above results suggest that the abundance of the complex organic molecules may vary along protostellar evolution. This may also be the case for massive star forming regions. In order to assess this possibility, it seems useful to observe nearby regions containing massive protostars with different evolutionary stages. NGC2264 is a famous star forming region lying at a distance of 760 pc. The brightest IR source in this region is IRS1, which is a 9.5 $M_\odot$ B2 ZAMS star (Allen 1972). IRS1 is associated with a dense molecular clump, as revealed by the CS $J = 7 - 6$ line and submillimeter-wave continuum observations (Schreyer et al. 1997; Ward-Thompson et al. 2000). In addition to IRS1, the NGC2264IRS1 region contains several submillimeter-wave continuum sources, MMS1–5, each of which is forming one high mass star or multiple intermediate stars (Ward-Thompson et al. 2000). In particular, MMS3 is a dense core associated with the small star cluster, and has a compact molecular outflow (Schreyer et al. 1997; Ward-Thompson et al. 2000). Ward-Thompson et al. (2000) postulated that MMS3 involves a high mass equivalent of a Class 0 protostar and is younger than IRS1 region.

We first observed IRS1 with NRO 45 m (Sakai et al. 2007). However, we were not able to detect any lines of the complex organic molecules (Sakai et al. 2007). Then, we focused on the CH_3OH peak position located in the MMS3 region, which is located at $30''$ southeast of IRS1 (Schreyer et al. 1997). Toward this position, we succeeded to detect two lines of the $8_{1,8}-7_{1,7}$ transition of $HCOOCH_3$ (Sakai et al. 2007). The MMS3 region is much less active than the

IRS1 region, indicating that the abundance of $HCOOCH_3$ is not only dependent on the luminosity. A similar trend is reported for low-mass star forming regions (Bottinelli et al. 2007). Instead, the chemical evolutionary effect described in 2.1 would also play an important role in the massive star forming regions. In addition, we may have to consider the dynamical evolutionary effect. As seen in our interferometric map of the 3.35 mm continuum emission (Fig. 3), the protostellar core around IRS1 seems to have mostly been disrupted, although the warm dense gas traced by the CS $J = 7 - 6$ line and the submillimeter-wave dust continuum emission still surrounds IRS1. If $HCOOCH_3$ is formed in or around the protostellar core, it is no longer produced newly in the IRS1 region. Hence, the $HCOOCH_3$ abundance becomes lower in the IRS1 region.

3 How the complex organic molecules distribute?

3.1 Difference between massive and low-mass star forming regions

We studied detailed distributions of the complex organic molecules in the NGC2264 region with NMA (Sakai et al. 2007). In spite of the limited S/N ratio of the $HCOOCH_3$ map, it can be seen in Fig. 3 that the $HCOOCH_3$ emission is rather faint toward the continuum peak, whereas it is relatively strong toward positions slightly offset from the continuum peak. Although the $HCOOCH_3$ emission seems to be

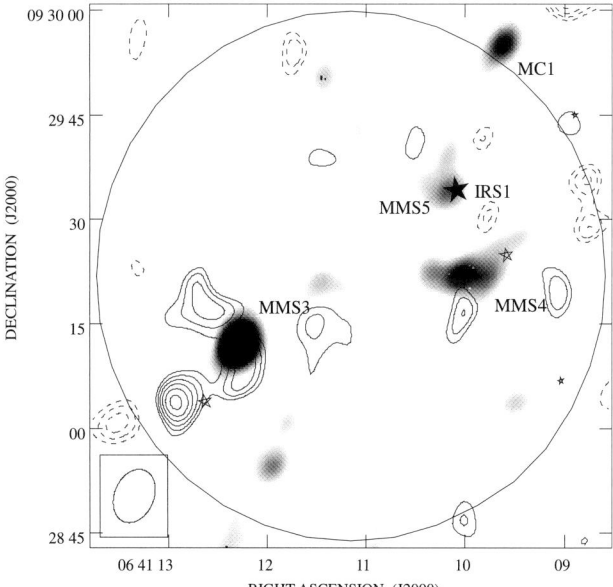

Fig. 3 Integrated intensity map of $HCOOCH_3$ (*contours*) observed with NMA, averaged over the A-state and the E-state. The rms is 8.5 mJy beam^{-1}. The lowest contour and the contour intervals are 2.5 σ and 0.5 σ, respectively. Negative contours are shown by dashed lines. The gray-scale image represents the 3.35 mm continuum image (Sakai et al. 2007)

distributed over the size of the star cluster, the HCOOCH$_3$ peaks are not associated with any known infrared sources. Particularly, it is strongest at the south-east peak, being apart from the continuum peak by 13$''$ (0.05 pc). This situation is similar to that seen in Orion KL. It is well known that HCOOCH$_3$ is more abundant toward Compact Ridge which is apart from the KL object and the dust continuum peak by about 10$''$ (e.g. Liu et al. 2002; Beuther et al. 2005). The effect of the outflow shock is suggested as an origin of Compact Ridge, but no direct evidences of the interaction with the outflow, such as velocity shifts and line broadenings, have been reported as far as we know. Therefore, the origin of this positional difference is still an open question. Very few regions have been studied so far with enough spatial resolution except Orion KL. Famous bright sources such as W3(OH) (Wyrowski et al. 1999) and G34.3+0.2 (Remijan et al. 2003) are too distant to be well resolved. Our result in NGC2264 MMS3 is the second well-resolved case, which would provide us with an important clue to understand the origin of Compact Ridge.

On the other hand, the distribution of the complex organic molecules including HCOOCH$_3$ is very compact around protostars in a low-mass star forming region, IRAS16293 (e.g. Kuan et al. 2004; Bottinelli et al. 2004b). In the low-mass case, complex organic molecules seems to be produced in the dense infalling material near the protostar. In contrast, these molecules distribute more globally over the size of star cluster in massive star forming regions. Therefore, the physical and chemical situations for production of complex organic molecules would be different between the low-mass and massive star forming regions.

3.2 Implication to production mechanisms

According to Nomura et al. (2004), the time scale for production of the complex organic molecules is about 10^4 yr after injection of parent molecules. However, our detection of HCOOCH$_3$ in a very young protostar IRAS4B suggests that HCOOCH$_3$ should be produced in a shorter timescale. In this relation, Bottinelli et al. pointed out that the crossing time of the infalling gas through the hot corino region is 10^3 yr which is much shorter than the gas-phase formation time scale. Because of these reasons, we should seriously consider a possibility that HCOOCH$_3$ is formed directly on grain surface and is released into the gas phase just after the onset of star formation. Recently, Garrot and Herbst examined grain surface reactions to produce HCOOCH$_3$ during the warming up phase (Garrod et al. 2006). This may resolve the time scale problem mentioned above.

Bottinelli et al. (2007) reported that the abundances of complex organic molecules in low mass star forming regions are higher than those in hot cores of massive star forming regions. According to the model by Garrod et al. (2006),

slower temperature raise results in higher abundances of the complex organic molecules. Since the physical evolution is slower for low mass star forming regions, higher abundances of complex organic molecules are expected, being consistent with the observational results. On the other hand, the difference in distribution between low-mass and massive star forming regions is hard to be interpreted. The difference may originate, for instance, from a difference in initial conditions, different speed of temperature raise, or different degree of molecular destructions in the later evolutionary stage.

4 Prospects with ALMA

For complete understanding of the chemical evolution from the protostellar core to protoplanetary disks, we still need more observations of complex organic molecules toward both of low-mass and massive star forming regions. However, the observation of complex organic molecules in low-mass star forming regions is very difficult because of weakness of the lines. Actually, the on-source integration time was as long as 18.5 hr and 22 hr in our observations toward IRAS4A and L1527, respectively. It is also difficult to resolve the distribution even for the massive star forming regions, because of the limited spatial resolution ($\sim$1$''$) of currently available instruments. ALMA will overcome these difficulties to deliver a clear view on the behaviors of the complex organic molecules.

References

Allen, D.A.: Infrared objects in H II Regions. Astrophys. J. **172**, L55 (1972)

Bachiller, R., et al.: Molecules in protostellar outflows. In: Complex Molecules in Space, Denmark (2006)

Beuther, H., et al.: Line imaging of Orion KL at 865 mum with the submillimeter array. Astrophys. J. **632**, 355 (2005)

Blake, G.A., et al.: Molecular abundances in OMC-1. Astrophys. J. **315**, 621 (1987)

Bottinelli, S., et al.: Complex molecules in the hot core of the low-mass protostar NGC 1333 IRAS 4A. Astrophys. J. **615**, 354 (2004a)

Bottinelli, S., et al.: Near-arcsecond resolution observations of the hot corino of the solar-type protostar IRAS 16293-2422. Astrophys. J. **617**, L69 (2004b)

Bottinelli, S., et al.: Hot corinos in NGC1333-IRAS4B and IRAS2A. Astron. Astrophys. **463**, 601 (2007)

Cazaux, S., et al.: The hot core around the low-mass protostar IRAS 16293-2422. Astrophys. J. **593**, L51 (2003)

Choi, M.: High-resolution observations of the molecular clouds in the NGC 1333 IRAS 4 region. Astrophys. J. **553**, 219 (2001)

Garrod, R.T., et al.: Formation of methyl formate and other organic species in the warm-up phase of hot molecular cores. Astron. Astrophys. **457**, 927 (2006)

Jørgensen, J.K., et al.: Physical structure and CO abundance of low-mass protostellar envelopes. Astron. Astrophys. **389**, 908 (2002)

Jørgensen, J.K., et al.: Probing the inner 200 AU of low-mass protostars with the submillimeter array. Astrophys. J. **632**, 973 (2005)

Kuan, Y.-J., et al.: Organic molecules in low-mass protostellar hot cores. Astrophys. J. **616**, L27 (2004)

Liu, S.-Y., et al.: Formic acid in Orion KL from 1 millimeter observations with the Berkeley-Illinois-Maryland Association Array. Astrophys. J. **576**, 255 (2002)

Maret, S., et al.: The H_2CO abundance in the inner warm regions of low mass protostellar envelopes. Astron. Astrophys. **416**, 577 (2004)

Millar, T.J., et al.: Gas phase reactions and rate coefficients for use in astrochemistry. Astron. Astrophys. Suppl. Ser. **87**, 585 (1991)

Nomura, H., et al.: The physical and chemical structure of hot molecular cores. Astron. Astrophys. **414**, 409 (2004)

Remijan, A., et al.: A survey of acetic acid toward hot molecular cores. Astrophys. J. **590**, 314 (2003)

Sakai, N., et al.: Detection of $HCOOCH_3$ toward a low-mass protostar, NGC 1333 IRAS 4B. Publ. Astron. Soc. Jpn. **58**, L15 (2006)

Sakai, N., et al.: Methyl formate in the NGC2264 IRS1 Region. Astrophys. J. **660**, 363 (2007)

Schreyer, K., et al.: A molecular line and infrared study of NGC 2264 IRS 1. Astron. Astrophys. **326**, 347 (1997)

Ward-Thompson, D., et al.: Dust emission from star-forming regions. VI. The submillimetre YSO cluster in NGC 2264. Astron. Astrophys. **355**, 1122 (2000)

Wyrowski, F., et al.: Hot gas and dust in a protostellar cluster near W3(OH). Astrophys. J. **514**, L43 (1999)

Revealing the "fingerprints" of the magnetic precursor of C-shocks

Izaskun Jiménez-Serra · Jesús Martín-Pintado ·
Arturo Rodríguez-Franco · Paola Caselli · Serena Viti ·
Tom Hartquist

Originally published in the journal Astrophysics and Space Science, Volume 313, Nos 1–3.
DOI: 10.1007/s10509-007-9645-y © Springer Science+Business Media B.V. 2007

Abstract We present the first C-shock and radiative transfer model that calculates the evolution of the line profiles of neutral and ion species like SiO, $H^{13}CO^+$ and $HN^{13}C$ for different flow times along the propagation of the shock through the unperturbed gas. We find that the line profiles of SiO characteristic of the magnetic precursor stage have very narrow linewidths and are centered at velocities close to the ambient cloud velocity, as observed toward the young shocks in the L1448-mm outflow. Consistently with previous works, our model also reproduces the broad SiO emission detected in the high velocity gas in this outflow, for the downstream postshock gas in the shock. This implies that the different velocity components observed in L1448-mm are due to the coexistence of different shocks at different evolutionary stages.

Keywords ISM: clouds · ISM: jets and outflows · Physical processes: shock waves · ISM: individual (L1448)

I. Jiménez-Serra (✉) · J. Martín-Pintado · A. Rodríguez-Franco
DAMIR-Instituto de Estructura de la Materia (CSIC),
C/ Serrano 121, 28006 Madrid, Spain
e-mail: izaskun@damir.iem.csic.es

P. Caselli
School of Physics and Astronomy, University of Leeds, LS2 9JT,
Leeds, UK

S. Viti
Department of Physics and Astronomy, University College
London, WC1E 6BT, London, UK

T. Hartquist
School of Physics and Astronomy, University of Leeds, LS2 9JT
Leeds, UK

1 Introduction

In star forming regions within dense molecular dark clouds, it is common to find the interaction of C-shock waves associated with bipolar outflows. The early stages of C-shocks are characterized by the interaction of the magnetic precursor, which accelerates, compresses and heats the ion fluid with respect to the neutral one (Draine 1980). The subsequent ion-neutral velocity drift leads to the sputtering of dust grains, injecting large amounts of molecular material into the gas phase (e.g. Draine et al. 1983).

Since silicon is heavily depleted onto dust grains ($\chi(SiO) \leq 10^{-12}$ in dark clouds; e.g. Ziurys et al. 1989), the large SiO abundances [$\chi(SiO) \sim 10^{-6}$] found in outflows makes this molecule an excellent tracer of the gas processed by C-shocks (Martín-Pintado et al. 1992). The typical SiO line profiles observed in these regions are very broad and centered at velocities very different from that of the ambient cloud (Martín-Pintado et al. 1992).

Surprisingly, Jiménez-Serra et al. (2004) reported the detection of very narrow SiO emission at almost ambient velocities (in the precursor component at 5.2 km s^{-1}, which implies a velocity shift of $\sim$0.5 km s^{-1} with respect to the ambient cloud at 4.7 km s^{-1}) toward the young shocks of the L1448-mm outflow. The derived SiO abundance is of $\sim 10^{-11}$. This emission, which appears together with the enhancement of the ions with respect to the neutrals, was proposed to be a signature of the magnetic precursor of C-shocks (Jiménez-Serra et al. 2004). The progressive enhancement of the abundances of shock tracers like SiO, CH_3OH and SO from the pre-shock gas to the precursor and post-shock components, is also consistent with an evolutionary sequence for the propagation of a C-shock through the ambient gas (Jiménez-Serra et al. 2005).

R. Bachiller, J. Cernicharo (eds.), *Science with the Atacama Large Millimeter Array*. DOI: 10.1007/978-1-4020-6935-2_30

While previous steady-state C-shock modelling has succeeded in reproducing the large abundances and the broad line profiles of SiO generated in the evolved post-shock gas (Schilke et al. 1997), there is not theoretical confirmation of the lower abundance and the narrow emission of SiO found in L1448-mm and expected to be produced by the precursor. Jiménez-Serra et al. (2007), who have recently calculated the evolution of the sputtering of dust grains in C-shocks, have predicted the increasing enhancement of the SiO and CH_3OH abundances as a result of the progressive erosion of dust grains as observed in L1448-mm (Jiménez-Serra et al. 2005). We present the first C-shock and radiative transfer model that allows to study the temporal evolution of the line profiles of SiO and of ion and neutral species like $H^{13}CO^+$ and $HN^{13}C$, for the different stages of the shock. We will compare these results with the molecular line profiles observed in the L1448-mm outflow.

2 The model

The steady-state profiles of the velocity, density and temperature of the ion and neutral fluids for a plane-parallel C-shock with shock velocity v_s, are approximated by the parametric model of Jiménez-Serra et al. (2007) (with z_n, z_i, z_T, z_0, a_T and b_T as the parameters that define the physical structure of the shock). As in Jiménez-Serra et al. (2007), we assume that silicon, which is mainly present in the olivine grain cores, is also a minor constituent of the water icy mantles. While the sputtering of the mantles is calculated following the procedure described in Caselli et al. (1997), the sputtering of the grain cores is determined by using the sputtering yields calculated by May et al. (2000) (see Jiménez-Serra et al. 2007 for details). As colliding particles, we have considered H_2 and He (the most abundant, but not the most efficient, sputtering agents; Jiménez-Serra et al. 2007), and the heavier species C, O, Si, Fe and CO.

The main differences with respect to the work of Jiménez-Serra et al. (2007) are that (i) we calculate the sputtering of grains as a function of the radial velocity, v_{LSR}, which is in the preshock frame (this will make it easier the comparison of the predicted SiO line profiles with the observed ones); and (ii) we let the model stop at certain flow times, t_{inst}, in the evolution of the shock. We refer *flow times* as the times associated with the neutral fluid (Jiménez-Serra et al. 2007).

The line profiles are calculated by means of a LVG model. We have used the H_2 collisional rates derived by Dayou and Balança (2006) for SiO, and by Schöier et al. (2005) for $H^{13}CO^+$ and $HN^{13}C$. The molecular level populations are determined in each plane-parallel layer of gas within the shock assuming that these layers are non-interacting. The linewidths are estimated by

considering only thermal and turbulent broadening (with $\Delta v_{tur} \sim 0.45$ km s^{-1}). The arising radiation is transfered through all the previous layers, and the final line profiles are calculated by adding up all the individual layer contributions.

3 The physical structure of the C-shock

In Fig. 1 we show the steady-state physical structure of the C-shock obtained with the parametric model of Jiménez-Serra et al. (2007) for $v_s = 70$ km s^{-1}, $n_0 = 10^5$ cm^{-3} and $z_0 = 0$ cm. This shock velocity is consistent with the terminal velocity of the broad SiO emission observed in L1448-mm (Martín-Pintado et al. 1992). Although this velocity could be considered as excessive for a C-shock (Draine et al. 1983), the recent results of Le Bourlot et al. (2002) show that the critical velocity of C-shocks can be increased to ~ 100 km s^{-1} for moderate H_2 densities and high magnetic fields.

The parameters z_n, z_i, z_T, a_T and b_T adopted for this model (see Table 1) have been chosen to match the intensity and central radial velocity of the narrow SiO emission reported for the magnetic precursor component in L1448-mm (see Sect. 5 and Jiménez-Serra et al. 2004). As expected in C-shocks (Draine et al. 1983), Fig. 1 shows that the ions are kinematically and thermally decoupled with respect to the neutrals in the precursor. According to Fig. 1, the precursor extends up to $\Delta z \sim 0.001$ pc $= 3 \times 10^{15}$ cm.

We note that the evolution of the physical parameters in the C-shock correspond to those typical of the steady-state (see e.g. Draine et al. 1983). Since the L1448-mm outflow is very young ($t_{dyn} \sim 3500$ yr), one may consider that the assumption of steadiness may not be valid for our case. However, although we cannot rule out the possibility that a J-type component may appear within the C-shock structure in the far downstream gas due to the high shock velocity, the time-scales required to attain the steady-state in a medium with $n_0 \sim 10^5$ cm^{-3} are of only ~ 1000 yr, which is of the same order of magnitude as the dynamical age of this outflow (Lesaffre et al. 2004).

4 The sputtering of dust grains

By including the physical structure of the C-shock into the sputtering equations of Jiménez-Serra et al. (2007), we can calculate the evolution of the gas phase abundances of Si/SiO ejected from the icy water mantles and the olivine grain cores. Consistently with the results of Jiménez-Serra et al. (2007), He is the most efficient sputtering agent for the water icy mantles at high shock velocities, and CO and O

Fig. 1 Physical structure of a C-shock calculated by using the parametric model of Jiménez-Serra et al. (2007) for $v_s = 70$ km s^{-1}, $n_0 = 10^5$ cm^{-3}, $v_0 = 4.7$ km s^{-1} and $T_0 = 10$ K. The subscripts n and i denote *neutrals* and *ions* (or *charged fluid*) respectively

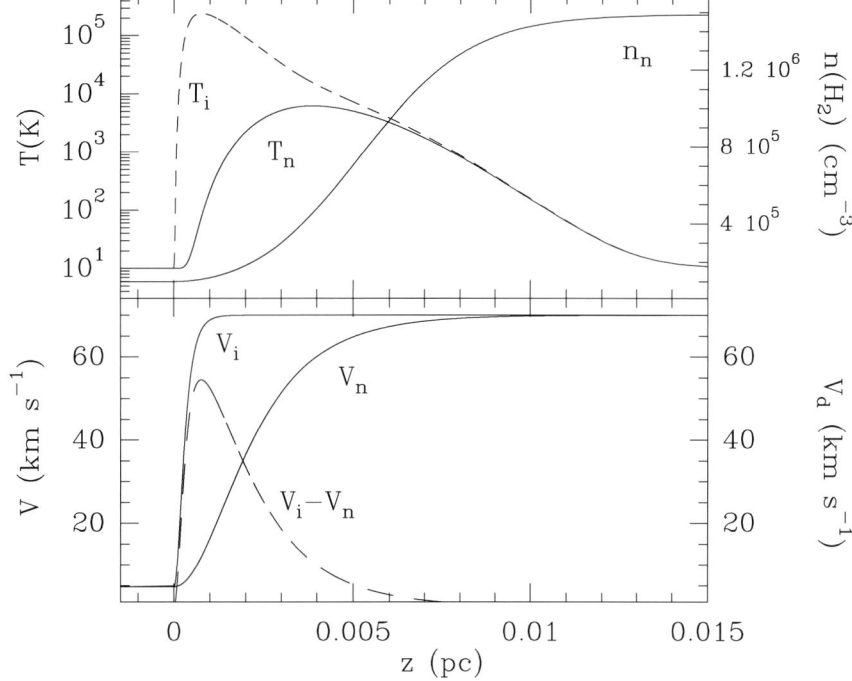

Table 1 Parameters z_n, z_i, z_T, a_T and b_T for the shock in Fig. 1

z_n (cm)	z_i (cm)	z_T (cm)	a_T (K$^{1/6}$ cm^{-1})	b_T	z_n/z_i
4.8×10^{15}	6.6×10^{14}	2.0×10^{15}	9.7×10^{-16}	6	7.3

eject all Si/SiO from the grain cores. The ion-neutral velocity drift, $v_d = |v_n - v_i|$, required to attain the *detectability* limit for the Si/SiO abundance [i.e. $\chi(\text{SiO}) \geq 10^{-12}$] is of $v_d \sim 6$ km s^{-1}. All material from the grain cores is completely injected into the gas phase for $v_d \sim 20$ km s^{-1}, which implies flow times of only $t_n \sim 10$ yr.

5 The evolution of the molecular line profiles

In Fig. 2, we show the evolution of the line profiles of SiO obtained for 10 different flow times in the shock. Although the charged dust grains are coupled to the ion fluid in the C-shock (Draine et al. 1983), we note that the SiO line profiles are plotted as a function of the neutral velocity. This is consistent with the fact that, once the SiO molecules are ejected from the grain surface, they are *instantaneously* slowed down to the velocity of the neutral fluid. The *slowed-down* time-scales, t_s, for this molecule in a medium with $n_0 = 10^5$ cm^{-3} and $T_k = 10$ K is of only $t_s \sim 21$ days (where t_s is estimated by doing $t_s^{-1} \approx 2 \times 10^{-12} T_k^{1/2} n_0$ s). From Fig. 2, we find that the SiO emission evolves from very narrow line profiles at almost ambient velocities, to broad line profiles centered at the shock velocity of $v_s = 70$ km s^{-1}. In contrast with previous shock modelling (e.g.

Schilke et al. 1997), Fig. 2 shows that the SiO line profiles characteristic of the magnetic precursor stage, have narrow linewidths ($\Delta v \sim 0.5$ km s^{-1}) and are centered at radial velocities ($v_{\text{LSR}} \sim 5.2$ km s^{-1}) which slightly differ from that of the ambient cloud ($v_0 = 4.7$ km s^{-1}). The velocity shift of the SiO line peak is naturally produced by the inefficiency of the grain mantle sputtering at the beginning of the shock (it only generates *detectable* SiO abundances for $v_d \geq 6$ km s^{-1}; Sect. 4). The narrow linewidth of the SiO profile is due to the low neutral temperature ($T_k \sim 10$ K) in the precursor.

As the shock evolves, the SiO emission develops two emission peaks at the moderate ($v_n \sim 20$ km s^{-1}) and high velocity regimes ($v_n \sim 60$–70 km s^{-1}; Fig. 2). The SiO line profile at $t_{\text{inst}} = 1520$ yr is consistent with those obtained by Schilke et al. (1997) by considering only the SiO emission generated in the evolved postshock gas.

We note that, not only the final time-scales of the shock ($t_{\text{inst}} = 1520$ yr) are similar to the dynamical age of L1448-mm (Sect. 3), but the flow times of the precursor stage and the moderate velocity gas ($t_{\text{inst}} = 3$–72 yr) are of the same order of magnitude as those found in the young shocks of this outflow ($t_{\text{dyn}} \sim 100$ yr; Girart and Acord 2001).

As expected for the ion and neutral fluids in the precursor, the differences between the line profiles of H^{13}CO$^+$ and HN^{13}C are evident at $t_{\text{inst}} = 3$ yr (see lower panels of Fig. 3). While the H^{13}CO$^+$ line peak is slightly (red)shifted with respect to the ambient gas, the neutrals show its maximum emission at this component. For larger values of t_{inst}, the terminal velocity of the ion emission is rapidly *saturated* (i.e.

Fig. 2 Evolution of the line profiles of SiO for different flow times in the C-shock. The flow times t_{inst}, for which we have stopped the model are indicated in the upper part of each plot

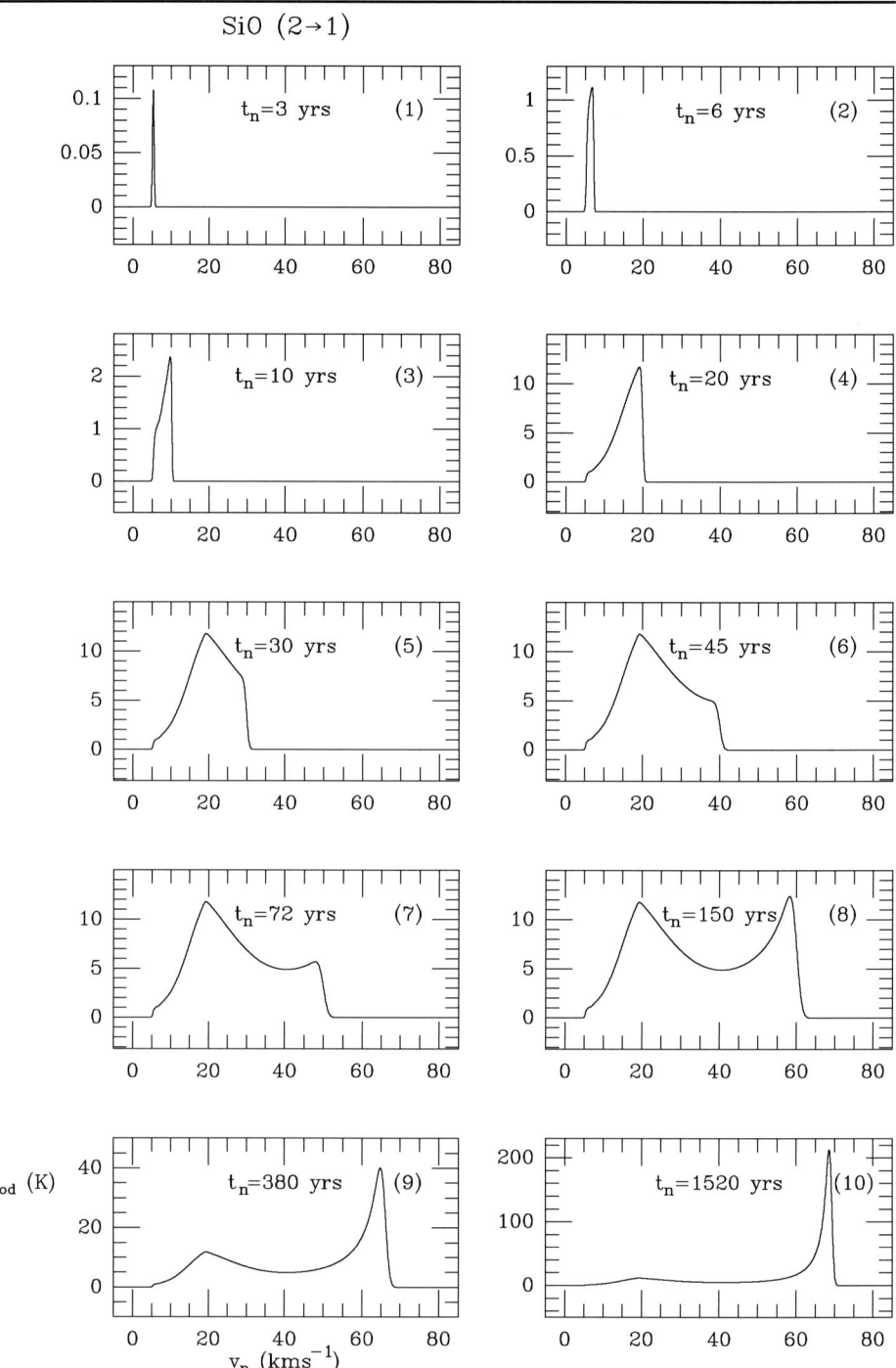

it rapidly reaches v_s), while the terminal neutral velocity is progressively increased with the evolution of the shock.

6 Comparison with observations

In Fig. 3, we present the comparison between the observed (upper panels) and the predicted (lower panels) line profiles of SiO, $H^{13}CO^+$ and $HN^{13}C$ for the magnetic precur-

sor stage at $t_{inst} = 3$ yr. We find that our model reproduces the narrow SiO emission detected in the precursor component of L1448-mm, and explains the velocity (red)shift (and consequently, the enhancement) of the ions with respect to the neutrals observed in the young shocks of this outflow (Jiménez-Serra et al. 2004). This confirms the idea that the narrow SiO line profiles are indeed signatures of the early interaction of C-shocks in young molecular outflows (Jiménez-Serra et al. 2004).

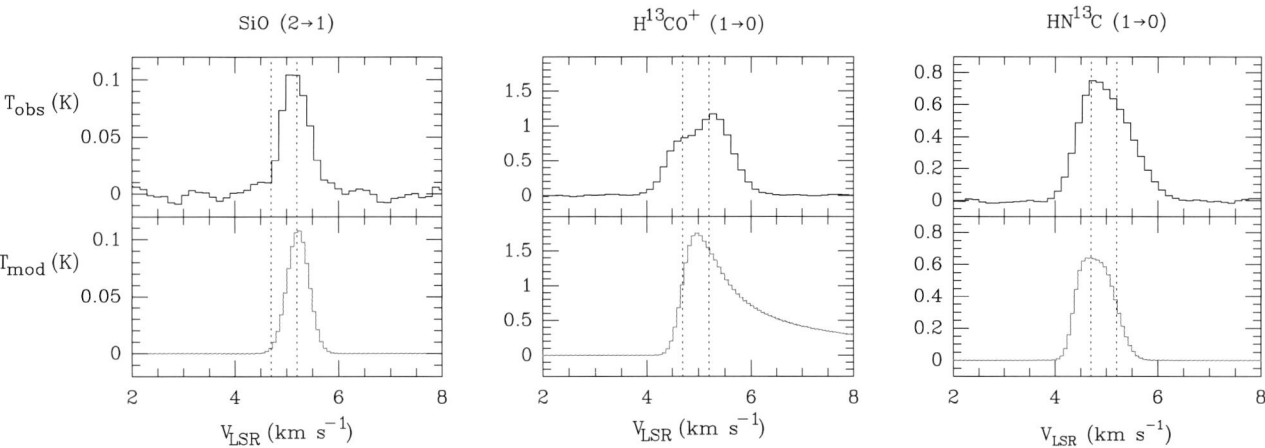

Fig. 3 Comparison of the line profiles of SiO, $H^{13}CO^+$ and $HN^{13}C$ predicted for the precursor at $t_{inst} = 3$ yr, and those observed in the young shocks of L1448-mm (Jiménez-Serra et al. 2004). *Vertical dotted lines* indicate the ambient (at 4.7 km s^{-1}) and precursor (at 5.2 km s^{-1}) components

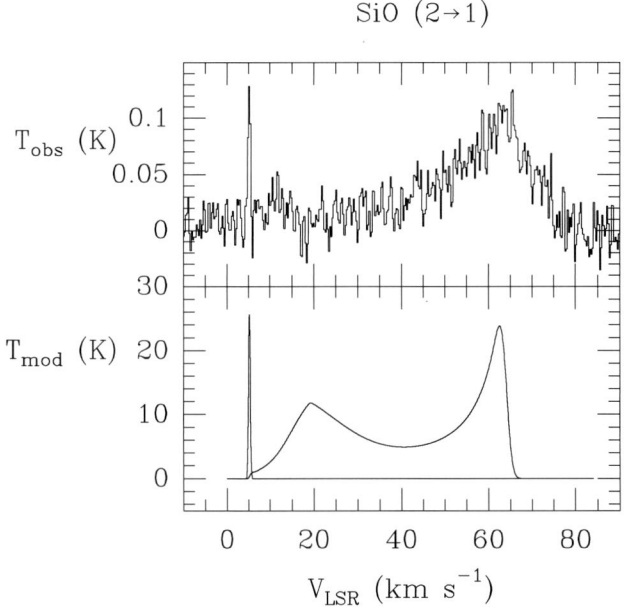

Fig. 4 Comparison between the SiO emission observed toward L1448-mm, and the superposition of the SiO line profiles obtained for $t_{inst} = 3$ and 250 yr. The line profile at $t_{inst} = 3$ yr has been multiplied by a factor of $\sim$240

Figure 4 shows the comparison of the SiO emission observed toward L1448-mm, with the superposition of the SiO line profiles obtained for two different stages in the shock (at $t_{inst} = 3$ and 250 yr). This explains the presence of different velocity regimes in this outflow (precursor, moderate velocity and high velocity gas; Jiménez-Serra et al. 2005) due to the coexistence of two different shocks at two different evolutionary stages within the same outflow region sampled by the sigle-dish beam observations of SiO (Jiménez-Serra et al. 2004).

In summary, we present the first C-shock and radiative transfer modelling that calculates the evolution of the line

profiles of molecules like SiO, $H^{13}CO^+$ and $HN^{13}C$ for different flow times in the shock. These results reveal that the narrow SiO emission and the enhancement of the ions found in the young L1448-mm outflow are signatures of the interaction of the magnetic precursor of C-shocks. The detection of the different velocity components (precursor, moderate velocity and high velocity gas) in this outflow suggests that we are observing the propagation of two different shocks at the magnetic precursor stage and at the late postshock gas. Due to the small dimensions of the precursor region (predicted to be of $\leq 1''$ at $\sim$300 pc for L1448-mm), interferometric observations with ALMA will be crucial to really understand the evolution of these shocks in very young molecular outflows.

Acknowledgement This work was supported by the Spanish MEC through projects number AYA2003-02785-E and ESP2004-00665 and by the "Comunidad de Madrid" Government under PRICIT project S-0505/ESP-0277 (ASTROCAM).

References

Caselli, P., et al.: Astron. Astrophys. **322**, 296 (1997)
Dayou, F., Balança, C.: Astron. Astrophys. **459**, 297 (2006)
Draine, B.T.: Astrophys. J. **241**, 1021 (1980)
Draine, B.T., et al.: Astrophys. J. **264**, 485 (1983)
Girart, J.M., Acord, J.M.P.: Astrophys. J. **552**, L63 (2001)
Jiménez-Serra, I., et al.: Astrophys. J. **603**, L49 (2004)
Jiménez-Serra, I., et al.: Astrophys. J. **627**, L121 (2005)
Jiménez-Serra, I., et al.: Astron. Astrophys. (2007, submitted)
Le Bourlot, J., et al.: Mon. Not. R. Astron. Soc. **332**, 985 (2002)
Lesaffre, P., et al.: Astron. Astrophys. **427**, 147 (2004)
Martín-Pintado, J., et al.: Astron. Astrophys. **254**, 315 (1992)
May, P.W., et al.: Mon. Not. R. Astron. Soc. **318**, 809 (2000)
Schilke, P., et al.: Astron. Astrophys. **321**, 293 (1997)
Schöier, F.L., et al.: Astron. Astrophys. **432**, 369 (2005)
Ziurys, L.M., et al.: Astrophys. J. **343**, 301 (1989)

A new evolutionary scenario of intermediate-mass star-formation revealed by multi-wavelength observations of OMC-2/3

Satoko Takahashi · Masao Saito · Shigehisa Takakuwa ·
Ryohei Kawabe

Originally published in the journal Astrophysics and Space Science, Volume 313, Nos 1–3.
DOI: 10.1007/s10509-007-9638-x © Springer Science+Business Media B.V. 2007

Abstract We have performed millimeter- and submillimeter-wave survey observations using the Nobeyama millimeter array (NMA) and the Atacama Submillimeter Telescope Experiment (ASTE) in one of the nearest intermediate-mass (IM) star-forming regions: Orion Molecular Cloud-2/3 (OMC-2/3). Using the high-resolution capabilities offered by the NMA (∼several arcsec), we observed dust continuum and $H^{13}CO^+(1-0)$ emission in 12 pre- and proto-stellar candidates identified previously in single-dish millimeter observations. We unveiled the evolutionary changes with variations of the morphology and velocity structure of the dense envelopes traced by the $H^{13}CO^+(1-0)$ emission. Furthermore, using the high-sensitivity capabilities offered by the ASTE, we searched for large-scale molecular outflows associated with these pre- and proto-stellar candidates observed with the NMA. As a result of the CO(3–2) observations, we detected six molecular outflows associated with the dense gas envelopes traced by $H^{13}CO^+(1-0)$ and 3.3 mm continuum emission. The estimated CO outflow momentum increases with the evolutionary sequence from early to late type of the protostellar cores. We also found that

the 24 μm flux increases as the dense gas evolutionary sequence. We propose that the enhancement of the 24 μm flux is caused by the growth of the cavity (i.e. the CO outflow destroys the envelope) as the evolutionary sequence. Our results show that the dissipation of the dense gas envelope plays an essential role in the evolution of the IM protostars. The extremely high-sensitivity and high-angular resolution offered by ALMA will reveal unprecedented details of the inner ∼50 AU of these protostars, which will provide us a break through in the classic scenario of IM star/disk formation.

Keywords IM protostars · Molecular outflows

1 Introduction

In the last two decades, the development of millimeter interferometers has enabled us to establish a standard scenario of sun-like ($M_* \sim 1 M_\odot$) star formation. However, details of formation and evolution of more massive stars ($\geq 2 M_\odot$), that is, intermediate- and high-mass protostars, remain poorly understood. Studies of the formation and evolution of IM protostars will allow us to understand whether the established low-mass star-formation scenario is applicable to a wide range of protostar masses. However, there is no systematic observations toward IM protostars. It is important to directly verify the accretion and dissipation (or destruction) processes of dense gas envelopes around IM protostars. For these purposes, we have performed multi-wavelength and multi-line survey observations toward IM pre- and proto-stellar candidates in the OMC-2/3 region.

The OMC-2/3 region ($d = 450$ pc; Genzel and Stutzki 1989), which is located at the northern part of Orion A

Current address:
S. Takahashi
Academia Sinica Institute of Astronomy & Astrophysics,
P.O. Box 23-141, Taipei, Taiwan 106 ROC

S. Takahashi (✉)
The Graduate University for Advanced Studies, National
Astronomical Observatory of Japan (NAOJ), Osawa 2-21-1,
Mitaka, Tokyo 181-8588, Japan
e-mail: satoko_t@asiaa.sinica.edu.tw

M. Saito · S. Takakuwa · R. Kawabe
ALMA Project Office, National Astronomical Observatory of
Japan (NAOJ), Osawa 2-21-1, Mitaka, Tokyo 181-8588, Japan

Table 1 NMA Observational Parameters

	H^{13}CO$^+$(1–0)	3.3 mm
Velocity resolution (km s^{-1})	0.1	–
Rms noise level (Jy beam^{-1})	0.1	0.001
Primary beam (HPBW in arcsec)	77 (3500 AU)	
Spatial resolution (arcsec)	3.5–6.0 (1600–2700 AU)	

giant molecular cloud, is one of the nearest active star-forming regions. There are 28 millimeter- and 33 submillimeter dust condensations in this region (Chini et al. 1997; Lis et al. 1998; Nielbock et al. 2003). Six condensations are associated with class 0-type SEDs by the L_{bol}/L_{smm} diagnose (Chini et al. 1997). In addition, the bolometric luminosity and core mass of these sources are at least one order of magnitude larger than low-mass counterparts, suggesting that these sources potentially form IM protostars (2–3 $M_\odot$ or a A0 star at the ZAMS).

2 Observations

We have observed the H^{13}CO$^+$ ($J = 1$–0; 86.754330 GHz) emission and the 3.3 mm continuum emission toward 12 millimeter sources in the OMC-3 region (SIMBA a and MMS 1 to MMS 10) and northern part of the OMC-2 region (FIR 2) with the six-elements NMA C+D configurations from 2004 November to 2006 May. The NMA observational parameters are summarized in Table 1. The CO(3–2; 345.795990 GHz) data have been taken with the ASTE 10 m telescope located at the Pampa la Bola, Chile during the period of 2005 September. The On-The Fly (OTF) mapping technique was employed to cover the entire OMC-2/3 region. The effective FWHM resolution is 26″ (corresponding to 0.06 pc). The typical rms noise level was 0.47 K in T_A^* with a velocity resolution of 1.08 km s^{-1}. In order to identify protostars along the OMC filament, we have also fetched archived 24 μm data obtained by infrared camera MIPS (Multiband Imaging Photometer) equipped in the infrared space telescope Spitzer.

3 Survey results

Figure 1b shows distribution of the CO(3–2) high-velocity emission in OMC-2/3. We totally identified 14 outflows and eight of them are newly identified in the present study (the detailed results of the CO(3–2) emission taken with the ASTE telescope will be in the forthcoming paper). Specifically, eight out of the fourteen outflows, SIMBA a and c, MMS 2, MMS 5, MMS 7, MMS 9, FIR 2, FIR 6b, and FIR 6c are accompanied with the 1.3 mm dust condensations

taken with the single-dish observations, and 24 μm Spitzer sources, suggesting the presence of the heating sources.

We have detected the compact 3.3 mm continuum emission (i.e., thermal dust emission) with a size scale of a few × 1000 AU toward the eight 1.3 mm sources, SIMBA a, MMS 1, MMS 2, MMS 5, MMS 6, MMS 7, MMS 9 and FIR 2 using the NMA. Seven out of the eight 3.3 mm sources (i.e., except MMS 1) are associated with the 24 μm Spitzer sources.

Figure 1c shows integrated intensity maps of the H^{13}CO$^+$ emission taken with the NMA. Significant H^{13}CO$^+$(1–0) emission has been detected toward nine out of the twelve 1.3 mm sources, SIMBA a, MMS 1, MMS2, MMS 3, MMS5, MMS6, MMS 7, MMS 8, and FIR 2, with a size and mass scales of 0.01–0.1 pc and 0.5–5$M_\odot$, respectively. The detected H^{13}CO$^+$ envelopes show various morphologies (centrally condensed, fan-shaped structure, etc.).

Velocity gradients along the major axis, implying rotating motion, are detected toward 5/12 sources. This velocity gradient was detected toward both pre- and proto-stellar candidates. In addition, along the minor axis in the H^{13}CO$^+$ envelope, we detected the gas dispersing motion along and/or perpendicular to the outflow axis toward 4/12 samples. This velocity gradient was only detected toward protostellar candidates. On the other hand, there is no clear evidence of gas accreting motion in the H^{13}CO$^+$ emission. From the intensive survey, we unveiled presence or absence of the molecular outflows and central protostars toward millimeter condensations in the OMC-2/3 region, and also revealed spatial- and velocity- structure of the associated dense gas envelopes traced by the H^{13}CO$^+$(1–0) emission.

4 Discussion

In order to discuss the evolutionary sequence of the IM protostars, we used six independent multi-wavelength data, our H^{13}CO$^+$(1–0) and 3.3 mm continuum emission, CO(3–2) emission and JHK_s images taken with the SIRIUS/IRSF, archive 24 μm Spitzer data, and published VLA-3.6 cm data from (Reipurth et al. 1999). H^{13}CO$^+$(1–0) emission traces a distribution and kinematics of the dense gas envelope. The size and momentum of the molecular outflows traced by the CO(3–2) emission reflect the age (or) activity of the CO outflow. 3.3 mm and 24 μm compact continuum sources trace a region, which is dense ($\geq 10^7$ cm^{-3}) and hot (> 150 K) inner envelope associated with the protostars. Reflection nebula (i.e., cavity-like structure along the molecular outflow) in JHK_s images is related to the dissipation of circumstellar material. Further free-free jet traced by the VLA-3.6 cm emission show an activity of central jet.

"Pre-stellar" sources, MMS 4, MMS 8 and MMS 10, have no 3.3 mm and 24 μm compact sources above 3σ sig-

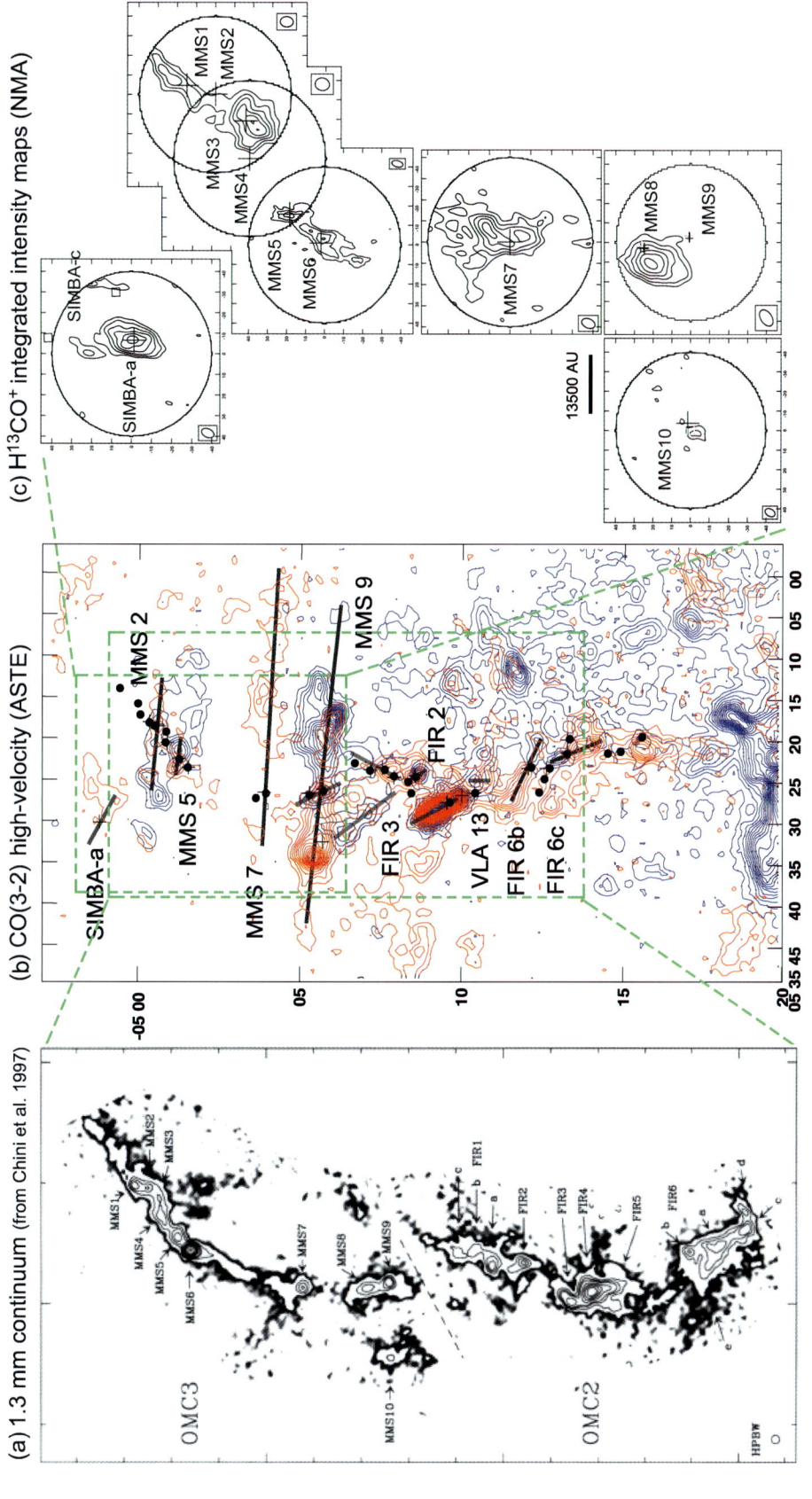

Fig. 1 **a** A 1.3 mm map of the OMC-2 and -3 region from Chini et al. (1997). **b** Survey results of molecular outflows traced by the high-velocity CO(3–2) emission taken with the ASTE with a velocity range of $V_{lsr} = -6.2$ to 7.8 km s^{-1} (*blue*) and $V_{lsr} = 14.3$ to 27.3 km s^{-1} (*red*), respectively. **c** Survey results of dense envelopes traced by the H^{13}CO$^+$(1–0) emission taken with the NMA. *Crosses* in (**b**) and (**c**) show the positions of the 1.3 mm source (from Chini et al. 1997). *Dots* in (**b**) show the positions of the 350 μm source (from Lis et al. 1998)

nal level, suggesting no signature of the dense and hot inner envelope. Upper-limit of average densities estimated by the 3.3 mm compact dusty component, several $\times 10^6$ cm^{-3}, is typically one order of magnitude less than that of protostellar samples in OMC-2/3 (a few $\times 10^5$ cm s^{-3}). In this phase, there is no signature of jet and outflow. These objects are associated with the relatively extended H^{13}CO$^+$(1–0) component(s).

"Class A" sources, SIMBA a, MMS 5 and FIR 2 are associated with 3.3 mm and faint 24 μm compact sources, suggesting the presence of the dense and inner hot envelope. Further 0.1 pc-scale CO outflow is detected toward each object. These results imply that protostars are already formed in central region and may begin mass accretion, although we did not detect mass accretion directly. Detected velocity gradient along the major axis implies the rotational motion of the envelope traced by the H^{13}CO$^+$ emission. We also found an evidence the dense gas traced by the H^{13}CO$^+$ emission entrained by the CO outflow. In this phase, we did not detect signature of free–free jet and reflection nebula.

"Class B" sources, MMS 2, MMS 7 and MMS 9 are associated with 3.3 mm and bright 24 μm compact sources, suggesting presence of the dense and hot envelope. Both of a 0.5–1.0 pc scale of CO outflow and free–free jet are associated with the objects. In addition, JHK_s images show reflection nebula around the heating source, suggesting the formation of the cavity along the outflow axis. Fan-shaped structure traced by the H^{13}CO$^+$ emission with a size scale of 0.1 pc have been observed in this phase. The peak position of the H^{13}CO$^+$ emission does not coincide with the 3.3 mm continuum emission.

From these results, we consider that the above classification with observational results suggest the formation scenario of the IM protostars with dissipating processes of circumstellar materials. Since protostars are formed in denser region ($\geq$ a few $\times 10^7$ cm^{-3}) traced by the 3.3 mm continuum emission taken with the NMA, inner hot envelope are observed by the 24 μm emission. Class A corresponds to the earlier evolutionary stage of the proto-stellar core with a rotating envelope traced by the H^{13}CO$^+$(1–0) emission. In this phase, small scale outflow with a size scale of 0.1 pc is observed. Class B corresponds to the later evolutionary stage of the proto-stellar cores. We detected dense gas dissipation in H^{13}CO$^+$(1–0) emission toward MMS 2 and MMS 7 (Detailed descriptions are in Takahashi et al. 2006). On the other hand, we did not detect significant H^{13}CO$^+$ emission toward MMS 9, implying dissipation of the dense gas. In Class B, large scale outflows (0.5–1 pc) are observed.

Furthermore we found the relation between the outflow growth and evolution of the circumstellar gas. Figure 2 shows a 24 μm flux densities plotted as a function of the outflow momenta derived from CO(3–2) observations. Our

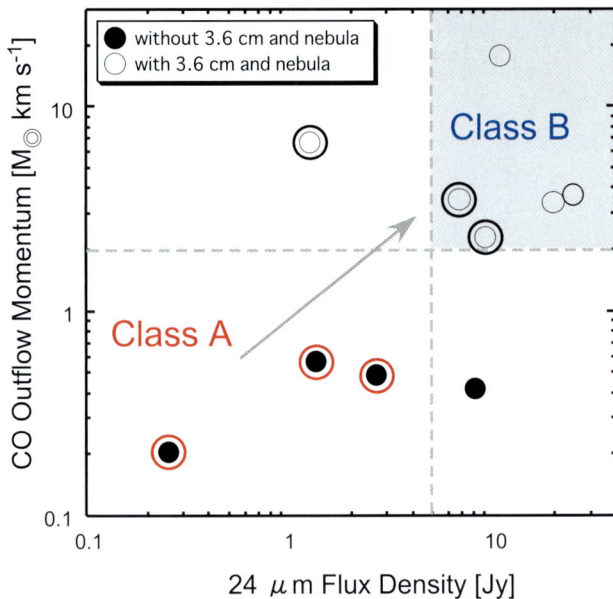

Fig. 2 The outflow momentum derived by CO(3–2) emission plotted as a function of the 24 μm flux

results show that the 24 μm flux has a positive correlation with the CO outflow momentum except for one object. In addition, Objects with the higher CO outflow momentum and 24 μm flux denoted by a shaded area in Fig. 2, are associated with a cm jet and bright reflection nebula in the NIR image. The presence of the cm jet and NIR nebula implies the formation of cavities along the outflow axis and the dissipation of the dense gas envelope. These phenomena are consistent with the dense core dissipation from Class A to ClassB denoted by Fig. 2. We consider that the dissipation of the dense gas plays an essential role in the evolution of the IM protostars.

Acknowledgements The authors acknowledge N. Kusakabe and A. Ishihara for helping us to reduct SIRIUS/IRSF data. S. Takahashi was financially supported by the Japan Society for the Promotion of Science (JSPS) for Young Scientists.

References

Chini, R., et al.: Dust filament and star formation in OMC-2 and OMC-3. Astrophys. J. **474**, L135 (1997)

Genzel, R., Stutzki, J.: The Orion Molecular Cloud and star-forming region. Annu. Rev. Astron. Astrophys. **27**, 41 (1989)

Lis, D.C., et al.: 350 μm continuum imaging of the Orion A Molecular Cloud with the submillimeter high angular camera. Astrophys. J. **509**, 299 (1998)

Nielbock, M., et al.: The stellar content of OMC-2/3. Astron. Astrophys. **408**, 245 (2003)

Reipurth, B., et al.: VLA detection of protostars in OMC-2/3. Astrophys. J. **118**, 983 (1999)

Takahashi, S., et al.: Millimeter- and submillimeter- wave observations of the OMC-2/3 region. I. Dispersing and rotating core around the intermediate-mass protostar MMS7. Astrophys. J. **651**, 933 (2006)

Scientific role of ACA for low-mass star-formation study

Shigehisa Takakuwa · Daisuke Iono · Baltasar Vila-Vilaro · Tomohiko Sekiguchi · Ryohei Kawabe

Originally published in the journal Astrophysics and Space Science, Volume 313, Nos 1–3.
DOI: 10.1007/s10509-007-9617-2 © Springer Science+Business Media B.V. 2007

Abstract We discuss the scientific role of the Atacama Compact Array (ACA), the Japanese contribution to the ALMA project, for low-mass star-formation study. Our recent observations of several low-mass protostellar envelopes in the submillimeter CS ($J = 7$–6) and HCN ($J = 4$–3) lines with the SMA and ASTE have revealed that these submillimeter emissions are more extended than $\sim$2000 AU and show different velocity structures from those traced by millimeter lines. These results suggest the importance of taking short-spacing informations the ACA can offer. Our comprehensive imaging simulations of these protostellar envelopes, as well as prestellar cores and debris disks, unprecedentedly demonstrate the scientific importance of ACA.

Keywords ACA · Imaging simulations · Low-mass star formation

1 Introduction

The Japanese ALMA group will construct the Atacama Compact Array (ACA), which consists of twelve 7-m antennas as an interferometric array (The ACA 7-m Array) and four 12-m antennas as a single-dish array (The ACA Total Power Array). The ACA can recover missing short-spacing information which the other part of ALMA, that is, fifty 12-m antennas as a large interferometric array (The 12-m array), cannot sample (0–12 m), and can significantly enhance the imaging capability of ALMA (Pety et al. 2001c; Tsutsumi et al. 2004; Morita and Holdaway 2005).

However, there is little study on the effect of the ACA from scientific point of view. In particular, recent high-sensitivity or high-resolution submillimeter observations of low-mass star-forming regions with the existing submillimeter telescopes such as Atacama Submillimeter Telescope Experiment (ASTE) (Ezawa et al. 2004) and the Submillimeter Array (SMA) (Ho et al. 2004) have revealed unexpected nature of low-mass star-forming regions in the submillimeter emissions. In order to verify the scientific role of the ACA in the field of low-mass star formation based on these new observational results, we have made comprehensive imaging simulations using the GILDAS imaging simulator (Pety et al. 2001a, 2001b), with the "real" astronomical images as an input model.

In this paper, first we will present our recent observational results of low-mass star-forming regions with the SMA and ASTE, as well as the millimeter telescopes at the Nobeyama Radio Observatory of Japan. Then, with these observational results as input models, we will show results of our imaging simulations with the ALMA and ACA, which unprecedentedly demonstrate the scientific importance of the ACA.

S. Takakuwa (✉) · D. Iono · B. Vila-Vilaro · T. Sekiguchi · R. Kawabe
ALMA Project Office, National Astronomical Observatory of Japan, Osawa 2-21-1, Mitaka, Tokyo 181-8588, Japan
e-mail: s.takakuwa@nao.ac.jp

Present address:
S. Takakuwa
Institute of Astronomy and Astrophysics, Academia Sinica, P.O. Box 23-141, Taipei 10617, Taiwan, China
e-mail: takakuwa@asiaa.sinica.edu.tw

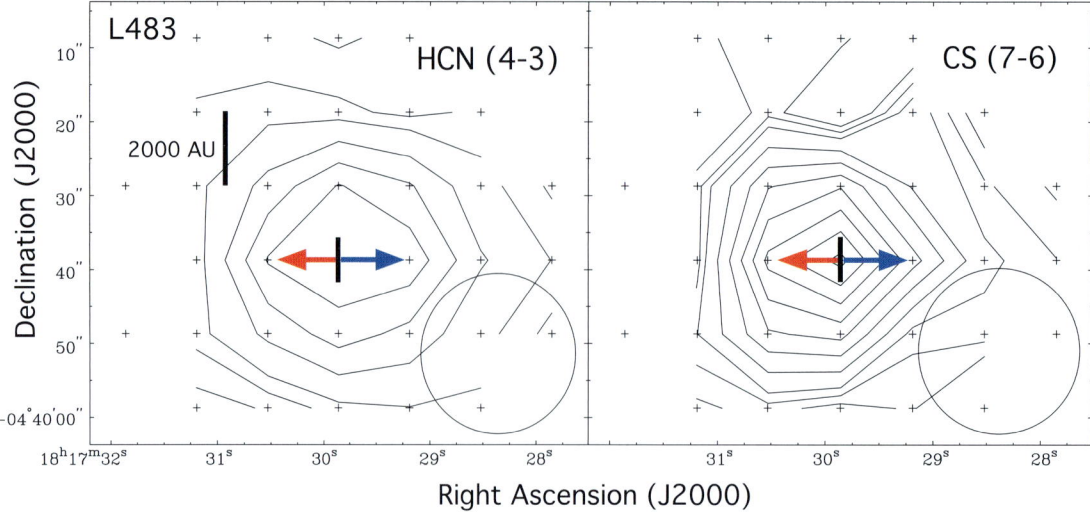

Fig. 1 Total integrated intensity maps (integrated velocity range 4.2–6.9 km s⁻¹) of the HCN (4–3) (*left*) and CS (7–6) (*right*) emission in L483, taken with ASTE. Contour levels are 2, 4, 6 σ, and then 10 σ in steps of 4 σ (1 σ = 0.0733 K km s⁻¹). The highest contour in the HCN map is 18 σ and that in the CS map 34 σ. Crosses indicate observed positions, and open circles at the bottom right corner beam sizes. Red and blue arrows show the direction of the redshifted and blueshifted molecular outflow, respectively, and the roots of the arrows indicate the protostellar position

2 Low-mass protostellar envelopes

2.1 ASTE results

In Fig. 1, we show total integrated intensity maps of the submillimeter HCN ($J = 4$–3; 354.505 GHz) and CS ($J = 7$–6; 342.883 GHz) lines toward one of the low-mass protostars, L483, observed with ASTE. Details of the ASTE observations as well as the scientific discussions are presented by Takakuwa et al. (2007a). There appears a western extension both in the HCN and CS emissions, and the structures traced by these submillimeter emissions are resolved with ASTE. The deconvolved size of the HCN emission measured from a 2-dimensional Gaussian fitting to the image is 5500×3700 (AU) (P.A. $= 78°$), while in the CS emission only the major axis is resolved ($\sim$2300 AU). This result suggest that these submillimeter emissions, which should trace gas temperatures above $\sim$40 (K) and densities above $\sim$10⁷ cm⁻³, can be more extended than $\sim$2000 AU in the low-mass protostellar envelope.

Figure 2 presents the velocity structure traced by the submillimeter CS line in L483. Along the axis of the associated molecular outflow (Tafalla et al. 2000), the CS (7–6) line is redshifted at the west of the protostar and blueshifted at the east. The same velocity gradient is also found in the HCN (4–3) line. Interestingly, this trend of the velocity gradient is opposite to that of the molecular outflow, and that in the 3-mm counterpart of the CS (2–1) line and other 3-mm lines such as N₂H⁺ (1–0) (Tafalla et al. 2000). The same results are also seen in another low-mass protostellar envelope around B335 (Takakuwa et al. 2007a). These results suggest

that the submillimeter emissions have different origin from that of the millimeter lines.

In summary, submillimeter molecular-line emissions can be more extended than expected, and trace different gas components from those by millimeter emissions in low-mass protostellar envelopes.

2.2 SMA+JCMT results and imaging simulations

The ASTE results shown in Sect. 2.1 have revealed the extended nature of submillimeter emissions in low-mass protostellar envelopes. We obtained a similar result in IRAS 16293-2422, a famous Class 0 protostar in ρ Oph, with SMA and JCMT observations. The upper-left panel in Fig. 3 shows a total integrated intensity map of the HCN (4–3) emission in IRAS 16293-2422 observed with the SMA and JCMT, which shows that the submillimeter HCN emission is more extended than 3000 AU (Takakuwa et al. 2007b). We have made GILDAS imaging simulations with this observational result as an input model, and the results are shown in other panels of Fig. 3. In our simulations, we included thermal noises and pointing errors within 0.6″, following the specification of ALMA. As shown in the upper panels, with the 12-m array only the extended envelope component can not be recovered, whereas with the total power array and the ACA, the extended envelope component is recovered properly. The lower panels show the images of the fidelity. Here, the fidelity ($\equiv F(i, j)$ for each image pixel i, j) is defined as;

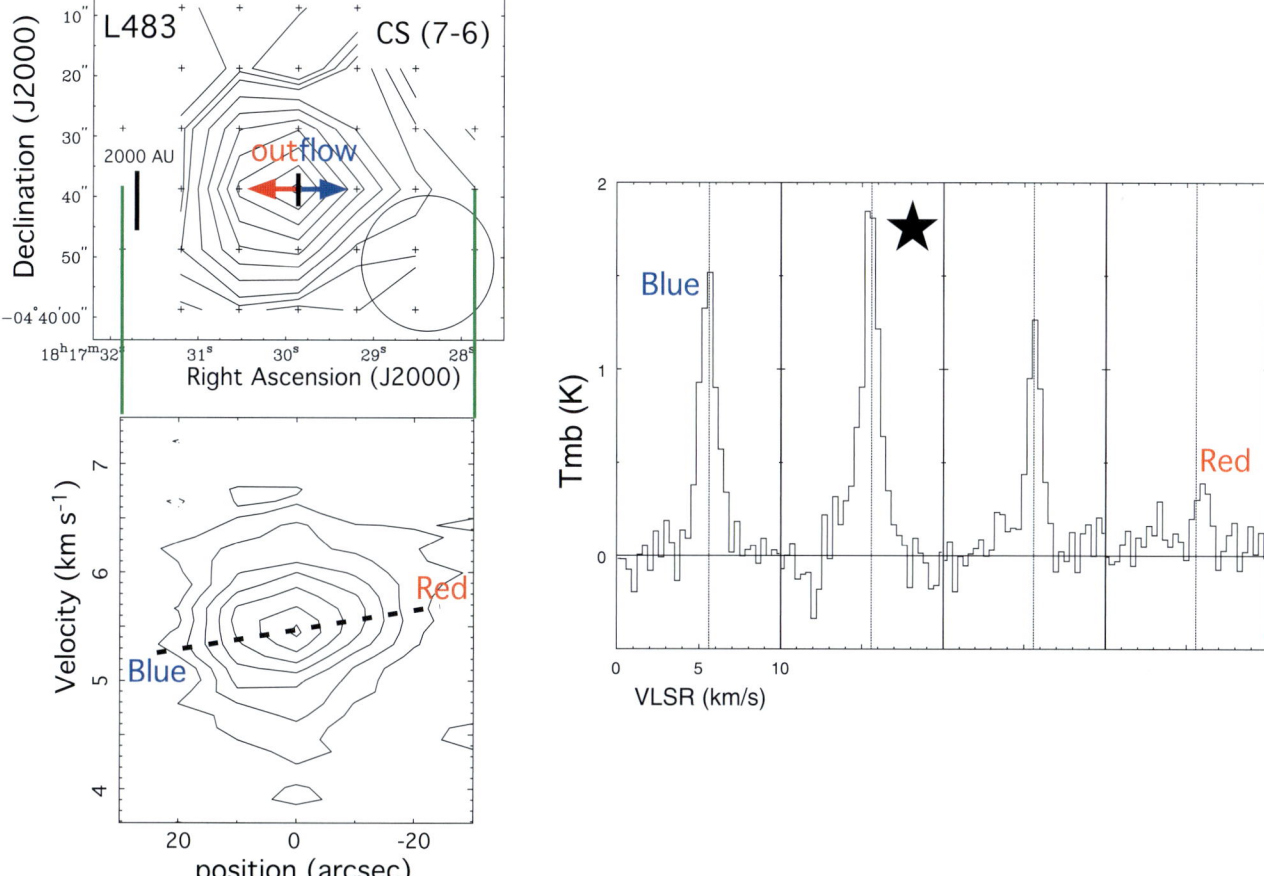

Fig. 2 ASTE results of the velocity structure traced by the submillimeter CS (7–6) line in L483. (*Upper left*) Total integrated intensity map of the CS (7–6) emission in L483, shown in Fig. 1. (*Lower left*) Position-velocity diagram of the CS (7–6) line along the axis of the associated molecular outflow in L483. Contour levels are from 2 σ in steps of 2 σ (1σ = 0.133 K). A *dashed line* delineates the detected velocity gradient. (*Right*) CS (7–6) line profile map at a grid spacing of 10″ along the outflow axis in L483. A *star mark* indicates the position of the protostar

$$F(i, j) = \frac{|\text{Model}(i, j)|}{\max(|\text{Diff}(i, j)|, 0.7 \times \text{rms}(\text{Diff}(i, j)))}, \quad (1)$$

$$\text{Diff}(i, j) = \text{Model}(i, j) - \text{Simulated}(i, j), \quad (2)$$

where Model(i, j) denotes "true" astronomical images as an input to the imaging simulator, Simulated(i, j) "observed" images or output images from the imaging simulator, and rms(Diff(i, j)) denotes the rms value of the difference image for the entire image pixel. The fidelity is a quantitative measure of the imaging quality, and higher fidelity values indicate better imaging observations (for detailed discussion on the fidelity, see Pety et al. 2001c). Figure 3 demonstrates that the inclusion of the ACA can achieve the maximum fidelity. In other words, the ACA is crucial for studies of low-mass protostellar envelopes where submillimeter emissions are extended (>2000 AU) and have different velocity structures from those in millimeter emissions.

3 Prestellar cores and clumps

Takakuwa et al. (2003) have found that there are ~2500 AU-scale "clumps" inside a CH$_3$OH-rich prestellar core (CH$_3$OH core 6) in the TMC-1C region. These small-scale clumps are also found in other prestellar cores (Lemme et al. 1995; Ohashi et al. 1999), and could be fundamental elements of structures and evolution of prestellar cores. Figure 4 shows the clumpy structure in CH$_3$OH core 6 of TMC-1C (upper panels), as well as results of the imaging simulations (middle and lower panels). It is clear that without the ACA the clumps in the prestellar core cannot be identified. These results suggest that the ACA is essential to study structures and evolution of prestellar cores.

4 Debris disk

Figure 5 shows results of imaging simulations with a SCUBA image of a debris disk around ϵ Eridani (Greaves

Astrophys Space Sci (2008) 313: 169–173

IRAS 16293-2422 HCN (J=4-3; 354.5 GHz)

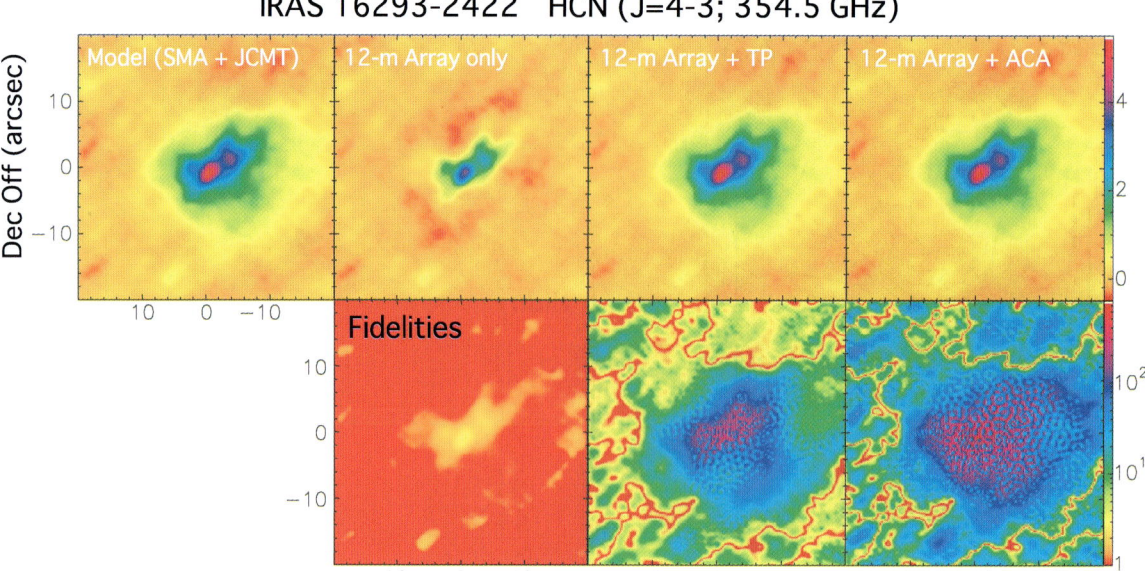

Fig. 3 Example of our ACA imaging simulations of low-mass protostellar envelopes. The input model (*upper left panel*) is from our observational result of the HCN (4–3) emission in IRAS 16293-2422 taken with the SMA and JCMT (Takakuwa et al. 2007b). *Other upper panels* show the simulated images with the 12-m array only, 12-m array + total power, and the 12-m array + ACA from *left* to *right panels*, respectively. *Lower panels* show images of the fidelity of *above panels*

Fig. 4 Results of our ACA imaging simulations of the prestellar core in the TMC-1C region. The input model (*upper panels*) is from our observational velocity channel maps of the CH_3OH (2_0–1_0 A^+) line in the TMC-1C core 6 taken with the NRO 45-m telescope and the Nobeyama Millimeter Array (Takakuwa et al. 2003). The LSR velocity is 5.5, 5.6, and 5.7 km s^{-1} from *left* to *right panels*. *Middle and lower panels* show the simulated images with the 12-m array only and the 12-m array + ACA, respectively

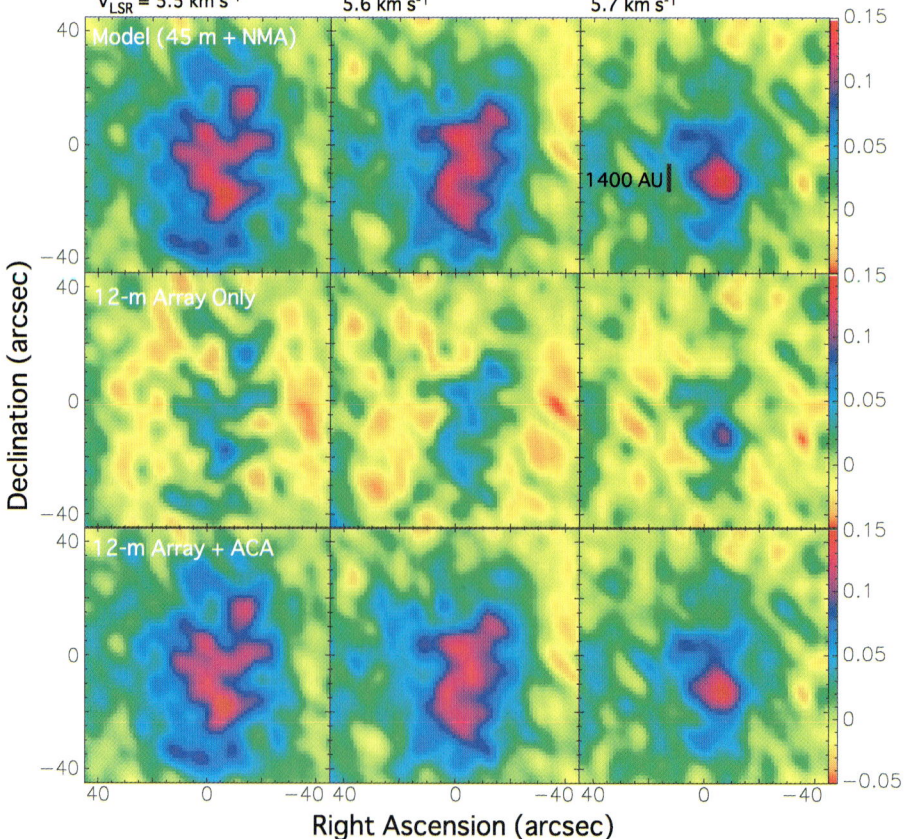

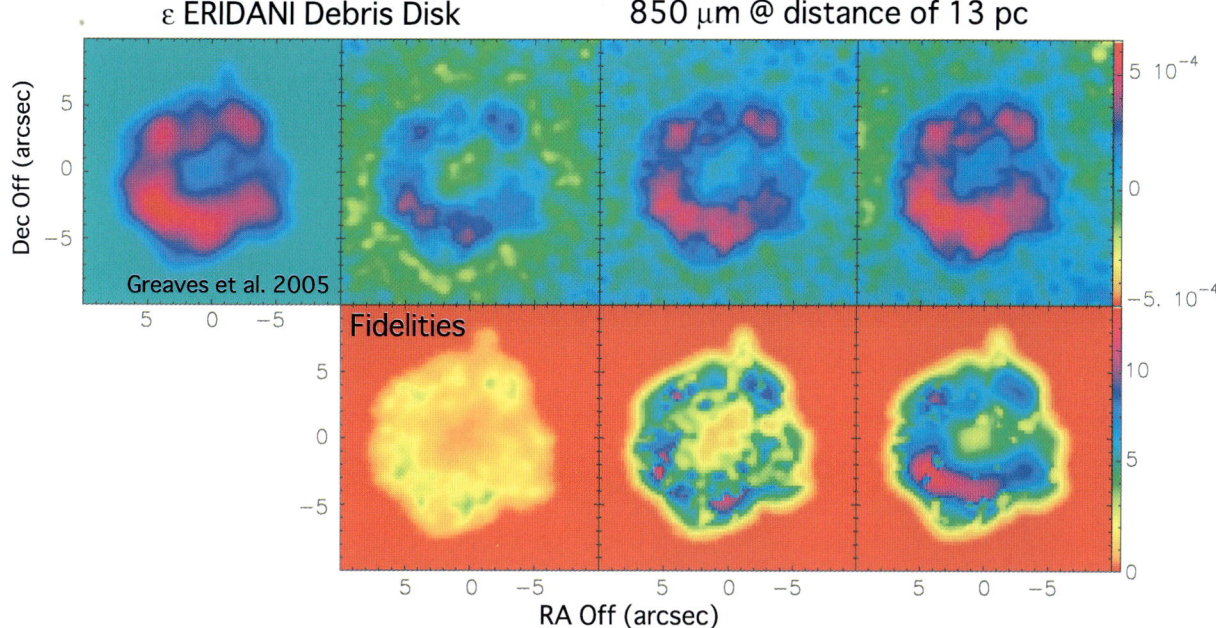

Fig. 5 Example of our ACA imaging simulations of debris disks. The input model (*upper left panel*) is from SCUBA observations of ϵ ERIDANI by Greaves et al. (2005). We put the target four times more distant than the original location (3.22 pc), in order to avoid huge mo- saicing and to save computational time. *Other upper panels* show the simulated images with the 12-m array only, 12-m array + total power, and the 12-m array + ACA from *left* to *right panels*, respectively. *Lower panels* show images of the fidelity of *above panels*

et al. 2005) as an input model. We put the source four times more distant than the original location, to avoid huge mo- saicing and to save computational time. It is obvious that the inclusion of the ACA significantly improves the imag- ing of the debris disk, suggesting the demand of the ACA for studies of debris disks and planet formation.

5 Summary

We have made ALMA/ACA imaging simulations with re- cent observational results as input models. Our simulations prove the scientific role of the ACA for low-mass star- formation study, ranging from prestellar cores, protostellar envelopes to debris disks.

Acknowledgements We are grateful to Drs. Greaves and Poulton for providing us their SCUBA data of ϵ ERIDANI as an input model to our imaging simulations. We also acknowledge the co-authors in Takakuwa et al. (2003, 2007a), and Takakuwa et al. (2007b), for pro- viding the observational images.

References

Ezawa, H., Kawabe, R., Kohno, K., Yamamoto, S.: The Atacama sub- millimeter telescope experiment (ASTE). Proc. SPIE **5489**, 763 (2004)

Greaves, J.S., Holland, W.S., et al.: Structure in the ϵ Eridani debris disk. Astrophys. J. **619**, L187 (2005)

Ho, P.T.P., Moran, J.M., Lo, K.-Y.: The submillimeter array. Astrophys. J. **616**, L1 (2004)

Lemme, C., Walmsley, C.M., Wilson, T.L., Muders, D.: A detailed study of an extremely quiescent core: L1498. Astron. Astrophys. **302**, 509 (1995)

Morita, K., Holdaway, M.A.: Array configuration design of the Ata- cama compact array. ALMA Memo, 538 (2005)

Ohashi, N., Lee, S.-W., Wilner, D.J., Hayashi, M.: CCS imaging of the starless core L1544: an envelope with infall and rotation. Astro- phys. J. **518**, L41 (1999)

Pety, J., Gueth, F., Guilloteau, S.: ALMA+ACA simulation tool. ALMA Memo, 386 (2001a)

Pety, J., Gueth, F., Guilloteau, S.: ALMA+ACA simulation results. ALMA Memo, 387 (2001b)

Pety, J., Gueth, F., Guilloteau, S.: Impact of ACA on the wide-field imaging capabilities of ALMA. ALMA Memo, 398 (2001c)

Tafalla, M., Myers, P.C., Mardones, D., Bachiller, R.: L483: a protostar in transition from Class 0 to Class I. Astron. Astrophys. **359**, 967 (2000)

Takakuwa, S., Kamazaki, T., Saito, M., Hirano, N.: $H^{13}CO^+$ and CH_3OH line observations of prestellar dense cores in the TMC- 1C region. II. Internal structure. Astrophys. J. **584**, 818 (2003)

Takakuwa, S., Kamazaki, T., Saito, M., Yamaguchi, N., Kohno, K.: ASTE observations of warm gas in low-mass protostellar en- velopes: different kinematics between submillimeter and millime- ter lines. Publ. Astron. Soc. Jpn. **59**, 1 (2007a)

Takakuwa, S., Ohashi, N., Bourke, T.L., Hirano, N., Ho, P.T.P., Jør- gensen, J.K., Kuan, Y.-J., Wilner, D.J., Yeh, S.C.C.: Arcsecond- resolution submillimeter HCN imaging of the binary protostar IRAS 16293-2422. Astrophys. J. **662**, 431 (2007b)

Tsutsumi, T., Morita, K., Hasegawa, T., Pety, J.: Wide-field imaging of ALMA with the Atacama compact array: imaging simulations. ALMA Memo, 488 (2004)

Planetary atmospheres with ALMA

Emmanuel Lellouch

Originally published in the journal Astrophysics and Space Science, Volume 313, Nos 1–3.
DOI: 10.1007/s10509-007-9637-y © Springer Science+Business Media B.V. 2007

Abstract Thanks to its sensitivity, spatial resolution and instantaneous *uv*-coverage, ALMA will permit many new studies related to the general topic of the couplings between chemistry and dynamics in planetary atmospheres. It will include: (1) three-dimensional mapping of composition, temperatures and winds in the atmospheres of Mars, Venus and Titan; (2) several aspects of Giant Planet composition and dynamics, such as the origin of oxygen, the evolution of Shoemaker–Levy 9 products in Jupiter's atmosphere, and the deep atmosphere structure and meteorology; (3) the study of tenuous and distant atmospheres (Io, Enceladus, Pluto, Triton and other Kuiper Belt objects).

Keywords Planets · Submillimeter · Interferometry

1 Introduction

Heterodyne spectroscopy at mm/submm wavelengths has proven over the last 25 years to be a powerful tool to study planetary atmospheres. Detection of minor species is often more sensitive than in any other spectral range. The high spectral resolving power allows the detailed investigation of line shapes, providing information on the atmospheric temperature structure and the vertical profiles of molecular abundances, and, when the source is resolved, allowing direct wind measurements from the determination of Doppler shift of line cores. The chemical, thermal and dynamical state of planetary atmospheres are usually intimately coupled: the thermal field drives the wind field, which itself affects the horizontal distribution of minor species. In return, the latter can impact the temperature field through their heating and cooling properties.

As of today, molecular rotational lines from Mars, Venus, the Giant Planets, Io and Titan have been succesfully observed. While large single-dish telescopes and current interferometers already permit a modest resolution of the largest of these bodies (Mars, Venus, Jupiter), ALMA will permit to resolve the planetary disks down to much finer scales. Even if the maximum spatial resolution offered by ALMA (e.g. $0.015''$ at 1 mm) will probably not be usable due to S/N limitations, realistic spatial resolutions will be 0.1–$0.2''$, i.e. 70–150 kilometers at 1 AU and 700–1500 km at 10 AU. The excellent *uv*-coverage will allow one to achieve "instantaneous imaging", an invaluable tool for studying time-variable phenomena in planetary atmospheres.

The other strength of ALMA will come from its unsurpassed sensitivity, which will gives access to weaker lines, more distant objects, and more tenuous atmospheres. Thanks to the high spatial resolution, limb sounding will be possible, which will improve the detection sensitivity (although vertical resolution at the limb will not be possible). In addition, the broad (2×8 GHz) bandwidth offered by the ALMA receivers will be favorable for the search for broad lines formed in deep atmospheres, e.g. Venus and the Giant Planets. Finally, the existence of a large facility in the Southern hemisphere will be valuable given that planets may go as far as $-27°$ in declination.

The general goal of ALMA in the field of planetary atmospheres will thus be the study of couplings between chemistry and dynamics, through 3-D measurements of abundances, temperatures and winds. In what follows, we review recent achievements from heterodyne spectroscopy

E. Lellouch (✉)
LESIA, Observatoire de Paris, 5 place Jules Janssen, 92190 Meudon Cedex, France
e-mail: emmanuel.lellouch@obspm.fr

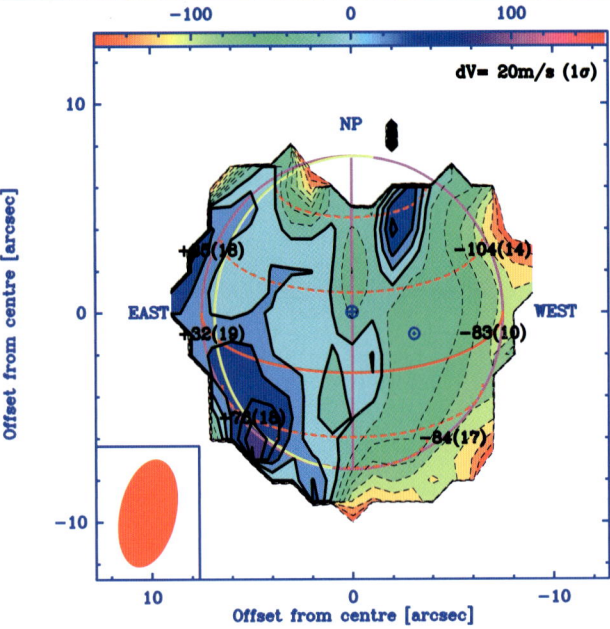

Fig. 1 Example of wind measurements in the Martian middle atmosphere (near 50 km altitude) from CO(1-0) observations. Data obtained at PdBI in May 1999. From (Moreno et al. 2006)

of planetary atmospheres and illustrate the potential of ALMA.

2 Venus and Mars

2.1 The dynamics of the middle atmospheres

Thermal and wind sounding of the middle atmospheres of Venus (70–120 km) and Mars (0–80 km) can be performed from CO line mapping. While temperature sounding is not a unique capability of heterodyne spectroscopy (and is actually performed at much higher spatial resolution using IR sounding from planetary orbiters), direct wind measurements can be obtained only by this technique. Current results from interferometry, performed mostly for the martian atmosphere, indicate a complex wind structure, variable with season and dust content, and generally at odds (except near solstice) with global climate model predictions (Fig. 1, Moreno et al. 2006). In Venus, the middle atmosphere circulation is characterized by the coexistence of (i) a retrograde zonal flow, (ii) a subsolar-to-antisolar flow, which a large body of temporal variability on all timescales from days to weeks (Clancy et al. 2005) and a strong coupling to the distribution of CO and other tracers.

ALMA will allow to study fine wind structures, including jets, waves, orographic winds and, thanks to its *uv*-coverage, to study temporal variability with unprecedented time resolution.

2.2 The martian water cycle

The martian and venusian atmospheres contain small amounts of water. On Mars, the vertically-integrated water column is well-known to vary wildly with season and latitude. This so-called "water cycle" reflects the exchanges of water between the atmospheric (gas phase, clouds) and surface reservoirs (the northern polar glacier and possibly the subsurface). However, the vertical profile of water is much less well characterized, although millimeter-wave observations (Clancy et al. 1996) indicates that on a planetary average, it also varies with season, being more confined to the surface near aphelion than near perihelion. This "varying hygropause" has important implications for the inter-hemispheric water transport and for atmospheric chemistry as a whole. Furthermore, a subtle fractionation effect is expected to occur on deuterated water. The HDO vapor pressure is slightly lower than that for H_2O. Thus, at cloud formation, the condensed phase is enriched in D and the gas phase is correspondingly depleted. When this effect is included in general circulation models, it is found that the D/H ratio is expected to show vertical and latitudinal variations (Fig. 2, Montmessin et al. 2005). Understanding this effect is not only interesting in itself, but also needed to properly interpret the D/H ratio, as measured from HDO/H_2O, in terms of the evolution of the martian atmosphere. ALMA will address this problem by obtaining seasonal maps of HDO and H_2O (as traced by $H_2^{18}O$).

2.3 Chemistry

Improving our knowledge of atmospheric chemistry is achieved by (i) monitoring known species and studying correlations between them, and (ii) searching for new species. In the martian atmosphere, the coupled photochemistry of CO_2 and water gives birth to several species, among which CO, O_2, O_3 and H_2O_2 have been detected. In Venus, the additional presence of sulfur- and chlorine-bearing species complicates the chemistry. On both planets, ALMA will be a sensitive tool to search for new species, including (not exclusively) NO, HO_2 and OH on Mars, and H_2S, OCS, O_2, and HCl above Venus clouds. Another subject of investigation will be the monitoring and mapping of the water and SO_2 content above Venus' clouds, which appears to be strongly variable, perhaps in relation with temperature fluctuations (Fig. 4, Gurwell et al. 2007). It is worth noting that the first detection of H_2O_2 on Mars and of SO_2 above Venus clouds were achieved at sub-millimeter wavelengths (Clancy et al. 2004, 2005; Fig. 3).

3 Titan

Titan's atmosphere is probably the one in which the couplings between chemistry and dynamics are the most marked.

Fig. 2 H$_2$O and HDO cycle in the martian atmosphere, calculated from general circulation model. *Top*: HDO distribution vs. latitude and season. *Bottom*: D/H ratio (given as 1/2 HDO/H$_2$O) vs. latitude and season.
From (Montmessin et al. 2005)

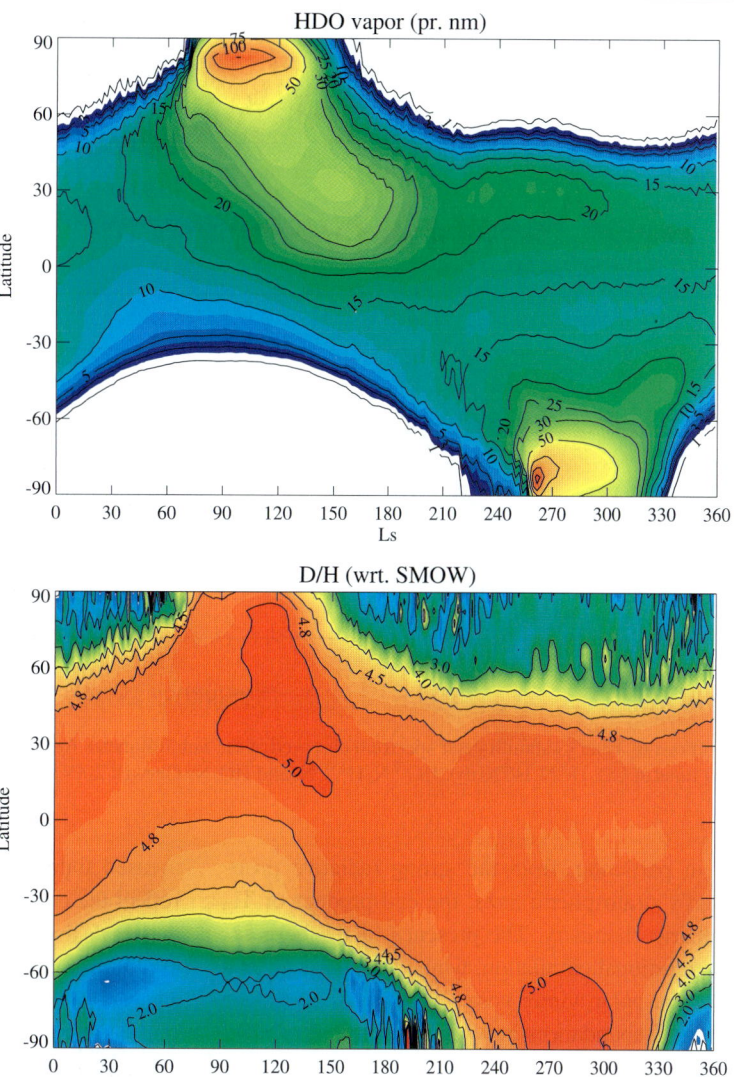

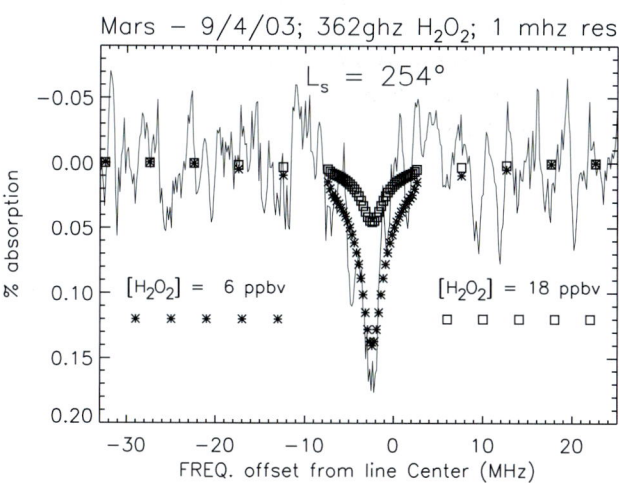

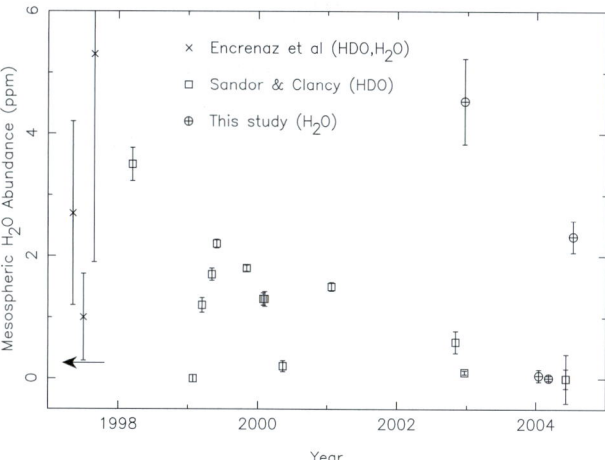

Fig. 3 The first detection of hydrogen peroxide in the martian atmosphere, obtained at JCMT. From (Clancy et al. 2005)

Fig. 4 The disk-average abundance of water above Venus clouds, measured from ground-based and SWAS millimeter/submillimeter observations. Note the erratic variability. From (Gurwell et al. 2007)

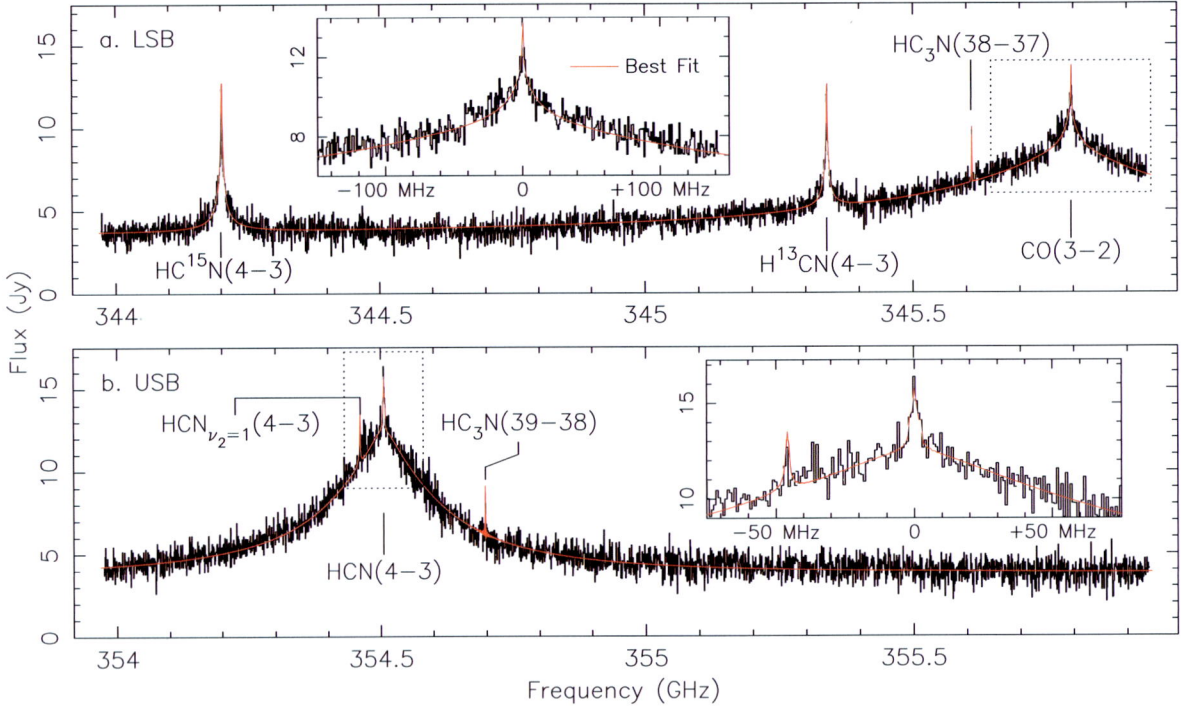

Fig. 5 Two portions of Titan's submillimeter spectrum observed at the Submillimeter Array, covering the 344–346 and 354–356 GHz range, and showing strong lines of CO, HCN, HCN(v_2), $HC^{15}N$, $H^{13}CN$, and HC_3N (Gurwell 2004)

This stems, in particular, from the strong radiative properties of Titan's stratospheric haze, the end-product of Titan's atmospheric chemistry. The presence of the haze reinforces the meridional circulation, which transports all minor species to the polar winter vortex in which they appear to have increased abundances and specific vertical profiles (Hourdin et al. 2004).

Titan's atmosphere is rich in hydrocarbon, nitrile and oxygen species, several of which (CO, HCN, HC_3N, CH_3CN, H_2O, and some isotopes) have been detected at mm/submm wavelengths. Titan submillimeter lines (Fig. 5) usually exhibits broad and narrow emission components, well suited to the versatility of the ALMA backends. PdBI observations (Moreno et al. 2005) have allowed to disk-resolve these lines, and to measure Doppler shifts indicative of stratospheric and mesospheric winds unreachable by any other technique (Figs. 6, 7).

The increased spatial resolution of ALMA will offer the possibility to study the latitudinal structure of these winds. ALMA will also map all the above detected species, hereby continuing the 3-D investigation of Titan's chemistry undertaken by *Cassini* which should last until 2010–2012. ALMA will also be able to study some of the trace species detected in Titan's thermosphere by *Cassini* (e.g. CH_3CCH, C_2H_3CN, NH_3) which revealed an organic chemistry of the upper atmosphere richer than anticipated. ALMA may also be able to determine rare isotope species, e.g. DCN or HDO. Deuterium in Titan is currently measured only in CH_4. Its in-

creased abundance with respect to the protosolar value must reflect the combination of the original composition of planetesimals that formed Titan with the subsequent atmospheric evolution, including hydrogen escape (Lunine et al. 1999). Measuring the D/H ratio in a species other than methane (e.g. DCN) would be of high interest.

Finally, continuum measurements, which probe the upper troposphere temperature, will provide new constraints about Titan's meteorology, especially through the determination of the latitudinal temperature contrasts near 30 km. All of the above phenomena should be monitored on a 30-year timescale, which corresponds to the annual period of Titan.

4 Giant Planets

Depending on their formation level, (sub)mm lines in Giant Planets may appear as broad (several GHz) tropospheric absorptions (e.g. PH_3, NH_3 in Jupiter and Saturn), relatively narrow (10–50 MHz) stratospheric emissions (e.g. CO, HCN, H_2O in Jupiter), or the combination of both (e.g. CO on Neptune (Fig. 8)). So far, observing these lines has required either low resolution broad-band spectroscopy (e.g. from FTS) or repeated tunings. The broad bandwidth and the multi-resolution mode offered by ALMA will greatly facilitate observations. A number of problems will be addressed:

Fig. 6 Example of line position fitting, enabling wind measurements. The example shows lines of CH_3CN and HC_3N observed on Titan with PdBI. From (Moreno et al. 2005)

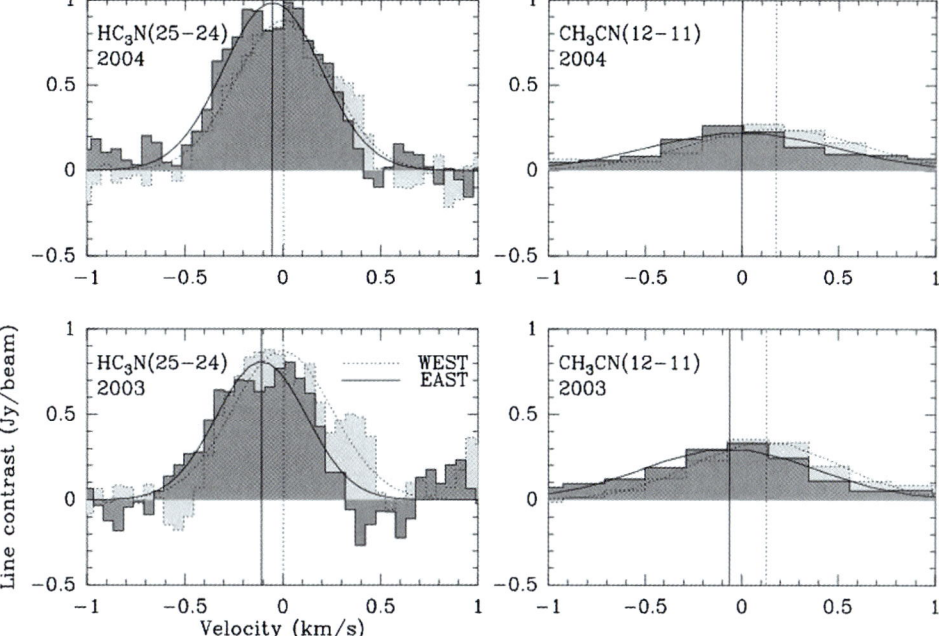

– The origin of external water in the outer planets. The presence of H_2O, CO and CO_2 in the upper atmospheres of the Giant Planets and Titan implies the existence of an external source of oxygen, whose nature is uncertain. While *Herschel* will address this question from extensive H_2O observations, ALMA will contribute to the problem by (i) establishing the CO vertical profile in Uranus and Saturn, (ii) determine its latitudinal distribution, (iii) attempt to measure the D/H ratio in this component. This latter measurement could be a clue to disentangling between cometary and local (rings, satellite) sources of water.

– The long-term evolution of the Shoemaker–Levy 9 impact products in Jupiter's atmosphere. The SL9 impacts in 1994 produced large amounts of CO, CS, and HCN, which in principle are chemically stable. Monitoring the evolution of their distribution with altitude and latitude will constrain the horizontal and vertical mixing in the atmosphere at a level where no tracers (such as clouds) exist to constrain the circulation.

– The tropospheric structure. As detailed in (Hosftadter 2006), brightness temperature maps of the Giant Planets (so far mostly performed at centimeter wavelengths) reflect the interplay between the tropospheric temperature field and its composition, especially the humidity of condensible species (NH_3, H_2S). Another important objective will be the search for phosphine, so far undetected, in Uranus and Neptune.

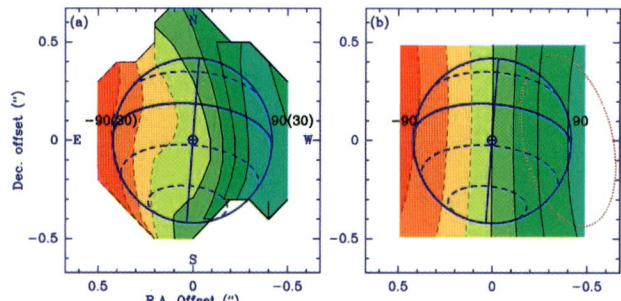

Fig. 7 Wind maps at 300 ± 150 km in Titan's atmosphere, measured from lineshifts of CH_3CN lines. From (Moreno et al. 2005)

5 Tenuous atmospheres

5.1 Io and Enceladus

Io's atmosphere is spatially heterogeneous, with local dayside pressure varying typically from 0.1 to several nanobars. It is mostly composed of SO_2 (∼90%), with traces of SO, NaCl, and S_2. The essential (and inter-related) questions regarding the nature of Io's atmosphere are: (i) the relative importance of sublimation equilibrium and direct volcanic output in maintaing the atmosphere, (ii) its vertical structure and dynamics, with the possible coexistence of plume dynamics with day-to-night, sublimation-driven, planetary-driven flows.

Disk-resolved observations of Io's atmosphere have been obtained recently at PdBI and SMA (Moullet et al. 2007; Gurwell et al. 2006), providing first indications on the spatial extent of the atmosphere and of its dynamics. With an anticipated spatial resolution of up to $0.1''$ (Io's disk is $1''$ in

Fig. 8 CO on Neptune. *Top*: the broad tropospheric absorption. *Bottom*: the narrow stratospheric emission. The combined observation of the two components indicates that CO is not uniform with altitude. *Insets* in the bottom panels show various CO vertical profiles and the associated eddy diffusion coefficients. From (Lellouch et al. 2005)

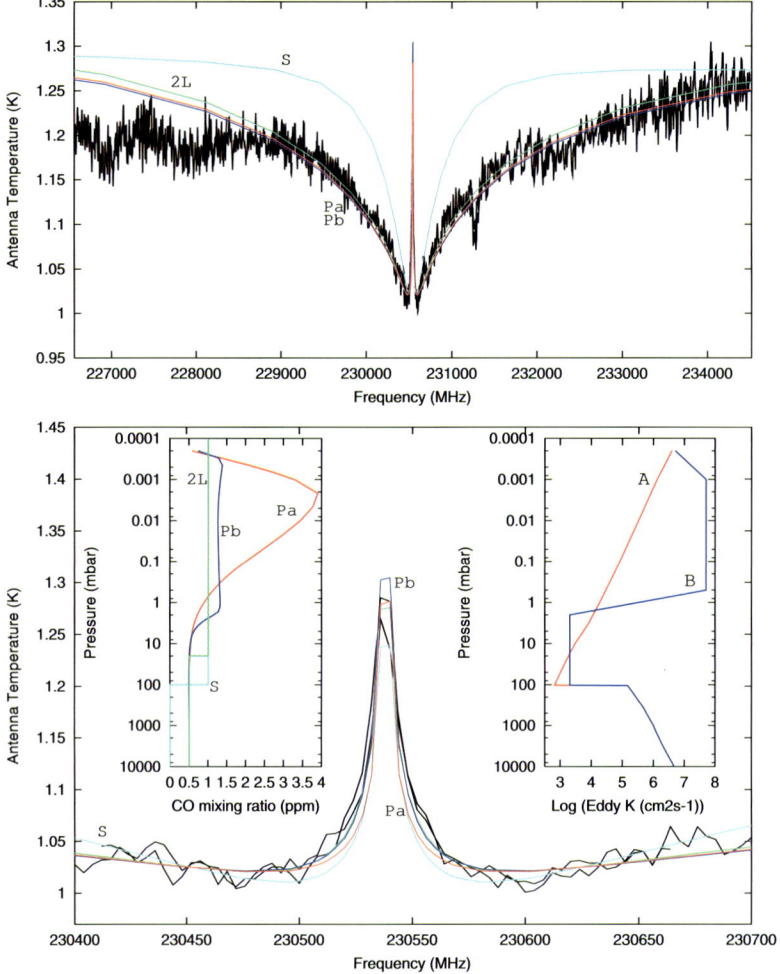

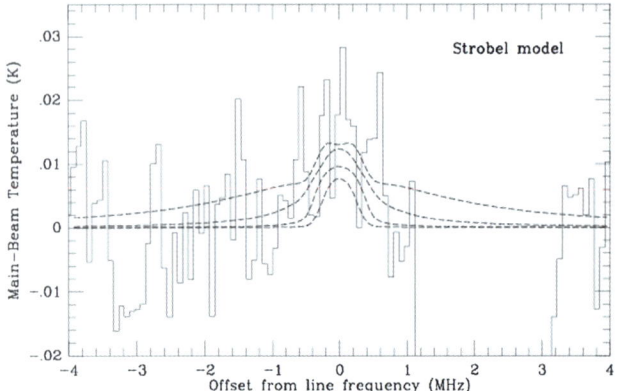

Fig. 9 The search for CO at 230 GHz in Pluto's atmosphere with IRAM-30 m. The *dashed lines* indicate models for various CO abundances. From (Bockelée-Morvan et al. 2001)

cluding S_2O, KCl, H_2S, ClO, SiO. Some of these species are purely volcanic in origin (i.e. have negligible vapor pressure at Io surface temperatures) and measuring their abundances would provide insight into the composition of volcanic magmas.

The atmosphere discovered by *Cassini* above the South Pole of Saturn's satellite Enceladus is in some respect similar to Io's, being apparently driven by a cryovolcanic plume. This atmosphere is primarily composed of water (typical column 10^{16} cm^{-2}), with CO, N_2 and CH_4 at the several percent level. Although a mapping seems difficult given the small apparent size of the target (about $0.15''$), ALMA could at least measure CO, thereby constraining the unknown atmospheric temperature.

5.2 Trans-neptunian objects

Several of the largest Trans-neptunian objects (Pluto, Eris, 2005 FY9) and Neptune's satellite and former Kuiper-Belt object Triton have surfaces covered by volatile ices, including N_2, CH_4 and CO. At least for Pluto and Triton, it is known that sublimation equilibrium of these ices gives rise

diameter), ALMA observations are expected to bring much new progress: in addition to the detailed localisation of the sources and the measurements of local and global winds, ALMA will be able to perform a detailed study of its composition by searching and mapping for several new species, into an atmosphere with a typical 10 µbar pressure (currently increasing for Pluto) dominated by N_2 and in which CH_4 has also been detected. Detecting CO (and perhaps other species, such as HCN) is an important goal, not only for extending the inventory of species in these atmospheres, but also because CO appears to be an important species controlling (through rotational cooling) their thermal budget. Attempts to detect CO have been performed at IRAM 30-m. Although they were unsuccessful (Bockelée-Morvan et al. 2001), comparison with models suggests that the detection of CO could be achieved with only a moderate increase of the sensitivity. ALMA should thus provide an easy detection of CO on Pluto and similarly on Triton, and the search could then be extended to other "dwarf planets" and large Kuiper Belt objects. Note also that the observation of several lines of CO would allow one to infer the atmospheric temperature structure, which is particularly unconstrained in the case of Pluto.

6 Specificity of planetary observations

We close by listing a few specificities of planetary and satellite observations which should be considered when preparing the observation and reduction softwares for ALMA.

– Planets and satellites are moving objects, which requires position tracking from ephemeris. While this will certainly be implemented for planets, it would also be important to achieve it for satellites, which as illustrated above, will represent a significant fraction of the science for planetary atmospheres.
– Velocity tracking is highly desirable, though admittedly velocity corrections could be done in the reduction stage. In any case, LSR velocity is not the appropriate velocity reference for Solar System objects, and the topocentric velocity is required.
– Some of the targets (Jupiter, Saturn, Venus, Mars at opposition) can be large compared to the primary beam of the

ALMA telescopes, which will require the use of ACA as well as mosaicking, along with accurate pointing.

References

Bockelée-Morvan, D., et al.: Search for CO gas in Pluto, Centaurs and Kuiper Belt objects at radio wavelengths. Astron. Astrophys. **377**, 343–349 (2001)

Clancy, R., et al.: Water vapor saturation at low altitudes around aphelion: a key to martian climate? Icarus **122**, 36–62 (1996)

Clancy, R.T., et al.: A measurement of the 362 GHz absorption line of Mars atmosphere H_2O_2. Icarus **168**, 116–121 (2004)

Clancy, R., et al.: Extreme variability in the middle atmosphere of Venus. In: AGU Fall Meeting, paper 33A-0223 (2005)

Gurwell, M.A.: Submillimeter observations of Titan: global measures of stratospheric temperature, CO, HCN, HN_3N and the isotopic ratios $^{12}C/^{13}C$ and $^{14}N/^{15}N$. Astrophys. J. **616**, L7–L10 (2004)

Gurwell, M.A., et al.: Solar system science with the submillimeter array. In: Science with ALMA: A New Era for Astrophysics, Madrid, 13–17 November 2006

Gurwell, M.A., et al.: SWAS observations of water vapor in the Venus mesosphere. Icarus **188**, 288–304 (2007)

Hosftadter, M.: Millimeter, submillimeter observations of Uranus and Neptune. In: Science with ALMA: A New Era for Astrophysics, Madrid, 13–17 November 2006

Hourdin, F., et al.: Titan's stratospheric composition driven by condensation and dynamics, J. Geophys. Res. **109**, E12004 (2004), doi: 10.1029/2004JE002282

Lellouch, E., et al.: A dual origin for Neptune's carbon monoxide? Astron. Astrophys. **430**, 37–41 (2005)

Lunine, J., et al.: On the volatile inventory of Titan from isotopic abundances of nitrogen and methane. Planet. Space Sci. **41**, 1291–1303 (1999)

Montmessin, F., et al.: Modelling the annual cycle of HDO in the martian atmosphere, J. Geophys. Res. **110**, E03006 (2005), doi: 10.1029/2004JE002357

Moreno, R., et al.: Interferometric measurements of zonal winds on Titan. Astron. Astrophys. **437**, 319–328 (2005)

Moreno, R., et al.: Wind measurements in Mars middle atmospheres at equinox and solstices: IRAM Plateau de Bure interferometric CO observations. In: Second Workshop on Mars Atmosphere Models and Observations, pp. 134–135, Granada, 27 February–3 March, 2006

Moullet, A., et al.: Io SO_2 atmosphere: first disk-resolved millimeter observations. Astrophys. Space Sci. (2007, this volume)

Cometary science with ALMA

Dominique Bockelée-Morvan

Originally published in the journal Astrophysics and Space Science, Volume 313, Nos 1–3.
DOI: 10.1007/s10509-007-9641-2 © Springer Science+Business Media B.V. 2007

Abstract This paper presents a prospect for the observations of comets with ALMA. Thanks to unprecedented sensitivity, angular resolution and instantaneous uv-coverage, key measurements on a number of topics related to the chemical and physical properties of the coma and the nucleus will be obtained. These include (1) the identification of new molecular species and measurements of key isotopic ratios, (2) measurements of the composition of short-period comets coming from the trans-Neptunian scattered disc, to investigate chemical diversity within the whole comet population, (3) imaging of gas jets and their relationship with dust features, (4) the study of extended sources of gas in the coma, and (5) the study of the physical and outgassing properties of the nucleus.

Keywords Comets · Submillimeter · Interferometry

1 Introduction

Comets contain potentially a lot of informations regarding the formation and early evolution of the Solar System. These remnants of the planetesimals that formed the outer planets contain a record of the physical and chemical conditions of the primitive Solar Nebula, in the region and at the time of their formation. The recent results obtained from astronomical observations, in situ exploration and laboratory analyses of samples returned by the Stardust mission illustrate the importance of comets for studying Solar System formation (Irvine et al. 2000; Brownlee et al. 2007).

D. Bockelée-Morvan (✉)
LESIA, Observatoire de Paris, 5 place Jules Janssen, 92190
Meudon Cedex, France
e-mail: dominique.bockelee@obspm.fr

Comets have been observed in the millimeter and submillimeter wavelength ranges for about 20 years. A wealth of information that cannot be obtained by other techniques has been retrieved regarding coma molecular and isotopic composition, and dust production. Interferometric observations have been performed in a very limited sample of comets, that were bright enough for the observations to be successful. These observations will be discussed here to illustrate the potential of ALMA for cometary science. We invite the reader to read the review paper of Biver (2005), for complementary information.

2 Molecular and isotopic composition

Millimeter and submillimeter spectroscopy is a very powerful tool for detecting parent molecules released from the nucleus, because of the cold environment (coma temperatures are typically 40–100 K) and radiative excitation which leads generally to low rotational temperatures. About two dozens molecules (not including isotopologues, molecular ions, atoms and radicals) have now been identified in cometary atmospheres, most of them using radio techniques (Fig. 1). Complex species, such as methyl formate and ethylene glycol, are detected in radio spectra (Fig. 2). The composition of cometary volatiles present striking similarities with that of star forming regions (interstellar ices, hot cores, in particular), suggesting that cometary molecules formed by similar processes (Irvine et al. 2000; Bockelée-Morvan et al. 2000).

2.1 New identifications

Abundances of detected cometary parent molecules range from less than 0.01% to 20% relative to water, and generally

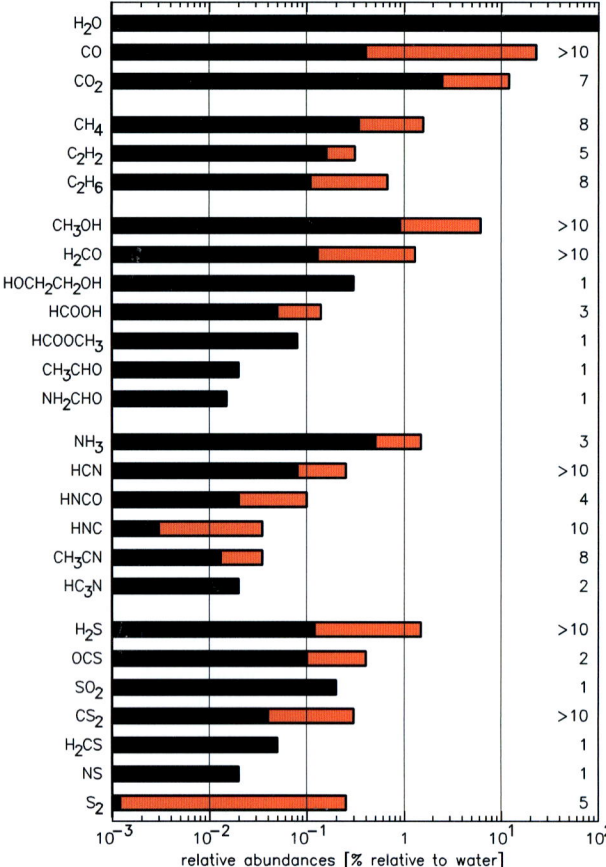

Fig. 1 Abundances relative to water of cometary parent molecules. The range of measured values is shown in the *gray portion*. The number of comets in which the molecule has been detected is given in the *right* (adapted from Bockelée-Morvan et al. 2004, courtesy of J. Crovisier)

decrease with increasing complexity, except for hydrocarbons. The cumulative histogram of the number of detected species N, as a function of their abundance X, suggests that observational biases affect the detection of species with abundances $<10^{-3}$ and that a large number of molecules remains to be identified in cometary atmospheres (Crovisier et al. 2004). With a gain in sensitivity of ~ 10 with respect to presently available instrumentation, ALMA should enable the detection of new cometary species in bright comets.

2.2 Chemical diversity

Surveys of parent volatile abundances show strong evidence for chemical diversity among comets. Depending on the molecule, abundances vary by a factor of a few to several tens (for CO) among comets (Fig. 1). Diversity is observed both within the population of Oort cloud comets (the dynamical class of nearly-isotropic comets) and within the population of ecliptic, short-period comets (dynamical class previously called Jupiter family comets), which originate from the trans-Neptunian scattered disc. Comets in these

two reservoirs were not formed in situ (they were expelled from inner regions by the giant planets), and possibly could have originated from distinct regions of the Solar Nebula. Presently, no obvious correlation is observed between the abundances of the hypervolatile species and the dynamical class. No trend between volatility and comet-to-comet variability is observed either (Biver et al. 2002; Gibb et al. 2003).

So far, ecliptic comets have been poorly investigated by radio spectroscopy, because of their relatively low level of activity. A number of molecules have only been detected in comet C/1995 O1 (Hale–Bopp) or in a small number of comets. ALMA will allow to assess the composition of a large sample of comets from the two reservoirs, leading to progress in the understanding of chemical diversity and relationships between cometary composition and formation region.

2.3 Isotopic ratios

Isotopic ratios are key diagnostics of the physical and chemical conditions that prevailed during the formation of cometary volatiles. So far, only a few isotopic ratios have been measured in cometary gases (see the reviews of Bockelée-Morvan et al. 2004; Altwegg and Bockelée-Morvan 2003). Measurements have been obtained for the $^{12}C/^{13}C$ ratio in C_2, CN and HCN, for the $^{32}S/^{34}S$ ratio in CS, S^+, H_2S, and for the $^{16}O/^{18}O$ ratio in water: they are consistent with terrestrial values. On the other hand, the $^{14}N/^{15}N$ ratio measured in CN in a dozen of comets is ~ 140, i.e., two times lower than the terrestrial value, whereas the $HC^{14}N/HC^{15}N$ measured in comet Hale–Bopp from submillimeter spectroscopy was found terrestrial. This suggests that HCN is not the only source of CN in cometary atmospheres. Stardust comet samples show refractory phases with low $^{14}N/^{15}N$ ratio, which are possibly progenitor of the CN radical. Surprisingly, the $^{14}CN/^{15}CN$ does not vary among comets, while they exhibit strongly different dust-to-gas-ratios. Thanks to ALMA, it will be possible to obtain additional measurements of $^{14}N/^{15}N$ in HCN, including for short-period comets. Measurements of $^{14}N/^{15}N$ in other N-bearing species, such as HNC, which origin in cometary atmospheres is debated (Rodgers and Charnley 2001), will be hopefully possible in very bright comets.

ALMA will provide important information regarding D/H ratios. HDO has been observed from submillimeter spectroscopy in the bright comets Hale–Bopp and C/1996 B2 (Hyakutake), from which a D/H ratio of 3×10^{-4} consistent with Giotto measurements in comet 1P/Halley has been derived. So far, these three comets are Oort cloud comets. Data on short-period comets will be obtained with HIFI aboard the Herschel Space Observatory. ALMA, which

Fig. 2 Detection of ethylene glycol (HOCH$_2$CH$_2$OH) and other species in comet Hale–Bopp (Crovisier et al. 2004)

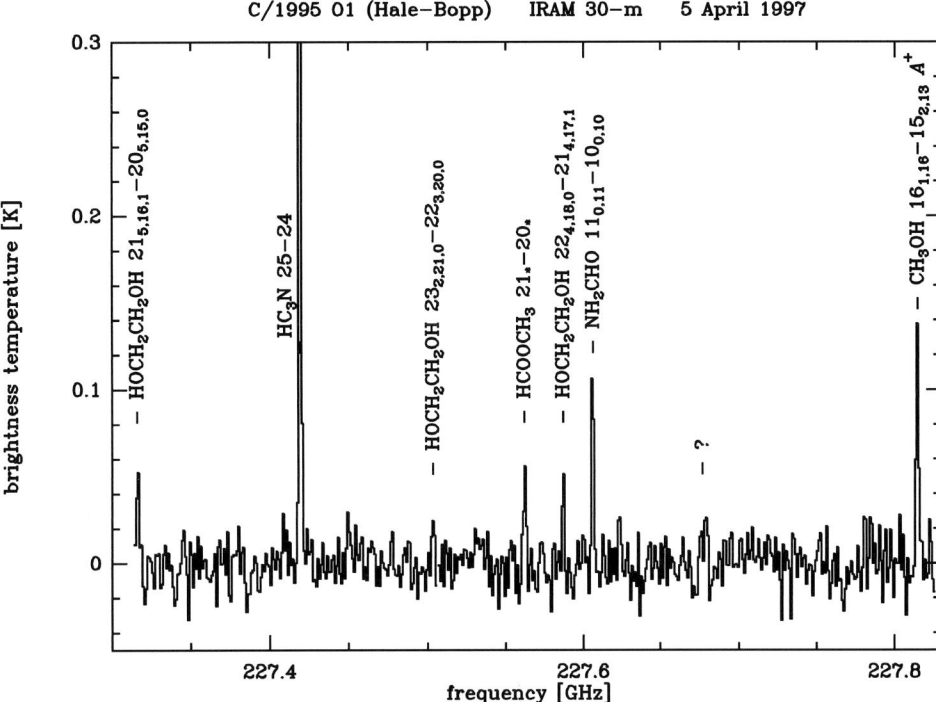

is competitive with Herschel providing the 464 or 894 GHz HDO lines can be observed, will allow further investigations in a large sample of objects. Because isotopic fractionation is molecule dependent, D/H measurements in other species are equally important. DCN has been detected in comet Hale–Bopp, yielding a D/H ratio in HCN seven times higher than that measured in cometary water. ALMA will have the sensitivity to detect DCN in typically one Oort cloud comet per year, and in short-period comets making a close approach to Earth (Biver 2005). Investigation of other D-bearing species and multi-deuterated molecules will be certainly fruitful for comparison with the extreme deuteration observed in the early phases of low-mass protostars formation (Ceccarelli et al. 2007). However, their detection may only be possible in very bright comets.

3 Comet activity

The high sensitivity of ALMA will make possible the observations of comets with low levels of activity. Comets with gas production Q_{H_2O} as low as 10^{27} s^{-1} will be detectable in 1 h integration time in the HCN J(4–3) line when at 1 AU from Earth. Since the HCN/H$_2$O relative abundance does not vary significantly from comet-to-comet, HCN can be used as a proxy for measuring the water production rate.

Measuring comet gas production is specially interesting if constraints on the nucleus size (Sect. 5) and dust production can be obtained in parallel. Based on optical observations, which are mostly sensitive to small micrometric particles, comets exhibit a wide range of dust-to-gas production ratios, which can indicate primordial differences in their dust-to-gas content, evolutionary effects at the surface or sub-surface layers, or differences in particle size distribution. Because most of the dust mass resides in large particles, ALMA observations of dust thermal emission will provide the total dust mass production rate and the dust-to-gas ratio in a statistically significant sample of comets. Combining constraints on the nucleus size (Sect. 5) and the gas production, information on the active area fraction of the nucleus surface can be obtained.

ALMA will have the sensitivity to monitor cometary activity over a wide range of heliocentric distances. Until now, monitoring observations along the cometary orbit were only possible for a few bright comets. Compared to other techniques, radio spectroscopy is powerful for monitoring gas production rates as the comet recedes from the Sun because the coma gets colder. In comet Hale–Bopp, the production rates of a number of parent molecules were measured up to 4–5 AU from the Sun, and up to 14 AU for CO using radio spectroscopy (Fig. 3, Biver et al. 2002). Similar studies will be possible in less productive comets with ALMA. Gas production curves are almost the only observational tool we have to obtain informations regarding the physical state of cometary ices, and their thermal properties and sublimation mechanisms. The study of distant cometary activity, which is governed by the released of hypervolatiles species such as CO, is important to understand how these gases are trapped in the nucleus. Gas and dust production curves complement

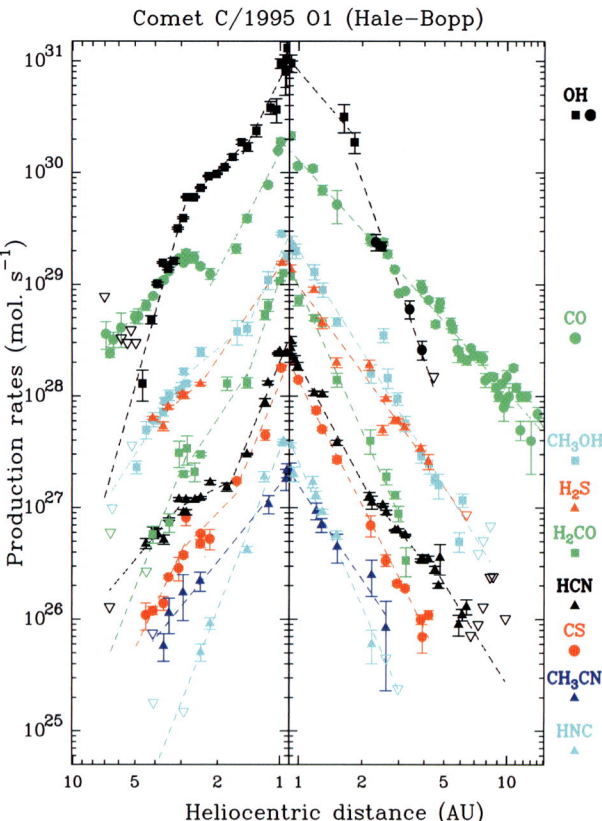

Fig. 3 The Christmas tree: gas production curves in comet Hale–Bopp from radio observations at IRAM, JCMT, CSO, SEST and Nançay telescopes (Biver et al. 2002)

visual light curves for the study of seasonal effects, dust mantling, etc.

4 Coma mapping

4.1 Coma structure

The study of the spatial distribution of gases and dust in the inner coma provides clues on the distribution of the outgassing at the surface of the nucleus, and on the gas dynamics processes occurring in the coma. Many comets exhibit day–night asymmetries and detailed, well defined structures which indicate that the surfaces of the nuclei are not uniformly active, with at least part of the material coming from isolated active areas. Structure in the coma can also result from outflowing material that experiences density enhancements due to the topology of the nucleus.

So far, structural information on the distribution of gases in the coma was mainly obtained from visible imaging of daughter products produced far out in the coma. With ALMA, it will be possible to map at high angular resolution the distribution in the inner coma of several key constituents released from the nucleus. Images of the dust thermal emission will be obtained simultaneously, enabling the

comparison of gas and dust features. For the gases, information on the spatial distribution along the line of sight will be extracted by analysing the line shapes.

This study will strongly benefit from the good instantaneous uv-coverage of ALMA. In contrast to most astronomical sources, comet activity varies on short timescales. In addition, the coma is shaped by the rotation of the nucleus (typically a few to a few tens hours).

The results obtained in comet Hale–Bopp from IRAM Plateau de Bure observations demonstrate the potential of ALMA for studying the inner cometary atmosphere and its temporal evolution. The spectral shift of the CO 230 GHz line recorded in autocorrelation mode and by the antenna pairs showed sinus-like variations with a period equal to the rotation period of 11.35 h (Fig. 4, Henry et al. 2002). Temporal modulations related to nucleus rotation were also observed in the interferometric visibilities. Maps derived from data subsets of $\sim$1 h duration showed that the centroid of the CO emission moved perpendicularly to the direction of the rotation axis (Fig. 4). These observations could be explained by the presence of a strong CO jet rotating with the nucleus (Henry et al. 2002; Boissier et al. 2007). The analysis of these data was hampered by the limited instantaneous uv-coverage of the Plateau de Bure interferometer, but they clearly showed that rotating comas can be detected by interferometric radio imaging, and that useful constraints on the rotation properties of cometary nuclei can also be obtained.

As detailed in Biver (2005), the most favorable lines for imaging are those of HCN, which could be mapped at medium resolution (500 km) with good time sampling in at least one comet per year. HCN imaging with a resolution of 50 km at the nucleus will be possible in brighter comets or in comets making close (0.1 AU) approach to Earth. The comparison of the spatial distribution of different molecules may reveal compositional inhomogeneities at the nucleus surface.

4.2 Extended sources

There are several observational pieces of evidence that some molecules observed in the coma are not directly released by the nucleus. The presence of distributed source of molecules in the coma was suggested for explaining the radial distributions of CO and H_2CO in comets Halley and Hale–Bopp, and the presence of CN, C_2 and C_3 gas jets. Such species could be produced by the degradation of organic grains (e.g., thermal degradation of polyoxymethylene is suggested for H_2CO, Fray et al. 2006). Some molecules could be the product of gas-phase chemistry. For example, isomerization of HCN from impact with fast hydrogen atoms was proposed for explaining the heliocentric evolution of the HNC/HCN ratio in comet Hale–Bopp (Rodgers and Charnley 1998).

Interferometric radio measurements already provided significant results in this field. The extended nature of H_2CO

Fig. 4 *Top*: maps of CO $J(2–1)$ obtained in comet Hale–Bopp at different times on 11 March 1997 using Plateau de Bure interferometer; the *arrows* show the displacement of the photometric centre with respect to the mean position. *Bottom left*: time evolution of the photometric centre; the arrow shows the spin axis orientation. *Bottom right*: time evolution of the velocity offset of the autocorrelation spectra of CO $J(2–1)$ obtained on 11 March. From Henry et al. (2002)

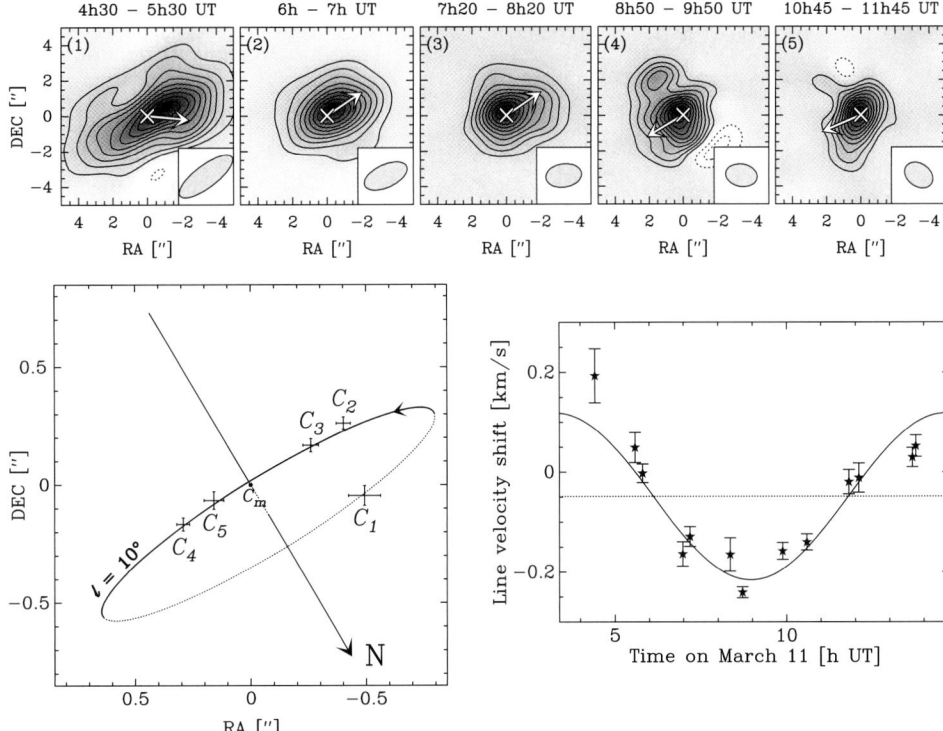

source was confirmed in comet Hale–Bopp (Wink et al. 1999). The radial distribution of HCN and H_2S was measured to be consistent with nucleus origin (Boissier et al. 2007; Snyder et al. 2001). Comparison of HNC and HCN $J(1–0)$ maps excludes HCN isomerization as the main process for producing HNC in Hale–Bopp coma (Fig. 5, Bockelée-Morvan et al. 2005). SO imaging showed that this radical is produced by the photolysis of SO_2 (Fig. 6, Boissier et al. 2007). Constraints on the photodissociation lifetime of CS were obtained (Boissier et al. 2007; Snyder et al. 2001).

With ALMA, it will be possible to study the radial distribution of cometary molecules in the coma and investigate how it varies with, e.g., the heliocentric distance, the comet activity, the dust-to-gas ratio. Important insights regarding the production mechanisms may be anticipated. Both autocorrelation measurements and interferometric mapping will be necessary, as the typical sizes of the distributed sources range from a few 100 km to several 10000 km (0.2–20", typically).

5 Nucleus properties

Cometary nuclei are among the most difficult Solar System objects to detect and characterize, suffering from the dual problem of being faint and immersed in a coma. So there is still limited information on their properties, such as size, shape, albedo, thermal inertia, for statistical studies (Lamy et al. 2004).

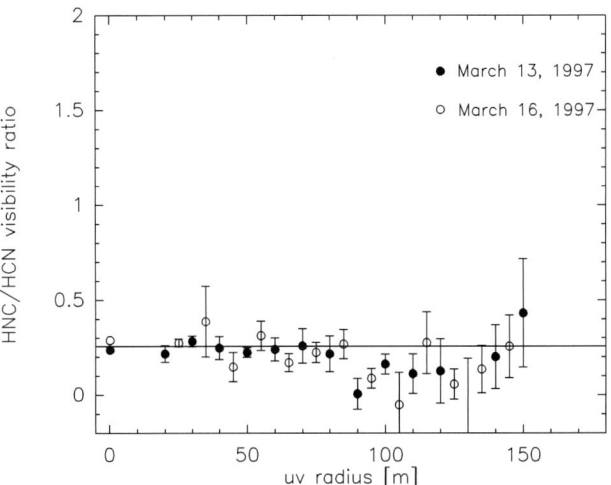

Fig. 5 HNC/HCN visibility ratio as a function of uv-radius (i.e., baseline length) from observations of comet Hale–Bopp at Plateau de Bure interferometer on March 9 (HCN $J(1–0)$) and March 13 and 16 (HNC $J(1–0)$) 1997 (Bockelée-Morvan et al. 2005). The zero baseline corresponds to autocorrelation measurements. The decrease of the HNC/HCN visibility ratio at long baselines may indicate production of HNC in the inner coma ($\ll 2000$ km). A much more extended HNC distribution is expected for HNC produced by the isomerization of HCN from H-impact or ion-molecule reactions (Bockelée-Morvan et al. 2005)

To first approximation (slow rotor, emissivity of 1), the total flux density emitted by a nucleus of diameter D is $F[\text{mJy}] = 0.03\lambda[\text{mm}]^{-2} \cdot D^2 \cdot r_h[\text{AU}]^{-0.5} \cdot \Delta[\text{AU}]^{-2}$, where r_h and Δ are the heliocentric and geocentric distances, re-

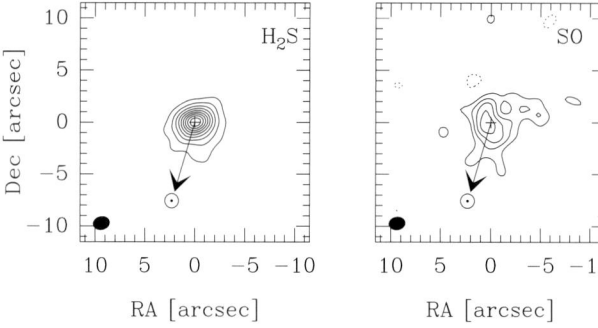

Fig. 6 Interferometric maps of H$_2$S 216.7 GHz and SO 219.9 GHz lines observed on 13 March 1997 at IRAM Plateau de Bure Interferometer. The synthesized beam is shown in the *left bottom* corner. The Sun direction is indicated by the *arrow*. These data are consistent with H$_2$S being released by the nucleus and SO being produced at least partly by the photolysis of SO$_2$ (Boissier et al. 2007)

spectively. In one hour integration time, a 2-km diameter body at $r_h = \Delta = 1$ AU can be detected with ALMA (5-σ at 0.8 mm). Observations have to be conducted with ALMA in extended configuration to warrant a weak contribution of the dust thermal emission in the synthesized beam. In intermediate configurations with 5-km baselines, a nucleus to dust contrast of typically 20 is expected for a 2-km diameter body observed at 0.8 mm (L. Jorda, private communication). Note that constraints on the size of Hale–Bopp's nucleus were obtained using the Plateau de Bure interferometer in compact configuration, but it required careful retrieval of the dust contribution in measured visibilities (Altenhoff et al. 1999). Thermal data obtained with ALMA, if complemented by measurements of reflected light, will allow the determination of the nucleus albedo. It can be anticipated that for large objects, or comets making close approaches to Earth, rotation light curves will be measured to retrieve information on their shape, rotation and thermal surface properties.

The maximum angular resolution of ALMA is 5 km for an object at 1 AU from Earth. Only large comet nuclei will thus be resolved with ALMA.

6 Specificity of comet observations

Astronomical observations of comets are generally far from being standard, and this will also be the case for ALMA. We list here a few specificities:

– Comets are moving objects. Accurate position tracking is required. This can be done either from user-supplied ephemeris or user-supplied comet orbital elements, if the ALMA software collection includes a code for computing cometary ephemeris. The later option is preferred. Observations have to be done using the most recent determined osculating orbital elements or ephemeris, implying close interaction between ALMA staff and cometary observers.

– Velocity tracking is highly desirable, though admittedly velocity corrections could be done afterwards from the reduction pipeline. In any case, LSR velocity is not the appropriate velocity reference for Solar System objects. Geocentric velocity is required.

– Analysis of cometary data may require specific reduction tools, e.g., for velocity corrections or analysis of short-term temporal variabilities.

– Autocorrelation measurements will provide the best sensitivity for detection of weak lines of long-lived cometary species. Autocorrelation measurements will usefully complement interferometric data for the study of the radial distribution of cometary species by providing the zero spacing.

– Comets are unpredictable objects to be often best observed over a fixed short (a week or so) time interval. The observations will have to be performed with the current ALMA antenna configuration. It is important to point out that ALMA will be useful for cometary studies, regardless its configuration. Extended configurations will be suitable for nucleus studies. In this case, coma compositional studies will be best performed from autocorrelation measurements; interferometric mapping of the coma could be attempted using antenna pairs with short spacings and the ACA array, if comet brightness permits it. Compact (<1 km) ALMA configurations will be optimum for interferometric mapping of the coma.

– Though automated sky survey programs for near-Earth asteroids and comet hunters are discovering many comets well before their perihelion, a significant number of comets are still to be observed as targets of opportunity (TOO), either because they were lately discovered or because they turned out to be much more active than anticipated. Therefore, unlike many other astronomical sources, comets will be often observed with ALMA as TOOs on Discretionary Director Time. Cometary science would greatly benefit from the implementation of an approved international TOO key program.

References

Altenhoff, W.J., et al.: Coordinated radio continuum observations of comets Hyakutake and Hale–Bopp from 22 to 860 GHz. Astron. Astrophys. **348**, 1020–1034 (1999)

Altwegg, K., Bockelée-Morvan, D.: Isotopic abundances in comets. Space Sci. Rev., 139–154 (2003)

Biver, N.: Comets with ALMA. In: The Dusty and Molecular Universe: A Prelude to Herschel and ALMA, ESA SP-577, pp. 151–156 (2005)

Biver, N., et al.: Chemical composition diversity among 24 comets observed at radio wavelengths. Earth Moon, Planets **90**, 323–333 (2002)

Biver, N., et al.: The 1995–2002 long-term monitoring of comet C/1995 O1 (Hale–Bopp) at radio wavelengths. Earth Moon, Planets **90**, 5–14 (2002)

Bockelée-Morvan, D., et al.: New constraints on the origin of HNC. Asteroids, Comets, Meteors, Buzios, 7–12 August 2005

Bockelée-Morvan, D., et al.: New molecules found in comet C/1995 O1 (Hale–Bopp). Investing the link between cometary and interstellar material. Astron. Astrophys. **353**, 1101–1114 (2000)

Bockelée-Morvan, D., et al.: The composition of cometary volatiles. In: Comets II, pp. 391–423. University of Arizona Press, Tucson (2004)

Boissier, J., et al.: Molecular spatial distributions observed in comet Hale–Bopp with IRAM Plateau de Bure interferometer. Astrophys. Space Sci. (2007, this volume)

Brownlee, D., et al.: Comet 81P/Wild 2 under a microscope. Science, **314**, 1711–1716 (2007)

Ceccarelli, C., et al.: Extreme deuteration and hot corinos: the earliest chemical signatures of low-mass star formation. In: Protostars and Planets V, pp. 47–62. University of Arizona Press, Tucson (2007)

Crovisier, J., et al.: Ethylene glycol in comet C/1995 O1 (Hale–Bopp). Astron. Astrophys. **418**, L35 (2004)

Crovisier, J., et al.: The composition of ices in comet C/1995 O1 (Hale–Bopp) from radio spectroscopy. Further results and upper limits on undetected species. Astron. Astrophys. **418**, 1141–1157 (2004)

Fray, N., et al.: Heliocentric evolution of the degradation of polyoxymethylene: application to the origin of formaldehyde (H_2CO)

extended source in comet C/1995 O1 (Hale–Bopp). Icarus **184**, 239–254 (2006)

Gibb, E., et al.: Methane in Oort cloud comets. Icarus **165**, 391–406 (2003)

Henry, F., et al.: Observations of rotating jets of carbon monoxide in comet Hale–Bopp with the IRAM Interferometer. Earth Moon, Planets **90**, 57–60 (2002)

Irvine, W.M., et al.: Comets: a link between interstellar and nebular chemistry. In: Protostars and Planets IV, pp. 1159–1200. University of Arizona Press, Tucson (2000)

Lamy, P., et al.: The sizes, shapes, albedos, and colors of cometary nuclei. In: Comets II, pp. 223–264. University of Arizona Press, Tucson (2004)

Rodgers, S.D., Charnley, S.B.: HNC and HCN in comets. Astrophys. J. **501**, L227 (1998)

Rodgers, S.D., Charnley, S.B.: On the origin of HNC in Comet Lee. Mon. Not. R. Astron. Soc. **323**, 84–92 (2001)

Snyder, L.E., et al.: BIMA array photodissociation measurements of HCN and CS in comet Hale–Bopp (C/1995 O1). Astron. J. **121**, 1147–1154 (2001)

Wink, J., et al.: Evidences for extended sources and temporal modulations in molecular observations of C/1995 O1(Hale–Bopp) at the IRAM interferometer. Earth Moon, Planets **78**, 63 (1999)

Observations of asteroids with ALMA

Amy J. Lovell

Originally published in the journal Astrophysics and Space Science, Volume 313, Nos 1–3.
DOI: 10.1007/s10509-007-9619-0 © Springer Science+Business Media B.V. 2007

Abstract Thermal observations of large asteroids at millimeter wavelengths have revealed high amplitude rotational lightcurves. Such lightcurves are important constraints on thermophysical models of asteroids, and provide unique insight into the nature of their surface and subsurface composition. A better understanding of asteroid surfaces provides insight into the composition, physical structures, and processing history of these surviving remnants from the formation of our solar system. In addition, detailed observations of the larger asteroids, accompanied by thermophysical models with appropriate temporal and spatial resolution, promise to decrease uncertainties in their flux predictions. Of particular interest are the near-Earth objects, which can be observed at large phase angles, permitting better assessment of the thermal response of their unilluminated surfaces. The high sensitivity of ALMA will enable us to detect many small bodies in all the major groups, to obtain lightcurves for a large sample of main-belt and near-Earth objects, to resolve the surfaces of some large objects, and to separate the emission from primary and secondary objects in binary pairs. In addition to the science goals of asteroid studies, these bodies may also prove useful operationally because those with known shapes and well-characterized lightcurves could be employed for flux calibration by ALMA and other high frequency instruments.

Keywords ALMA · Asteroids · Thermal emission · Radio astronomy · Flux calibration

A.J. Lovell (✉)
Agnes Scott College, Decatur, GA 30030, USA
e-mail: alovell@agnesscott.edu

Present address:
A.J. Lovell
Arecibo Observatory, Arecibo, PR 00612, USA

1 Introduction

The study of asteroids with sensitive, modern telescopes and instrumentation provides an opportunity to investigate these remnants of solar system formation in unprecedented detail. The largest of the asteroids are surviving protoplanets, while smaller asteroids may reveal diverse interiors of former protoplanets and provide clues to the collision history of small-body populations. While asteroids and other categories of heliocentric small bodies may be found throughout the solar system, from a few solar radii to over tens of thousands of Astronomical Units (AU), the largest of the known small-body populations are in five major groups: Near Earth Objects (NEOs), Main Belt Asteroids (MBAs), Jupiter Trojans, Centaurs, and Kuiper Belt Objects (KBOs). We will consider here the importance each of these populations and what contributions ALMA can make to scientific investigation of asteroids in each group.

For each population of bodies, the conditions under which they originally formed in the solar nebula are different in density, temperature, volatile composition, and dust-to-gas ratios. Compositional differences, isotope ratios, and other observable properties provide clues that will enhance our understanding of the protoplanetary disk and the early solar system in which asteroids formed. Which populations are most likely primordial? Do asteroids still contain the ices with which they formed? Which bodies are rich in water? How does the heating history differ between groups of asteroids? What does the study of asteroids tell us about the radial mixing of nebular materials? These questions illustrate how asteroids provide an opportunity for detailed, close-range study of star and planet formation in general, and the protosolar nebula in particular.

Observational studies of asteroids in the optical and near-infrared are based the detection of sunlight reflected from the surfaces. Unresolved photometric investigations have revealed colors and albedos for many bodies, while lightcurves sampling over the several-hour rotation of the bodies (Pravec et al. 2002) led to spin period estimates, and constraints on asteroid shapes (see for example Ďurech and Kaasalainen 2003; Kaasalainen et al. 2002; Lacerda and Luu 2006). Spectroscopic studies have led to numerous classification schemes (Bus et al. 2002 and references therein) and the ability to compare asteroid surfaces with meteorite spectra (Burbine et al. 2002). Ground- and space-based infrared radiometry has provided estimates of asteroid diameters (see Tedesco et al. 2002 and references therein).

Beginning with the first spacecraft images of asteroid Gaspra in 1991 (Veverka et al. 1994), space-based observations have revealed that asteroids are incredibly diverse and fascinating. Direct imaging illustrates the characteristics of surface regolith and roughness, and gives an indication of recent impact history (Chapman et al. 2002). Realistic, detailed shape models that have resulted from spacecraft images, in combination with mass estimates made possible from flybys or satellite orbits, enable reliable bulk density estimates. Geological studies of asteroids have only been possible because of high-resolution space-based observations (Sullivan et al. 2002; Binzel et al. 1997). Asteroid surfaces imaged to date display regional diversity in mineralogy, dramatic topographical features, large-scale concavities, and a variety of apparent surface ages.

Earth-based radar observations have been carried out on over 120 different asteroids. Observations carried out since the 1997 upgrade of the Arecibo Observatory system have made many new discoveries about the nature of asteroids (Magri et al. 2007), including shapes, spin states, and in some cases the discovery of satellites (Merline et al. 2002). This unique approach beams a high-power radio frequency pulse at the asteroid, and receives an echo at different times, depending on the range to various points on the surface. Radar echoes also provide information on the Doppler shifts arising from rotation, and the degree of polarization, related to surface roughness at centimeter scales. Such observations provide constraints on the sizes, spin rates, and physical characteristics of many more asteroids than can be visited by spacecraft (Ostro et al. 2002).

2 Thermal emission observations

While much has been learned about asteroid surfaces from the approaches described above, much more remains unknown about specific physical properties of these bodies. Studies of thermal emission provide insight into physical properties such as the porosity and depth of loose, powdery regolith, the thermal inertia of the body, and whether or not the asteroid includes metals, ices, or water. Millimeter-wavelength observations of asteroids (Altenhoff et al. 1994;

Redman et al. 1992) are dominated by thermal emission originating from temperatures at depth, not from the immediate surface like observations of reflected light. Typically, the depth probed by thermal emission is several times the observation wavelength; thus, observations at 1 mm characterize on average the top 1 cm of the surface. Since the top centimeters of airless bodies are typically covered with loose regolith (Johnston et al. 1989), thermal observations at several different wavelengths (Redman et al. 1998) provide a means of estimating the regolith depth. Asteroid temperatures, ranging from tens to hundreds of Kelvin, yield higher fluxes at the sub-millimeter wavelengths than at longer millimeter or centimeter wavelengths.

3 Potential for ALMA observations

The important capabilities of ALMA for advancing asteroid science will be the sensitivity, resolution, and multifrequency capacity of the array. As of early 2007, approximately 148,000 asteroids have sufficiently well-established orbits to have designated minor planet numbers, and over 200,000 additional asteroids have preliminary orbits. A sizable fraction of known asteroids should be detectable down to a fraction of a mJy and hundreds will be resolvable at sizes larger than 30 mas. Using ALMA, we can move beyond observations of the largest individual asteroids and into a broader investigation of each population. A broader, statistically significant approach will enable a better understanding of the relationship between the current physical characteristics of asteroids and their source regions in the solar nebula.

Table 1 shows the numbers of bodies in each asteroid population detectable above a flux of 0.1 mJy. Values are based on the entire population, except for NEOs, where the estimates are based on close approaches (<0.1 AU) only in the last five years. The flux estimates were made assuming blackbody radiation, using the geocentric distance at opposition (or closest approach for NEOs), heliocentric distance r, a temperature $T = 300/\sqrt{r}$, albedo of 0.11 and an estimate of diameter based on absolute magnitude H: $D = 4000 \times 10^{-0.2H}$. Bodies with lower albedo would have slightly larger diameters, and thus larger fluxes. For the outer solar system, numbers in this table should be considered lower limits. Also tabulated are the number of bodies at each flux level that are large enough in angular size to resolve.

Table 1 Objects above flux and angular size threshold

	NEOs	MBAs	Trojans	KBOs
>0.1 mJy	900	135000	600	880
>30 mas	22	695	5	10

Clearly, ALMA has the potential to add significantly to our knowledge of these populations of bodies, each of which will be treated in greater detail below.

3.1 Near Earth objects

Near Earth Objects (NEOs) are those bodies, including both asteroids and comets, with orbital paths that approach or cross that of Earth. This population is of political interest because of possible impact hazards and natural resource potential. NEOs can be observed at closer range than other bodies, and under a much wider variety of illumination conditions. Because of a high probability for gravitational interactions with planets or the Sun, and because of a variety of other dynamical effects, the NEO population is short-lived and must be replenished from other sources of small bodies. Some fraction of the near-Earth asteroids (NEAs) may be extinct comets, but the balance have likely entered their current orbits after passing through a resonance in the Main Asteroid Belt.

Physical characteristics, such as density and porosity, are of particular interest for NEOs for assessing impact hazards. Thermal emission studies of these bodies have greater potential for constraining thermal inertia, because, unlike superior objects, they can be observed at large phase angles, revealing thermal emission from the non-illuminated portion of their surfaces. Sensitive thermal observations at millimeter and sub-millimeter wavelengths will provide an important complement to the increasing volume of optical, infrared, and radar data on NEOs. Figure 1 shows the close approach distances and fluxes of objects studied to date with radar, with highlighted points for those which appear larger than 30 mas at close approach.

Some NEO observations may be challenging due to uncertain ephemerides, rapid motions on the sky, and the close

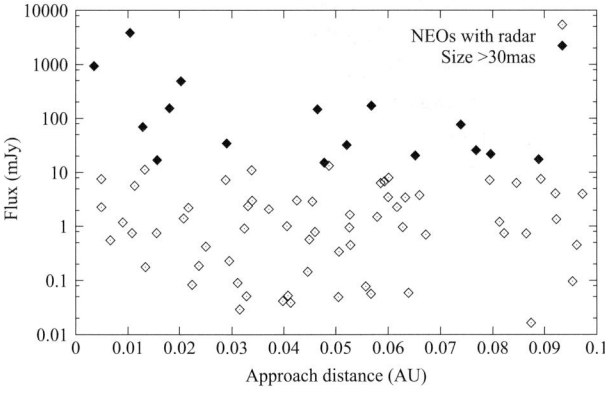

Fig. 1 Close approaches of NEOs that have been observed with ground-based radar. Flux estimates in mJy show that nearly all of the objects accessible to radar investigation are observable to ALMA at their close approach to Earth. If the angular size at the time of close approach is predicted to be greater than 30 mas, the fluxes are highlighted as *filled squares*

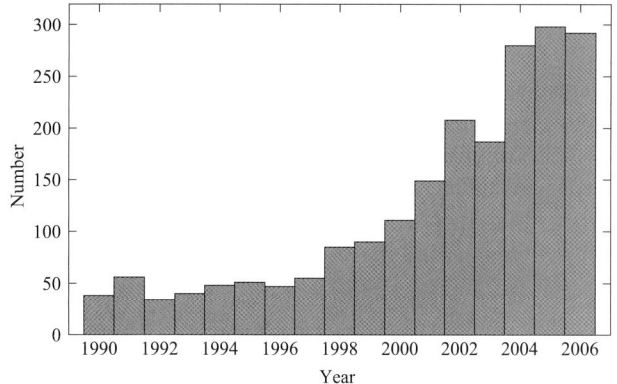

Fig. 2 A histogram showing the number of known NEO approaches within 0.1 AU of Earth each year from 1990–2006. Active sky surveys have dramatically increased the number of observable objects in the last few years

proximity to the Sun; however, the potential for scientific gain is very high. Fortunately, many of the objects in this group approach close enough to the Earth to be detectable, or even resolvable, by ALMA.

A host of survey programs (Stokes et al. 2002) have recently detected large numbers of new bodies, not only in the near-Earth population, but throughout the solar system. A number of surveys increased their detection efficiency with facilities upgrades and expansions in the last several years. As a result of these efforts, the number of known NEOs has increased by an order of magnitude over the last decade. Figure 2 shows a histogram by year since 1990 of the number of known objects that approached within 0.1 AU of Earth. New survey instruments currently under development promise to increase the NEO database by another order of magnitude in the coming decade. In early January, 2007, over 4400 NEOs were known (Chamberlain 2007), 700 of those over 1 km in diameter. By the time ALMA is fully operational, the inventory of near Earth asteroids larger than 1 km should be nearly complete.

3.2 Main belt asteroids

The main belt of asteroids lies between the orbits of Mars and Jupiter, and was the first major population of asteroids to be observed in our solar system. This group contains several of the larger known minor planets, and the overwhelming majority (over 300,000) of the known asteroids of any size. The largest of these asteroids are protoplanets which survived the epoch of solar system formation essentially intact. Collisions played a large role in the evolution of asteroids in this region of the solar system, and most smaller Main Belt Asteroids (MBAs) are fragments of catastrophic impacts.

Recent Hubble Space Telescope (HST) imaging at high resolution has improved understanding of the large-scale properties and processes acting on the largest asteroids

(Zellner et al. 1997, 2005; Storrs et al. 2005; Parker et al. 2002; Thomas et al. 1997), while some spacecraft *in situ* imaging has revealed smaller scale processes and characteristics (Saito et al. 2006). The abundance of asteroid imaging data enhances the contribution that ALMA thermal observations can make to a better understanding of these bodies. In combination, thermal and imaging data constrain physical composition and processing history of the asteroids, leading to better constrains on the conditions under which they formed. Over 100,000 known MBAs will be detectable with ALMA, with almost 700 of those larger than 30 mas.

The Main Belt Asteroids 1 Ceres, 2 Pallas, and 4 Vesta are the largest surviving protoplanets from the formation of the solar system. Observations of 345 GHz thermal emission from these asteroids have revealed lightcurves which vary well over 10% with rotation (Barerra-Pineda et al. 2005). Ceres, the largest asteroid, has a very low amplitude visual lightcurve and appears largely homogeneous in HST images (Li et al. 2006). Visible lightcurves of Vesta and disk-resolved images with adaptive optics (Drummond et al. 1998) and HST (Binzel et al. 1997) observations indicate a heterogeneous surface. These observations imply different evolution of these bodies.

Figure 3 shows a comparison between lightcurves of Vesta, using the Heinrich Hertz Submillimeter Telescope (SMT) at 345 GHz (above) and the Very Large Array (VLA) at 23 GHz (K band, below). The variations present in each of these lightcurves demonstrate that the thermal characteristics of this body vary across its surface, and these variations persist to depths of several centimeters. Multi-frequency ob-

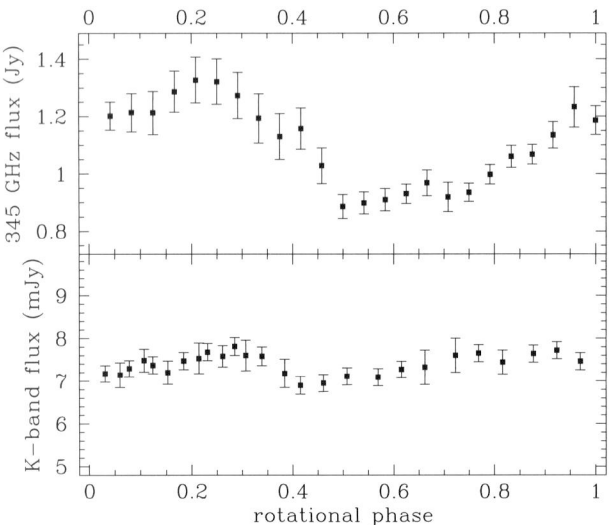

Fig. 3 Lightcurves of Vesta at 345 GHz and K-band. The *upper panel* shows the continuum lightcurve of Vesta at 345 GHz, from January 2003 (Chamberlain et al. 2007). The *lower panel* shows the VLA K-band lightcurve of Vesta in November 2004. Though plotted to different scales, the vertical scaling is such that the relative amplitude is preserved. The physical parameters influencing the lightcurve variations appear to persist several centimeters into the sub-surface of Vesta

servations with ALMA can probe the response of Vesta at intermediate wavelengths to see how lightcurve variations vary in amplitude with depth. Vesta and other large asteroids will benefit from the high-resolution imaging capabilities of ALMA with larger baselines, providing tighter constraints on thermal properties by region across their surfaces.

3.3 Trojan asteroids

Trojan asteroids orbit the Sun at the L_4 and L_5 Lagrangian points, 60° ahead of and behind the planets in their orbits. The Jovian Trojans are by far the largest population, with over 2000 known objects. Because Trojan orbits do not easily capture new objects, these bodies were more likely trapped by Jupiter's migration very early in the history of the solar system (Morbidelli et al. 2005), and have been preserved at cold temperatures ever since. For these reasons, Trojan Asteroids may be primitive objects, remnant planetesimals of those which formed into Jupiter (Marjari et al. 2002). Because of their distance from Earth and relatively small sizes (up to about 100 km), very few thermal studies have done of Jovian Trojans (Emery et al. 2006; Fernández et al. 2003). Five Neptunian Trojans have been discovered at L_4, but even the largest of these is unlikely to be detectable above 0.1 mJy.

It is an open question what degree of compositional diversity there may be among the Trojans. Thermal studies of Trojan asteroids, even with ALMA, will primarily be unresolved. However, such detections still provide valuable constraints on the characteristics and thus potential formation circumstances of these bodies. 600 known objects in this population should be detectable at opposition with ALMA above 0.1 mJy, opening an opportunity to compare them to similar-sized bodies in the other asteroid populations to investigate trends with distance in the protosolar nebula.

3.4 Centaurs and Kuiper Belt objects

Centaurs are a population of small bodies with semi-major axes between the orbits of Jupiter and Neptune. Objects in the giant planets region have a high probability of gravitational perturbation, so Centaurs are a small transient population, but may hold important clues to dynamical evolution of cold, outer solar system planetesimals. Transneptunian bodies, with semi-major axes outside that of Neptune, fall into the Kuiper Belt. The Kuiper Belt Objects (KBOs) are subdivided into several categories: classical objects with low inclination and eccentricity, resonant objects populating the 3 : 2 mean motion resonance with Neptune, and scattered objects with higher eccentricities (Chiang et al. 2007). Much of this population either formed at large heliocentric distances or were scattered by Uranus and Neptune during solar system formation. Thus, these are possibly the least thermally processed objects remaining from the era of planetary formation, and are interesting as tracers.

ALMA should be able to detect 880 of the known KBOs with fluxes above 0.1 mJy. Depending on the geocentric distance at the time of opposition, bodies with diameters larger than 700–800 km should be able to be resolved at 30 mas. Current size estimates place 10–12 objects at this level, though it is likely that more large bodies will be detected before ALMA is fully operational.

3.5 Minor body satellites

A number of asteroids, across the different populations, are now known to have satellites. The orbital parameters of these satellites provide a means of constraining asteroid masses, which, in combination with existing size estimates, enables an estimate of bulk density. The density of 243 Ida, the first asteroid to be seen with a satellite (Belton et al. 1996), is ~3 g cm^{-3} (Belton et al. 1996), while the density of binary Trojan asteroid 617 Patroclus is considerably lower, ~0.8 g cm^{-3} (Marchis 2006). Since density is an important influence in the propagation of thermal emission, as well as a constraint on the formation circumstances and history of solar system bodies, ALMA thermal observations of binary asteroids will offer another means of probing conditions that existed in the nebula.

Search programs are ongoing and many new satellites are being discovered annually (Stokes et al. 2002). Figure 4 shows the separation of binary pairs known throughout the solar system. Bodies separated by more than 30 mas may be seen separately by ALMA, in the cases where the secondary object is also stronger than 0.1 mJy (filled symbols). It is not expected that ALMA would discover new satellites, but instead would enhance our understanding of those discovered by other means.

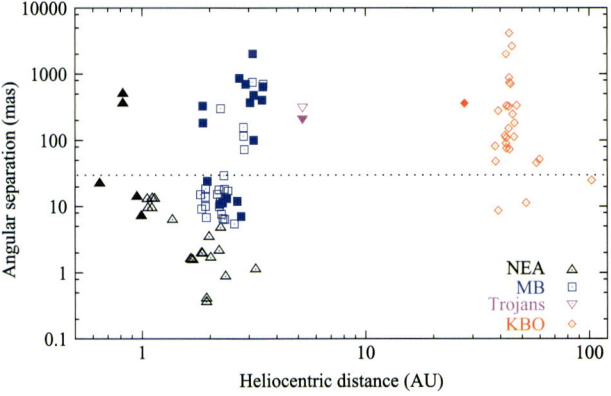

Fig. 4 Distribution of binary asteroids at all heliocentric distances. Maximum angular separations between the primary and secondary can be quite large, enabling a number of these objects to be detected separately by ALMA. Those systems for which the secondary object would appear brighter than 0.1 mJy at 1 mm are highlighted as *filled points*

Among binary asteroids, primary to secondary mass ratios vary widely, as do the separations between the two bodies. These observed characteristics provide valuable constraints on formation, collisions and subsequent gravitational interactions in each population. For example, a sizable fraction, 16%, of NEAs appear to be binary, compared to ~1% in the KBOs, suggesting a different satellite formation mechanism for each of these, distinct from that acting in the main belt. Mechanisms such as rotational disruption, cratering ejecta, and mutual capture are being investigated for satellite formation in the near-Earth, main belt, and outer solar system populations, respectively (Richardson and Walsh 2006). Each of these formation mechanisms would predict a different set of relative thermal characteristics for binary pairs, and a number of such pairs are separated widely enough to be observed individually by ALMA, further constraining the formation models.

4 Note on flux calibration

Thermal variations observed for large asteroids may not simply follow predicted projected area changes as the body rotates. For this reason, observers should use caution in employing asteroids as flux calibration standards without prior knowledge of lightcurves at or near the same observation frequency. Nonetheless, a multi-frequency study of lightcurves has the potential to characterize the fluxes of the largest asteroids and could be used as a basis for a predictive flux model. Such a model, taking into account the real shape of the body, the illumination in terms of phase as well as sub-solar and sub-earth latitudes, could produce a satisfactory prediction of fluxes for several large asteroids. This type of study would be a beneficial early science project for ALMA, in order to provide supplementary flux calibrators that are well-distributed across the sky, and easily observed from Atacama.

5 Conclusions

Recent advances in asteroid science have opened a wide field of investigations, both of current populations, and for the window they offer into formation and subsequent dynamical history of the solar system. Thousands of known objects will be observable by ALMA, and many more will be discovered within a decade. As a result, ALMA is poised with the capacity to increase dramatically the number of objects detected and characterized. Such thermal studies of entire asteroid populations will enable bulk comparisons of bodies formed at different heliocentric distances, and will be a dramatic improvement beyond the few individual objects and their characterizations to date. Resolved temperature maps of the largest bodies promise to solve the question of what lies behind large sub-millimeter thermal heterogeneity seen on some asteroids. Resolution of binary pairs further constrains the formation environments as well as the satellite creation mechanisms. In conclusion, ALMA thermal observations will lead to better constraints on conditions throughout the protoplanetary disk, as reflected in the current populations of asteroids.

Acknowledgements   I am grateful to the organizing committee of the ALMA 2006 Symposium for inviting me to present a case for asteroid science with ALMA. I would like to thank the members of the Arecibo Observatory staff for helpful conversations regarding the content of this paper. In particular, I thank Mike Nolan for providing radar observation statistics.

References

Altenhoff, W.J., et al.: Millimeter-wavelength observations of minor planets. Astron. Astrophys. **287**, 641–646 (1994)

Barerra-Pineda, P., et al.: Asteroid and minor bodies science with the Large Millimetric Telescope. Rev. Mex. Astron. Astrofís. (Ser. Conf.) **24**, 188–191 (2005)

Belton, M.J.S., et al.: Bulk density of asteroid 243 Ida from the orbit of its satellite Dactyl. Nature **374**, 785–788 (1996)

Belton, M.J.S., et al.: The discovery and orbit of 1993 (243)1 Dactyl. Icarus **120**, 185–199 (1996)

Binzel, R.P., et al.: Geologic mapping of Vesta from 1994 Hubble Space Telescope images. Icarus **128**, 95–103 (1997)

Burbine, T.H. et al.: Meteoritic parent bodies: their number and identification. In: Bottke, et al. (eds.) Asteroids III, pp. 653–667. University of Arizona Press (2002)

Bus, S.J., et al.: Visible-wavelength spectroscopy of asteroids. In: Bottke, et al. (eds.) Asteroids III, pp. 169–182. University of Arizona Press (2002)

Chamberlain, A.: NEO Discovery Statistics. http://neo.jpl.nasa.gov/stats. Cited 26 Jan 2007 (2007)

Chamberlain, A., et al.: Submillimeter lightcurves of Vesta. Icarus (2007, in press)

Chapman, C.R., et al.: Cratering on asteroids from Galileo and NEAR Shoemaker. In: Bottke, et al. (eds.) Asteroids III, pp. 315–330. University of Arizona Press (2002)

Chiang, E., et al.: A brief history of transneptunian space. In: Reipurth, et al. (eds.) Protostars and Planets V, pp. 895–911. University of Arizona Press (2007)

Drummond, J.D., et al.: Full adaptive optics images of asteroids Ceres and Vesta; Rotational poles and triaxial ellipsoid dimensions. Icarus **132**, 80–99 (1998)

Ďurech, J., Kaasalainen, M.: Photometric signatures of highly nonconvex and binary asteroids. Astron. Astrophys. **404**, 709–714 (2003)

Emery, J., et al.: Thermal emission spectroscopy (5.2–38 μm) of three Trojan asteroids with the Spitzer Space Telescope: Detection of fine-grained silicates. Icarus **182**, 496–512 (2006)

Fernández, Y.R., et al.: The Albedo distribution of Jovian Trojan asteroids. Astron. J. **126**, 1563–1574 (2003)

Johnston, K.J., et al.: The microwave spectra of the asteroids Pallas. Vesta, and Hygiea. Astron. J. **98**, 335–340 (1989)

Kaasalainen, M., et al.: Models of twenty asteroids from photometric data. Icarus **159**, 369–395 (2002)

Lacerda, P., Luu, J.: Analysis of the rotational properties of Kuiper Belt objects. Astron. J. **131**, 2314–2326 (2006)

Li, J.-Y., et al.: Photometric analysis of 1 Ceres and surface mapping from HST observations. Icarus **182**, 143–160 (2006)

Magri, C., et al.: A radar survey of main-belt asteroids: Arecibo observations of 55 objects during 1999–2003. Icarus **186**, 126–151 (2007)

Marchis, F.: A low density of 0.8 g cm^{-3} for the Trojan binary asteroid 617 Patroclus. Nature **439**, 565–567 (2006)

Marjari, F., et al.: Origin and evolution of trojan asteroids. In: Bottke, et al. (eds.) Asteroids III, pp. 725–738. University of Arizona Press (2002)

Merline, W., et al.: Asteroids do have satellites. In: Bottke, et al. (eds.) Asteroids III, pp. 289–312. University of Arizona Press (2002)

Morbidelli, A., et al.: Chaotic capture of Jupiters Trojan asteroids in the early Solar System. Nature **435**, 462–465 (2005)

Ostro, S.J., et al.: Asteroid Radar Astronomy. In: Bottke, et al. (eds.) Asteroids III, pp. 151–168. University of Arizona Press (2002)

Parker, J.W., et al.: Analysis of the first disk-resolved images of Ceres from ultraviolet observations with the Hubble Space Telescope. Astron. J. **123**, 549–557 (2002)

Pravec, P., et al.: Asteroid Rotations, In: Bottke, et al. (eds.) Asteroids III, pp. 113–122. University of Arizona Press (2002)

Redman, R.O., et al.: Millimeter and submillimeter observations of the asteroid 4 Vesta. Astron. J. **104**, 405–411 (1992)

Redman, R.O., et al.: High-quality photometry of asteroids at millimeter and submillimeter Wavelengths. Astron. J. **116**, 1478–1490 (1998)

Richardson, D., Walsh, K.: Binary minor planets. Annu. Rev. Earth Planet. Sci. **34**, 47–81 (2006)

Saito, J., et al.: Detailed images of asteroid 25143 Itokawa from Hayabusa. Science **312**, 1341–1344 (2006)

Stokes, G.H., et al.: Near-Earth asteroid search programs. In: Bottke, et al. (eds.) Asteroids III, pp. 45–54. University of Arizona Press (2002)

Storrs, A.D., et al.: A closer look at main belt asteroids 1: WF/PC images. Icarus **173**, 409–416 (2005)

Sullivan, R.J., et al.: Asteroid geology from Galileo and NEAR Shoemaker data. In: Bottke, et al. (eds.) Asteroids III, pp. 331–350. University of Arizona Press (2002)

Tedesco, E.F., et al.: The supplemental IRAS minor planet survey. Astron. J. **123**, 1056–1085 (2002)

Thomas, P.C., et al.: Vesta: spin pole, size, and shape from HST images. Icarus **128**, 88–94 (1997)

Veverka, J., et al.: Galileo's encounter with 951 Gaspra: overview. Icarus **107**, 2–17 (1994)

Zellner, B.H., et al.: Hubble Space Telescope images of asteroid 4 Vesta in 1994. Icarus **128**, 83–87 (1997)

Zellner, N.E.B., et al.: Near-IR imaging of asteroid 4 Vesta. Icarus **177**, 190–195 (2005)

ALMA as the ideal probe of the solar chromosphere

Maria A. Loukitcheva · Sami K. Solanki ·
Stephen White

Originally published in the journal Astrophysics and Space Science, Volume 313, Nos 1–3.
DOI: 10.1007/s10509-007-9626-1 © Springer Science+Business Media B.V. 2007

Abstract The very nature of the solar chromosphere, its structuring and dynamics, remains far from being properly understood, in spite of intensive research. Here we point out the potential of chromospheric observations at millimeter wavelengths to resolve this long-standing problem. Computations carried out with a sophisticated dynamic model of the solar chromosphere due to Carlsson and Stein demonstrate that millimeter emission is extremely sensitive to dynamic processes in the chromosphere and the appropriate wavelengths to look for dynamic signatures are in the range 0.8–5.0 mm. The model also suggests that high resolution observations at mm wavelengths, as will be provided by ALMA, will have the unique property of reacting to both the hot and the cool gas, and thus will have the potential of distinguishing between rival models of the solar atmosphere. Thus, initial results obtained from the observations of the quiet Sun at 3.5 mm with the BIMA array (resolution of 12″) reveal significant oscillations with amplitudes of 50–150 K and frequencies of 1.5–8 mHz with a tendency toward short-period oscillations in internetwork and longer periods in network regions. However higher spatial resolution, such as that provided by ALMA, is required for a clean separation between the features within the solar atmosphere and for an adequate comparison with the output of the comprehensive dynamic simulations.

Keywords Sun · Solar chromosphere · Millimeter observations

M.A. Loukitcheva (✉) · S.K. Solanki
Max-Planck-Institut für Sonnensystemforschung,
37191 Katlenburg-Lindau, Germany
e-mail: lukicheva@mps.mpg.de

M.A. Loukitcheva
Astronomical Institute, St. Petersburg University,
Universitetskii pr. 28, Peterhof, 198504 St. Petersburg, Russia

S. White
Astronomy Department, University of Maryland, College Park,
MD 20742, USA

1 Introduction

The chromosphere remains the least understood layer of the solar atmosphere, with the very basics of its structure being hotly debated: is it better described by the classical picture of a steady temperature rise as a function of height, with superposed weak oscillations (e.g. semi empirical models of Vernazza et al. (1981); Fontenla et al. (1990)), or does the temperature keep dropping outwards, with very hot shocks producing strong localized heating (radiation hydrodynamic simulations of Carlsson and Stein (1995, 2002), and Wedemeyer et al. (2004))? The latter concept is consistent with the IR observations of carbon monoxide, which require cool gas to be present at chromospheric heights (see, e.g. Ayres 2002).

Thus, existing models cannot provide a complete description of the solar chromosphere. Consequently nowadays two alternative pictures of the chromosphere co-exist and the role played by chromospheric dynamics in the structuring of this atmospheric layer is a subject of intense scientific debate.

One reason for conflicting models is that they are based either on atomic chromospheric lines and continua in the UV or on molecular lines in the IR, since UV observations are practically blind to cool gas in a dynamic chromosphere, while the IR observations sample only the cool part of the chromosphere. Improved and more sensitive diagnostics of the chromospheric structure and dynamics, that sample both the hot and the cool gas and should distinguish between the rival models, are provided by observations at millimeter wavelengths with an acceptable spatial resolution as was proposed by Loukitcheva et al. (2004). In this contribution we review the unique chromospheric observations at 3.5 mm with the Berkeley-Illinois-Maryland Array and the analysis of the intensity variations expected from the model of Carlsson and Stein for mm wavelengths. We postulate the requirements for mm observations with the future instruments, with emphasis on spatial and temporal resolution. Finally we discuss the prospects for chromospheric studies with ALMA.

2 Results

2.1 Analysis of the BIMA observations at 3.5 mm

The Berkeley-Illinois-Maryland Array (BIMA) operating at a wavelength of 3.5 mm (frequency of 85 GHz) has been the only interferometer in the mm range frequently used for solar observations. The BIMA telescopes are now part of the CARMA array which will also carry out such observations. With the BIMA data obtained in the years 2003 and 2004 we have constructed two-dimensional maps of the solar chromosphere with a resolution of 12″, which represents the highest spatial resolution achieved so far at this wavelength for non-flare solar observations. The BIMA images have led to new insights in to chromospheric structure and to the detection of spatially-resolved chromospheric oscillations at mm wavelengths. The details of the restoration procedure and extensive tests of the sensitivity of the BIMA data to the detection of dynamic signatures can be found in White et al. (2006).

With the currently available resolution the contrast of the brightness structures is evaluated to be up to 30% of the quiet-sun brightness at 3.5 mm (White et al. 2006). However, the similarity of brightness structures, derived from the mm images and seen in other chromospheric emissions (Fig. 1), in spite of the difference in resolution of the images (1–2″ resolution of the UV images), implies that the BIMA resolution is not enough to resolve the millimeter fine structure and observations with spatial resolution much higher than 12″ are required. A detailed analysis of the relations between the millimeter emission, magnetic field and other chromospheric diagnostics is in preparation.

In the millimeter brightness we detected intensity oscillations with typical amplitudes of 50–150 K in the range of periods from 120 to 700 seconds (frequency range 1.5–8 mHz). We found a tendency toward short period oscillations in internetwork and longer periods in network regions in the quiet Sun, which is in good agreement with the results obtained at other wavelengths. At 3 mm the inner parts of

the chromospheric cells exhibit a behavior typical of the internetwork with the maximum of the Fourier power in the 3-minute range, however, most of the oscillations are quasi-periodic, showing up in wave trains of finite duration lasting for typically 1–3 wave periods (see also Loukitcheva et al. 2006).

2.2 Analysis of the CS model millimeter spectrum

The response of the submillimeter and millimeter radiation to a time-series generated by Carlsson and Stein (CS) was computed under the assumption of thermal free-free radiation by Loukitcheva et al. (2004). The results are depicted in Fig. 2 as the excess intensity as a function of wavelength and time.

Wave periods of approximately 3 min can be clearly distinguished in the intensity at all considered wavelengths. Though the dominant frequency of the oscillations changes slightly with wavelength, for all mm wavelengths it lies in the range of 3 minutes. The difference from one period of time to another can be explained by the presence of merging shocks during certain time intervals. The differences in the light curves at different wavelengths are caused primarily by the difference in the formation heights of the emitted radiation. In general the amplitudes of the oscillations compared to the radiation temperature are large, in this sense mm wavelength radiation combines the advantages of the CO lines, which mainly see the cool gas, with those of atomic lines and UV continua, which mainly sample the hot gas.

On the whole, the brightness temperatures are extremely time-dependent at millimeter wavelengths, following changes in the atmospheric parameters. With increasing wavelength the amplitude of the brightness oscillations grows significantly, reaches its maximum value at 2.2 mm (expected to be 15% of the quiet-Sun brightness temperature), and decreases rapidly towards longer wavelengths. Thus we can identify the range 0.8–5.0 mm as the appropriate range of mm wavelengths at which one can expect the clearest signatures of dynamic effects. A careful look at the mm brightness spectrum as a function of time (see Fig. 2) reveals a time delay between the oscillations at long and short millimeter wavelengths. Hence, it is possible to study wave modes traveling in the chromosphere by comparing sub-mm with mm observations.

3 Discussion

The CS model predicts that spatially and temporally resolved observations should clearly exhibit the signatures of the strong shock waves. However, a direct comparison of the observational data products (RMS values, histogram skewness, Fourier and wavelet spectra, etc.), referring to regions

Fig. 1 Portrait of the solar chromosphere at the center of the Sun's disk at 4 different wavelengths on May 18, 2004. From *top left* to *bottom right*: MDI longitudinal photospheric magnetogram, UV 1600 A image from TRACE, CaII K line center image from BBSO and BIMA image at 3.5 mm

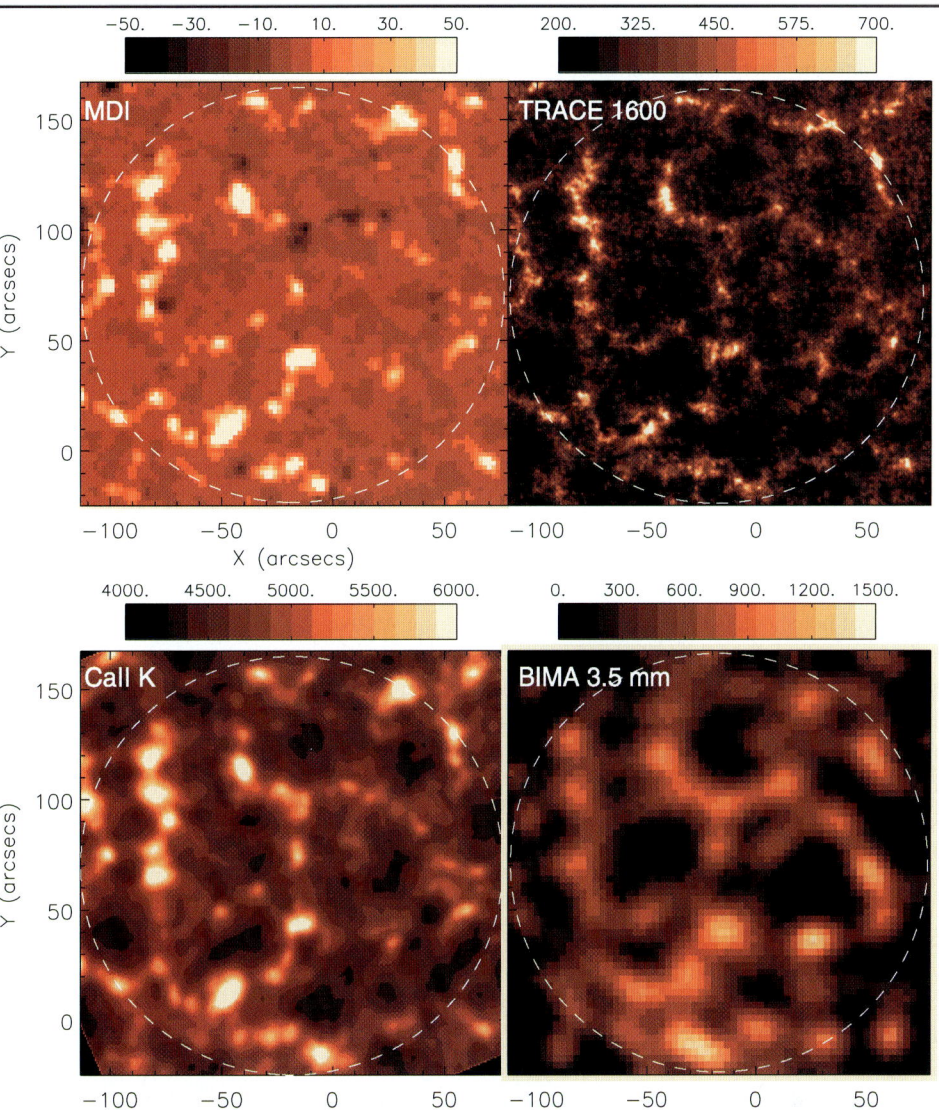

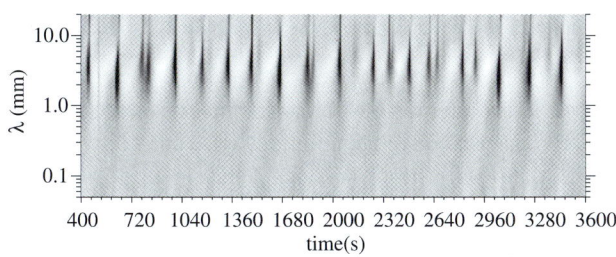

Fig. 2 Evolution of the Carlsson and Stein model millimeter spectrum with time. Negative grey scale representing excess intensity as a function of time and wavelength

with weak magnetic field like the quiet Sun internetwork, with the corresponding products expected from the simulations of Carlsson and Stein exhibits large differences. In particular, the RMS of the brightness temperature is nearly an order of magnitude larger in the model (800 K at 3 mm)

than in the observations (100 K). Another difference is the absence of longer periods in the model power spectrum. But these discrepancies do not rule out the CS models. On the one hand the model is one dimensional and hence does not predict a coherence length of the oscillations, while on the other hand we are not able to resolve individual oscillating elements due to the limited spatial resolution of the observations.

Consequently we estimated the influence of the spatial smearing on the model parameters of chromospheric dynamics and on the observed oscillatory power. Thus we confirmed that the very limited spatial resolution currently available hinders a clean separation between cells and network and typically both network and internetwork areas contribute to the recorded BIMA radiation. From the analysis of the observational data it was found that power in all frequency ranges increases significantly with improving resolution. Consistency between the power predicted by the CS model and the observed power is obtained if the coherence length of oscillating elements is on the order of $1''$.

Our results are consistent with Wedemeyer-Böhm et al. (2005), who computed the millimeter wave signature resulting from the 3-D simulations of Wedemeyer et al. (2004). Although the 3-D simulations suffer from the fact that the radiative transfer of energy is computed entirely in LTE, which becomes a poor assumption at chromospheric heights, the authors believe that the chromospheric pattern and its temporal evolution is representative of the non-magnetic internetwork regions of the solar chromosphere. The simulations display a complex 3D structure of the chromospheric layers, which is highly dynamical on temporal scales of 20–25 s and on spatial scales comparable to solar granulation, which is in good agreement with the $1''$ size of oscillating elements that we deduced. According to Wedemeyer et al. (2004) the chromospheric temperature structure is characterized by a pattern of hot shock waves, which originate from convective motions, and cool gas lying between the shocks. The intensity distribution at mm wavelengths follows the pattern of the shocks in the chromosphere with a sub arcsecond size of the features associated with the shocks. All this complex and dynamic 3D structure can be deduced from observations at mm wavelengths with a sufficiently high spatial resolution of better than $1''$.

4 Summary

Simultaneous mm-submm observations at different wavelengths can be used for the tomography of the solar atmosphere, as radiation at the different wavelengths originates from different layers, with the average formation height increasing with wavelength. Such observations also provide a strong test of present and future models. However, observations that might be able to uncover the nature of the chromosphere should meet the following requirements:

- multiband observations in mm-submm domain (0.8–5.0 mm) to address shock waves and chromospheric oscillation modes
- arcsecond spatial resolution to resolve fine structure
- temporal resolution better than a few seconds to follow its evolution in time
- FOV size of order of $1'$
- accurate absolute calibration of the observations (Bastian 2002).

These requirements look very similar to the technical specification of the continuum observations with the Atacama Large Millimeter Array (ALMA), which represents an enormous advance over existing instrumentation operating at mm-submm wavelengths. ALMA will produce images of the highest resolution available for the foreseeable future (although the technical problem of sampling both large and small spatial scales simultaneously, required for high-quality imaging of the chromosphere, will remain a challenge) and will be the most sensitive instrument operating at submm-mm wavelengths. To summarize, ALMA will be an extraordinarily powerful instrument for studying the solar chromosphere. It will finally allow the mapping of the three-dimensional thermal structure of the solar chromosphere which will be a real breakthrough in solar studies.

Acknowledgements The use of BIMA for scientific research carried out at the University of Maryland is supported by NSF grant AST-0028963. Solar research at the University of Maryland is supported by NSF grant ATM 99-90809 and NASA grants NAG 5-8192, NAG 5-10175, NAG 5-12860 and NAG 5-11872.

References

Ayres, T.R.: Does the Sun have a full-time COmosphere? Astrophys. J. **575**, 1104–1115 (2002)

Bastian, T.S.: ALMA and the Sun. Astronomische Nachrichten **323**, 271–276 (2002)

Carlsson, M., Stein, R.F.: Does a nonmagnetic solar chromosphere exist? Astrophys. J. **440**, L29–L32 (1995)

Carlsson, M., Stein, R.F.: Dynamic hydrogen ionization. Astrophys. J. **572**, 626–635 (2002)

Fontenla, J.M., Avrett, E.H., Loeser, R.: Energy balance in the solar transition region. III—Helium emission in hydrostatic, constant-abundance models with diffusion. Astrophys. J. **406**, 319–345 (1990)

Loukitcheva, M., Solanki, S.K., Carlsson, M., Stein, R.F.: Millimeter observations and chromospheric dynamics. Astron. Astrophys. **419**, 747–756 (2004)

Loukitcheva, M., Solanki, S.K., White, S.: The dynamics of the solar chromosphere: comparison of model predictions with millimeter-interferometer observations. Astron. Astrophys. **456**, 713–723 (2006)

Vernazza, J.E., Avrett, E.H., Loeser, R.: Structure of the solar chromosphere. III—Models of the EUV brightness components of the quiet-sun. Astrophys. J. Suppl. Ser. **45**, 635–725 (1981)

Wedemeyer, S., Freytag, B., Steffen, M., Ludwig, H.-G., Holweger, H.: Numerical simulation of the three-dimensional structure and dynamics of the non-magnetic solar chromosphere. Astron. Astrophys. **414**, 1121–1137 (2004)

Wedemeyer-Böhm, S., Ludwig, H.-G., Steffen, M., Freytag, B., Holweger, H.: The shock-patterned solar chromosphere in the light of ALMA. In: Favata, et al. (eds.) Proceedings of the 13th Cambridge Workshop on Cool Stars, Stellar Systems and the Sun, Hamburg, Germany, ESA SP-560, pp. 1035–1038 (2005)

White, S., Loukitcheva, M., Solanki, S.K.: High-resolution millimeter-interferometer observations of the solar chromosphere. Astron. Astrophys. **456**, 697–711 (2006)

The study of evolved stars with ALMA

Hans Olofsson

Originally published in the journal Astrophysics and Space Science, Volume 313, Nos 1–3.
DOI: 10.1007/s10509-007-9640-3 © Springer Science+Business Media B.V. 2007

Abstract Intense mass loss occurs for low- and intermediate-mass stars on the asymptotic giant branch (AGB), and for the higher mass ($\gtrsim 8$ M$_\odot$) stars during their red supergiant evolution. These winds affect the evolution of the stars profoundly, creates circumstellar envelopes of gas and dust, as well as enrich the interstellar medium with heavy elements and grain particles. The mass loss characteristics are well-studied, but the basic processes are still not understood in detail, and the mass-loss rate of an individual star cannot be derived from first principles. These objects also provide us with fascinating systems, in which intricate interplays between various physical and chemical processes take place, and their relative simplicity in terms of geometry, density distribution, and kinematics makes them excellent astrophysical laboratories. The review concentrates on the aspects of AGB stars and their mass loss which are of particular interest in connection with ALMA.

Keywords Asymptotic giant branch stars · Circumstellar envelopes · Isotope ratios · Mass loss · Molecular abundances

1 AGB stars

1.1 The AGB star phenomenon

We are concerned with red giants on the asymptotic giant branch (AGB), although much of what will be said on the circumstellar medium applies also to red supergiants.

H. Olofsson (✉)
Onsala Space Observatory, 43992 Onsala, Sweden
e-mail: hans.olofsson@chalmers.se

An AGB star has effectively divided itself into three parts: a small, dense, and very hot core that is strongly gravitationally bound, a large, tenuous, and cooler stellar envelope where the external parts are only weakly gravitationally bound, and a huge circumstellar envelope (CSE) of gas and dust, formed through mass loss and decoupled from the star. The AGB is the final stellar evolutionary phase for all stars in the range about 0.8–8 M$_\odot$ (the borders depend mainly on the treatment of convection). This means that a large fraction of all stars that have died in our universe have done this as AGB-stars.

The AGB stars are more or less regularly time-variable, e.g., the Mira variability where regular pulsations occur on a time scale of about a year with an amplitude in luminosity of about a factor of two. Of considerably more astrophysical importance is the He-shell-flash process (or thermal pulsing, a phenomenon unique to AGB stars), which is believed to be the process that leads to the dredge-up of nuclear-enriched matter to the surface (and among other things leads to the formation of carbon stars). This phenomenon occurs on a time scale of about 10^3 years and it repeats itself (relatively) regularly on a time scale of 10^{4-5} years.

Of particular interest is the termination of the AGB evolution and the subsequent metamorphosis when the core becomes a white dwarf that for a brief moment (on an astronomical time scale) lights up the escaping CSE and a planetary nebula (PN) is formed. This is one of the most spectacular events on the astronomical sky, where an ordinary star creates an extra-ordinary event.

The single most important process for the evolution of an AGB star is the surface mass loss. It may reach values above 10^{-4} M$_\odot$ yr^{-1}, and it is the process that determines it all: the lifetime on the AGB, the luminosity reached, the gas/dust return, the chemical composition of the returned gas, etc. Remarkably, this driving engine for the stellar evolution is directly accessible to observations (rather than being hidden in the core as the nuclear burning processes). It is of particular importance to determine the temporal behavior of the mass-loss rate, and how it depends on the stellar mass, metallicity, pulsational behavior, etc. It must be emphasized that even if this phenomenon is well established, our knowledge of its details is limited, and its magnitude certainly cannot be calculated from first principles.

It is clear that an AGB star and its descendants are intricate objects, where a full description requires a complex interplay between different physical/chemical processes with different time scales. ALMA may provide crucial information for our understanding of these objects.

1.2 The importance of AGB stars

The very high mass-loss rates reached over time scales in excess of 10^4 years mean that AGB stars have a large mass return and so are important for the cosmic gas/dust cycle: they produce heavy elements (e.g., through the 3α- and s-processes), they produce dust particles that become the core of interstellar dust particles, and they produce complex molecules (e.g., PAHs). The mass return is dominated by those objects that reach a mass-loss rate of 10^{-5} $M_\odot$ yr^{-1} and more. AGB stars are very luminous, up to 5×10^4 $L_\odot$, and old, and hence are excellent probes of the structure, kinematics, and starformation history of galaxies. Finally, the reasonably simple geometry and kinematics of the CSEs make them excellent astrophysical and astrochemical laboratories.

1.3 ALMA and AGB stars

In the next sections we present various aspects of the research on AGB stars where ALMA can provide crucial insight. We present a number of simple formulae that are useful, in the context of ALMA, to estimate flux densities, observational spaces, angular sizes etc. for both the central stars and the CSEs (distances are given in kpc, frequencies in GHz, luminosities in $L_\odot$, masses in $M_\odot$, mass-loss rates in $M_\odot$ yr^{-1}, radial distances in cm, time scales in years, and velocities in km s^{-1}).

1.4 Stellar archeology

AGB stars and their CSEs have the interesting property that the temporal behavior of the central star is imprinted spatially in the expanding CSE, although not necessarily in a straightforward manner. Interestingly, the angular resolution of ALMA fits well the important time scales (t) of AGB stars,

$$\theta_t \approx \frac{v_e t}{D} \approx 0.3 \left[\frac{t}{100} \right] \left[\frac{1}{D} \right] \text{arcs},\tag{1}$$

where D is the distance to the source, and v_e the envelope expansion velocity (assumed to be 10 km s^{-1}). Examples are stellar pulsation (≈ 1 yr), super-period effects ($\lesssim 5$ yr), clump ejection (≈ 1 yr), gas-dust interaction ($\lesssim 100$ yr), thermal pulsing ($10^{2–3}$ yr repeated on a time scale of $10^{4–5}$ yr), and the termination of the AGB ($\lesssim 100$ yr). In addition, post-AGB objects are very dynamical objects: fast winds, bipolar outflows, jets, shocks, ionization fronts, and dissociation fronts are frequent phenomena.

2 ALMA and the central stars

The huge collecting area and resolving power of ALMA will make it a very important instrument for the study of stellar physics.

Assuming that the central stars are perfect blackbody emitters, the expected flux densities are given by,

$$S_* \approx 2 \left[\frac{L}{10^4} \right] \left[\frac{2500}{T_{\text{eff}}} \right]^3 \left[\frac{\nu}{230} \right]^2 \left[\frac{1}{D} \right]^2 \text{mJy}\tag{2}$$

where L is the stellar luminosity, T_{eff} the stellar effective temperature, and ν the observing frequency. This results in an observational space of the order of

$$D_{*,230} \approx 5 \left[\frac{L}{10^4} \right]^{0.5} \left[\frac{2500}{T_{\text{eff}}} \right]^{1.5} \text{kpc}\tag{3}$$

in 1 hour of observing time at 230 GHz ($5\sigma = 85$ μJy). Note that $S_{350} \approx 2S_{230}$, and $N_{350} \approx 2N_{230}$ (where N is the noise figure) so that the S/N-ratio will hardly increase with frequency. Thus, a large number of AGB-stars of various types are easily detected by ALMA (in addition, they can be important as calibration sources).

The angular size of an AGB star is

$$\theta_* \approx 0.05 \left[\frac{L}{10^4} \right]^{0.5} \left[\frac{2500}{T_{\text{eff}}} \right]^2 \left[\frac{0.1}{D} \right] \text{arcs}.\tag{4}$$

Since the most nearby AGB stars lie at ≈ 50 pc, we expect $\theta_* \lesssim 0.''1$. With a resolution of $0.''02$ ALMA reaches $5\sigma \approx 5$ K in 1 hour of observing time at 230 GHz, and $5\sigma \approx 10$ K in 1 hour of observing time at 350 GHz at a resolution of $0.''01$. Actually, AGB stars have convective outer layers, and we expect to see of the order of tens of convection cells per star. Hence, the characteristic angular scales are a factor of a few lower than those given above. Thus, a moderate number of AGB stars and supergiants may be imaged by ALMA at a reasonable resolution.

The situation may be more complicated, since, as shown by Reid and Menten (1997), AGB stars appear to have a radio photosphere of about twice the size of the optical disk and it is not clear how its structure is related to that of the convective photosphere. As the only example of a (partly)

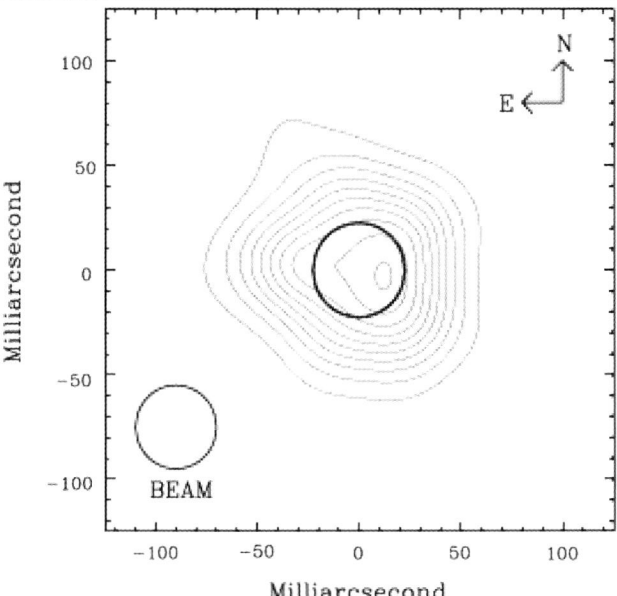

Fig. 1 A 7 mm radio continuum image towards the red supergiant α Ori obtained with the VLA (Lim et al. 1998). The estimated photospheric size is indicated

resolved stellar image at radio wavelengths we show the VLA 7 mm image of the supergiant α Ori in Fig. 1.

3 ALMA and the CSEs

ALMA observations of dust continuum and molecular line emission from AGB CSEs will be crucial for our understanding of stellar evolution on the AGB (and also that of red supergiants). Of particular importance is the estimate of the mass-loss rate, and estimates based on circumstellar emission in some form appear to be the most reliable, i.e., $\dot{M} \approx v_{\mathrm{e}} M_{\mathrm{em}}/R_{\mathrm{em}}$, where M_{em} is the mass inside the radius of the emitting region R_{em} that expands with the velocity v_{e}. They also have the widest applicability, and cover essentially the entire AGB and beyond. The most reliable mass-loss-rate estimates are obtained by modeling circumstellar dust continuum and CO radio line emission, and we therefore calculate the approximate observational spaces for dust continuum and CO line emission.

3.1 The dusty CSE

Assuming optically thin emission (which is reasonable for even high mass-loss rates at the observing frequencies of ALMA) the flux density of a dusty CSE is (using a dust-to-gas mass ratio of 0.005, and a dust emissivity of $100\,\mathrm{cm^2\,g^{-1}}$ at 60 µm which scales linearly with frequency)

$$S_{\mathrm{CSE,d}} \approx 15 \left[\frac{\dot{M}}{10^{-6}} \right] \left[\frac{\nu}{230} \right]^3 \left[\frac{1}{D} \right]^2 \mathrm{mJy}. \qquad (5)$$

This results in an observational space in 1 hour of observing time at 230 GHz of the order of ($5\sigma = 85$ µJy)

$$D_{\mathrm{CSE,d,230}} \approx 13 \left[\frac{\dot{M}}{10^{-6}} \right]^{0.5} \mathrm{kpc}. \qquad (6)$$

This means that ALMA may reach 10^{-7} M$_\odot$ yr^{-1} objects in the Galactic Centre in 4 hours, and 10^{-5} M$_\odot$ yr^{-1} objects in the LMC in 2 hours. Note that $S_{350} \approx 4S_{230}$, while $N_{350} \approx 2N_{230}$ so that there is a gain in sensitivity by using a higher observing frequency.

Note also that $S_* \propto \nu^2$ while $S_{\mathrm{CSE,d}} \propto \nu^3$ so that eventually the CSE emission becomes stronger than the stellar emission at some frequency that depends on the mass-loss rate. However, the surface brightness of the circumstellar emission is, of course, much lower than that of the stellar emission.

The mm/sub-mm dust emission is sharply peaked at the centre, but it extends much further than any molecular line emission (and hence potentially probes the mass-loss rate over much longer time scales), and it is optically thin even at high mass-loss rates.

3.2 The gaseous CSE

CO radio line emission is the dominant molecular-line mass-loss-rate estimator. We therefore give an estimate of the expected CO($J = 2 \to 1$) line flux density (based on detailed radiative transfer modeling by Ramstedt et al., in prep.)

$$S_{\mathrm{CO(2-1)}} \approx 6 \left[\frac{\dot{M}}{10^{-6}} \right]^{1.2} \left[\frac{15}{v_{\mathrm{e}}} \right]^{1.6} \left[\frac{f_{\mathrm{CO}}}{10^{-3}} \right]^{0.7} \left[\frac{1}{D} \right]^2 \mathrm{Jy} \qquad (7)$$

where f_{CO} is the abundance of CO with respect to H$_2$. Note that $S_{\mathrm{CO(3-2)}} \gtrsim 2S_{\mathrm{CO(2-1)}}$ at relatively low mass-loss rates and $S_{\mathrm{CO(3-2)}} \approx S_{\mathrm{CO(2-1)}}$ at high mass-loss rates, while $N_{350} \approx 2N_{230}$ so that there is little, if any, gain in sensitivity by going to higher-J lines.

The observational space for the CO($J = 2 \to 1$) line in 1 hour is of the order of ($5\sigma = 6$ mJy at 2 km s^{-1} resolution)

$$D_{\mathrm{CO(2-1)}} \approx 30 \left[\frac{\dot{M}}{10^{-6}} \right]^{0.6} \left[\frac{15}{v_{\mathrm{e}}} \right]^{0.8} \left[\frac{f_{\mathrm{CO}}}{10^{-3}} \right]^{0.4} \mathrm{kpc}. \qquad (8)$$

Thus, ALMA may reach 10^{-7} M$_\odot$ yr^{-1} objects in the Galactic Centre in 1 hour, 10^{-6} M$_\odot$ yr^{-1} objects in the LMC in 4 hours, but for M31 we require 40 hours to detect 10^{-5} M$_\odot$ yr^{-1} objects (assuming $v_{\mathrm{e}} = 15$ km s^{-1}, $f_{\mathrm{CO}} = 10^{-3}$).

The size of the emitting region is dependent on the transition observed, but an upper limit is given by the size of the CO envelope (determined by photodissociation, Mamon et al. 1988)

$$\theta_{\mathrm{CO}} \approx 5 \left[0.1 + 0.9 \left(\frac{\dot{M}}{10^{-6}} \right)^{0.7} \right] \left[\frac{1}{D} \right] \mathrm{arcs}. \qquad (9)$$

Thus, $\theta_{CO}(8\text{ kpc}, 10^{-6}\text{ M}_\odot\text{ yr}^{-1}) \approx 0''.5$ and the same size is obtained for 50 kpc and $10^{-5}\text{ M}_\odot\text{ yr}^{-1}$, meaning that properly resolved CO CSEs is possible out to several kpc, while in the LMC we are essentially restricted to detections.

3.3 Circumstellar molecules

Presently, 69 molecular species are detected in AGB CSEs. A large fraction are unique to the circumstellar medium (compared to the interstellar medium), Table 1. About 80% are detected at radio wavelengths, and the rest (except C_2) are detected at IR wavelengths. This is impressive considering that AGB CSEs are low-mass objects, but about half of them have been detected in only IRC+10216 (probably the most nearby C-star, ≈ 120 pc, and it has a very high mass-loss rate, $\approx 2 \times 10^{-5}\text{ M}_\odot\text{ yr}^{-1}$). This is far from satisfactory if we aim for an understanding of circumstellar chemistry. Measured brightness distributions are limited to a dozen species, in a single line each, observed towards IRC+10216.

The first more detailed study of circumstellar abundances in a larger sample of sources were performed by González Delgado et al. (2003); SiO in about 40 M-stars. This was extended by Schöier et al. (2006b) to include also 20 C-stars. The results are a large spread in estimated circumstellar SiO abundances (although it is unclear how much of this is real, due to the difficulty in establishing a proper circumstellar model for each star) and no apparent difference between the two chemistries, indicating that non-LTE chemistry plays a

role. Recent interferometric results show that a more complicated SiO abundance distribution is required to explain the data, an inner component of high abundance, and an outer, extended component of low abundance (Schöier et al. 2004, 2006a), indicating that adsorption onto grains plays a role. This shows that our knowledge of stellar/circumstellar chemistry is not complete by studying only one object, and observations with ALMA may considerably increase the observational data base.

ALMA may also lead to discoveries of new species. Crude estimates of what is detectable is obtained using this formula

$$S \approx 6g_u A_{ul} \left[\frac{f_X}{10^{-8}}\right]\left[\frac{\dot{M}}{10^{-6}}\right]\left[\frac{R_e}{10^{16}}\right]\left[\frac{15}{v_e}\right]^2$$
$$\times \frac{e^{-E_l/kT_x}}{Q(T_x)}\left[\frac{1}{D}\right]^2 \text{ Jy} \tag{10}$$

which is based on the assumption of optically thin emission, a Boltzmann population distribution at an assumed temperature T_x, as well as an assumption of the size of the emitting region R_e (g_u is the degeneracy of the upper level, A_{ul} the Einstein A-coefficient, f_X the abundance with respect to H_2, E_l the energy of the lower level, and Q the partition function). As an example, a low-abundance species like H_2CO ($f = 10^{-8}$ in IRC+10216) is detectable with ALMA in the $3_{12}-2_{11}$ line at 226 GHz in 1 hour of observing time out to about 1 kpc in a $10^{-5}\text{ M}_\odot\text{ yr}^{-1}$ object ($5\sigma = 4$ mJy at 4 km s^{-1} resolution; assuming $T_x = 25$ K, $R_e = 10^{16}$ cm, and $v_e = 15$ km s^{-1}).

The problem of line crowding is not as bad as for star formation sources, but it cannot be ignored for the more nearby, high-mass-loss rate, C-rich objects. In particular, lines from vibrationally excited states of the more complex species may be abundant. Methods for interpreting spectra of many molecular lines may have to be used.

3.4 Circumstellar isotopic ratios

Another, and very important, side of abundance estimates is the isotopic ratios of various elements. Only for IRC+10216 there exists a reasonable setup of isotope ratios; see Table 2 where various isotope ratios are compared to the corresponding solar values (Kahane et al. 1992, 2000; Wannier et al. 1991). These data show the potential, but also the problem that the weakness of the lines severely limits the number of sources that has been possible to study so far.

4 The circumstellar structure

The geometry, i.e., the three-dimensional density distribution, and the kinematics of AGB CSEs have implications for

Table 1 Molecules detected in AGB CSEs

2-atoms:	AlCl	CN	NaCl	SiN
	AlF	CP	OH	SiO
	C_2	CS	PN	SiS
	CO	KCl	SiC	SO
3-atoms:	AlNC	CO_2	HNC	SiC_2
	C_3	HCN	MgCN	SiCN
	C_2H	H_2O	MgNC	SiNC
	C_2S	H_2S	NaCN	SO_2
4-atoms:	ℓ-C_3H	C_3S	H_2CO	NH_3
	C_3N	C_2H_2	H_2CS	SiC_3
	C_3O	HC_2N		
5-atoms:	C_5	c-C_3H_2	HC_3N	HNC_3
	C_4H	CH_2CN	HC_2NC	SiH_4
	C_4Si	CH_4	H_2C_3	
6-atoms:	C_5H	C_2H_4	HC_4N	H_2C_4
	C_5N	CH_3CN		
$\geq$7-atoms:	C_6H	C_8H	HC_5N	HC_9N
	C_7H	CH_2CHCN	HC_7N	H_2C_6
Ions:	HCO^+	C_6H^-		

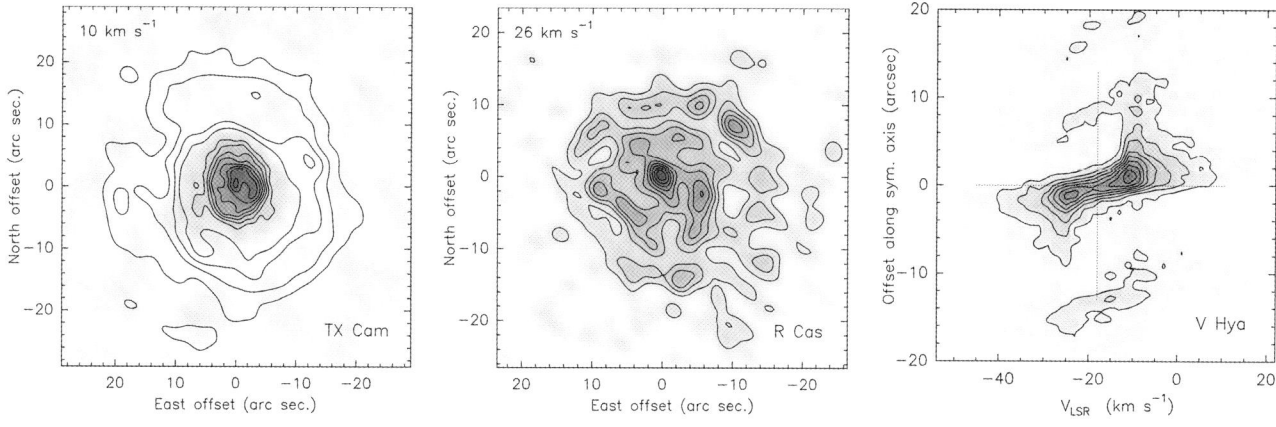

Fig. 2 CO images (at the systemic velocities) of the CSEs around TX Cam (*left*; outer parts in $J = 1 \rightarrow 0$ and central part in $J = 2 \rightarrow 1$) and R Cas (*middle*; $J = 1 \rightarrow 0$). Position-velocity diagram of the CO($J = 2 \rightarrow 1$) emission towards V Hya (*right*). IRAM PdB interferometer data (Castro-Carrizo et al. 2007)

Table 2 Isotope ratios in the CSE of IRC+10216

Isotope ratio	Ratio	IRC/Solar	Species
$^{12}C/^{13}C$	45	0.5	CS, SiC$_2$
$^{12}C/^{14}C$	>63 000		CO
$^{14}N/^{15}N$	5300	20	HCN
$^{16}O/^{17}O$	840	0.3	CO
$^{16}O/^{18}O$	1300	3	CO
$^{28}Si/^{29}Si$	>15	>0.8	SiS, SiC$_2$
$^{28}Si/^{30}Si$	>20	>0.7	SiS, SiC$_2$
$^{32}S/^{33}S$	>60	>0.5	SiS
$^{32}S/^{34}S$	22	1	CS, SiS
$^{32}S/^{36}S$	2400	2.7	CS, SiS
$^{35}Cl/^{37}Cl$	2	0.7	AlCl, NaCl, KCl
$^{24}Mg/^{25}Mg$	8	1	MgNC
$^{24}Mg/^{26}Mg$	7	1	MgNC

our understanding of the mechanism that produces the mass loss, and for the geometry of the successors, the PNe. The determination of the geometry is not a simple observational task, since the conversion from a 2D brightness distribution to a 3D density distribution can be quite complicated. Here, radiative (e.g., saturation, maser action), excitation, chemistry, and observational (e.g., the lack of interferometer sensitivity to extended emission) effects play an important role. Indeed, these complications in combination with a lack of observational data with sufficiently high angular resolution have the effect that our knowledge of the geometry of AGB CSEs is limited.

Spectacular results have been obtained on the small scales using SiO masers that show roughly circular ring structures of diameters a few stellar radii (Boboltz et al. 1997), H$_2$O masers that can estimate the magnetic field as a distance from the star (Vlemmings et al. 2002), and imaging

of dust emission which show proper motions of individual dust clumps (Tuthill et al. 2000; Weigelt et al. 2002). On larger scales, (Fong et al. 2006) recently published an analysis of existing CO radio line imaging. The on-going large program on the IRAM PdB interferometer to map the CSEs of objects on and beyond the AGB in the CO($J = 1 \rightarrow 0$ and $2 \rightarrow 1$) lines will provide the best pre-ALMA data base (Castro-Carrizo et al. 2007). Initial results show a wide variety of AGB CSEs from essentially spherical to highly elongated, essentially smooth to highly clumpy, etc., Fig. 2.

Whether the circumstellar medium is smooth or clumpy is important not only for the understanding of the mass-loss mechanism, but also for the excitation of atoms and molecules, the radiative transfer of line as well as continuum emission, and the ability of the molecules to survive the dissociating radiation. Thus, it may have a profound effect on our modeling of the CSE emission.

The following formulae give simple estimates of expected continuum and line flux densities from individual circumstellar clumps. Guided by results from SiO maser emission we assume an initial clump size, r_{cl}, of 10^{13} cm (a density of 10^{11} cm^{-3} results in a clump mass, M_{cl}, of 10^{-6} M$_\odot$). The size of such a clump will be 0.01 [0.1/D] arcs. We assume initial dust and kinetic temperatures of 1000 K. This results in a continuum flux density of (the same assumptions on the dust as above)

$$S_{d,cl} \approx 0.006 \left[\frac{M_{cl}}{10^{-6}} \right] \left[\frac{T_d}{1000} \right] \left[\frac{\nu}{230} \right]^3 \left[\frac{1}{D} \right]^2 \text{mJy}, \qquad (11)$$

and a continuum optical depth of

$$\tau_{d,cl} \approx 0, 1 \left[\frac{M_{cl}}{10^{-6}} \right] \left[\frac{10^{13}}{r_{cl}} \right]^2 \left[\frac{\nu}{230} \right]. \qquad (12)$$

The corresponding results for the CO($J = 2 - 1$) line are (assuming thermal excitation, and $f_{CO} = 10^{-3}$)

$$S_{CO(2-1),cl} \approx 0.03 \left[\frac{M_{cl}}{10^{-6}} \right] \left[\frac{1000}{T_x} \right] \left[\frac{1}{D} \right]^2 \text{ mJy}, \quad (13)$$

$$\tau_{CO(2-1),cl} \approx 100 \left[\frac{M_{cl}}{10^{-6}} \right] \left[\frac{10^{13}}{r_{cl}} \right]^2 \left[\frac{1000}{T_x} \right]^2. \quad (14)$$

As an example, for IRC+10216, where $\theta_{cl} \gtrsim 0.''01$ at $D = 120$ pc, the results are $S_{d,cl} \approx 0.4$ mJy at 230 GHz and $S_{CO(2-1),cl} \approx 2$ mJy. That is, the dust clumps are easily detected, but just about resolved, in 1 hour (85 μJy at $0.''02$ resolution), while the gas clumps are more difficult to detect ($5\sigma = 4$ mJy in 1 hour at $0.''02$ and 4 km s^{-1} resolutions). For dust the situation improves with frequency, and we can also expect the clumps to expand as they recede from the star. Hence, ALMA provides the possibility to study simultaneously the gas and dust distributions in some detail for nearby objects.

5 Termination of the AGB phase

A considerable increase in mass-loss rate as the end of the AGB is approached, often called the superwind phase, is reasonably well established (Delfosse et al. 1997; Fong et al. 2002; Justtanont et al. 1996), but the details are far from understood. However, at the end of the AGB the star must decrease its mass-loss rate substantially on a short time scale. The termination of the AGB is reached when the stellar envelope mass becomes very low, say ≈ 0.001 M$_\odot$. If the mass-loss rate is as high as 10^{-4} M$_\odot$ yr^{-1}, it must decrease with orders of magnitude within decades, a phenomenon for which there is very little observational information (Lewis 2002).

Perhaps, already at this point the deviations from spherical symmetry, which will become so apparent during the evolution towards a PN, are manifested close to the star (Meixner et al. 1999; Ueta et al. 2001). Interesting examples are V Hya (Sahai et al. 2003), IRC+10216 (Weigelt et al. 2002), RV Boo (Bergman and Kerschbaum 2000), CIT6 (Schmidt et al. 2002), and IRC+10011 (Vinković et al. 2004). A spectacular example is provided by W43A, where H$_2$O masers show a (magnetically) collimated, precessing jet (Imai et al. 2002; Vlemmings et al. 2006) emanating through a biconically expanding envelope which is traced by SiO maser emission (Imai et al. 2005) (although admittedly the evolutionary nature of this object is uncertain). To this we should add a handful of AGB and post-AGB objects for which the evidence for the presence of a disk is relatively convincing, e.g., the carbon stars EU And and BM Gem, and the post-AGB objects the Red Rectangle and AC Her. (Jura and Kahane 1999) argued that the narrow CO radio line features observed towards these four objects could be interpreted as signatures of long-lived reservoirs of orbiting gas, and the Red Rectangle has a disk with Keplerian rotation (Bujarrabal et al. 2005). During this period the star is completely obscured in the visual, and ALMA can therefore contribute significantly to our understanding of stellar evolution during this period.

6 Post-AGB circumstellar chemistry

An increasing UV flux and the presence of shocks in post-AGB objects will have an effect on the chemistry (Woods et al. 2003). This is verified by the detections of polyacetylenes in the C-rich proto-PNe AFGL618 and AFGL2688, and methylpolyynes and the benzene ring in AFGL618 (Cernicharo et al. 2001a, 2001b). A number of ionic species have been detected in the CSE around a young PN, NGC7027 (Latter et al. 1993). In addition, the abundant H$_2$ molecule is readily detectable in PNe of the bipolar type (Cox et al. 2002). The molecular species detected only in post-AGB CSEs are CH, CH$^+$, CO$^+$, H$_2$, N$_2$H$^+$, OCS, HC$_4$H, HC$_6$H, CH$_3$C$_2$H, CH$_3$C$_4$H, and C$_6$H$_6$. ALMA will provide the angular resolution required to follow the temporal evolution of the chemistry.

References

Bergman, P., Kerschbaum, F., Olofsson, H.: Astron. Astrophys. **353**, 257 (2000)

Boboltz, D.A., Diamond, P.J., Kemball, A.J.: Astrophys. J. **487**, L147 (1997)

Bujarrabal, V., Castro-Carrizo, A., Alcolea, J., Neri, R.: Astron. Astrophys. **441**, 1031 (2005)

Castro-Carrizo, A., Neri, R., Winters, J., et al.: In: Kerschbaum, F., Charbonnel, C., Wing, R. (eds.), ASP Conf. Ser. Why Galaxies Care About AGB Stars. ASP, San Francisco (2007, in press)

Cernicharo, J., Heras, A.M., Pardo, J.R., et al.: Astrophys. J. **546**, L127 (2001a)

Cernicharo, J., Heras, A.M., Tielens, A.G.G.M., et al.: Astrophys. J. **546**, L123 (2001b)

Cox, P., Huggins, P.J., Maillard, J.-P., et al.: Astron. Astrophys. **384**, 603 (2002)

Delfosse, X., Kahane, C., Forveille, T.: Astron. Astrophys. **320**, 249 (1997)

Fong, D., Justtanont, K., Meixner, M., Campbell, M.T.: Astron. Astrophys. **396**, 581 (2002)

Fong, D., Meixner, M., Sutton, E.C., Zalucha, A., Welch, W.J.: Astrophys. J. **652**, 1626 (2006)

González Delgado, D., Olofsson, H., Kerschbaum, F., et al.: Astron. Astrophys. **411**, 123 (2003)

Imai, H., Obara, K., Diamond, P.J., Omodaka, T., Sasao, T.: Nature **417**, 829 (2002)

Imai, H., Nakashima, J.-i., Diamond, P.J., Miyazaki, A., Deguchi, S.: Astrophys. J. **622**, L125 (2005)

Jura, M., Kahane, C.: Astrophys. J. **521**, 302 (1999)

Justtanont, K., Skinner, C.J., Tielens, A.G.G.M., Meixner, M., Baas, F.: Astrophys. J. **456**, 337 (1996)

Kahane, C., Cernicharo, J., Gomez-Gonzalez, J., Guelin, M.: Astron. Astrophys. **256**, 235 (1992)

Kahane, C., Dufour, E., Busso, M., et al.: Astron. Astrophys. **357**, 669 (2000)

Latter, W.B., Walker, C.K., Maloney, P.R.: Astrophys. J. **419**, L97 (1993)

Lewis, B.M.: Mon. Not. R. Astron. Soc. **576**, 445 (2002)

Lim, J., Carilli, C.L., White, S.M., Beasley, A.J., Marson, R.G.: Nature **392**, 575 (1998)

Mamon, G.A., Glassgold, A.E., Huggins, P.J.: Astrophys. J. **328**, 797 (1988)

Meixner, M., Ueta, T., Dayal, A., et al.: Astrophys. J. Suppl. Ser. **122**, 221 (1999)

Reid, M.J., Menten, K.M.: Astrophys. J. **476**, 327 (1997)

Sahai, R., Morris, M., Knapp, G.R., Young, K., Barnbaum, C.: Nature **426**, 261 (2003)

Schmidt, G.D., Hines, D.C., Swift, S.: Astrophys. J. **576**, 429 (2002)

Schöier, F.L., Olofsson, H., Wong, T., Lindqvist, M., Kerschbaum, F.: Astron. Astrophys. **422**, 651 (2004)

Schöier, F.L., Fong, D., Olofsson, H., Zhang, Q., Patel, N.: Astrophys. J. **649**, 965 (2006a)

Schöier, F.L., Olofsson, H., Lundgren, A.A.: Astron. Astrophys. **454**, 247 (2006b)

Tuthill, P.G., Monnier, J.D., Danchi, W.C., Lopez, B.: Astrophys. J. **543**, 284 (2000)

Ueta, T., Meixner, M., Hinz, P.M., et al.: Astrophys. J. **557**, 831 (2001)

Vinković, D., Blöcker, T., Hofmann, K.-H., Elitzur, M., Weigelt, G.: Mon. Not. R. Astron. Soc. **352**, 852 (2004)

Vlemmings, W.H.T., Diamond, P.J., van Langevelde, H.J.: Astron. Astrophys. **394**, 589 (2002)

Vlemmings, W.H.T., Diamond, P.J., Imai, H.: Nature **440**, 58 (2006)

Wannier, P.G., Andersson, B.-G., Olofsson, H., Ukita, N., Young, K.: Astrophys. J. **380**, 593 (1991)

Weigelt, G., Balega, Y.Y., Blöcker, T., et al.: Astron. Astrophys. **392**, 131 (2002)

Woods, P.M., Millar, T.J., Herbst, E., Zijlstra, A.A.: Astron. Astrophys. **402**, 189 (2003)

Molecular lines from protoplanetary nebulae: observations with ALMA

Valentín Bujarrabal

Originally published in the journal Astrophysics and Space Science, Volume 313, Nos 1–3.
DOI: 10.1007/s10509-007-9494-8 © Springer Science+Business Media B.V. 2007

Abstract Planetary nebulae (PNe) are formed in a very fast process. In just about 1000 years, the nebula evolves from a spherical and slowly expanding AGB envelope to a PN, with usually axial symmetry and high axial velocities. Molecular lines are known to probe most of the nebular material in young PNe and protoplanetary nebulae (PPNe), and are therefore very useful to study such an impressive evolution. Many quantitative results on these objects have been so obtained, including general structure, total mass and density distribution, kinetic temperatures, velocity fields, etc. Existing observations probe both the gas accelerated by post-AGB shocks and the quiescent components. But the study of crucial regions to understand PN formation (recently shocked shells, regions heated by the stellar UV and inner rotating disks) requires observations at higher frequency and with better spatial resolution.

Keywords (Stars:) circumstellar matter · Stars: AGB and post-AGB · Stars: mass loss · Radio lines: stars

PACS 97.10.Fy · 97.10.Me · 98.38.Ly · 98.58.Li

1 Introduction

The most remarkable property of AGB stars is, probably, that they are losing significant amounts of material. At the tip of the AGB, this mass ejection process becomes very strong, with a mass-loss rate as high as 10^{-4} $M_\odot$ yr^{-1}, even

V. Bujarrabal (✉)
Observatorio Astronómico Nacional (IGN), Apdo. 112, E-28803, Alcalá de Henares, Spain
e-mail: v.bujarrabal@oan.es

up to 10^{-3} $M_\odot$ yr^{-1}, at the very end of the AGB evolution. Of course, the star cannot maintain such a high mass-loss rate during a long time: after a few thousand years, the star has practically ejected itself. Then, in a transformation as short as $\sim$ 1000 yr, it leaves the AGB. The surrounding shell is then very massive, $\sim$ 1 $M_\odot$.

The central core now becomes visible, because all layers outside it have been ejected; the new star is a very hot and small blue dwarf. This hot star is able to illuminate and ionize the envelope, that becomes a planetary nebula (PN), through the transition phase of protoplanetary nebula (PPN). During this very short phase, $\sim$ 1000 yr, the nebular morphology and kinematics dramatically change: the spherical, slowly expanding AGB envelope becomes a nebula with, usually, axial symmetry and high axial velocities. Typically, gas masses of about 0.1–0.5 $M_\odot$ are accelerated to reach velocities $\sim$ 30–100 km s^{-1}, up to 400 km s^{-1}. It is usually assumed that this spectacular metamorphosis results from the impact on the slow AGB envelope of a fast and highly collimated (jet-like) wind, ejected in the late-AGB or early post-AGB phases. But, as we will see, the basic phenomena are still not well understood.

Molecular lines are a very useful tool to study the nebulae around post-AGB stars, particularly young planetary nebulae and protoplanetary nebulae. As we will see, many important quantitative results are obtained from these observations. In evolved PNe, however, the diffuse gas and very hot central star yield strong photoionization, and molecules become very rare. In intermediate-evolution sources, massive PDRs are present, surrounded by warm molecule-rich gas.

The most useful molecular species to study PPNe is carbon monoxide. CO lines provide important observational advantages: they are intense, due to the high abundance of this molecule, and can be observed with present radiotele-scopes, yielding high spectral and spatial resolution.

The excitation of the low-energy rotational CO lines is particularly easy to describe. For instance, the most usually observed transitions are the $J = 1$–0 and 2–1 lines, at mm-wavelengths, and the first rotational level ($J = 1$) is at just 5.5 K from the ground state. These lines are moreover easily thermalized, due to their particularly low Einstein co-efficients. Note that optical depth of CO lines in PPNe is often not negligible, but it can be treated, using some escape probability formalism or, if possible, observing lines of the rare isotope ^{13}CO. These mm-wave lines do not require high-excitation conditions, so they are very well suited to study most nebular gas, which is relatively cold. On the other hand, it is readily shown that the intensity of these CO lines tends to decrease as soon as the temperature reaches ~ 50 K, see e.g. Bujarrabal et al. (1997), due to the population of higher-J levels.

Finally, the chemistry of this molecule is very simple, with a high and quite constant abundance; except when photodissociation is important, in this case, the molecule is rapidly destroyed. So, CO lines are a good tracer of molecule-rich gas.

1.1 Protoplanetary dynamics

As we have mentioned, the evolution of protoplanetary neb-ulae during their short life is really fast and deep. The in-volved dynamics is particularly impressive.

The dynamical phenomena that are believed to be at work in our case are:

(1) During the AGB phase, the slow (~ 10 km s^{-1}) wind is thought to be driven by shock propagation in the inner circumstellar layers, whose origin is the important pul-sation activity of these stars, and by radiation pressure acting onto grains, once they are formed at a certain dis-tance from the photosphere.

(2) In the the post-AGB phase, the star ejects outflows that are very collimated along an axis. They are very fast (up to 1000 km s^{-1}) but carry a small amount of mass. The origin of these jets is not well understood. The most popular theories propose a magnetocentrifugal launch-ing mechanism, similar to that at work in protostar out-flows, due to reaccretion of circumstellar material by the star or a companion from a rotating disk. To form such a disk, we need that the previously ejected gas win an-gular momentum, thanks to interaction with a stellar or substellar companion.

(3) These fast outflows must interact with the fossil AGB envelope, yielding bow shocks. Assuming that the fast jets were very energetic at the beginning of the phase, the shocks will significantly accelerate the polar caps

of the massive AGB shells. So, the bipolar lobes ob-served in PNe and PPNe would be formed; slow equato-rial disks, less affected by the shock interaction, would subsist. However, we will see later that we do not see traces of massive shocked gas in some objects, in which the post-AGB acceleration could take place following a different process.

We note that mechanisms of this kind, probably involv-ing magnetic forces and differential rotation, are necessary to explain post-AGB dynamics, because of the clear axial symmetry of the massive bipolar outflows and the very high linear momentum carried by them, much higher than that provided by radiation pressure.

See, as general references, Habing (1996), Balick and Frank (2002), Soker (2002), Frank and Blackman (2004).

2 Molecular data on PPNe

2.1 Statistical studies

The observation of molecular lines has been particularly useful to study the general properties of PPNe. Many im-portant quantitative results have been obtained, mainly from CO lines. See general properties of molecular gas in PPNe in Bujarrabal et al. (2001), and references therein.

A total of 30 or 40 PPNe (i.e. young PNe with stellar temperatures not higher than 30000 K) have been relatively well studied in CO emission. In more than 2/3 of these, the mass of the molecular gas is large, between about 0.1 and 2 $M_\odot$. For them, we can conclude that CO probes the bulk of the nebular material. In some objects, we can say that most of the total material is detected by observations of molecular lines: most of the stellar mass has been ejected and is now in the form of molecular cool gas. We can also estimate the ejection time from the extent and velocity of the CO emitting gas, i.e. the time during which the mass was ejected by the star (presumably in the late AGB phases). We usually find short times, less than about a thousand years: the ejection by the star of most of the material that forms the nebulae also seems to be a very fast phenomenon.

Only in $\sim 15\%$ of the PPNe with useful CO data we do not detect fast outflows (with axial velocities between 30 and 400 km s^{-1}), down to reasonable detection limits. Of course, most of these cases could be due to projection effects (jet directions close to the sky plane), high photodissociation in extended components, or peculiar structures (in fact, sources with low initial mass or probably containing rotating disks are among them).

When the fast bipolar outflows are mapped, they are found to systematically show a linear increase of the velocity with the distance to the star ('Hubble-like law'). This seems to correspond to ballistic movements, due to a sudden acceleration in the past plus free expansion since then. In well studied objects, it is also possible to estimate the time during which the dynamics was active (i.e. the forces accelerating the outflows were not negligible), values smaller than 300 yr are found.

In more than 80% of the sources with useful CO data, the momentum carried by the high-velocity flows is too high to be driven by radiation pressure, at least under expected conditions (even taking into account the effects of multiple scattering in opaque shells). Another mechanism must be invoked (Sect. 1.1).

In well mapped sources, a clear axial symmetry is almost always found; the only well studied PPN that appears to be spherical after accurate mapping is, to my knowledge, IRAS 22272+5435. In general, the structure is composite, showing the bipolar fast flow plus a disk or torus, usually with low expansion velocity comparable to that of AGB shells. In some cases, a slow halo is also detected.

The existence of disks in rotation has been proposed to be a common phenomenon in PPNe, to explain the protoplanetary dynamics as well as some observational facts, but up to date they have been well identified only in one or two sources (Sect. 2.3).

Finally we note that in many sources the measured temperature from CO mm-wave lines is very low, $\sim$ 10–30 K. But a strong selection effect may be present here, because these low-J transition tend to select cold gas and are not adequate to study the warm components expected in PPNe (Sect. 1).

2.2 High-resolution mapping of PPNe

Several objects have been very carefully studied in molecular line emission. At this respect, mm-wave interferometers have been extremely useful, because the total size of these objects, 5–15 arcseconds, is well suited to instruments like PdBI and OVRO. I will mention a few relevant studies.

CRL 618 is a very well studied PPN. Accurate mapping of this source in CO emission has been performed (Sánchez Contreras et al. 2004), as well as a model that is probably the most detailed one developed for a PPN. The various components of the model correspond to features actually detected in the maps. There are two slow components: a halo corresponding to a mass-loss process more or less similar to those typical of obscured AGB stars, plus a very compact and dense one, corresponding to a mass-loss rate as high as 2×10^{-4} $M_\odot$ yr^{-1}, but lasting only about 500 yr. Two fast axial outflows are found: a double shell, plus a very fast inner outflow, running inside the cavity left by the shells. The temperature of these two components is anomalously high, several hundred K.

The two cavity walls seem to be in fact two bow shocks, because of their velocity field, shape and high temperature.

This is confirmed by the comparison of their extent and shape with images of Hα emission (Sánchez Contreras et al. 2004). Hα comes from a series of shocked knots and apparently traces a post-AGB collimated jet. This jet extends almost exactly up to the tips of the double cavity, suggesting that they are impinging onto it. It seems very probable that shock acceleration is active in this source.

M 1-92 is also very well studied. Maps and the corresponding modelling (Bujarrabal et al. 1998) show a clearly axial structure, with two fast lobes plus a slow equatorial disk. These data showed a clear ballistic expansion in the bipolar lobes. Very recent results (Alcolea et al. 2007) confirm these previous results, but the very high resolution in these data (better than $0''.5$) reveals a surprising result. The linear velocity gradient applies even to the central disk, where no intensity minimum is seen at the systemic velocity. Moreover, even with this high resolution, the brightness remains relatively low, not showing signs of shock heating in the lobes. It is therefore probable (Alcolea et al. 2007) that the whole nebula was formed by a sudden process of anisotropic ejection, about 1000 yr ago, and, at least in this object, shock interaction may have had negligible effects on the overall nebular dynamics. See all details on these results in the contribution by Alcolea et al. to this symposium.

Other interesting nebulae, showing a variety of phenomena, have been also well studied. OH 231.8+4.2 (Alcolea et al. 2001; Bujarrabal et al. 2001) has a massive, cool component that is very fast and elongated, and shows an almost exactly linear relation between velocity and position, again due to ballistic movements. In the optical, it shows a remarkable bow-like shock, detected in Hα emission, resulting from the interaction of the fast massive component with outer, more diffuse shells.

We can also mention IRAS 21282+5050 (included in the new PdBI survey of AGB and post-AGB sources, Castro-Carrizo et al., in preparation), in which most of the gas is concentrated in a cylinder-like structure in moderate expansion; HD 101584 and Hen 3-1475 (Olofsson 1999; Huggins et al. 2004), showing high amounts of mass and linear momentum; etc. Other examples are presented in Bujarrabal (2006).

2.3 Rotating disks in PPNe

As we have seen, rotating disks in PPNe have been proposed to explain the very energetic and collimated jets found in these objects (Sect. 1). Rotating disks could also explain certain properties sometimes found in post-AGB objects, like the detection of hot dust in the NIR (which should require a stable reservoir of grains close to the star), the relative lack of refractory elements in the stellar atmospheres (due possibly to reaccretion of nebular material), and the peculiar profiles found in a few PPNe (difficult to explain

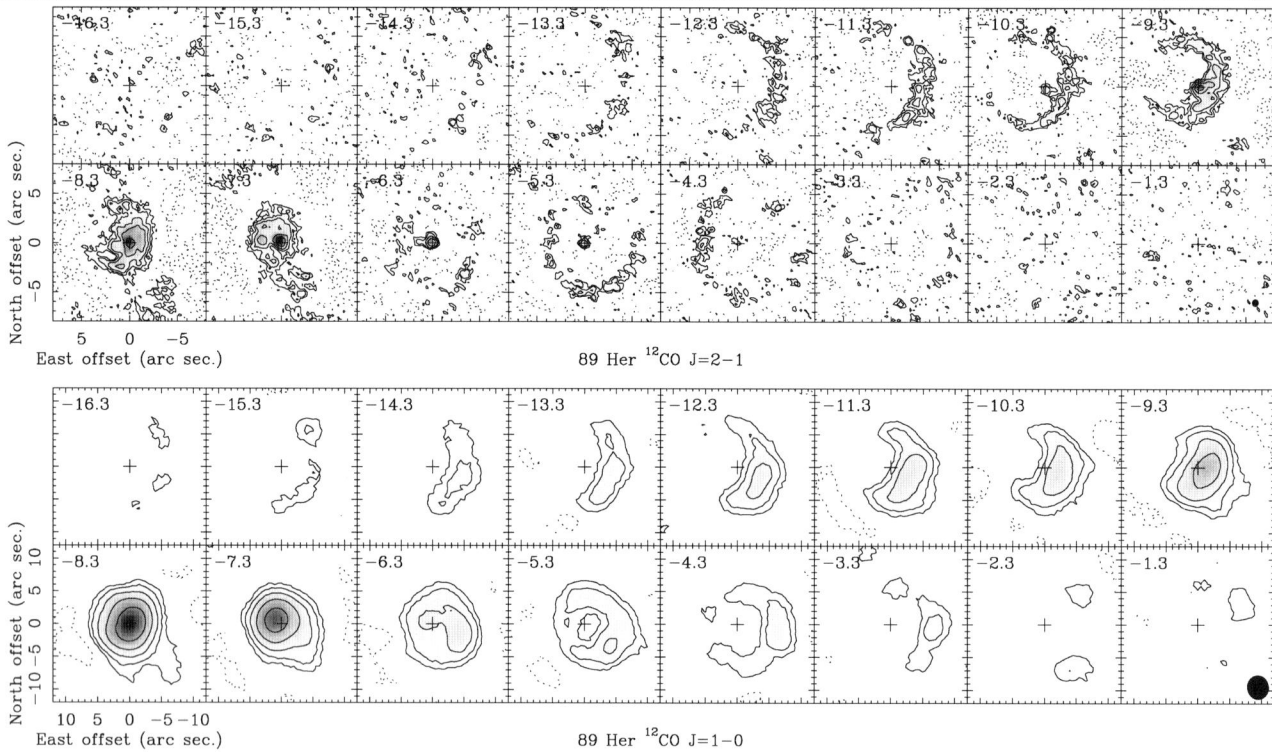

Fig. 1 Channel maps of CO $J = 2-1$ and 1–0 from 89 Her. The first contours are, respectively, 15 and 30 mJy/beam; the contours are separated by a factor 2 and the negative contours (at -15 and -30 mJy, respectively) are indicated by *dashed lines*. The LSR velocity in $km\,s^{-1}$ for the center of each channel is indicated in the *upper left corner*. The *black ellipses* in the *last panels* indicate the beam half-maximum sizes

for standard expanding nebulae). These properties are systematically found to be associated to binarity, as expected (Sect. 1.1). See as general reference van Winckel (2003).

However, rotation has been directly confirmed in only one PPN. CO $J = 2-1$ and $J = 1-0$ maps of the Red Rectangle (Bujarrabal et al. 2005) show a structure elongated along the equatorial plane of the nebula, in which a thick dust disk was known to be present. The CO velocity-position diagrams in the equatorial direction show clearly the signature of keplerian rotation. Models of CO emission confirm that the disk is rotating, and that the rotation is keplerian at least in the inner $\sim 8 \times 10^{15}$ cm. This dynamics corresponds to a central mass (probably the mass of the binary central star) of $\sim 1.5\ M_\odot$. This keplerian disk is relatively hot, with temperatures $\gtrsim 100$ K.

Outside that radius, the disk is still rotating, but a very slow expansion (at less than 1 $km\,s^{-1}$) also appears. This low expansion velocity, comparable to the thermal velocity dispersion, suggests that the expansion is just due to pressure gradients or to some kind of evaporation. This is supported by the fact that the models tend to predict a significant decrease of the density at the radius at which the dynamics change, by about a factor three.

The interest of these results depends a lot on whether such structures are prevalent in the inner regions of PPNe, such that they could explain the ubiquitous post-AGB axial

jets. Therefore, we would like very much to know whether other rotating disks appear in PPNe. 89 Her is a well known post-AGB star that presents (as the Red Rectangle) the properties mentioned above, which probably indicate the presence of a keplerian disk.

Recent CO line maps of the nebula around 89 Her (Bujarrabal et al. 2007) show a peculiar brightness distribution (Fig. 1). The nebula model explaining these observations is depicted in Fig. 2. The nebula shows two components: an extended hourglass-like one in expansion and a compact one, of which the size has not been resolved (represented in Fig. 2 by the black circle). Most of the molecular mass, $\sim 10^{-2}$ $M_\odot$, is found in the compact component.

We only know the total velocity dispersion of the compact component, ~ 5 $km\,s^{-1}$, and an upper limit to its size, $\sim 0''.4$ (6×10^{15} cm at 1 kp, the distance of 89 Her). If we assume that it is an expanding disk, we can calculate a typical lifetime, obtaining a value $\lesssim 100$ yr. This is a very short time, in particular compared with the typical lifetime of the hourglass, ~ 3500 yr. It is also a very short time compared to the typical scales of the evolution of this star: we do not expect that a post-AGB source ejects such a massive component and the star is not evolving so fast, because its effective temperature has not changed since 1950.

On the other hand, assuming that the central compact component is a keplerian (stable) disk, we obtain a total

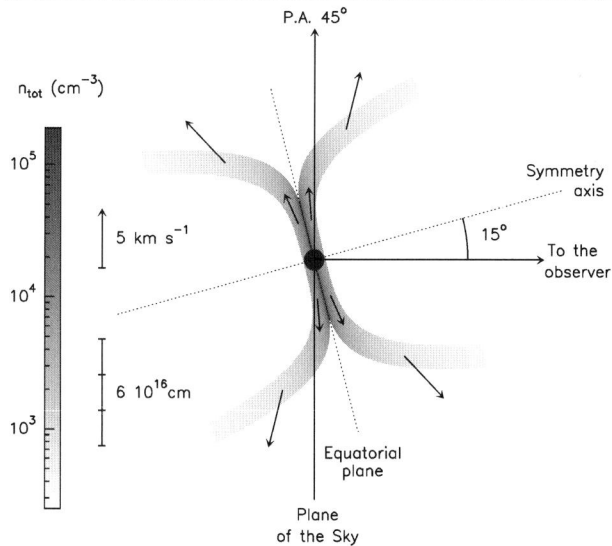

Fig. 2 Geometry and distribution of density and velocity in the model nebula for 89 Her. Note that only limits were obtained for the size and physical conditions of the central, compact component, represented by the *black circle*

diameter of the disk (for reasonable stellar masses) quite comparable with the small sizes derived for the central hot-dust component from NIR interferometry (Bujarrabal et al. 2007). Therefore, 89 Her likely harbors another rotating disk.

Several other PPNe around binary stars share with the Red Rectangle and 89 Her the NIR excess that could be due to the presence of disks. Therefore, these CO data support the identification of inner stable reservoirs of material in keplerian rotation in these objects. However, the small angular size of the proposed keplerian disk in 89 Her can be a common property of such candidate disks, in view of the high dust temperature measured in them (and that they are not closer than 89 Her). These small regions cannot be resolved by present day radiointerferometers; therefore a direct confirmation of the keplerian dynamics will require the use of future instruments.

3 ALMA: higher frequency, higher resolution

We have mentioned (Sect. 1) that the commonly imaged mm-wave lines cannot select warm regions. Since these components tend to be compact, their mm-wave emission is dominated by the contribution of cool regions, and the warm gas is hardly identified in the data. However, very interesting compact, warm regions appear in PPNe and, as we will see, are crucial to understand their evolution.

3.1 Warm, chemically active regions

The chemical evolution of PPNe is dominated by photo-induced reactions, particularly, by photodissociation of

molecules. For stellar temperatures larger than ~ 10000 K, detectable PDRs appear due to molecule dissociation (Castro-Carrizo et al. 2001; Fong et al. 2001). A thin warm layer of molecules should appear around the PDR. In these warm regions we also expect a very active photoinduced chemistry, see (Pardo et al. 2005) and references therein, due to the photodissociation of stable molecules and the subsequent fast reactions including radicals.

However, such photon-dominated regions have been scarcely studied, mainly due to the lack of resolution and of high-frequency observations. The properties of the molecule-rich warm gas heated by stellar photons remain up to now practically unknown. In particular, the stratification of molecules, at expected scales smaller than $1''$ in PPNe (Pardo et al. 2005), has not been yet mapped.

Shocks may also trigger the chemical evolution of PPNe, also destroying stable molecules. Such a mechanism could be the dominant agent of the chemical evolution in some objects (Sánchez Contreras et al. 2000). However, the difficult observation of shocked layers (Sect. 3.3) hampers such studies.

3.2 Inner rotating disks

The systematic detection of rotating disks in PPNe requires good mapping of relatively warm and compact regions. Typically, maps with a resolution $\sim 0''.1$ (a few 10^{15} cm) are required (Sect. 2.3). We stress the interest of systematic detection of disks in PPNe, in order to conclude on their effects on the protoplanetary dynamics.

Even in particularly developed and nearby disks, like that in the Red Rectangle, the inner regions of the disk were not observed. These inner regions (at less or about 10^{15} cm) are particularly important, because only at such short distances we could see the effects of the companion on the disk and the possible infall onto it or the primary.

3.3 Recently shocked layers

The recently shocked (still warm) layers in the massive components of PPNe are particularly important, because shocks are thought to be the main mechanism driving the spectacular dynamics of PPNe. We see the effects of these shocks (wide massive bubbles, strong axial acceleration), but the shocked regions themselves remain very poorly observed. Existing maps have only detected them in one or two objects, and, even in those sources, the shocked layers have been not resolved.

From existing data, we estimate that the width of the shocked layers is $\lesssim 0''.2$. So with present-day resolutions, hardly better than $1''$, these layers can only be scarcely studied. We do not know the distributions of the velocity, density and temperature in such layers: parameters basic to understand the shock process and their relevance in the PPNe evolution.

As we have seen (Sect. 2.2), we are not even sure of how often such shocks appear and are relevant in the nebular dynamics. Their identification requires observations with subarcsecond resolution of temperature-depending emission. It is probable that many (a majority of?) sources show a lack of shock features, like in M 1-92. If this is the case, the formation of highly axisymmetric nebulae, with bipolar lobes and high axial velocities, would not result from wind interaction, but appear naturally because the nebula ejection has been a strongly anisotropic and very sudden process.

3.4 Observations of young planetary nebulae with ALMA

ALMA will allow observations with high resolution (in particular thanks to the high telescope surface) and at high frequency. Therefore, we can expect accurate mapping with this instrument of that warm, compact regions mentioned above. We recall that these regions are crucial to understand the impressive evolution of PPNe, in particular the involved dynamics. Other aspects of these rapidly changing objects, like the chemical evolution, will also be addressed.

We can check that ALMA will allow systematic observations at subarcsecond resolution. For instance, we will map the CO $J = 3$–2 and $J = 6$–5 lines with a resolution of $0''.05$. A noise $\sigma \sim 4$ K (in brightness units) will be obtained, which means that we will be able to carefully study with such a resolution regions hotter than about 50 K. We also will observe, for instance, the CO $J = 6$–5 transition with a resolution of $0''.02$, attaining $\sigma \sim 25$ K, suitable to observe regions warmer than 300 K: the inner rotating disks, PDRs and surrounding regions, shocks, ... Note that these S/N ratio estimates are generous, so it will be also possible to observe a variety of molecular lines (rarer species, optically thin lines like those of ^{13}CO, ...) in a variety of sources (i.e. not only in the most remarkable, intense ones, which are not always the most interesting ones).

In summary, ALMA is expected to provide observations of a large number of molecular lines in a wide variety of post-AGB sources, with a resolution more than ten times better than those currently obtained with present telescopes. These observations will yield a wide range of (perhaps unexpected) results, addressing in particular the main phenomena driving the evolution of these objects.

Acknowledgements   This work has been supported by the *Spanish Ministry of Education & Science*, project numbers AYA2003-7584 and ESP2003-04957.

References

Alcolea, J., Bujarrabal, V., Sánchez Contreras, C., Neri, R., Zweigle, J.: Astron. Astrophys. **373**, 932 (2001)

Alcolea, J., Neri, R., Bujarrabal, V.: Astron. Astrophys., in press (2007)

Balick, B., Frank, A.: Annu. Rev. Astron. Astrophys. **40**, 439 (2002)

Bujarrabal, V.: In: Barlow, M.J., Méndez, R.H. (eds.) IAU Symp. No 234, Planetary Nebulae in our Galaxy and Beyond, p. 193. Cambridge University Press, Cambridge (2006)

Bujarrabal, V., Alcolea, J., Neri, R., Grewing, M.: Astron. Astrophys. **320**, 540 (1997)

Bujarrabal, V., Alcolea, J., Neri, R.: Astrophys. J. **504**, 915 (1998)

Bujarrabal, V., Castro-Carrizo, A., Alcolea, J., Sánchez Contreras, C.: Astron. Astrophys. **377**, 868 (2001)

Bujarrabal, V., Castro-Carrizo, A., Alcolea, J., Neri, R.: Astron. Astrophys. **441**, 1031 (2005)

Bujarrabal, V., van Winckel, H., Neri, R., et al.: Astron. Astrophys., in press (2007)

Castro-Carrizo, A., Bujarrabal, V., Fong, D., et al.: Astron. Astrophys. **367**, 674 (2001)

Fong, D., Meixner, M., Castro-Carrizo, A., et al.: Astrophys. J. **367**, 652 (2001)

Frank, A., Blackman, E.G.: Astrophys. J. **614**, 737 (2004)

Habing, H.R.: Astron. Astrophys. Rev. **7**, 97 (1996)

Huggins, P.J., Muthu, C., Bachiller, R., Forveille, T., Cox, P.: Astron. Astrophys. **414**, 581 (2004)

Olofsson, H., Nyman, L.-Å.: Astron. Astrophys. **347**, 194 (1999)

Pardo, J.R., Cernicharo, J., Goicoechea, J.R.: Astrophys. J. **628**, 275 (2005)

Soker, N.: Astrophys. J. **568**, 726 (2002)

Sánchez Contreras, C., Bujarrabal, V., Neri, R., Alcolea, J.: Astron. Astrophys. **357**, 651 (2000)

Sánchez Contreras, C., Bujarrabal, V., Castro-Carrizo, A., Alcolea, J., Sargent, A.: Astrophys. J. **617**, 1142 (2004)

van Winckel, H.: Annu. Rev. Astron. Astrophys. **41**, 391 (2003)

Planetary nebulae and ALMA

P.J. Huggins

Originally published in the journal Astrophysics and Space Science, Volume 313, Nos 1–3.
DOI: 10.1007/s10509-007-9612-7 © Springer Science+Business Media B.V. 2007

Abstract Our understanding of the late evolution of intermediate mass stars ($\sim$1–8M$_\odot$) through the planetary nebula phase is undergoing major developments. Observations at infrared and millimeter wavelengths have revealed important components of neutral gas and dust in the nebulae that directly trace their formation from mass-loss on the Asymptotic Giant Branch. At the same time, high resolution imaging, especially with the Hubble Space Telescope, has revealed a surprising array of structures in the nebulae: multiple arcs, tori, jets, and myriads of small scale fragments. None of these are fully understood, and all involve the neutral gas component. This paper highlights recent observations of these structures and discusses the open questions, with an emphasis on those areas where observations with ALMA are likely to make important contributions.

Keywords Planetary nebulae · Molecular lines

1 Introduction

Planetary Nebulae (PNe) play a central role in our picture of the late stages of evolution of intermediate mass stars ($\sim$1–8M$_\odot$). They are formed in the transition from Asymptotic Giant Branch (AGB) stars to white dwarfs, and their study constitutes an important area of stellar and Galactic astronomy, e.g., (Barlow and Mendez 2006). The ionized nebulae are easily identified and have been observed for more than two centuries, so one might think that they form a well understood class of object, but this turns out not to be the case.

P.J. Huggins (✉)
Physics Department, New York University, New York, NY 10003, USA
e-mail: patrick.huggins@nyu.edu

Basic aspects of their origin and evolution are uncertain and even controversial.

This paper reviews our current view of the nebulae with an emphasis on the directions in which observations with ALMA are likely have a significant impact. The paper is divided into three parts: first, a reminder about the properties of PNe; second, an outline of the outstanding scientific challenges in the field and the limitations of our current observations; and third, a discussion of three specific areas—the large scale structure, the small scale structure, and the chemistry—where high resolution observations are especially important for future progress.

2 The properties of PNe

It is well established that PNe originate in the neutral mass-loss of stars on the AGB, but the exact mode of the final ejection is uncertain (see Sect. 4). The subsequent evolution of the central stellar component is relatively well understood: it evolves to high temperature at nearly constant luminosity, and then turns onto a white dwarf cooling track, e.g., (Bloecker 1995). An important characteristic of this evolution is the extremely rapid time scale of a few 100 yr to a few 1000 yr (depending on the core mass), much faster than the evolution on the AGB. In the early post-AGB evolution—the proto-PN phase—the ejected circumstellar gas remains mostly neutral (Bujarrabal, this volume), but when the central star reaches a temperature of $\sim$30,000 K, there is a sudden onset of photo-ionization, and a bona fide PN is born.

Important advances in characterizing the morphology of the ionized nebulae have been made at optical wavelengths using high resolution imaging with HST. The increase in angular resolution by a factor $\sim$10 in going from observations with ground-based telescopes to HST (resolution is $\sim$0.1$''$)

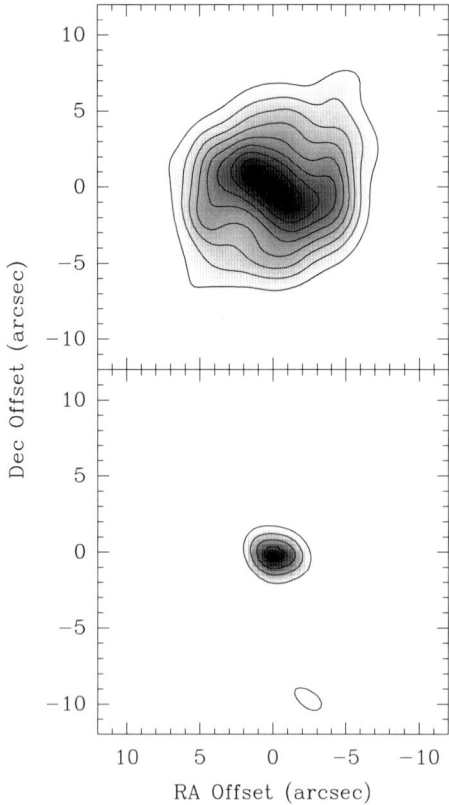

Fig. 1 Millimeter observations of the young PN M1-16. *Top*: CO (2–1) emission in integrated intensity. *Bottom*: the millimeter continuum. Data from the IRAM interferometer. Figure adapted from (Huggins et al. 2000)

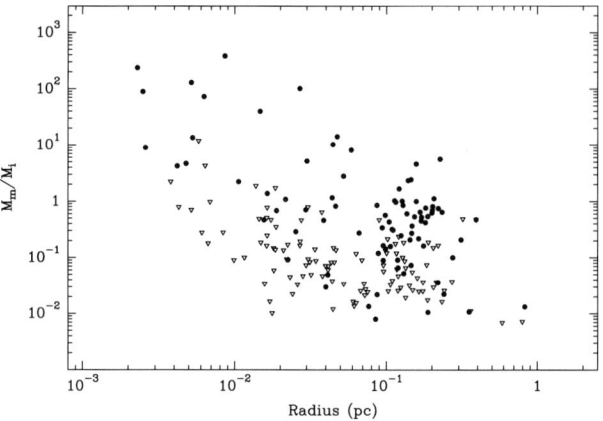

Fig. 2 Mass ratio of molecular/ionized components of PNe vs. nebula radius. Data based on CO observations; *filled circles* are detections, *triangles* are upper limits (Huggins et al. 2007)

has revealed striking and varied structure in the nebulae on both large and small size scales. PNe with smooth, spherically symmetric morphologies are rare.

The physical properties of the ionized gas are well characterized. The temperature is $\sim$10,000 K. The typical densities decrease from 10^5 to 10^2 cm^{-3} as the nebulae expand, and the FUV and EUV radiation fields are intense in the early phases. The total mass ranges up to $\sim$1M$_\odot$. The typical expansion velocity is $\sim$25 km s^{-1}, and the nebulae disperse in the ISM on a time scale of a few $\times$ 10^4 yr, leaving the remnant core as a newly formed white dwarf.

In spite of the hostile conditions, many PNe show a component of neutral gas, which can be studied in CO, H$_2$, and other molecular species, as well as neutral atoms (e.g., Huggins et al. 1996, 2005; Kastner et al. 1996). At millimeter wavelengths only the relatively nearby objects can be usefully mapped. A classic example is the young PN M1-16 shown in Fig. 1. The millimeter continuum shows the newly formed ionized nebula in free–free emission, and CO reveals the surrounding molecular envelope that is being photo-ionized from the inside. Other well studied examples observed in CO and H$_2$ (which traces highly excited gas) include NGC 7027, e.g., (Cox et al. 2002; Fong et al. 2006), and the very extended PN NGC 7293 (the

Helix Nebula) where the molecular gas is imaged in some detail (Young et al. 1999; Cox et al. 1998; Speck et al. 2002; Hora et al. 2006). Well resolved examples of this type demonstrate that the structure in the neutral gas is largely responsible for the structure observed in the ionized nebulae.

It is to be expected that the amount of neutral gas in PNe shows evolution and population effects, and this is observed to be the case. This is illustrated in Fig. 2, which shows results based on survey observations in CO. Flux measurements have been used to estimate the mass ratio of molecular to ionized gas (assuming a nominal CO/H$_2$ ratio) and this is plotted against the nebula size, which roughly corresponds to the expansion age. The detected PNe show that the mass ratio decreases with nebula age, as expected from dissociation and photo-ionization as the nebulae expand. There are also great differences between PNe of the same age, and in many the molecular component has been destroyed, even in the early PNe or proto-PNe phases. This is likely related to the amount and mode of the final ejection (Sect. 4) as well as the evolution time of the central star.

3 Current limitations and scientific challenges

The main limitation of current observations of the neutral gas in PNe is that we can study only the nearest examples in any degree of detail, so we are unable to explore the full extent of the phenomenon. The situation is illustrated in Table 1, which gives the angular sizes of the ionized nebulae for three nearby archetypes, separated in expansion age by factors of $\sim$10. For a young PN like NGC 7027 ($\tau_{\exp} \sim 1000$ yr), we can determine the structure with a moderate degree of detail at a resolution of $1''$. However, for younger systems like AFGL 618 (see Bujarrabal, this

Table 1 Angular sizes of nearby PN archetypes

PN	age (yr)	dist (pc)	θ_{ion} ($''$)	θ_{ion} at GC ($''$)
AFGL 618	100	900	0.3	0.03
NGC 7027	1000	700	10	0.9
NGC 7293	10,000	200	600	15

volume) the structure close to the ionized region is completely unresolved. On the other hand, for the nearest highly evolved PNe like NGC 7293, the angular size is very large. However, even in this case, it turns out that the structure of most interest is still at sub arc second size scales (see Sect. 5).

Thus a factor of >10 increase in resolution at millimeter wavelengths is likely to produce a revolution in our perception of the neutral gas in PNe, comparable to the change brought about by HST at optical wavelengths. It would completely transform our view of the nearest objects, and extend our vision to a significant population in the Galaxy.

The need for improved observations is directly driven by the current scientific challenges in the field. The most pressing open questions related to the neutral component in PNe are:

– The origin of the large scale structure.
– The origin of the small scale structure.
– The possible role of magnetic fields.
– The survival of molecules and the associated chemistry.

These issues are all inter-connected, and relate directly or indirectly to our basic lack of understanding of the PN formation process. The study of the neutral component is therefore central to the understanding of PNe.

4 Large scale structure

The origin of the large scale structure in PNe is the most widely debated of the current open questions, prompted by the bewildering range of morphologies revealed by high resolution observations with HST. Concentric arcs, e.g., (Corradi et al. 2004), which surround some PNe can be traced to effects in the mass loss on the AGB (Mauron and Huggins 1999, 2000), but their origin is unknown. Much more widely observed are asymmetries in the large scale structure of the nebulae, and these can be broadly classified into two categories: equatorial enhancements variously described as disks or tori; and bi-polar, multi-polar, or point symmetric structures produced by collimated outflows, or jets. These features are usually somewhat smoothed out in evolved PNe, because the thermal velocity of the ionized gas is comparable to the expansion velocity, but they are more prominent in younger objects. Jets are common or perhaps ubiquitous in the youngest objects (Sahai and Trauger 1998).

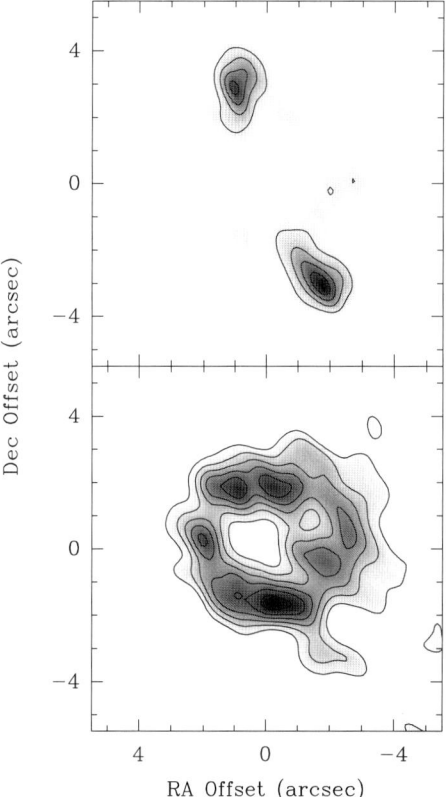

Fig. 3 Millimeter observations of the young PN BD+30°3639. *Top*: CO (2–1) emission in integrated intensity. *Bottom*: the millimeter continuum. Data from the IRAM interferometer. Figure adapted from (Bachiller et al. 2000)

The early development of PN structure in the neutral gas can be seen in proto-PNe (Bujarrabal, this volume) and even in late AGB stars (Mauron and Huggins 2006), and it can be traced through into fully developed PNe. In some cases the jets are contained within the neutral envelope. One example of this is the young elliptical PN BD+30°3639, shown in Fig. 3. In this case there is no signature at all of jets in optical images. High resolution millimeter observations in the continuum show the elliptical ionized nebula in free-free emission, but in CO there are two molecular bullets, seen on either side of the central star, with equal and opposite ejection velocities of 50 km s^{-1}, projected on the line of sight (Bachiller et al. 2000). The interpretation is that there are invisible jets from the central star system that impact the neutral gas at the periphery of the ionized nebula and create the dense knots seen in CO.

In the archetype NGC 7027, observations in H$_2$ at $1''$ resolution reveal holes in the PDR gas which interfaces between the ionized nebula and the surrounding envelope. The holes are symmetric in position and velocity about the central star, indicating the action of multiple jet pairs (Cox et al. 2002). In the optical, HST images of NGC 7027 show multiple bipolar structures composed of numerous shrapnel-like components, and at X-ray wavelengths Chandra reveals

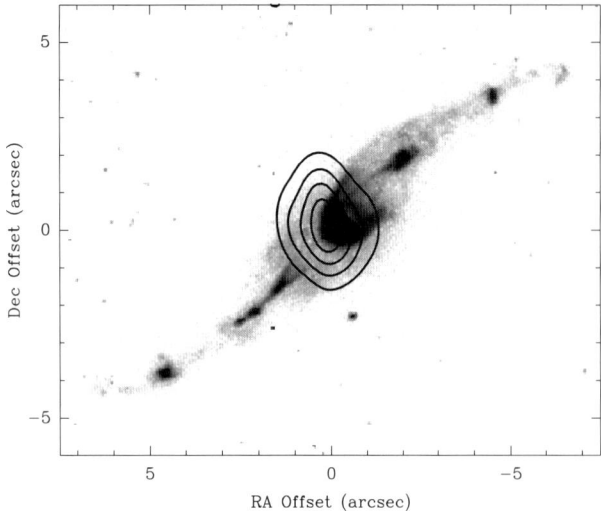

Fig. 4 The jets and torus of the young PN He 3-1475. CO (2–1) *contours* in integrated intensity, superposed on HST image in Hα+N II. CO data from the IRAM interferometer (Huggins et al. 2004)

X-ray emission along the most prominent bi-polar axis in the molecular gas (Kastner et al. 2001).

In other cases, the jets have broken through the dense molecular envelope and are more prominent at optical wavelengths. An example of an evolved PN system is KjPn 8, which shows at least two jet systems. The most recent has a velocity of 300 km s^{-1} and extends over 240$''$ (Lopez et al. 1997). Around the origin of these jets there is a slowly expanding 25$''$ molecular torus seen in CO which lies orthogonal to the jet axis (Forveille et al. 1998). In the younger object M1-16, there are episodic jets at velocities of ∼300 km s^{-1} extending over 100$''$ (Schwarz 1992) which have pierced the molecular envelope and entrained the molecular gas; an equatorial enhancement and the extension of molecular gas along the jet axis can be seen in Fig. 1. Probably the most spectacular system is the very young PN He 3-1475 (Fig. 4). HST observations reveal episodic jets with velocities >1000 km s^{-1}, which have changed direction, forming point symmetric S-shaped curves, e.g., (Borkowski and Harrington 2001). The jets have burst through a dense molecular torus around the center, and entrained gas along the bi-conical sides of the outflow.

The current theoretical view of the jet and torus ejection is uncertain. There is a widespread, but not universal view that binaries are involved in determining the symmetries, e.g., (Morris 1987; Soker and Livio 1994). There is some consensus that magnetic fields collimate the jets, but the physical circumstances that produce the tori and the jets are not known. Possibilities include: a disk around a binary companion which captures the mass loss of the primary; and various direct interaction scenarios including common envelopes with a magnetic wind or a disk

around the primary formed by the break-up of a secondary.

The highest possible resolution observations of the neutral gas in young PNe, proto-PNe, and the cores of AGB stars will be crucial for sorting out this question. The prospects are excellent: for example a young 500 yr old torus-jet system with a torus expansion velocity of 10 km s^{-1} and at jet velocity of 100 km s^{-1} has dimensions of 2$''$ × 20$''$ at a distance of 1 kpc and 0.3$''$ × 3$''$ at the Galactic center. At a resolution better than ∼0.1$''$ our view of the nearest cases (at ∼1 kpc) will be greatly enhanced, and the range of the phenomena observed can be greatly expanded. The putative disks from which jets might be launched are likely to be very small (∼1 AU or smaller) and a challenge even in the nearest cases, but the somewhat larger scale environment in which the jets and tori form is likely to produce vital clues on the formation mechanism.

The role of magnetic fields in helping to form the large scale structure in PNe is uncertain and controversial (e.g, Soker 2006), although it is probable that they are relevant on local scales (Huggins and Manley 2005). There are several interesting developments on the observational side. The first magnetic fields (at kG levels) have recently been detected in the central stars of PNe (Jordan et al. 2005). Strong fields are detected in the localized maser spots in AGB envelopes, and these extend into the young PN phase, e.g., (Miranda et al. 2001). In addition, the polarization of dust emission, which dominates the infrared and sub-millimeter spectra of proto-PNe and young PNe, has been detected in the PN NGC 7027, NGC 6537, and NGC 6302 using JCMT (Sabin et al. 2007). It will be important to develop this approach at high resolution in AGB stars, proto-PNe and PNe as a probe of the field geometry.

5 Small scale structure

When we turn to the small scale structure of PNe we again see a range of phenomena we do not clearly understand. The focus here is on cometary globules, but there is a range of other structure including ansae and filaments, whose origins are also uncertain.

The globules are best seen in NGC 7293, the nearest PN ($d \sim 200$ pc) with a significant molecular component. They were first observed more than half a century ago, but their detailed morphology only became clear as a result of high resolution imaging with HST (O'Dell and Handron 1996). The globules have compact heads, 1$''$–2$''$ in size, with tails extending radially away from the star, and in NGC 7293 they number about 10^4 (Meixner et al. 2005). The similarity of the structures and the large numbers immediately prompt the questions: What are the globules? And how do they form?

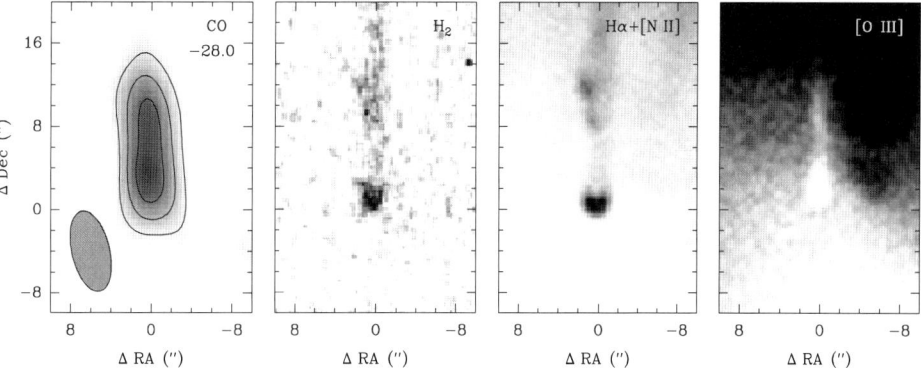

Fig. 5 Cometary globule in the Helix Nebula. *Right*: dust absorption, seen against the nebula emission in [O III] 5007Å. *Center right*: Hα+[N II] 6584 Å. *Center left*: H$_2$ $v = 1 - 0$ $S(1)$. *Left*: CO 1–0 central channel. Data from the NTT and IRAM interferometer (Huggins et al. 2002)

We know the answer to the first question: the globules are dense, neutral condensations embedded in the ionized gas. The neutral component of an individual globule in NGC 7293 has been resolved in CO and H$_2$ (see Fig. 5). The H$_2$ component can also be seen in large numbers of globules in HST and Spitzer images in the near infrared (Meixner et al. 2005; Hora et al. 2006). Based on CO measurements and dust absorption, the globules are found to have planet-like masses of $\sim 10^{-5} M_\odot$, and their large numbers constitute most or all of the mass of the neutral gas in the envelope. The CO observations (Huggins et al. 1992, 2002) show that the gas is quiescent. There are mini-PDRs on the surfaces facing the central star, and there are molecules not only in the globule heads, but also in the tails.

Globules are also seen in other nearby PNe at high resolution with HST, e.g., in NGC 6853 (the Dumbbell Nebula) and NGC 6720 (the Ring Nebula) (O'Dell et al. 2002). The globules in NGC 6720 have recently been observed in CO (Josselin et al. 2007) using the IRAM interferometer. The largest structures are just resolved, and a long filament at the periphery of the nebula is seen to consist of a string of globules, which likely has a bearing on the formation process.

Although observations show that the present evolution of globules is dominated by photo-ionization effects, the current theoretical view of the origin of globules is uncertain. There are two basic scenarios for the origin of the heads: an instability in the atmosphere of the precursor AGB star (Dyson et al. 1989), but also see (Huggins and Mauron 2002); and formation in the circumstellar gas from an instability during nebula formation, e.g., (Capriotti 1973). Similarly for the tails, it is not certain whether they form from material swept from the heads by a wind, or if they form at the same time as the heads; in addition, shadowing of the stellar radiation field by the heads is also likely to play a role.

The observational opportunities for very high resolution millimeter observations are clear. For the nearby archetypes, spatio-kinematic imaging of heads and tails, including proper motions of the flows (possible in the Helix over sev-

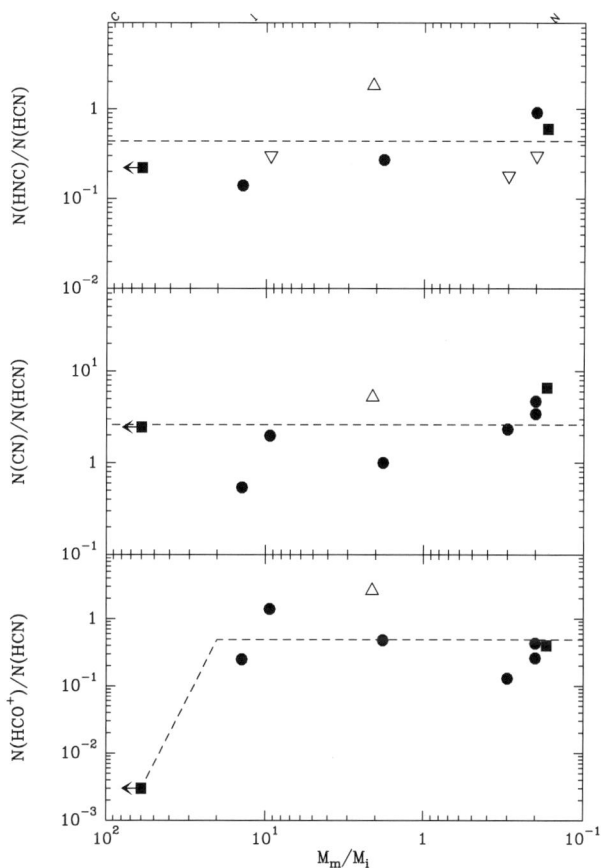

Fig. 6 Abundance ratios in PNe vs. molecular-to-ionized mass ratio. From left to right the data points are for: AFGL 2688, M1-17, NGC 7027, IC 5117, M1-13, BV5-1, K3-94, K3-24, NGC 7293. Adapted from (Josselin and Bachiller 2003)

eral years) will be important for testing hydrodynamic simulations of the various formation scenarios. Increased resolution will also allow the possibility of observing globules in more distant PNe at earlier stages of evolution, to pin down the formation. It seems very likely that these kind of observations could solve all the current open questions in this area.

6 PDRs and chemistry

The physical conditions in the neutral gas change rapidly during PN formation, providing novel environments for the study of PDRs and molecular chemistry, although their study so far is at an early stage of development. The environments include the dense inner regions of the AGB envelopes, slowly expanding tori and possibly long-lived disks, molecular shocks (both O-rich and C-rich) at jet-envelope interfaces, dense PDRs with intense radiation fields as the central star turns on, and globule-tail environments far from the star.

There is a rich chemistry in the very earliest PN phases, exemplified by the spectral line scans of AFGL 618, e.g., (Pardo et al. 2005). One interesting result seen here is increased abundances of cyano-polyyne and poly-acetylene chain molecules, which are attributed to polymerization induced by the chemistry in the rapidly increasing ultraviolet radiation field.

The chemistry in more fully developed PNe has not been extensively studied, but over a dozen molecular species have been detected—see recent inventories by (Hasegawa 2005; Ziurys 2006). Sub-millimeter C I emission has also been detected, e.g., (Bachiller et al. 1994; Young et al. 1997), and is potentially important for PDR studies. The characteristics of the observed chemistry are strongly enhanced ions and radicals (e.g., HCO^+ and CN), which can in general terms be attributed to the radiation field. The strongest source is NGC 7027, where the millimeter lines of species such as HCN and HCO^+ are a few degrees with JCMT. It is possible to construct (carbon rich) PDR models that generate the enhanced species (Hasegawa et al. 2000) but the thickness of the PDR ($<1''$) is not spatially resolved by current observations, so there is a serious lack of geometrical constraints on the chemical structure.

It is striking that lines of several species can be observed in PNe over a large range in evolutionary age, even in highly evolved objects like NGC 7293 at an offset of $\sim 400''$ from the central star (Bachiller et al. 1997; Josselin and Bachiller 2003). A surprising result is that the abundances of the radicals and ions observed is relatively constant for all PNe (Fig. 6). Models are not easily able to account for this (Redman et al. 2003). In the evolved PNe the molecules observed are localized in (unresolved) globules. The head-tail structures are likely to produce quite complex spatially varying chemistry. To make significant progress the need is to resolve the physical structure, so again the need is for high resolution capabilities.

7 Conclusions

Millimeter observations have already made important contributions to our current understanding of PNe, and the prospects for future observations are excellent. There are important open questions concerning the development of the neutral gas that relate directly to key uncertainties in how the nebulae form. These questions can be addressed with the high resolution capabilities of ALMA, and the results promise to provide important contributions to our understanding of this basic phase in the evolution of ordinary stars.

Acknowledgements This work was supported in part by the NSF (AST 03-07277).

References

Bachiller, R., Huggins, P.J., Cox, P., Forveille, T.: Astron. Astrophys. **281**, L93 (1994)

Bachiller, R., Forveille, T., Huggins, P.J., Cox, P.: Astron. Astrophys. **324**, 1123 (1997)

Bachiller, R., Forveille, T., Huggins, P.J., Cox, P., Maillard, J.P.: Astron. Astrophys. **353**, L5 (2000)

Barlow, M.J., Mendez, R.H.: Planetary Nebulae in our Galaxy and Beyond, IAU Symp. No. 234. Cambridge University Press, Cambridge (2006)

Bloecker, T.: Astron. Astrophys. **297**, 727 (1995)

Borkowski, K.J., Harrington, J.P.: Astrophys. J. **550**, 778 (2001)

Capriotti, E.R.: Astrophys. J. **179**, 495 (1973)

Corradi, R.L.M., Sánchez-Blázquez, P., Mellema, G., Giammanco, C., Schwarz, H.E.: Astron. Astrophys. **417**, 637 (2004)

Cox, P., et al.: Astrophys. J. **495**, L23 (1998)

Cox, P., Huggins, P.J., Maillard, J.-P., Habart, E., Morisset, C., Bachiller, R., Forveille, T.: Astron. Astrophys. **384**, 603 (2002)

Dyson, J.E., Hartquist, T.W., Pettini, M., Smith, L.J.: Mon. Not. R. Astron. Soc. **241**, 625 (1989)

Fong, D., Meixner, M., Sutton, E.C., Zalucha, A., Welch, W.J.: Astrophys. J. **652**, 1626 (2006)

Forveille, T., Huggins, P.J., Bachiller, R., Cox, P.: Astrophys. J. **495**, L111 (1998)

Hasegawa, T., Volk, K., Kwok, S.: Astrophys. J. **532**, 994 (2000)

Hasegawa, T.: AIP Conf. Proc. **804**, 218 (2005)

Hora, J.L., Latter, W.B., Smith, H.A., Marengo, M.: Astrophys. J. **652**, 426 (2006)

Huggins, P.J., Manley, S.P.: Publ. Astron. Soc. Pac. **117**, 665 (2005)

Huggins, P.J., Mauron, N.: Astron. Astrophys. **393**, 273 (2002)

Huggins, P.J., Bachiller, R., Cox, P., Forveille, T.: Astrophys. J. **401**, L43 (1992)

Huggins, P.J., Bachiller, R., Cox, P., Forveille, T.: Astron. Astrophys. **315**, 284 (1996)

Huggins, P.J., Forveille, T., Bachiller, R., Cox, P.: Astrophys. J. **544**, 889 (2000)

Huggins, P.J., Forveille, T., Bachiller, R., Cox, P., Ageorges, N., Walsh, J.R.: Astrophys. J. **573**, L55 (2002)

Huggins, P.J., Muthu, C., Bachiller, R., Forveille, T., Cox, P.: Astron. Astrophys. **414**, 581 (2004)

Huggins, P.J., Bachiller, R., Planesas, P., Forveille, T., Cox, P.: Astrophys. J. Suppl. Ser. **160**, 272 (2005)

Huggins, P.J., et al.: (2007, in preparation)

Jordan, S., Werner, K., O'Toole, S.J.: Astron. Astrophys. **432**, 273 (2005)

Josselin, E., Bachiller, R.: Astron. Astrophys. **397**, 659 (2003)

Josselin, E., et al.: (2007, in preparation)

Kastner, J.H., Weintraub, D.A., Gatley, I., Merrill, K.M., Probst, R.G.: Astrophys. J. **462**, 777 (1996)

Kastner, J.H., Vrtilek, S.D., Soker, N.: Astrophys. J. **550**, L189 (2001)

Lopez, J.A., Meaburn, J., Bryce, M., Rodriguez, L.F.: Astrophys. J. **475**, 705 (1997)

Mauron, N., Huggins, P.J.: Astron. Astrophys. **349**, 203 (1999)

Mauron, N., Huggins, P.J.: Astron. Astrophys. **359**, 707 (2000)

Mauron, N., Huggins, P.J.: Astron. Astrophys. **452**, 257 (2006)

Meixner, M., McCullough, P., Hartman, J., Son, M., Speck, A.: Astron. J. **130**, 1784 (2005)

Miranda, L.F., Gómez, Y., Anglada, G., Torrelles, J.M.: Nature **414**, 284 (2001)

Morris, M.: Publ. Astron. Soc. Pac. **99**, 1115 (1987)

O'Dell, C.R., Handron, K.D.: Astron. J. **111**, 1630 (1996)

O'Dell, C.R., Balick, B., Hajian, A.R., Henney, W.J., Burkert, A.: Astron. J. **123**, 3329 (2002)

Pardo, J.R., Cernicharo, J., Goicoechea, J.R.: Astrophys. J. **628**, 275 (2005)

Redman, M.P., Viti, S., Cau, P., Williams, D.A.: Mon. Not. R. Astron. Soc. **345**, 1291 (2003)

Sabin, L., Zijlstra, A.A., Greaves, J.S.: Mon. Not. R. Astron. Soc. **376**, 378 (2007)

Sahai, R., Trauger, J.T.: Astron. J. **116**, 1357 (1998)

Schwarz, H.E.: Astron. Astrophys. **264**, L1 (1992)

Soker, N., Livio, M.: Astrophys. J. **421**, 219 (1994)

Soker, N.: Publ. Astron. Soc. Pac. **118**, 260 (2006)

Speck, A.K., Meixner, M., Fong, D., McCullough, P.R., Moser, D.E., Ueta, T.: Astron. J. **123**, 346 (2002)

Young, K., Cox, P., Huggins, P.J., Forveille, T., Bachiller, R.: Astrophys. J. **482**, L101 (1997)

Young, K., Cox, P., Huggins, P.J., Forveille, T., Bachiller, R.: Astrophys. J. **522**, 387 (1999)

Ziurys, L.M.: Proc. NAS **103**, 12274 (2006)

Chemistry in the circumstellar medium

Unveiling the dust formation zone

T.J. Millar

Originally published in the journal Astrophysics and Space Science, Volume 313, Nos 1–3.
DOI: 10.1007/s10509-007-9636-z © Springer Science+Business Media B.V. 2007

Abstract The growth of dust grains in the inner regions of late-type stars is shrouded in mystery due to the difficulty of understanding the growth of heterogeneous particles from simple atoms and molecules and the lack of observational data. This article reviews the molecular processes important in circumstellar envelopes and discusses how ALMA might be used to probe the dust formation zone either directly or indirectly.

Keywords Circumstellar envelopes · AGB stars · Astrochemistry

1 Introduction

Significant amounts of interstellar dust are known to be formed in the inner regions of the circumstellar envelopes of asymptotic stars. These grains are intimately linked to the mass-loss process as they absorb stellar radiation and share the momentum they gain with the gas, dragging gas away from the stellar surface and initiating a mass-loss which leads to a the creation of a circumstellar envelope (CSE) which might contain up to a solar mass of gas. These CSEs incorporate a huge range in physical conditions, from very dense ($n(H_2) \sim 10^{15}$ cm^{-3}), hot ($T \sim 2000$ K) molecular gas just above the photosphere, to gas with properties similar to that found in dark molecular clouds, to diffuse regions dominated by the external UV radiation field which produces atomic gas far from the central star. Overall the gas

density follows an r^{-2} density distribution, while the dust extinction, which is proportional to the dust column density to infinity and controls the effects of the external UV radiation field, is proportional to r^{-1}. These overall distributions break down close to the photosphere where stellar pulsations give rise to periodic shock waves which compress, heat and alter the chemical composition of the gas. In the following, I will make a few comments about each of the chemically active regions in the CSE with particular attention on C-rich envelopes.

2 CSE chemistry

As mentioned above, the wide range in physical conditions experienced by a parcel of gas and dust as it traverses the envelope of the star on their way to the interstellar medium makes it appropriate to consider individual regions of activity.

2.1 LTE chemistry

Local Thermodynamic Equilibrium (LTE) holds in regions of high temperature and density close to the photosphere and molecular abundances are determined by elemental composition of the gas and pressure. Molecules form which minimise free-energy with the result that in carbon stars, the most abundant species are H_2, CO, N_2, C_2H_2, and HCN, while in oxygen-rich stars, H_2, H_2O and CO dominate. Models for LTE chemistry have a long history and have been extensively developed by Tsuji and collaborators (e.g. Tsuji 1973). These species are very stable and act as parents for the rich non-LTE chemistry which occurs further out in the envelope.

T.J. Millar (✉)
Astrophysics Research Centre, School of Mathematics and Physics, Queen's University Belfast, Belfast BT7 1NN, UK
e-mail: Tom.Millar@qub.ac.uk

In this region, molecules such as CS are a natural target for ALMA, particularly if they are pumped by IR radiation in which case maser emission may ensue. The CS abundance itself depends very sensitively on the C/O ratio and may be a useful tracer around C/O $\sim$ 1 and in monitoring the evolution of the AGB to protoplanetary nebulae (PPN) to planetary nebula (PN) phase. Infrared pumping often leaves molecules in vibrationally excited states and Fronfría et al. (2006) have recently detected vibrationally excited HCN up to $v = 4$ in IRC+10216. Such emission is limited to within a few arcsec of the star.

2.2 Shocks and dust formation

AGB stars are a prolific source of dust particles but the process is difficult to model quantitatively. In the absence of pulsations, the r^{-2} density distribution means that gas does not necessarily spend enough time in a high density zone where collisions can be frequent enough to allow dust particles to grow efficiently. Dust nucleation requires a cooling flow, high collisional rates and time. Periodic shock waves, caused by the pulsations of the central star allow the gas to be heated and compressed thus reducing collisional time scales and allowing particles to nucleate in the cooling post-shock gas. Moreover, the mass-loss process is not efficient until the grains have grown to the point where they can absorb enough of the stellar radiative momentum. Before this stage is reached, an individual parcel of gas may experience several shock heating/cooling cycles each of which can affect its chemical composition, and in particular aid the growth of the seed molecules from which the dust grains grow. To give an idea of the parameter space, a 20 km s^{-1} shock at 1.2 stellar radii increases the gas temperature immediately to 4500 K in a region with gas density $\sim 6 \times 10^{15}$ cm^{-3}.

The spatial scale on which dust nucleates is around 5 stellar radii, a region which is again exposed to IR pumping. Vibrationally excited and masing molecules are natural probes. For example, Fronfría et al. (2006) have detected the thermal $v = 1$ transitions of SiS while observations of the $J = 5–4$ transition of SiO with the SMA have enabled Schöier et al. (2006) to show that the abundance of SiO is much more abundant at 3–8 stellar radii than its LTE abundance, indicating that it can be formed in the shock chemistry; at larger radii the SiO abundance decreases, most likely due to incorporation of SiO in to the grains, although other processes may be at work here.

Willacy and Cherchneff (1998) first investigated chemistry in periodic shocks and found that the process could be broken down into three major events. First, immediately post-shock where conditions are so extreme that H$_2$ and other molecules produced in LTE are completely destroyed. Second, the cooling post-shock gas in which H$_2$ and other

molecules can form. Third, a 'freeze-out' region in which the dynamical time is less than the chemical time so that chemical abundances get frozen in. Abundances at the end of one pulsation, when the gas has essentially relaxed back to its pre-shock conditions, can be much different from their LTE values. Pulsational shocks have been studied to see if they are the cause of some of the abundance anomalies, including the detection of water in the CSE of the carbon-rich star IRC+10216. Although no significant abundance of water is formed, other abundances can be different from LTE values by many orders of magnitude. Willacy and Cherchneff (1998) found that HCN and CS are destroyed and SiO enhanced by periodic shocks. Oxygen-rich stars are expected to form relatively few carbon-bearing molecules if LTE holds, CO being the notable exception. Duari et al. (1999) discussed pulsational shock chemistry in O-rich stars and found that HCN, by eight orders of magnitude, and CS and CO$_2$, both by three orders of magnitude, are enhanced over their LTE values. For S-type stars, that is C/O $\sim$ 1, Duari and Hatchell (2000) have shown that HCN is enhanced by some four orders of magnitude. Such a result is consistent with the detection of vibrationally excited HCN, pumped by IR radiation within 33 stellar radii in χ Cygni (Duari and Hatchell 2000).

In carbon-rich stars the growth of carbonaceous grains is thought to proceed through the build up of PAH particles with the abundant acetylene molecule acting as the feedstock for this growth. Cau (2002) has used the model developed by Frenklach and Fiegelson (1989) to discuss the growth of PAH molecules following a pulsational shock. Cau's model used the same chemical synthesis which produces the first ring molecule, benzene, to build bigger cyclic species, and found that a molecule with seven rings was formed very efficiently during one pulsational period. His model also allowed these ring molecules to associate with themselves to form dimers, again finding that dimer growth is efficient during one period. However, the yield of dimers produced, that is the abundance of carbon in dimers compared to the total abundance of all carbon, is not very large, certainly not large enough to provide seed particles for the growth of the carbonaceous grains observed in the CSE.

3 The photochemical region

The fact that AGB stars are cool means that the UV photons which affect the chemistry are external rather than internal. Since dust extinction is inversely proportional to radial distance from the star, UV photons photodissociate parent molecules at distances which are about 100 times larger than the pulsational/dust formation region. Little is known observationally about the intermediate region although there is some evidence, from the observations of SiO (Schöier

Fig. 1 The radial distribution of hydrocarbon chains

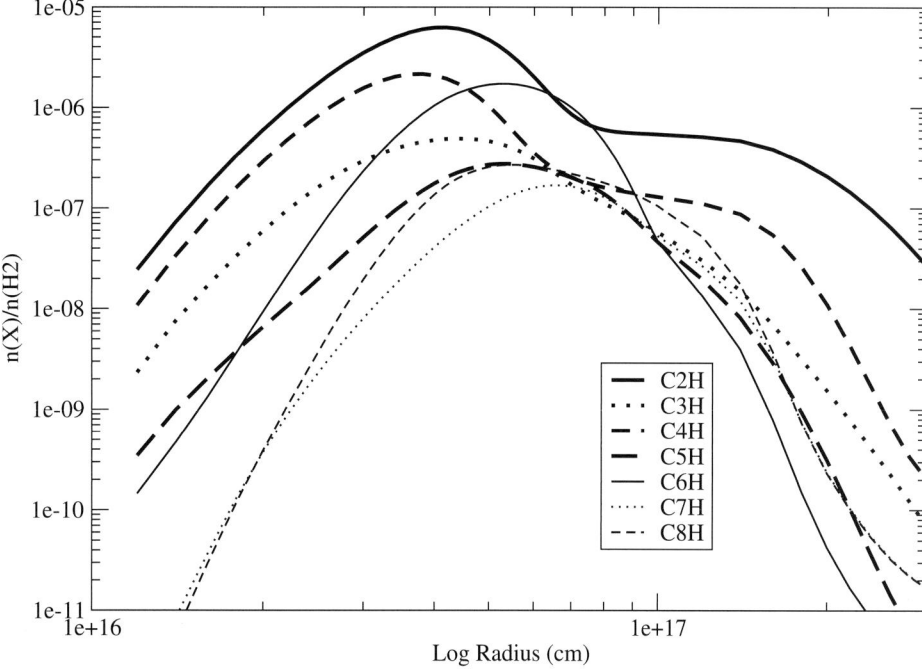

Fig. 2 The radial distribution of cyanopolyynes

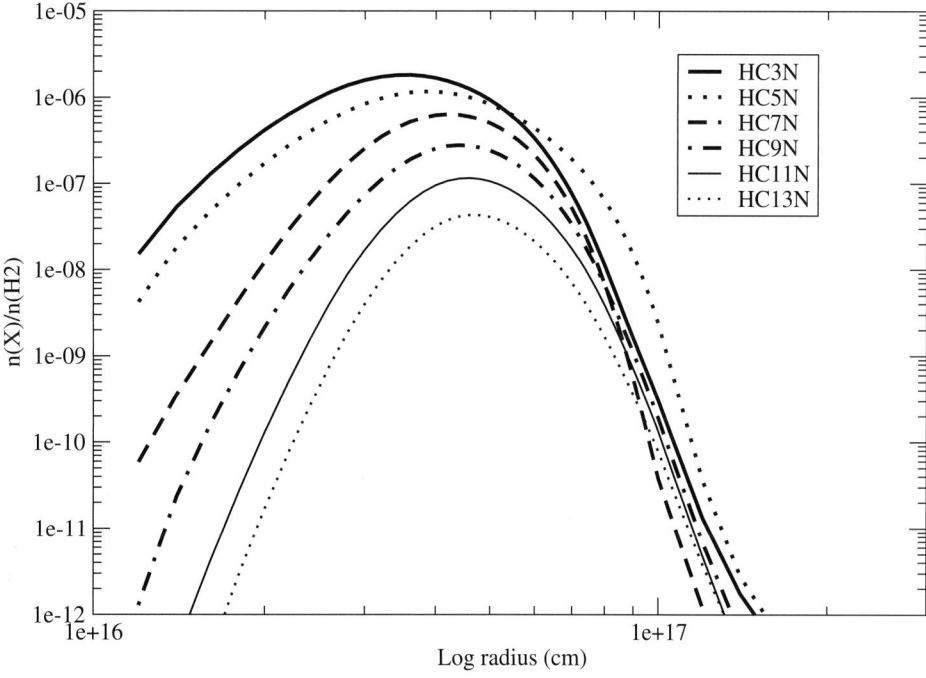

et al. 2006) discussed above, that the gas-grain interaction is important. This interaction may result in both the depletion of molecules, such as SiO, and to formation, particularly through the formation of hydrogenated species such as NH_3, CH_4, and C_2H_4. The most stable species produced in the LTE and dust-forming regions, H_2 and CO, are unaffected by the radiation field so that it is the photodissociation and photoionization of acetylene which produces the ions and radicals which drive an extensive organic chem-

istry through both ion-neutral and neutral-neutral reactions. Figure 1, which presents the radial distributions of some hydrocarbon species, shows that species with an even number of carbon atoms tend to be more abundant than those with an odd number, due to the importance of reactions involving C_2H and $C_2H_2^+$. Figure 2 shows the radial distributions of the cyanopolyyne species which are formed mainly in neutral-neutral reactions. We see that their abundances decrease as the chains get longer while there is a shift in the radius at which the peak abundances occur. Such a shift could be observed through interferometric observations; current observations show that the distributions of HC$_3$N and HC$_5$N are very similar (Guélin et al. 2000). ALMA will be able to investigate this with greater spatial resolution and extend the study to larger cyanopolyynes.

The recent detection of C$_6$H$^-$ in IRC+10216 by Mc-Carthy et al. (2006), was predicted implicitly by the calculations of Millar et al. (2000) who predicted anion/neutral ratios of between 0.01 and 0.4 based on efficient formation through radiative electron attachment, which occurs when the neutral has six or more atoms, and destruction mainly through photodissociation. Detailed calculations by Millar et al. (2007) show that anions as small as C$_4$H$^-$ should be detectable in IRC+10216. Despite the fact that the radiative attachment of electrons is only about 1% efficient, the large column density of C$_4$H in IRC+10216 means that the column density of C$_4$H$^-$ is appreciable.

4 Metals

A large number of metal-containing molecules have been detected in IRC+10216 and other more evolved objects. The LTE calculations predict metal halides should be present and NaCl, KCl, AlCl and AlF have been observed with distributions which peak on the star and have spatial extents of a few arcsec. Fractional abundances can be large—Highberger et al. (2001) and Cernicharo and Guélin (1987) find AlF/H$_2$ and AlCl/H$_2$ $\sim 10^{-7}$, which implies that a large fraction of elemental aluminum, which is depleted by several orders of magnitude in diffuse interstellar clouds, presumably in refractory grain cores, is not incorporated into dust grains.

ALMA will be able to probe this central region very effectively and test LTE models, test the pulsational shock models and the growth of carbon chains, rings and PAH molecules in the dust-forming region through study of their high frequency ro-vibrational transitions.

Other metal molecules have a shell-like distribution on a scale of about 15 arcsec, very similar to those found for the cyanopolyynes, indicating that they are formed by chemistry, either in the gas-phase or as the result of some desorption process from the dust grains. The molecules tend to be cyanides or isocyanides and include MgCN, MgNC, AlNC, SiCN and SiNC. The thickness of the emission is only a few arcsec and ALMA will be ideally suited to their study. Careful analysis should enable us to determine the fraction of metals which have never been incorporated into the dust grains, and thereby constrain models of grain nucleation and growth, or the fraction of metals which have been released from dust grains in the outer envelope, which may impact our understanding of grain erosion in the interstellar medium.

In order to see if the metal cyanides might be formed by gas-phase chemistry rather than release from dust grains, we have performed some test calculations using the chemical scheme proposed by Dunbar and Petrie (2002) in which MgNC is formed by the radiative association of Mg$^+$ with the cyanopolyynes:

$$Mg^+ + HC_nN \longrightarrow HC_nNMg^+ + h\nu$$

where $n = 3, 5, 7, \ldots$ followed by dissociative recombination with electrons. Figure 3 shows the fractional abundance of MgNC which results. For an initial fractional abundance

Fig. 3 The radial distribution of MgNC

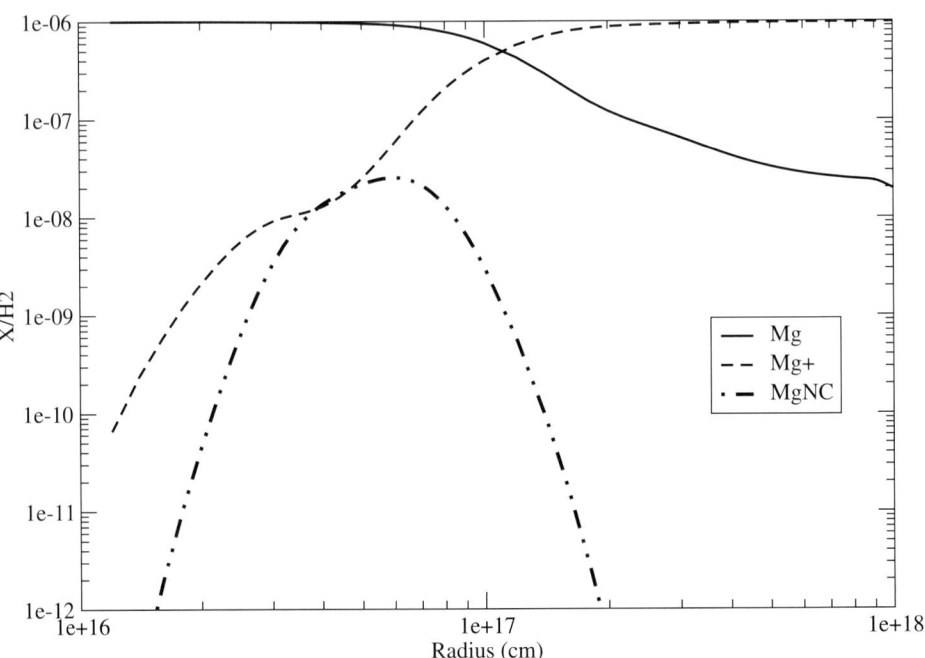

of Mg of 10^{-6}, the MgNC column density calculated is 1.1×10^{13} cm^{-2}, compared to $\sim 2 \times 10^{13}$ cm^{-2} observed for each of MgNC and MgCN. The dominant reaction is actually with HC$_7$N which has an association rate about four orders of magnitude larger than that of HC$_3$N at low temperatures. The initial Mg abundance adopted is very large, about 3% of the cosmic abundance and may be unrealistic but the result, since it depends on the presence of Mg$^+$ in the outer envelope, will hold as long as this fractional abundance of Mg is contained in parent gas-phase species.

Much of the molecular emission detected in the outer envelope appears to correlate well with the dust arcs and rings detected by Mauron and Huggins (2000); ALMA will give a definitive answer. These arcs have density enhancements on the order of three and are thin with widths ranging form 0.5 to 3 arcsec, corresponding to time-scales of a few tens of years with the total radial extent of the arcs corresponding to around a thousand years. These structures may play an important role in the chemistry in the outer CSE since they not only provide additional extinction against the destructive radiation field but they also provide a higher collisional rate under which chemistry can proceed. These arcs of higher than average density are thought to have been formed through helium shell flashes. ALMA will be able to probe the isotopic abundance of these metal cyanides and therefore trace the nuclear chemistry in the star as a function of time.

5 Conclusions

ALMA will greatly aid the study of AGB stars. In particular, its sensitivity will allow us to investigate objects other than IRC+10216 in some depth and to look for similarities and differences. IRC+10216 will remain important as the closest AGB star to earth and we can expect a host of data on small-scale structure, including some hints as to how very non-spherical objects such as PPN and PN evolve from spherical AGB stars. Studies of the LTE region and pulsational shocks will enable us to tie down both the role of pulsations in creating the conditions under which dust grains nucleate and grow and the particular chemical intermediaries which promote this growth. While we expect

this to demonstrate the importance of hydrocarbon species in C-rich objects, we do not know the molecular precursors to dust in O-rich stars; ALMA will be able to tell us. A very exciting opportunity exists to probe nucleosynthesis in AGB stars as a function of time through observations of isotopic species as a function of radial distance in CSEs. Finally, the importance of the gas-grain interaction in the outer CSE is poorly understood. ALMA should be able to determine whether grains act as a source or sink of molecules, a role which may change with position in the envelope. If the interaction of grains with the external radiation field and/or cosmic-ray particles does provide a source of gas-phase species, it will have profound implications for the transport of complex organic molecules from AGB stars to interstellar clouds.

Acknowledgements Astrophysics at Queen's University Belfast is supported by a grant from PPARC.

References

Cau, P.: Astron. Astrophys. **392**, 203 (2002)

Cernicharo, J., Guélin, M.: Astron. Astrophys. **183**, L10 (1987)

Duari, D., Hatchell, J.: Astron. Astrophys. **358**, L25 (2000)

Duari, D., Cherchneff, I., Willacy, K.: Astron. Astrophys. **341**, L47 (1999)

Dunbar, R.C., Petrie, S.: Astrophys. J. **564**, 792 (2002)

Frenklach, M., Fiegelson, D.: Astrophys. J. **341**, 372 (1989)

Fronfría Expósito, J.P., Agúndez, M., Tercero, B., Pardo, J.P., Cernicharo, J.: Astrophys. J. **646**, L127 (2006)

Guélin, M., Lucas, R., Neri, R., Bremer, M., Broguiere, D.: In: Minh, Y.C., van Dishoeck, E.F. (eds.) Astrochemistry: From Molecular Clouds to Planetary Systems, p. 365. Kluwer, Dordrecht (2000)

Highberger, J.L., Savage, C., Bieging, J.H., Ziurys, L.M.: Astrophys. J. **562**, 790 (2001)

Mauron, N., Huggins, P.J.: Astron. Astrophys. **359**, 707 (2000)

McCarthy, M.C., Gottleib, C.A., Gupta, H., Thaddeus, P.: Astrophys. J. **652**, L141 (2006)

Millar, T.J., Herbst, E., Bettens, R.P.A.: Mon. Not. R. Astron. Soc. **316**, 195 (2000)

Millar, T.J., Walsh, C., Cordiner, M., Ní Chuímin, R., Herbst, E.: Astrophys. J. **662**, L87 (2007)

Schöier, F.L., Fong, D., Olofsson, H., Zhang, Q., Patel, N.: Astrophys. J. **649**, 965 (2006)

Tsuji, T.: Astron. Astrophys. **23**, 411 (1973)

Willacy, K., Cherchneff, I.: Astron. Astrophys. **330**, 676 (1998)

Understanding the chemical complexity in Circumstellar Envelopes of C-Rich AGB stars: the case of IRC +10216

M. Agúndez · J. Cernicharo · J.R. Pardo ·
J.P. Fonfría Expósito · M. Guélin · E.D. Tenenbaum ·
L.M. Ziurys · A.J. Apponi

Originally published in the journal Astrophysics and Space Science, Volume 313, Nos 1–3.
DOI: 10.1007/s10509-007-9495-7 © Springer Science+Business Media B.V. 2007

Abstract The circumstellar envelopes of carbon-rich AGB stars show a chemical complexity that is exemplified by the prototypical object IRC +10216, in which about 60 different molecules have been detected to date. Most of these species are carbon chains of the type C_nH, C_nH_2, C_nN, HC_nN. We present the detection of new species (CH_2CHCN, CH_2CN, H_2CS, CH_3CCH and C_3O) achieved thanks to the systematic observation of the full 3 mm window with the IRAM 30m telescope plus some ARO 12m observations. All these species, known to exist in the interstellar medium, are detected for the first time in a circumstellar envelope around an AGB star. These five molecules are most likely formed in the outer expanding envelope rather than in the stellar photosphere. A pure gas phase chemical model of the circumstellar envelope is reasonably successful in explaining the derived abundances, and additionally allows to elucidate the chemical formation routes and to predict the spatial distribution of the detected species.

Keywords Astrochemistry · Circumstellar matter · Molecular processes · AGB stars · IRC +10216

M. Agúndez (✉) · J. Cernicharo · J.R. Pardo ·
J.P. Fonfría Expósito
Departamento de Astrofísica Molecular e Infrarroja, Instituto de Estructura de la Materia, CSIC, Serrano 121, 28006 Madrid, Spain
e-mail: marce@damir.iem.csic.es

M. Guélin
Institut de Radioastronomie Millimétrique, 300 rue de la Piscine, 38406 St. Martin d'Heres, France

E.D. Tenenbaum · L.M. Ziurys · A.J. Apponi
Departments of Chemistry and Astronomy, University of Arizona, 933 North Cherry Avenue, Tucson, AZ 85721, USA

1 Introduction

IRC +10216 was discovered in the late sixties as an extremely bright object in the mid infrared (Becklin et al. 1969). Since then and with the development of radioastronomy during the seventies it was recognized as one of the richest molecular sources in the sky together with some others such as the Orion nebula, Sagittarius-B2 and the Taurus molecular cloud complex.

IRC +10216 consists of a central carbon-rich AGB star, i.e. C/O>1 in the photosphere, losing mass at a rate of $2-4 \times 10^{-5}$ $M_\odot$ yr^{-1} in the form of a quasi-spherical wind that produces an extended circumstellar envelope (CSE) from which most of the molecular emission arises. To date, some 60 different molecules have been detected in this source, most of which are organic molecules consisting of a linear and highly unsaturated backbone of carbon atoms. Among these species there are cyanopolyynes ($HC_{2n+1}N$) and their radicals ($C_{2n+1}N$), polyyne radicals (C_nH), carbenes (H_2C_n), radicals ($HC_{2n}N$) as well as S-bearing (C_nS) and Si-bearing (SiC_n) species (Glassgold 1996; Cernicharo et al. 2000).

It is nowadays accepted that the formation of these molecules occurs either under chemical equilibrium in the dense ($n > 10^{10}$ cm^{-3}) and hot ($T_k \sim 2000$ K) vicinity of the stellar photosphere or in the colder and less dense outer envelope when the interstellar UV field photodissociate/ionize the molecules flowing out from the star producing radicals/ions which undergo rapid neutral-neutral and ion-molecule reactions (Lafont et al. 1982; Millar et al. 2000). This picture of circumstellar photochemistry resembles that occurring in cold dense clouds ($T_k = 10$ K, $n \sim 10^4$ cm^{-3}) such as TMC-1 (Kaifu et al. 2004). In both places organic molecules are mostly unsaturated, which is typical of low temperature non-equilibrium chemistry and simply reflects the trend

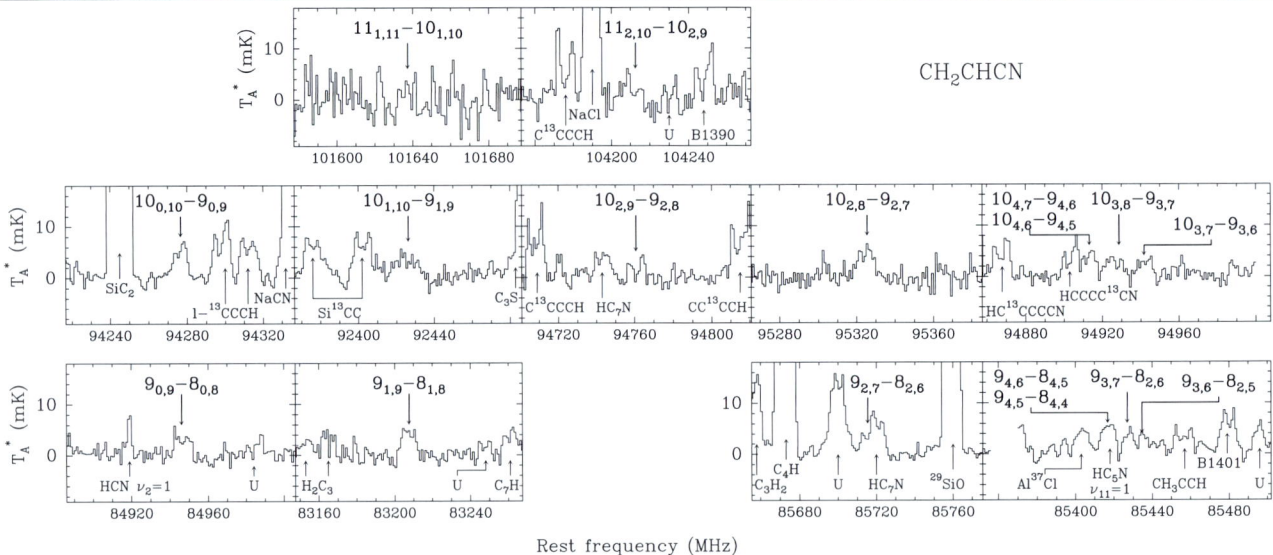

Fig. 1 Rotational lines of CH$_2$CHCN detected toward IRC +10216 with the IRAM 30m telescope

of gas phase bimolecular reactions in ejecting an hydrogen atom and the low reactivity of H$_2$ with most hydrocarbons.

In this contribution we report on the detection toward IRC +10216 of the partially saturated C-bearing species CH$_2$CHCN, CH$_2$CN and CH$_3$CCH; the S-bearing molecule H$_2$CS and the oxygen-carbon chain C$_3$O. All these species are known to exist in cold dense clouds (Matthews and Sears 1983; Irvine et al. 1988, 1981, 1989; Brown et al. 1985). Thus, their detection in IRC +10216 stress the similarity between the chemistry taking place in cold dense clouds and in the CSEs of C-rich AGB stars.

2 Observations

The observations of CH$_2$CHCN, CH$_2$CN, H$_2$CS and CH$_3$CCH were achieved with the IRAM 30m telescope (see e.g. Fig. 1) while C$_3$O was observed with both the IRAM 30m and ARO 12m telescopes.

The IRAM 30m observations were carried out during several sessions from 1990 to 2005, most of them after 2002 in the context of a λ 3 mm line survey of IRC+10216 from 80 to 115.75 GHz (Cernicharo et al., in preparation). Two 3 mm SIS receivers with orthogonal polarizations were used in single-sideband mode with image sideband rejections >20 dB. The standard wobbler switching mode was used with an offset of 4′. The back end was a 512 two-pole filter with half-power widths and spacings equal to 1.0 MHz.

The ARO 12m observations were done in several runs between 2003 and 2005. The receivers were dual-channel cooled SIS mixers at 2 and 3 mm, operated in single-sideband mode with ∼18 dB rejection of the image sideband. The back ends were two 256 channel filter banks

with 1 and 2 MHz resolutions, configured in parallel mode (2 × 128 channels) for the two receiver channels. A millimeter auto correlator spectrometer with 2048 channels of 768 kHz resolution was operated simultaneously to confirm features seen in the filter banks. Data were taken in beam switching mode with ±2′ subreflector throw.

3 Molecular column densities

The number of CH$_2$CHCN, CH$_2$CN and C$_3$O lines detected was large enough to allow us to construct rotational temperature diagrams (see e.g. Fig. 2). The rotational temperatures derived (see Table 1) are within the range of other shell distributed molecules in IRC +10216: 20–50 K (Cernicharo et al. 2000). For CH$_3$CCH and H$_2$CS we observed only a few transitions with similar upper level energies and it was not possible to constrain the rotational temperature which was assumed to be 30 K.

In Table 1 we give the column densities[1] (averaged over the IRAM 30m beam, $\theta_{MB} = 21$–$31''$ at λ 3 mm) of the species detected for the first time (in bold) as well as values or upper limits derived for other related species. The column densities of the new species are in the range 10^{12}–10^{13} cm^{-2}. Note for example that both CH$_3$CN and CH$_3$CCH have very similar column densities although the lines of CH$_3$CCH are some 30 times weaker than those of CH$_3$CN mostly because the different electric dipole moments (0.780 D vs. 3.925 D). The two related species

---

[1] For CH$_2$CN and H$_2$CS, the two molecules which have two interchangeable nuclei with non zero spin, we have assumed an ortho-to-para ratio 3:1 when deriving their column densities.

Table 1 Column densities of some molecules in IRC +10216

Molecule	T_{rot} (K)	N_{tot} (cm^{-2}) Observed		Calculated This work	MI00
CH$_2$CHCN	44	5.1(12)		1.4(13)	2.2(11)
CH$_3$CN	40	3.0(13)		6.3(12)	6.8(12)
CH$_2$CN	49	8.4(12)		4.6(12)	1.4(13)
CH$_3$C$_3$N	30^a	<1.3(12)		1.2(11)	1.4(12)
CH$_3$CCH	30^a	1.6(13)		1.1(12)	8.0(12)
CH$_3$C$_4$H	30^a	<9.7(12)		8.2(12)	9.0(12)
H$_2$CO	28	5.4(12)	(Ford et al. 2004)	2.8(12)	–
H$_2$CS	30^a	1.0(13)		1.3(12)	4.4(11)
C$_2$O	30^a	<7.0(12)	(Tenenbaum et al. 2006)	7.0(11)	–
C$_3$O	27	2.6(12)	(Tenenbaum et al. 2006)	2.8(12)	–
C$_4$O	30^a	<6.0(12)	(Tenenbaum et al. 2006)	–	–
C$_5$O	30^a	<3.0(12)	(Tenenbaum et al. 2006)	–	–

Notes: $a(b)$ refers to $a \times 10^b$. The total column density across the source N_{tot} is twice the radial column density N_{rad}. A superscript "a" indicates an assumed value.

References: (MI00) Chemical model of Millar et al. (2000)

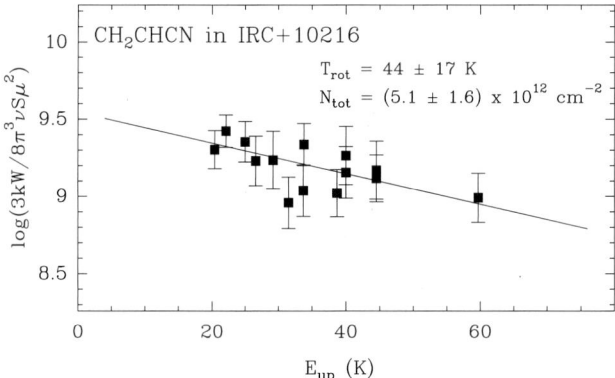

Fig. 2 Rotational temperature diagram of CH$_2$CHCN

CH$_3$CN and CH$_2$CN have similar column densities which may indicate a common chemical origin (see Sect. 4). The carbon-rich nature of IRC +10216 makes oxygen-bearing species to have a low abundance. For example thioformaldehyde is more abundant than formaldehyde despite the cosmic abundance of oxygen being 50 times larger than that of sulphur.

4 Molecular synthesis in the outer envelope

In order to explain how the detected species are formed we have performed a detailed chemical model of the outer envelope. The chemical network consists of 385 gas phase species linked by 6547 reactions, whose rate constants have been taken from the astrochemical databases UMIST99 (Le Teuff et al. 2000) and osu.2003 (Smith et al. 2004), re-

cently revised.[2] The temperature and density radial profiles as well as other physical parameters are taken from (Agúndez and Cernicharo 2006). The resulting abundance radial profiles for CH$_2$CHCN, CH$_2$CN, CH$_3$CCH, H$_2$CS and related species are plotted in Fig. 3. The model predicts that these four molecules together with C$_3$O (not plotted in Fig. 3, see Tenenbaum et al. (2006) for a detailed discussion) are formed with an extended shell-type distribution ($r \sim 20''$) via several gas phase reactions.

Vinyl cyanide (CH$_2$CHCN) is formed by the reaction between CN and ethylene (C$_2$H$_4$), which is most likely the main formation route in dark clouds (Herbst and Leung 1990). Fortunately, the reaction has been studied in the laboratory and has been found to be very rapid at low temperature (Sims et al. 1993) and to produce vinyl cyanide (Choi et al. 2004). The predicted column density agrees reasonably well with the observational value.

In our model both CH$_2$CN and CH$_3$CN are mostly formed (>90%) through the dissociative recombination (DR) of CH$_3$CNH$^+$

$$CH^+ \xrightarrow{H_2} CH_2^+ \xrightarrow{H_2} CH_3^+ \xrightarrow{HCN} CH_3CNH^+ \xrightarrow{e^-} CH_2CN$$
$$\xrightarrow{e^-} CH_3CN$$

whereas the major destruction process (>90 %) for both species is photodissociation. The branching ratios in the DR of CH$_3$CNH$^+$ are not known and are assumed to be equal. Assuming, as we do, that the photodissociation rate of CH$_3$CN and CH$_2$CN are equal, the observed CH$_3$CN/CH$_2$CN ratio suggests branching ratios in the DR

[2]See http://www.udfa.net and http://www.physics.ohio-state.edu/~eric/research.html.

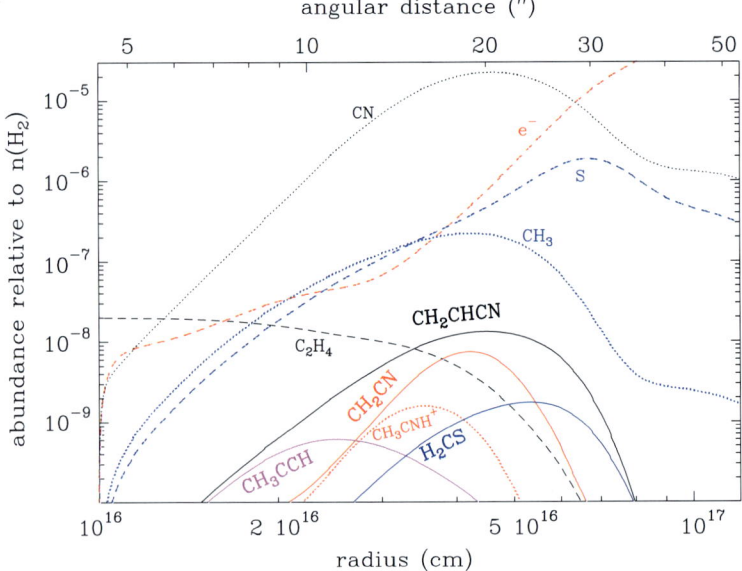

Fig. 3 Abundances of CH$_2$CHCN, CH$_2$CN, CH$_3$CCH and H$_2$CS (*solid lines*) and related species (*dotted and dashed lines*) given by the chemical model, as a function of radius (*bottom axis*) and angular distance (*top axis*) for an assumed stellar distance of 150 pc

of CH$_3$CNH$^+$ of 0.8 for the (CH$_3$CN + H) channel and 0.2 for (CH$_2$CN + H$_2$ or 2H). This estimate will be strongly affected if the photodissociation rates of CH$_3$CN and CH$_2$CN are very different but not if, as has been suggested (Herbst and Leung 1990; Turner et al. 1990), CH$_2$CN does indeed react with atomic oxygen, the abundance of which is too low at the radius where CH$_2$CN is present.

The synthesis of CH$_3$CCH involves ion-molecule reactions with the dissociative recombination of the C$_3$H$_5^+$ and C$_4$H$_5^+$ ions as the last step. The model underproduces it by an order of magnitude, probably due to uncertainties and/or incompleteness in the chemical network, which affect the formation rate of the last step species C$_3$H$_5^+$ and C$_4$H$_5^+$. The heavier chain CH$_3$C$_4$H is also predicted with a column density even higher than that of CH$_3$CCH. The larger rotational partition function works against lines being detectable, but the higher dipole moment (1.21 D vs. 0.78 D) could result in line intensities similar to those of CH$_3$CCH.

Thioformaldehyde is formed by the reaction S + CH$_3$ and in less extent through the DR of H$_3$CS$^+$ (see Agúndez and Cernicharo 2006 for a detailed discussion). The order of magnitude of discrepancy between model and observations reduces to a factor 4 with further non-local non-LTE radiative transfer calculations. We note, however, that a significant fraction of both H$_2$CO and H$_2$CS could be formed in grain surfaces by hydrogenation of CO and CS respectively.

The detection of C$_3$O in IRC +10216 (Tenenbaum et al. 2006), only previously detected in the dark clouds TMC-1 (Brown et al. 1985) and Elias 18 (Trigilio et al. 2007), stress both the similarity between dark clouds and C-rich CSEs chemistries and also the non-negligible oxygen chemistry taking place in C-rich CSEs. Actually astrochemical networks consider that C$_3$O is formed through DR of the molecular ions H$_n$C$_m$O$^+$ (see Fig. 4). However, the C$_3$O

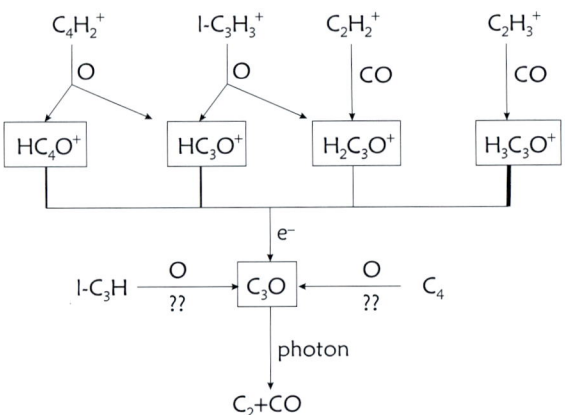

Fig. 4 Scheme showing the main chemical formation routes to C$_3$O. From Tenenbaum et al. (2006)

column density derived in IRC +10216 is an order of magnitude higher than calculated which could imply additional chemical routes for its formation, e.g. neutral-neutral reactions of atomic oxygen with carbon chain radicals such as those suggested in Fig. 4.

Acknowledgements We would like to thank the IRAM 30m telescope staff for their assistance during the observations. MA acknowledges a grant from Spanish MEC: AP2003-4619.

References

Agúndez, M., Cernicharo, J.: Oxygen chemistry in the circumstellar envelope of the carbon-rich star IRC +10216. Astrophys. J. **650**, 374 (2006)

Becklin, E.E., Frogel, J.A., Hyland, A.R., Kristian, J., Neugebauer, G.: The unusual infrared object IRC +10216. Astrophys. J. Lett. **158**, L133 (1969)

Brown, R.D., et al.: Tricarbon monoxide in TMC-1. Astrophys. J. **297**, 302 (1985)

Cernicharo, J., Guélin, M., Kahane, C.: A λ 2 mm molecular line survey of the C-star envelope IRC+10216. Astron. Astrophys. Suppl. Ser. **142**, 181 (2000)

Choi, N., Blitz, M.A., McKee, K., Pilling, M.J., Seakins, P.W.: H atom branching ratios from the reactions of CN radicals with C_2H_2 and C_2H_4. Chem. Phys. Lett. **384**, 68 (2004)

Ford, K.E.S., Neufeld, D.A., Schilke, P., Melnick, G.J.: Detection of formaldehyde toward the extreme carbon star IRC +10216. Astrophys. J. **614**, 990 (2004)

Glassgold, A.E.: Circumstellar photochemistry. Annu. Rev. Astron. Astrophys. **34**, 241 (1996)

Herbst, E., Leung, C.M.: The gas phase production of CH_2CN and other organo-nitrogen species in dense interstellar clouds. Astron. Astrophys. **233**, 177 (1990)

Irvine, W.M., et al.: The increasing chemical complexity of the Taurus dark clouds—detection of CH_3CCH and C_4H. Astrophys. J. Lett. **248**, 113 (1981)

Irvine, W.M., et al.: Identification of the interstellar cyanomethyl radical (CH_2CN) in the molecular clouds TMC-1 and Sagittarius B2. Astrophys. J. Lett. **334**, 107 (1988)

Irvine, W.M., et al.: Observations of some oxygen-containing and sulfur-containing organic molecules in cold dark clouds. Astrophys. J. **342**, 871 (1989)

Kaifu, N., et al.: A 8.8-50 GHz complete spectral line survey toward TMC-1 I. Survey data. Publ. Astron. Soc. Jpn. **56**, 69 (2004)

Lafont, S., Lucas, R., Omont, A.: Molecular abundances in IRC +10216. Astron. Astrophys. **106**, 201 (1982)

Le Teuff, Y.H., Millar, T.J., Marckwick, A.J.: The UMIST database for astrochemistry 1999. Astron. Astrophys. Suppl. Ser. **146**, 157 (2000)

Matthews, H.E., Sears, T.J.: The detection of vinyl cyanide in TMC-1. Astrophys. J. **272**, 149 (1983)

Millar, T.J., Herbst, E., Bettens, R.P.A.: Large molecules in the envelope surrounding IRC +10216. Mon. Not. Roy. Astron. Soc. **316**, 195 (2000)

Sims, I.R., Queffelec, J.-L., Travers, D., Rowe, B.R., Herbert, L.B., Karthauser, J., Smith, I.W.M.: Rate constants for the reactions of CN with hydrocarbons at low and ultra-low temperatures. Chem. Phys. Lett. **211**, 461 (1993)

Smith, I.W.M., Herbst, E., Chang, Q.: Rapid neutral-neutral reactions at low temperatures: a new network and first results for TMC-1. Mon. Not. Roy. Astron. Soc. **350**, 323 (2004)

Tenenbaum, E.D., Apponi, A.J., Ziurys, L.M., Agúndez, M., Cernicharo, J., Pardo, J.R., Guélin, M.: Detection of C_3O in IRC +10216: oxygen-carbon chemistry in the outer envelope. Astrophys. J. Lett. **649**, 17 (2006)

Trigilio, C., Palumbo, M.E., Siringo, C., Leto, P.: Science with ALMA: a new era for astrophysics. Springer Astrophysics and Space Science, Madrid (2007)

Turner, B.E., Friberg, P., Irvine, W.M., Saito, S., Yamamoto, S.: Interstellar cyanomethane. Astrophys. J. **355**, 546 (1990)

New Plateau de Bure observations of M 1–92; unveiling the core

Javier Alcolea · Valentín Bujarrabal · Roberto Neri

Originally published in the journal Astrophysics and Space Science, Volume 313, Nos 1–3.
DOI: 10.1007/s10509-007-9621-6 © Springer Science+Business Media B.V. 2007

Abstract M 1–92 is a very well studied bipolar pPN that can be considered an archetype of this type of sources; it shows a clear axial symmetry, along with the kinematics characteristic of this class of envelopes around post-AGB stars. We performed sub-arcsecond resolution observations of the $J = 2-1$ rotational line of ^{13}CO in M 1–92 with the new extended configurations of the IRAM Plateau de Bure interferometer, for studying the morphology and velocity field of the molecular gas better in the nebula, particularly in its central parts. We found that the equatorial structure dividing the two lobes is a thin flat disk, which expands radially with a velocity proportional to the distance to the central stellar system. The kinetic age of this equatorial flow is very similar to that measured in the two lobes, suggesting that the whole structure was formed as a result of a single event some 1200 yr ago, after which the nebula reached an expansion velocity field with axial symmetry. The small widths and velocity dispersion in the gas forming the lobe walls confirm

Based on observations carried out with the IRAM Plateau de Bure Interferometer. IRAM is supported by INSU/CNRS (France), MPG (Germany) and IGN (Spain).

J. Alcolea (✉)
Observatorio Astronómico Nacional (OAN-IGN), C/ Alfonso XII No. 3, 28014 Madrid, Spain
e-mail: j.alcolea@oan.es

V. Bujarrabal
Observatorio Astronómico Nacional (OAN-IGN), Apartado 112, 28803 Alcalá de Henares, Spain
e-mail: v.bujarrabal@oan.es

R. Neri
Institut de Radio Astronomie Millimétrique (IRAM), 300 Rue de la Piscine, 38406 St.-Martin d'Hères, France
e-mail: neri@iram.fr

that the acceleration responsible for the nebular shape could not last more than 100–120 yr. In view of the similarity to η Car, we speculate on the possibility that the whole nebula was formed as a result of a magneto-rotational explosion in a common-envelope system. The study of the possible importance of this mechanism in the context of global PNe and pPNe reshaping should be one on the fields in which future ALMA observations will make a crucial contribution.

Keywords Circumstellar matter · Post-AGB stars · M 1–92

PACS 97.10.Fy · 97.10.Me · 95.75.Kk

1 Introduction

Planetary nebulae (PNe) are among the most bizarre structures found in space; in practice no two are alike (Balick and Frank 2002). Yet these nebulae are the result of the post-AGB evolution of the circumstellar envelopes (CSEs) that surround giant stars while they are at the AGB (see Olofsson, this volume). How the variety of complex and intricate shapes of PNe can arise out of the nicely spherical CSEs is one of the great mysteries of modern Astrophysics. To better understand this metamorphosis we should have a look at the so called pre-planetary nebulae (pPNe), envelopes around post-AGB stars that are in their way to become a PN but have not yet reach that phase. Today we know that at these early stages in the post-AGB evolution, the envelopes show clear departures from sphericity and very energetic dynamics (Bujarrabal 2007).

M 1–92, also known as Minkowski's Footprint, is a bipolar pPNe that has been very well studied in the optical, infrared and radio wavelengths. At optical wavelengths, it consists in a two-lobe reflection nebula. These two lobes define a clear axis of symmetry oriented at a position angle

(PA) of 311°, and are divided by a dusty obscuring equatorial structure that partially hides the central star and the SE lobe, the one that is receding from us (Minkowski 1946). In the middle of each of the two lobes, atomic line emission in Hα and some forbidden lines reveal the presence of shocks (Solf 1994; Bujarrabal et al. 1998b), which propagate along a fast and tenuous post-AGB bipolar jet that runs along the symmetry axis of the nebula (Solf 1994; Arrieta et al. 2005). The star is not directly seen but its dust reflected spectrum shows a composite nature that has been attributed to the presence of a binary system (Arrieta et al. 2005). The distance to M 1–92 has been estimated as 2.5 kpc, adopting a standard luminosity of 10^4 $L_\odot$ for a post-AGB star (Cohen and Kuhi 1977).

However, the best tracer of the bulk of the nebula is CO since most of the material is still in the form of molecular gas. This emission has been previously mapped in detail by us. In particular we observed the optically thin line ^{13}CO $J = 2$–1 using the IRAM Plateau de Bure Interferometer, obtaining a spatial resolution of nearly $1''$ (Bujarrabal et al. 1998a). The total molecular gas detected in M 1–92 is $0.9\,M_\odot$. These observations revealed that the two lobes are in fact two cavities enclosed by a thin layer of gas (and dust, where the stellar light is scattered), divided by a slightly denser equatorial structure. As for the velocity field, we concluded that in the lobes the gas presents a linear velocity gradient, i.e. the farther from the center the gas is located the higher the expansion velocity it shows. Assuming that this Hubble-like velocity field is the result of a very brief acceleration, we derived a kinetic age for these structures of about 1000 yr. For the equatorial component, the maps were not very detailed because the spatial resolution was comparable to its diameter of $2''$, and we just concluded that the observations were compatible with a disk with a CSE-like constant velocity expansion.

In this contribution we report on new Plateau de Bure observations of higher spatial resolution aimed to better study the structure of M 1–92, especially in its central parts.

2 Observations and results

The new observations of ^{13}CO $J = 2$–1 in M 1–92 were conducted in Winter 2006, using the new Plateau de Bure six-antenna layouts code-named A & B (the most extended ones), which provided projected baselines up to 760 m. The correlator was configured to observe the spectral line with a velocity resolution of $0.2\,\mathrm{km\,s}^{-1}$. In addition to this line, we simultaneously observed the $J = 1$–0 transitions of ^{13}CO and C^{18}O, and the continuum emission at 2.6 and 1.3 mm, but these results will be discussed elsewhere.

The ^{13}CO $J = 2$–1 visibilities were calibrated and the images were generated in the standard way for Plateau de

Bure observations, see Alcolea et al. for additional information (Alcolea et al. 2007). The integrated spectrum was compared to those previously obtained by us with the Plateau de Bure and Pico de Veleta telescopes (Bujarrabal et al. 1998a). We found that in the new images there is a significant fraction of lost flux (over-resolved by the instrument), as high as 60% at some velocities. Therefore, to obtain images containing all the emission but still having better spatial resolution, we decided to merge the new data with those from our 1995–1997 observations (Bujarrabal et al. 1998a), using robust weighting. In the resulting maps we have recovered all the flux as before but achieving a spatial resolution much better than in our previous paper: the resulting clean beam has a HPBW of $0''\!.51 \times 0''\!.35$ (major axis at P.A. = 36°). The data set was also re-sampled to a resolution of $3.25\,\mathrm{km\,s}^{-1}$ to increase the S/N and to ease the comparison to our previously published maps of this source (Bujarrabal et al. 1998a).

The new velocity maps and position vs. velocity diagrams confirmed our previous model for the molecular gas. The two emptied bubbles divided by an equatorial thin disk can be identified even more clearly (see Fig. 1 and Figs. 1 and 2 in Alcolea et al. 2007). Thanks to the new improved resolution, we are able to measure for the first time the width and velocity dispersion in the walls of the two emptied lobes. We find that these figures are typically $0''\!.25$ (9×10^{15} cm for the assumed distance to the source) and less than $3\,\mathrm{km\,s}^{-1}$, respectively. These numbers are about 10%–20% of the values of the diameter of the lobes and the typical expansion velocity of the gas, indicating that whatever process was responsible for the acceleration of the nebula along its symmetry axis, it should have lasted much less than the kinetic age of the lobes themselves; i.e. we have an upper limit for the duration of the forces of 100–200 yr. We also note that the boundaries of the lobes are much better defined in their inner side than outside, where a sort of halo is detected, particularly at low expansion velocities.

However the most important result is that with the new improved resolution, the kinematics of the equatorial disk is not compatible with our previous assumptions. The observed maps do not correspond to a thin disk, radially expanding at constant velocity, as it should happen for a equatorial undisturb remnant of a standard CSE. For example the maximum emission occurs at the center of nebula, and at the systemic velocity. This rather surprising finding is confirmed in images with still higher spatial resolution, that were obtained using only data from the most extended configuration. In these images we have achieved a spatial resolution of $0''\!.46 \times 0''\!.26$ (major axis at P.A. = 25°), and a velocity resolution of just $1\,\mathrm{km\,s}^{-1}$. We find that for the central structure, the emission at the different velocities traces a set of parallel strips, perpendicular to the symmetry axis of the nebula (Alcolea et al. 2007). These strips show a linear gradient too, but of opposite sign with respect to the

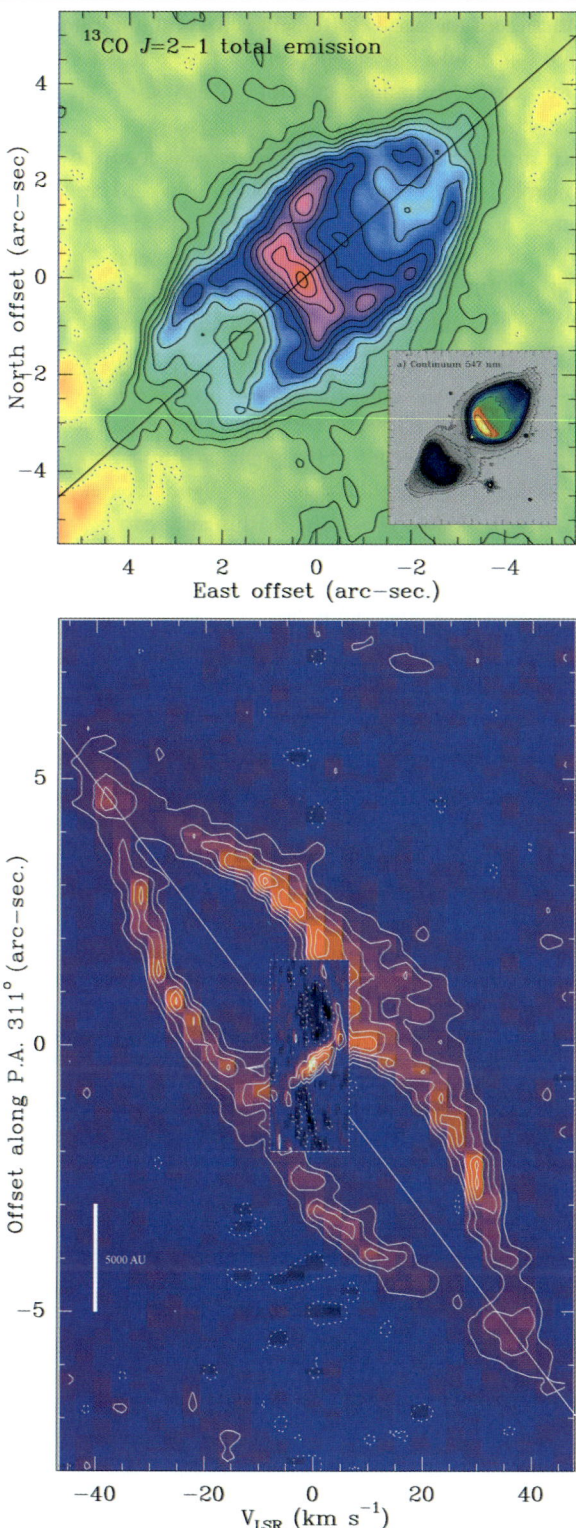

Fig. 1 *Top panel.* Integrated intensity map of ^{13}CO $J = 2$–1 in M 1–92 (new + old data). The *inset* shows the continuum HST image at 547 nm (Bujarrabal et al. 1998b). *Tick marks* are the same in the two plots. *Bottom panel.* Position vs. velocity diagram (old + new data) for a cut along the symmetry axis of the nebula. In the central part we have used data from the A configuration only. This plot also represents the structure of the nebula, for which the linear scale is given by the vertical bar

gradient found in the lobes. These are the signatures of a radially expanding disk, in which the velocity modulus is not constant but proportional to the distance to the center. We find another Hubble-like velocity law for the equatorial disk. In fact, if we assume an inclination of symmetry axis of nebula with respect to the plane of the sky of 38°, instead of the standard value of 32°–35° (but still within the errors estimated for this parameter Solf 1994; Bujarrabal et al. 1998a), the kinetic ages, t_{kin}, for both the two lobes and the equatorial disk become the same, 1200 yr. If this hypothesis is correct we should conclude that in all the molecular gas the velocity is just proportional to the distance, $\vec{V} = \vec{R}/t_{\mathrm{kin}}$, as one would expect if all the nebula was formed as a result of a sudden acceleration event, i.e. an explosion. In this case we can directly use the position vs. velocity diagram along the symmetry axis for roughly deriving a model of the real structure of M 1–92. Finally, at the maximum linear resolution achieved of 9×10^{15} cm, we have not found signs of rotation whatsoever.

3 Discussion

In general, all theories aiming to explain the shapes of non spherical PNe and pPNe assume that these asymmetries are the result of a two-wind interaction. Originally it was thought that at the end of the AGB the mass loss turned into non isotropic but enhanced along a certain equatorial plane, resulting in the formation of a collimating disk, which would channel further mass loss along its perpendicular axis (Mastrodemos and Morris 1998, 1999). More recently, the discovery that such disks would be unable to produce the highly collimated flows often seen in the post-AGB phase, has raised the idea that the interaction of these bipolar flows with the formerly spherical CSE would result in the complex observed structures (Balick and Frank 2002). The origin of these bipolar post-AGB ejections is also a very debated issue between theories that necessarily involve the presence of stellar companions and those which do not (Bujarrabal et al. 2000). However the results we have found in M 1–92 really challenge all these scenarios: we have two lobes that are axially expanding, and a disk also in expansion but radially. In fact we can describe the whole nebula as a structure formed out of a single acceleration event, after which it continues freely expanding. The present shape and kinematics of M 1–92 would be just the result of the (axially symmetric) angular dependence of the forces during such an explosive-like event.

As far as we know there is only another example among nebulae around evolved stars in which a similar object has been ejected during a short period of time. *The Homunculus* is a very massive dusty envelope surrounding a binary system including one of the brightest luminous blue variables (LBVs), η Car. This nebula also consist of a two emptied bubbles plus a flat thin torus. Using three HST images, Morse et al. measured the proper motions in the sky of the dust in this structure, finding that the expansion in all directions follows a single Hubble-like velocity law (Morse et al. 2001). They reach the conclusion that all the structure was ejected 160 yr ago, during the great eruption of η Car by the mid of the XIX Century, an event that lasted only about ten years. In spite of the differences between η Car and the central star(s) of M 1–92, the similitude between both nebulae is obvious. (Note that the mass of star(s) is much larger in the case of η Car, but the same also applies to the energy involved in the ejection process since both the expansion velocity and mass of its envelope are probably ten times higher than in our case.)

For the case of *The Homunculus* Matt et al. have developed a model in which a magneto-rotational explosion results in the ejection of part of the convective mantle of the LBV (Matt et al. 2006). This explosion is caused by the winding of the stellar magnetic field as consequence of a differential rotation between the core and the mantle. When the magnetic pressure overcomes the weight of the upper layers a copious mass loss is driven along the polar directions of the magnetic field. At the same time the magnetic forces squeeze the material towards the equatorial plane and an ejection in this plane is driven too. More recently, Nordhaus and Blackman have concluded that such explosion can be produced when, in a common-envelope system, the companion supplies enough angular momentum to the mantle to sustain the increasing magnetic energy (Nordhaus and Blackman 2006).

In view of the similitude between the envelopes of η Car and of M 1–92, and since the central star in our case is also a binary system, we suggest that a scaled down version of the above described mechanism could explain the observed morphology and velocity field. In fact, Matt & Balick also suggested that the same phenomenon could explain the shapes of other pPNe like IRAS 17106–3046, in which the two-lobe plus disk structure is evident (Matt and Balick 2004). (Note however that in this latter case we do not have information on the velocity field.) In this scenario we should conclude that the molecular gas now forming the M 1–92 was directly ejected from the stellar mantle, and did not encounter an AGB CSE with significant mass in its way out. Otherwise it would very difficult to explain how the present nebula could display such accurate ballistic movement. Yet, if all the nebula we see now was expelled in 100 yr or less, we should also conclude that the post-AGB evolution of our source was indeed initiated by the explosive event occurred 1200 yr ago. This scenario might seem very intricate but note that the latest statistics on the PN population suggest that most of them, if not all

of them, are the result of the envelopes around binary systems, that could have undergone a phase of strong interaction during a common-envelope phase (De Marco 2007; Soker 2006).

Even if the proposed mechanism is indeed the one responsible for the formation of M 1–92, it would be very premature to conclude on its possible importance for the problem of the shaping of pPNe and PNe in general. On the one hand M 1–92 is the only pPN in which a single Hubble-law can account for the velocity field in all the molecular gas, but on the other hand there are just a few sources for which such spatial, and spectral high-resolution images are available. In CLR 2688 a single Hubble law is present in the CO jets detected in the polar directions, but this law might not hold along the equatorial plane (Cox et al. 2000; Ueta et al. 2006). However, in this case these post-AGB flows are clearly interacting with the CSE formed while at the AGB, and therefore free expansion of the ejected material is not expected. In the Red Rectangle the equatorial disk is dominated by Keplerian rotation (Bujarrabal et al. 2005), and this could also be the case of 89 Her (Bujarrabal et al. 2007). Interestingly, these two sources are both low-mass binary systems. Clearly, to conclude on these questions we should wait until high-quality images of a statistically significant number of sources become available. Of course this will happen as soon as the Atacama Large Millimeter Array becomes fully operational in the next years.

Acknowledgements J.A. and V.B. acknowledge their partial support from the *Spanish Ministry of Education and Science* under projects AYA2003-7584 and ESP2003-04957. Data reduction and plots have been performed using the Gildas software package. J.A. also thanks the Debian/Gnu developer community for its efforts in supporting a free working environment for computers.

References

Alcolea, J., Neri, R., Bujarrabal, V.: Astron. Astrophys. **468**, L41–L44 (2007)

Arrieta, A., Torres-Peimbert, S., Georgiev, L.: Astrophys. J. **623**, 252–268 (2005)

Balick, B., Frank, A.: Annu. Rev. Astron. Astrophys. **40**, 439–486 (2002)

Bujarrabal, V.: Astrophys. Space Sci., doi:10.1007/s10509-007-9494-8 (2007)

Bujarrabal, V., Alcolea, J., Neri, R.: Astrophys. J. **504**, 915 (1998a)

Bujarrabal, V., Alcolea, J., Sahai, R., Zamorano, J., Zijlstra, A.A.: Astron. Astrophys. **331**, 361–371 (1998b)

Bujarrabal, V., García-Segura, G., Morris, M., Soker, N., Terzian, Y.: In: J.H. Kastner, N. Soker, S. Rappaport (eds.) Asymmetrical Planetary Nebulae II: From Origins to Microstructures. ASP Conf. Ser., vol. 199, p. 201 (2000)

Bujarrabal, V., Castro-Carrizo, A., Alcolea, J., Neri, R.: Astron. Astrophys. **441**, 1031–1038 (2005)

Bujarrabal, V., van Winckel, H., Neri, R., Alcolea, J., Castro-Carrizo, A., Deroo, P.: Astron. Astrophys. **468**, L45–L48 (2007)

Cohen, M., Kuhi, L.V.: Astrophys. J. **213**, 79–92 (1977)

Cox, P., Lucas, R., Huggins, P.J., Forveille, T., Bachiller, R., Guilloteau, S., Maillard, J.P., Omont, A.: Astron. Astrophys. **353**, L25–L28 (2000)

De Marco, O.: In: M.J. Barlow, R.H. Méndez (eds.) IAU Symp. 234: Planetary Nebulae in the Milky Way and Beyond, p. 111 (2007), astro-ph/0605626

Mastrodemos, N., Morris, M.: Astrophys. J. **497**, 303 (1998)

Mastrodemos, N., Morris, M.: Astrophys. J. **523**, 357–380 (1999)

Matt, S., Balick, B.: Astrophys. J. **615**, 921–933 (2004)

Matt, S., Frank, A., Blackman, E.G.: Astrophys. J. Lett. **647**, L45–L48 (2006)

Minkowski, R.: Publ. Astron. Soc. Pac. **58**, 305 (1946)

Morse, J.A., Kellogg, J.R., Bally, J., Davidson, K., Balick, B., Ebbets, D.: Astrophys. J. Lett. **548**, L207–L211 (2001)

Nordhaus, J., Blackman, E.G.: Mon. Not. R. Astron. Soc. **370**, 2004–2012 (2006)

Soker, N.: Astrophys. J. Lett. **645**, L57–L60 (2006)

Solf, J.: Astron. Astrophys. **282**, 567–585 (1994)

Ueta, T., Murakawa, K., Meixner, M.: Astrophys. J. **641**, 1113–1121 (2006)

A massive, dusty toroid with large grains in the pre-planetary nebula IRAS22036+5306

R. Sahai · K. Young · N. Patel · C. Sánchez Contreras · M. Morris

Originally published in the journal Astrophysics and Space Science, Volume 313, Nos 1–3.
DOI: 10.1007/s10509-007-9644-z © Springer Science+Business Media B.V. 2007

Abstract Using the Submillimeter Array (SMA), we have obtained high angular-resolution ($\sim 1''$) interferometric maps of the submillimeter (0.88 mm) continuum and CO $J = 3$–2 line from IRAS 22036+5306 (I 22036), a bipolar pre-planetary nebula (PPN) with knotty jets discovered in our HST SNAPshot survey of young PPNe. In addition, we have obtained supporting lower-resolution ($\sim 10''$) 2.6 mm continuum and CO, ^{13}CO $J = 1$–0 observations with the Owens Valley Radio Observatory (OVRO) interferometer. We find an unresolved source of submillimeter (and millimeter-wave) continuum emission in I 22036, implying a very substantial mass (0.02–$0.04 M_\odot$) of large (i.e., radius $\gtrsim 1$ mm), cold ($\lesssim 50$ K) dust grains associated with I 22036's toroidal waist. The CO $J = 3$–2 observations show the presence of a very fast (~ 220 km s^{-1}), highly collimated, massive ($0.03 M_\odot$) bipolar outflow with a very large scalar momentum (about 10^{39} g cm s^{-1}), and the characteristic spatio-kinematic structure of bow-shocks at the tips of this outflow. The fast outflow in I 22036, as in most PPNe, cannot be driven by radiation pressure. The large mass of the torus suggests that it has most likely resulted from common-envelope evolution in a binary, however it remains to be seen whether or not the time-scales required for the growth of grains to millimeter sizes in the torus are commensurate with such a formation scenario. The presence of the torus should facilitate the formation of the accretion disk needed to launch the jet. We also find that the ^{13}C/^{12}C ratio in I 22036 is very high (0.16), close to the maximum value achieved in equilibrium CNO-nucleosynthesis (0.33). The combination of the high circumstellar mass (i.e., in the torus and an extended dust shell inferred from ISO far-infrared spectra) and the high ^{13}C/^{12}C ratio in I 22036 provides strong support for this object having evolved from a massive ($\gtrsim 4 M_\odot$) progenitor in which hot-bottom-burning has occurred.

Keywords Circumstellar matter · Planetary nebulae: individual (IRAS 22036+5306) · Reflection nebulae · AGB and post-AGB stars · Mass loss · Winds & outflows · Interferometry

R. Sahai (✉)
Jet Propulsion Laboratory, Caltech, 4800 Oak Grove Drive, Pasadena, CA 91109, USA
e-mail: sahai@jpl.nasa.gov

K. Young · N. Patel
Harvard-Smithsonian Center for Astrophysics, 60 Garden St., Cambridge, MA 02138, USA

C. Sánchez Contreras
Instituto de Estructura de la Materia, CSIC, Serrano 121, 28006 Madrid, Spain

M. Morris
Department of Physics and Astrophysics, University of California, Los Angeles, CA 90095, USA

1 Introduction

Pre-Planetary nebulae (PPNe)—objects in transition between the AGB and planetary nebula (PN) evolutionary phases—hold the key to one of the most vexing and long-standing problems in our understanding of these very late stages of evolution for low and intermediate mass stars. Current observational evidence shows that during the PPN phase, the dense, isotropic mass-loss which generally marks the late AGB phase changes to high-velocity, bipolar or multipolar mass-loss. The emergence of fast collimated outflows or jets during the PPNe or the very late AGB phase has been hypothesized as the primary mechanism for this dramatic change in the geometry and dynamics of the mass-loss (Sahai 2004; Sahai and Trauger 1998). However, the physical mechanism for producing the fast outflows remains unknown. (Sub)millimeter-wave interferometric observations are the best probes of the dynamics and energetics of the shock acceleration process (which transfers a substantial amount of directed momentum to large parts of the dense AGB wind), as well as the mass of cold material in different parts of the nebula.

We have therefore carried out observations of the submillimeter continuum and CO $J = 3$–2 observations of the PPN IRAS22036+5306 (I 22036) with the Submillimeter Array (SMA); supporting interferometric observations at 2.6 mm were also obtained with the Owens Valley Radio Observatory (OVRO) interferometer. I 22036 is a bipolar PPN with knotty jets discovered in our Hubble Space Telescope (HST) snapshot survey of young PPNe, with an estimated distance and luminosity of 2 kpc and $2300 L_\odot$, respectively (Sahai et al. 2003).

2 Observations

The submillimeter data were obtained with the SMA, using two tracks, on July 20 (UT) and Aug. 29 (UT), 2004, with the array in the compact and extended configurations, respectively. The receivers were tuned to center the 345.796 GHz rest frequency of the CO(3–2) line in the upper sideband, after correcting for the source's -45 km s^{-1} LSR velocity. For both tracks the correlator was configured to have the highest available uniform resolution, 406.25 kHz channel (0.35 km s^{-1}), over a total bandwidth of 2 GHz (further details in Sahai et al. 2006). The continuum map was made using the lower sideband (LSB) data, and uniform weighting of Tsys-weighted visibilities. These data were supplemented by lower-resolution observations of the CO and ^{13}CO $J = 1$–0 line at OVRO on 2002 September 21 as part of a snapshot survey of PPNe (details in Sánchez Contreras and Sahai 2004).

3 Results

The major results of our study are as follows (details of the scientific analysis are given in Sahai et al. 2006):

(i) The CO $J = 3$–2 spatially integrated line profile covers a total velocity extent (FWZI) of $\sim$500 km s^{-1}. The CO and ^{13}CO $J = 1$–0 also show large velocity widths, although with a smaller spread due to the lower sensitivity of these data (Fig. 1).

(ii) The CO $J = 3$–2 emission comes from a massive high-velocity bipolar outflow aligned with the major axis of the nebula. The characteristic position-velocity signature of a bow-shock resulting from a high-velocity bipolar collimated outflow/jet interacting with dense ambient material can be seen in the CO $J = 3$–2 data in both the red and blue-shifted components of the outflow. The P–V structure has a high degree of symmetry around the center, indicating that the bipolar outflow and surrounding medium are correspondingly symmetric. The bow-shocks do not lie at the tips of the optical lobes but are associated with the bright optical regions S1, S2, and their obscured counterparts on the far side of the nebula (Fig. 2).

(iii) A central, unresolved, continuum source with a flux of 0.29 ± 0.04 Jy (8.4 ± 1.7 mJy) at $\lambda = 0.88$ mm (2.6 mm) is seen in the SMA (OVRO) data.

(iv) We have estimated the masses of the high- and low-velocity molecular components using LVG models to constrain the excitation temperature and optical depths. The mass in the high-velocity outflow, i.e., at velocities offset more than ± 15 km s^{-1} from the systemic velocity, is $0.03 M_\odot$, and the scalar momentum is 1.1×10^{39} g cm s^{-1}. The large value of the scalar momentum, together with the small age of the high-velocity outflow, implies that it cannot be driven by radiation pressure. The total molecular mass derived from integrating the flux of the CO $J = 3$–2 line over its full velocity range, is $\sim$0.065$M_\odot$, much lower than the value ($\sim$4.7$M_\odot$) estimated previously from a dust-shell model of I 22036's near to far-infrared fluxes

Fig. 1 CO spectra of I 22036 **a** CO $J = 3$–2 line integrated over the entire emission region (from SMA data), **b** CO and ^{13}CO(1–0) peak intensity in the unresolved emission source for each line (from OVRO data)

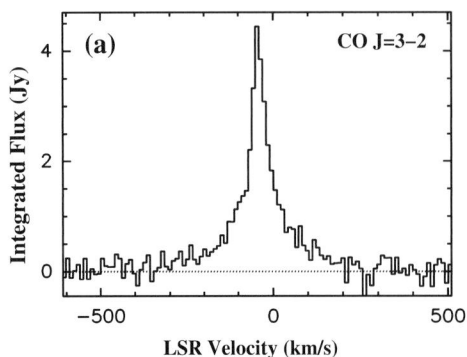
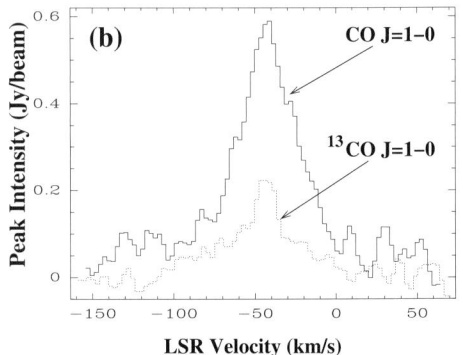

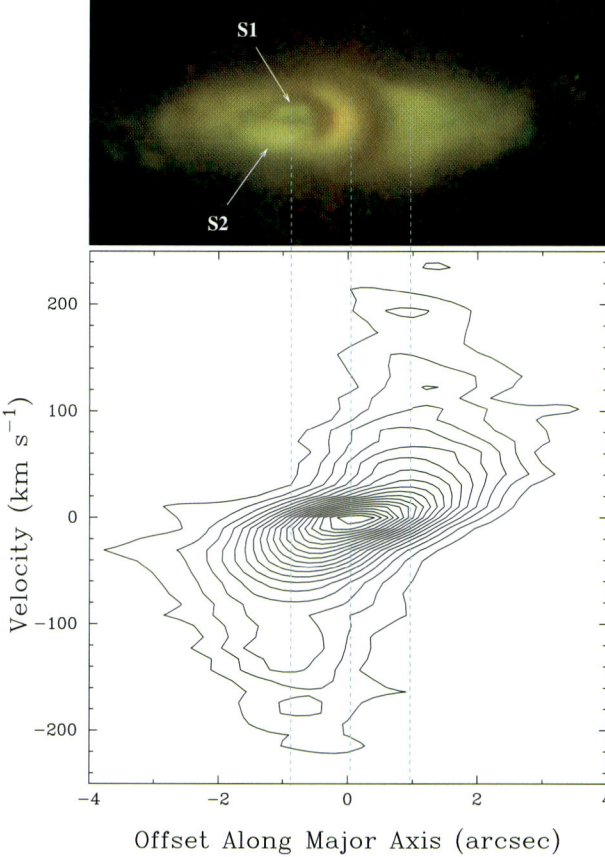

Fig. 2 Position–velocity plot of the CO $J = 3$–2 emission along the major axis of I 22036. Contour levels are 5 to 95% of the peak in steps of 5%. A color HST image of the nebula (with the 0.6 μm image in *green* and the 0.8 μm image in *red*) is shown, with its major axis aligned along the spatial axis in the P–V plot, for comparison

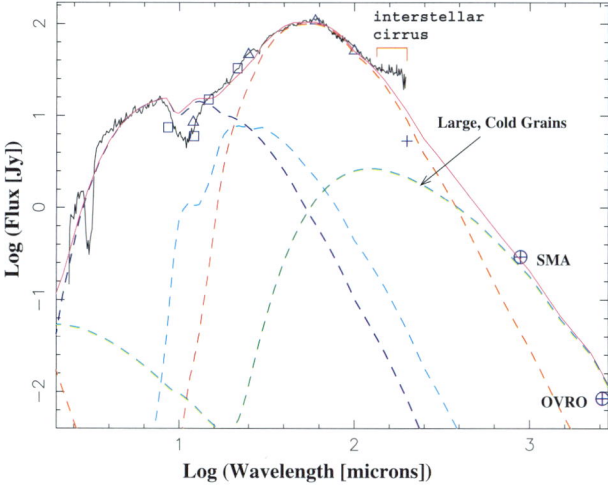

Fig. 3 Observations [*black curve*: ISO spectra, *blue symbols*: photometric data—SMA 0.88 mm and OVRO 2.6 mm: *circles w/crosses*, MSX: *squares*, IRAS: *triangles*, ISO/PHOT: *cross*] and a model spectrum (*magenta curve*) of I 22036. Individual components (*dashed curves*) of the model are also shown: cool (*red*) & warm (*cyan*) shells, hot inner disk (*blue*) and a component with large, cold grains (*green*), probably associated with I 22036's dusty waist

(Fig. 3), assuming a gas-to-dust ratio, $\delta = 200$. First, the interferometric $J = 3$–2 observations resolve out any emission from the large shell (inner and outer radii of 1.4×10^{17} and 5.3×10^{17} cm) in which the cool dust resides, and second, the gas in this shell may not be sufficiently excited.

(v) The small size of the continuum source indicates that it is associated with the dusty toroidal waist of I 22036. The spectral index of the continuum from 0.88 mm to 2.6 mm, α (defined as $F_\nu \sim \nu^\alpha$) is 3.3 is too steep to be a result of thermal brehmstrahhlung. The 0.88 and 2.6 mm continuum fluxes are significantly in excess of the SED resulting from the dust model by Sarkar and Sahai (2006), and must come from a substantial mass ($\sim 0.04 M_\odot$) of large (of radius $\gtrsim 1$ mm) cold ($\lesssim 50$ K) grains in this waist (Fig. 3).

(vi) The ^{13}C/^{12}C ratio in I 22036 is 0.16, close to the maximum value achieved in equilibrium CNO-nucleosynthesis (0.33). The enhancement is most likely due to the operation of hot-bottom-burning, which operates in stars with initial masses $\gtrsim 4 M_\odot$.

4 The shaping of bipolar planetary nebulae

The dynamical evolution and shaping of PPNe is believed to result from the shock interaction between a fast, collimated post-AGB wind and the slowly expanding, dense wind ejected during the previous AGB phase. A significant amount of momentum is believed to be transferred to the AGB shell by means of bow-like shocks at two localized regions where the bipolar post-AGB wind impinges on, and thus accelerates and heats the AGB wind. Although morphological identifications of bow-shocks have been made in PPNe, their characteristic spatio-kinematic structure has seldom been identified. Thus, the bow-shock structures inferred kinematically from our CO $J = 3$–2 imaging of I 22036 provide direct support for, and are an important input for quantitative modelling of the physics of, the interaction of collimated fast winds with AGB envelopes as a production mechanism for bipolar PPNe (see, e.g., Lee and Sahai 2003). The launching mechanism for such jets remains unknown, but given the similarity between the fast outflow speeds in PPNe and jets in low-mass pre-main sequence stars with accretion disks, it is plausible that the mechanisms which drive jets in PPNe and YSOs are very similar. In this scenario, the accretion disk in a PPN would be produced around a relatively close companion low-mass star, which accretes material from the dense wind of the AGB primary (Morris 1987; Mastrodemos and Morris 1998).

5 Dusty waists and large grains

Although we can now state with some confidence that the formation of bipolar and multipolar lobes in PPNe is initiated by collimated outflows, the formation of the dense, dusty waists in these objects remains a puzzle. The presence of large grains with a very substantial mass associated with I 22036's toroidal waist, is a surprising result, but it is not unprecedented. Evidence for comparable masses of large grains has been uncovered for other well-studied PPNe such as the Egg Nebula ($\sim 0.01 M_\odot$, Jura et al. 2000) and the Red Rectangle (up to $0.01 M_\odot$, Men'shchikov et al. 2002). In the quadrupolar PPN IRAS 19475+3119, a comparison of a detailed model of its full SED (from optical wavelengths to 200 μm) to the observed 0.85 mm and 2.6 mm fluxes, that there is a substantial 0.85 mm excess flux which also requires the presence of large, cool grains (Sarkar and Sahai 2006; Sahai et al. 2007). We speculate that the formation of dusty waists and the presence of large grains in PPNe are due to intimately-linked physical processes, thus a resolution of the issue of waist formation may well lead to a resolution of how the large grains are made or vice versa. A possible mechanism is the destruction of volatile cometary-debris disks within several hundred to a thousand AU (Stern et al. 1990) by an intermediate-mass star during its luminous post-main-sequence evolution, providing a source of large solid particles which could form a dusty torus. However, the total mass of large grains which we have found in I 22036 appears far too large compared to the mass estimates for our Kuiper Belt ($\sim 0.1 M_\oplus$) or the inner parts of the Oort Cloud ($\sim 40 M_\oplus$), and argues against such a process, unless of course the Kuiper Belt/Oort Cloud analogs of an intermediate mass star like that in I 22036 are significantly more massive than those in our Solar system. The large mass of the torus suggests that it has most likely resulted from common-envelope evolution in a binary, however it remains to be seen whether or not the time-scales required for the growth of grains to millimeter sizes in the torus are commensurate with such a formation scenario. The presence of the torus would certainly facilitate the formation of the accretion disk needed to launch the jet by providing a dense reservoir of material for building the disk.

6 The future: ALMA observations of pre-planetary & planetary nebulae

The unprecedented sensitivity and resolution ($\sim 0.1''$) of ALMA will revolutionize the study of PPNe and PNe, by enabling us to map the outflows in a statistical sample of bipolar and multipolar objects, cleanly separating the emission from individual lobes. We will be able to determine expansion time-scales for each lobe, thereby distinguishing between simultaneous-formation versus sequential-formation models of multilobe structure. ALMA will help us isolate the mysterious toroidal waist component from the polar outflows, and determine its kinematics. Measurements of the submillimeter and millimeter-wave continuum fluxes, together with gas tracers such as CO, will allow us to determine the mass of gas and dust in the torus independently. The mass of the torus may provide a discriminant between different formation theories for the torus such as common-envelope evolution or wind accretion in a binary system.

Acknowledgements   RS is thankful to NASA for partial financial support for this work from ADP award (399.20.00.08), LTSA award (399.20.40.06), and STSCI HST award (GO 09463.01).

References

Jura, M., Turner, J.L., Van Dyk, S., Knapp, G.R.: Astrophys. J. **528**, L105 (2000)

Lee, C.-F., Sahai, R.: Astrophys. J. **586**, 319 (2003)

Mastrodemos, N., Morris, M.: Astrophys. J. **497**, 303 (1998)

Men'shchikov, A.B., Schertl, D., Tuthill, P.G., Weigelt, G., Yungelson, L.R.: Astron. Astrophys. **393**, 867 (2002)

Morris, M.: Publ. Astron. Soc. Pac. **99**, 1115 (1987)

Sahai, R.: In: Asymmetrical Planetary Nebulae III. ASP Conf. Ser., vol. 313, p. 141 (2004)

Sahai, R., Trauger, J.T.: Astron. J. **116**, 1357 (1998)

Sahai, R., Zijlstra, A., Sánchez Contreras, C., Morris, M.: Astrophys. J. **586**, L81 (2003)

Sahai, R., Young, K., Patel, N.A., Sánchez Contreras, C., Morris, M.: Astrophys. J. **653**, 1241 (2006)

Sahai, R., Sánchez Contreras, C., Morris, M., Claussen, M.: Astrophys. J. **658**, 410 (2007)

Sánchez Contreras, C., Sahai, R.: In: Asymmetrical Planetary Nebulae III. ASP Conf. Ser., vol. 313, p. 377 (2004)

Sarkar, G., Sahai, R.: Astrophys. J. **644**, 1171 (2006)

Stern, S.A., Shull, J.M., Brandt, J.C.: Nature **345**, 305 (1990)

Gas dynamics and structure of galaxies

ALMA targets and capabilities

Kazushi Sakamoto

Originally published in the journal Astrophysics and Space Science, Volume 313, Nos 1–3.
DOI: 10.1007/s10509-007-9624-3 © Springer Science+Business Media B.V. 2007

Abstract Between gas dynamics and structure of galaxies is a two-way relation. On one hand, gas dynamics in a galaxy is largely determined by the structure of the galaxy, and on the other hand, gas dynamics can gradually alter the galaxy structure through redistribution of mass and angular momentum within the galaxy. The first half of this relation should mostly determine gas distribution and regulate star formation in undisturbed spirals, and the second half has been suggested to cause secular evolution of spiral galaxies—a slow mode of galaxy evolution in the absence of major mergers. Our knowledge on this relation is going to be greatly deepened by the ALMA. Focusing on the galaxy evolution through gas dynamics, I briefly review what we know about the subject. Then I try to look out what the ALMA can do to answer open questions in the field. It is pointed out that the ALMA will be able to fully map all the spiral galaxies between 1 and 25 Mpc at $1''$ resolution in 1000 hours.

Keywords ALMA · Galactic gas dynamics · Galactic structure

1 Introduction

The 'gas dynamics and structure of galaxies' is a vast subject on which the ALMA will certainly make breakthroughs. It covers almost everything between the neighbor subjects of galaxy formation and star formation in galaxies. Given the formidable task to review such a broad subject, I choose to concentrate on a particular area around the bar and secular evolution of disk galaxies. Even this subject has a rich history and remarkable recent progress. Kormendy and Kennicutt (2004, hereafter KK04) extensively reviewed much of them. So, while outlining the subject, I try to cover some of the latest developments in the field, look things more from a viewpoint of gas, and discuss the (sub-)millimeter observations that the ALMA can do for the subject.

2 Bar-driven gas transport in spiral galaxies

In their 1977 paper, Matsuda and Nelson (1977) nicely summed up an outcome of gas dynamics in barred galaxies. They wrote, "*almost all the gas originally occupying the region within the corotation-points* [i.e., the region that the bar sweeps] *is swept into a dense nucleus at the center of the bar*". This is because gas in a bar tends to have a shock in the bar's leading edge and loses energy and angular momentum there (Sørensen et al. 1976). In addition, the bar exerts negative gravitational torque on the gas piled up at the bar shock lanes (Fig. 1).

Observationally, this bar-driven gas transport has been confirmed, among other observations, by measuring the central concentration of molecular gas in galaxy disks (Sakamoto et al. 1999b; Sheth et al. 2005). Figure 2 shows histograms of the degree of gas concentration in barred and unbarred galaxies. They show that barred galaxies tend to have higher degrees of gas concentration than unbarred galaxies do. On average, the concentration index is a factor of ~ 3 higher in barred spirals than in unbarred counterparts, suggesting that two-thirds of the gas in the central kiloparsec of barred galaxies has been transported there from outside by the bars. The typical mass of molecular gas in the central kpc of these barred galaxies is a little more than $10^8 M_\odot$ for

K. Sakamoto (✉)
National Astronomical Observatory, Mitaka, Tokyo, Japan
e-mail: sakamoto.kazushi@nao.ac.jp

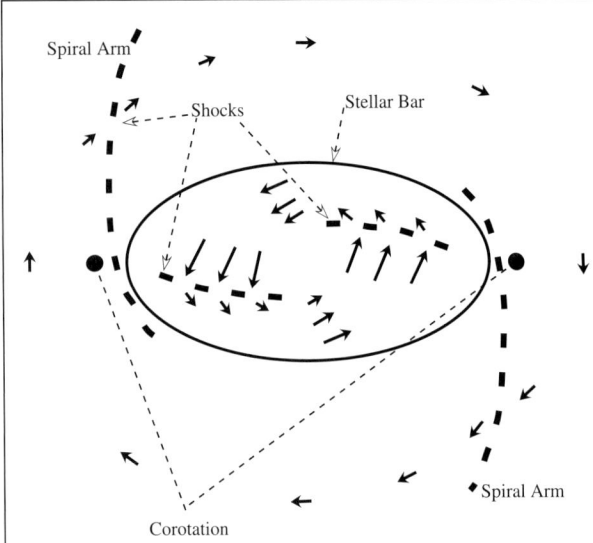

Fig. 1 An illustration of gas motion in a barred galaxy, based on Fig. 1 of Matsuda and Nelson (1977). *Filled arrows* indicate gas velocities in the frame rotating with the bar, the *thick dashed lines* show gas density peaks, and the *pair of black dots* are at the corotation radius

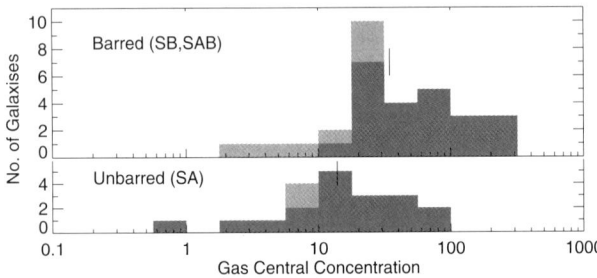

Fig. 2 Degrees of gas concentration in barred and unbarred spiral galaxies. The *abscissa* is the mean surface density of molecular gas in the central kpc divided by the mean surface density of molecular gas averaged over the optical disk of each galaxy. *Light-gray* data are upper limits, and the *vertical bars* are medians. Data are from Young et al. (1995), Sakamoto et al. (1999b), and Sheth et al. (2005)

the classical conversion factor. The two thirds of the large amount of mass is there because of the bars, although the mass could be reduced by the choice the CO to H_2 conversion factor.

3 Bar-driven galaxy evolution

The addition of a large amount of gas to the centers of barred galaxies has consequences. An immediate result of the gas supply is active star formation (e.g., Sérsic and Pastoriza 1967). It adds stars to the central kpc, which is usually the bulge region. Another possible outcome of the bar-driven gas transport is that the structure of the galaxies may change because of the redistribution of mass and angular momentum within each galaxy. Specifically, the bar that transported

the gas may be destroyed. This bar-dissolution has been a focus of galaxy simulations since the 90's, and has also been a target of many observational studies (see KK04).

The bar-dissolution is important because it closes a loop of secular evolution for spiral galaxies. The loop goes like the following. First, we know that the majority of galaxies have a bar, and that the bars transport gas toward galactic centers as we have seen. Active star formation results. It has been pointed out that bulges of late type spirals are disky pseudo-bulges (e.g., Kormendy et al. 2006). Hence, forming stars in the nuclear gas concentrations can let those pseudo-bulges grow and move the galaxies toward earlier types. In the mean time, a new bar can form in disk galaxies either spontaneously or triggered by galaxy interactions, according to numerical modeling. Thus, if there is the self-destruction of bars due to the bar-driven gas transport, then the loop of bar formation and destruction closes and at least some galaxies evolve across the Hubble's tuning-fork diagram between the barred and unbarred branches and from later types to earlier types. Various lines of evidence suggest that the relative importance of the slow-building bulge compared to the classical bulge (made earlier through merger) is larger in later type spiral galaxies than in earlier type spirals (KK04).

3.1 Bar-dissolution theories

Recent theories on bar-dissolution have suggested two things. Firstly, the bar-driven central mass concentrations in real galaxies are *not* massive enough to destroy bars according to numerical simulations of stellar dynamics (Shen and Sellwood 2004; Athanassoula et al. 2005). Secondly, it has been pointed out that it is the gas-to-bar transfer of angular momentum that destroys bars and that the process takes only a couple of Gyrs for a typical late-type spiral galaxy (Bournaud et al. 2005).

In the latter, Bournaud et al. (2005) reported that the bar strength in their barred galaxy simulations dumped in a Gyr or two and *before* a large amount of gas accumulated at the galactic center. They took it as evidence that the gas-to-bar transfer of angular momentum, not the central mass concentration, destroyed the bar. There appears to be no inconsistency between this result and the negative results of bar dissolution from mass concentration, because the latter simulations added a growing central mass at the galactic center rather than moving gas in the galaxy. It appears therefore that the bar-dissolution and secular evolution scenario is still alive (at least theoretically) and that gas plays a key role there.

3.2 Bar statistics as a function of redshift

While theorists were simulating barred galaxies in computers, observers looked at the sky to tell how old bars are and when they formed. There have been much progress in the last few years in the observations, thanks to the look-back studies using the Hubble Space Telescope and also to ground-based surveys of nearby galaxies.

It now appears that bars have been in spiral galaxies for the last 10 Gyr or so. Elmegreen et al. (2004) showed that the fraction of large bars as a function of redshift is roughly constant out to $z = 1$ or 2, or for the last 8 or 10 Gyr. Jogee et al. (2004) and Marinova and Jogee (2007) also obtained a similar fraction using larger samples from $z = 0$ to 1; the fraction of strong bars is about $1/3$ and appears unchanged. The rarity of bars at high redshifts in earlier studies has been attributed to band shifting (Sheth et al. 2003) and insufficient spatial resolution.

The nearly constant bar fraction means that bars must be either long-lived despite the bar-dissolution theories, or that bars are short-lived but are recurrent. If at least some bars are recurrent, as some theories predict, then there must be galaxies with young bars.

3.3 Young and old bars

Are there young bars, and have we seen them? Hüttemeister et al. (1999, 2000) pointed out that gas-rich bars are plausibly young. Their idea is that if a bar is young and has started sweeping gas only recently, then much gas must be still in the bar on its way to the galactic center. They showed two cases of such gas-filled bars in UGC 2855 and NGC 7479. Similar bars have been observed in other galaxies too, in NGC 3627 (Helfer et al. 2003) and M83 (Fig. 3) to name a few. Interestingly, UGC 2855 and NGC 3627 are undergoing galaxy interaction, which may have induced the bars.

In addition to the possible young bars, we may have already seen old bars. Figure 4 shows barred galaxies of

two types, each possibly representing galaxies with young bars and those with old bars. On the left is again M83, whose Hubble type is SAB(s)c. The bar in the galaxy has much gas and dust seen as optical absorption, CO emission, and mid-infrared emission. The gas-loaded bar is possibly young. On the right is NGC 6744 whose classification is SAB(r)bc. The (r) means that the galaxy has an inner ring, unlike the (s)-type (spiral-type) M83. Sandage pointed out long ago that intermediate-type barred galaxies with an inner ring do not show leading-edge dust lanes in their bars (Sandage 1961). NGC 6744 has such a classification and indeed lacks a strong dust lane in its bar in the optical and mid-IR.[1] KK04 suggested that the difference is because the bars in r-type galaxies are old and have already swept all the gas they can sweep to the centers. Their interpretation, if correct, is consistent with the prediction of Matsuda and Nelson that a bar will sweep almost all the gas within the bar radius ($\approx$ corotation radius) to the nucleus.

There remains many things to be studied in this bar-dating problem. For example, the difference in gas distribution within various types of bars may reflect differences in gas dynamics or amount of available gas rather than different bar ages. Also, it has not been clear whether the

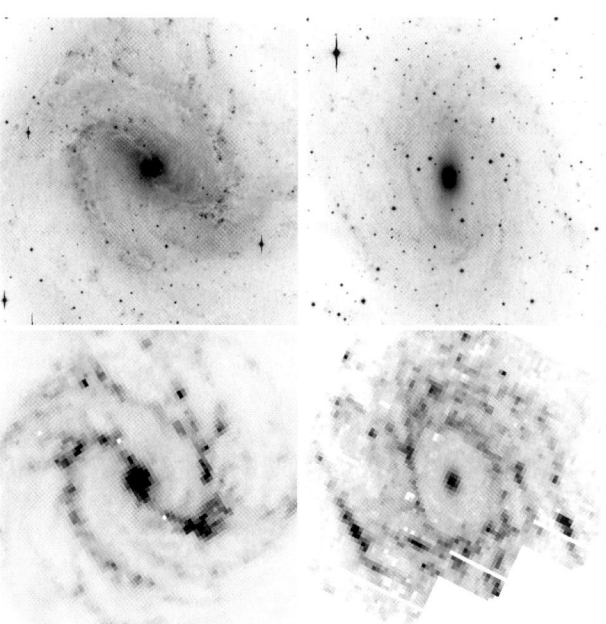

Fig. 4 Spiral galaxies with gas-loaded and gas-deficient bars. The bar may be older in the latter than in the former. (*left*) M83. SAB(s)c. (*right*) NGC 6744. SAB(r)bc. Its inner ring encircles the bar. (*top row*) *R*-band images from Larsen and Richtler (1999). (*bottom row*) ISO 7 μm images from Roussel et al. (2001). Each panel covers $(7.5')^2$

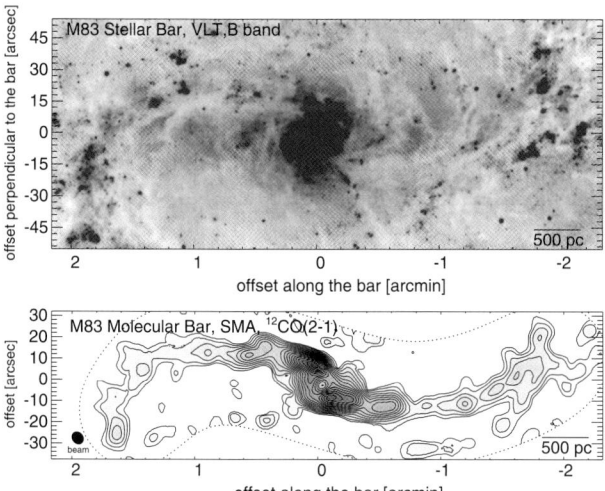

Fig. 3 A bar that has conspicuous gas/dust lanes in its leading edge (Sakamoto et al. in preparation). It appears that much gas is still being funneled to the galactic center though the bar

---

[1] The point here is the weakness of any dust lane in the bar. A bar always has some gas to funnel because it is continuously fed by evolved stars in the disk. The galaxy's NED images do hint at vary faint dust features in the bar.

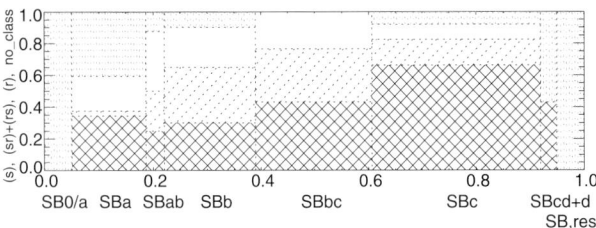

Fig. 5 Distribution of inner ring/spiral morphologies among the barred galaxies in the RSA (Sandage and Tammann 1981). Of all the barred spirals, 17, 42, and 19% are assigned (r), (s), and (rs) or (sr), respectively; the remaining 22% have no classification about inner spiral/ring. The fraction of (r)-type (= blank area) increases from later types to earlier types (i.e., from SBd to SBa). Sandage (1961) noted the lack of dust lanes in the early-type barred galaxies (~SBa) in addition to the lack of dust lanes in the intermediate-type inner-ring galaxies. Do these mean that the bars in later types last less and are younger than the bars in earlier types? The ALMA will hopefully answer the question

recurrent-bar model and the presence of young bars are reconcilable to the metallicity-gradient observations, which should reflect the history of bar-driven gas mixing and star formation in the galaxies (Dutil and Roy 1999). If the gas-rich and gas-poor bars are younger and older bars, then their distribution (or relative fraction) as a function of various galaxy parameters is of interest.[2] The Hubble classification may give us a crude answer (Fig. 5). However, the optical classification is not based on gas distribution, and it was manually assigned. To some extent, satellite imaging in the mid-IR can be a substitute of CO mapping (Regan et al. 2006, and in this conference). The ISO images in Roussel et al. (2001) do support the weakness of dust lanes in the bars in the r-type galaxies than in the s- (and rs-) type galaxies. However, there are only a small number of spatially resolved bars (~20), in which only a few are r-type. The analysis of Spitzer and AKARI data is therefore eagerly waited. In any case, imaging surveys of gas in the disks of spiral galaxies are highly needed both for better understanding of gas dynamics and for better statistics of gas distribution in spiral galaxies. They let us infer the bar ages and the significance of the secular evolution in galaxies.

4 Current status of gas observations

As we have seen, the existing pieces of evidence imply (though have not proven) that there are young and old bars. The presence of young bars along with the nearly constant bar fraction over the last ~8 Gyr implies that some galaxies have recurrent bars and for them the secular evolution has been ongoing. Numerical models suggest that gas dynamics

---

[2]Dr. F. Combes asked about this in the question time for this talk.

Table 1 Mapping surveys of molecular gas in nearby galaxies

Name	Resolution	No. of galaxies	Year	Ref.[b]
PdBI NUGA	$1''$	25	2007[a]	1
NRO Virgo	$3''$	15	2003	2
NRO-OVRO	$4''$	20	1999	3
NRO Seyfert-HII	$4''$	25	2007[a]	4
BIMA SONG	$6''$	44	2003	5
NRO 45m	$15''$	40	2007[a]	6
FCRAO	$45''$	300	1995	7

[a]Presented at this conference

[b]References: 1. García-Burillo et al. (2007), 2. Sofue et al. (2003), 3. Sakamoto et al. (1999a), 4. Kohno (2007), 5. Helfer et al. (2003), 6. Kuno et al. (2007), 7. Young et al. (1995)

plays a key role throughout this process. We need gas observations of spiral galaxies in order to test and refine this scenario from gas distribution, kinematics, and their statistical properties. Molecular gas is the main component of the ISM in the inner disks of galaxies where bars are and most of star formation occurs. Thus the gas observations that we need most are the observations of molecular gas, although HI provides important complementary information.

Table 1 lists major imaging or mapping surveys of molecular gas (via CO) in nearby spiral galaxies, excluding those on mergers. These are our main source of information for the statistical properties linked to the gas dynamical problems in this article. For example, the gas concentration statistics in Fig. 2 are from three of these surveys. In addition to these surveys, there have been a large number of works on individual or smaller samples of nearby galaxies, usually addressing more specific subjects than the larger surveys do.

Even though the listed surveys took long observing times at among the most powerful millimeter radio telescopes, the data currently in our hands have some weakness. For example, higher resolution surveys tend to have a smaller sample and a smaller imaging area ($\lesssim 1'$) in each galaxy. A small sample of a few dozens of galaxies makes statistical analysis difficult. It was fine to divide the sample in half using a single parameter, barred or unbarred, and compare the two using a single index of gas concentration, but it gets difficult to add more parameters such as earlier or later Hubble types, and r- or s-types. It is clear, however, that we need more elaborate parameterization of galaxy structure, gas distribution, and kinematics. We need it for multivariable statistics to address the secular evolution problem on a firm statistical ground. Thus ALMA's power is much needed here.

5 ALMA capabilities

As an example of the ALMA's power, I note that the ALMA enables us to fully map a volume limited sample of sev-

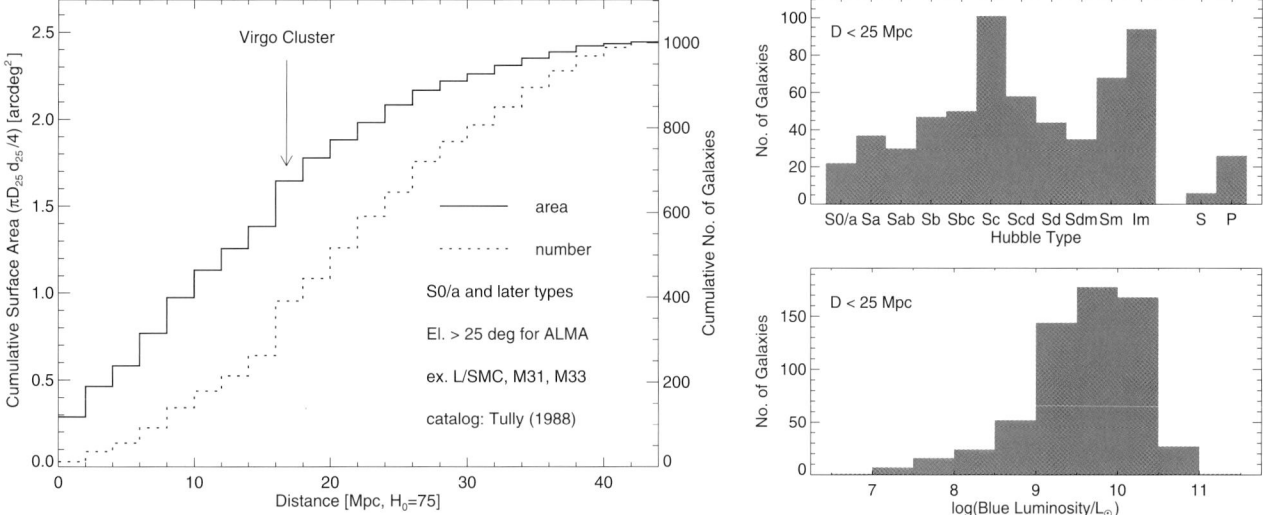

Fig. 6 (*left*) Cumulative number and surface area of nearby spiral galaxies suitable for ALMA observations. The galaxies are from Tully's Nearby Galaxies Catalog (Tully 1988). The de Vaucouleurs radius (i.e., the isophote radius at 25 mag arcsec^{-2} in the B band) is used for the surface area. Four of the nearest galaxies, LMC, SMC, M31, and M33, are excluded from this plot because they are dis-

tinctively large on the sky, having surface areas of 77, 9.7, 2.6, and 0.4 arcdeg^2, respectively. There are 618 potential target galaxies within 25 Mpc according to the catalog, and their total surface area is 2.0 arcdeg^2. (*right*) Distributions of Hubble types and blue luminosities of the spiral galaxies within 25 Mpc

eral hundred galaxies at an arcsecond resolution with high sensitivity in a realistic amount of time. A survey like this overcomes most of the weak points in the existing data, allowing us to address not only the evolution issues but also a wide range of problems in the general areas of gas dynamics, galaxy structure, and star formation.

Figure 6 shows the cumulative number and surface area of nearby spiral galaxies as a function of distance. I used the Nearby Galaxies Catalog (Tully 1988) and counted only the galaxies that rise above an elevation of 25° at Chajnantor. There are 622 spiral (i.e., S0/a or later) galaxies within 25 Mpc for ALMA observations. The total surface area of these galaxies within their de Vaucouleurs radii (R_{25}) is 2.0 arcdeg^2 when the four largest galaxies are excluded. (Excluded are LMC, SMC, M31, and M33. They are within 0.7 Mpc.) The cumulative surface area should be a linear function of distance for a complete catalog in a flat and uniform universe. The slower-than-linear increase at larger distances in the plot suggests that the catalog is incomplete beyond ∼20 Mpc, presumably missing distant dwarfs and low-surface-brightness galaxies. The uncatalogued sources aside, these galaxies are the ALMA's prime targets to study the relation between gas dynamics and galaxy structure in nearby galaxies. The figure also shows distributions of morphology and luminosity of the galaxies.

The ALMA will be able to map the entire 2 arcdeg^2 surface of the galaxies at an arcsecond resolution with a reasonable sensitivity in 1000 hours. The mapping is fastest in the 1 mm window containing the CO(2–1) line. The 50-element ALMA can map at 7 arcmin^2 hr^{-1} in 1 mm to achieve a

target sensitivity of $\sigma = 0.1$ K (T_b) for $\Delta V = 10$ km s^{-1}, assuming a system temperature of 170 K (for a precipitable water vapor (pwv) of 2.3 mm) from the ESO sensitivity calculator.[3] About 20 s are spent on each point if the Nyquist sampling is employed. The mapping is faster by a factor of 4 than in 3 mm CO(1–0) for the same pwv, and also faster by a factor of 1.1 than in CO(3–2) even if a better weather (pwv = 1.2 mm) is assumed for 345 GHz.[4]

Scientifically, the value of the data set will be enormous. The area that the ALMA can map in 1000 hrs contains bars, which are shorter than D_{25} ($= 2R_{25}$), spiral arms, most of the galaxy disks, and the galaxy nuclei. Hence the items mentioned in the previous sections—distribution of the gas concentration degrees and the fraction of the gas-loaded and gas-deficient bars across the galaxy types—can be studied with accurate statistics. Even a quantitative morphological-classification of galaxies solely based on molecular gas distribution will be possible. The high-resolution velocity information across the galaxy disks can be directly compared to those from numerical modeling of galaxy gas dynamics to infer the gravitational potential and mass distribution in the galaxies. The target sensitivity used above is such that the 3 σ molecular gas surface density is $10 M_\odot$ pc^{-2} in a

---

[3]http://www.eso.org/projects/alma/science/bin/sensitivity.html

[4]The total on-source time to map a large area with a fixed velocity resolution, a fixed target sensitivity in temperature, and a fixed spatial resolution depends on the frequency f and the system temperature T_{sys} and the aperture efficiency η_a at the frequency as $t_{on} \propto \frac{(T_{sys}/\eta_a)^2}{f^3}$.

$10 \ \mathrm{km\,s^{-1}}$ channel if $N(\mathrm{H_2})/I_{\mathrm{CO(2-1)}}$ is $2 \times 10^{20} \ \mathrm{cm^{-2}}$ $(\mathrm{K\,km\,s^{-1}})^{-1}$. A $10^5 M_\odot$ GMC in a 10 Mpc galaxy would be detected at 10 σ. The target sensitivity and linear resolution for 10 Mpc galaxies match those in the M33 observations of Engargiola et al. (2003). Their mosaic covering the most in D_{25} detected 148 GMCs in the sub-L^* spiral. Thus numerous GMCs and GMAs (maybe $\gtrsim 10^5$) should be detected across the over 600 galaxy disks observed at the 10–100 pc resolutions. Their dynamical responses to galaxy structures such as bars and spiral arms will be an important subject to study in the context of the cloud life-cycle and the regulation of star formation in the disks of spiral galaxies. In addition to the ^{12}CO line, the 2×5.5 GHz bandwidth in 1 mm provides us with continuum and lines of other molecules (e.g., CO isotopologues) useful to diagnose ISM properties. The data let us find interesting regions for further study, and the ALMA has capabilities to image the selected regions in higher resolution and sensitivity and in multiple wavelengths. Finally, the high-quality data of $z = 0$ galaxies is a necessity for comparison with high-z observations to study galaxy evolution.[5]

Technically, the main challenges for observations like these are in the massive mosaicking toward extended sources. It is vital to fully develop mosaic techniques such as the on-the-fly (OTF) mosaicking, to build a robust calibration procedure and an efficient reduction pipeline, and to integrate the Atacama Compact Array (ACA) for the zero-spacing flux. The OTF is needed because each mosaic point needs only 20 seconds of integration while each of the 200 largest galaxies takes more than an hour to map (and the largest dozen takes more than 10 hours for each). Without OTF, each point only has an instantaneous u-v coverage and the resulting synthesized beam varies within the galaxy. A (semi-)automated reduction pipeline is absolutely needed to cope with the overwhelming data rate. On the other hand, the kind of observations envisioned here are not demanding in terms of weather conditions. Since a few 100 nearby galaxies are in the sky at any moment, this may be a convenient fall-back project that can be invoked 75% of time (for the pwv ≤ 2.3 mm) during compact array configurations.

It seems inevitable that a data set of this magnitude is collected with the ALMA whether in a coherent manner or from numerous independent projects. It may not take much more than a few years once the array starts full operation, considering the fraction of time spent to map nearby galaxies at the existing (sub-)millimeter telescopes. The resulting

ALMA atlas of nearby galaxies will be a great asset in astronomy. In addition to the aforementioned scientific outputs from the data themselves, there will be synergy between the ALMA observations, theoretical modeling, and observations in other wavelengths. Combined together, they will bring us comprehensive understanding of the galactic gas dynamics, galaxy structure, and galaxy evolution.

6 Further beyond

The ALMA will enable us, among other things, a full mapping of molecular gas in several hundred local galaxies at an arcsecond resolution. It is revolutionary for the research of gas dynamics and structure of galaxies, and the array will keep us happily busy for many years. In preparation for the years after that, I note that the ALMA will *not* allow the same full-imaging survey (1) in several different wavelengths to better diagnose gas properties, (2) at a 10 times higher sensitivity to map more tenuous gas or weaker lines, or (3) at a 0.1″ resolution to resolve individual molecular clouds in over 500 galaxies, according to the calculation in the previous section. The last one would take 10^7 hr ($=10^3$ yr) in 1 mm if the sensitivity requirement is the same. Although these projects can be done in smaller scales and the need for a full scale project is questionable at this point, the power of the ALMA does have a limit. We may need array receivers for multi-beam interferometry in a future ALMA upgrade to enhance its mapping capability.

Acknowledgements I thank the ALMA-J and the conference organizer for their financial supports to attend this stimulating meeting. The VLT image of M83 is kindly provided by Drs. F. Cameron and L. Tacconi-Garman through Dr. A.J. Baker. This contribution made use of the NASA ADS, NED, and the NAOJ ADAC.

References

Athanassoula, E., Lambert, J.C., Dehnen, W.: Mon. Not. R. Astron. Soc. **363**, 496 (2005)

Bournaud, F., Combes, F., Semelin, B.: Mon. Not. R. Astron. Soc. **364**, L18 (2005)

Dutil, Y., Roy, J.-R.: Astrophys. J. **516**, 62 (1999)

Elmegreen, B.G., Elmegreen, D.M., Hirst, A.C.: Astrophys. J. **612**, 191 (2004)

Engargiola, G., Plambeck, R.L., Rosolowsky, E., Blitz, L.: Astrophys. J. Suppl. Ser. **149**, 343 (2003)

García-Burillo, S., et al.: Astrophys. Space Sci. (2007). doi:10.1007/s10509-007-9627-0

Helfer, T., et al.: Astrophys. J. Suppl. Ser. **145**, 259 (2003)

Hüttemeister, S., Aalto, S., Wall, W.F.: Astron. Astrophys. **346**, 45 (1999)

Hüttemeister, S., Aalto, S., Das, M., Wall, W.F.: Astron. Astrophys. **363**, 93 (2000)

Jogee, S., et al.: Astrophys. J. **615**, L105 (2004)

Kohno, K.: Astrophys. Space Sci. (2007). doi:10.1007/s10509-007-9695-1

---

[5]Dr. C. Wilson asked in the session whether gas observations with the ALMA would help us detect bars at high-z. A gas morphology similar to the one in M83 certainly suggests a bar even if the bar is not detected in the optical/IR. On the other hand, gas observations may well miss old bars that do not have much gas in their leading-edge lanes. An interesting possibility, however, is that the fraction of such aged bars (or the r-type gas distribution) may be smaller at higher redshifts.

Kormendy, J., Kennicutt, R.C. Jr.: Annu. Rev. Astron. Astrophys. **42**, 603 (2004)

Kormendy, J., Cornell, M.E., Block, D.L., Knapen, J.H., Allard, E.L.: Astrophys. J. **642**, 745 (2006)

Kuno, N., et al.: Publ. Astron. Soc. Jpn. **59**, 117 (2007)

Larsen, S.S., Richtler, T.: Astron. Astrophys. **345**, 59 (1999)

Marinova, I., Jogee, S.: Astrophys. J. **659**, 1176 (2007)

Matsuda, T., Nelson, A.H.: Nature **266**, 607 (1977)

Regan, M., et al.: Astrophys. J. **652**, 1112 (2006)

Roussel, H., et al.: Astron. Astrophys. **369**, 473 (2001)

Sakamoto, K., Okumura, S.K., Ishizuki, S., Scoville, N.Z.: Astrophys. J. Suppl. Ser. **124**, 403 (1999a)

Sakamoto, K., Okumura, S.K., Ishizuki, S., Scoville, N.Z.: Astrophys. J. **525**, 691 (1999b)

Sakamoto, K., et al.: (in preparation)

Sandage, A.: The Hubble Atlas of Galaxies. Carnegie Institution, Washington (1961)

Sandage, A., Tammann, G.A.: A Revised Shapley Ames Catalog of Bright Galaxies (RSA). Carnegie Institution, Washington (1981)

Sérsic, J.L., Pastoriza, M.: Publ. Astron. Soc. Pac. **79**, 152 (1967)

Shen, J., Sellwood, J.A.: Astrophys. J. **604**, 614 (2004)

Sheth, K., Regan, M.W., Scoville, N.Z., Strubbe, L.E.: Astrophys. J. **592**, (2003)

Sheth, K., et al.: Astrophys. J. **632**, 217 (2005)

Sofue, Y., et al.: Publ. Astron. Soc. Jpn. **55**, 17 (2003)

Sørensen, S.-A., Matsuda, T., Fujimoto, M.: Astrophys. Space Sci. **43**, 491 (1976)

Tully, R.B.: Nearby Galaxies Catalog. Cambridge Univ. Press, Cambridge (1988)

Young, J.S., et al.: Astrophys. J. Suppl. Ser. **98**, 219 (1995)

Investigations of star formation in galaxies using ALMA

Min S. Yun

Originally published in the journal Astrophysics and Space Science, Volume 313, Nos 1–3.
DOI: 10.1007/s10509-007-9642-1 © Springer Science+Business Media B.V. 2007

Abstract ALMA provides unprecedented sensitivity and resolution to study gas and dust emission in the millimeter and submillimeter bands. The magnitude of the improvement is such that not only conventional studies can be done much better but entirely new tools and research fields should also become accessible. In this article, I examine several specific areas where new capabilities of ALMA will bring significant quantitative improvements to the determination of star formation rate and properties of the gas fueling the activities. I propose a survey of nearby galaxies with well measured metallicity gradient during the early phase of the ALMA operation as one of the key science projects.

Keywords Star formation · Molecular gas · Dust · Recombination lines

1 Introduction

In one of the earliest investigations of a physical link between star formation activities and gas distribution in external galaxies, Scoville and Young (1983) found the radial distribution of far-IR, radio continuum, blue light, and Hα in M51 closely resembling that of CO (and presumably H$_2$), with an approximately exponential distribution. In contrast, the λ21 cm neutral hydrogen (HI) emission does not. Such observations suggest that star formation is fueled by molecular gas, following a power law dependence of star formation rate (SFR) on gas density—so called "Schmidt Law",

SFR $\propto \Sigma_{\rm gas}^n$ with $n = 1$–2 (Schmidt 1959). Later, Kennicutt (1989, 1998a) has shown that the power-law index is in the range of $n = 1.3$–1.5 by analyzing a large ensemble of published global gas density and SFR measurements.

Noting the formation of molecular gas as a key step in the star formation process, Blitz and Rosolowsky (2006) have proposed a modified star formation prescription based on pressure determining the degree to which the ISM is molecular: $\Sigma_{\rm SFR} \propto \Sigma_g P_{\rm ext}^{0.92}$, where the external pressure is determined by total gas surface density Σ_g, vertical velocity dispersion v_g, and mid-plane gas and stellar mass density ρ_g and ρ_* ($P_{\rm ext} = (2G)^{0.5} \Sigma_g v_g [\rho_*^{1/2} + (\frac{\pi}{4} \rho_g)^{1/2}]$). This prescription offers a better description of star formation activity in the Milky Way disk and is supported by the remarkable uniform trend in the observed molecular fraction as a function of external pressure in a large sample of nearby galaxies.

Another intriguing idea is a radiation pressure-supported starburst. Motivated by the suggestion of Scoville (2003) that radiation pressure may support a gas disk against gravity in the nuclear starburst regions of ULIRGs, Thompson et al. (2005) produced a detailed model for such a system and proposed a star formation density dependence on gas density, $\Sigma_{\rm SFR} \propto \Sigma_{\rm gas}/\kappa$, where κ is mean opacity of the gas disk. Assuming a characteristic starburst region size of 100 pc (Condon et al. 1991), typical radiation density in these nuclear starburst regions is $F \sim 10^{13}$ $L_\odot$/kpc^2 with a limiting SFR density of $\Sigma_{\rm SFR} \sim 10^3$ $M_\odot$ yr^{-1} kpc^{-2}. The inferred molecular gas disk sizes from the existing measurements are generally significantly larger, $R_{\rm CO} \sim 100$–1000 pc (Scoville et al. 1997; Downes and Solomon 1998) [also C. Wilson in this volume], but the existing data lack sufficient resolution to offer a real insight into this problem. Detailed morphological and dynamical structures of massive gas disks will

M.S. Yun (✉)
Department of Astronomy, University of Massachusetts, Amherst, MA 01106, USA
e-mail: myun@astro.umass.edu

Fig. 1 Greyscale images of the Cygnus HII complex at 60 μm (*left*) and 1.4 GHz (*right*; from Yun et al. 2007). HII regions and HII complexes appear bright at both wavelengths. Each image is about 10 degrees on a side, which corresponds to 350 pc at a distance of 2 kpc

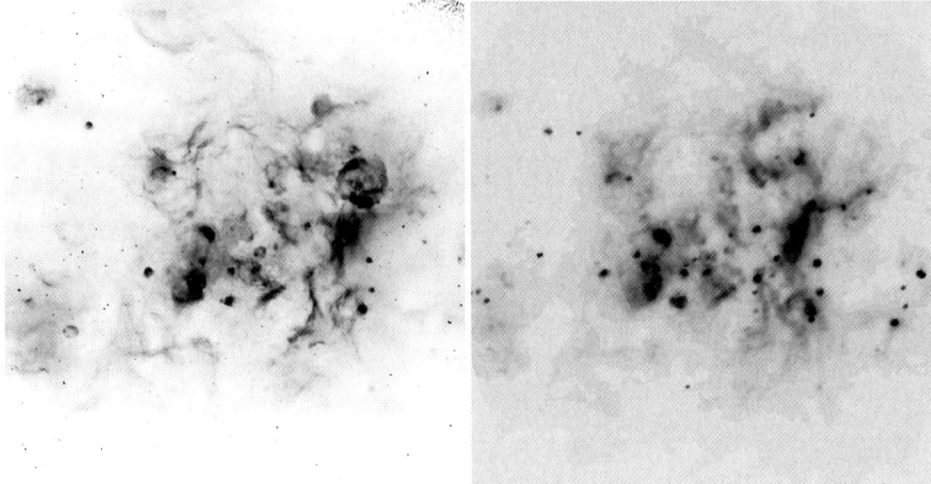

be revealed by future ALMA observations. This is a good example of how ALMA will play a transformational role in the study of star formation in external galaxies.

In this brief review, I explore several specific areas where the superb sensitivity and angular resolution of ALMA will bring significant quantitative improvements to the determination of star formation rate and the properties of the gas fueling the star formation activities.

2 Improved determination of star formation rates

Over the years tracers at all wavelengths, from thermal plasma emission in the X-ray to non-thermal synchrotron emission in the radio, have been explored as quantitative SFR indicators. Extinction and optical depth effects limit the usefulness of traditional tracers such as UV and Hα since massive young stars form inside dusty cocoons located within giant molecular clouds. Nuclear starburst regions in ULIRGs may be optically thick even at wavelengths as long as 100 μm (Scoville et al. 1991; Downes et al. 1993). Spitzer and its IR instruments have yielded remarkable new insights into massive star formation in nearby galaxies (e.g., Gordon et al. 2004). However, the Spitzer angular resolution ($6''$ at 24 μm) is still modest at best.

Star formation rate derived from the far-IR is not immune from complications either because FIR luminosity is an integral property over the past starburst history and as A-type stars dominate the luminosity. Also, not all UV photons are processed by dust, and some escape along the superwinds traced in X-ray and Hα. Tracers of massive star formation in the mm/submm wavelength window accessible by ALMA are particularly appealing because of low optical depth and the sub-arcsec angular resolution achievable by ALMA. The two particularly attractive SFR tracers are thermal free-free emission and hydrogen recombination lines as they reflect the *current* massive star formation rate.

2.1 Thermal free-free emission

The tight correlation between radio continuum and far-IR has been known for a long time, and indeed it is one of the tightest known in astrophysics (see review by Condon 1992). The broad utility of radio continuum as an effective tracer of star formation activity is demonstrated by the fact that >98% of FIR selected IRAS 2 Jy sample galaxies by Yun et al. (2001) follow this correlation, independent of luminosity. Radio AGN contamination rate is less than 1% in the same sample.

The global radio-FIR correlation is generally well explained by the massive star formation process which generates copious amounts of UV radiation that is then reprocessed to FIR by dust, along with the cosmic-ray injection and acceleration by Type-II supernovae. What is surprising is the close correlation between these two tracers even at parsec scales as shown in Fig. 1, as non-thermal synchrotron emission is thought to arise from a kpc scale cosmic ray processing volume. A detailed analysis of spectral energy distribution (SED) for individual emitting regions suggests that both non-thermal synchrotron and thermal free-free emission contribute to the observed spatial correlation.

At wavelengths shorter than 1 cm, radio continuum emission is expected to be optically thin, with significant or even dominant contribution arising from thermal free-free component. At $\lambda \sim 3$ mm ($\nu \sim 100$ GHz), continuum emission may be almost entirely optically thin free-free emission, except for some of the densest ionized regions such as "ultra-compact" HI regions where optical depth may exceed unity.

Thermal free-free emission from an ionized gas cloud with temperature T_e and optical depth τ_{ff} has brightness temperature of $T_b = T_e(1 - e^{-\tau_{ff}})$, where optical depth τ_{ff} is given by $\tau_{ff} = 0.082 T_e^{-1.35} \nu_{\text{GHz}}^{-2.1} \text{EM}$ and emission measure is $\text{EM} \equiv \int n_e^2 dl$. Then for $\tau_{ff} \ll 1$,

$$T_b = \tau_{ff} T_e = 0.082 T_e^{-0.35} \nu_{\text{GHz}}^{-2.1} \text{EM}. \tag{1}$$

The thermal spectral luminosity L_T of an HII region photoionized by hot stars is proportional to the Lyman continuum photon production rate N_{Ly} (see Condon 1992) as

$$N_{Ly} \geq 6.3 \times 10^{52} \left(\frac{T_e}{10^4 \text{ K}} \right) \left(\frac{\nu}{\text{GHz}} \right)^{-0.1} \left(\frac{L_T}{10^{20} \text{ W Hz}^{-1}} \right). \tag{2}$$

2.2 Millimeter/submillimeter hydrogen recombination lines

Hydrogen recombination lines in the optical and IR bands (Hα, Hβ, Pβ, Brα, and Brγ) have been broadly used as a probe of instantaneous measure of SFR for years (see review by Kennicutt 1998b). Recombination lines at meter and centimeter wavelengths are far less useful because they are often weak and are subject to stimulated emission (see review by Brown et al., 1978). Recent studies of millimeter radio recombination lines (RRLs) in Galactic HII regions (Gordon and Walmsley 1990) and starburst galaxies (Seaquist et al. 1994, 1996; Puxley et al. 1997) have revealed surprisingly high line fluxes, and the analysis of line brightness suggests non-LTE and optically thin spontaneous emissions ($S_\nu \propto \nu^2$). High electron density is needed to produce recombination line emission at millimeter wavelengths, and they may serve as a useful indicator of ionized gas density (Seaquist et al. 1996). The line-to-continuum ratio is also a direct probe of electron temperature (Roberts et al. 1991).

A general expressions for an optically thin hydrogen recombination line is $T_L = T_L^*[b(1 - \frac{1}{2}\tau_c\beta)]$ (see Rohlfs and Wilson 1996), which is

$$T_L = 1920 T_e^{-1.5} \text{EM} \left(\frac{\Delta\nu}{\text{kHz}} \right)^{-1} \left[b \left(1 - \frac{1}{2}\tau_c\beta \right) \right] \tag{3}$$

where T_L^* is the LTE line brightness temperature, and the departure coefficient b and the non-LTE factor β can be computed analytically (Salem and Brockehurst 1979; Walmsley 1990). In the Rayleigh-Jeans limit, the line integral is then $S_L \Delta V = 2k \frac{T_L \Omega_b}{\lambda^2} \Delta V$ and

$$S_L \Delta V = 4.8 \times 10^{-4} T_e^{-1.5} \text{EM} \theta^2$$
$$\times \nu_{\text{GHz}} \left[b \left(1 - \frac{1}{2}\tau_c\beta \right) \right] \text{Jy km s}^{-1} \tag{4}$$

where θ is the source angular size in arcsec. For optically thin emission at mm wavelengths, the non-LTE term involving β becomes negligible, and the relative line flux ratio becomes essentially the LTE value. The observed recombination line strengths can then be used to infer T_e and n_e in a straightforward way.

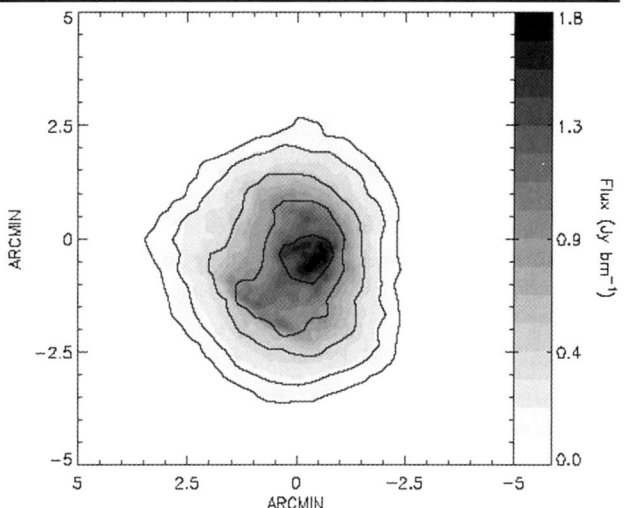

Fig. 2 Derived EM images in contours over the 8.4 GHz continuum image from the VLA+GBT observations by Shepherd et al. (2001) in greyscales. The contours correspond to EM of 0.1, 0.2, 0.5, 1.0, and 1.5×10^6 cm^{-6} pc

2.3 Test case: M42 (Orion Nebula)

As a first test of the utility of mm recombination lines for ionized gas pressure and Lyman continuum photon production rate, we observed and analyzed the recombination line strength and distribution in the prototypical galactic HII region M42 (Orion Nebula). M42 is a pocket of ionized gas that has been described as a "blister" on the surface of the Orion A molecular cloud (O'Dell 2001). The nebula, at a distance of 480 ± 80 pc (Genzel et al. 1981), has a diameter of approximately 8 pc (Hillenbrand 1997) and hosts the Orion Nebula Cluster, which is the nearest region of active high mass star formation, with an extent of about 3 pc (20′). There are about 3500 stars within the 2.5 pc of the cluster center, but the dynamics and evolution of the nebula are primarily shaped by the massive O- and B-type stars near the center ("the Trapezium"). The most massive star θ^1 Ori C is a type O7 star (25 $M_\odot$) and is the dominant source of ionization in this HII region.

We have imaged the central $15' \times 12'$ region of M42 in H41α (92.018 GHz) and H42α (85.673 GHz) at 50″ resolution using the FCRAO 14-m telescope (Fischer et al., in preparation). Using the VLA+GBT 8.4 GHz radio continuum image as a model for the thermal free-free continuum (Brown et al. 1978), these recombination line images are used to compute average EM = 5×10^5 pc cm^{-6} (see Fig. 2) and $n_e = 10^3$ cm^{-3}. The Lyman continuum photon production rate $N_{Ly} = 3$–4×10^{48} sec^{-1} is in good agreement with the expected rate for an O7 star, 4.2×10^{48} sec^{-1}.

2.4 Test case: Arp 220

The prototypical ultraluminous infrared galaxy Arp 220 (D = 72 Mpc) has an FIR luminosity of 1.5×10^{12} $L_\odot$, which translates to a star formation rate of 260 $M_\odot$/yr using the conversion relation given by Kennicutt (1998b). Arp 220 represents an extreme example of a star forming system, both in terms of its high SFR and the total luminosity fraction that emerges in the FIR (>99%). Therefore, Arp 220 is an interesting test case that also contrasts nicely with the Orion Nebula example above. There are important questions for which the new mm recombination line observations may offer a new insight. What is the current star formation rate? Where is the location of the current star formation activity? What are the physical properties of gas in the starburst regions? What is the mass of the neutral and ionized gas fueling the starburst?

Analyzing H92α emission in Arp 220 observed using the VLA, Zhao et al. (1996) found a wide range of physical conditions that could produce the observed line and continuum emission at 8.4 GHz: (1) a cluster of 110 H II regions with $n_e = 10^4$ cm^{-3} and $M_{HII} = 2 \times 10^5$ $M_\odot$; (2) a uniform density slab with $n_e = 10^4$ cm^{-3} and $M_{HII} = 3 \times 10^7$ $M_\odot$; and (3) a uniform density slab with $n_e = 100$ cm^{-3} and $M_{HII} = 6 \times 10^8$ $M_\odot$. Adding IRAM 30-m telescope observations of H31α, H40α, and H42α lines, Anantharamaiah et al. (2000) found it necessary to invoke three *separate* components of ionized gas to account for the observed recombination line strengths, with $n_e = 10^{3-5}$ cm^{-3}, EM $= 10^{5-9}$ pc cm^{-6}, and $M_{HII} = 10^{3-7}$ $M_\odot$.

The H41α spectrum we obtained with a 25 hour long integration using the OVRO millimeter interferometer, shown in Fig. 3, is much narrower in linewidth than the CO lines and is asymmetrical with a pronounced redshifted wing. The high density gas tracer HCN (4–3) line also shows a similar asymmetry, as does the H92α emission. In contrast, the IRAM 30-m telescope H40α and H42α spectra reported by Anantharamaiah et al. (2000) are more symmetric and much brighter, but these spectra are also noisier, obtained with just 3 hours of integration time.

Disregarding the uncertain IRAM 30-m measurements for the moment, the VLA H92α and our OVRO H41α emission can be explained using a single component non-LTE model with $T_e = 7500$ K and EM $= 10^8$ cm^{-6} pc assuming a source size of $1'' \times 0.5''$ (Anantharamaiah et al. 2000). If the emitting region consists of two 100 pc diameter spherical clouds surrounding the two merging nuclei (Sakamoto et al. 1999), then the average density and mass of the ionized gas are $n_e = 10^3$ cm^{-3} and $M_{HII} = 3 \times 10^7$ $M_\odot$. If the emitting source is made of two 100 pc disks with a scaling height of 10 pc, $n_e = 3 \times 10^3$ cm^{-3} and $M_{HII} = 8 \times 10^6$ $M_\odot$. Lastly, if the emission arises in two 100 pc diameter sheets with a thickness of 0.1 pc, then $n_e = 3 \times 10^4$ cm^{-3} and $M_{HII} = 4 \times 10^5$ $M_\odot$. Therefore, we conclude that the mean density of the ionized gas is lower than that of the molecular gas ($n_e = 10^{3-4}$ cm^{-3}), and the total ionized gas mass is only a few percent of the total molecular gas mass.

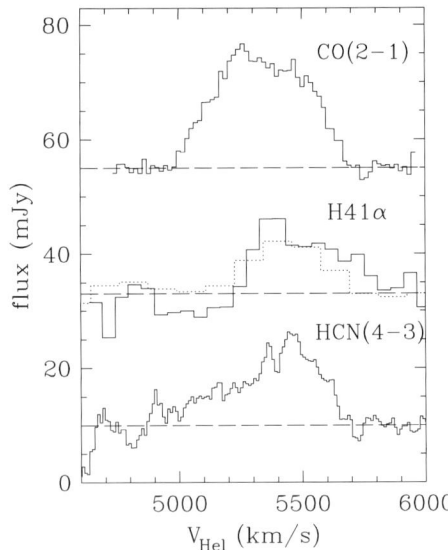

Fig. 3 Our OVRO mm array H41α spectrum is shown in comparison with the CO (2–1), H92α, and HCN (4–3) spectra Yun et al. (2007). The H92α spectrum (Anantharamaiah et al. 2000), shown in *dotted line*, is only that of the eastern nucleus, multiplied by a factor of 30 for an easier comparison

Assuming EM $= 10^8$ cm^{-6} pc and $T_e = 7500$ K with a geometry of two 100 pc diameter clouds for the ionized gas, we compute $N_{Ly} \geq 1.6 \times 10^{55}$ s^{-1}. Using the conversion relation given by Kennicutt (1998b), SFR $= 1.1 \times 10^{-53} N_{Ly}$ $M_\odot$ yr^{-1}, we derive a SFR of ≥ 180 $M_\odot$ per year, which is in good agreement with the SFR derived from the FIR luminosity.

3 Improved determination of gas and dust mass

The flip side of an improved estimation of star formation rate is a better quantitative understanding of gas and dust mass that fuels the star formation activity. All proposed star formation "laws" formulate a star formation rate with a power-law dependence on gas density. Exploring these hypothesis beyond the order of magnitude estimation requires accurate molecular gas mass determination, leading to critical tests of the threshold density for the "Kennicutt law", the HI-to-H$_2$ phase transition in Blitz-Rosoloski relation, and hydrostatic equilibrium in the radiation-supported starburst disks.

Since molecular hydrogen H$_2$ does not have a permanent dipole moment, observations of trace molecules such as CO have been used to estimate total H$_2$ masses of molecular gas clouds. The CO-to-H$_2$ conversion factor ("X-factor") is calibrated specifically by comparing the CO luminosity and virial masses of spatially resolved molecular clouds in the Milky Way—see Fig. 4 and a review by Young and Scoville (1991). This conversion relation has been shown to be robust against variations in physical properties such as gas

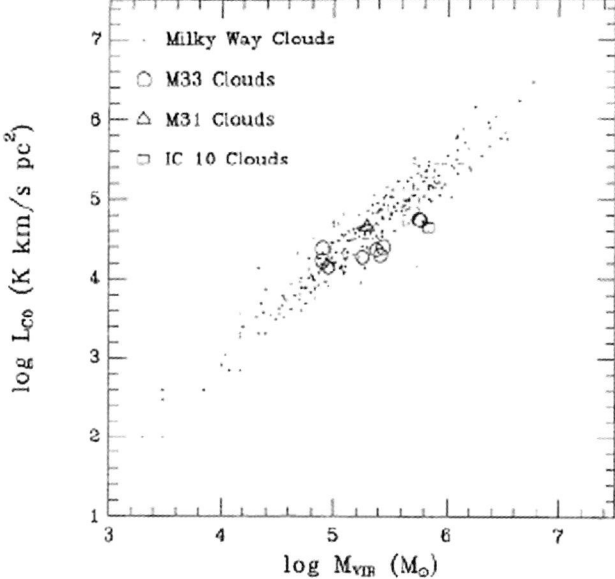

Fig. 4 Comparison of CO luminosity and virial masses of spatially resolved molecular clouds in the Milky Way and nearby galaxies (from Young and Scoville 1991)

density, temperature, and even metallicity to a degree. However, the range of properties of the gas fueling astrophysical phenomena in external galaxies is much broader than commonly found in the Milky Way disk or in a few nearby galaxies. A significant impact of chemical evolution as well as excitation dependence on density, UV, X-ray, and shock heating have been well documented in recent studies of ISM in nearby galaxies (see discussions by J. Turner, C. Wilson, and R. Regan in this volume). For example, presence of substantial warm, dense gas in the inner parsec of our Galactic Center is revealed from the synthesis of measurements from the radio to IR (see Shukla et al. 2004), despite the inference of an apparent neutral gas gap from earlier lower density and temperature tracer observations. Similarly, a joint analysis of new, multiple submm molecular transitions by Iono et al. (2007) has led to a significantly higher mean gas density and temperature for the gas in the nuclear starburst regions of NGC 6240 ($n \sim 10^5$ cm^{-3}, $T \sim 100$ K) than previously reported.

Superb angular resolution and sensitivity of ALMA should provide a definitive calibration of the X-factor from the observations of thousands of suitable galaxies. For the Galactic X-factor determination, the CfA/Columbia CO survey (Dame et al. 2001) mapped the Milky Way disk with a $7.5'$ resolution (about 25 pc at a 10 kpc distance), while the FCRAO CO Galactic Plane Surveys (Heyer et al. 1998) mapped the Galactic disk with a $45''$ resolution (2.5 pc at 10 kpc distance). ALMA can image CO emission in external galaxies with a $0.1'' = 1.5$ pc resolution at the distance of M81 (3.3 Mpc) and with a 10 pc resolution at the distance of Virgo cluster (18 Mpc). An $L_{CO} - M_{VIR}$ plot like

the one shown in Fig. 4 can be constructed for *every galaxy observed with ALMA out to a* $\sim$20 Mpc *distance*. Such an analysis not only offers the X-factor appropriate for that particular galaxy and an accurate total gas mass, but it also should allow the determination of any radial and azimuthal trends within the galaxy as well. A survey of nearby galaxies with well known metallicity gradient (e.g., M101, NGC628, NGC3344 (Zaritsky et al. 1994)) should be conducted during the early phase of the ALMA operation as a key science project.

One of the persistent puzzles in the study of gas and dust in external galaxies is that the gas-to-dust ratios derived are significantly larger than the Galactic value of $\sim$150 (Spitzer 1978). This discrepancy is thought to arise from the far-IR ($\lambda \leq 100$ µm) measurements used to derive the dust masses in external galaxies are not sensitive to the cold (10–20 K) dust component, and the derived gas-to-dust mass ratio is expected to approach the Galactic value with the inclusion of longer wavelength measurements. Therefore it is somewhat surprising that the average M_{HI+H_2}/M_d ratio derived for the Spitzer 160 µm sample is still $\sim$ 1000 despite the inclusion of measurements at nearly twice the wavelength (Yun et al. 2007). The observed SEDs indeed demonstrate that the same "warm" dust responsible for the IRAS 60 and 100 µm emission can also largely account for the Spitzer 160 µm flux as well (also see Devereux et al. 1990). Dunne et al. (2001) have shown that the gas-to-dust ratio can be lowered to a mean value of $\sim$250 if their 450 and 850 µm measurements of a sample of nearby galaxies are interpreted as arising from a distinct cold (17–24 K) dust component with optically thin emission ($\beta = 2$). Some degeneracy between dust temperature and emissivity (lower T_d for a larger value of β) is inherent to these single temperature dust emission models, and the difference in the dust mass arises mainly from the difference in the assumed dust emissivity. New high quality measurements of a larger sample of galaxies with a broader wavelength coverage is needed to achieve a significant advancement on this matter, and this is another area where ALMA enables a transformational science by yielding the requisite sensitive and accurate multi-wavelength measurements.

4 Discussion and summary

Deriving a reliable star formation rate for a heavily obscured systems such as young super stellar clusters in the Antennae (2006) or the starburst nuclei of ULIRGs like Arp 220 has to rely on optically tracers at longer wavelengths. Free-free emission and hydrogen recombination lines in the millimeter and submillimeter wavelengths accessible by ALMA are potentially powerful new tools. The analyses of free-free and recombination lines at millimeter wavelengths in Orion

Nebula and Arp 220 show them to be quite promising, and the Lyman continuum photon production rates derived are in excellent agreement with previous estimates.

The presence of dust within HII regions that may quench the ionizing photons is a potentially serious concern for using free-free continuum and recombination lines as a direct probe of the massive star formation rate. This is the reason why the Lyman continuum photon production rate in (3) is given as a lower limit. The excellent agreement between the new and the previous estimates for Orion Nebula and Arp 220 suggests some robustness for the millimeter free-free and recombination line methods. Interpreting the continuum at λ 3–10 mm as purely free-free emission may be problematic as SED analysis often show significant thermal dust and synchrotron contribution as well (see Fig. 1 by Condon 1992). Observations at centimeter wavelengths suggests that the extent of dense ionized gas is more compact than the extent of the nuclear molecular gas disks traced in CO (Zhao et al. 1996, 1997), and the recombination line emission may be biased to the most intense star forming regions.

A particularly attractive transformational capability of ALMA is its ability to resolve individual GMCs in external galaxies and determine their virial masses. Applying the same methods used for the Galactic GMC calibration, it should be possible to determine the conversion relation from CO luminosity to H_2 mass for 1000 or more galaxies, and the use of *in situ* determination of gas mass will greatly improve the tests and calibrations of various star formation "laws". Therefore, I propose a survey of nearby galaxies with well measured metallicity gradient during the early phase of the ALMA operation as one of its key science projects.

Acknowledgements   I acknowledge the support of the American Astronomical Society and the National Science Foundation in the form of an International Travel Grant, which enabled me to attend this conference.

References

Anantharamaiah, K.R., et al.: Starburst in the ultraluminous galaxy Arp 220: Constraints from observations of radio recombination lines and continuum. Astrophys. J. **537**, 613 (2000)

Blitz, L., Rosolowsky, E.: The role of pressure in GMC formation II: The H_2-pressure relation. Astrophys. J. **650**, 933 (2006)

Brandl, B.R., et al.: Deep near-infrared imaging and photometry of the antennae galaxies with WIRC. Astrophys. J. **635**, 280 (2006)

Brown, R.L., Lockman, F.J., Knapp, G.R.: Radio recombination lines. Annu. Rev. Astron. Astrophys. **16**, 445 (1978)

Condon, J.J.: Radio emission from normal galaxies. Annu. Rev. Astron. Astrophys. **30**, 575 (1992)

Condon, J.J., et al.: Compact starbursts in ultraluminous infrared galaxies. Astrophys. J. **378**, 65 (1991)

Dame, T.M., Hartmann, D., Thaddeus, P.: The Milky Way in molecular clouds: A new complete CO survey. Astrophys. J. **547**, 792 (2001)

Devereux, N.A., Young, J.S.: The gas/dust ratio in spiral galaxies. Astrophys. J. **359**, 42 (1990)

Downes, D., Solomon, P.M.: Rotating nuclear rings and extreme starbursts in ultraluminous galaxies. Astrophys. J. **507**, 615 (1998)

Downes, D., Solomon, P.M., Radford, S.J.E.: Molecular gas mass and far-infrared emission from distant luminous galaxies. Astrophys. J. **414**, L13 (1993)

Dunne, L., Eales, S.A.: The SCUBA local universe galaxy survey-II. 450-μm data: Evidence for cold dust in bright IRAS galaxies. Mon. Not. R. Astron. Soc. **327**, 697 (2001)

Genzel, R., et al.: Proper motions and distances of H_2O maser sources I—The outflow in Orion-KL. Astrophys. J. **244**, 884 (1981)

Gordon, M.A., Walmsley, C.M.: An observational study of millimeter-wave recombination lines. Astrophys. J. **365**, 606 (1990)

Gordon, K.D., et al.: Spatially resolved ultraviolet, Hα, infrared, and radio star formation in M81. Astrophys. J. Suppl. Ser. **154**, 215 (2004)

Heyer, M.H., et al.: The five college radio astronomy observatory CO survey of the outer galaxy. Astrophys. J. Suppl. Ser. **115**, 241 (1998)

Hillenbrand, L.A.: On the stellar population and star-forming history of the Orion Nebula cluster. Astron. J. **113**, 1733 (1997)

Iono, D., et al.: High resolution imaging of warm and dense molecular gas in the nuclear region of the luminous infrared galaxy NGC6240. Astrophys. J. **659**, 283 (2007)

Kennicutt, R.C.: The star formation law in galactic disks. Astrophys. J. **344**, 685 (1989)

Kennicutt, R.C.: The global Schmidt law in star-forming galaxies. Astrophys. J. **498**, 541 (1998a)

Kennicutt, R.C.: Star formation in galaxies along the Hubble sequence. Annu. Rev. Astron. Astrophys. **36**, 189 (1998b)

O'Dell, C.R.: The Orion Nebula and its associated population. Annu. Rev. Astron. Astrophys. **39**, 990 (2001)

Puxley, P.J., et al.: Observations of millimeter-wavelength hydrogen recombination lines in the galaxy NGC 253. Astrophys. J. **485**, 143 (1997)

Roberts, D.A., et al.: VLA radio recombination line observations of Sagittarius A West. Astrophys. J. **366**, L15 (1991)

Rohlfs, K., Wilson, T.L.: In: Tools of Radio Astronomy, 3rd edn., pp. 327–349. Springer, Berlin (1996)

Sakamoto, K., et al.: Counterrotating nuclear disks in Arp 220. Astrophys. J. **514**, 68 (1999)

Salem, M., Brockehurst, M.: A table of departure coefficients from thermodynamic equilibrium b_N factor for hydrogenic ions. Astrophys. J. Suppl. Ser. **39**, 633 (1979)

Scoville, N.: Starburst and AGN connections and models. J. Korean Astron. Soc. **36**, 167 (2003)

Scoville, N., Young, J.S.: The molecular gas distribution in M51. Astrophys. J. **265**, 148 (1983)

Scoville, N.Z., et al.: Dust and gas in the core of Arp 220 (IC 4553). Astrophys. J. **366**, L5 (1991)

Scoville, N.Z., Yun, M.S., Bryant, P.M.: Arcsecond imaging of CO emission in the nucleus of Arp 220. Astrophys. J. **484**, 702 (1997)

Schmidt, M.: The rate of star formation. Astrophys. J. **129**, 243 (1959)

Seaquist, E.R., Kerton, C.R., Bell, M.B.: Millimeter recombination lines and the state of ionized gas in M82. Astrophys. J. **429**, 612 (1994)

Seaquist, E.R., et al.: Millimeter recombination line emission in the starburst galaxy M82. Astrophys. J. **465**, 691 (1996)

Shepherd, D.S., Maddalena, R., McMullin, J.P.: The Orion Nebula in 3.6 cm continuum emission: The first combination of VLA and GBT data. Bull. Am. Astron. Soc. **33**, 1502 (2001)

Shukla, H., Yun, M.S., Scoville, N.Z.: Dense, ionized, and neutral gas surrounding Sagittarius A*. Astrophys. J. **616**, 231 (2004)

Spitzer, L.: In: Physical Processes in the Interstellar Medium, p. 333. Wiley, New York (1978)

Thompson, T.A., Quataert, E., Murray, N.: Radiation pressure-supported starburst disks and active galactic nucleus fueling. Astrophys. J. **630**, 167 (2005)

Walmsley, C.M.: Level populations for millimeter recombination lines. Astron. Astrophys. Suppl. Ser. **82**, 201 (1990)

Young, J.S., Scoville, N.Z.: Molecular gas in galaxies. Annu. Rev. Astron. Astrophys. **29**, 581 (1991)

Yun, M.S., Reddy, N.A., Condon, J.J.: Radio properties of infrared-selected galaxies in the IRAS 2 Jy sample. Astrophys. J. **554**, 803 (2001)

Yun, M.S. et al.: Cold gas and dust properties of the brightest 160 μm sources in the Spitzer first look survey field. Astrophys. J. (2007, submitted)

Zaritsky, D., Kennicutt, R.C., Huchra, J.P.: HII regions and the abundance properties of spiral galaxies. Astrophys. J. **420**, 87 (1994)

Zhao, J.H., et al.: Radio recombination lines from the nuclear regions of starburst galaxies. Astrophys. J. **472**, 54 (1996)

Zhao, J.H., et al.: High-density, compact HII regions in the starburst galaxies NGC 3628 and IC 694: High-resolution VLA observations of the H92 alpha radio recombination line. Astrophys. J. **482**, 186 (1997)

Probing the feeding and feedback of activity near and far

S. García-Burillo · F. Combes · J. Graciá-Carpio ·
A. Usero · M. Guélin

Originally published in the journal Astrophysics and Space Science, Volume 313, Nos 1–3.
DOI: 10.1007/s10509-007-9627-0 © Springer Science+Business Media B.V. 2007

Abstract High-resolution CO maps are an essential tool to
search for observational evidence of AGN fueling in galaxy
nuclei. While their capabilities will be surpassed by ALMA,
current mm-interferometers can already provide relevant information on scales which are critical for the process of angular momentum transfer in fueling the AGN. In this context we present the latest results issued from the NUclei of
GAlaxies (NUGA) project, a high-resolution (0.5″–1″) CO
survey of low luminosity AGNs conducted with the IRAM
Plateau de Bure interferometer (PdBI). The use of more specific molecular tracers of dense gas can probe the feedback
influence of activity on the chemistry and energy balance
in the interstellar medium of nearby galaxies, a prerequisite to understanding how feedback operate at higher redshift galaxies. We discuss the results obtained in an ongoing
study devoted to probe the feedback of activity from nearby
Seyferts to high-redshift QSO.

S. García-Burillo (✉) · J. Graciá-Carpio · A. Usero
Observatorio Astronómico Nacional, Alfonso XII, 3,
28014 Madrid, Spain
e-mail: s.gburillo@oan.es

J. Graciá-Carpio
e-mail: j.gracia@oan.es

A. Usero
e-mail: a.usero@oan.es

F. Combes
Observatoire de Paris, LERMA, 61, Av. de l'Observatoire,
75014 Paris, France
e-mail: Francoise.combes@obspm.fr

M. Guélin
IRAM, 300, Rue de la Piscine, 38406 St Mt d'Hères, France
e-mail: guelin@iram.fr

Keywords Galaxies: active · Galaxies: ISM · Galaxies:
starburst · Infrared: galaxies · ISM: molecules · Radio lines:
galaxies

1 AGN fueling: the NUGA survey

While it is widely accepted that the nuclei of most galaxies
host supermassive black holes, the fact that only a fraction
of galaxies in the Local Universe are active indicates that
AGN fueling *at present* is an episodic phenomenon. This
intermittency suggests that AGN fueling is a finely tuned
self-regulated process in nearby galaxies. For high luminosity AGNs (HLAGNs) the presence of ∼kpc-scale non-
axisymmetric perturbations and the onset of activity seem
to be linked. However the quest of a *universal* mechanism
for AGN feeding in LLAGNs is complicated by the fact that
the AGN duty cycle is bound to be short ($\sim 10^{7-8}$ yrs) (e.g.,
see review by Combes 2003). This time-scale problem translates into a spatial-scale problem: only by probing the scales
which are critical for removing angular momentum during
the very last stages of the process (≤ 10–100 pc), the fueling
problem can be addressed in LLAGNs. With this objective,
we have conducted a high-resolution ($\sim 0.5''$–$1''$) CO survey
of a sample of 25 LLAGNs carried out with the PdBI.

The CO line maps provided by the NUclei of GAlaxies
(NUGA) project (fully described by García-Burillo et al.
2003a and García-Burillo et al. 2003b) can probe molecular gas in the circumnuclear disks of these galaxies with
unprecedented resolution and sensitivity. This allows us
to search for observational evidences of ongoing feeding
through a detailed study of the distribution and kinematics of molecular gas. Furthermore with the information on
the stellar potentials at hand, obtained through NIR images
for most of the NUGA galaxies, gravitational torque maps

NGC4579

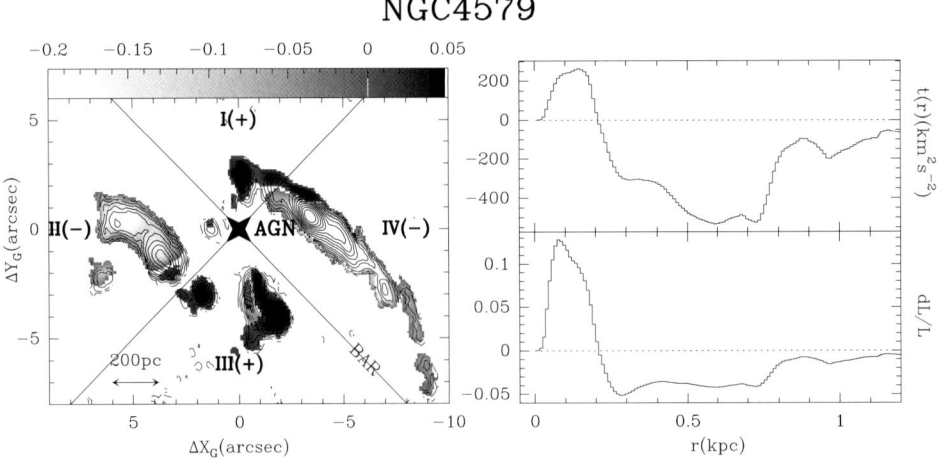

Fig. 1 (*left*) We overlay the $^{12}CO(1-0)$ contours with the map of the effective angular momentum variation in the nucleus of NGC 4579. The grey scale is normalized to the maximum absolute value in the map. The torques change sign as expected if the *butterfly* diagram, defined by the orientation of quadrants I-to-IV, can be attributed to the action of the large-scale bar of NGC 4579. (*right*) Torques are systematically strong and negative for the bulk of the molecular gas in NGC 4579, from $r = 200$ pc out to $r = 1200$ pc. In the vicinity of the AGN, however, torques become positive and AGN feeding is not presently favored

and the ensuing gas flow time-scales can be derived. We can thus investigate on a quantitative basis whether gravity torques *alone* are able explain the low level of nuclear activity in these galaxies. The first results obtained from the stellar torque analysis in four representative LLAGNs have been published by García-Burillo et al. (2005). Gravity torques exerted by the stellar potential on the gas disks of these LLAGNs seem to fail in accounting for the feeding of the AGN. In a first step, gravity torques help driving the gas inwards and feeding a nuclear starburst on scales of ~a few 100 pc. On smaller scales, however, stellar torques have virtually no role in AGN fueling in the current epoch: torques on the gas are not negative all the way to the center, but on the contrary they become positive and would paradoxically halt the feeding (see Fig. 1 adapted from García-Burillo et al. 2005). In a second step, a mechanism complementary to the present configuration of gravity torques is needed to drive gas inflow on smaller scales.

Different scenarios can be envisaged to solve the puzzle. An extremely short-lived ($<10^7$ yrs) perturbation in the stellar potential (e.g., nuclear bar or $m = 1$ instability) could be responsible for the fueling, whereas it would be very difficult to have a smoking gun evidence of its action (see however the case of NGC2782 discussed by Hunt et al., in preparation). Alternatively, gravity torques may need to be assisted to drive the gas to the center by other means independent of the intervention of the stellar potential. García-Burillo et al. (2005) have discussed the efficiency of viscosity as a potential driver of AGN fueling in the inner circumnuclear disks of the four LLAGNs where the gravity torque failure was established. In their paper they proposed a new evolutionary scenario in which the onset of nuclear activ-

ity can be understood as a recurrent phase during the typical lifetime of any galaxy. In this scenario, where viscosity could be efficient at dissolving nuclear rings at the Inner Lindblad Resonance (ILR) region of barred AGN hosts, the recurrence of activity in galaxies is indirectly related to that of the bar instabilities although the various active phases are not necessarily coincident with the maximum strength of a single bar episode. These activity episodes may appear at different evolutionary stages of the bar potential, depending on the balance between gravity torques and viscosity. Viscosity can indeed counter-balance positive gravity torques inside the ILR if the bar strength is lowered due to secular evolution. Several (~3–4) episodes of activity may appear per bar cycle. The wide variety of morphologies identified in the CO maps of NUGA targets corroborates that there might not be a universal pattern associated with LLAGNs (García-Burillo et al. 2003a).

We are currently extending the gravity torque analysis to the whole NUGA sample by including also the information provided by HI maps of the galaxies. These HI maps obtained at the VLA (see Haan et al., in preparation) are crucial to complete the picture of the gas flow time-scales from the outer region down to the inner playground of the galaxy disks; this will test the validity of our fueling scenario.

2 The feedback of activity

The onset of activity in galaxies, sometimes accompanied by a nuclear starburst (SB) episode, can have a profound influence on the properties of the AGN host interstellar medium. In particular, activity can change the physical and chemical properties of molecular gas in galactic nuclei through the action of strong radiation fields (UV photons and X-rays) and violent mass flows (potential drivers of shock chemistry). The use of chemical diagnostic tools is essential in order to trace the evolution of the SB/AGN phenomenon. But first and foremost, quantifying the feedback of activity is a prerequisite to deriving an accurate estimate of the mass of dense molecular gas which is more directly involved in the episodes of star formation and/or AGN fueling. This information is required to characterize the star formation efficiency (SFE) law which should be based on a reliable estimate of the dense molecular gas content in galaxies.

As fully discussed in Sect. 2.1, there is observational evidence of *overluminous* $J = 1$–0 lines of HCN in the nuclei of nearby Seyferts (e.g., Tacconi et al. 1994; Kohno et al. 2001; Usero et al. 2004), as well as in some luminous and ultraluminous infrared galaxies (LIRGs and ULIRGs) (e.g., Gao and Solomon 2004; Graciá-Carpio et al. 2006). In these galaxies, the HCN(1–0) line is overluminous with respect to CO(1–0), but also with respect to other tracers more adapted to probe the dense molecular gas phase ($n > a$ few 10^4 cm^{-3}), such as the HCO$^+$(1–0) line. With no other constraints at hand (e.g., observations in higher-J transitions), overluminous HCN(1–0) lines can be simply explained by a scenario invoking a higher fraction of very dense ($n > 10^5$ cm^{-3}) molecular gas in these galaxies (Gao and Solomon 2004). However, the fact that this result can also be explained by an enhancement of HCN abundances (Tacconi et al. 1994; Usero et al. 2004; Graciá-Carpio et al. 2006) or also by non-collisional excitation of the lines (García-Burillo et al. 2006) casts doubts on the reliability of HCN lines *alone* as a *true quantitative* tracer of dense molecular gas. To elucidate the question we need multiline and multispecies observations, aimed at disentangling the effects of excitation from those of chemistry. Examples of this effort are the surveys presented by Graciá-Carpio et al. (2007) and Papadopoulos et al. (2007).

2.1 Probing AGN feedback near and far

Molecular gas can be exposed to strong X-rays close to the central engine of AGN. X-rays can extend their power deep into the clouds by their ability to penetrate huge gas column densities out to $A_v = 100$–1000. X-ray dominated regions (XDR) can produce an increase of the gas phase abundances of a certain set of ions, radicals and molecular species (e.g., HCN) (Lepp and Dalgarno 1996). Moreover, AGNs can be strong MIR emitters; this complicates the interpretation of observations (e.g., HCN lines) if IR fluorescent excitation does play a role for certain molecular lines. The role of these two typical AGN ingredients (XDR chemistry and IR pumping) and their possible allies (SB-driven chemistry: hot cores?) must be assessed before converting luminosities into masses.

The large HCN/CO abundance ratio derived in the nucleus of the Seyfert 2 galaxy NGC 1068 by Tacconi et al. 1994 was the first observational evidence that molecular gas chemistry can be shaped by activity. Usero et al. (2004) completed observations of the CND of NGC 1068 using the 30m telescope for eight molecular species. The global analysis of the combined survey suggests that the bulk of the molecular gas in the circumnuclear disk of NGC 1068 has become a giant XDR.

Graciá-Carpio et al. (2006) have recently completed observations with the IRAM 30 m telescope in the 1–0 line of HCO$^+$ of a sample of 16 galaxies including 10 LIRGs and 6 ULIRGs. The results of the first HCO$^+$ survey in LIRGs and ULIRGs show that the HCN/HCO$^+$ luminosity ratio increases with L_{IR}. This a priori unexpected trend, confirmed by new HCN and HCO$^+$ observations (see Graciá-Carpio et al., this volume) provides evidence that HCN may not be a fair tracer of dense gas in the most extreme LIRGs. A plausible scenario accounting for the observed trends implies that X-rays may shape the chemistry of molecular gas at $L_{IR} > 10^{12} L_\odot$. Alternatively, it has also been argued that the abundance of HCN can be enhanced in the molecular gas closely associated with high-mass star forming regions. In either case the reliability of HCN as a straightforward tracer of dense molecular gas in ULIRGs should be put on hold (Graciá-Carpio et al. 2006). If we assume that HCN(1–0) line luminosities are roughly proportional to the dense mole-

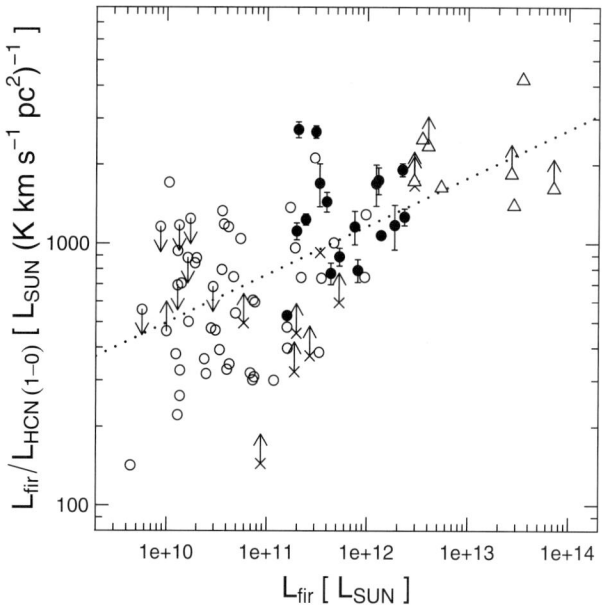

Fig. 2 New HCN(1–0) observations obtained by Graciá-Carpio et al. (2007) (*filled circles*) show that the star formation efficiency (SFE) measured with respect to the dense gas mass derived from HCN is not constant as a function of FIR luminosity. Data from Gao and Solomon (2004) appear as open circles; data of QSO-hosts are taken from Evans et al. (2006) (*crosses*) and data from high-redshift galaxies are taken from Wagg et al. (2005) and Carilli et al. (2005) (*open triangles*)

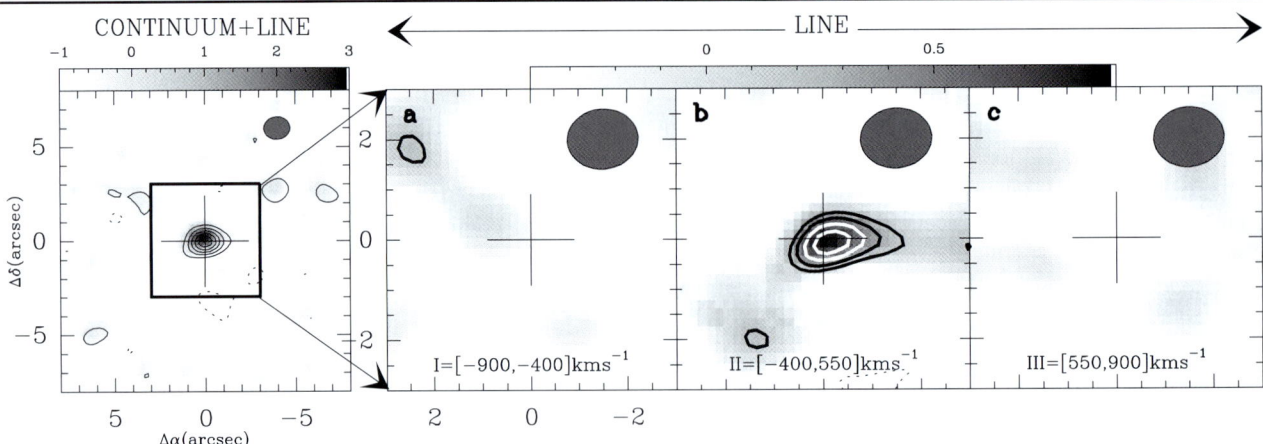

Fig. 3 The first detection of HCO$^+$ emission at redshift $z \sim 4$ in the QSO APM 08279 (García-Burillo et al. 2006). We show the continuum+line emission (*left panel*) and HCO$^+$(5–4) line emission in the three velocity intervals shown (*right panels*)

cular gas, the new HCN data of Graciá-Carpio et al. (2007) included in Fig. 2 shows evidence that the SFE is not constant as a function of L_{FIR}, contrary to the claim of Gao and Solomon (2004). It is tempting to foresee that a similar plot derived from HCO$^+$ line data would expand differences in the SFE among galaxies as a function of L_{FIR}, according to the findings of Graciá-Carpio et al. (2006). These results illustrate the use of chemical diagnostics to address questions such as the variation of SFE in galaxies.

Motivated by our results in ULIRGs, we have recently made observations with the PdBI of the broad absorption line quasar APM 08279 at $z = 3.9$. This source has the enormous potential of being a powerful IR emitter, with a clear dominant contribution from the AGN; taken together, these ingredients make of this source an ideal testbed where to probe the feedback of activity in the terms discussed above. Emission of high-J CO lines ($J = 9$–8 and 4–3), mapped by Downes et al. (1999), revealed the presence of a circumnuclear disk of hot and dense molecular gas. Wagg et al. (2005) reported the detection of HCN (5–4) emission in this quasar using the IRAM PdBI. García-Burillo et al. (2006) have recently reported the detection of HCO$^+$ (5–4) emission in APM 08279 (Fig. 3). The HCN/HCO luminosity ratio measured by García-Burillo et al. (2006) is $\sim$3 times larger than that predicted by simple radiative transfer models which assume that excitation of the two lines is collisional and that chemical abundances for the two molecular species are comparable. More recently we were able to detect HNC(5–4) emission and tentatively detect CN through the 4–3 line in this source using the PdBI (See Fig. 4 adapted from Guélin et al. 2007). If we assume that the excitation of these lines is mainly collisional, the inferred abundance of HCN in APM 08279 would be *anomalously* high with respect to CO, but, most notably, also with respect to HCO$^+$ and CN: [HCN]/[CO] $\sim 10^{-2}$–10^{-3},

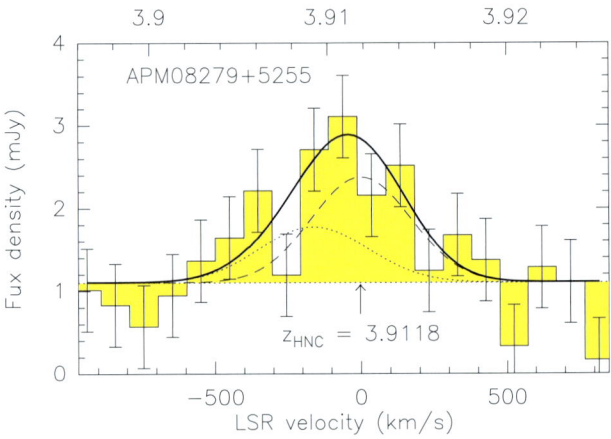

Fig. 4 Spectrum of the HNC(5–4) and CN(4–3) emissions from APM 08279 (Guélin et al. 2007). The *thick line* represents the best fit spectrum and the *dotted and dashed lines* the contributions from CN and HNC

[HCN]/[HCO$^+$] $\sim$ 10 and [HCN]/[CN] $\geq$ 10. Of particular note, the inferred abundances of HNC and HCN are comparable: [HNC] $\sim 0.6 \times$ [HCN]. While this result casts doubts on HCN as an unbiased tracer of the dense gas in AGNs, it should be emphasized that, to date, there is no simple chemical scheme (XDR?, PDR?, hot core-like chemistry?) able to account for all the abundance ratios derived in APM 08279.

Alternatively (García-Burillo et al. 2006) and (Guélin et al. 2007) argue that the excitation of HCN, HCO$^+$ and HNC lines in APM 08279 may be not only collisional, but also radiative. A confirmation of that possibility would require the observation of lower-J lines of these molecular species, together with a full treatment of IR pumping in the radiative transfer models. These scenarios have completely different but equally relevant implications for the interpretation of high-J molecular line observations of dense gas in other high-redshift galaxies.

References

Carilli, C.L., Solomon, P., Vanden Bout, P., et al.: Astrophys. J. **618**, 586 (2005)

Combes, F.: In: Collin, S., Combes, F., Shlosman, I. (eds.) Active Galactic Nuclei: from Central Engine to Host Galaxy. ASP Conference Series, vol. 290, p. 411 (2003)

Downes, D., Neri, R., Wiklind, T., Wilner, D.J., Shaver, P.A.: Astrophys. J. Lett. **513**, L1 (1999)

Evans, A.S., Solomon, P.M., Tacconi, L.J., Vavilkin, T., Downes, D.: Astron. J. **132**, 2398 (2006)

Gao, Y., Solomon, P.M.: Astrophys. J. Suppl. Ser. **152**, 63 (2004)

García-Burillo, S., Combes, F., Eckart, A., et al.: In: Collin, S., Combes, F., Shlosman, I. (eds.) Active Galactic Nuclei: from Central Engine to Host Galaxy. ASP Conference Series, vol. 290, p. 423 (2003a)

García-Burillo, S., Combes, F., Hunt, L.K., et al.: Astron. Astrophys. **407**, 485 (2003b)

García-Burillo, S., Combes, F., Schinnerer, E., Boone, F., Hunt, L.K.: Astron. Astrophys. **441**, 1011 (2005)

García-Burillo, S., Graciá-Carpio, J., Guélin, M., et al.: Astrophys. J. Lett. **645**, L17 (2006)

Graciá-Carpio, J., García-Burillo, S., Planesas, P., Colina, L.: Astrophys. J. Lett. **640**, L135 (2006)

Graciá-Carpio, J., García-Burillo, S., Planesas, P.: Dense molecular gas in a sample of LIRGs and ULIRGs: the low-redshift connection to the huge high-redshift starbursts and AGNs. Astrophys. Space Sci. (2007) doi:10.1007/s10509-007-9629-y

Guélin, M., Salomé, P., Neri, R., et al.: Astron. Astrophys. **462**, L45 (2007)

Kohno, K., Matsushita, S., Vila-Vilaró, B., et al.: The central kiloparsec of starbursts and AGN: the La palma connection. ASP Conf. Ser. **249**, 672 (2001)

Lepp, S., Dalgarno, A.: Astron. Astrophys. **306**, L21 (1996)

Papadopoulos, P.P., Greve, T.R., van der Werf, P., Müehle, S., Isaak, K., Gao, Y.: A large CO and HCN line survey of luminous infrared galaxies. astro-ph/0701829v1 (2007)

Tacconi, L.J., Genzel, R., Blietz, M., et al.: Astrophys. J. **426**, L77 (1994)

Usero, A., García-Burillo, S., Fuente, A., Martín-Pintado, J., Rodríguez-Fernández, N.J.: Astron. Astrophys. **419**, 897 (2004)

Wagg, J., Wilner, D.J., Neri, R., Downes, D., Wiklind, T.: Astrophys. J. Lett. **634**, L13 (2005)

Chemical complexity in galaxies

Jean L. Turner · David S. Meier

Originally published in the journal Astrophysics and Space Science, Volume 313, Nos 1–3.
DOI: 10.1007/s10509-007-9633-2 © Springer Science+Business Media B.V. 2007

Abstract ALMA will be able to detect a broad spectrum of molecular lines in galaxies. Current observations indicate that the molecular line emission from galaxies is remarkably variable, even on kpc scales. Imaging spectroscopy at resolutions of an arcsecond or better will reduce the chemical complexity by allowing regions of physical conditions to be defined and classified.

Keywords Galaxies: ISM · Astrochemistry · Galaxies: individual (IC 342, M 82)

1 Introduction

A rich spectrum of molecular lines is available at millimeter and submillimeter wavelengths for the study of molecular clouds in galaxies. A recent single dish survey in the $\lambda = 2$ mm band of the starburst Sc galaxy NGC 253 revealed lines from 25 (Martín et al. 2006) of the over 35 molecular species detected to date in external galaxies (see contribution by S. Martín in this volume). Within the reach of ALMA will be thousands of transitions of over a hundred molecules, bringing a vast set of potential diagnostics of molecular gas and physical conditions within galaxies.

With such chemical riches comes complexity. Existing observations of the most commonly used molecular tracers, such as CO, HCN, H_2, and ammonia, often give conflicting results for the properties of molecular gas in galaxies. Mean densities inferred over decaparsec scales of $\sim 10^3$ cm^{-3} to $\sim 10^6$ cm^{-5}, depending on which molecular tracer is used. Temperatures inferred from molecular lines can also vary, from $\sim$10–900 K. While CO emission, particularly in spiral arms, appears to be optically thick, as expected from observations of Galactic clouds, there appears to be a widespread diffuse and emissive optically thin CO component in the interarm regions of spiral galaxies, mixed in with the thick gas (Wiklind et al. 1990; Crosthwaite et al. 2002).

Spatial resolution can provide key information to resolve the complexity and confusion in molecular spectra of galaxies, by isolating areas of common physical conditions. Arcsecond imaging of molecular line emission with millimeter arrays gives resolutions comparable to individual giant molecular clouds in the nearest galaxies. Dense cloud tracers identify where the dense cloud cores are (Downes et al. 1992), and their gas properties should be very different from those of diffuse molecular gas found between spiral arms (Wright et al. 1993). Nuclear activity can affect relative molecular abundances (Helfer and Blitz 1997) on these sizescales (see the contribution by S. García-Burillo in this volume). These small-scale variations in molecular emission properties are of interest to astrophysicists, since these are potential diagnostics of physical processes such as shocks or irradiation. The variations are also of interest to astrochemists, since where molecules are found within galaxies may provide clues toward solving the mystery of how they form. ALMA will be a tool for both astrophysicists and astrochemists, with exquisite sensitivity to many different molecules and transitions.

An example of the imaging spectroscopy of a nearby normal spiral galaxy, IC 342, with data from the Owens Valley Millimeter Array, is presented in the next section.

J.L. Turner (✉)
Department of Physics and Astronomy, UCLA, Los Angeles, CA 90095-1547, USA
e-mail: turner@astro.ucla.edu

D.S. Meier
Jansky Fellow, NRAO, Socorro, NM 87801, USA
e-mail: dmeier@nrao.edu

Fig. 1 (*Left*) CO, 3 mm continuum, and HST F555 (V) image of the central 300 pc of IC 342. (*Right*) Schematic of bar orbits and other influences on the molecular clouds in the center of IC 342

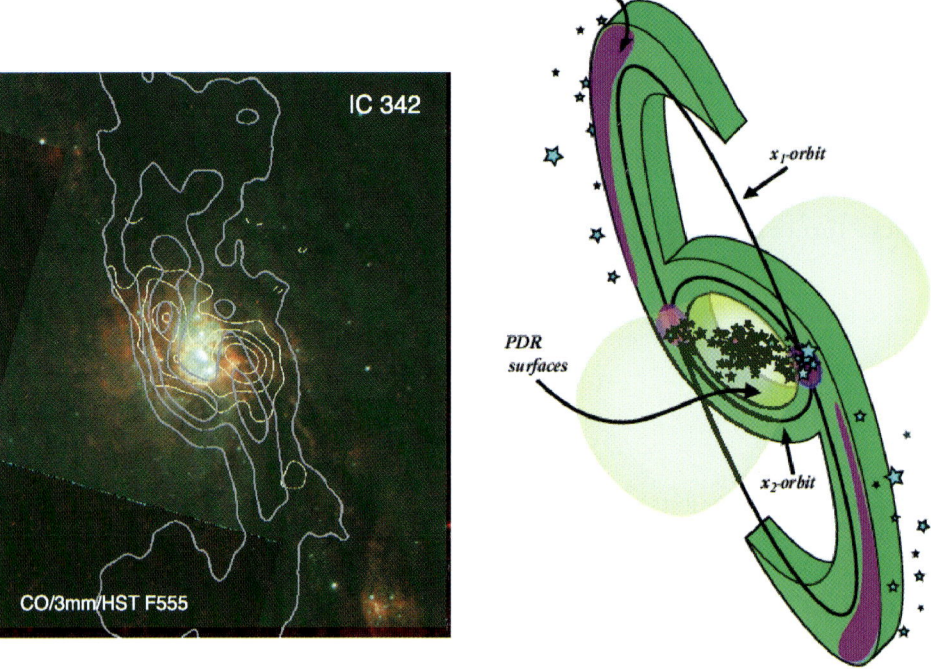

This example illustrates that with a sufficiently large number of spectral lines and the appropriate statistical analysis, the imaging of the line emission of many molecules can reduce the complexity of the interpretation of the physical and chemical processes in the molecular clouds.

2 Principal component analysis of the center of a normal spiral galaxy: imaging chemistry in IC 342

IC 342 is a nearby Scd galaxy, nearly face-on, and at a distance of 3 Mpc ($1'' = 15$ pc). An intense episode of nuclear star formation is indicated by a bright infrared and radio continuum source of luminosity $L_{IR} \sim 10^8$ L$_\odot$ (Becklin et al. 1980; Turner and Ho 1983), which is offset by $\sim$50 pc from the dynamical center and nuclear star cluster (Böker et al. 1997). Molecular gas in the center of IC 342 forms a barlike structure (Lo et al. 1984; Ishizuki et al. 1990) within the central 300 pc. A schematic of the stellar x_1 and x_2 bar orbits and other dynamical features of this central molecular bar are shown in Fig. 1 (Meier and Turner 2005). Streaming of the gas along the bar (Lo et al. 1984) leads to strong shearing motion, corresponding to a velocity differential of 50 km s^{-1} (Turner and Hurt 1992) across the arms. The current burst of star formation is found at the southwestern intersection of the x_1 and x_2 bar orbits, where the stellar orbits change from parallel to perpendicular to the bar (Meier and Turner 2001, 2005).

IC 342 has many faces, depending on which molecules are used to observe it. Careful analysis of CO and its isotopologues indicate that the bulk of the molecular gas in the center of IC 342 tends to be cool, 20 K or less (Meier and Turner 2001). Dust temperatures measured to 160 μm with the KAO are $\sim$40 K (Rickard and Harvey 1984). There is evidence for an extremely warm component of molecular gas, ranging from 50–150 K as indicated by the lower inversion lines of ammonia, CO(7–6) and the H$_2$ rotational lines (Martin and Ho 1986; Harris et al. 1991; Rigopoulou et al. 2002; Mauersberger et al. 2003). Gas temperatures as high as 800–900 K are inferred from the higher inversion lines of ammonia (Mauersberger et al. 2003) within the same general region as the cooler CO gas. However, the velocity of the CO(7–6) emission suggests that it originates in the clouds to the northeast of the dynamical center (Harris et al. 1991), a preliminary indication that CO alone gives an incomplete view of the spatial distribution of molecular line emission in IC 342.

How can higher resolution and more molecules refine our understanding of the molecular gas in the center of IC 342? Images were made of the central kpc of IC 342 in eight molecules with the Owens Valley Millimeter Array: C$_2$H, C^{34}S, N$_2$H$^+$, CH$_3$OH, HNCO, HNC, HC$_3$N, and SO by Meier and Turner (2005). Results were combined with existing observations of the CO and CO isotopologues of the 2–1 and 1–0 transitions (Meier and Turner 2001) and HCN (Downes et al. 1992) maps. The seven images (SO was not detected) are shown in Fig. 2. The $5''$ resolution maps reveal a remarkable degree of chemical variation across the central kpc. These lines arise from molecules with similar upper level energies and critical densities; the differences appear to be due to molecular abundance variations within the nuclear region (Meier and Turner 2005).

Fig. 2 Images of 3 mm
transitions of molecules in the
center of IC 342. Data from the
Owens Valley Millimeter Array
at a resolution of 4″. Contours
of the 3 mm emission of $^{12}C^{16}O$
are shown in each panel,
showing where the molecular
clouds are found. From Meier
and Turner (2005)

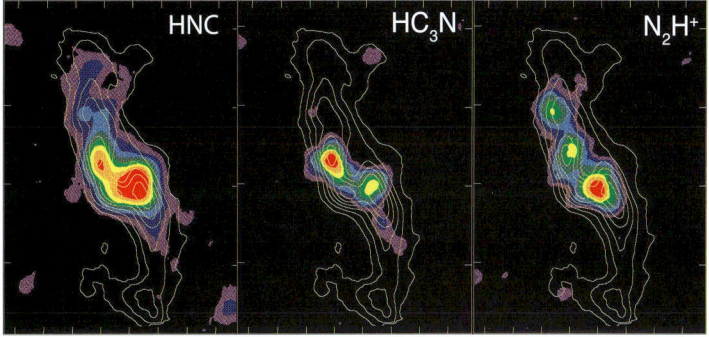

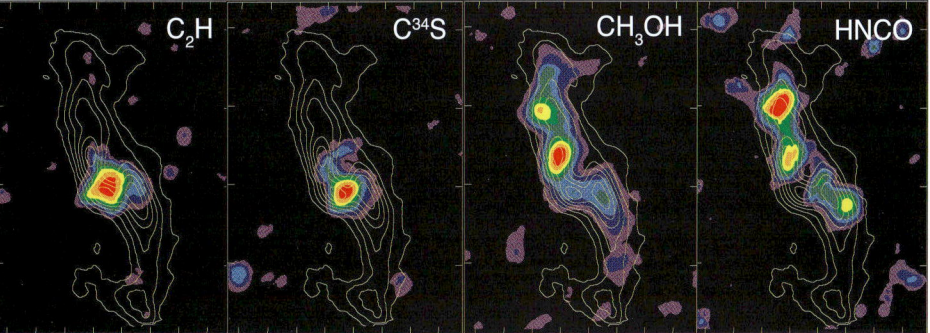

The combination of nearly a dozen different molecular lines with an image containing ∼220 independent resolution elements is a dataset sufficiently large that one can study the statistical correlations among the molecules as a function of location. Following the methods of Ungerechts et al. for the Orion Ridge (Ungerechts et al. 1997), a principal component analysis was done of the molecular line intensities across the nucleus of IC 342. The principal component analysis extracts an unbiased set of correlations from the data by choosing independent axes that maximize variance. Strong correlations are apparent in the first two principal components; a third component seems to indicate weak radial variations.

The spatial map of principal component axis 1 is shown in Fig. 3. With a resemblance to the CO map of Fig. 1, it represents the best "average" map for this set of molecules, which trace a slightly denser gas component than CO. This first axis simply tells us that the molecular lines arise in molecular clouds. More precisely, principal component axis 1 appears to be the density-weighted mean column density. The molecules with the highest projections (i.e., best correlation with density-weighted mean column density) along PC axis 1 are $C^{18}O$, N_2H^+, HNC, and HCN. In fact the PC 1 map appears to be an effective average of the $C^{18}O$ and HNC maps. These 3 mm lines appear to be the best general tracers of quiescent clouds and the overall denser molecular gas component characteristic of the nuclear region of a spiral galaxy.

Principal component axis 2 distinguishes two further major correlations, which are roughly anticorrelated here (mod-

ulo the requirement of PC axis 1 that molecules are found in molecular clouds). With positive projections along PC axis 2 are the emission of HNCO and CH_3OH (methanol). These molecules are found preferentially along the molecular bar arms. Observed Galactic abundances of methanol are very difficult to produce through gas-phase chemistry, so methanol is believed to be a tracer of grain chemistry. That methanol lies along the bar arm in IC 342 suggests that it is produced by the processing of grain mantles in the shocks along the bar arms. The SiO image of IC 342 of Usero et al. (2006) looks very much like the CH_3OH image (Usero et al. 2006), which would also support this interpretation. The chemistry of HNCO has been less certain from Galactic studies; both gas or grain chemistry have been suggested for it. Based on the IC 342 image, and the very good correlation with methanol and SiO, we suggest that it is also a product of grain chemistry (the excellent correlation of HNCO and methanol is also seen in the barred galaxy Maffei 2, Meier et al. in prep.). It is interesting that these putative shock tracers are found most strongly along the northern bar arm, which also has less active star formation.

The second correlation evident in PC axis 2 of Fig. 3 is that of C_2H and $C^{34}S$. Both of these molecules are restricted to the central 100 pc of IC 342. The emission from these lines is distinctive in that they are not found along the bar arms where the other species are found nor are they coincident with the bulk of the youngest embedded star formation. These molecules are likely to be tracers of highly irradiated molecular gas on the inner faces of the central ring,

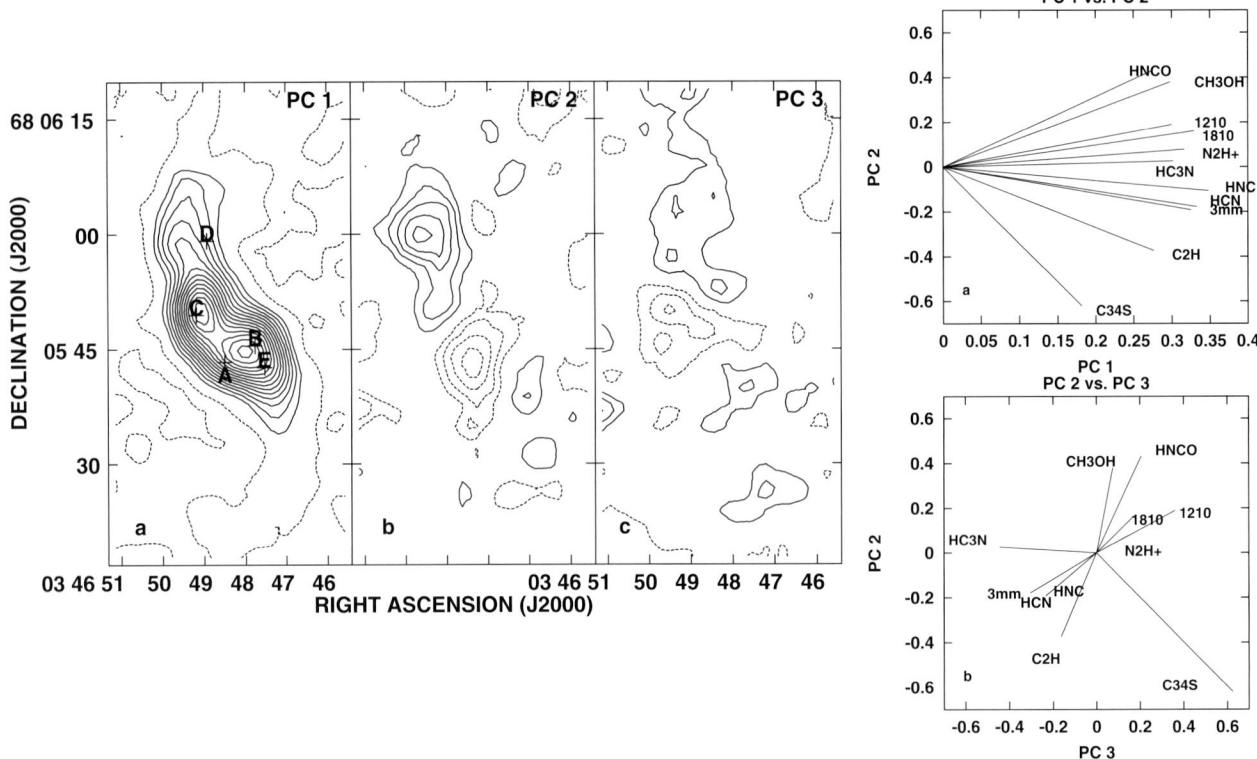

Fig. 3 Principal component maps of 3 mm lines in IC 342 and the projections of the molecules along the principal axes. From Meier and Turner (2005). *Letters* indicate individual molecular clouds identified in higher resolution maps (Downes et al. 1992)

Table 1 Correlation matrix for 3 mm molecules in the nucleus of IC 342

	^{12}CO	C^{18}O	3 mm	C$_2$H	C^{34}S	CH$_3$OH	HC$_3$N	HCN	HNC	HNCO
^{12}CO	1.0	…	…	…	…	…	…	…	…	…
C^{18}O	0.82	1.0	…	…	…	…	…	…	…	…
3 mm	0.65	0.76	1.0	…	…	…	…	…	…	…
C$_2$H	0.53	0.62	0.76	1.0	…	…	…	…	…	…
C^{34}S	0.38	0.39	0.48	0.50	1.0	…	…	…	…	…
CH$_3$OH	0.75	0.80	0.67	0.49	0.21	1.0	…	…	…	…
HC$_3$N	0.60	0.71	0.85	0.57	0.30	0.68	1.0	…	…	…
HCN	0.65	0.75	0.90	0.74	0.49	0.64	0.77	1.0	…	…
HNC	0.76	0.85	0.94	0.79	0.49	0.72	0.81	0.91	1.0	…
HNCO	0.67	0.75	0.58	0.42	0.19	0.78	0.57	0.57	0.66	1.0
N$_2$H$^+$	0.73	0.81	0.75	0.59	0.42	0.71	0.68	0.74	0.82	0.69

which contains a bright nuclear star cluster, presumably located at the dynamical center of the galaxy (or a combination of many star clusters, C. Max private communication). The fact that these two lines appear to be more closely associated with the visible nuclear cluster, estimated to be 60 Myr in age (Böker et al. 1997), than with the IR and radio source of the current starburst, suggests that this gas may be more closely correlated with a B star population than the young O

stars. A similar result has been found for PAHs in galaxies (Peeters et al. 2004).

Another view of the results of the principal component analysis is the correlation matrix, shown in Table 1. This matrix represents the correlations between individual pairs of molecules and the 3 mm continuum. Of particular interest are the unusually low correlations and high correlations. The spatial anti-correlation of the molecular bar arm molecules

Fig. 4 Imaging of 3 mm molecular lines in M82, made using the Berkeley-Illinois-Maryland Association Array (Meier and Turner in prep.)

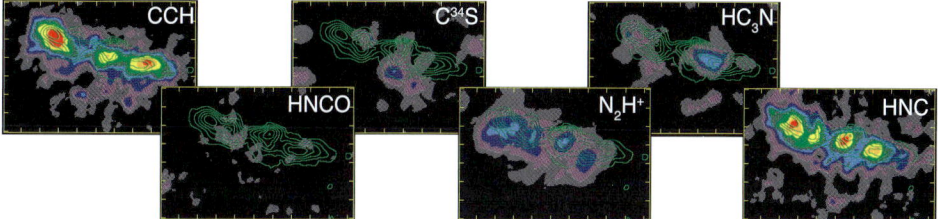

and the nuclear "PDR" molecules of PC axis 2 described above is evident in unusually low values of the correlation coefficient. The correlation between HNC and HCN is very high, and perhaps this is not surprising since they are isomers. But the very highest correlation, and it is remarkably high at 0.94, is between HNC and the 3 mm continuum. This is closely followed by a correlation between HCN and 3 mm continuum, which at 0.91 is nearly as tight. The 3 mm continuum in IC 342 is free-free emission, with little contribution from dust (Meier and Turner 2001). Thus the dense gas tracers HNC and HCN are extremely well-correlated with free-free emission. This result suggests that the correlation found by Gao and Solomon (2004) based on global HCN fluxes (mostly) holds down to the scales of individual giant molecular clouds.

While it is encouraging that the ability to resolve features such as bar arms, orbital intersections, and nuclear star cluster appears to simplify the chemical analysis, there is still more to learn about the molecular clouds in IC 342. The transitions here all trace the cool and relatively dense molecular gas component. While there are indications that the warmer molecular cloud component at 50–150 K traced by ammonia, H_2, and upper J levels of CO, and the hot component above 500 K traced by the upper ammonia inversion levels, are found in different parts of the nucleus than the lower energy gas traced by CO (Harris et al. 1991; Ho et al. 1990), the hot molecular gas chemistry remains as yet relatively unexplored in the imaging domain.

3 A different kind of chemistry: M82

A different situation is represented by the dwarf starburst galaxy, M82. At $L_{IR} \sim 3–4 \times 10^9$ $L_\odot$, M82 is one of the most energetic local starbursts. Since it is a dwarf galaxy, processes associated with the starburst might be expected to dominate the physical processes in the molecular gas of this galaxy, including the chemistry. These processes would include high radiation fields and shocks associated with the many supernova remnants within the nucleus.

In Fig. 4 are shown the same set of molecules mapped in M82 as were mapped in IC 342, observed with the Berkeley-Illinois-Maryland Association array. Emission from C_2H is particularly bright in M82, which reflects chemistry in the presence of strong radiation fields. HNC is also bright. Based on our analysis of IC 342, we infer this is because HNC traces the regions of free-free emission within the starburst, which are extensive in M82. Methanol and HNCO, tracers of grain chemistry, are relatively weak in M82 compared to the dense gas tracer molecules and the PDR molecules. This suggests that either the many supernova remnants observed in M82 either do not have as great an impact on the molecular gas as do the bar arms in IC 342 or that the harsh interstellar radiation field has chemically processed these species even further. This set of images is consistent with the characterization of M82 as a "giant PDR" in terms of its chemistry (Mauersberger et al. 1991; García-Burillo et al. 2002).

4 Summary

High spatial resolution observations of millimeter-wave emission from molecules indicates that the chemical structure of nearby galaxies is variable and complex, but that it is possible to spatially isolate clouds with particular characteristics, such as clouds experiencing shocks within spiral arms, or clouds near high radiation fields. However, this work is at the limits of the feasibility of today's interferometers, requiring long integration times and multiple spectral tunings. ALMA will revolutionize the study of chemistry in external galaxies due to its dramatic increase in resolution, sensitivity, and instantaneous bandwidth. At Band 6, ALMA will be able to achieve similar sensitivities to the current IC 342 dataset (~20 mK) in a few minutes over 8 GHz simultaneous bandwidth, with six times the spatial resolution. With this degree of improvement, mapping significant fractions of nearby galaxies at resolutions of a few pc in an extensive set of molecular species becomes possible, permitting the study of astrochemistry in different local galactic environments in much the fashion as large scale surveys do in our Galaxy today.

Pushing out beyond the local neighborhood of galaxies, ALMA will be able to detect IC 342-like GMCs in the brighter lines (e.g. HNC and CH_3OH, (Aalto et al. 2002); see contribution by S. Aalto in this volume) out to 75 Mpc in 1 hour on source; the fainter lines to Virgo distances at least. Arp 220-like systems will be detectable in species like

HNC and HC$_3$N to $z \sim 0.1$ in an 8 hr track. This will open up an array of galaxy types—dwarfs, ULIRGs, mergers, and potentially even ellipticals—to astrochemical scrutiny. The recent detections of HCN(3-2) and HCO$^+$(1-0) towards the Cloverleaf galaxy at $z = 2.56$ (Solomon et al. 2003; Riechers et al. 2006) demonstrate that for the most luminous systems chemistry is within reach across the observable universe.

Acknowledgements   This research is supported by the U.S. National Science Foundation.

References

Aalto, S., Polatidis, A.G., Hüttemeister, S., Curran, S.J.: Astron. Astrophys. **381**, 783 (2002)

Becklin, E.E., Gatley, I., Matthews, K., Neugebauer, G., Sellgren, K., Werner, M.W., Wynn-Williams, C.G.: Astrophys. J. **236**, 441 (1980)

Böker, T., Förster-Schreiber, N.M., Genzel, R.: Astron. J. **114**, 1883 (1997)

Crosthwaite, L.P., Turner, J.L., Buchholz, L., Ho, P.T.P., Martin, R.N.: Astron. J. **123**, 1892 (2002)

Downes, D., Radford, S.J.E., Guilloteau, S., Guelin, M., Greve, A., Morris, D.: Astron. Astrophys. **262**, 424 (1992)

Gao, Y., Solomon, P.: Astrophys. J. **606**, 271 (2004)

García-Burillo, S., Martín-Pintado, J., Fuente, A., Usero, A., Neri, R.: Astrophys. J. **575**, 55 (2002)

Harris, A.I., Stutzki, J., Graf, U.U., Russell, A.P.G., Genzel, R., Hills, R.: Astrophys. J. **382**, 75 (1991)

Helfer, T.T., Blitz, L.: Astrophys. J. **478**, 162 (1997)

Ho, P.T.P., Martin, R.N., Turner, J.L., Jackson, J.M.: Astrophys. J. **355**, 19 (1990)

Ishizuki, S., Kawabe, R., Ishiguro, M., Okumura, S.K., Morita, K.-I., Chikkada, Y., Kasuga, T.: Nature **334**, 224 (1990)

Lo, K.Y., et al.: Astrophys. J. **282**, L59 (1984)

Martin, R.N., Ho, P.T.P.: Astrophys. J. **308**, 7 (1986)

Martín, S., Mauersberger, R., Martín-Pintado, J., Henkel, C., García-Burillo, S.: Astrophys. J. Suppl. Ser. **164**, 450 (2006)

Mauersberger, R., Henkel, C., Walmsley, C.M., Sage, L.J., Wiklind, T.: Astron. Astrophys. **247**, 307 (1991)

Mauersberger, R., Henkel, C., Weiss, A., Peck, A.B., Hagiwara, Y.: Astron. Astrophys. **403**, 561 (2003)

Meier, D.S., Turner, J.L.: Astrophys. J. **551**, 687 (2001)

Meier, D.S., Turner, J.L.: Astrophys. J. **618**, 259 (2005)

Peeters, E., Spoon, H.W.W., Tielens, A.G.G.M.: Astrophys. J. **613**, 986 (2004)

Rickard, L.J., Harvey, P.M.: Astron. J. **89**, 1520 (1984)

Riechers, D.A., Walter, F., Carilli, C.L., Weiss, A., Bertoldi, F., Menten, K.M., Knudsen, K.K., Cox, P.: Astrophys. J. **645**, 13 (2006)

Rigopoulou, D., Kunze, D., Lutz, D., Genzel, R., Moorwood, A.F.M.: Astron. Astrophys. **389**, 374 (2002)

Solomon, P., Vanden Bout, P., Carilli, C., Guelin, M.: Nature **426**, 636 (2003)

Turner, J.L., Ho, P.T.P.: Astrophys. J. **268**, L79 (1983)

Turner, J.L., Hurt, R.L.: Astrophys. J. **384**, 72 (1992)

Ungerechts, H., Bergin, E., Goldsmith, P.F., Irvine, W.M., Schloeb, F.P., Snell, R.L.: Astrophys. J. **482**, 245 (1997)

Usero, A., García-Burillo, S., Martin-Pintado, J., Fuente, A., Neri, R.: Astron. Astrophys. **448**, 457 (2006)

Wiklind, T., Rydbeck, G., Hjalmarson, A., Bergman, P.: Astron. Astrophys. **232**, 11 (1990)

Wright, M.C.H., Ishizuki, S., Turner, J.L., Ho, P.T.P., Lo, K.Y.: Astrophys. J. **406**, 470 (1993)

Chemistry in luminous AGN and starburst galaxies

Susanne Aalto

Originally published in the journal Astrophysics and Space Science, Volume 313, Nos 1–3.
DOI: 10.1007/s10509-007-9643-0 © Springer Science+Business Media B.V. 2007

Abstract Molecular line emission is a useful tool for probing the highly obscured inner kpc of starburst galaxies and buried AGNs. Molecular line ratios serve as diagnostic tools of the physical conditions of the gas—but also of its chemical properties. Both provide important clues to the type and evolutionary stage of the nuclear activity. While CO emission remains the main tracer for molecular distribution and dynamics, molecules such as HCN, HNC, HCO^+, CN and HC_3N are useful for probing the properties of the denser $(n \gtrsim 10^4$ $cm^{-3})$, star-forming gas. Here I discuss current views on how line emission from these species can be interpreted in luminous galaxies. HNC, HCO^+ and CN are all species that can be associated both with photon dominated regions (PDRs) in starbursts—as well as X-ray dominated regions (XDRs) associated with AGN activity. HC_3N line emission may identify galaxies where the starburst is in the early stage of its evolution.

Keywords Galaxies: starburst · Galaxies: active · Radio lines: ISM · ISM: molecules

1 Introduction

Single dish and high resolution studies of CO and ^{13}CO have shown that the CO/^{13}CO 1–0 and 2–1 line intensity ratios are efficient diagnostic tools for large scale ISM properties in galaxies. In particular the ratio between diffuse and self-gravitating gas - as well as effects of temperature and density

S. Aalto (✉)
Department of Radio Astronomy and Space Science, Onsala
Space Observatory, Chalmers University of Technology, 439 92
Onsala, Sweden
e-mail: susanne@oso.chalmers.se

in the moderately dense $(n = 10^2–10^4$ $cm^{-3})$ phase of the molecular gas (e.g. Aalto et al. 1991; Young and Sanders 1986; Casoli et al. 1992; Wall et al. 1993; Aalto et al. 1995; Paglione et al. 2001; Glenn and Hunter 2001; Tosaki et al. 2002).

Observing the denser $(n > 10^4$ $cm^{-3})$ gas phase of the molecular gas requires molecules with higher dipole moments than CO. A useful dense gas tracer is HCN which has been used for extensive surveys of external galaxies (e.g. Solomon et al. 1992; Helfer and Blitz 1993; Paglione et al. 1995; Curran et al. 2000; Gao and Solomon 2004). Abundance and excitation differences among galaxies may, however, cause both the CO and HCN molecular $J = 1–0$ emission to trace gas differently from galaxy to galaxy. The line ratio may still be very useful in classifying galaxies: Galaxies with Seyfert nuclei often have low CO/HCN 1–0 line ratios—ranging from 3 to 10 (e.g. Bryant 1996; Kohno et al. 2001). It is unclear whether this is due to excitation, chemistry or both. Globally, the ratios for Seyfert galaxies approach those for more normal starburst galaxies (e.g. Curran et al. 2000). Starburst galaxies more typically have higher ratios, ranging from 10 to 30 (e.g. Aalto et al. 1995; Kohno et al. 2001; Gao and Solomon 2004).

The type and evolutionary stage of the dense gas traced by HCN can also be addressed through observing and modeling the chemistry and excitation of the gas. Efforts to study the excitation of HCN through observing higher transitions emission are now in progress (e.g. Greve et al. 2006). Additional high density tracer molecules and their excitation may be combined to build models of the properties of the nuclear activity. Studies of nearby, less luminous, galaxies are now showing spectacular and interesting results. One example is IC 342 where Meier et al. (2005) have probed the properties of the dense star forming gas through studying a large number of molecules at high resolution. Martín and collaborators (Martín et al. 2005) have performed an extensive spectral scan of the nearby starburst galaxy NGC 253. In IC 342, Montero-Castano et al. (2006) have mapped ammonia (NH_3) at high resolution at 25 GHz to probe the hot and dense gas. More detailed discussions of the chemistry of nearby galaxies can be found elsewhere in this volume.

Below, I discuss four high density tracer molecules: HNC, HCO^+, CN and HC_3N and interpretations of their line emission from luminous ($L_{IR} > 10^{11} L_\odot$, galaxies.

2 The HCN/HCO^+ 1–0 line ratio

HCO^+ is suggested by e.g. Kohno et al. (2001) to be a tracer of the fraction of the dense gas which is involved in star formation (see also Imanishi et al. 2004). They compared the HCN/HCO^+ 1–0 line ratio in a selection of Seyfert and starburst galaxies and find that the relative HCO^+ 1–0 luminosity is significantly higher in starbursts. According to models by Maloney et al. (1996) a deficiency of HCO^+ is expected near a hard X-ray source—in an X-ray dominated region (XDR) resulting in an elevated HCN/HCO^+ line intensity ratio. Interestingly, elevated HCN/HCO^+ 1–0 line ratios are found in ULIRGs by Graciá-Carpio et al. (2006) and in general the line ratio appears to correlate with FIR luminosity. The authors attribute this to the presence of an AGN where the X-rays affect the chemistry to impact the HCN/HCO^+ abundance ratio. Usero and collaborators (Usero et al. 2004) studied the circum nuclear disk (CND) of the Seyfert galaxy NGC 1068 in several molecular lines, and presented the first extragalactic detection of the reactive ion HOC^+. They suggest that the CND is an XDR based on several lines, including an elevated HCN/HCO^+ 1–0 line ratio.

2.1 XDRs?

The X-ray irradiation of molecular gas leads to a so-called X-ray dominated region (e.g., Maloney et al. 1996; Lepp and Dalgarno 1996; Meijerink and Spaans 2005; Meijerink et al. 2006; Meijerink et al. 2007) similar to PDRs associated with bright UV sources (Tielens and Hollenbach 1985). The more energetic (1–100 keV) X-ray photons penetrate large columns (10^{22}–10^{24} cm^{-2}) of gas and lead to a different ion-molecule chemistry. Species like C, C^+ and CO co-exist (unlike the stratified PDR) and the abundances of H^+ and He^+ are much larger.

However, Meijerink and Spaans (2005) and Meijerink et al. (2007) suggest that an expected AGN signature is not an elevated HCN/HCO^+ line ratio, instead a slightly more luminous HCO^+ 1–0 line than HCN 1–0 is expected from most of their XDR models. According to Meijerink et al. HCN/HCO^+ 1–0 line ratios >1 may be possible at the edges of a moderate density ($n = 10^4$–10^5 cm^{-3}) XDR. Line emissivities are however quite modest in such a region. A high X-ray flux leads to more destruction of HCO^+ at the edge of an XDR, through dissociative recombination, but the large degree of ionization, and the associated ion-molecule chemistry, more than compensates for this, in the models by Meijerink and Spaans, at hydrogen columns in excess of 10^{23} cm^{-2} and/or flux-to-density ratios of less than 10^{-3} erg s^{-1} cm (Meijerink and Spaans 2005). In Fig. 1 model HCN/HCO^+ 1–0 line ratios for XDRs and PDRs are presented showing that the HCN/HCO^+ 1–0 line ratios are below unity for the model-XDR. PDRs are the interface regions between young massive stars and the molecular clouds and should be abundant in active, somewhat evolved, starbursts. Perhaps the key to the issue is that XDRs should be divided into low- and high-column density XDRs,

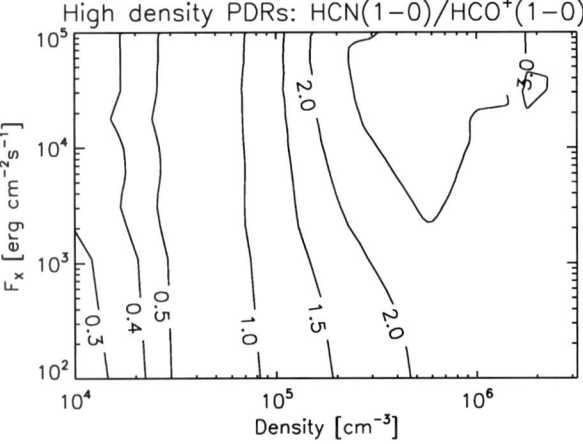
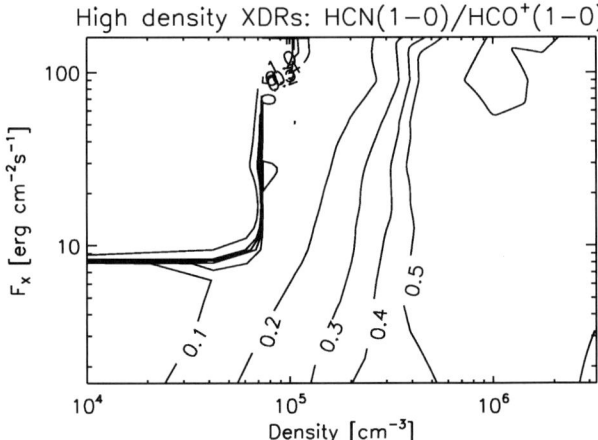

Fig. 1 A grid of one dimensional single-sided PDR/XDR slab models in density and FUV/X-ray irradiation for HCN/HCO^+ 1–0 line ratios (Meijerink et al. 2007). PDRs (*left panel*) and XDRs (*right panel*). The model clouds have a size of 1 pc, but choice of cloud size does not influence the computed line intensity ratios. Individual cloud velocity line width is assumed to be 5 km s^{-1}

where the low column density XDRs are the ones with the HCN/HCO$^+$ 1–0 line ratios exceeding unity.

A more unambiguous XDR-diagnostic may be highly excited ($J > 10$) rotational lines of CO (Meijerink et al. 2007). XDRs more easily produce very warm CO compared to PDRs (also PDRs with high cosmic ray rates). This is because CO is present already at the warm XDR edge, initiated through C$^+$ + OH route above, while the higher UV flux at the edge of PDRs, for the same density and total energy input, suppresses CO.

2.2 Warm, dense gas in young starbursts?

Relatively low HCO$^+$ abundances may be expected in young, synchrotron-deficient, starbursts which would also result in HCN/HCO$^+$ 1–0 line intensity ratios exceeding unity. An alternative interpretation to an elevated HCN/HCO$^+$ 1–0 line ratio may therefore be that some Seyfert nuclei are surrounded by regions of very young star formation. If the starburst is young enough not to have produced a significant number of supernovae, then the relative HCO$^+$ 1–0 luminosity may be low compared to that of HCN 1–0. An interesting alternative to this is if it is not the age of the starburst that results in the observed line ratio—but its structure. In the section on CN further down we discuss the possibility of dust-bound H II regions, where the UV photons quickly become absorbed by dust. This may affect the chemistry of other species, for instance HCO$^+$. Thus, the impact of ISM structure and gas-to-dust ratio on line ratios should be explored further.

Further observational and modeling efforts are necessary to resolve this issue. Since HCO$^+$ becomes excited at lower densities than HCN it is possible that the HCO$^+$ and HCN line emission do not sample the same regions. Furthermore, it is essential to take excitation and optical depth effects into account before interpreting line ratios in terms of abundances.

3 CN: tracing both PDRs and XDRs

The radical CN is another tracer of dense gas, with a somewhat lower (by a factor of 5) critical density than HCN. The abundance of the CN radical becomes enhanced at the inner edge of a PDR at an A_V of about 2 magnitudes. At larger depths into the cloud the CN abundance radically declines and the $\frac{[\text{HCN}]}{[\text{CN}]}$ abundance ratio increases (e.g. Jansen et al. 1995). Observations by Rodríguez-Franco and collaborators of the CN emission towards the Orion A molecular complex (Rodríguez-Franco et al. 1998) show that the morphology of the CN emission is dominated by the ionization fronts of the H II regions. The authors conclude that this molecule is an excellent tracer of regions affected by UV radiation. Thus,

the emission from the CN molecule should serve as a measure of the relative importance of gas in PDRs. In an XDR (where the X-rays from an AGN impacts the surrounding ISM) one also expects an enhancement of CN abundances over a wide range of ionization rates (e.g. Lepp and Dalgarno 1996). In contrast to PDRs, the CN abundances may remain large in XDRs also when hydrogen column densities are large. CN abundances remain greater than those of HCN and HCO$^+$ throughout the XDR.

3.1 Faint CN 1–0 line emission in luminous galaxies?

In this context, it is therefore surprising that the relative CN 1–0 line intensities appear to be moderate to low in luminous galaxies (Aalto et al. 2002), although the sample of 13 galaxies in the study is too small to base firm conclusions on. More extragalactic surveys of CN would be welcome. It is interesting to speculate whether the apparent CN-deficiency is related to the [C II] 158 μm fine structure line being found to be faint in ULIRGs compared to less luminous starburst galaxies. Malhotra et al. (1997) report a decreasing trend in $\frac{F_{[\text{CII}]}}{F_{\text{FIR}}}$ with increasing $\frac{f(60)}{f(100)}$ μm flux ratio. One potential explanation to this is that the PDRs may be quenched in the high pressure, high density environment in the deep potentials of the ULIRGs and the HII regions exist in forms of small-volume, ultracompact HII regions that are dust-bounded. This may also result in faint CN line emission. However, without more reliable statistical information it is difficult to speculate further.

4 Luminous HNC line emission

Aalto et al. (2002, 2007a) find surprisingly bright HNC 1–0 and 3–2 emission towards luminous galaxies. Particularly bright HNC 3–2 emission is found towards the IR luminous and deeply obscured galaxies Arp 220 and NGC 4418 (Aalto et al. 2007b). Intense HNC line emission is unexpected in galaxies where the gas has high kinetic temperatures. In warm Galactic star forming regions, the emission is weak due to temperature-dependent neutral-neutral chemical reactions which to the destruction of HNC (Schilke et al. 1992; Hirota et al. 1998). In luminous galaxies, however, this appears not to be the case. When $I(\text{HNC}) = I(\text{HCN})$ it can be explained through ion-neutral chemistry in photon dominated regions (PDRs). When HNC line emission is *overluminous* ($I(\text{HNC}) > I(\text{HCN})$) however, then either radiative excitation of HNC occurs through mid-infrared pumping—or, the chemistry is being affected by X-rays so that the HNC abundance can exceed the HCN abundance (Aalto et al. 2007b; Meijerink et al. 2007). It has also been suggested that shocks may enhance HNC abundances (Arce and Sargent 2004).

4.1 Pumping or XDRs?

Mid-IR pumping: Both HCN and HNC have degenerate bending modes in the IR. The molecule absorbs IR-photons to the bending mode (its first vibrational state) and then it decays back to the ground state via its P branch ($v = 1\text{–}0$, $\Delta J = +1$) or R-branch ($v = 1\text{–}0$, $\Delta J = -1$). In this way, a vibrational excitation may produce a change in the rotational state in the ground level and can be treated (effectively) as a collisional excitation in the statistical equations. Thus, IR pumping excites the molecule to the higher rotational level by a selection rule $\Delta J = 2$ (Fig. 2). For HNC, the bending mode occurs at $\lambda = 21.5$ μm (464.2 cm^{-1}) with an energy level $h\nu/k = 669$ K and an A-coefficient of $A_{IR} = 5.2$ s^{-1}. For HCN the mode occurs at $\lambda = 14$ μm (713.5 cm^{-1}), energy level $h\nu/k = 1027$ K and $A_{IR} = 1.7$ s^{-1}. *It is therefore significantly easier to pump HNC, than HCN.* The pumping of HNC may start to become effective when the IR background reaches an optically thick brightness temperature of $T_B \approx 50$ K and the gas densities are below critical.

X-ray chemistry: Alternatively, models of XDRs (Meijerink and Spaans 2005; Meijerink et al. 2006; Meijerink et al. 2007) indicate that the HNC/HCN column density ratio is elevated, and reaches a value of ~ 2, for gas densities around 10^5 cm^{-3}. In XDRs, the degree of ionization is much larger than in PDRs. Hence, reactions with, e.g., He$^+$, H$^+$, C$^+$ and many others, that lead to the formation and destruction of HCN and HNC in an asymmetric way (with HCNH$^+$ and H$_2$NC$^+$ as intermediates), are much more important. It is interesting to note that HNC 1–0 emission in the edge-on Seyfert galaxy NGC 4945 seems to be more nuclear than HCN 1–0 (Cunningham and Whiteoak 2005), and

it is tempting to suggest that this may be due to the chemical impact of the Seyfert nucleus.

Regardless of whether it is mid-IR pumping or X-ray induced chemistry (or a combination of both) that is behind the overluminous HNC 3–2 emission, it is clear that the *prevailing ISM conditions are unusual compared to similar scales in more normal galaxies.* The buried nuclear activity affects the properties of the ISM on scales large enough to create global effects on observed molecular line ratios.

4.2 HNC at high resolution in the ULIRG Arp 220

Arp 220 is an ultraluminous gas-rich merger of two galaxies where the two nuclei are separated by ~ 300 pc. Preliminary results from a high resolution study of HNC 3–2 with the SMA (Aalto et al. in preparation), reveal bright line emission emerging from the western nucleus—as well as fainter, broader emission north and south of the western nucleus. Only faint emission is detected in the eastern nucleus—which also appears to be the case for CN 2–1 and HCN 4–3 (Wiedner et al. in preparation). Thus, since the CO 2–1 luminosity is similar in the two nuclei (Sakamoto et al. 1999), there appears to be a real difference in gas properties between them.

Averaged in a region of 200 pc, the brightness temperature of the nuclear HNC 3–2 emission appears to be 25–40 K in a velocity range of a few hundred km s^{-1}. If the emission is clumpy, the brightness temperature is a lower limit to the gas temperature and it shows that luminous HNC line emission is indeed emerging from warm, dense gas—unlike the situation in Milky Way clouds (see Sect. 4). Interestingly, the HNC 3–2 brightness temperature is similar to those found for CO 2–1 in the western nucleus of Arp 220 by Sakamoto et al. (1999). An east-west rotational major axis of the western nucleus is observed for the HNC 3–2 emission, consistent with what is found for CO and HI.

5 HC$_3$N: young star formation?

Meier et al. (2005) discuss HC$_3$N chemistry in the nearby galaxy IC 342. It may be formed through the reaction C$_2$H$_2$ + CN $\rightarrow$ HC$_3$N + H, and destroyed through photodissociation and collisions with C$^+$. In the Orion region, the emission from HC$_3$N is bright toward hot, dense cores, while the HC$_3$N/CN abundance ratio is only 10^{-3} in PDRs (Rodríguez-Franco et al. 1998). Since C$^+$ is abundant in XDRs, HC$_3$N should not be thriving in this an X-ray affected environment (although the behavior of HC$_3$N abundances in XDRs have not yet been modeled). Bright HC$_3$N line emission should therefore indicate neither PDRs nor XDRs. Instead, it implies the presence of dense, warm gas which is not being strongly radiatively impacted upon—potentially indicative of young star formation.

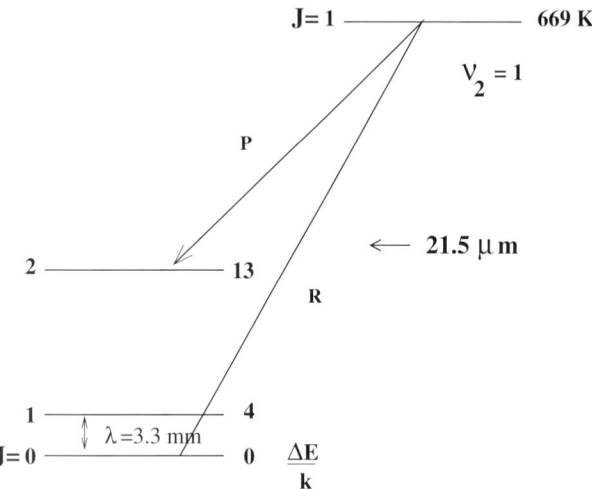

Fig. 2 A schematic picture of the pumping of the HNC rotational levels via the mid IR bending transitions (Aalto et al. 2007b). The figure shows how the rotational $J = 2$ level may become populated through the $\Delta J = 2$ selection rule through mid-IR pumping. The principle is the same for higher J levels

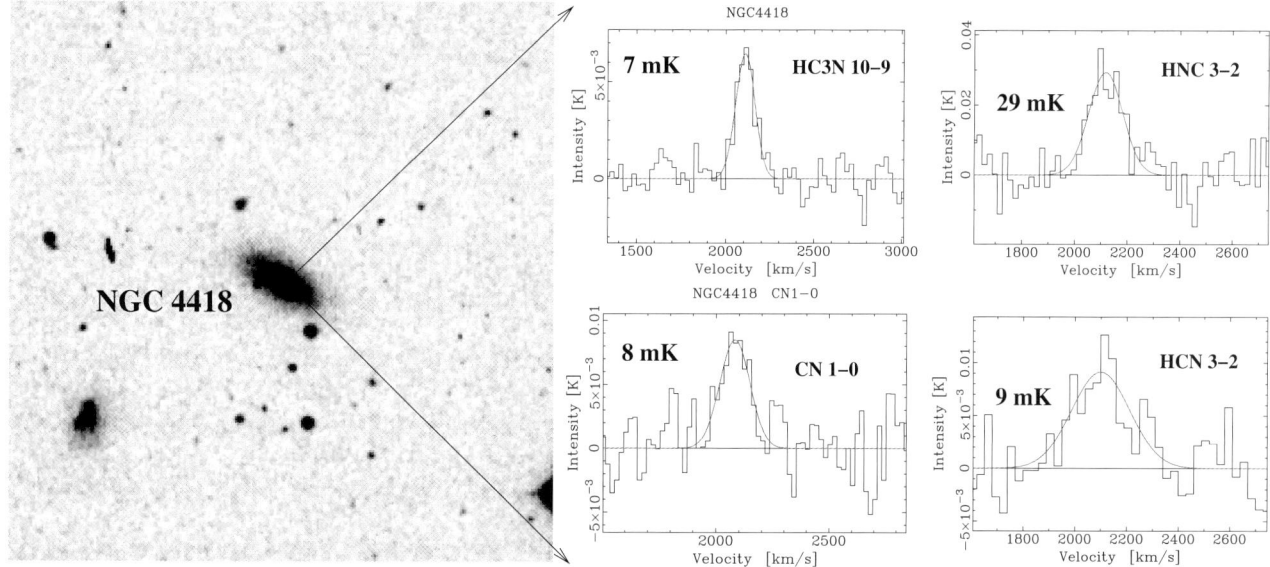

Fig. 3 *Left panel*: DSS image of the luminous infrared galaxy NGC 4418. *Right panel*: The *upper left figure* shows the HC$_3$N 10–9 line emission and the *lower panel* the CN 1–0. The *upper right figure* shows the HNC 3–2 line emission from the central region of the galaxy

and the *lower right panel* shows the HCN 3–2 emission from the same region (Aalto et al. 2007a, 2007b). It is evident that the HNC 3–2 line intensity is more than a factor of two greater than that of HCN towards NGC 4418

In a survey of HNC 1–0 in a sample of 13 luminous galaxies (Aalto et al. 2002), HC$_3$N 10–9 line emission was detected in the same frequency band in two galaxies: Arp 220 (strong emission) and NGC 1808 (weak). It appears that significant HC$_3$N line emission is not a common feature of luminous galaxies. A high resolution study of IC 342 (Meier et al. 2005) shows that the emission is strongly associated with 3 mm continuum emission. Multiple transitions have been observed in NGC 253 (Mauersberger et al. 1990) where the starburst is suggested to be young. In M 82, for comparison, HC$_3$N line emission is weak.

The two galaxies with (so far) the most luminous HC$_3$N emission are Arp 220 (Aalto et al. 2002) and NGC 4418 (Aalto et al. 2007a). Despite Arp 220 being more than an order of magnitude more luminous, the two galaxies have several mid-IR and molecular properties in common. Both Arp 220 and NGC 4418 show deep silicate absorption in their mid-IR spectra (Spoon et al. 2001) suggesting visual extinction exceeding 50 magnitudes and both galaxies have overluminous (see Sect. 4) HNC (I(HNC) > I(HCN)) $J =$ 3–2 emission (Aalto et al. 2007b).

5.1 NGC 4418

NGC 4418 is an interacting IR luminous early type spiral with a high infrared surface brightness (Evans et al. 2003) and most of its molecular gas is concentrated in the inner kpc (Dale et al. 2005). It is unclear what is driving the IR luminosity: NGC 4418 is a high-q galaxy where $q =$ the logarithmic ratio of far-infrared to radio flux densities and

this FIR-excess is suggested to be caused by a nascent starburst (Lisenfeld et al. 1996; Roussel et al. 2003). The relatively faint radio emission is then due to a synchrotron deficiency in the young ($\lesssim 2 \times 10^6$ yrs) starburst. Imanishi et al. (2004) find, however, that the estimated star formation luminosity from the observed PAH emission can account for only a small fraction of the infrared luminosity thus suggesting that NGC 4418 may be largely AGN-driven. They furthermore report an elevated HCN/HCO$^+$ 1–0 line intensity ratio which they suggest is due to the presence of an AGN-powered XDR-chemistry, although depth-dependent models (Meijerink and Spaans 2005; Meijerink et al. 2007) seem to suggest an opposite trend, with HCN/HCO$^+$ 1–0 < 1 under X-ray irradiation (see Sect. 2.1).

NGC 4418 shows very bright HC$_3$N 10–9, 16–15 and 25–24 line emission (Fig. 3) and the question is how that should be interpreted in terms of starburst and/or AGN activity. If the line emission coexists with other line emission from high density tracers—which is very compact ($\lesssim 2''$)—then both a starburst and an AGN has limited impact on the surrounding molecular ISM. Either through the activity being in their early stage of evolution (pre-PDR starburst/young- or weak AGN) or it is due to the ISM structure of a compact distribution (see Sect. 2.2).

6 Outlook

ALMA will offer cloud-scale resolution and sensitivity for many ULIRGs and semi-distant galaxies. For Arp 220, for example, a full 8 hour 345 GHz track will result in 1.5 K sensitivity at 17 pc resolution ($10\,km\,s^{-1}$ velocity resolution). This will be enough to image the properties of the nuclear molecular clouds and search for gradients and correlations on small scales. Brightness temperatures can be used to constrain physical conditions and radiative transport models.

The high sensitivity of ALMA also enables broad spectral scans of starburst and active galaxies. The rich chemistry of NGC 253 revealed in the spectral scan by Martín et al. (2005) serves as an illustration for this. They compare the chemistry of NGC 253 and M 82 and confirm previous suggestions that M 82 is in a more evolved, PDR-dominated stage than NGC 253. Such spectral fingerprints may help classify various types of starbursts in terms of evolution and intrinsic properties. Even if the analogy to optical/UV spectral classification models should not be taken too far it should be possible to produce models of ensembles of clouds to help classify the overall properties of the starburst. This opens up new possibilities of using ALMA to its fullest potential when studying chemistry of luminous galaxies.

Acknowledgements The author would like to thank the Swedisch Research Counsil.

References

Aalto, S., Johansson, L.E.B., Booth, R.S., Black, J.H.: Astron. Astrophys. **249**, 323 (1991)

Aalto, S., Booth, R., Black, J.H.: Astron. Astrophys. **300**, 369 (1995)

Aalto, S., Polatidis, A.G., Hüttemeister, S., Curran, S.J.: Astron. Astrophys. **381**, 783 (2002)

Aalto, S., Monje, R., Martin, S.: Astron. Astrophys. (2007a, in press)

Aalto, S., Spaans, M., Wiedner, M., Hüttemeister, S.: Astron. Astrophys. (2007b, in press)

Arce, H.G., Sargent, A.: Astrophys. J. **612**, 342 (2004)

Bryant, P.M.: PhD thesis, Caltech (1996)

Casoli, F., Dupraz, C., Combes, F.: Astron. Astrophys. **264**, 55 (1992)

Cunningham, M.R., Whiteoak, J.B.: Mon. Not. R. Astron. Soc. **364**, 37 (2005)

Curran, S.J., Aalto, S., Booth, R.S.: Astron. Astrophys. Suppl. Ser. **141**, 193 (2000)

Dale, D.A., Sheth, K., Helou, G., Regan, M.W., Hüttemeister, S.: Astron. J. **129**, 2197 (2005)

Evans, A.S., Becklin, E.E., Scoville, N.Z., Neugebauer, G., Soifer, B.T., Matthews, K., Ressler, M., Werner, M., Rieke, M.: Astron. J. **125**, 2341 (2003)

Gao, Y., Solomon, P.M.: Astrophys. J. Suppl. Ser. **152**, 63 (2004)

Glenn, J., Hunter, T.R.: Astrophys. J. Suppl. Ser. **135**, 177 (2001)

Graciá-Carpio, J., García-Burillo, S., Planesas, P., Colina, L.: Astrophys. J. **640**, 135 (2006)

Greve, T., et al.: 2006, In: Charmandaris, V., Rigopoulou, D., Kylafis, N. (eds.) Studying Galaxy Evolution with Spitzer and Herschel. CUP Conf. Ser.

Helfer, T.T., Blitz, L.: Astrophys. J. **419**, 86 (1993)

Hirota, T., Yamamoto, S., Mikami, H., Ohishi, M.: Astrophys. J. **503**, 717 (1998)

Imanishi, M., Nakanishi, K., Kuno, N., Kohno, K.: Astron. J. **128**, 2037 (2004)

Jansen, D.J., van Dishoeck, E.F., Black, J.H., Spaans, M., Sosin, C.: Astron. Astrophys. **302**, 223 (1995)

Kohno, K., Matsushita, S., Vila-Vilaro, B., Okumura, S.K., Shibatsuka, T., Okiura, M., Ishizuki, S., Kawabe, R.: 2001, In: Knapen, Beckman, Shlosman, Mahoney (eds.) The Central Kiloparsec of Starbursts and AGN: The La Palma Connection. ASP Conf. Ser., vol. 249, p. 672

Lepp, S., Dalgarno, A.: Astron. Astrophys. **306**, L21 (1996)

Lisenfeld, U., Völk, H.J., Xu, C.: Astron. Astrophys. **314**, 745 (1996)

Malhotra, S., Helou, G., Stacey, G., et al.: Astrophys. J. Lett. **491**, 27 (1997)

Maloney, P.R., Hollenbach, D.J., Tielens, A.G.G.M.: Astrophys. J. **466**, 561 (1996)

Martín, S., Mauersberger, R., Martín-Pintado, J., Henkel, C., García-Burillo, S.: 2005 In: Lis, Blake, Herbst (eds.) Astrochemistry Throughout the Universe: Recent Successes and Current Challenges. IAU Symposium no. 231, p. 219

Mauersberger, R., Henkel, C., Sage, L.J.: Astron. Astrophys. **236**, 63 (1990)

Meier, D.S., Turner, J.L.: Astrophys. J. **618**, 259 (2005)

Meijerink, R., Spaans, M.: Astron. Astrophys. **436**, 397 (2005)

Meijerink, R., Spaans, M., Israel, F.P.: Astrophys. J. **650**, L103 (2006)

Meijerink, R., Spaans, M., Israel, F.P.: Astron. Astrophys. (2007, in press), astro-ph/0610360

Montero-Castano, M., Herrnstein, Robeson, M., Ho, P.T.P.: Astrophys. J. **646**, 919 (2006)

Paglione, T.A.D., Tosaki, T., Jackson, J.M.: Astrophys. J. **454**, L117 (1995)

Paglione, T.A.D., Wall, W.F., Young, J.S., Heyer, M.H., Richard, M., Goldstein, M., Kaufman, Z., Nantais, J., Perry, G.: Astrophys. J. **135**, 183 (2001)

Rodríguez-Franco, A., Martin-Pintado, J., Fuente, A.: Astron. Astrophys. **329**, 1097 (1998)

Roussel, H., et al.: Astrophys. J. **593**, 733 (2003)

Sakamoto, K., Scoville, N.Z., Yun, M.S., Crosas, M., Genzel, R., Tacconi, L.J.: Astrophys. J. **514**, 68 (1999)

Schilke, P., Walmsley, C.M., Pineau de Forest, G., et al.: Astron. Astrophys. **256**, 595 (1992)

Solomon, P.M., Downes, D., Radford, S.J.E.: Astrophys. J. **387**, L55 (1992)

Spoon, H.W.W., Keane, J.V., Tielens, A.G.G.M., Lutz, D., Moorwood, A.F.M.: Astron. Astrophys. **365**, 353 (2001)

Tielens, A.G.G.M., Hollenbach, D.: Astrophys. J. **291**, 722 (1985)

Tosaki, et al.: Publ. Astron. Soc. Jpn. **54**, 209 (2002)

Wall, W.F., Jaffe, D.T., Bash, F.N., et al.: Astrophys. J. **414**, 98 (1993)

Usero, A., García-Burillo, S., Fuente, A., Martín-Pintado, J., Rodríguez-Fernández, N.J.: Astron. Astrophys. **419**, 897 (2004)

Young, J.S., Sanders, D.B.: Astrophys. J. **302**, 680 (1986)

Dense gas in normal and active galaxies

Kotaro Kohno · Koichiro Nakanishi · Tomoka Tosaki ·
Kazuyuki Muraoka · Rie Miura · Hajime Ezawa ·
Ryohei Kawabe

Originally published in the journal Astrophysics and Space Science, Volume 313, Nos 1–3.
DOI: 10.1007/s10509-007-9695-1 © Springer Science+Business Media B.V. 2007

Abstract Dense molecular medium plays essential roles in galaxies. As demonstrated by the tight and linear correlation between HCN(1–0) and FIR luminosities among starforming galaxies, from very nearby to high-z ones, the observation of a dense molecular component is indispensable to understand the star formation laws in galaxies. In order to obtain a general picture of the global distributions of dense molecular medium in normal star-forming galaxies, we have conducted an extragalactic CO(3–2) imaging survey of nearby spiral galaxies using the Atacama Submillimeter Telescope Experiment (ASTE). From the survey (ADIoS; ASTE Dense gas Imaging of Star-forming galaxies), CO(3–2) images of M 83 and NGC 986 are presented. Emphasis is placed on the correlation between the CO(3–2)/CO(1–0) ratio and the star formation efficiency in galaxies. In the central regions of some active galaxies, on the other hand, we often find enhanced or overluminous HCN(1–0) emission. The HCN(1–0)/CO(1–0) and HCN(1–0)/HCO$^+$(1–0) intensities are often enhanced up to ~ 0.2–0.3 and ~ 2–3, respectively. Such elevated ratios have never been observed in the nuclear starburst regions. One possible explanation for these high HCN(1–0)/CO(1–0) and HCN(1–0)/HCO$^+$(1–0) ratios is X-ray induced chemistry in X-ray dominated regions (XDRs), i.e., the overabundance of the HCN molecule in the X-ray irradiated dense molecular tori. If this view is true, the known tight correlation between HCN(1–0) and the star-formation rate breaks in the vicinity of active nuclei. Although the interpretation of these ratios is still an open question, these ratios have a great potential for a new diagnostic tool for the energy sources of dusty galaxies in the ALMA era because these molecular lines are free from dust extinction.

Keywords ISM · Starburst · Seyfert · Abundance

K. Kohno (✉) · K. Muraoka
Institute of Astronomy, University of Tokyo, 2-21-1 Osawa,
Mitaka, Tokyo, 181-0015, Japan
e-mail: kkohno@ioa.s.u-tokyo.ac.jp

K. Nakanishi · T. Tosaki
Nobeyama Radio Observatory, 462-2 Minamimaki, Minamisaku,
Nagano, 384-1305, Japan

R. Miura · H. Ezawa · R. Kawabe
National Astronomical Observatory of Japan, 2-21-1 Osawa,
Mitaka, Tokyo, 181-8588, Japan

1 Introduction

Dense molecular medium is one of the indispensable components to understand the star-formation laws in galaxies. This is because stars are formed from dense molecular cores and not from diffuse envelopes of giant molecular clouds (GMCs); in fact, millimeter-wave observations of HCN(1–0) emission, which requires dense ($n_{H_2} > 10^4$ cm^{-3}) environments for its collisional excitation due to its high permanent dipole moment ($\mu = 3.0$ Debye), show a tight and linear correlation between HCN(1–0) and far-infrared (FIR) luminosities over ~ 8 orders of magnitude, i.e., from Galactic GMC scales to distant extremely luminous galaxies (Solomon et al. 1992; Gao and Solomon 2004; Cariili et al. 2005; Wu et al. 2005).

However, the weakness of HCN emission (typically 1/10 in T_b of CO in the central regions of galaxies, and 1/30–1/50 of CO in the disk regions of the Galaxy (Helfer and Blitz 1997) and galaxies (Helfer and Blitz 1993; Kohno et al. 1996, 1999, 2002, 2003)) often prevents us from producing large-scale maps of dense molecular medium through HCN observations.

The first issue to be addressed in this paper is then a use of high-J CO lines as another tracer of dense molecular medium in galaxies; previous observational studies of CO(3–2) emission from nearby galaxies suggest that it is indeed detectable even in the disk regions of galaxies (Wielebinski et al. 1999; Dumke et al. 2001; Walsh et al. 2002). We are therefore motivated to conduct a large-scale extragalactic survey of nearby star-forming galaxies in the CO(3–2) line. This topic will be discussed in Sect. 2.

The other issue of this paper is the dense molecular gas at the very centers of Seyfert galaxies. Dense molecular medium also plays various roles in the vicinity of active galactic nuclei (AGNs). The presence of dense and dusty interstellar matter (ISM), which obscures the broad line regions in the AGNs, is inevitable at <1 pc–a few 10 pc scales according to the proposed unified model of Seyfert galaxies. This circumnuclear dense ISM could be a reservoir of fuel for the active nuclei and a site for massive star formation. In fact, strong HCN (1–0) emission has been detected in the prototypical type-2 Seyfert NGC 1068 (Jackson et al. 1993; Tacconi et al. 1994; Helfer and Blitz 1995). Similar enhancements in the low-luminosity Seyfert galaxies such as NGC 5194 (Kohno et al. 1996), NGC 1097 (Kohno et al. 2003), NGC 5033 (Kohno 2005), and NGC 6951 (Krips et al. 2007) have been reported. In these Seyfert nuclei, the HCN (1–0) to CO (1–0) integrated intensity ratios in the brightness temperature scale, $R_{\mathrm{HCN/CO}}$, are enhanced up to approximately 0.4–0.6, and the kinematics of the HCN line indicates that this dense molecular medium could be the outer envelope of the obscuring material (Jackson et al. 1993; Tacconi et al. 1994; Kohno et al. 1996; Krips et al. 2007).

However, if we recall that HCN emission exhibits a tight correlation with massive star formation, one should wonder whether massive star formation occurs at the very centers of these Seyfert galaxies. Interestingly, this enhanced HCN emission is often overluminous with respect to HCO$^+$ emission (Kohno et al. 2001; Kohno 2005). What is happening at the very centers of Seyfert galaxies (such as NGC 1068), where very enhanced HCN(1–0) emission is often observed? Is HCN emission still a tracer of massive star formation there? These issues will be addressed in Sect. 3.

2 Dense gas imaging survey of spiral galaxies using the ASTE

Submillimeter-wave CO(3–2) emission is another tracer of dense gas because the Einstein A coefficient is scaled as $\propto \nu^3$; hence, the critical density of CO(3–2) emission is higher than that of CO(1–0) by a factor of $\sim 3^3$, i.e., $n_{\mathrm{H_2}} \sim 10^4$ cm^{-3}.

The Atacama Submillimeter Telescope Experiment (ASTE) (Ezawa et al. 2004), a new 10 m telescope in the

Fig. 1 A view of the ASTE observatory at Pampa la Bola (4860 m) in the Atacama desert, Chile

Table 1 The Atacama Submillimeter Telescope Experiment (ASTE)

Diameter of the main reflector	10 m
Diameter of the sub reflector	60 cm
Surface accuracy of the main reflector	19 μm
Observing frequency range	320–370 GHz
Main beam efficiency at 350 GHz	0.6–0.7 (winter nights)
Radio pointing accuracy	2″ r.m.s.
System noise temperature	150–250 K in DSB
Spectrometer band width	512 MHz
Spectrometer channels	1024 channels

Atacama desert in northern Chile, provides us with an ideal opportunity to make large-scale maps of CO(3–2) emission because of the high beam efficiency of the telescope and the low system noise temperature owing to the good receiver system and atmospheric conditions at the site, as well as its efficient On-The-Fly (OTF) mapping capability. Figure 1 shows the photograph of the ASTE observatory; the parameters of the ASTE are listed in Table 1.

In order to understand the global distributions of dense molecular medium in galaxies, we have conducted an extragalactic CO(3–2) imaging survey of nearby spiral galaxies, ADIoS (ASTE Dense gas Imaging of Spiral galaxies). Current sample galaxies of the ADIoS project are listed in Table 2, and some recent selected highlights from ADIoS, M 83 (Muraoka et al. 2007) and NGC 986 (Kohno et al. submitted), are displayed below. Observations of another intriguing target of ADIoS, NGC 604 (the most luminous giant HII region in the spiral galaxy M 33), are also reported

Table 2 Current sample list of ADIoS (ASTE Dense gas Imaging of Star-forming galaxies)

Name	Area	CO(3–2) reference	Complementary CO(1–0) data
M 83	$5' \times 5'$	Muraoka et al. (2007)	NRO 45 m (Kuno et al. 2007)
a GMA in M 31	$1'.5 \times 1'.5$	Tosaki et al. (2007b)	NRO 45 m (Tosaki et al. 2007b)
NGC 986	$3' \times 3'$	Kohno et al. (submitted)	–
NGC 253	$9' \times 3'$	Nakanishi et al. (in prep.)	NRO 45 m (Sorai et al. 2000)
NGC 604 (in M 33)	$5' \times 5'$	Tosaki et al. (2007a)	NRO 45 m (Tosaki et al. 2007a)

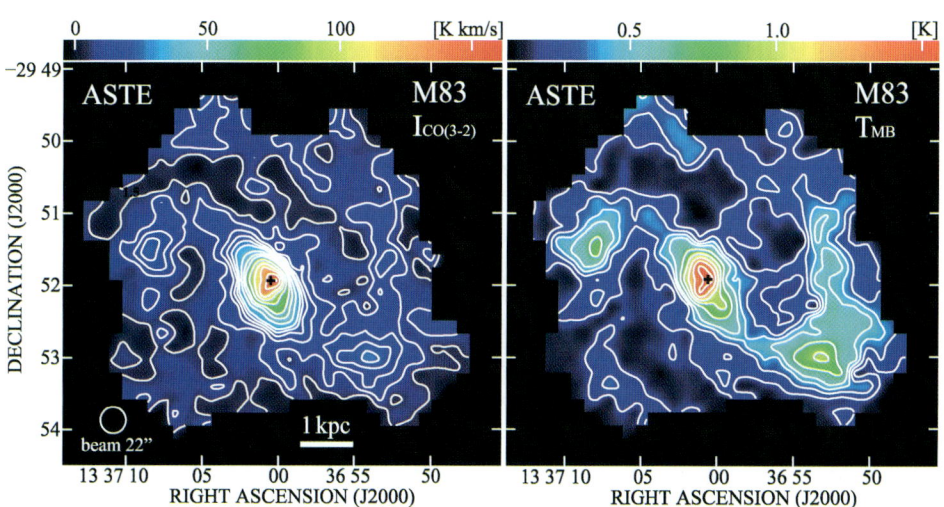

Fig. 2 (*Left*) A contour map of the CO(3–2) integrated intensity of M 83. The *central cross* indicates the central reference position of the map. Contours are at 5, 10, 15, 20, 25, 30, 40, 50, 60, 80, 100, 120, 140, and 160 K km s^{-1}, and the peak is 161.2 K km s^{-1}. The 1 σ noise level of the map is 2 K km s^{-1}. (*Right*) A contour map of the CO(3–2) peak brightness temperature. Contours are at 0.2, 0.3, 0.4, 0.5, 0.7, 0.9, 1.1, 1.3, and 1.5 K, and the peak is 1.53 K

(Miura et al. submitted; Tosaki et al. 2007a). Further recent scientific results from ASTE can be found in Takami et al. (2006), Takahashi et al. (2006), Takakuwa et al. (2007), Oka et al. (2007), Nagai et al. (2007), Tosaki et al. (2007b), Komugi et al. (2007), Nakanishi et al. (2007), Hatsukade et al. (2007), Tanaka et al. (2007).

2.1 M 83

This is one of the best-studied barred spiral galaxies due to its proximity and face-on view, together with the richness of the ISM and star formation. Several CO(3-2) observations of M 83 have already been reported (Israel and Baas 2001; Dumke et al. 2001; Bayet et al. 2006), yet the sensitivity and size of these maps are not sufficient to depict the entire structure of dense molecular medium in M 83. We observed CO(3-2) emission in the central $5' \times 5'$ (or 6.6×6.6 kpc at $D = 4.5$ Mpc) region of the nearby barred spiral galaxy M 83. The integrated intensity map, along with the peak temperature map, are shown in Fig. 2. We successfully resolved the major structures, i.e., the nuclear starburst region, bar, and inner spiral arms in CO(3-2) emission at a resolution of $22''$ or 480 pc. We found a good spatial coincidence between the CO(3-2) and 6 cm continuum emissions. The global CO(3-2) luminosity was measured as $L'_{\mathrm{CO(3-2)}} = 5.1 \times 10^8$ K km s^{-1} pc^2 within the observed region. More

than half of the $L'_{\mathrm{CO(3-2)}}$ originates from the disk region ($r > 0.5$ kpc), indicating that the CO(3-2) emission in the disk region significantly contributes to the global CO luminosity. From a comparison of the CO(3-2) data with CO(1-0) intensities measured with the Nobeyama 45-m telescope, we determined the radial profile of the CO(3-2)/CO(1-0) integrated intensity ratio, $R_{3-2/1-0}$. The ratio is almost unity in the central region ($r < 0.25$ kpc), whereas it drops to a constant value, 0.6–0.7, in the disk region. The radial profile of the star formation efficiencies (SFEs), determined from the 6 cm radio continuum and CO(1-0) emission (star formation rate per unit gas mass), shows the same trend as that of $R_{3-2/1-0}$ (Fig. 3). This implies that the spatial variation in the dense gas fraction traced by $R_{3-2/1-0}$ may govern the spatial variation in SFE in M 83.

2.2 NGC 986

Our CO(3-2) image of the $3' \times 3'$ (or 20×20 kpc at a distance of 23.2 Mpc) region of the southern barred spiral galaxy NGC 986 in Fig. 4 revealed the presence of a large (the major axis of 14 kpc in total length) gaseous bar filled with a dense molecular medium along the dark lanes seen in the optical images. This is the largest "dense-gas rich bar" known. The dense gas bar discovered in NGC 986 could be in an early phase (Hüttemeister et al. 1999; Sakamoto 2007) and also could be a huge reservoir of possible "fuel" for future starbursts in the central region, implying that the star formation in the central region of NGC 986 could still be in the growing stage. The spatial coincidence between the overall distributions of dense molecular gas traced by CO(3–2) and the massive star formation depicted by Hα is also remarkable in Fig. 4c. This is what is expected from the recent CO(3–2) observations that demonstrate a tight correlation between CO(3–2) and Hα luminosities (Komugi et al. 2007), indicating an intimate association between the dense molecular medium and massive star formation (see also Yao et al. 2003).

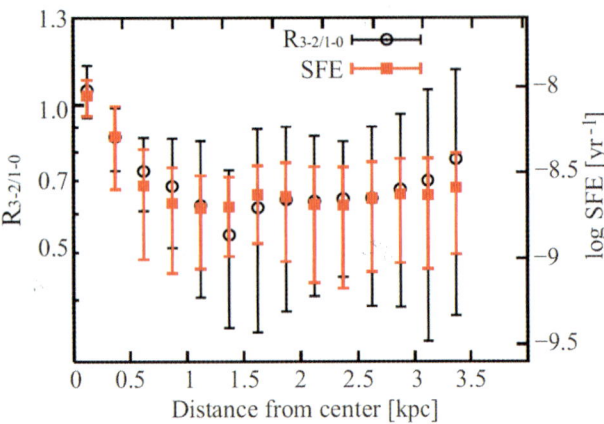

Fig. 3 The azimuthally averaged SFE and CO(3–2)/CO(1–0) integrated intensity ratio ($R_{3-2/1-0}$) as a function of the galactocentric radius of M 83. Both profiles show a strong peak at the center, whereas no significant enhancement is visible at the bar-end, despite the fact that there is a secondary peak in the CO intensities and SFRs

3 Dense molecular gas at the centers of Seyfert and starburst galaxies

In order to understand the nature of enhanced or overluminous HCN emission in some Seyfert nuclei, we have conducted an extensive "3D" imaging survey of the CO(1–0), HCN(1–0), and HCO$^+$(1–0) lines in the local Seyfert and starburst galaxies using the Nobeyama Millimeter Array (NMA) and the RAINBOW interferometer. The majority of the Seyfert sample galaxies are from the Palomar Northern Seyfert sample (Ho and Ulvestad 2001). A few southern Seyfert galaxies are also included in the sample. It should be noted that the HCN(1–0) and HCO$^+$(1–0) lines are observed simultaneously by using the Ultra Wide Band Correlator (Okumura et al. 2000). This is essential to obtain accurate HCN/HCO$^+$ integrated intensity ratios (R_{HCN/HCO^+} hereafter). We obtained CO(1–0) images of 18 Seyfert galaxies (6 type-1s and 12 type-2s) and 7 starburst/HII galaxies with a typical resolution of ∼ 4″. Simultaneous HCN(1–0) and HCO$^+$(1–0) observations were conducted toward 17 Seyferts and 10 starburst galaxies with typical resolutions of ∼ 2″ to 6″ and sensitivities of a few mJy beam^{-1} for a ∼ 50 km s^{-1} velocity channel.

Some of the Seyfert galaxies, including NGC 1068 (Fig. 5), NGC 1097 (Kohno et al. 2007), NGC 5033 (Kohno 2005), and NGC 5194 (Fig. 6), show enhanced or overluminous HCN(1–0) emission. The $R_{HCN/CO}$ and R_{HCN/HCO^+} ratios in these Seyfert galaxies are enhanced up to ∼0.2–0.5 and ∼2–3, respectively. Such elevated ratios are never observed in our nuclear starburst sample. Note that Seyfert galaxies accompanied with a compact nuclear starburst (e.g.,

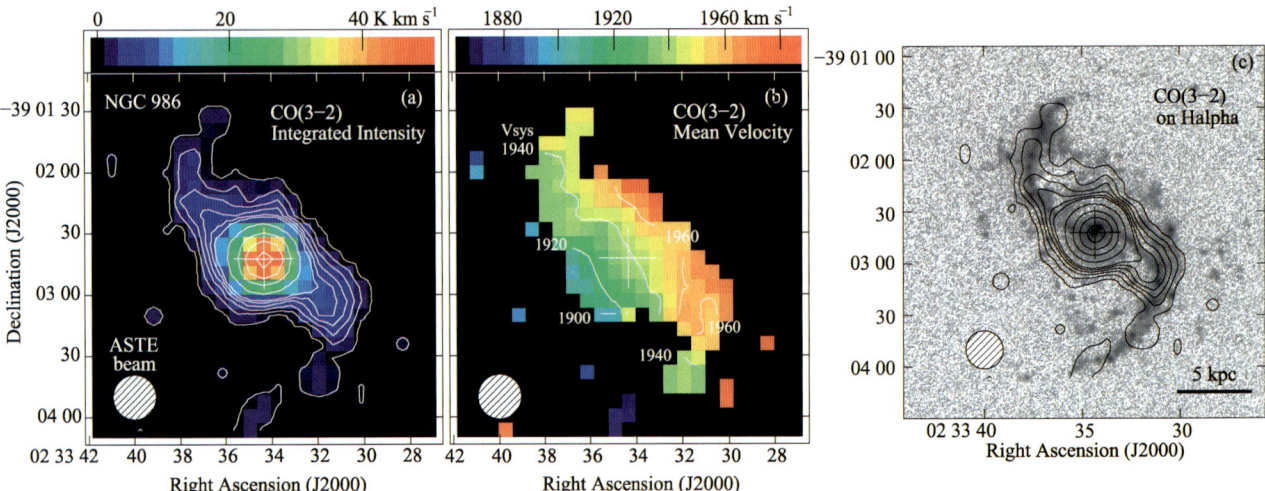

Fig. 4 (**a**) Velocity-integrated CO(3–2) intensity map of NGC 986, obtained with the OTF mode using the SC345 receiver mounted on ASTE. Image size is $3' \times 3'$ or 20×20 kpc at a distance of 23 Mpc. Contour levels are 1, 4, 7, 10, 13, 16, 30, 50, 70, & 90% of the peak intensity, 55.2 K km s^{-1} or 3.35×10^3 Jy beam^{-1} km s^{-1}. The *central* *cross* indicates the position of the nucleus (the peak of NIR continuum from 2MASS/NED). (**b**) Intensity-weighted mean velocity map of CO(3–2) in NGC 986. Contours are labeled by the LSR velocity. (**c**) Contour map of the CO(3–2) integrated intensity superposed on the continuum-subtracted Hα image (Hameed and Devereux 1999)

Fig. 5 Molecular line images of NGC 1068. (*Top left*) CO(1–0) image obtained with BIMA (Helfer and Blitz 1995). (*Bottom left*) HCN(1–0) image obtained using NMA. (*Bottom right*) HCO$^+$(1–0) image. (*Top right*) HCN and HCO$^+$ spectra in the $6''.0 \times 3''.9$ aperture centered on the nucleus. Overluminous HCN(1–0) emission at the center of NGC 1068 is remarkable; $R_{HCN/HCO^+} = 2.1 \pm 0.13$. In contrast, R_{HCN/HCO^+} in the circumnuclear ring/arms is approximately unity, which is typical for starburst regions

Fig. 6 Molecular line images of NGC 5194. (*Top left*) CO(1–0) image obtained with NMA (Sakamoto et al. 1999). (*Bottom left*) HCN(1–0) image obtained using NMA. (*Bottom right*) HCO$^+$(1–0) image. (*Top right*) HCN and HCO$^+$ spectra in the $7'' \times 6''$ aperture centered on the nucleus. The R_{HCN/HCO^+} value at the center is extremely high: 2.5 ± 0.43

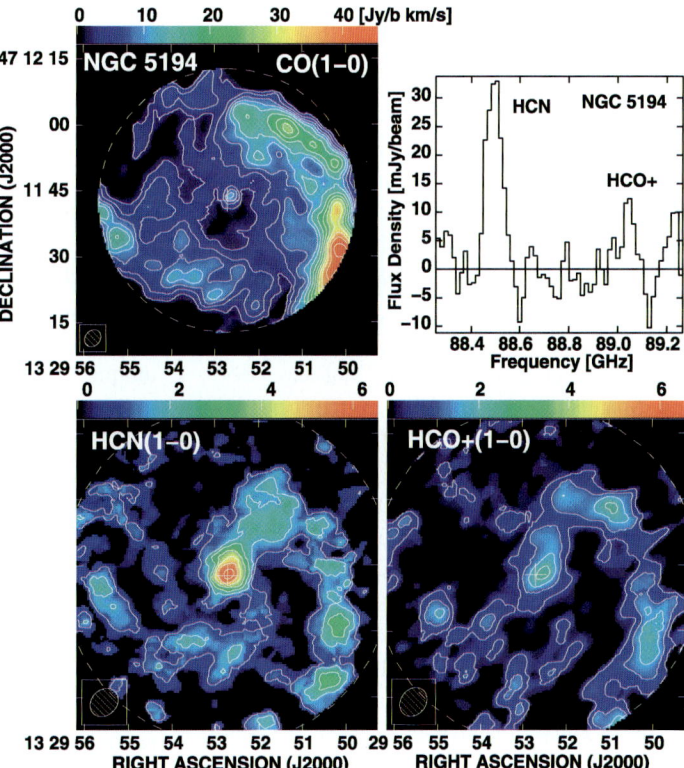

NGC 6764, Fig. 7) exhibit line ratios similar to those of nuclear starburst galaxies. One possible explanation for these high $R_{HCN/CO}$ and R_{HCN/HCO^+} ratios is the chemistry due to X-ray dominated regions (XDRs, Maloney et al. 1996), i.e., the overabundance of the HCN molecule in the X-ray irradiated dense molecular tori (Lepp and Dalgarno 1996;

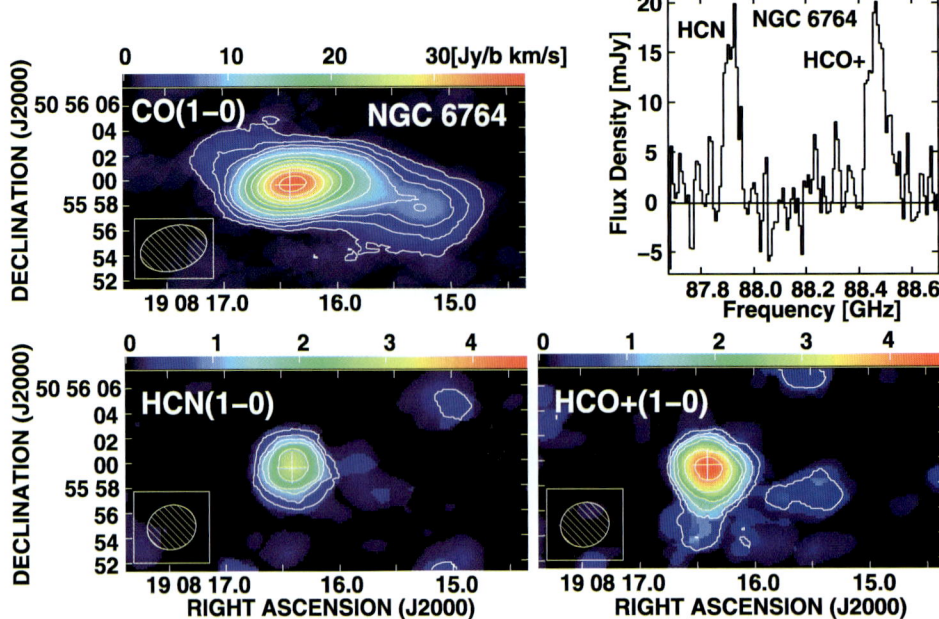

Fig. 7 Molecular line images of NGC 6764. (*Top left*) CO(1–0), (*bottom left*) HCN(1–0), and (*bottom right*) HCO$^+$(1–0) images of NGC 6764 using NMA. (*Top right*) HCN and HCO$^+$ spectra in the $5''.6 \times 3''.6$ aperture centered on the nucleus. No significant enhancement of HCN(1–0) emission with respect to CO nor HCO$^+$ can be seen; $R_{HCN/HCO^+} = 0.63 \pm 0.09$, which suggests a composite nature

(i.e., AGN accompanied with a compact nuclear starburst within the observing beam). In fact, the strong Wolf-Rayet feature (Conti 1991) and large-scale outflow (Hota and Saikia 2006) indicate that young starburst dominates the central region of this galaxy, although the presence of an active nucleus is evident according to the time variability of the nuclear X-ray source (Schinnerer et al. 2000)

Maloney 1999; Meijerink and Spaans 2005). One of the keys is the temperature of molecular clouds in the XDRs; in contrast to photo dissociation regions (PDRs) where UV photons are blocked at the surface of molecular clouds, high-energy photons can penetrate into deep inside molecular clouds. Moreover, heating due to photo-ionization in XDRs is considerably more efficient than that of photoelectric heating in PDRs. Consequently, the temperature of molecular clouds in XDRs becomes very high compared with molecular clouds in PDRs (Maloney 1999; Meijerink and Spaans 2005). In fact, at the center of NGC 5194, a host of low-luminosity AGN (see Refs. in Kohno et al. 1996), very high kinetic temperature of the molecular gas has been suggested (Matsushita et al. 1998, 2004), and this nucleus is a representative one showing the overluminous HCN(1–0) emission in our sample (Fig. 6).

If high $R_{HCN/CO}$ and R_{HCN/HCO^+} values are signatures of the overabundance of HCN molecule due to XDR chemistry, the known tight correlation between HCN(1–0) and star formation rate reported in star-forming galaxies breaks in the vicinity of active nuclei; the overluminous HCN(1–0) emission does not imply an elevated massive star formation rate there. Although the interpretation of these ratios is still an open question (Meijerink and Spaans 2005; Papadopoulos 2007; Papadopoulos et al. submitted; Aalto et al. 2007; Aalto 2007; Garcia-Burillo et al. 2007; Meijerink et al. 2007), HCN/HCO$^+$ ratios (especially for

high-J transitions, Graciá-Carpio et al. 2006; Garcia-Burillo et al. 2006) have a great potential as a new diagnostic tool for the energy sources of distant dusty galaxies in the ALMA era because these molecular lines are free from dust extinction. HCN(1–0) and HCO$^+$(1–0) observations of LIRGs and ULIRGs suggest the presence of hidden active nuclei in some of these IR luminous objects (Imanishi et al. 2004, 2006, 2007; Graciá-Carpio et al. 2006; Garcia-Burillo et al. 2007) although the conclusions are not unique yet (e.g., Aalto et al. 2007; Aalto 2007).

Acknowledgements We appreciate the feedback of an anonymous referee whose comments improved the manuscript. We would like to acknowledge all the members involved with the ASTE team for their great efforts in the ASTE project. The ASTE is the joint project between National Astronomical Observatory of Japan (NAOJ) and universities including the University of Tokyo, University of Chile, Nagoya University, Osaka Prefecture University, and Ibaragi University. We are also grateful to the NMA staff for their continuous efforts to operate the array. This study was financially supported by MEXT Grant-in-Aid for Scientific Research on Priority Areas No. 15071202. Observations with the ASTE were carried out remotely from Japan by using NTT's GEMnet2 and its partner R&E (Research and Education) networks, which are based on AccessNova collaboration of University of Chile, NTT Laboratories, and NAOJ.

References

Aalto, S.: Astrophys. Space Sci. (2007), doi:10.1007/s10509-007-9643-0

Aalto, S., et al.: Astron. Astrophys. **464**, 193 (2007)

Bayet, E., et al.: Astron. Astrophys. **460**, 467 (2006)

Cariili, C.L., et al.: Astrophys. J. **618**, 586 (2005)

Conti, P.S.: Astrophys. J. **377**, 115 (1991)

Dumke, M., et al.: Astron. Astrophys. **373**, 853 (2001)

Ezawa, H., et al.: Proc. SPIE **5489**, 763 (2004)

Gao, Y., Solomon, P.M.: Astrophys. J. **606**, 271 (2004)

Garcia-Burillo, S., et al.: Astrophys. J. **645**, L17 (2006)

Garcia-Burillo, S., et al.: New Astron. Rev. **51**, 160 (2007)

Graciá-Carpio, J., et al.: Astrophys. J. **640**, L135 (2006)

Hameed, S., Devereux, N.: Astron. J. **118**, 730 (1999)

Hatsukade, B., et al.: Publ. Astron. Soc. Jpn. **59**, 67 (2007)

Helfer, T.T., Blitz, L.: Astrophys. J. **419**, 86 (1993)

Helfer, T.T., Blitz, L.: Astrophys. J. **450**, 90 (1995)

Helfer, T.T., Blitz, L.: Astrophys. J. **478**, 233 (1997)

Ho, L.C., Ulvestad, J.S.: Astrophys. J. Suppl. Ser. **133**, 77 (2001)

Hota, A., Saikia, D.J.: Mon. Not. R. Astron. Soc. **371**, 945 (2006)

Hüttemeister, S., et al.: Astron. Astrophys. **346**, 45 (1999)

Imanishi, M., et al.: Astron. J. **128**, 2037 (2004)

Imanishi, M., et al.: Astron. J. **131**, 2888 (2006)

Imanishi, M. et al.: Astron. J. (2007, in press)

Israel, F.P., Baas, F.: Astron. Astrophys. **371**, 433 (2001)

Jackson, J.M., et al.: Astrophys. J. **418**, L13 (1993)

Kohno, K.: AIPC **783**, 203 (2005)

Kohno, K., et al.: Astrophys. J. **461**, L29 (1996)

Kohno, K., et al.: Astrophys. J. **511**, 157 (1999)

Kohno, K., et al.: Astron. Soc. Pac. Conf. Ser. **249**, 672 (2001)

Kohno, K., et al.: Publ. Astron. Soc. Jpn. **54**, 541 (2002)

Kohno, K., et al.: Publ. Astron. Soc. Jpn. **55**, L1 (2003)

Kohno, K., et al.: Astron. Soc. Pac. Conf. Ser. **373**, 647 (2007)

Kohno, K., et al.: Publ. Astron. Soc. Jpn. (submitted)

Komugi, S., et al.: Publ. Astron. Soc. Jpn. **59**, 55 (2007)

Krips, M., et al.: Astron. Astrophys. **468**, 63 (2007)

Kuno, N., et al.: Publ. Astron. Soc. Jpn. **59**, 117 (2007)

Lepp, S., Dalgarno, A.: Astron. Astrophys. **306**, L21 (1996)

Maloney, P.R.: Astrophys. Space Sci. **266**, 207 (1999)

Maloney, P.R., et al.: Astrophys. J. **466**, 561 (1996)

Matsushita, S., et al.: Astrophys. J. **495**, 267 (1998)

Matsushita, S., et al.: Astrophys. J. **616**, L55 (2004)

Meijerink, R., Spaans, M.: Astron. Astrophys. **436**, 397 (2005)

Meijerink, R., et al.: Astron. Astrophys. **461**, 793 (2007)

Miura, R. et al.: Astrophys. Space Sci. (submitted)

Muraoka, K., et al.: Publ. Astron. Soc. Jpn. **59**, 43 (2007)

Nagai, M., et al.: Publ. Astron. Soc. Jpn. **59**, 25 (2007)

Nakanishi, H., et al.: Publ. Astron. Soc. Jpn. **59**, 61 (2007)

Oka, T., et al.: Publ. Astron. Soc. Jpn. **59**, 15 (2007)

Okumura, S.K., et al.: Publ. Astron. Soc. Jpn. **52**, 393 (2000)

Papadopoulos, P.P.: Astrophys. J. **656**, 792 (2007)

Papadopoulos, P.P. et al.: Astrophys. Space Sci. (submitted)

Sakamoto, K.: Astrophys. Space Sci. (2007), doi:10.1007/s10509-007-9624-3

Sakamoto, K., et al.: Astrophys. J. Suppl. Ser. **124**, 403 (1999)

Schinnerer, E., et al.: Astrophys. J. **545**, 205 (2000)

Solomon, P.M., et al.: Astrophys. J. **387**, L55 (1992)

Sorai, K., et al.: Publ. Astron. Soc. Jpn. **52**, 785 (2000)

Tacconi, L.J., et al.: Astrophys. J. **426**, L77 (1994)

Takahashi, S., et al.: Astrophys. J. **651**, 933 (2006)

Takakuwa, S., et al.: Publ. Astron. Soc. Jpn. **59**, 1 (2007)

Takami, M., et al.: Publ. Astron. Soc. Jpn. **58**, 563 (2006)

Tanaka, K., et al.: Publ. Astron. Soc. Jpn. **59**, 323 (2007)

Tosaki, T., et al.: Astrophys. J. **664**, L27 (2007a)

Tosaki, T., et al.: Publ. Astron. Soc. Jpn. **59**, 33 (2007b)

Walsh, W., et al.: Astron. Astrophys. **388**, 7 (2002)

Wielebinski, R., et al.: Astron. Astrophys. **347**, 634 (1999)

Wu, J., et al.: Astrophys. J. **635**, L173 (2005)

Yao, L., et al.: Astrophys. J. **588**, 771 (2003)

Spectroscopic surveys of cosmic evolution

Nick Scoville

Originally published in the journal Astrophysics and Space Science, Volume 313, Nos 1–3.
DOI: 10.1007/s10509-007-9685-3 © Springer Science+Business Media B.V. 2007

Abstract ALMA will be the premier instrument for the study of galaxy evolution in the early universe—enabling studies of the gas content, dynamics and dynamical masses, and star formation with unparalleled resolution and sensitivity. Galaxy evolution and AGN growth in the early universe are believed to be strongly driven by merging and dynamical interactions. Thus, a full exploration of the environmental influence is absolutely essential. The Cosmic Evolution Survey (COSMOS) is specifically designed to probe the correlated coevolution of galaxies, star formation, active galactic nuclei (AGN) and dark matter (DM) large-scale structure (LSS) over the redshift range $z > 0.5$ to 3. In this contribution I review the characteristics of the COSMOS survey and very exciting initial results on mapping large scale structure in galaxies and dark matter. The survey includes multi-wavelength imaging and spectroscopy from X-ray to radio wavelengths covering a 2 square degree equatorial field. Given the very high sensitivity and resolution of these datasets, COSMOS will provide unprecedented samples of objects at $z > 3$ for followup studies wit ALMA.

Keywords Surveys · Galaxy evolution · Cosmology

1 Introduction

Our understanding of the formation and evolution of galaxies and their large-scale structures (LSS) has advanced enormously over the last decade—a result of a phenomenal synergy between theoretical and observational efforts. Deep observational studies using the Hubble Space Telescope (HST) and the largest ground based telescopes have probed galaxy and AGN populations back to redshift $z = 6$ when the universe had aged less than 1 billion of its current 13 billion years. Just as remarkable is the enormous success of numerical simulations for Λ CDM models in reproducing many of the current LSS characteristics, all starting from an initial, nearly uniform, hot universe!

1.1 The cosmic evolution survey—COSMOS

The COSMOS survey (Scoville et al. 2007a) is the first survey encompassing a sufficiently large area that it can address the coupled evolution of LSS and galaxies, star formation and AGN. COSMOS is the largest HST survey ever undertaken—imaging an equatorial, 2 square degree field with single-orbit I-band exposures to a depth of $I_{AB} = 28$ mag (5σ). Extensive multi-λ ground and space-based observations of this field have been gathered spanning the entire spectrum from X-ray, UV, optical/IR, mid-infrared, mm/submm and to radio with extremely high sensitivity imaging and spectroscopy. This full spectrum approach is required to probe the coupled evolution of young and old stellar populations, starbursts, the ISM (molecular and ionized components), AGN and dark matter. Each of these cosmic components may be best probed quite differently. The multi-λ approach is also required due to the differential redshifting of cosmic history and the presence of dust obscuration in many of the most rapidly evolving galactic regions. The large area coverage of COSMOS is motivated to sample the largest structures existing in the local universe—smaller area coverage can lead to severe cosmic variance problems.

COSMOS detects $\simeq 2 \times 10^6$ galaxies and AGN sampling a volume in the high redshift universe ($z > 0.5$ to 4) ap-

N. Scoville (✉)
Astronomy Department, Caltech, 105-24, Pasadena, CA 91125, USA
e-mail: nzs@astro.caltech.edu

proaching that sampled locally by the Sloan Digital Sky Survey (SDSS). Subaru optical imaging and photometric redshifts have been determined for approximately 800,000 galaxies. The COSMOS spectroscopic surveys (VLT and Magellan) will yield 45,000 galaxies with accurate redshifts at $z = 0.5$–2.5, all having 0.05 arcsec HST imaging. Here I provide a brief overview of the scientific goals of the COS-MOS survey and an overall summary of the survey observational program since this will very likely become the major survey field for early universe imaging with ALMA.

2 COSMOS science goals

The COSMOS survey addresses nearly every aspect of observational cosmology over the majority of the Hubble time at $z \geq 0.5$ to 6:

- The assembly of galaxies, clusters and dark matter up to $2 \times 10^{14} M_\odot$;
- Reconstruction of the dark matter distributions and characteristics out to $z \sim 1$ using weak gravitational lensing at $z < 1.5$;
- The evolution of galaxy morphology, galactic merger rates and star formation as a function of LSS environment and redshift;
- Evolution of AGN and the dependence of black hole growth on galaxy morphology and environment; and
- The mass and luminosity distribution of the earliest galaxies, AGN and intergalactic gas at $z = 3$ to 6 and their clustering.

The growth of galaxies, AGN and dark matter structure is traced in COSMOS over a period corresponding to $\sim 75\%$ of the age of the universe. Galaxies in the early universe are built up by two major processes: dissipational collapse and merging of lower mass protogalactic and galactic components. Their intrinsic evolution is then driven by the conversion of primordial and interstellar gas into stars, with galactic merging and interactions triggering star formation and starbursts.

The need to sample very large scales arises from the fact that structure occurs on mass scales up to $\geq 10^{14} M_\odot$ and existing smaller surveys are likely to be unrepresentative at $z \sim 1$. Earlier projects, such as GOODS and GEMS, adequately sample masses up to $3 \times 10^{13} M_\odot$, whereas COS-MOS samples the largest known structures at $\sim 2 \times 10^{14} M_\odot$ (dark and luminous matter). COSMOS will measure all galaxy populations in their large-scale context as a function of redshift, providing essential guidelines for the next generation of theoretical models.

3 Major observational components

In this section, we briefly review the major ingredients of the COSMOS survey.

3.1 Galaxy evolution: HST imaging and SEDs

The evolutionary status of galaxies can be analyzed from either their morphologies or their spectral energy distributions (SED, characterizing the stellar population).

Morphological parameters for the galaxies are obtained from the HST imaging (e.g. bulge/disk ratios, concentration, asymmetry, size, multiplicity, clumpiness). The COSMOS I-band HST images have sufficient depth and resolution to allow classical bulge-disk decomposition for L^* galaxies at $z \leq 2$, while less detailed structural parameters such as compactness, asymmetry, clumpiness and size can be measured for all galaxies down to the spectroscopic limit ($I_{AB} \sim 25$), out to $z \sim 5$.

In COSMOS, deep imaging (from Subaru, GALEX, UKIRT and NOAO, and SPITZER-IRAC) provides SEDs to characterize the integrated stellar populations of the 2 million galaxies detected with HST. The SEDs are derived self-consistently with the photometric redshift determinations.

3.2 Galaxy redshifts: photometric and spectroscopic

Determining the redshifts or lookback time of individual galaxies is clearly one of the most difficult and time consuming aspects of any cosmological evolution survey. In COSMOS this is even more difficult since the redshifts are needed with sufficient precision not just to determine the cosmic epoch, but also to place the galaxies within or outside of LSS appearing along the line of sight. Without high precision, structures become 'blurred' due to scattering of galaxies to different distances in the line of sight and for specific galaxies, their environment cannot be determined.

In COSMOS, photometric redshifts are obtained from deep (mostly ground-based) imaging—from Subaru, CFHT, UKIRT, and NOAO (Taniguchi et al. 2007; Capak et al. 2007a; Mobasher et al. 2007). At present the photometric-redshift accuracy is $\sigma_z/(1 + z) \sim 0.02$ for approximately 4×10^5 galaxies at $z < 2.5$ (Mobasher et al. 2007; Ilbert et al. 2007). These photometric redshifts are based on 20 bands of optical/IR data and the accuracy is unprecedented for photo-z's. It is sufficient to enable initial definition of the LSS, especially for the denser environments (Scoville et al. 2007b).

Very large spectroscopic surveys are now ongoing as part of COSMOS at the VLT and Magellan telescopes (Lilly 2007; Impey et al. 2007). The spectroscopic sample will eventually include approximately 45,000 galaxies and 1000 AGN down to limits of $I_{AB} = 24.5$ and $I_B = 25.5$ mag.

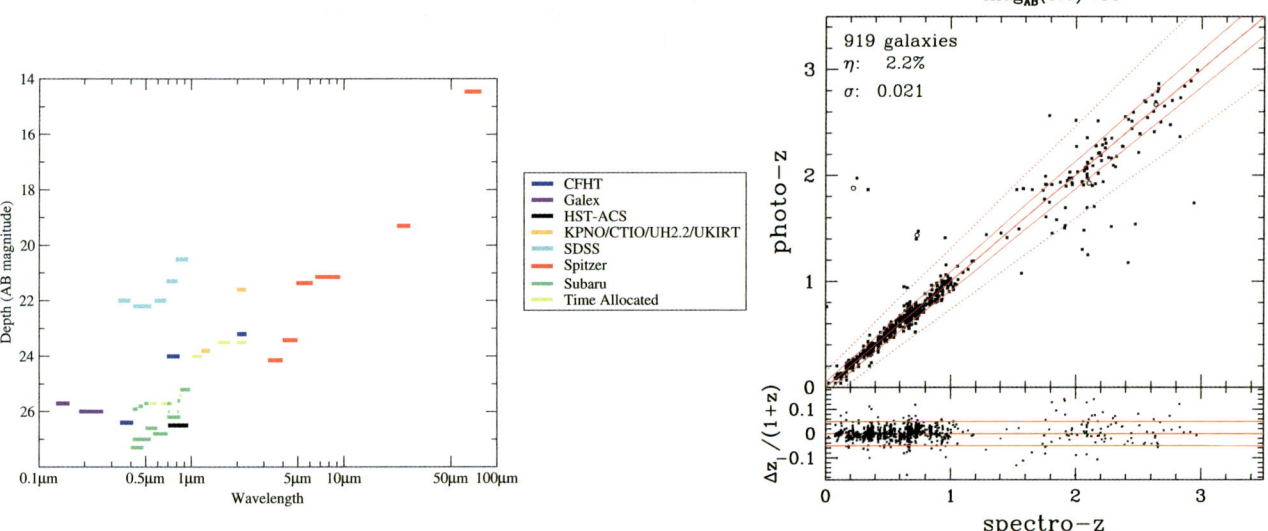

Fig. 1 *Left*: the 5σ sensitivities are shown for the UV-optical and IR bands in COSMOS (3 arcsec apertures except ACS 0.15 arcsec, Capak et al. 2007b). *Right*: comparison of the newly derived COSMOS photo-z's using 20 bands (Ilbert et al. 2007) with the spectroscopic redshifts (Lilly 2007) for 2100 galaxies with $3.6\,\mu m < 24\,mag_{AB}$ (some with I_{AB} down to 26 mag) indicates $\sigma_z/(1+z) < 0.02$. For the 65 X-ray AGN in our 24 μm flux-limited IRS sample, there are optical spectroscopic redshifts from Magellan and VLT

These spectroscopic samples will provide very precise definition of the environment, albeit for smaller subsets of the overall COSMOS galaxy population.

3.3 Galaxy overdensities and weak lensing

The environment or LSS in which a given galaxy resides is defined from the local number density of galaxies or from the DM density as determined from weak lensing or the local galaxy velocity dispersion. The COSMOS HST imaging provides measures of the close-in environment (from galaxy multiplicity and merger indicators such as tidal distortions) and larger-scale DM environment. As noted in Sect. 3.2, definition of the environment is critically dependent on moderately high accuracy spectroscopic (or photometric) redshifts; the integrated, multi-wavelength approach adopted for COSMOS is intended to maximize the impact and utility of each component. Having multiple approaches to environmental determination also provides added confidence.

The wide-area, uniform ACS and Subaru COSMOS surveys allow determination of spatial correlation functions as a function of type (morphological and SED) and luminosity and their evolution with redshift and environment. The enormous sizes of the samples which become available in COSMOS will enable precision approaching that of SDSS but at much higher redshift. COSMOS provides ~ 40 slices of the universe back to $z \sim 4$ to reveal the spatial distribution and shapes of hundreds of thousands of galaxies sampling the full range of cosmic structure.

A major goal of COSMOS has recently been realized and was a highlight of the Jan'07 AAS meeting and a Nature cover (see Fig. 1)—the first imaging of large scale structures (38 LSS on scales of 3–30 Mpc at $z = 0.2$–1.1) in both the dark matter from weak lensing (Massey et al. 2007) and in the baryons from galaxy overdensities (Scoville et al. 2007b; Guzzo et al. 2007) and diffuse X-ray emission (Hasinger et al. 2007; Finoguenov et al. 2007). The dependence of morphology and galactic spectral energy distribution (SED) on environment has been followed in these structures and into the field population (Scarlata et al. 2007; Capak et al. 2007b; Scoville et al. 2007b). Large structures have also been seen in COSMOS at $z \sim 3$–4.5 in Lyman Break Galaxies (Giavalisco et al. 2007) and at $z = 5.7$ in Lyα emitting sources (Murayama et al. 2007).

3.4 Activity: starbursts and AGN

The COSMOS survey samples $\sim 45{,}000$ galaxies spectroscopically—providing an enormous sample of emission line tracers of both starbursts and AGN over a broad range of redshift. In addition, complete very high sensitivity radio continuum (Schinnerer et al. 2007) and X-ray (Hasinger et al. 2007) coverage directly probes the population of AGN; the radio sensitivity can detect the starburst population at $z \sim 1$–2. Perhaps most importantly, the coverage with Spitzer will detect dust embedded ultraluminous starbursts and AGN out to $z \sim 2$–3. Thus, COSMOS encompasses enormous samples of galaxies with multiple, independent tracers of luminous activity—which can then be analyzed as a function of both redshift and environment—opening up fundamental investigations of starburst and AGN fueling in the early universe.

Fig. 2 Comparison of the weak lensing dark matter mass map (*contours*) with galaxy LSS (*blue*) and diffuse X-ray emission (*red*) (Massey et al. 2007; Scoville et al. 2007b). This is the projected density distribution *obtained from the full 3-d distribution* derived using photometric redshifts for line of sight discrimination. Major structures shown here occur at $z = 0.25, 0.5, 0.73$ and 1.0 with sizes up to ~40 Mpc and total masses up to $10^{15} M_\odot$, similar to that of the COMA cluster

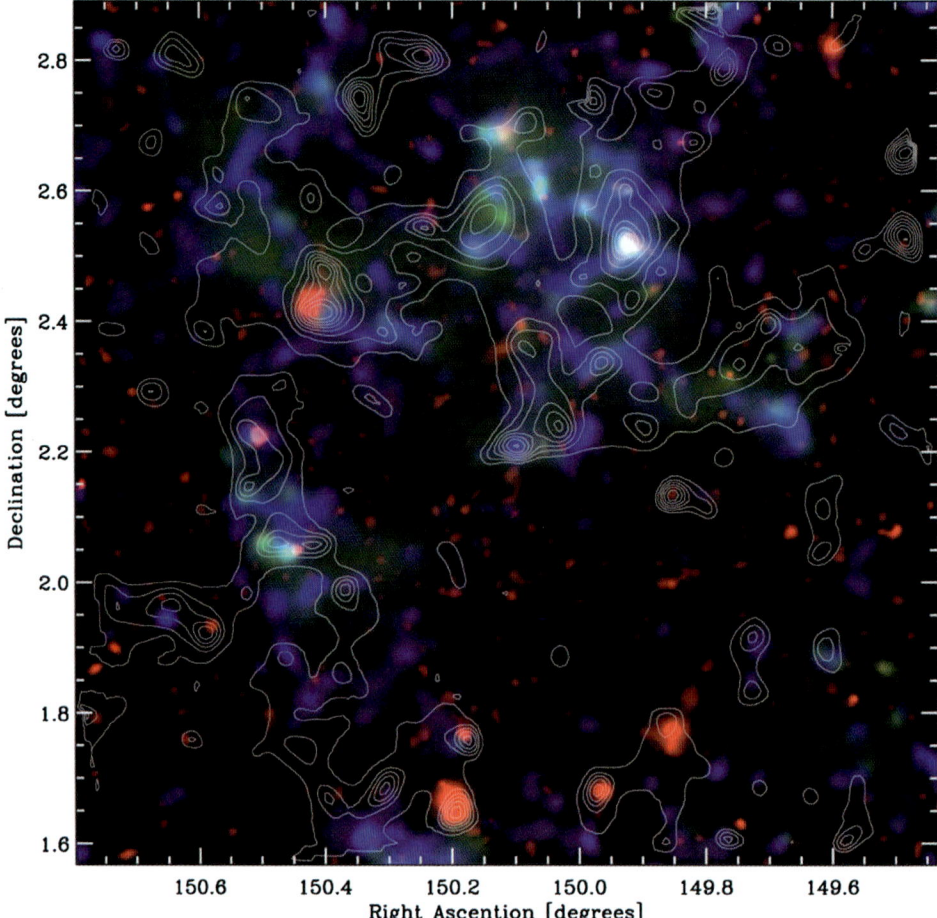

Submm-λ surveys of the COSMOS field have been initiated to identify the most luminous starbursts at $z > 1$. In the long term, high resolution imaging with ALMA will be a vital capability—providing resolved images of the neutral ISM, luminosity distribution and dynamical masses for virtually all COSMOS galaxies having ISMs equivalent to the Galaxy.

3.5 $z = 3$–6: high redshift galaxies, LSS and IGM

The large areal coverage and high sensitivity of the COSMOS survey will result in significant samples of $z > 3$ objects, selected by multi-band color criteria (e.g. BzK selection), the Lyman-break method (Giavalisco et al. 2007), or by direct detection of Lyα emission lines (Murayama et al. 2007). At these higher redshifts, the field subtends over 200 Mpc (comoving) and samples a volume similar to that sampled locally by SDSS.

4 COSMOS multi-wavelength surveys

The COSMOS field is located near the celestial equator ($\delta = 2°$) to ensure visibility by all astronomical facilities,

especially unique instruments such as the next generation 20–30 m optical/IR telescope(s). The time requirements for deep imaging and spectroscopy over a total area of 2 square degrees, containing over a million galaxies makes it strategically imperative that the field be readily observable by all large optical/IR telescopes. For radio studies, high-declination fields such as Lockman Hole, HDF-North, Groth strip and CDF-South are ruled out—they can not be easily observed by *both* (E)VLA in the north and ALMA in the south.

The COSMOS field is accessible to essentially all astronomical facilities, enabling complete multi-λ datasets (X-ray, UV, optical/IR, FIR/submm to radio). The status of these observational programs is continuously updated on the COSMOS website: http://www.astro.caltech.edu/~cosmos/.

The extensive allocations on Subaru, CFHT, UKIRT and NOAO have providing extremely deep photometry for 22 bands from U to K_s, enabling accurate photo-z's, integrated colors and color selection of populations (e.g. LBGs, EROs, AGN, etc.) for essentially all objects detected in the 2 square degree ACS field. XMM has devoted 1.4 Ms to a complete X-ray survey of the field (Hasinger et al. 2007), and COSMOS was one of the deep-GALEX fields

Table 1 Multi-λ COSMOS data

Data	Bands, λ, Res.	AB mag 5σ pt. src	Investigators/Time
HST-ACS	814I	28.8	C12-13 581 o
HST-ACS	475g	28.15	C12 9 o
HST-NIC3	160W	25.6 (6% area)	C12-13 590 o
HST-WFPC2	300W	25.4	C12-13 590 o
Subaru-SCam	B, V, r′, i, z′, g′	27–26	Taniguchi et al. 10 n
Subaru-SCam	10 IB filters	26	Taniguchi et al. 16 n, Scoville 3 n
Subaru-SCam	NB816	25	Taniguchi et al. 2.5 n
CFHT-Megacam	u*	27	Sanders et al. 33 hr
CFHT-Megacam	u, i*	26	LeFevre et al. 12 hr
CFHT-LS	u–z		Deep LS Survey
LBT	Ugr	27	Impey/Schinnerer/Giallongo et al. 48 hr
NOAO/CTIO	K_s	21	Mobasher et al. 18 n
CFHT/UKIRT	J, H, K	24.5–23.5	Sanders et al. 25 n
UH-88	J	21	Sanders et al. 22 n
GALEX	FUV, NUV	26.1, 25.8	Schminovich et al. 200 ks
XMM-EPIC	0.5–10 keV	10^{-15} cgs	Hasinger et al. 1.4 Ms
CXO	0.5–7 keV	0.9°	Elvis et al. 1.8 Ms
VLT-VIMOS sp.	($R = 200$)	# = 3000 ($I_{AB} < 23$)	Kneib et al. 20 hr
VLT-VIMOS sp.	($R = 200, 600$)	# = 45000 ($I_{AB} < 25, z \geq 0.8$)	Lilly et al. 540 hr
Mag.-IMAX sp.	($R = 3000$)	# = 2000	Impey, McCarthy, Elvis 30 n
Keck/GEMINI sp.	($R = 5,000$)	# = 4000 ($I < 24$)	Team Members 10 n
Spitzer-MIPS	160, 70, 24 μm	17, 1, 0.15 mJy (5σ)	Sanders et al. 450 hr
Spitzer-IRAC	8, 6, 4.5, 3 μm	11, 9, 3, 2 μJy (5σ)	Sanders et al. 170 hr
IRAM-MAMBO	1.2 mm	1 mJy (20 × 20 arcmin)	Bertoldi et al. 90 hr
CSO-Bolocam	1.1 mm	3 mJy	Aquirre et al. 40 n
JCMT-Aztec	1.1 mm	0.9 mJy (1σ)	Sanders et al. 5 n
VLA-A	20 cm	25 μJy (1σ)	Schinnerer et al. 60 hr
VLA-A/C	20 cm	10 μJy (1σ)	Schinnerer et al. 275 hr
VLA-A-1°	20 cm	7 μJy (1σ)	Schinnerer et al. 75 hr
SZA (full field)	9 mm	S–Z to $2 \times 10^{14} M_\odot$	Carlstrom et al. 2 mth

for UV imaging (Schiminovich et al. 2007). The COSMOS VLA survey was allocated 320 hrs for the largest, deep radio survey every done (Schinnerer et al. 2007). The XMM, GALEX and VLA surveys are all now complete. Deep mid-infrared observations (IRAC) and far-infrared observations (MIPS) of the full COSMOS field have been carried out with Spitzer in GO-2&3 (Sanders et al. 2007). At submm-wavelengths, partial surveys of COSMOS are on-going at the CSO, IRAM-30 m and JCMT telescopes. The COSMOS field will also be a prime survey field for Herschel—providing longer wavelength coverage in the far infrared.

Acknowledgements The major COSMOS datasets become publicly available in staged releases (following calibration and validation) through the web site for IPAC/IRSA: http://irsa.ipac.caltech.edu/data/COSMOS/. I gratefully acknowledge the contributions of the entire COSMOS collaboration consisting of more than 70 scientists. The HST COSMOS Treasury program was supported through NASA grant HST-GO-09822.

References

Capak, P., et al.: Astrophys. J. Suppl. Ser. **172**, 99 (2007a)
Capak, P., et al.: Astrophys. J. Suppl. Ser. **172**, 284 (2007b)
Finoguenov, A., et al.: Astrophys. J. Suppl. Ser. **172**, 182 (2007)
Giavalisco, M., et al.: Astrophys. J. Suppl. Ser. (2007, in press)
Guzzo, L., et al.: Astrophys. J. Suppl. Ser. **172**, 254 (2007)
Hasinger, G., et al.: Astrophys. J. Suppl. Ser. **172**, 29 (2007)
Ilbert, O., et al.: Astrophys. J. (2007, in press)
Impey, C.D., et al.: Astrophys. J. Suppl. Ser. (2007, in press)
Lilly, S., et al.: Astrophys. J. Suppl. Ser. **172**, 70 (2007)
Massey, R., et al.: Nature **445**, 286 (2007)
Mobasher, B., et al.: Astrophys. J. Suppl. Ser. **172**, 117 (2007)
Murayama, T., et al.: Astrophys. J. Suppl. Ser. **172**, 523 (2007)
Sanders, D.B., et al.: Astrophys. J. Suppl. Ser. **172**, 86 (2007)
Scarlata, C., et al.: Astrophys. J. Suppl. Ser. **172**, 406 (2007)
Schinnerer, A., et al.: Astrophys. J. Suppl. Ser. **172**, 46 (2007)
Schiminovich, D., et al.: Astrophys. J. Suppl. Ser. (2007, in press)
Scoville, N., et al.: Astrophys. J. Suppl. Ser. **172**, 1 (2007a)
Scoville, N., et al.: Astrophys. J. Suppl. Ser. **172**, 150 (2007b)
Taniguchi, Y., et al.: Astrophys. J. Suppl. Ser. **172**, 9 (2007)

Observations of molecular clouds in nearby galaxies with ALMA

Nario Kuno · Akihiko Hirota · Tomoka Tosaki · Rie Miura

Originally published in the journal Astrophysics and Space Science, Volume 313, Nos 1–3.
DOI: 10.1007/s10509-007-9620-7 © Springer Science+Business Media B.V. 2007

Abstract We present recent results of the observations of giant molecular clouds in nearby galaxies with the Nobeyama 45 m telescope and Millimeter Array. We give some brief comments about observations of GMCs in nearby galaxies with ALMA.

Keywords Galaxies · Star formation · Molecular clouds

1 Nobeyama CO atlas of nearby spiral galaxies

We made a CO mapping survey of 40 nearby spiral galaxies with the multi-beam receiver BEARS (25-Beam Array Receiver System) mounted on the 45 m telescope (Kuno et al. 2007). It is often said that spiral density waves trigger star formation, while bars suppress star formation by large shear. So, it is interesting to investigate the relation between SFE and strength of arm and bar. Using the CO data, we are investigating the spatial variation of star formation efficiency (SFE) in some galaxies to study the influence of spiral and bar structures on star formation. We defined arm, interarm, bar, and central regions using 2MASS image (Fig. 1). SFR in each region was derived from Hα data and total gas mass was derived from HI and CO data. Figure 2 shows SFE in arm, interarm, bar, and central region in NGC 4321. In NGC 4321, SFE in bar is lower than that in the arms, although the

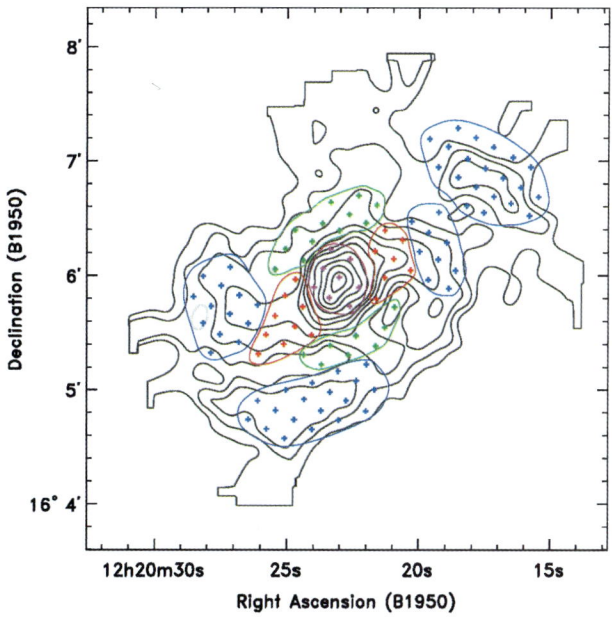

Fig. 1 CO integrated intensity map of NGC 4321 (Kuno et al. 2007). The arm, interarm, bar, and central regions are divided by *solid line*. The *crosses* indicate the observing grid

surface density of total gas is comparable. SFE in the interarm regions is also lower than that in the arms. These are preliminary results and we are making same comparison for other galaxies to see the relation between SFE and strength of arm and bar.

2 Observations of GMCs in nearby galaxies

To understand the reason for the spatial variation of SFE, it is important to investigate the difference of the properties of

N. Kuno (✉) · T. Tosaki
Nobeyama Radio Observatory, Minamimaki-mura,
Minamisaku-gun, Nagano 384-1305, Japan
e-mail: kuno@nro.nao.ac.jp

A. Hirota · R. Miura
University of Tokyo, Bunkyo-ku, Tokyo 113-033, Japan

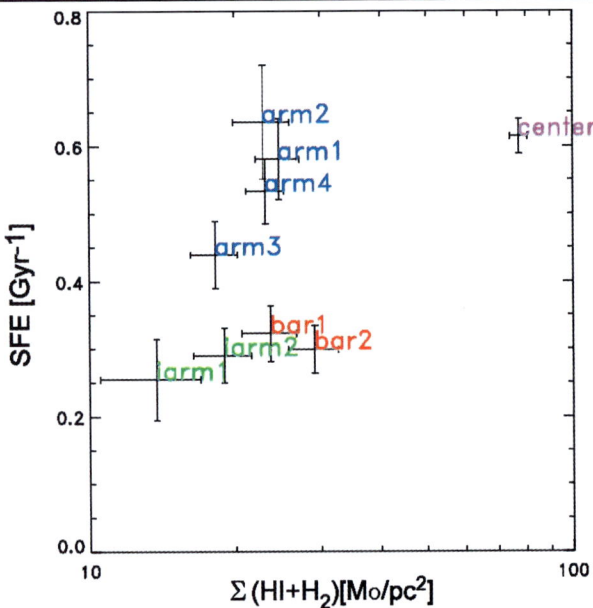

Fig. 2 SFE in arm, interarm, bar, and central regions in NGC 4321 derived from Hα, HI and CO data (Hirota et al. 2007)

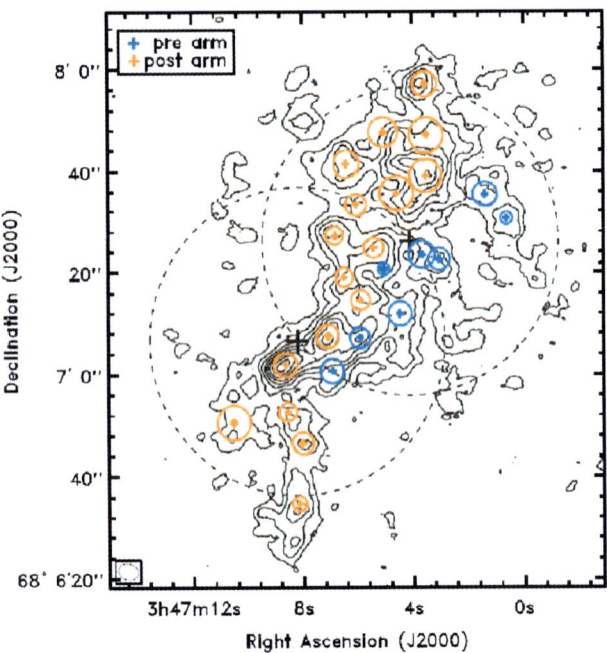

Fig. 3 CO integrated intensity map of IC342 obtained with the 45 m telescope and NMA (Hirota et al. 2007). The *circles* indicate identified molecular clouds

molecular clouds, such as mass function and dense gas fraction. For such studies, we have to resolve individual molecular cloud in nearby galaxies. We present some observations of giant molecular clouds in nearby galaxies performed with NMA.

Fig. 4 Comparison between virial mass and CO luminosity mass of the molecular clouds identified in the arm region (Hirota et al. 2007)

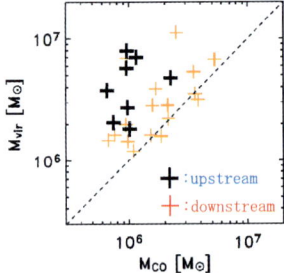

2.1 IC 342

We observed the spiral arm and bar regions with the 45 m telescope and NMA to investigate how these structures change the properties of molecular clouds. Here, we present the results of the arm region. Figure 3 is the CO map obtained by combining the data with the 45 m telescope and NMA. We identified 25 clouds in the arm region. We found the velocity change across the arm as often seen in many spiral galaxies. So, we separated these clouds into two groups, namely, clouds located upstream and downstream of the velocity change.

Figure 4 is the comparison of virial mass and CO luminosity mass of the clouds. We can see the property of molecular clouds changes across the arm. That is, upstream diffuse clouds become gravitationally bound downstream. Or the upstream clouds may be simply assembly of small clouds which are not gravitationally bound. If we can get higher spatial resolution, we may be able to distinguish them and see the formation process of GMCs directly.

2.2 M31

In many spiral galaxies, complexes of star forming regions whose size is comparable with GMAs (Giant Molecular Associations) are often seen. In order to study how star formation proceeds in GMAs, we made observations of a GMA in M31 to resolve its internal structure. Figures 5 and 6 are the CO map obtained with the 45 m telescope and NMA, respectively. Within the GMA, we can see many clumps whose size is similar to GMC in Milky Way. The missing flux is estimated to be about 50%. These results suggest that the GMA consists of many GMCs and diffuse envelope. We are searching where dense gas is formed in the GMA by observing with dense gas tracers.

2.3 M33

We observed molecular clouds around the giant HII region NGC 604. Figure 7 (right) is the CO map of NGC 604 obtained with the 45 m telescope. We found some complexes of molecular clouds that were not detected in the previous

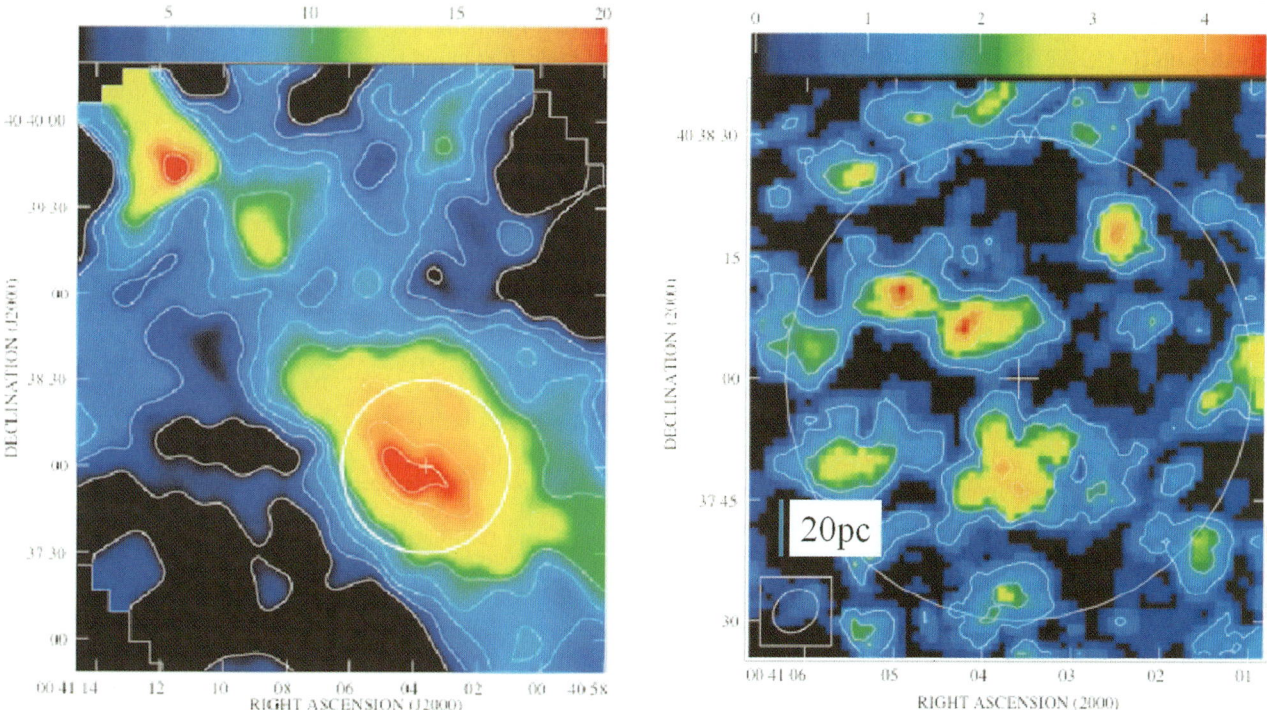

Fig. 5 CO map of M31 obtained with NRO 45 m telescope. The *circle* indicates the field of view of NMA (Tosaki et al. 2007)

Fig. 6 CO map of M31 obtained with NMA. The *circle* indicates the field of view of NMA (Tosaki et al. 2007)

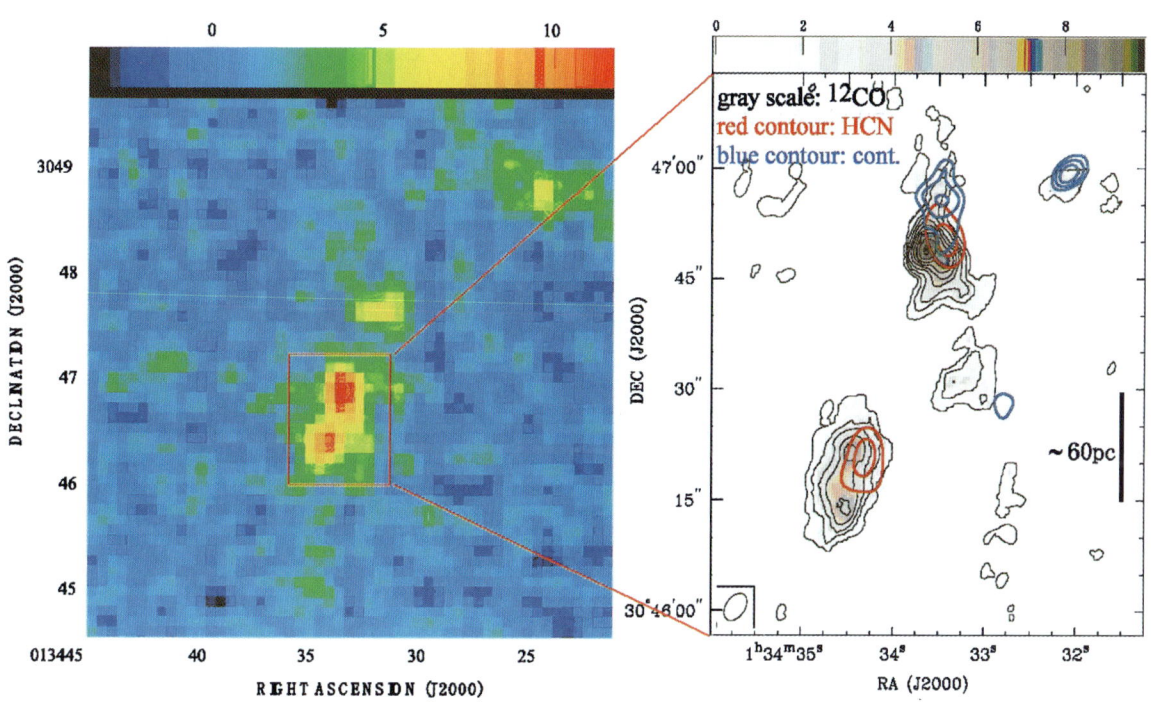

Fig. 7 CO map of M33 obtained with NRO 45 m telescope (*right*) and NMA (*left*). The *circle* indicates the field of view of NMA (Miura et al. 2007)

observations. Figure 7 (left) is the CO map of the two complexes in Fig. 1 obtained with NMA. We identified 8 molecular clouds in the complexes whose size and mass are comparable with those of GMCs in our Galaxy. Furthermore, we made observations of HCN and radio continuum and found that these clouds are separated into four types, although the mass of the molecular clouds are almost the same; molecular clouds with (1) CO, (2) CO + HCN, (3) CO + HCN + radio continuum, (4) CO + radio continuum. In this region, sequential star formation seems to be triggered by NGC 604. The difference of the properties of molecular clouds may mean the difference of the stage of star formation, that is, from early stage clouds which have dense gas traced by HCN to late stage clouds which have star forming regions traced by radio continuum.

3 Observations of GMCs in nearby galaxies with ALMA

We can resolve only giant molecular clouds in nearby galaxies with present instruments. For example, however, if we can get $0.1''$ resolution with ALMA, it corresponds to 5 pc at a distance of 10 Mpc. So, we can expect to resolve even much smaller molecular clouds in nearby galaxies with ALMA. Furthermore, we can observe dense gas in individual molecular cloud with high sensitivity of ALMA. So, we can study the relation between star formation efficiency and the properties of molecular clouds in detail. And it must give us some clues for understanding of massive star formation.

However, even for ALMA, it takes long observing time to map entire disk of nearby galaxies with such high angular resolution and high sensitivity. So, we have to make observations with present instruments which can be used to select target area for ALMA observations.

References

Hirota, et al.: (2007, in preparation)
Kuno, et al.: Publ. Astron. Soc. Jpn. **59**, 117 (2007)
Miura, et al.: (2007, in preparation)
Tosaki, et al.: (2007, in preparation)

Luminous infrared galaxies with the submillimeter array: probing the extremes of star formation

C.D. Wilson · G.R. Petitpas · D. Iono · A. Peck · M. Krips · B.E. Warren · A.J. Baker · M.S. Yun · Y. Pihlstrom · C. Mihos · S. Matsushita · M. Juvela · P.T.P. Ho · T.J. Cox · L. Armus

Originally published in the journal Astrophysics and Space Science, Volume 313, Nos 1–3.
DOI: 10.1007/s10509-007-9618-1 © Springer Science+Business Media B.V. 2007

Abstract Luminous and Ultraluminous infrared galaxies (ULIRGs) contain the most intense regions of star formation in the local universe. Because molecular gas is the fuel for current and future star formation, the physical properties and distribution of the warm, dense molecular gas are key components for understanding the processes and timescales controlling star formation in these merger and merger remnant galaxies. We present new results from a legacy project on the Submillimeter Array which is producing high resolution images of a representative sample of galaxies with $\log L_{FIR} > 11.4$ and $D < 200$ Mpc.

Keywords Galaxies: infrared · Galaxies: ISM · Galaxies: individual (Mrk231, Mrk273, UGC5101, IRAS10565+2448)

1 Introduction

Ultra-luminous infrared galaxies (ULIRGs) contain the regions of most intense star formation in the local universe. Although their high rates of star formation and accretion appear to be triggered by the merger of two gas-rich galaxies (Sanders et al. 1988; Veilleux et al. 2002), the detailed physical connection between galaxy mergers and star formation and, in particular, the time evolution of this process, is not well understood. Relating numerical hydrodynamical models (Mihos and Hernquist 1996; Cox et al. 2004) to observations is complicated by the difficulty in identifying the precise stage of the merger (Murphy et al. 2001). In addition, while high resolution imaging has found that most ULIRGs have nuclear separation from <0.3 kpc to 48 kpc (Murphy et al. 1996), other merging galaxies with these nuclear separations which are *not* ULIRGs have also been found (Braine et al. 2004). These observations suggest that the onset of the intense star formation and accretion which produces a ULIRG is not a simple function of the age of the merger and leaves open the question of whether all luminous infrared

C.D. Wilson (✉) · B.E. Warren
Dept. of Physics & Astronomy, McMaster University, Hamilton, Ontario L8S 4M1, Canada
e-mail: wilson@physics.mcmaster.ca

G.R. Petitpas · A. Peck · M. Krips
Smithsonian Astrophysical Observatory, Hilo, USA

D. Iono
National Astronomical Observatories of Japan, Tokyo, Japan

A.J. Baker
State University of New Jersey, Piscataway, USA

M.S. Yun
University of Massachusetts, Amherst, USA

Y. Pihlstrom
University of New Mexico, Albuquerque, USA

C. Mihos
Case Western Reserve University, Cleveland, USA

S. Matsushita · P.T.P. Ho
Institute of Astronomy & Astrophysics, Academia Sinica, Taipei, Taiwan

M. Juvela
University of Helsinki Observatory, Helsinki, Finland

T.J. Cox
Center for Astrophysics, Cambridge, USA

L. Armus
Infrared Processing and Analysis Center, Pasadena, USA

galaxies (LIRGs; $11 \leq \log(L_{FIR}/L_{\odot}) < 12$) will undergo a ULIRG phase ($12 \geq \log(L_{FIR}/L_{\odot})$) at some point in their evolution.

Local ULIRGs are also important as the closest analogs to the high-redshift submillimeter galaxies (Blain et al. 2002): both populations have high infrared luminosities, large amounts of molecular gas (Frayer et al. 1998, 1999; Neri et al. 2003; Greve et al. 2005; Tacconi et al. 2006) and display morphological evidence of past or ongoing mergers (Veilleux et al. 2002; Conselice et al. 2003). Since galaxy merger rates are substantially higher in the early universe (Le Fèvre et al. 2000; Gottlöber et al. 2001), understanding the physical and dynamical properties of nearby ULIRGs is also important for understanding the processes in the early universe which give rise to the very luminous submillimeter galaxy population.

In this paper, we present some first results obtained with the Submillimeter Array (SMA) for a sample of 14 luminous and ultraluminous infrared galaxies. The galaxies have all been observed in the CO J = 3-2 lines and in continuum emission at 880 µm, which allows us to compare the dust and gas properties in the central kiloparsec.

2 Sample and observations

For this survey, we selected a sample of luminous and ultraluminous infrared galaxies with redshifts $z < 0.045$ (distances $D_L < 200$ Mpc, adopting $H_o = 70$ km s^{-1} Mpc^{-1}, $\Omega_M = 0.3$, $\Omega_\Lambda = 0.7$) and far-infrared luminosities $\log L_{FIR} > 11.4$. The distance limit was chosen so as to be able to achieve spatial resolutions of order 1 kpc or better in all the target galaxies. The luminosity cutoff was chosen to allow us to span a wide range of merger properties and luminosities while still concentrating on the most infrared luminous nearby galaxies. Out of a total of 39 galaxies above declination $-20°$ (Sanders et al. 2003) which meet these two criteria, we selected 14 galaxies with previous interferometric observations in the CO J = 1-0 transition to be observed in this survey.

The observations discussed here were obtained between 2005 May 16 and 2006 April 22 using either the compact or extended configuration of the SMA. The configuration was chosen so as to achieve a spatial resolution of 0.7–1 kpc in the CO J = 3-2 line; the typical angular resolution in the extended configuration is 0.7″. The correlator was configured to have a spectral resolution of 0.8125 MHz (~0.7 km s^{-1} for CO J = 3-2) with a bandwidth of 2 GHz in each of the upper and lower sidebands. Each galaxy was observed with the ^{12}CO J = 3-2 transition in the lower sideband of the receiver and the adjacent continuum in the upper sideband, 10 GHz away. All galaxies were detected strongly in the CO J = 3-2 line, and also in the continuum with a signal to noise

of four or better; examples of the CO J = 3-2 integrated emission are shown in Figs. 1 and 2.

3 The gas to dust mass ratio

We can use our new CO J = 3-2 and 880 µm data to estimate the gas to dust mass ratio in the central regions of these luminous infrared galaxies. The gas to dust ratio is interesting because it allows us to probe the physical properties of the interstellar medium on kiloparsec scales in regions of galaxies with intense heating from starburst activity and possibly an active galactic nucleus.

We have calculated the dust mass from the 880 µm flux assuming a dust emissivity at 880 µm, κ, of 1 cm^2 g^{-1} (Henning et al. 1995; Johnstone et al. 2000). With this assumption, the dust mass is given by

$$M_{dust} = 74220 S_{880} D_L^2 (\exp(17/T_D) - 1)/\kappa \ (M_{\odot})$$

where S_{880} is the 880 µm flux in Jy and D_L is the luminosity distance in Mpc. We calculated the dust temperature, T_D, from the 60 and 100 µm fluxes (Sanders et al. 2003) assuming the emission at these wavelengths is optically thick (Solomon et al. 1997).

To calculate the H$_2$ gas mass, we adopt the revised CO-to H$_2$ conversion factor advocated by (Downes and Solomon 1998),

$$M_{H_2} = 0.8 L'_{CO}(1 - 0)$$

where M_{H_2} is the H$_2$ gas mass in M$_{\odot}$ and $L_{CO}(1-0)$ is the luminosity of the CO J = 1-0 line in K km s^{-1} pc^2. (This equation corresponds to a CO-to-H$_2$ conversion factor of 0.5×10^{20} H$_2$ cm^{-2} (K km s^{-1})$^{-1}$.) We adopt an average CO J = 3-2/J = 1-0 line ratio of 0.5, which is consistent with single dish and interferometric measurements of Mrk231 and Mrk273 (Wilson et al. 2007) in calculating the H$_2$ gas mass.

We focus here on the gas-to-dust mass ratio determined in the central beam to probe the extreme central region of each galaxy or galaxy component. Thus, we compare the H$_2$ mass calculated from the integrated CO J = 3-2 intensity in a single beam at the peak of the emission with the dust mass calculated from the peak 880 µm continuum intensity. This analysis gives an average gas to dust ratio including all 13 galaxies or galaxy components analyzed to date of 207 ± 49 (standard deviation 177). Thus, our new SMA data suggest that the gas to dust mass ratio in the central kiloparsec of these luminous and ultraluminous infrared galaxies is very similar to the gas to dust ratio measured in the Milky Way. This result is somewhat surprising given that the dust in these nuclear regions is subject to more intense heating (as well as perhaps processing due to shocks) compared to typical regions in the Milky Way.

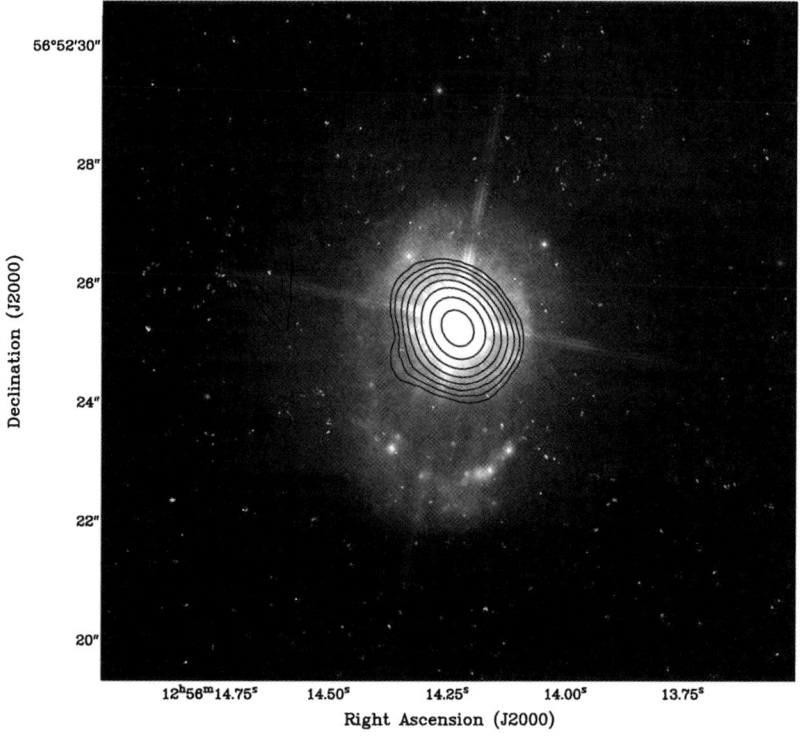

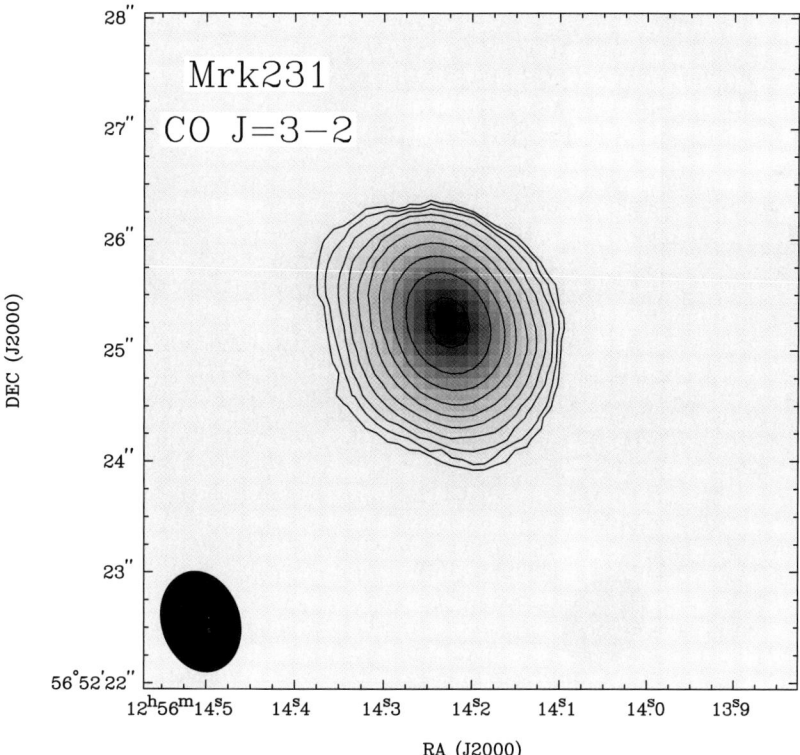

Fig. 1 (*Top*) CO J = 3-2 integrated intensity map for Mrk 231 overlaid on an I band image from the Hubble Space Telescope. Note that the coordinate alignment between the optical and radio images is only approximate. (*Bottom*) Closer view of the CO J = 3-2 integrated intensity map. The lowest contour is 2σ (5.2 Jy/beam km/s) contours increase by factors of 1.5. The $\sim 0.8''$ beam is indicated in the lower left corner

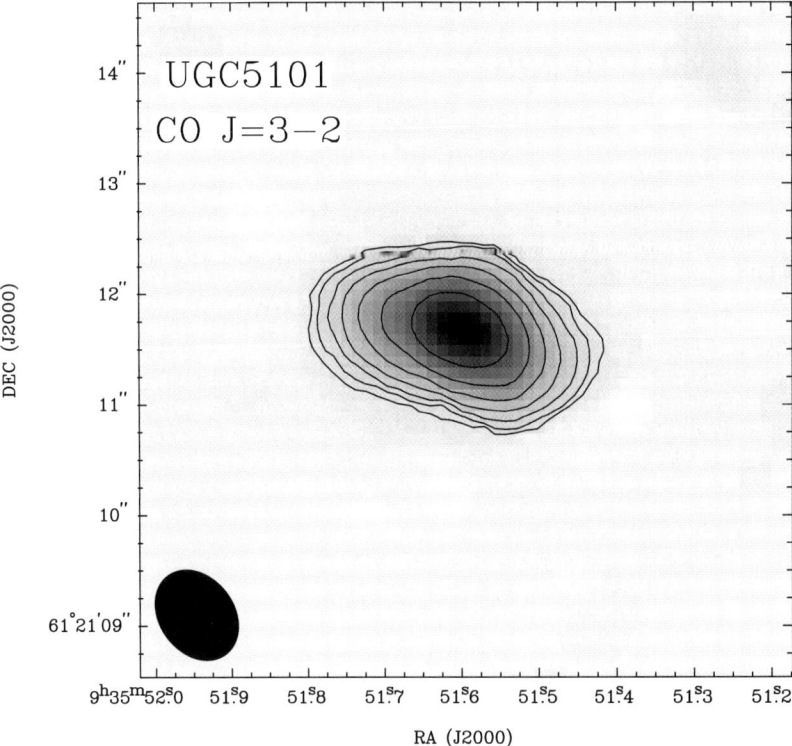

Fig. 2 (*Top*) CO J = 3-2 integrated intensity map for UGC 5101 overlaid on an I band image from the Hubble Space Telescope. Note that the coordinate alignment between the optical and radio images is only approximate. (*Bottom*) Closer view of the CO J = 3-2 integrated intensity map. The lowest contour is 2σ (7.6 Jy/beam km/s) contours increase by factors of 1.5. The $\sim 0.8''$ beam is indicated in the lower left corner

4 Comparing ULIRGs at low and high redshift

We have compared four of the brightest and most compact galaxies in our sample (Mrk 231, Mrk 273, UGC 5101, and IRAS 10565+2448) with eight high-redshift submillimeter galaxies that have been observed in the CO J = 3-2 line (Downes and Solomon 2003; Genzel et al. 2003; Weiss et al. 2003; Tacconi et al. 2006). By choosing this high-redshift sample that was observed in exactly the same CO line, we do not have to worry that CO excitation effects may bias our results.

Compared to the local ULIRGs, the high-redshift galaxies are at least an order of magnitude more luminous in the CO J = 3-2 line; this result was already known from single dish observations of both sets of galaxies. We also see that the high redshift galaxies have somewhat broader line widths on average (560 ± 90 km/s) compared to the local ULIRGs (370 ± 90 km/s), which suggests that the high redshift galaxies are more massive systems. Finally, the high redshift galaxies that have been spatially resolved have much larger diameters (5 kpc full width half maximum) compared to 0.6 kpc for the local galaxies. If this result is robust, it suggests that the CO surface brightness (or equivalently, the molecular gas surface density) may be up to an order of magnitude *lower* in the high redshift galaxies than in the local ULIRG sample. However, this conclusion remains to be confirmed by an analysis of our complete local sample (Iono et al. 2007) and perhaps higher resolution observations of the high redshift galaxies to confirm their diameters.

5 Future work

This SMA legacy survey aims to address five broad scientific questions:

1. *What are the distributions, kinematics, and physical conditions of dense molecular gas in U/LIRGs?* By combining our new SMA data with published and archival CO J = 1-0 and J = 2-1 data and radiative transfer models (Juvela 1997), we will be able to determine the physical properties of the cool, warm, and dense molecular gas components which feed the starburst and any AGN activity.
2. *What is the distribution of the dust in U/LIRGs?* High-resolution 340 GHz continuum images from the SMA can be combined with Spitzer IRS spectra to estimate the dust temperature via the mid-infrared to submillimeter spectral energy distribution (SED) and the dust mass (including both small and large grains).
3. *Do the gas properties change as the interaction progresses?* This data set covering a range of mergers from mid to late stages will allow us to determine how the distribution and kinematics of the gas change as a function of physical conditions such as density and temperature, and to correlate those changes with the stage of the merger.
4. *How do the properties of the dense gas in local U/LIRGs compare to those of the high-redshift submillimeter galaxies?* Armed with a robust local sample of 14 U/LIRGs, we can make a rigorous comparison of the properties of the gas with those in higher redshift galaxies (Greve et al. 2005; Tacconi et al. 2006).
5. *What is the origin of nuclear OH megamasers?* Our sample, which contains galaxies with and without OH megamasers, is well suited for identifying any unique nuclear conditions (physical, chemical, kinematical) that produce OH megamasers.

This survey should give us a good understanding of the kiloparsec-scale properties of the interstellar medium in these very luminous and active galaxies. In the future, with ALMA we will be able to probe much smaller spatial and mass scales in many of these same galaxies. For example, ALMA will be able to detect giant molecular clouds with masses of 5×10^6 M$_\odot$ (3σ) out to the distance limit of our survey (200 Mpc) in just one hour of integration. Thus, ALMA will let us search for and study giant molecular clouds, the fundamental star forming unit in galaxies, in a much wider range of environments than is possible with current telescopes.

References

Blain, A.W., Smail, I., Ivison, R.J., Kneib, J.-P., Frayer, D.T.: Submillimeter galaxies. Phys. Rep. **369**, 111–176 (2002)

Braine, J., Lisenfeld, U., Duc, P.-A., et al.: Colliding molecular clouds in head-on galaxy collisions. Astron. Astrophys. **418**, 419–428 (2004)

Conselice, C.J., Chapman, S.C., Windhorst, R.A.: Evidence for a major merger origin of high-redshift submillimeter galaxies. Astrophys. J. **596**, L5–L8 (2003)

Cox, T.J., Primack, J., Jonsson, P., Somerville, R.S.: Generating hot gas in simulations of disk-galaxy major mergers. Astrophys. J. **607**, L87–L90 (2004)

Downes, D., Solomon, P.M.: Rotating nuclear rings and extreme starbursts in ultraluminous galaxies. Astrophys. J. **507**, 615–654 (1998)

Downes, D., Solomon, P.M.: Molecular gas and dust at z = 2.6 in SMM J14011+0252: a strongly lensed ultraluminous galaxy, not a huge massive disk. Astrophys. J. **582**, 37–48 (2003)

Frayer, D.T., Ivison, R.J., Scoville, N.Z., et al.: Molecular gas in the Z = 2.8 submillimeter galaxy SMM 02399-0136. Astrophys. J. **506**, L7–L10 (1998)

Frayer, D.T., Ivison, R.J., Scoville, N.Z., et al.: Molecular gas in the Z = 2.565 submillimeter galaxy SMM J14011+0252. Astrophys. J. **514**, L13–L16 (1999)

Genzel, R., et al.: Spatially resolved millimeter interferometry of SMM J02399-0136: a very massive galaxy at z = 2.8. Astrophys. J. **584**, 633–642 (2003)

Gottlöber, S., Klypin, A., Kravtsov, A.V.: Merging history as a function of halo environment. Astrophys. J. **546**, 223–233 (2001)

Greve, T.R., Bertoldi, F., Smail, I., et al.: An interferometric CO survey of luminous submillimetre galaxies. Mon. Not. Roy. Astron. Soc. **359**, 1165–1183 (2005)

Henning, Th., Michel, B., Stognienko, R.: Dust opacities in dense regions. Planet. Space Sci. **43**, 1333–1343 (1995)

Iono, D., et al.: in preparation (2007)

Johnstone, D., Wilson, C.D., Moriarty-Schieven, G., Joncas, G., Smith, G., Fich, M.: Large-area mapping at 850 microns. II. Analysis of the clump distribution in the ρ ophiuchi molecular cloud. Astrophys. J. **545**, 327–339 (2000)

Juvela, M.: Non-LTE radiative transfer in clumpy molecular clouds. Astron. Astrophys. **322**, 943–961 (1997)

Le Fèvre, O., Abraham, R., Lilly, S.J., et al.: Hubble space telescope imaging of the CFRS and LDSS redshift surveys—IV. Influence of mergers in the evolution of faint field galaxies from z 1. Mon. Not. Roy. Astron. Soc. **311**, 565–575 (2000)

Mihos, J.C., Hernquist, L.: Gasdynamics and starbursts in major mergers. Astrophys. J. **464**, 641–663 (1996)

Murphy, T.W. Jr., Armus, L., Matthews, K., et al.: Visual and near-infrared imaging of ultraluminous infrared galaxies: the IRAS 2 Jy sample. Astron. J. **111**, 1025–1052 (1996)

Murphy, T.W. Jr., Soifer, B.T., Matthews, K., Armus, L.: Age dating ultraluminous infrared galaxies along the merger sequence. Astrophys. J. **559**, 201–224 (2001)

Neri, R., et al.: Interferometric observations of powerful co emission from three submillimeter galaxies at $z = 2.39$, 2.51, and 3.35. Astrophys. J. **597**, L113–L116 (2003)

Sanders, D.B., Soifer, B.T., Elias, J.H., Madore, B.F., Matthews, K., Neugebauer, G., Scoville, N.Z.: Ultraluminous infrared galaxies and the origin of quasars. Astrophys. J. **325**, 74–91 (1988)

Sanders, D.B., Mazzarella, J.M., Kim, D.-C., Surace, J.A., Soifer, B.T.: The IRAS revised bright galaxy sample. Astron. J. **126**, 1607–1664 (2003)

Solomon, P.M., Downes, D., Radford, S.J.E., Barrett, J.W.: The molecular interstellar medium in ultraluminous infrared galaxies. Astrophys. J. **478**, 144–161 (1997)

Tacconi, L., et al.: High-resolution millimeter imaging of submillimeter galaxies. Astrophys. J. **640**, 228–240 (2006)

Veilleux, S., Kim, D.-C., Sanders, D.B.: Optical and near-infrared imaging of the IRAS 1 Jy sample of ultraluminous infrared galaxies. II. The analysis. Astrophys. J. Suppl. Ser. **143**, 315–376 (2002)

Weiss, A., Henkel, C., Downes, D., Walter, F.: Gas and dust in the Cloverleaf quasar at redshift 2.5. Astron. Astrophys. **409**, L41–L45 (2003)

Wilson, C.D., et al.: in preparation (2007)

The Galactic Center as nearby extragalactic chemical laboratory

Sergio Martín · Miguel Angel Requena-Torres ·
Jesús Martín-Pintado · Rainer Mauersberger

Originally published in the journal Astrophysics and Space Science, Volume 313, Nos 1–3.
DOI: 10.1007/s10509-007-9616-3 © Springer Science+Business Media B.V. 2007

Abstract The availability of wideband receivers altogether with the high collecting area of ALMA will easily cover large spectral regions, allowing us to create full molecular line surveys of a wide variety of extragalactic objects, far from our present reach. This observations will provide a full chemical description of a large number of galaxies with different morphologies and evolutionary stages. The recently published 2 mm line survey towards the starburst galaxy NGC 253 has shown the need of chemical templates within the Galaxy in order to fully understand molecular emission within the nuclear region of galaxies. It will be shown how the study of the chemical complexity of molecular cloud prototypes within the Galactic Center region may provide a key information to determine the heating mechanisms of molecular material in the nuclei of galaxies. In the nearby future this chemical studies will be further extended to the high-z universe and will be particularly important in sources such as submillimeter galaxies, not easily observable in other frequency ranges.

Keywords Galaxies: ISM · Galaxy: center · ISM: molecules

S. Martín (✉)
CfA, 60 Garden st., Cambridge, 02141 MA, USA
e-mail: smartin@cfa.harvard.edu

M.A. Requena-Torres · J. Martín-Pintado
DAMIR-IEM-CSIC, Serrano 121, 28006 Madrid, Spain

R. Mauersberger
IRAM, Avda. Divina Pastora 7, Local 20, 18012 Granada, Spain

1 Introduction

The heating of the interstellar medium within the nuclei of the diverse galaxy types, and therefore the chemistry of the molecular material, is expected to be dominated by different physical processes. Two are the main mechanisms that are thought to dominate the chemistry of the molecular material in the nuclei of starburst galaxies: the UV radiation stemming from massive stars which create large photodissociation regions (PDRs), and the presence of strong shock waves as a result of large-scale cloud collisions due to orbital motions within the galactic potential, the expanding bubbles from supernovae events and/or strong stellar winds from massive Wolf-Rayet stars.

The recently published 2 mm spectral line survey on NGC 253 (Martín et al. 2006) opened up the way for a spectral classification of the molecular content of the nuclear environment in starburst galaxies. However, for the clear emergence of a detailed classification, comparative studies have to be carried out, not only towards other prototypical galaxies, but towards galactic chemical prototypes that will help us understanding the origin of the emission in distant galaxies.

In this context, the center of our Galaxy appears to be an excellent local chemical laboratory where we can find examples of molecular gas heated by the same mechanisms expected to play an important role within starburst galaxies. The detailed study of the chemical composition in this region will allow us to find the best molecular tracers to differentiate and classify galactic nuclei.

2 Tracing the heating in galaxies

The observation of selected chemical tracers has been proved to be a powerful tool to identify the effects of the

Fig. 1 Methanol detection in two positions around the nuclear region of M 82 (Martín et al. 2006). The measured abundances of a few 10^{-9} were not expected for M 82, which is assumed to be the prototype of giant PDR outside the Galaxy. This detection points out the presence of an dense and warm additional molecular gas component, hidden from the pervasive UV radiation dominating the region

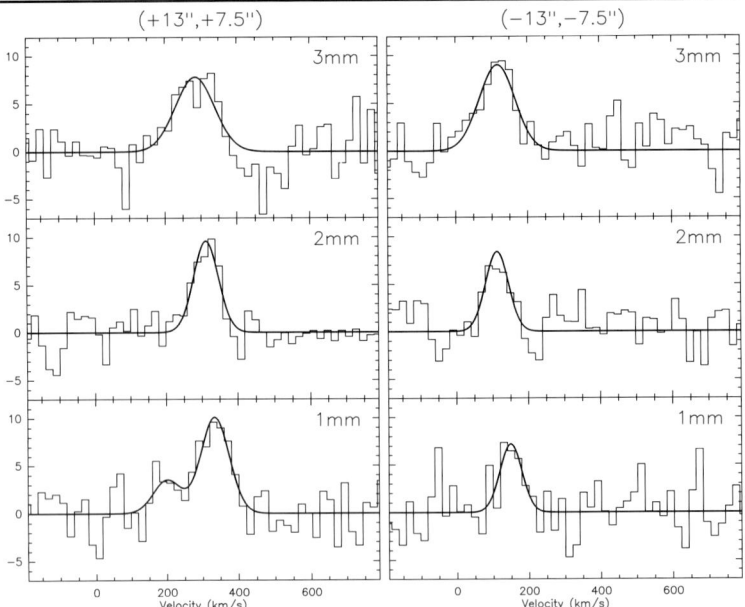

different heating mechanisms in the extragalactic ISM. With a resolution of 5–6″, the OVRO interferometric maps of IC 342 (Meier and Turner 2005) clearly shows the trend followed by the different tracers. Molecules such as C_2H and $C^{34}S$ peak towards the very nuclear region of IC 342 where the gas is highly pervaded by the UV radiation from the central star cluster, while species such as CH_3OH and HNCO, whose formation is known to be enhanced in surface of dust grains, are found to outline the arms of the nuclear minispiral where large scale cloud collisions would be responsible for their injection into gas phase.

Interesting chemical cases are those of the starburst galaxies M 82 and NGC 253. The detection in M 82 of high abundances of PDR tracer species such as CO^+, HOC^+, C_2H, HCO (Fuente et al. 2006) distributed throughout its central region (García-Burillo et al. 2002) has turned M 82 into the prototype of giant photodominated region outside the Galaxy. On the other hand, NGC 253 shows a molecular emission closely resembling that of the molecular envelope surrounding the Sgr B2 massive star forming complex (Martín et al. 2005, 2006), which implies a similar heating process (low velocity shocks) leading the chemistry in both objects. The differences observed in the chemistry of M 82 and NGC 253 are interpreted as differences in the evolutionary states of their nuclear starbursts (Wang et al. 2004). In this scenario, M 82 is regarded as the prototype of evolved starburst where the radiation from the massive stars recently formed in the burst is affecting the surrounding molecular material, while NGC 253 represents the early stage of the starburst where the gas is mainly affected by the shocks. In an UV dominated environment such as that in M 82, low abundances of easily photodissociated species such as methanol would be expected. However, methanol

is clearly detected in several positions within M 82 nuclear region (Martín et al. 2006, see Fig. 1) with abundances of a few 10^{-9}. These abundances are significantly higher than what is expected to be formed in gas-phase (Lee et al. 1996) and well above the abundances observed in PDRs (Jansen et al. 1995). Methanol is known to be efficiently formed in dust grains and injected into gas-phase via evaporation or disruption of ice mantles (Charnley et al. 1995). This suggest that within the central region of M 82, even if photodissociation dominates the heating, there is a reservoir of dense and warm molecular gas, similar to that found in other starburst like NGC 253 and Maffei 2, which is likely to represent the fuel of recent or ongoing star formation in the nuclear region of this galaxy.

The examples presented in this section clearly illustrate the complexity we find in the central regions of galaxies, where different gas phases with different physical conditions coexist, and how the chemical studies altogether with the high spatial resolution provided by interferometers are a powerful tool to disentangle each of their components.

3 Galactic Center templates

The 2 mm spectra of NGC 253 (Martín et al. 2006) has shown the potential of chemical studies on extragalactic object as a classification tool for galaxies. However, it has become evident the need of gathering information on Galactic templates of chemical prototypes in order to explain and understand the emission from galactic nuclei.

With this aim, we have carried out observations on a set of 13 positions distributed throughout the Galactic Center of our Galaxy (Martín et al. 2007). These sources were carefully selected to be a representative sample of the chemical

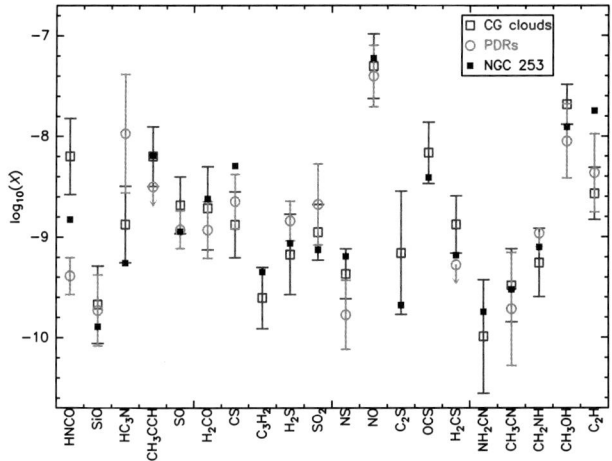

Fig. 2 Molecular abundances of the 20 brightest observed species in the 2 mm line survey of NGC 253 (*filled black squares*), compared with measured abundances observed in most Galactic Center molecular clouds (*open squares*) and in those sources strongly affected by UV radiation from nearby star clusters (*open circles*). The chemistry in NGC 253 are significantly well reproduced by those observed in typical GC molecular clouds

variety we find in this region. By observing a similar frequency range of the 2 mm atmospheric window to that observed in NGC 253, we were able to make a direct comparison between the spectra. Abundances of the 20 most intense species were calculated and compared for each source. Surprisingly, the observed abundances of all species are similar within a factor of 3 in all sources, which is in agreement with constant relative abundances of complex organic molecules observed in 40 positions within the Galactic Center region (Requena-Torres et al. 2006). This observed chemical composition homogeneity is claimed to be the result of a common grain mantle composition along the Galactic Center and the disruption of these grains by frequent low velocity shocks.

However, molecules such as HNCO and, in a less degree, CH_3OH, appear to be clearly underabundant in the sources which are affected by photodissociation, as observed for the positions in the sample close to the Central and Quintuplet Cluster. Figure 2 shows the average abundances derived from the those position observed not to be affected by strong UV radiation, designated as CG clouds, and the mean abundance value of the two positions observed close to the massive GC star clusters as example of UV-driven chemistry. Only CS appears to be slightly overabundant in PDRs. Therefore, the molecule of HNCO, which practically disappears in PDRs, is observed to be the molecular species showing the higher contrast between the material heated by shocks and that pervaded by strong UV radiation from stars.

Therefore, as mentioned above, these set of sources are representative of the two main physical processes leading the chemistry in starburst galaxies, namely large scale low velocity shocks and photodissociation.

4 Extragalactic comparison

On top of the abundances in the Galactic Center, Fig. 2 shows the abundances measured in the starburst galaxy NGC 253. The molecular composition of this galaxy is significantly similar to that observed in GC molecular clouds (Martín et al. 2006). When comparing the observations within our Galaxy with those in extragalactic objects it is important to take into account that single dish observations with available telescopes, such as the IRAM 30 m, the beam of $\sim 17''$ is covering a 1 pc region in the Galactic Center, while on NGC 253 an average of the emission within a region of ~ 200 pc is observed. However, observations show how prevailing heating mechanisms reveal clear chemical traces even over such extended regions (Mauersberger and Henkel 1993).

Figure 3 shows a comparison between the spectra of NGC 253 and two GC position, one in the envelope around the Sgr B2 complex as an example of GC molecular cloud affected by low velocity shocks, and one position in the circunnuclear disk (CND) around Sgr A* as example of PDR chemistry affected by the UV radiation from the central cluster. This comparison shows at a glance the results discussed from Fig. 2, proving the potential of galactic templates to understand extragalactic emission.

As far a HNCO is concerned as a chemical discriminator, we observe how its abundance in NGC 253 is clearly above that observed in PDRs, but not as high as the average value derived in GC clouds. Observations have been carried out towards M 82 at different transitions of HNCO (Martín et al. 2007) to check the suitability of this molecule as an extragalactic discriminator. Its non detection in M 82 points out a difference of more than a factor of 20 in the HNCO abundance with respect to that in NGC 253, supporting the idea that M 82 and NGC 253 are starburst galaxies in extreme evolutionary states (Wang et al. 2004).

5 From current instruments to ALMA

Current available interferometers such as SMA or PdBI easily reach angular resolutions of 1 arcsecond. At the distance of the nearest starbust galaxies (~ 2–3 Mpc) this turns into an spatial resolution of ~ 10–15 pc, which is of the same order as the typical sizes of Molecular Clouds in the GC (Martín-Pintado et al. 1987). Thus, we are now able to study the physical properties of individual molecular clouds in nearby galaxies.

The observing capabilities of ALMA, both in sensitivity and angular resolution will drastically change our view of extragalactic ISM. At the nearest galaxies, we will have a resolution equivalent to that we currently have with single-dish telescopes in the Galactic Center. It will be possible to

Fig. 3 Comparison between the 2 mm spectra of NGC 253 and those of two positions within the Galactic Center. *Middle* spectrum is a typical GC molecular cloud in a position in the Sgr B2 molecular envelope. *Bottom* spectrum corresponds to observations in the CND around Sgr A*, strongly affected by UV radiation. A significant resemblance is observed between the NGC 253 emission and that in GC molecular clouds.

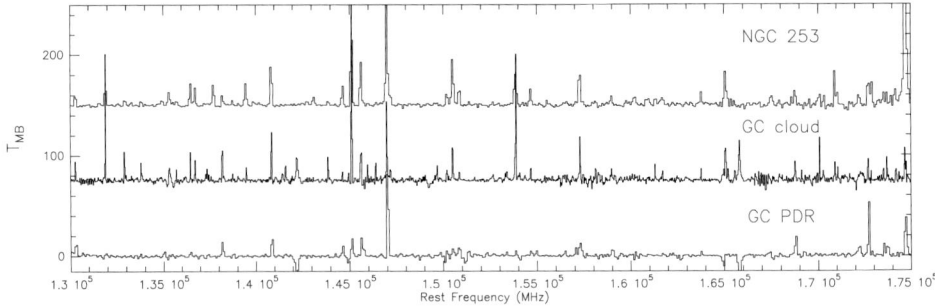

resolve the structure in single clouds within galaxies as well as resolve the circumnuclear material around their power sources, as we currently do within our own Galaxy.

Our understanding of the high redshift universe will also be strongly enhanced. The detailed study of the nuclei of ultraluminous galaxies such as Arp 220 will be a cornerstone in the chemical understanding of the high-z galaxies. The observation of molecular tracers such as HNCO, CH_3OH and CS, will be extended to a much wider range of sources up to the most distant galaxies where currently tens of hours must be devoted to scarcely resolve their CO structure (Walter et al. 2004). Particularly interesting will be the high-sensitivity observations of submillimeter galaxies, so elusive at other frequencies, where we might be able to establish a chemical classification by determining the ratios of these selected tracers which are out of our grasp with current instrumentation.

References

Charnley, S.B., Kress, M.E., Tielens, A.G.G.M., Millar, T.J.: Interstellar alcohols. Astrophys. J. **448**, 232–239 (1995)

Fuente, A., García-Burillo, S., Gerin, M., et al.: Detection of CO^+ in the nucleus of M 82. Astrophys. J. **641**, L105–L108 (2006)

García-Burillo, S., Martín-Pintado, J., Fuente, A., Usero, A.: Widespread HCO emission in the nuclear starburst of M 82. Astrophys. J. **575**, L55–L58 (2002)

Jansen, D.J., Spaans, M., Hogerheijde, M.R., van Dishoeck, E.F.: Millimeter and submillimeter observations of the Orion Bar. II. Chemical models. Astron. Astrophys. **303**, 541–553 (1995)

Lee, H.-H., Bettens, R.P.A., Herbst, E.: Fractional abundances of molecules in dense interstellar clouds: A compendium of recent model results. Astron. Astrophys. Suppl. Ser. **119**, 111 (1996)

Martín, S., Martín-Pintado, J., Mauersberger, R., Henkel, C., García-Burillo, S.: Sulfur chemistry and isotopic ratios in the starburst galaxy NGC 253. Astrophys. J. **620**, 210 (2005)

Martín, S., Martín-Pintado, J., Mauersberger, R.: Methanol detection in M 82. Astron. Astrophys. **450**, L13–L16 (2006)

Martín, S., Mauersberger, R., Martín-Pintado, J., Henkel, C., García-Burillo, S.: A 2 mm spectral line survey of the starburst galaxy NGC 253. Astrophys. J. Suppl. Ser. **164**, 450–476 (2006)

Martín-Pintado, J., de Vicente, P., Fuente, A., Planesas, P.: SiO emission from the galactic center molecular clouds. Astrophys. J. **482**, L45 (1987)

Mauersberger, R., Henkel, C.: Dense Gas in Galactic Nuclei. Rev. Mod. Astron. **6**, 69–102 (1993)

Meier, D.S., Turner, J.L.: Spatially resolved chemistry in nearby galaxies. I. The center of IC 342. Astrophys. J. **618**, 259–280 (2005)

Requena-Torres, M.A., Martín-Pintado, J., Rodríguez-Franco, A., et al.: Organic molecules in the Galactic center. Hot core chemistry without hot cores. Astron. Astrophys. **455**, 971–985 (2006)

Walter, F., Carilli, C., Bertoldi, F., et al.: Resolved Molecular Gas in a Quasar Host Galaxy at Redshift $z = 6.42$. Astrophys. J. **615**, L17–L20 (2004)

Wang, M., Henkel, C., Chin, Y.-N., et al.: Dense gas in nearby galaxies. XVI. The nuclear starburst environment in NGC 4945. Astron. Astrophys. **422**, 883–905 (2004)

Martín, S., Martín-Pintado, J., Requena-Torres, M.A., Mauersberger, R.: (2007, in preparation)

Studying the first galaxies with ALMA

C.L. Carilli · F. Walter · R. Wang · A. Wootten ·
K. Menten · F. Bertoldi · E. Schinnerer · P. Cox ·
A. Beelen · A. Omont

Originally published in the journal Astrophysics and Space Science, Volume 313, Nos 1–3.
DOI: 10.1007/s10509-007-9647-9 © Springer Science+Business Media B.V. 2007

Abstract We discuss observations of the first galaxies, within cosmic reionization, at centimeter and millimeter wavelengths. We present a summary of current observations of the host galaxies of the most distant QSOs ($z \sim 6$). These observations reveal the gas, dust, and star formation in the host galaxies on kpc-scales. These data imply an enriched ISM in the QSO host galaxies within 1 Gyr of the big bang, and are consistent with models of coeval supermassive black hole and spheroidal galaxy formation in major mergers at high redshift. Current instruments are limited to studying truly pathologic objects at these redshifts, meaning hyperluminous infrared galaxies ($L_{FIR} \sim 10^{13} L_{\odot}$). ALMA will provide the one to two orders of magnitude improvement in millimeter astronomy required to study normal star forming galaxies (i.e. Ly-α emitters) at $z \sim 6$. ALMA will reveal, at sub-kpc spatial resolution, the thermal gas and dust—the fundamental fuel for star formation—in galaxies into cosmic reionization.

C.L. Carilli (✉) · R. Wang · A. Wootten
National Radio Astronomy Observatory, 1003 Lopezville Road,
Socorro, N 87801, USA
e-mail: ccarilli@nrao.edu

F. Walter · E. Schinnerer
Max Planck Insitut für Astronomie, Heidelberg, Germany

K. Menten · A. Beelen
Max-Planck Institute for Radio Astronomy, Bonn, Germany

F. Bertoldi
Bonn University, Bonn, Germany

P. Cox
IRAM, Grenoble, France

A. Omont
Institute de Astrophysique, Paris, France

Keywords Radio astronomy · Extragalactic astronomy ·
Cosmology · Galaxy evolution and formation

1 Introduction: Cosmic reionization and the first galaxies

Observations of the first generation of galaxies and supermassive black holes (SMBH) provide the greatest leverage into theories of cosmic structure formation, and are a principle science driver for all future large area telescopes, from meter to X-ray wavelengths. The recent discovery of the Gunn-Peterson effect, i.e. Ly-α absorption by a partially neutral intergalactic medium (IGM), toward the most distant ($z \sim 6$) QSOs indicates that we have finally probed into the near-edge of cosmic reionization (Fan et al. 2006a). Reionization sets a fundamental benchmark in cosmic structure formation, indicating the formation of the first luminous objects which act to reionize the IGM. Detection of large scale polarization of the CMB, corresponding to Thomson scattering of the CMB by the IGM during reionization, suggests a significant ionization fraction extending to $z \sim 11 \pm 3$ (Page et al. 2006). Overall, current data indicate that cosmic reionization is a complex process, with significant variance in space and time (Fan et al. 2006b). The on-set of Gunn-Peterson absorption at $z \geq 6$ implies that the IGM becomes opaque at observed wavelengths ≤ 1 µm, such that observations of the first luminous objects will be limited to radio through near IR wavelengths.

In this contribution we will discuss the current status of centimeter and millimeter observations of the most distant objects. We will then discuss the revolution afforded by ALMA in this area of research, taking the field from the current study of truly pathologic, rare objects, to the study of the first generation of normal star forming galaxies. In this

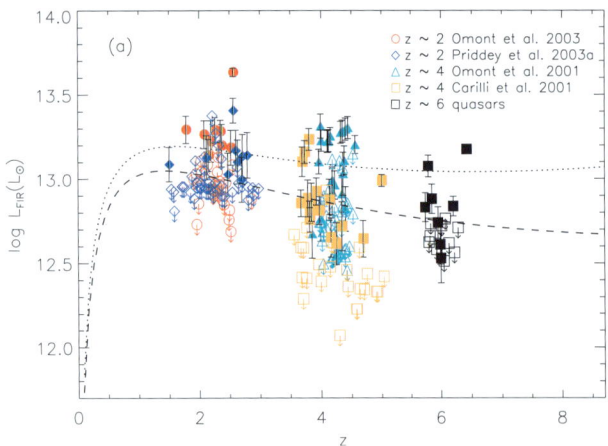

Fig. 1 The logarithm of the FIR luminosity versus redshift for different QSO samples observed at (sub)mm wavelengths (Wang et al. 2007). The *open symbols* with *arrows* denote upper limits. The *dashed* and *dotted lines* represent the typical 3σ detection limits of MAMBO at 250 GHz and SCUBA at 350 GHz, namely $S_{250} = 2.4$ mJy, and $S_{350} = 8.5$ mJy, respectively

contribution we concentrate on molecular line and dust continuum emission. Walter and Carilli (this volume) discuss the exciting prospects for studying the fine structure PDR cooling lines, such as [C(II)], from the first galaxies.

2 Current centimeter and millimeter observations of $z \sim 6$ objects

2.1 The host galaxies of $z \sim 6$ SDSS QSOs

Given the sensitivities of current instruments, observations of sources at $z \sim 6$ are restricted to Hyper Luminous Infrared galaxies (i.e. $L_{FIR} > 10^{13} L_\odot$). At these extreme redshifts, the samples remain limited to the host galaxies of optically luminous QSOs selected from the SDSS. The study of such systems has become paramount since the discovery of the bulge mass – black hole mass correlation in nearby galaxies, a result which suggests a fundamental relationship between black hole and spheroidal galaxy formation (Gebhardt et al. 2000). Our millimeter surveys of the $z \sim 6$ QSOs show that roughly 1/3 of optically selected QSOs are also hyper-luminous infrared galaxies ($L_{FIR} \geq 10^{13} L_\odot$), emitting copious thermal radiation from warm dust (Fig. 1; Wang et al. 2007). This corresponds to roughly 10% of the bolometric luminosity of the QSO (which is dominated by the AGN 'big blue bump'), and the question remains open as to the dominant dust heating mechanism: star formation or the AGN?

The best studied of the $z \sim 6$ QSOs is the most distant QSO known, J1148+5251, at $z = 6.419$. This galaxy been detected in thermal dust, non-thermal radio continuum, and CO line emission (Walter et al. 2003; Bertoldi et

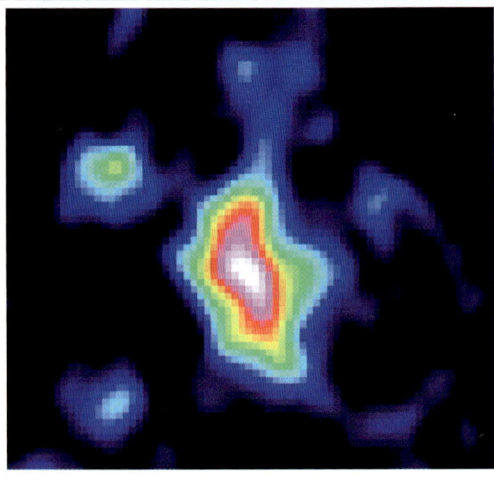

Fig. 2 The velocity integrated CO 3-2 emission from J1148+5251 imaged by the VLA at $0.4''$ resolution (Walter et al. 2004). The total flux is 0.15 Jy km s^{-1}, and the source is clearly resolved, with a full extend of about $1''$. The figure is about $3''$ on a side

al. 2003; Carilli et al. 2004), with an implied dust mass of $7 \times 10^8 M_\odot$, and a molecular gas mass of $2 \times 10^{10} M_\odot$. The molecular gas is extended over $\sim 1''$, or ~ 5.5 kpc (Fig. 2). High resolution VLA imaging of the molecular gas distribution in J1148+5251 provides the only direct measure of the host galaxy dynamical mass, resulting in a value of $\sim 4 \times 10^{10} M_\odot$ within 3 kpc of the galaxy center (Walter et al. 2004; Walter and Carilli, this volume). This mass is comparable to the gas mass, suggesting a baryon-dominated potential for the inner few kpc of the galaxy (Lintott et al. 2006), as is true in nearby spheroidal galaxies, and for ULIRGs (Downes and Solomon 1998). The dynamical mass is also more than an order of magnitude lower than expected based on the bulge mass–black hole mass correlation, suggesting a departure from this fundamental relationship at the highest redshifts, with the SMBH forming prior to the spheroidal galaxy (Walter et al. 2004).

The radio through near-IR SED of J1148+5251 is shown in Fig. 3 (Beelen et al. 2006). The Spitzer bands are consistent with the standard QSO optical through mid-IR SED, including a hot dust component (~ 1000 K), presumably heated by the AGN. However, the observed (sub)mm data reveal a clear rest-frame FIR excess. The rest frame FIR through radio SED is reasonably fit by a template that follows the radio through FIR correlation for star forming galaxies (Yun et al. 2000), with a dust temperature of 55 K. The implied star formation rate is of order $3000 M_\odot$ year^{-1} (Bertoldi et al. 2003). Most recently, IRAM 30 m observations of J1148+5251 have yielded the first detection of the fine structure line of [C(II)] at cosmologically significant redshifts (Maiolino et al. 2005). This line is thought to dominate ISM cooling in photon-dominated regions, i.e. the interface regions between giant molecular clouds and H(II) regions.

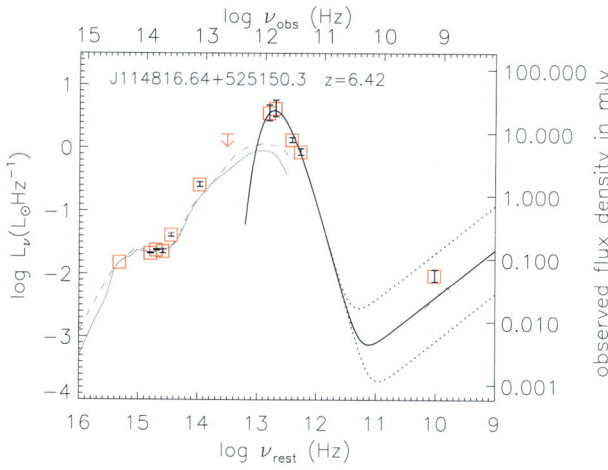

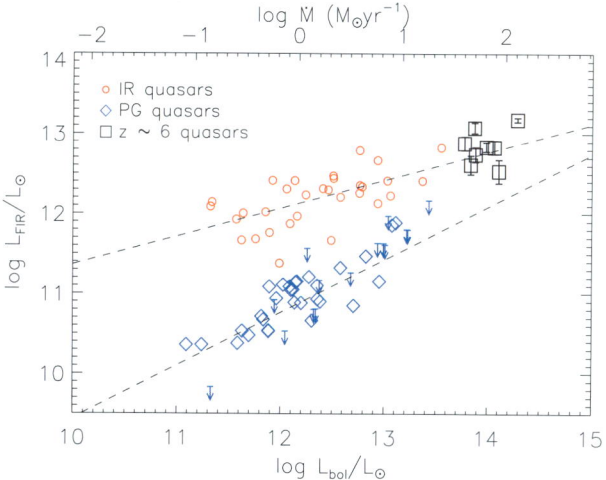

Fig. 3 The SED of J1148+5251 from the rest frame near-IR to the radio (Jiang et al. 2006; Wang et al. 2007; Beelen et al. 2006). The models at rest frame frequencies, $v > 10^{13}$ Hz are two standard QSO SEDs including emission from hot ($\sim$1000 K) dust. The rest-frame far-IR through radio model entails a 55 K modified black body, plus synchrotron radio emission that follows the radio-FIR correlation for star forming galaxies (Beelen et al. 2006; Wang et al. 2007). The *dashed lines* indicate the range defined by star forming galaxies (Yun et al. 2000)

Fig. 4 Correlation of L_{FIR}–L_{bol} for QSO host galaxies. The (sub)mm detected $z \sim 6$ quasars are plotted as *black squares* with error bars denoting 1σ r.m.s. The local IR and PG quasars from Hao et al. (2005) are plotted as *red circles* and *blue diamonds* with *arrows* denoting upper limits in L_{FIR}. The *dashed lines* represent the linear regression results for the two local quasar samples

These observations of J1148+5251 demonstrate that large reservoirs of dust and metal enriched atomic and molecular gas can exist in the most distant galaxies, within 870 Myr of the big bang. The molecular gas and [C(II)] emission suggest a substantially enriched ISM on kpc-scales. The molecular gas represents the requisite fuel for star formation.

The mere existence of such a large dust mass so early in the universe raises the interesting question: how does a galaxy form so much dust so early in the universe? The standard mechanism of dust formation in the cool winds from low mass (AGB) stars takes a factor two or so too long. Maiolino et al. (2004) and Strata et al. (2007) suggest dust formation associated with massive stars in these distant galaxies. They show that the reddening toward the most distant objects is consistent with different dust properties (i.e. silicates and amorphous carbon grains), as expected for dust formed in type-II SNe (although cf. Venkatesan et al. 2006).

Overall, we conclude that J1148+5251 is a likely candidate for the co-eval formation of a SMBH through Eddington-limited accretion, and a large spheroidal galaxy in a spectacular starburst, within 1 Gyr of the big bang. This conclusion is consistent with the general notion of 'downsizing' in both galaxy and supermassive black hole formation (Cowie et al. 1996; Heckman et al. 2004), meaning that the most massive black holes ($>10^9 M_\odot$) and galaxies ($>10^{12} M_\odot$) may form at high redshift in extreme, gas rich mergers/accretion events.

Li et al. (2006) and Robertson et al. (2007) have performed detailed modeling of a system like J1148+5251, in-

cluding feedback from the AGN to regulate star formation. They show that is plausible to form both the galaxy and the SMBH in rare peaks in the cosmic density field (comoving density $\sim 10^{-9}$ Mpc^{-3}), through a series of major mergers of gas rich galaxies, starting at $z \sim 14$, resulting in a SMBH of $\sim 10^9 M_\odot$, and a galaxy of total stellar mass $\sim 10^{12} M_\odot$ by $z \sim 6$. The system will eventually evolve into a rare, extreme mass cluster ($\sim 10^{15} M_\odot$) today. The ISM abundance in the inner few kpc will quickly rise to $\sim$ solar, although the dust formation mechanism remains uncertain.

In support of this conclusion, Fig. 4 shows the relationship between FIR luminosity and bolometric luminosity for the (sub)mm detected $z \sim$ QSOs, as well as a low redshift sample of optically selected QSOs (e.g. PG sample), and an IRAS selected sample of QSOs (Wang et al. 2007; Hao et al. 2005). The $z \sim 6$ sources fall at the extreme luminosity end of the sample, but interestingly, they also follow the trend in FIR to bolometric luminosity set by the low redshift IRAS QSOs, and as opposed to the trend set by the optically selected QSOs. The optically selected QSOs typically have early-type host galaxies, while the IRAS selected QSOs reside in major mergers, with co-eval starbursts (Hao et al. 2005).

A final interesting aspect of the molecular line studies of the most distant QSO host galaxies is the derivation of the sizes of the cosmic Stromgren spheres (Walter et al. 2003; Fan et al. 2006a). The size of the ionized region around the QSO, presumably formed by the radiation from the QSO, can be derived from the difference between the redshift of the host galaxy and the redshift of the on-set of the Gunn-Peterson trough. For J1148+5251, this redshift difference is

$\Delta z \sim 0.1$, implying a physical radius for the cosmic Stromgren sphere of $R = 4.7$ Mpc. This radius can be related to the cosmic neutral fraction using the QSO ionizing luminosity, and the mean baryon density, through the equation: $t_q = 10^5 R^3 f(HI)$, where t_q is the QSO lifetime, and $f(HI)$ is the IGM neutral fraction (White et al. 2005). For J1148+5251, the implied QSO lifetime is $\sim 10^7 f(HI)$ years. A number of authors (Wyithe et al. 2005; Fan et al. 2006a; Kurk et al. 2007) have inverted this equation in order to derive the IGM neutral fraction. Using the J1148+5251 CO host galaxy redshift, plus the redshifts for other $z \sim 6$ QSO host galaxies derived from low ionization broad lines (e.g. Mg(II)), they derive a mean neutral fraction at $z \sim 6.2$ of $f(HI) > 0.1$, assuming a fiducial QSO lifetime $\geq 10^6$ years.

2.2 Limits on normal galaxies: the Cosmos field

J1148+5251 is an extremely rare and pathologically luminous object, unlike anything seen nearby. For instance, there are only some 50 or so of these SDSS $z \sim 6$ QSOs on the entire sky!

We have recently investigated the properties of more normal star forming galaxies at $z \sim 6$ using the Ly-α emitting galaxies (LAEs) selected through a wide field, narrow band search of the Cosmos field (Murayama et al. 2007). The sensitivity to the Ly-α line is such that one can detect galaxies with star formation rates of $\sim 10 M_\odot$ year^{-1} into cosmic reionization. These galaxies are numerous, with roughly 100 deg^{-2} in a narrow redshift search range of $z = 5.7 \pm 0.05$. Extrapolation of the luminosity function to dwarf star forming galaxies could provide enough photons to reionization the universe (Fan et al. 2006a).

We have taken the sample of ~ 100 LAEs from the Cosmos field and searched for radio and millimeter emission using MAMBO (Bertoldi et al. 2007) and the VLA (Schinnerer et al. 2007). We do not detect any individual source down to 3σ limits of ~ 30 μJy beam^{-1} at 1.4 GHz, nor do we detect a source in a stacking analysis, to a 2σ limit of 2.5 μJy beam^{-1} (Carilli et al. 2007). At 250 GHz we do not detect any of the 10 LAEs that are located within the central regions of the COSMOS field covered by MAMBO (20' × 20') to a typical 2σ limit of $S_{250} < 2$ mJy. The radio data imply that there are no low luminosity radio AGN with $L_{1.4} > 6 \times 10^{24}$ W Hz^{-1} in the LAE sample.

These radio and millimeter observations rule out any highly obscured, extreme starbursts in the sample, i.e. any galaxies with massive star formation rates $> 1500 M_\odot$ year^{-1} in the full sample (based on the radio data), or $500 M_\odot$ year^{-1} for the 10% of the LAE sample that fall in the central MAMBO field. The stacking analysis implies an upper limit to the mean massive star formation rate of $\sim 100 M_\odot$ year^{-1}.

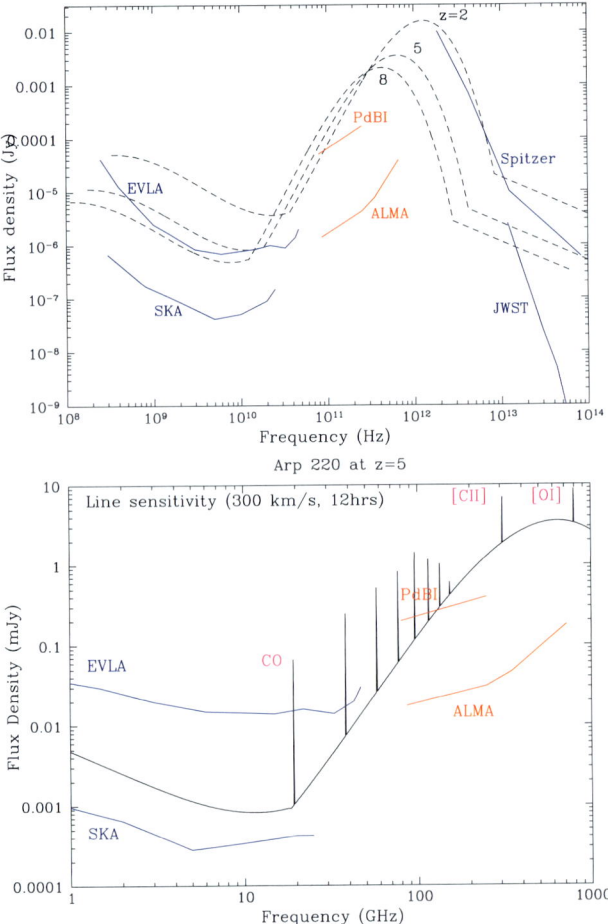

Fig. 5 *Top*: Continuum spectrum of an active star forming galaxy with a star formation rate $\sim 100 M_\odot$ year^{-1}, at $z = 2, 5,$ and 8. The curves show the continuum sensitivities of various telescopes in 12 hours. *Bottom*: Line spectrum of the same galaxy, but only at $z = 5$. The line sensitivities were derived assuming a line width of 300 km s^{-1}

While this study represents the most sensitive, widest field radio and mm study of $z \sim 6$ LAEs to date, it also accentuates the relatively poor limits that can be reached in the radio and mm for star forming galaxies at the highest redshifts, when compared to studies using the Ly-α line.

3 The ALMA revolution

ALMA will be able to detect the thermal emission from warm dust from a source like J1148+5251 in 1 second. Moreover, it will detect the more normal galaxy population ($S_{250} \sim 20$ μJy for SFR $\sim 10 M_\odot$ year^{-1}), in a few hours.

Figure 5 shows the sensitivity of current and future telescopes from the radio through the near-IR, along with the spectrum of an active star forming galaxy, like Arp 220 (SFR $\sim 100 M_\odot$ year^{-1} or $L_{FIR} \sim 10^{12} L_\odot$). The EVLA, and eventually the SKA, will study the non-thermal (and

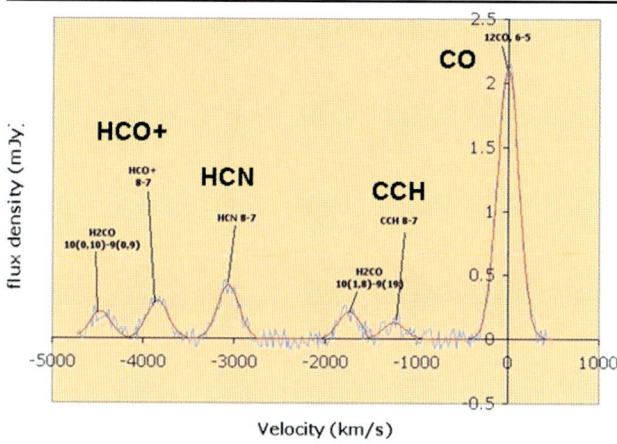

Fig. 6 A simulation of an ALMA spectrum of J1148+5251 at $z = 6.42$. The spectrum shows the 8 GHz bandpass, with an integration time of 24 hours, centered around 93 GHz

possibly free-free thermal) emission associated with star formation, and possibly AGN, from these distant galaxies, as well as the low order transitions from molecular gas. The JWST will observe the stars, the AGN, and the ionized gas. ALMA reveals the thermal emission from dust and gas, including high order molecular line transitions, and fine structure ISM cooling lines (Walter and Carilli, this volume), from the first galaxies—the basic fuel for galaxy formation. ALMA provides the more than an order of magnitude increase in sensitivity and resolution to both detect, and image at sub-kpc resolution, the gas and dust in normal star forming galaxies (e.g. Ly-α galaxies, with star formation rates $\sim 10 M_\odot$ year^{-1}) back to the first generation of galaxies during cosmic reionization.

As an example, Fig. 6 shows the calculated spectrum for J1148+5251 for the 90 GHz band of ALMA in 24 hours. The CO lines will be detected with essentially infinite signal-to-noise, allowing detailed imaging and dynamical studies on sub-kpc scales. Moreover, in a given 8 GHz bandwidth for ALMA, we will detect transitions from numerous astrochemically interesting molecules, such as the dense, pre-star forming gas tracers, HCN and HCO$^+$ (Gao et al. 2007).

Galaxy formation is a complex process, and proper studies require a panchromatic approach. ALMA represents the more than an order of magnitude increase in sensitivity required to probe normal galaxies into cosmic reionization.

Acknowledgements We acknowledge support from the Max-Planck Society and the Alexander von Humboldt Foundation through the Max-Planck Forschungspreis 2005. The National Radio Astronomy Observatory is a facility of the National Science Foundation, operated by Associated Universities, Inc.

References

Bertoldi, F., Carilli, C.L., Cox, P., Fan, X., et al.: Astron. Astrophys. **406**, L55 (2003)

Bertoldi, F., et al.: Astrophys. J. (2007, in press)

Beelen, A., et al.: Astrophys. J. **642**, 694 (2006)

Carilli, C., et al.: Astron. J. **128**, 997 (2004)

Carilli, C., et al.: Astrophys. J. (2007, in press), astro-ph/0612346

Cowie, L.L., et al.: Astron. J. **112**, 839 (1996)

Downes, D., Solomon, P.: Astrophys. J. **507**, 615 (1998)

Fan, X., et al.: Astron. J. **132**, 117 (2006a)

Fan, X., Carilli, C.L., Keating, B.: Annu. Rev. Astron. Astrophys. **44**, 415 (2006b)

Gao, Y., et al.: Astrophys. J. (2007, in press), astro-ph/703548

Gebhardt, K., et al.: Astrophys. J. **543**, L5 (2000)

Hao, C.N., Xia, X.Y., Mao, S., et al.: Astrophys. J. **625**, 78 (2005)

Heckman, T., et al.: Astrophys. J. **613**, 109 (2004)

Jiang, L., et al.: Astron. J. **132**, 2127 (2006)

Kurk, J., et al.: Astron. Astrophys. (2007, submitted)

Li, Y., et al.: Astrophys. J. (2006, in press), astro-ph/0608190

Lintott, C., Ferreras, I., Lahav, O.: Astrophys. J. **648**, 826 (2006)

Maiolino, R., et al.: Nature **431**, 533 (2004)

Maiolino, R., et al.: Astron. Astrophys. **440**, L51 (2005)

Murayama, T., Taniguchi, Y., Scoville, N.Z., et al.: Astrophys. J. Suppl. Ser. (2007), astro-ph/0702458

Page, L., et al.: Astrophys. J. (2006, in press), astro-ph/0603450

Robertson, B., et al.: Astrophys. J. (2007, in press), astro-ph/073456

Schinnerer, E., et al.: Astrophys. J. (2007, in press), astro-ph/0612314

Strata, G., et al.: Astrophys. J. Lett. (2007, submitted), astro-ph/0703349

Venkatesan, A., Nath, B., Shull, M.: Astrophys. J. **640**, 31 (2006)

Walter, F., Bertoldi, F., Carilli, C.L., et al.: Nature **424**, 406 (2003)

Walter, F., Carilli, C., Bertoldi, F., et al.: Astrophys. J. **615**, L17 (2004)

Wang, R., et al.: Astrophys. J. (2007, submitted)

White, R., et al.: Astron. J. **129**, 2102 (2005)

Wyithe, S., Loeb, A., Carilli, C.: Astrophys. J. **628**, 575 (2005)

Yun, M.S., et al.: Astrophys. J. **528**, 171 (2000)

Detecting the most distant (z > 7) objects with ALMA

Fabian Walter · Chris Carilli

Originally published in the journal Astrophysics and Space Science, Volume 313, Nos 1–3.
DOI: 10.1007/s10509-007-9634-1 © Springer Science+Business Media B.V. 2007

Abstract Detecting and studying objects at the highest redshifts, out to the end of Cosmic Reionization at $z > 7$, is clearly a key science goal of ALMA. ALMA will in principle be able to detect objects in this redshift range both from high-J ($J > 7$) CO transitions and emission from ionized carbon, [CII], which is one of the main cooling lines of the ISM. ALMA will even be able to resolve this emission for individual targets, which will be one of the few ways to determine dynamical masses for systems in the Epoch of Reionization. We discuss some of the current problems regarding the detection and characterization of objects at high redshifts and how ALMA will eliminate most (but not all) of them.

Keywords Radio lines: ISM · Galaxies: high-redshifts · ISM: lines and bands

1 Introduction: the highest redshift galaxies

In recent years, deep narrow band surveys have revealed a major population of Lyman Alpha Emitters (LAE) out to very high redshifts (e.g. Hu et al. 2002; Kurk et al. 2004; Stern et al. 2005; Murayama et al. 2007). In particular, Taniguchi et al. (2005) report the detection of 9 spectroscopically confirmed LAE at redshifts of $z \sim 6.6$ in the Subaru Deep Field (currently, the published LAE redshift record

F. Walter (✉)
Max Planck Insitut für Astronomie, Heidelberg, Germany
e-mail: walter@mpia.de

C. Carilli
National Radio Astronomy Observatory, Charlottesville, USA
e-mail: ccarilli@nrao.edu

holder is at $z = 6.98$, Iye et al. 2006). The mere presence of Lyman alpha emission in these sources provides strong evidence that they are undergoing bursts of star formation: the star formation rates of individual objects are $\sim 10 \, M_\odot \, yr^{-1}$ (based on their FUV luminosities) and their redshifts place them well within the end of cosmic reionization. They also appear to be very numerous: Taniguchi et al. (2005) find ~ 30 LAEs in only a quarter degree field (e.g., compared to ~ 10 QSOs at $z > 6$ which are distributed over a quarter of the sky! Fan et al. 2004). This implies that LAEs may play an important role in reionizing the universe at $z > 6$ (for a review see Fan et al. 2006). Investigating the physical properties of these sources are thus of great interest and ALMA will play a critical role in studying these objects, as discussed in the following.

2 Interstellar medium: CO vs. [CII] emission

2.1 Carbon monoxide (CO)

Constraining the properties of the molecular gas in objects at the end of cosmic reionization is clearly of key importance as such observations (1) will measure the available 'fuel' for star formation, (2) will help to constrain the dynamical mass of the system and will thus (3) allow to put these objects in an evolutionary context for early galaxy formation. Typically, at low and high redshifts, CO emission is used as a tracer for the molecular gas phase (e.g. review by Solomon and Vanden Bout 2005). It is important to keep in mind though, that, at the highest z, only the very high rotational lines of CO will be observable with ALMA. E.g. even in the lowest (currently funded) frequency band of ALMA (band 3, 84–119 GHz), only CO transitions with $J > 7$ (i.e., CO(7–6), CO(8–7), etc.) will be observable at $z > 7$. This is

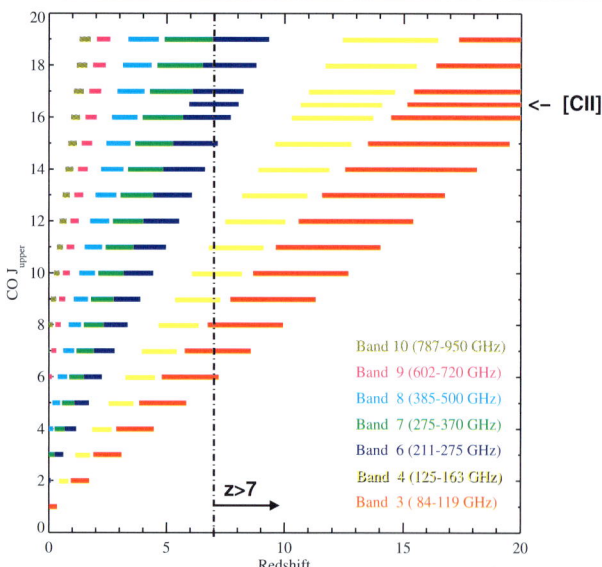

Fig. 1 ALMA CO 'discovery space': the *horizontal lines* indicate which CO transition (plotted on the *y*-axis) can be observed with which ALMA band as a function of redshift (plotted on the *x*-axis). For objects with $z > 7$, only the higher-J CO transitions can be observed with ALMA. The [CII] 'discovery space' is also indicated

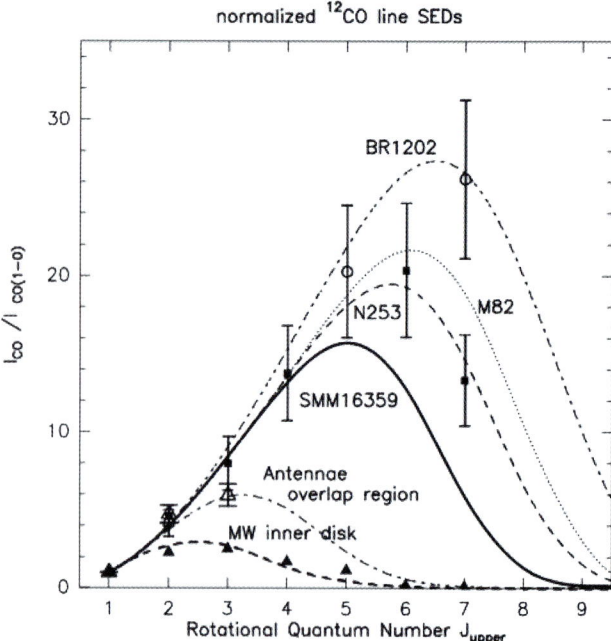

Fig. 2 Comparison of various normalized (by their CO(1–0) flux density) CO line SEDs at low and high redshift (figure taken from Weiss et al. 2007). The CO line SEDs decline rapidly beyond $J = 6$–8

graphically illustrated in Fig. 1 where we plot ALMA's 'CO discovery space' (i.e., which line can be observed at which redshift using which ALMA band). The high-J transitions correspond to highly excited gas (either due to high kinetic temperatures, high densities, or both) which may not be excited in normal starforming environments. This is shown in Fig. 2 (taken from Weiss et al. 2007) where measured CO line strengths (as a function of J, this is sometimes referred to as CO line ladders/SEDs) are plotted for a number of key sources. What is immediately obvious from this plot is that most objects have sharply decreasing CO line strengths beyond $J > 8$, in particular starforming systems such as NGC 253, or the sub-millimeter galaxy plotted in this diagram (the quasars appear to be more excited, but their CO line SED still turns over at $J \sim 7$, for an exceptional object see APM 07279, Weiss et al. 2007). This comparison immediately implies that emission from the CO molecule will typically be very difficult to observe with ALMA at $z > 7$ as the observable lines will simply not be excited.

2.2 Ionized carbon ([CII]) to the rescue!

An alternative tracer of the interstellar medium is one of the main cooling line of the ISM, the $^2P_{3/2} \rightarrow {}^2P_{1/2}$ fine-structure line of C^+ (or [CII]). In brief, the [CII] line is expected to be much stronger than any of the CO lines. Given its high frequency (157.74 µm, corresponding to 1900.54 GHz) [CII] studies in the local universe are limited to airborne or satellite missions (e.g. Stacey et al. 1991; Malhotra et al. 1997; Madden et al. 1997). These studies

have demonstrated that this single line can indeed carry a good fraction of the total infrared luminosity (L_{FIR}) of an entire galaxy. In the local universe, the ratio L_{CII}/L_{FIR} has been found to be $2–5 \times 10^{-4}$ in the case of ULIRGS (e.g. Gerin and Phillips 2000), but is more like $5–10 \times 10^{-3}$ in more typical starforming galaxies (for a discussion on possible reasons for the suppressed ratio in ULIRGs see, e.g., Luhman et al. 1998). Notably, the ratio has been found to be 1% or even higher in low metallicity environments. E.g., in the low-metallicity galaxy IC10, L_{CII}/L_{FIR} reaches values as high as 4%, with an average value of 2% (Madden et al. 1997; see Israel et al. 1996 for a similar result for the LMC). This is the reason why it has long been argued (e.g., Stark 1997) that observation of the [CII] line of pristine systems at the highest redshifts will likely be the key to study molecular gas in the earliest starforming systems, in particular in the era of ALMA. The ALMA [CII] 'discovery space' is also indicated in Fig. 1.

3 Expected [CII] line strengths

At the redshifts of the LAEs, the [CII] line is shifted to the 1 mm band of ALMA (band 6, 211–275 GHz). [CII] emission has recently been successfully detected using the IRAM 30 m in the highest redshift quasar J1148+5251 at $z = 6.42$ (Maiolino et al. 2005, see Fig. 3). The notable difference between J1148+5251 and the $z > 6$ LAEs is that the ratio L_{CII}/L_{FIR} has been found to be very low ($\sim 5 \times 10^{-4}$) in

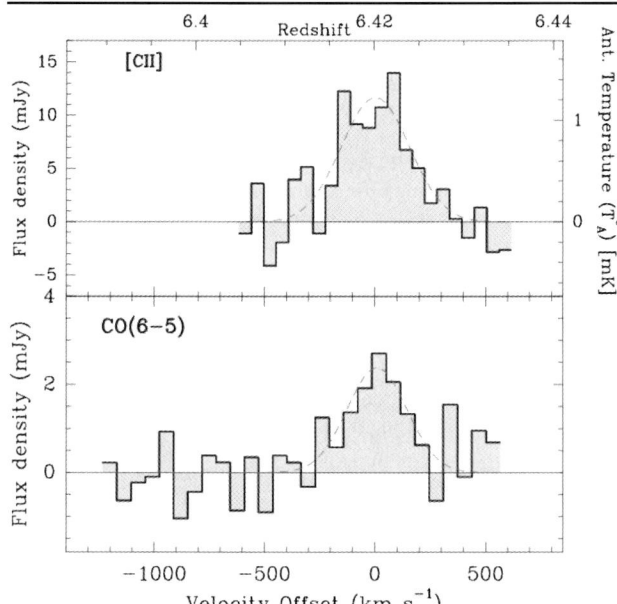

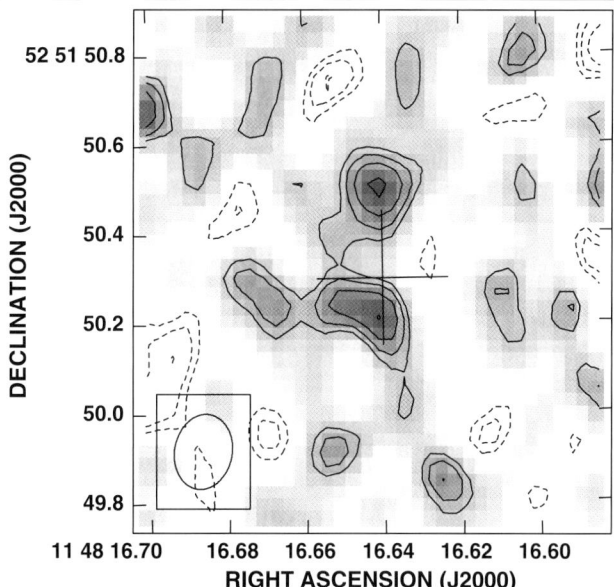

Fig. 3 *Top:* first detection of [CII] at high redshift in the $z = 6.42$ QSO J1148+5251 (Maiolino et al. 2005). *Bottom:* brightest CO transition ($J = 6$) in the same source (Bertoldi et al. 2003; Walter et al. 2003). Note that the [CII] line is brighter by a factor of ~ 5

Fig. 4 High–resolution CO image of the $z = 6.42$ QSO J1148+5251 obtained at the VLA (Walter et al. 2004). The resolution achieved in these observations ($0.15''$, corresponding to ~ 1 kpc) will be routinely reached with ALMA

J1148+5251, i.e. in perfect agreement with studies of low redshift ULIRGs that show a central AGN. On the contrary, the LAE are presumably pure starbursts (no evidence for an AGN is found, Taniguchi et al. 2005) and they likely have lower metallicities compared to the highly overdense regions in which the luminous quasars are supposedly present. All these arguments point towards a L_{CII}/L_{FIR} ratio in LAE that is close to what is found for nearby normal galaxies, or perhaps even for the metal-poor dwarf galaxies (i.e., around 1% or even higher). In other words, the [CII] luminosity of the LAE may well be an order of magnitude stronger (for a given IR luminosity) than what has been found in the $z = 6.4$ QSO. In the following we present a quick back-of-the envelope calculation based on the detected [CII] line strength in J1148+5251 (~ 10 mJy) which has a SFR of a few 1000 $M_\odot$ yr^{-1}. This SFR is more than two order of magnitudes higher than the SFR found in a typical LAE, but as the L_{CII}/L_{FIR} may be higher by an order of magnitude in the LAEs, the expected [CII] line strength of the LAE may be as high as 1 mJy. Such a line should be easily detectable with ALMA at high significance in a few hours.

4 Resolving the ISM

Detecting the [CII] (or CO) emission is critical to estimate the reservoir of the (molecular) gas in these early systems. A second step is then to spatially resolve the molecular gas distribution. In particular, given the typical diameters of galaxies of many kpc, a linear resolution of ~ 1 kpc is needed

to resolve the structure of the underlying galaxy. Such measurements are needed (1) to get an estimate for the size of the galaxy (and thus a better estimate for the dynamical mass), (2) to resolve potentially merging systems, and (3) to better constrain the physical properties of the gas (e.g., by measuring the brightness temperature of the hosts). A linear resolution of 1 kpc corresponds to a resolution of $0.15''$ at the redshifts under consideration ($1'' \sim 5.8$ kpc at $z = 6$). Such observations can then in turn be used to constrain the predictions by CDM simulations of early galaxy formation, and, if a large sample was available, put limits on the frequency of mergers at high redshift. In addition, such studies can be used to constrain the possible redshift-evolution of the M_{BH}–σ_v relation in high-z quasars. Such observations will clearly be feasible with ALMA in the extended arrays. High-resolution CO imaging is already possible with the current generation of telescopes: we have used the VLA to resolve the molecular gas in the host galaxy of the $z = 6.42$ QSO J1148+5251 (see Fig. 4, Walter et al. 2004).

5 The case for ALMA band 5

As a technical note: the ALMA redshift coverage for the [CII] line is not ideal as its frequency lies between the CO(17–16) and CO(16–15) transition (see Fig. 1). One concern is that the critical redshift range ($8 < z < 10.5$) is currently not fully covered: This frequency range corresponds to the ALMA band 5 which is only partly funded by the European Union as part of the Sixth Framework Programme

(FP6) for up to 8 antennas. Clearly, it would be highly desirable to equip as many ALMA antennas with band 5 receivers as possible.

6 Concluding remarks

ALMA observations of the [CII] line will play a fundamental role in studying the youngest galaxies in the Epoch of Cosmic Reionization at $z > 7$. Given the expected line strengths it should be possible to resolve these galaxies in the [CII] line emission on kpc scales. Such measurements would not only constrain the sizes but would also help to derive the dynamical masses in these early starforming systems. Given the typical CO excitation in starforming galaxies (i.e. drop in excitation around the $J \sim 6$ transitions), ALMA will likely act as a [CII]—rather than a CO—machine for objects at these extreme redshifts.

Acknowledgements   It is our pleasure to thank our collaborators on this project: Frank Bertoldi, Dominik Riechers, Pierre Cox, Roberto Maiolino and Axel Weiß.

References

Bertoldi, F., Carilli, C.L., Cox, P., Fan, X., et al.: Astron. Astrophys. **406**, L55 (2003)
Fan, X., Hennawi, J.F., Richards, G.T., et al.: Astron. J. **128**, 515 (2004)
Fan, X., Carilli, C.L., Keating, B.: Annu. Rev. Astron. Astrophys. **44**, 415 (2006)
Gerin, M., Phillips, T.G.: Astrophys. J. **537**, 644 (2000)
Hu, E.M., Cowie, L.L., McMahon, R.G., et al.: Astrophys. J. **568**, L75 (2002)
Israel, F.P., Maloney, P.R., Geis, N., Herrmann, F., Madden, S.C., Poglitsch, A., Stacey, G.J.: Astrophys. J. **465**, 738 (1996)
Iye, M., Ota, K., Kashikawa, N., Furusawa, H., Hashimoto, T., Hattori, T., Matsuda, Y., Morokuma, T., Ouchi, M., Shimasaku, K.: Nature **443**, 186 (2006)
Kurk, J.D., Cimatti, A., di Serego Alighieri, et al.: Astron. Astrophys. **422**, L13 (2004)
Luhman, M.L., et al.: Astrophys. J. Lett. **504**, L11 (1998)
Madden, S.C., Poglitsch, A., Geis, N., Stacey, G.J., Townes, C.H.: Astrophys. J. **483**, 200 (1997)
Maiolino, R., Cox, P., Caelli, P., et al.: astro-ph/0508064 (2005)
Malhotra, S., et al.: Astrophys. J. Lett. **491**, L27 (1997)
Murayama, T., Taniguchi, Y., Scoville, N.Z., et al.: 2007, Astrophys. J. Suppl. Ser., astro-ph/0702458 (2007)
Solomon, P.M., Vanden Bout, P.A.: Annu. Rev. Astron. Astrophys. **43**, 677 (2005)
Stacey, G.J., Geis, N., Genzel, R., Lugten, J.B., Poglitsch, A., Sternberg, A., Townes, C.H.: Astrophys. J. **373**, 423 (1991)
Stark, A.A.: Astrophys. J. **481**, 587 (1997)
Stern, D., Yost, S.A., Eckart, M.E., Harrison, F.A., Helfand, D.J., Djorgovski, S.G., Malhotra, S., Rhoads, J.E.: Astrophys. J. **619**, 12 (2005)
Taniguchi, Y., Ajiki, M., Nagao, T., et al.: Publ. Astron. Soc. Jpn. **57**, 165 (2005)
Walter, F., Bertoldi, F., Carilli, C.L., et al.: Nature **424**, 406 (2003)
Walter, F., Carilli, C., Bertoldi, F., et al.: Astrophys. J. **615**, L17 (2004)
Weiss, A., Downes, D., Neri, R., Walter, F., Henkel, C., Wilner, D.J., Wagg, J., Wiklind, T.: Astron. Astrophys. **467**, 955 (2007)

Redshift distribution of the submillimeter extragalactic background light

Is ALMA going to see many high-redshift galaxies?

Wei-Hao Wang · Lennox L. Cowie · Amy J. Barger

Originally published in the journal Astrophysics and Space Science, Volume 313, Nos 1–3.
DOI: 10.1007/s10509-007-9615-4 © Springer Science+Business Media B.V. 2007

Abstract The submillimeter (submm) extragalactic background light (EBL) traces the integrated star formation history throughout the cosmic time. Deep blank-field 850 μm and 1.4 GHz surveys and optical follow-up have been only able to determine the redshift of ∼20% of the submm EBL. The majority (80%) of the submm EBL is still below the confusion and sensitivity limits of current submm and radio instruments. We break through these limits with stacking analyses on our deep 850 μm image in the GOODS-N and find that the submm EBL mostly comes from galaxies at redshifts around 1.0. This redshift is much lower than the redshift of $z = 2$–3 previously implied from radio identified submm sources. This result significantly decreases the number of high redshift galaxies that may be seen by ALMA.

Keywords Submillimeter · Extragalactic background · Galaxy · ALMA

1 Introduction

The extragalactic background light (EBL) is an integrated measure of the history of the luminous energy production of the universe from both star formation and blackhole accretion. Directly emitted light is seen in the X-ray, UV, and optical, whereas dust reradiated energy appears in the far-infrared (FIR) and submillimeter (submm). The FIR and submm EBL measured by *COBE* is comparable to the optical EBL, showing that the submm wavelength is extremely important for understanding the evolution and formation of galaxies and active galactic nuclei. Further observations of the resolved submm EBL sources and X-ray observations (Alexander et al. 2003) suggest that these sources are mostly star forming galaxies and active galactic nucleus contribution is relatively small. In addition, the strong negative K-correction of the submm thermal spectra makes this waveband a very sensitive probe for high-redshift dust emission, although this negative K-correction alone does not necessarily make the observed submm sources a high-redshift population.

ALMA, as the most important next generation instrument in the submm wavelength, will provide the sensitivity and resolution for studying galaxy evolution at a great depth by observing the submm EBL sources. It will also have the potential of discovering high-redshift ($z > 4$) dusty galaxies that were previously missed by optical and near-infrared (NIR) surveys, if such a high-redshift population exists. In this paper we briefly summarize the previous understanding of the redshift distribution of the submm population and present our recent work on this topic. We also discuss the implication of our work to future ALMA surveys of high-redshift galaxies.

2 Redshift distribution of submillimeter sources

2.1 Bright SCUBA sources

Confusion limited blank-field SCUBA surveys have resolved ∼20%–30% of the submm EBL into point sources brighter than ∼2 mJy at 850 μm (hereafter bright SCUBA

W.-H. Wang (✉)
NRAO, 1003 Lopezville Rd., Socorro, NM 87801, USA
e-mail: whwang@aoc.nrao.edu

L.L. Cowie
Institute for Astronomy, University of Hawaii, Honolulu, USA

A.J. Barger
Department of Astronomy, University of Wisconsin-Madison,
Madison, USA

sources). Approximately 60% of these bright sources have radio counterparts at 1.4 GHz and their locations can be accurately determined by radio interferometry. Optical and near-infrared spectroscopy of radio identified bright SCUBA sources shows a redshift distribution of $z \sim 2$–3 (Chapman et al. 2003; Swinbank et al. 2004; Chapman et al. 2005). Radio-submm photometric redshifts of radio identified SCUBA sources generally also provide a similar redshift range (Barger et al. 2000; Ivison et al. 2002). These results show that the bright SCUBA sources are a high-redshift population. Furthermore, a common interpretation of the bright sources without radio counterparts is even higher redshifts, because of the positive K-correction in the radio.

Nevertheless, it is important to realize that the radio identified SCUBA sources only contribute at most $30\% \times 60\% = 20\%$ to the total submm EBL. The majority 80% of the submm EBL is either below the confusion limit of current submm instruments, or below the sensitivity limit of current radio interferometers. The redshift distribution of the fainter submm EBL sources does not necessarily match that of the bright sources, and is essentially unknown.

2.2 Stacking analyses on faint submm sources

Submm sources fainter than the confusion limit of SCUBA can be statistically detected with a stacking technique if there is prior information about their locations. In our most recent work (Wang et al. 2006, 2007) we performed stacking analyses on our 110 arcmin^2 GOODS-N SCUBA map (Wang et al. 2004) using galaxy samples selected from ground-based deep NIR images and ultradeep *Spitzer* IRAC images (GOODS *Spitzer* Legacy Science Program). The galaxy samples contain $\sim$3000 sources with very complete redshift information (spectroscopic and photometric). We found that a combination of >2 μJy K_s band sources and >2 μJy 8.0 μm sources picks up the largest amount of 850 μm flux. By averaging the measured 850 μm fluxes at the locations of the $K_s + 8.0$ μm galaxies, we detected a surface brightness of 29.3 ± 2.6 Jy deg^{-2} at 850 μm, corresponding to $\sim$67%–95% of the total 850 μm EBL measured by *COBE* (31 Jy deg^{-2} Puget et al. 1996, or 44 Jy deg^{-2} Fixsen et al. 1998). This large fraction of detected submm EBL is a substantial improvement over the result from just bright SCUBA sources. We also detected a 1.4 GHz EBL of 1.117 ± 0.026 Jy deg^{-2} and a 24 μm EBL of 6.42 ± 0.12 Jy deg^{-2} with the same stacking technique.

Moreover, by grouping the NIR galaxy sample with their redshifts and optical/NIR spectral energy distributions (SEDs), we found that most of the detected submm EBL comes from galaxies with intermediate class SEDs (Wang et al. 2006). Among the detected 29.3 Jy deg^{-2} 850 μm EBL, 24.7 Jy deg^{-2} comes from redshift identified sources. The unidentified sources contribute 4.6 ± 0.9 and $0.35 \pm$

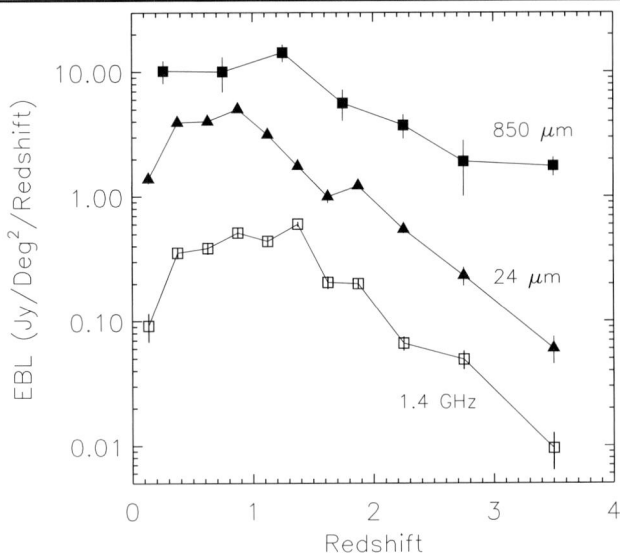

Fig. 1 Contributions to the 850 μm, 24 μm, and 1.4 GHz EBLs (in Jy deg^{-2} *per redshift*) vs. redshift. All the three EBLs are measured from the 2 mJy $K_s + 8.0$ μm sample. The total amounts of redshift identified EBLs are 24.7 ± 2.4, 5.8 ± 0.1, and 0.763 ± 0.024 Jy deg^{-2} at 850 μm, 24 μm, and 1.4 GHz, respectively. Note that all EBLs peak at $z \sim 1.0$. The slope of the submm EBL at $z > 1.5$ is much shallower because of the negative K-correction in the submm

0.01 Jy deg^{-2} to the 850 μm and 1.4 GHz EBL, respectively, and are dominated by just couples of radio bright (>300 μJy) but optically faint sources. We show the EBL contributions vs redshift in Fig. 1, derived from the redshift identified sources. As shown in the figure, most of the submm EBL comes from redshifts around 1.0. This result suggests that the faint submm EBL sources and the bright SCUBA sources are two different populations at different redshifts. (Further analyses in their 850/24 μm colors also show that their dust temperature properties are different, Wang et al. 2007.) While this result fits into the popular scenario of "cosmic downsizing," it is still somewhat surprising, and it has to be tested with the current dataset and with future observations.

2.3 Tests on the stacking results

To ensure that the stacking results are unbiased, we performed various tests on our data. Among the most important ones, we measured 850 μm fluxes at random positions and found the stacking fluxes are consistent with zero. This zero sum is a direct result of the two negative 50% sidelobes of the SCUBA jiggle map. It shows that the low angular resolution of the SCUBA map does not produce a confusing flux that biases the stacking result and only a real correlation between the NIR sample and the submm sources can provide a non-zero stacking flux. We also repeated the same stacking analyses on the 1.4 GHz radio image (Richards 2000) and 24 μm image of GOODS-N, and found that the detected radio and 24 µm EBL shows consistent SED class distribution (Wang et al. 2006) and redshift distribution (Fig. 1). These radio and 24 µm measurements provide a strong support to the above submm result since they come from maps of totally different angular resolutions and noise properties.

Although our stacking results seem to be robust, further tests on the low-redshift origin of the submm EBL have to be carried out with new and independent observations before ALMA comes online in order to provide critical inputs to ALMA observations. One possible test is to measure the redshift distribution of radio-identified faint submm sources lensed by clusters. These lensed, faint sources are below the normal confusion limit of SCUBA. They can provide a direct comparison with the statistically detected faint submm sources in the stacking analyses. Although lensing surveys may favor high redshift objects, good lensing geometry models can help to account for this bias effect.

2.4 Cosmic star formation history

With the measured redshift distribution of the submm and radio EBL, we can infer the cosmic star formation history (i.e., comoving star formation rate density, SFRD, as a function of redshift) using the standard star formation rate formula $\dot{M} = 1.7 \times 10^{-10} L_{\mathrm{IR}}/L_\odot$ (Kennicutt 1998). The infrared luminosities can be converted from either 850 µm fluxes or radio fluxes. The 850 µm conversion is probably relatively robust only for bright SCUBA sources since they have similar dust properties. The radio conversion is more reliable for the entire sample, given the tight correlation between radio power and infrared luminosity in star forming galaxies. The results are shown in Fig. 2. It is clear that although bright SCUBA sources have a redshift distribution strongly peaked at $z = 2$–3, the entire submm/infrared SFRD is relatively flat at $z > 1$.

3 Discussion

Our successful stacking detection of the 850 µm EBL can be attributed to the properties of the $K_s + 8.0$ µm sample. The rest-frame K_s band has relatively less extinction and is just slightly redder than 1.6 µm, where there is an opacity minimum in stellar atmosphere. Therefore the K_s band is very sensitive to slightly redshifted and dust extinguished stellar emission. Moreover, the 8.0 µm band directly picks up the blue end of the thermal dust emission and the strong PAH features from star forming galaxies. Consequently, sources selected at these two wavebands are highly correlated to the star forming, dusty submm population. On the other hand, these two wavebands may be biased against high redshift sources, as they start to miss the peaks of the stellar and dust spectra at $z > 1$. It is thus possible that a significant portion

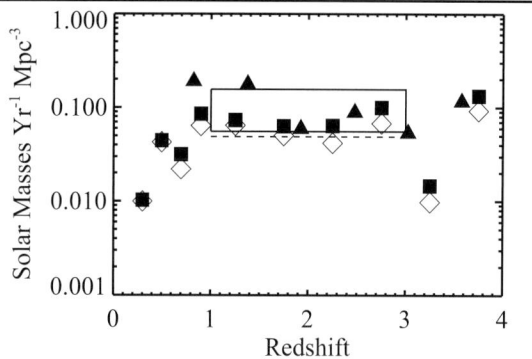

Fig. 2 SFRD vs. redshift (Wang et al. 2006). The *filled squares* show the SFRD derived from the radio EBL. The *open diamonds* show the same results when sources with 2–8 keV luminosities $> 10^{42}$ erg s^{-1} (sources containing active galactic nuclei) are excluded. The *filled triangles* show the SFRD computed using the 850 µm EBL. The *rectangular region* denotes the SFRD from the missing submm EBL that is not accounted for by our NIR sample, assuming that it lies in the redshift interval $z = 1$–3. The range corresponds to the uncertainty in the total 850 µm EBL

of the redshift unidentified submm EBL comes from redshifts greater or much greater than 1.0.

Nevertheless, our stacking analyses place approximately 17 Jy deg^{-2} of the submm EBL at $z < 1.5$, and leave 7 or 20 Jy deg^{-2} of the submm EBL redshift unidentified, depending on which of the *COBE* EBL measurements is adopted (Puget et al. 1996; Fixsen et al. 1998). While the exact amount of unidentified submm EBL and its redshift distribution is still open, the above results already dramatically lower the number of high-redshift submm emitting sources that ALMA may see. Because of this, any ALMA survey that targets on discovering high-redshift ($z > 4$) sources has to increase the area coverage. For example, if we assume the maximum missing EBL of 20 Jy deg^{-2} and place 20% of it at $z > 4$, we find a source density of 8×10^3 deg^{-2} for a typical source flux of 0.5 mJy (Cowie et al. 2002). This corresponds to 0.15 sources per ALMA field of view. To detect a significant sample of 100 of such sources at 850 µm at 10 σ, approximately 300 hour of integration is needed.

The above simple and optimistic integration time estimate does not yet include any effort that is needed to identify these high redshift submm sources. (As an example, multi-wavelength submm color selection of high redshift dust emission will require at least a few times more observing time.) The great difficulty of identifying them is hinted by the simple fact that they are still missed by our stacking analyses. This missing EBL was not picked up by even the deepest *Spitzer* 3.6–24 µm imaging and 10-m class ground-based optical imaging. After the submm EBL is fully resolved by ALMA, a significant amount of the ALMA sources will not have optical and NIR counterparts, or the counterparts will be too faint for spectroscopic follow-up with current ground-based and space-based instruments.

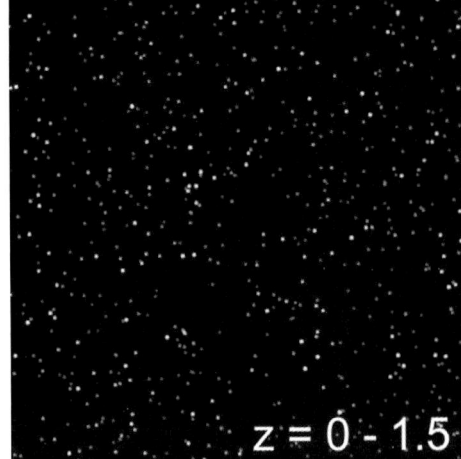

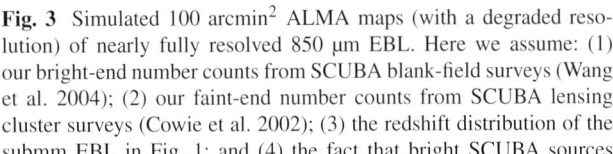

Fig. 3 Simulated 100 arcmin² ALMA maps (with a degraded resolution) of nearly fully resolved 850 µm EBL. Here we assume: (1) our bright-end number counts from SCUBA blank-field surveys (Wang et al. 2004); (2) our faint-end number counts from SCUBA lensing cluster surveys (Cowie et al. 2002); (3) the redshift distribution of the submm EBL in Fig. 1; and (4) the fact that bright SCUBA sources

almost only appear at $z > 1.5$ (Chapman et al. 2003, 2005). Under the above assumptions, sources at $z < 1.5$ dominate the total submm source number and the total submm EBL. At $z > 1.5$, the submm population is dominated by a relatively small number of bright SCUBA sources. This picture does not account for the EBL still missed by our stacking analyses, which may be at $z > 1.5$

They may be as well too faint for the EVLA to detect their synchrotron radiation, and for wide-band receivers on large single-dish telescopes to detect the redshifted CO lines. It is almost certain that in order to fully understand the submm EBL sources, next generation instruments in all wavebands from optical to millimeter are critically needed.

4 Summary

Bright submm sources identified in the radio contribute ~20% to the total submm EBL, and are at redshifts between 2 and 3. Fainter submm sources, which account for the majority of the submm EBL and cosmic star formation, can be detected with a stacking technique and appear to have a redshift distribution peaked at 1.0. This implies that most of the submm EBL sources resolve by ALMA will be low-redshift sources. To summarize this paper, in Fig. 3 we present simulated ALMA maps of submm EBL sources at $z < 1.5$ and $z = 1.5$–4.

References

Alexander, D.M., et al.: Astron. J. **125**, 383 (2003)

Barger, A.J., Cowie, L.L., Richards, E.A.: Astron. J. **119**, 2092 (2000)

Chapman, S.C., Blain, A.W., Ivison, R.J., Smail, I.R.: Nature **422**, 695 (2003)

Chapman, S.C., Blain, A.W., Smail, I., Ivison, R.J.: Astrophys. J. **622**, 772 (2005)

Cowie, L.L., Barger, A.J., Kneib, J.-P.: Astron. J. **123**, 2197 (2002)

Fixsen, D.J., Dwek, E., Mather, J.C., Bennett, C.L., Shafer, R.A.: Astrophys. J. **108**, 123 (1998)

Ivison, R.J., et al.: Mon. Not. R. Astron. Soc. **337**, 1 (2002)

Kennicutt, R.C.: Annu. Rev. Astron. Astrophys. **36**, 189 (1998)

Puget, J.-L., Abergel, A., Bernard, J.-P., Boulanger, F., Burton, W.B., Désert, F.-X., Hartmann, D.: Astron. Astrophys. **308**, L5 (1996)

Richards, E.A.: Astrophys. J. **533**, 611 (2000)

Swinbank, A.M., Smail, I., Chapman, S.C., Blain, A.W., Ivison, R.J., Keel, W.C.: Astrophys. J. **617**, 64 (2004)

Wang, W.-H., Cowie, L.L., Barger, A.J.: Astrophys. J. **613**, 655 (2004)

Wang, W.-H., Cowie, L.L., Barger, A.J.: Astrophys. J. **647**, 74 (2006)

Wang, W.-H., Cowie, L.L., Barger, A.J.: (2007, in preparation)

Molecular absorptions in high-z objects

F. Combes

Originally published in the journal Astrophysics and Space Science, Volume 313, Nos 1–3.
DOI: 10.1007/s10509-007-9632-3 © Springer Science+Business Media B.V. 2007

Abstract Molecular absorption lines measured along the line of sight of distant quasars are important probes of the gas evolution in galaxies as a function of redshift. A review is made of the handful of molecular absorbing systems studied so far, with the present sensitivity of mm instruments. They produce information on the chemistry of the ISM at $z \sim 1$, the physical state of the gas, in terms of clumpiness, density and temperature. The CMB temperature can be derived as a function of z, and also any possible variations of fundamental constants can be constrained. With the sensitivity of ALMA, many more absorbing systems can be studied, for which some predictions and perspectives are described.

Keywords Galaxies: interstellar medium · Quasars: absorption lines · Galaxies: evolution · Galaxies: high-redshift

1 Introduction

Molecular absorptions at intermediate redshift began to be studied more than a decade ago, after the discovery of CO absorption in front of the BL Lac object PKS1413+135 at $z = 0.25$ (Wiklind and Combes 1994). Although many groups undertook active searches, there are still now only 5 molecular absorbing systems detected at high z: PKS1413+135 and B3 1504+377, which are self-absorbing systems, and 3 gravitational lens systems B0218+357, PKS1830-211, PMN J0134-0931 (with OH

only). Table 1 summarises the properties of these systems, together with the few local extra-galactic ones.

With respect to emission, absorption measurements are quite sensitive, even to a small amount of molecular gas along the line of sight. The detection depends mainly on the background source intensity, and the rarity of the detections until now is due to that of strong millimetric radio sources. Due to its sensitivity increase, there could be $\sim$30–100 times more sources detected with ALMA.

1.1 Scientific goals

The study of molecular absorbing systems at intermediate and high redshift allows to reach several goals:

- to detect molecules at high z with much more sensitivity (down to 1 $M_\odot$) than with emission searches and with complementary insight in physical conditions
- to study the evolution with z of chemical abondances: not only CO lines are detectable, but molecular surveys are possible
- to measure the CMB temperature as a function of redshift, to independently estimate the Hubble constant, through the time delay between two gravitational lens images
- to probe the variation of fundamental constants (α, g_p, $\mu = m_e/m_p$). Several theories based on superstrings, Kaluza-Klein theory, or compactified extra-dimensions, predict spatio-temporal variations of the fundamental constants (Uzan 2003; Murphy et al. 2003; Chand et al. 2006).

1.2 New local absorptions

The Centaurus A (NGC 5128) dust lane is well known to absorb in front of the strong internal radio source, and the absorption is diluted in the emission for the CO lines (the same

F. Combes (✉)
LERMA, Observatoire de Paris, 61 Av. de l'Observatoire, 75014, Paris, France
e-mail: francoise.combes@obspm.fr

Table 1 Brief census of molecular absorbers in radio

Source	$z_a{}^1$	$z_e{}^2$	$N_c{}^3$	$N(H_2)^4$ cm^{-2}	ΔV^5 km/s	Molecules
Cen-A	0.0018	0.0018	17	2.0×10^{20}	80.	CO, HCN, HCO$^+$, N$_2$H$^+$, CS...
3C 293	0.045	0.045	3	1.5×10^{19}	40.	CO, HCN, HCO$^+$
4C 31.04	0.06	0.06	2	1.0×10^{19}	120.	CO, HCN, HCO$^+$
PKS1413+135	0.247	0.247	2	4.6×10^{20}	2.	CO, HCN, HCO$^+$, HNC
B3 1504+377	0.673	0.673	2	1.2×10^{21}	75.	CO, HCN, HCO$^+$, HNC
B 0218+357	0.685	0.94	1	4.0×10^{23}	20.	CO, HCN, HCO$^+$, H$_2$O, NH$_3$, H$_2$CO
PMN J0134-0931	0.765	2.22	3	–	100.	OH
PKS1830-211	0.885	2.51	2	4.0×10^{22}	40.	CO, HCN, HCO$^+$, N$_2$H$^+$, CS...

[1] Redshift of absorption lines

[2] Redshift of background continuum source

[3] Number of components in absorption

[4] Maximum H$_2$ column density over components

[5] Maximum velocity width

phenomenon is occurring also for M82 in a lesser extent). The absorption is however completely detached for the high density tracers, like HCO$^+$ and HCN (Wiklind and Combes 1997a). The absorption extends over very broad wings, suggesting perturbed kinematics, or outflows.

Recently, very broad absorption extending to the blue wing was observed in the HI line towards 3C 293 by Morganti et al. (2003): the total width of 1400 km/s absorption implies neutral gas entrained in the radio jet towards the observer. This observation is confirming theoretical expectations of AGN feedback on the interstellar medium of the host galaxies. The 3C 293 host galaxy has already been observed in the CO line, and found quite rich in molecular gas (Evans et al. 1999). Garcia-Burillo et al. (2006) have observed this strong radio source at 1 mm and 3 mm with the IRAM interferometer, and found several absorption components, in CO, HCO$^+$ and HCN. The high resolution helps to disentangle absorption from emission in the nucleus. The shape of the CO emission map suggests an interaction between the jet and the ISM, able to redirect the jet, and produce the HI outflow. The molecular lines are however not as broad as the HI line. The HCO$^+$ absorption has not only a component in front of the nucleus, but also in front of the radio jet. Strong HCO$^+$ and CO absorptions are also detected in front of 4C 31.04, clearly on the blue-side of the total spectrum, delineated by emission.

2 Higher redshift absorptions

After the first system PKS 1413+135, another internal absorption was detected with several components in B3

1504+377 (Wiklind and Combes 1996a). Then the absorption was detected in intervening systems, which amplifies the background quasar by lensing effects (B0218 +357, Wiklind and Combes 1995; Menten and Reid 1996; Gerin et al. 1997); in front of PKS1830-211, the redshift of the lens was found by sweeping the band over 14 GHz. This meant observing with 14 tunings, before detecting 2 absorption lines and determining unambiguously the redshift (Wiklind and Combes 1996b). The third gravitationally lensed quasar is PMN J0134-0931, detected only in the OH lines, but not in CO or HCO$^+$ (Kanekar et al. 2005).

The absorbing redshifts range up to $z \sim 1$ (the background quasar up to $z \sim 2$), and it becomes difficult to find higher redshift radio sources, that are strong enough in the millimeter domain. The synchroton spectrum is frequently steep, and the intensity fades at high frequency. This means that the K-correction plays a very negative role here. In addition, the number of quasars per comoving volume is expected to decrease after $z = 2$. A solution is to follow the high-z quasars and their emission in red-shifting our observations, down to the centimeter domain.

With the present instrumentation at IRAM, a full search was undertaken with selection of candidates as:

1—strong mm source (at least 0.15 Jy at 3 mm), about 150 sources, searching at the host redshift when known,

2—or at a different z, if already an absorption is detected in HI-21 cm, or DLAs, or MgII or CaII (e.g. Carilli et al. 1993),

3—all mm-strong radio-source with a known gravitational lens (VLBI) (Webster et al. 1995; Stickel and Kühr 1993; Jackson and Browne 2007).

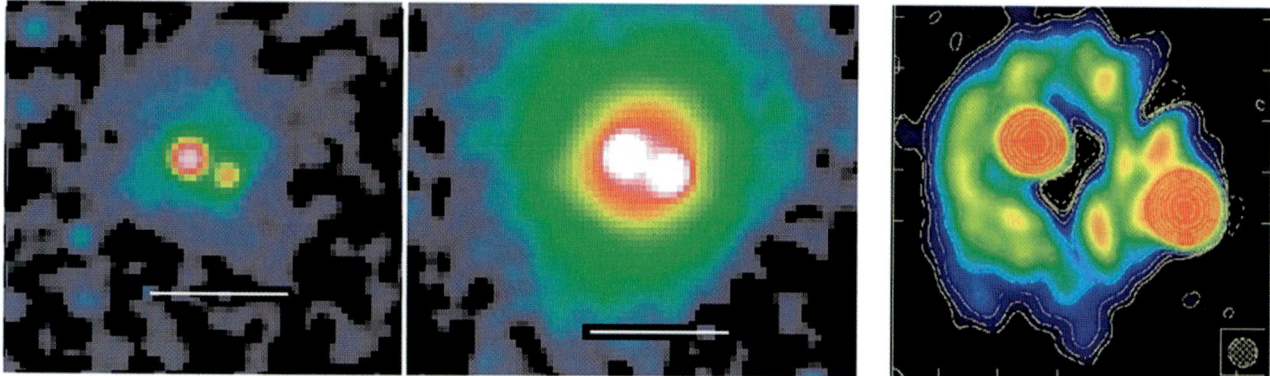

Fig. 1 The B0218+357 gravitational lens imaged by the HST in 2 bands, V (*left*) and H (*middle*, Jackson et al. 2000 and the CASTLES collaboration; in the *left* and *middle panels*, the *white bar* is 1 arcsec), and in radio with JIVE (*right*, Biggs et al. 1999). The distance between the two gravitational images is 0.335 arcsec

This survey led to mostly negative results, meaning that a much larger sample of radio-sources, and much more sensitivity is required to find more molecular absorption systems.

2.1 Absorption in the quasar host

The BL Lac PKS1413+135 at $z = 0.247$ is an edge-on galaxy, and the nucleus is obscured by Av > 30 mag (McHardy et al. 1994). On the line of sight, a very narrow absorption <1 km/s has been found: since the continuum source is highly variable, it was possible to probe the small scale structure of the interstellar medium (Wiklind and Combes 1997b).

Towards B3 1504+377 at $z = 0.672$, 7 different molecular absorption lines are detected, with a large separation 330 km/s, which could be explained by a highly noncircular motions in the center, with a more regular spiral arm in the outer parts. The observed HNC/HCN absorption ratio implies thermalization of the gas, with the excitation temperature equal to the kinetic one. As is frequently observed in molecular absorptions, the HCO^+ is enhanced by 10–100, which can only be explained by a combination of a diffuse and a clumpy medium (e.g. Lucas and Liszt 1994, 1998).

2.2 Absorption in the intervening lens galaxy

2.2.1 B0218+357

B0218+357 is amplified by a gravitational lens at $z = 0.685$: the source is split in 2 main images A and B, with an Einstein ring (cf Fig. 1). In VLBI, the A and B components reveal a detailed structure, with two bright cores and extended radio jet components (Biggs et al. 2003). It is the absorber with the largest column density around 10^{24} cm^{-2} at maximum. All three CO isotopic lines up to $C^{18}O$ are optically thick (Combes and Wiklind 1995).

This has allowed the search of many molecules, and in particular important ones undetected in our Galaxy due to atmospheric absorption at $z = 0$. Search for O_2 lines at 56, 119, 368 and 424 GHz in the rest frame have led to upper limits O_2/CO $< 2 \times 10^{-3}$ (Combes and Wiklind 1995; Combes et al. 1997), suggesting that most of the oxygen should be in the form of OI. The H_2O molecule at 557 GHz has been detected, and tentatively LiH at 444 GHz in the rest frame (Combes and Wiklind 1997, 1998), with $H_2O/H_2 = 10^{-5}$ and LiH/$H_2 \sim 3 \times 10^{-12}$. NH$_3$ has been detected at 2 cm (Henkel et al. 2005).

Recent deep HST imagery reveals the lensing galaxy, almost face-on, with spiral arms complicating the lens analysis (York et al. 2005). Due to extinction, the distance between the two images A and B, is 317 mas in optical, while 335 mas in radio. Monitoring the time-delay, together with a lensing model, taking into account the spiral arms, leads to an estimation of the Hubble constant of H$-0 = 70$ km/s/Mpc (while 61 km/s/Mpc if spiral arms are masked out).

2.2.2 PKS1830-211

Towards PKS1830-211, the lensing galaxy at $z = 0.88582$ splits the background source in two images A and B, each absorbed by a different velocity component (Frye et al. 1997; Wiklind and Combes 1998). The intrinsic temporal variability allows to monitor the time delay between the two components.

Even without resolving spatially the two images, it is possible to follow the intensity ratio between the two, since they are absorbing at two different velocities. The IRAM monitoring during 3 years (1 h per week) led to a time delay of 24 ± 5 days, and estimation of H$_0 = 69 \pm 12$ km/s/Mpc (Wiklind and Combes 1999).

The study of a large variety of molecules allows to tackle the evolution of chemical conditions. There does not seem to

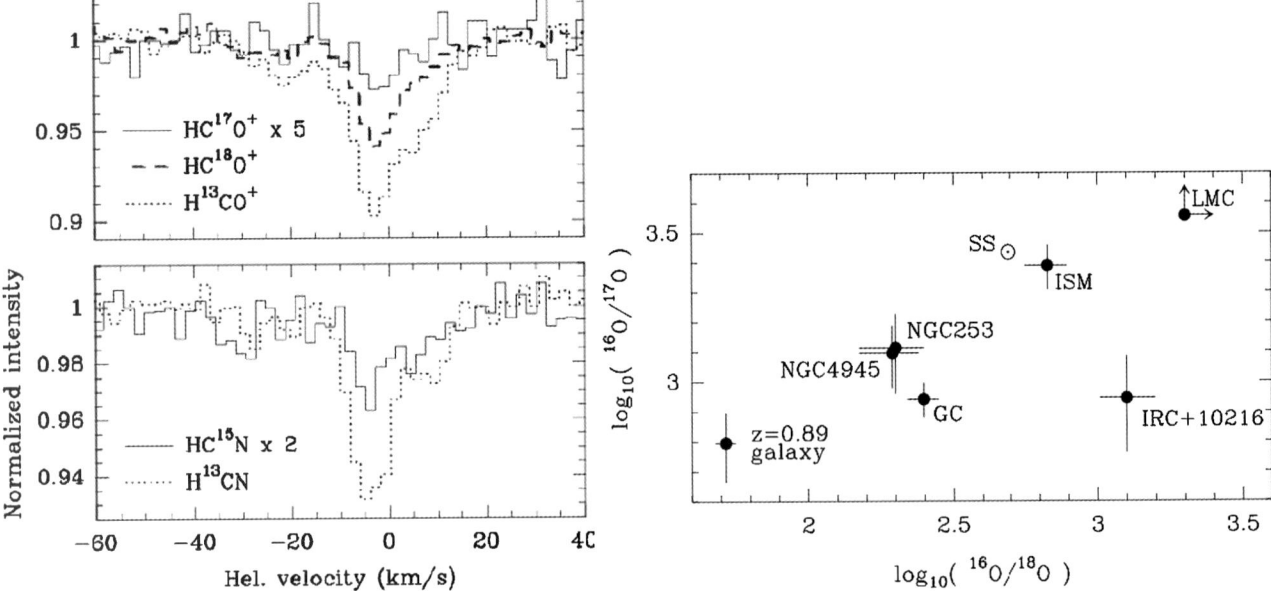

Fig. 2 (*Left*) Molecular absorption lines including isotopes, like ^{15}N, ^{17}O or ^{18}O, in PKS1830-211 by Müller et al. (2006). (*Right*) Oxygen isotopic ratios ^{16}O/^{17}O versus ^{16}O/^{18}O for different sources. The rel- ative abundances of the oxygen isotopes at $z = 0.89$ suggests that en- richment by low mass stars had not yet time to dominate in this young lens galaxy, in front of PKS1830-211

be variations at high z in comparison with $z = 0$ absorptions, but there is a large scatter, even locally (Lucas and Liszt 1994; Liszt et al. 2006).

Upper limits were reported for deuterated molecules (Shah et al. 1999) and for CI (Gerin et al. 1997). A recent survey with the IRAM interferometer of several isotopes (C, N, O or S) begins to find evidence for abundance evolution (Müller et al. 2006, Fig. 2).

Only low excitation diffuse gas is observed on the line of sight of PKS1830-211, the volumic density is so low that $T_{ex} \sim T_{CMB}$. The observation of several lines of the ro- tational ladder of the same molecule (HCN, HNC, N_2H^+, $H^{13}CO^+$, CS...) can then lead to a measure of T_{CMB}. Mil- limeter absorptions can then complement the measurement of $T_{CMB}(z)$ obtained from UV H_2 lines (Srianand et al. 2000; Reimers et al. 2003; Cui et al. 2005).

2.2.3 PMN J0134-0931

This recent absorber has been detected in HI and OH lines at GBT (Kanekar et al. 2005), from the $z = 0.7645$ lens in front of the background quasar at $z = 2.22$. The latter is split in 6 apparent images. Surprisingly, only upper limits of HCO$^+$ or H_2CO lines were obtained on this source, probably due to small-scale structure of the ISM, and very different contin- uum source extent across the radio spectrum. The absorption system provides a good probe of the fundamental constant variation.

2.3 Variations of constants

Although laboratory measurements and solar system obser- vations (e.g. Olive et al. 2002; Uzan 2003) do not show ev- idence for time variations of the fundamental constant α, high-z observations have revealed a possible variation over larger time-scales and also with space (Webb et al. 2001; Murphy et al. 2003) with some controversy (Chand et al. 2006; Tzanavaris et al. 2006).

While from Alkali Doublet (CIV, SiII, SiIV, MgII, AlIII, ...) on 22 absorbing systems and the "many-multiplet" method on 143 systems, a positive result $\Delta\alpha/\alpha = (-0.6 \pm 0.1) \times 10^{-5}$ has been claimed (Murphy et al. 2003), this has not been confirmed with the same method, $\Delta\alpha/\alpha = (-0.05 \pm 0.2) \times 10^{-5}$ by Chand et al. (2006), but see Mur- phy et al. (2006). Independent methods with radio lines are then welcome to better understand the systematics of the various techniques.

The radio domain has the big advantage of heterodyne techniques, with a spectral resolution of 10^6 or more, and dealing with cold gas and narrow lines. Also different con- stants can be probed, while comparing the optical lines with the HI 21 cm, the OH 18 cm and CO or HCO$^+$ rotational lines, which depend very differently on α, the electron- proton mass ratio $\mu = m_e/m_p$, or the proton gyromagnetic ratio g_p. Note that Ubachs and Reinhold (2004) and Rein- hold et al. (2006) respectively put bounds and report an in- dication of cosmological variation of μ based on labora- tory measurement and reanalysis of H_2 spectra of $\Delta\mu/\mu = (2.0 \pm 0.6) \times 10^{-5}$.

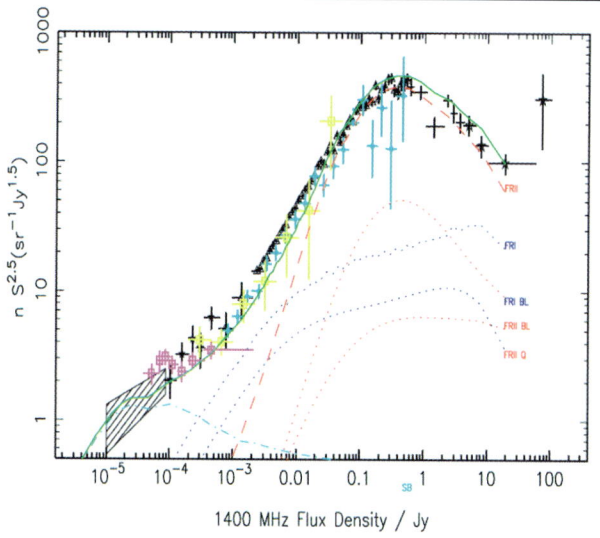

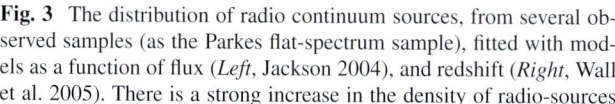

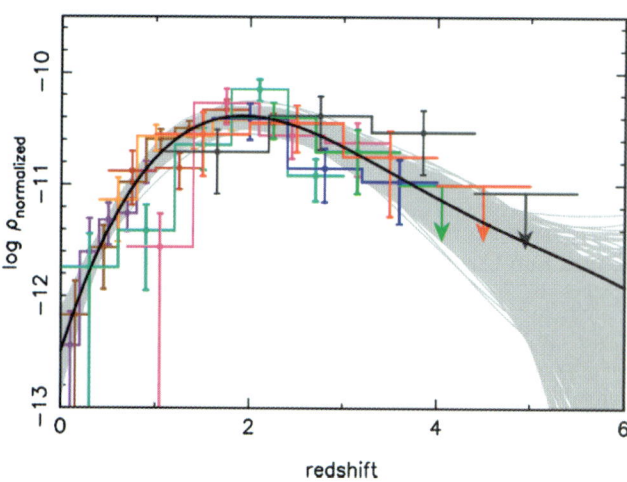

Fig. 3 The distribution of radio continuum sources, from several observed samples (as the Parkes flat-spectrum sample), fitted with models as a function of flux (*Left*, Jackson 2004), and redshift (*Right*, Wall et al. 2005). There is a strong increase in the density of radio-sources until $z = 2$, which translates into a strong increase in density of sources with decreasing flux, above what is expected for the euclidean count (falloff dn/dS as $S^{-5/2}$)

In PKS1413+135, a resolution of 40 m/s is required to resolve the lines, and the obtained upper limits for variations on $y = \alpha^2 g_p \mu$ are $\Delta y/y = (-0.20 \pm 0.20) \times 10^{-5}$ and $\Delta y/y = (-0.16 \pm 0.36) \times 10^{-5}$ for B0218+357 (Murphy et al. 2001). The main systematics is the kinematical bias, i.e. that the different lines do not come exactly from the same material along the line of sight, with the same velocity. Statistics with absorptions of HI and HCO$^+$ in our own Galaxy in front of remote quasars (Lucas and Liszt 1998) have measured a dispersion of about 1.2 km/s, corresponding to $\Delta y/y = 0.4 \times 10^{-5}$. The results combining lines in PMN J0134-0931 and B0218+357 on $F = g_p[\alpha^2/\mu]^{1.57}$ are $\Delta F/F = (0.44 \pm 0.36^{stat} \pm 1.0^{syst}) \times 10^{-5}$ for $0 < z < 0.7$ (with statistical and systematical errors separated). No variation is detected, while the sensitivity at 2σ on the α variation is $\Delta\alpha/\alpha \sim 6.7 \times 10^{-6}$, and on the mass ratio $\Delta\mu/\mu \sim 1.4 \times 10^{-5}$ over half of the age of the universe (Kanekar et al. 2005). It is then needed to find much more sources with ALMA.

3 Perspectives

The number of molecular absorptions so far (5 at high-z) and also the number of HI-21 cm absorbers (about 50 for $z > 0.1$) is surprisingly low. Why so few radio absorbers? One explanation, at least for the molecular absorptions, is that the high column density expected obscures the background quasars, introducing a strong bias against the optical detection of these remote sources. At least some could be known only in radio, but with no redshift available. Curran

et al. (2006) have noticed a strong correlation between the molecular fraction and the red colors of the quasars. Those with molecular absorptions are in general compact flat spectrum sources, where most of the emission is covered. Future searches should concentrate on sub-DLA systems, where the H_2 fraction is higher, as well as metals (Khare et al. 2006; Kulkarni et al. 2006).

Typical first projects with ALMA could be (included in the DRSP):

1— Molecular survey of PKS1413, PKS1830, CenA, in 7 wide priority bands, with spectral resolution of 1–4 km/s;

2— Search for new systems, towards 60 selected radio loud AGNs with mm cont flux >50 mJy, with criteria of obscuration, gravitational lensing and/or suppressed soft X-ray flux. When no redshift is known, the search could be over the entire redshift range using the technique of frequency scanning.

As shown in Fig. 3, it is now well-known that the volumic density of radio quasars peaked around $z = 2$ (Shaver et al. 1996; Wall et al. 2005), and there is a cutoff after $z = 3$. Optical quasars follow the same curve, in a similar way to the star formation history. In parallel, the number of sources as a function of flux N(S) increases well above the euclidean curve in $S^{-1.5}$, and we could expect to detect 1 or 2 orders of magnitude more quasars with ALMA. However at high-z, their millimeter flux is weakened by the non-favorable K-correction (compact and flat-spectrum sources being rare). In this domain, it is interesting to search 3 mm systems at cm wavelengths, with Band 1 and 2 of ALMA in the future.

References

Biggs, A.D., Browne, I.W.A., Helbig, P., et al.: Mon. Not. R. Astron. Soc. **304**, 349 (1999)

Biggs, A.D., Wucknitz, O., Porcas, R.W., et al.: Mon. Not. R. Astron. Soc. **338**, 599 (2003)

Carilli, C.L., Rupen, M.P., Yanny, B.: Astrophys. J. **412**, L59 (1993)

Chand, H., Srianand, R., Petitjean, P., et al.: Astron. Astrophys. **451**, 45 (2006)

Combes, F., Wiklind, T.: Astron. Astrophys. **303**, L61 (1995)

Combes, F., Wiklind, T.: Astrophys. J. **486**, L79 (1997)

Combes, F., Wiklind, T.: Astron. Astrophys. **334**, L81 (1998)

Combes, F., Wiklind, T., Nakai, N.: Astron. Astrophys. **327**, L17 (1997)

Cui, J., Bechtold, J., Ge, J., Meyer, D.M.: Astrophys. J. **633**, 649 (2005)

Curran, S.J., Whiting, M.T., Murphy, M.T., et al.: Mon. Not. R. Astron. Soc. **371**, 431 (2006)

Evans, A., Sanders, D., Surace, J., Mazzarella, J.: Astrophys. J. **511**, 730 (1999)

Frye, B., Welch, W.J., Broadhurst, T.: Astrophys. J. **478**, L25 (1997)

Garcia-Burillo, S., Combes, F., Usero, A., et al.: Astron. Astrophys. (2006, in preparation)

Gerin, M., Phillips, T.G., Benford, D.J., et al.: Astrophys. J. **488**, L31 (1997)

Henkel, C., Jethava, N., Kraus, A., et al.: Astron. Astrophys. **440**, 893 (2005)

Jackson, C.: New Astron. Rev. **48**, 1187 (2004)

Jackson, N., Browne, I.W.A.: Mon. Not. R. Astron. Soc. **374**, 168 (2007)

Jackson, N., Xanthopoulos, E., Browne, I.W.A.: Mon. Not. R. Astron. Soc. **311**, 389 (2000)

Kanekar, N., Carilli, C.L., Langston, G.I., et al.: Phys. Rev. Lett. **95**, z1301 (2005)

Khare, P., Kulkarni, V.P., Peroux, C., et al.: Astron. Astrophys. (astro-ph/0608127) (2006)

Kulkarni, V.P., Khare, P., Peroux, C., et al.: Astrophys. J. Lett. (astro-ph/0608126) (2006)

Lucas, R., Liszt, H.: Astron. Astrophys. **282**, L5 (1994)

Lucas, R., Liszt, H.: Astron. Astrophys. **337**, 246 (1998)

Liszt, H., Lucas, R., Pety, J.: Astron. Astrophys. **448**, 253 (2006)

McHardy, I., Merrifield, M., Abraham, R., Crawford, C.: Mon. Not. R. Astron. Soc. **268**, 681 (1994)

Menten, K.M., Reid, M.J.: Astrophys. J. **465**, L99 (1996)

Morganti, R., Oosterloo, T.A., Emonts, B.H.C., et al.: Astrophys. J. **593**, L69 (2003)

Müller, S., Guélin, M., Dumke, M., Lucas, R., Combes, F.: Astron. Astrophys. **458**, 417 (2006)

Murphy, M.T., Webb, J.K., Flambaum, V.V., et al.: Mon. Not. R. Astron. Soc. **327**, 1244 (2001)

Murphy, M.T., Webb, J.K., Flambaum, V.V.: Mon. Not. R. Astron. Soc. **345**, 609 (2003)

Murphy, M.T., Webb, J.K., Flambaum, V.V.: astro-ph/0612407 (2006, submitted)

Olive, K.A., Pospelov, M., Qian, Y.-Z., et al.: Phys. Rev. D **66**, d5022 (2002)

Reimers, D., Baade, R., Quast, R., Levshakov, S.A.: Astron. Astrophys. **410**, 785 (2003)

Reinhold, E., Buning, R., Hollenstein, U., Ivanchik, A., Petitjean, P., Ubachs, W.: Phys. Rev. Lett. **96**, 151101 (2006)

Shah, R.Y., Wootten, A., Mangum, J.G., et al.: Astron. Soc. Pacific. Conf. Ser. **156**, 233 (1999)

Shaver, P.A., Wall, J.V., Kellermann, K.I., et al.: Nature **384**, 439 (1996)

Srianand, R., Petitjean, P., Ledoux, C.: Nature **408**, 931 (2000)

Stickel, M., Kühr, H.: Astron. Astrophys. Suppl. Ser. **101**, 521 (1993)

Tzanavaris, P., Murphy, M.T., Webb, J.K., et al.: Mon. Not. R. Astron. Soc. **374**, 634 (2006)

Ubachs, W., Reinhold, E.: Phys. Rev. Lett. **92**, 101302 (2004)

Uzan, J.-P.: Rev. Mod. Phys. **75**, 403 (2003)

Wall, J., Jackson, C., Shaver, P., Hook, I., Kellermann, K.: Astron. Astrophys. **434**, 133 (2005)

Webb, J.K., Murphy, M.T., Flambaum, V.V., et al.: Phys. Rev. Lett. **87**, i1301 (2001)

Webster, R.L., Francis, P.J., Peterson, B.A., et al.: Nature **375**, 469 (1995)

Wiklind, T., Combes, F.: Astron. Astrophys. **286**, L9 (1994)

Wiklind, T., Combes, F.: Astron. Astrophys. **299**, 382 (1995)

Wiklind, T., Combes, F.: Astron. Astrophys. **315**, 86 (1996a)

Wiklind, T., Combes, F.: Nature **379**, 139 (1996b)

Wiklind, T., Combes, F.: Astron. Astrophys. **324**, 51 (1997a)

Wiklind, T., Combes, F.: Astron. Astrophys. **328**, 48 (1997b)

Wiklind, T., Combes, F.: Astrophys. J. **500**, 129 (1998)

Wiklind, T., Combes, F.: astro-ph/9909314 (1999)

York, T., Jackson, N., Browne, I.W.A., et al.: Mon. Not. R. Astron. Soc. **357**, 124 (2005)

Molecular signature of star formation at high redshifts

Serena Viti · Chris J. Lintott

Originally published in the journal Astrophysics and Space Science, Volume 313, Nos 1–3.
DOI: 10.1007/s10509-007-9631-4 © Springer Science+Business Media B.V. 2007

Abstract In recent years there has been much debate, both observational and theoretical, about the nature of star formation at high redshift. In particular, there seems to be strong evidence of a greatly enhanced star formation rate early in the Universe's evolution. Simulations investigating the nature of the first stars indicate that these were large, with masses in excess of 100 solar masses. By the use of a chemical model, we have simulated the molecular signature of massive star formation for a range of redshifts, using different input models of metallicity in the early Universe. We find that, as long as the number of massive stars exceeds that in the Milky Way by factor of at least 1000, then several 'hot-core' like molecules should have detectable emission. Although we predict that such signatures should already be partly detectable with current instruments (e.g. with the VLA), facilities such as ALMA will make this kind of observation possible at the highest redshifts.

Keywords Astrochemistry · Cosmology: early universe · Stars: formation

1 Introduction

There is currently great interest among astronomers in the nature of star formation in the early Universe. Observations of high redshift galaxies ($z \sim 2$ or so) suggest that the star formation rate was much larger than it is now (Smail et al. 2002). Simulations (e.g. Abel et al. 2002) of the formation of the first generation of stars from the cold neutral atomic hydrogen/helium gas in the post-recombination pre-stellar Universe suggest that the earliest stars were very massive ($\sim$100 solar masses) and evolved very rapidly. These first stars would have had two important effects in the early Universe: (i) they partially ionised the surrounding gas, and (ii) they led to a metal enrichment. Therefore, second generation stars formed in gas that was enriched in metals affected significantly the thermodynamic properties of the gas in the star-forming regions, and hence, the size, distribution and rate of formation of those stars.

It is of course currently nearly impossible to obtain direct observational information about the crucial initial step. However, the second generation of stars should be observable in galaxies at high redshift and observations of the second generation stars in distant galaxies may be used to infer the nature of the first generation of stars.

Indeed, it may be possible to approach the problem of star formation at high redshift by using our current understanding of the formation of massive stars in our own Milky Way and of their impact on the star forming regions in which they are embedded. This understanding has been gained from observations of molecular line emissions from such regions, and demonstrates that we can infer both the history of the star-formation process and details of the local physical conditions from the molecular line observations (e.g. Beuther et al. 2005; Hatchell et al. 1998).

In this article we briefly summarize recent work on the investigation of extragalactic massive star formation and show that the study of massive star formation in high redshift galaxies may be a novel tool for the exploration of the phys-

S. Viti (✉)
Physics and Astronomy Department, University College London, Gower Street, London, WC1E 6BT, UK
e-mail: sv@star.ucl.ac.uk

C.J. Lintott
Oxford Astrophysics, Denys Wilkinson Building, Keble Road, Oxford, OX2 6HD, UK
e-mail: cjl@astro.ox.ac.uk

ical conditions in those galaxies and of the nature of the formation of the first generation of stars.

2 Massive star formation beyond the Milky Way

The formation of stars occurs within dense cores inside molecular clouds. Due to the presence of dust, the presence of newly formed massive stars can only be detected by radio emissions from ultra-compact HII regions (UCHIIs), dense regions of less than a light year across immediately surrounding the newly formed star, by infrared and sub-millimetre emissions from dust close to the star outside the UCHII, and by the mm and sub-mm line emissions from a variety of molecular species that have been warmed by the nearby star. The dense ($\sim 10^7$ cm^{-3}), warm (~ 300 K), transient ($\sim 10^5$ years) compact (~ 0.03 pc) regions of molecular line emissions are called 'hot cores'. It is particularly through the molecular line emissions that the history of the star formation process and the description of the current physical conditions in the hot core can be obtained (e.g. Beuther et al. 2005; Hatchell et al. 1998; Viti et al. 2004).

The picture of the formation of a massive star in the Galaxy that emerges from such studies of hot cores is as follows. During the collapse of a star forming core, the increased gas number density promotes a more rapid chemistry in the gas phase (driven mainly by cosmic ray ionisation) and an enhanced interaction of gas phase species with dust grain surfaces. As long as the core temperature remains low, most species are deposited on dust grains and molecular ices accumulate. By the end of the collapse phase nearly all the gas phase material, other than H$_2$ and He, is in the ice (Brown et al. 1988; Caselli et al. 1998). Since the gas phase chemical composition is time dependent, the ice composition represents an integrated chemical tracer of the collapse history.

Once the star turns on, the dust temperature begins to rise and thermal desorption of molecules from the ices begins; thermal activation of solid-state chemistry may also occur. With an appropriate model of the dynamical, thermal and chemical processes as well as multi-species molecular observations of hot cores, then a description of the collapse and of the local physical conditions can be inferred.

Over the last decade, sophisticated models of hot cores, including both the preceding collapse phase and the subsequent warm-up phase have been developed (e.g. Brown et al. 1988; Caselli et al. 1998; Viti et al. 2004; Garrod and Herbst 2006). We adapted one such model of hot cores in our own Galaxy (Viti et al. 2004) to simulate the chemical evolution of high mass star forming regions at higher redshifts to address the following question: can molecular emission associated with massive star forming regions be detectable in distant systems (Lintott et al. 2005)?

The motivation for that study came from the recent molecular observations at high redshift. in particular, the CO detection at $z = 6.4$ in J1148+5251 (Walter et al. 2003; Bertoldi et al. 2003). Such emission indicates a large mass of molecular gas already at $z = 6$ and implies that the star formation rate is high (~ 3000 M$_\odot$ yr^{-1}). In Lintott et al. (2005) we investigated whether the signatures of hot cores in low metallicity environments may be strong enough to be detectable and we predicted the main chemical tracers of such activity. In contrast to galactic massive star formation simulations, these models used as initial elemental abundances the yields from the progenitor (1st generation) stars. Lintott et al. (2005) used three different models of yields from a zero-metallicity star, namely, the Chieffi and Limongi (2002) model of a progenitor of 80 M$_\odot$, the Umeda and Nomoto (2002) models of progenitors of pair instability supernovae with masses from 150 to 270 M$_\odot$ and the Heger and Woosley (2002) model a 80 M$_\odot$ helium core. The metallicity and the gas to dust ratios were also reduced appropriately and a range of values were used. The output of these chemical models gave the fractional abundances of a large number of molecules as a function of time. An example of the resulting chemical abundances is given in Table 1 for a range of metallicities for *one* hot core. Using the chemical abundances from Table 1 we inferred that if 10^7–10^8 hot cores are present in an unresolved high redshift source then many of the molecular species reach significant column densities. Hence one conclusion of that study was that observational searches for multiple molecular lines associated with massive star formation should be done and that comparisons of such observations with extragalactic star formation models may help constraining models of the first stars. It is clear that future facilities such as ALMA will provide a huge increase in our capacity to carry out such observations.

At the moment, however, it is certainly true that, observationally, it is a challenge to detect molecules other than CO, and possibly HCN, at high redshift. Hence, local starbursts may provide us with good local analogues to distance sources. In fact, M82, the closest starburst galaxy, has been extensively observed in several molecular lines (e.g. Mao et al. 2000; Weiß et al. 2001; García-Burillo et al. 2002; Martín et al. 2006) and, although most of the emission can be explained as arising from photon-dominated regions within the galaxy, some of the molecules, such as CH$_3$OH, can only be accounted for by the presence of dense gas.

2.1 Can models of star formation from near-primordial gas account for the observed HCN abundance?

Another important question that can be tackled by the use of chemical models is the exploration of the effect of AGNs on high-redshift star-forming systems (Lintott and Viti 2006). There is ample evidence of a strong correlation between the

Fig. 1 Change in abundance for HCN for different models varying in cosmic-ray ionization rates. Taken from Lintott and Viti (2006)

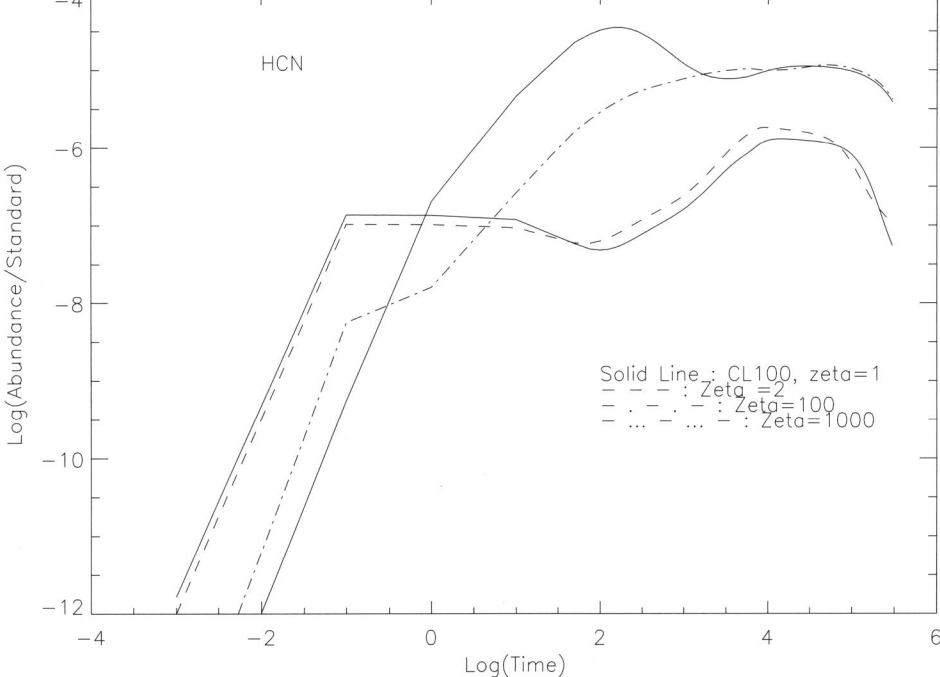

luminosity of the HCN (1–0) transition and the star formation rate in the Milky Way as well as external galaxies (Gao and Solomon 2004; Wu et al. 2005). However, this correlation seems to break down at high redshift, as the HCN luminosity appears to be lower than expected. The conventional explanation for this is that observed infrared excess is due to AGN activity. It is possible that star formation activity could equally mimic this effect. In fact, Lintott and Viti (2006) show that AGN activity should lead to an *increase* in the HCN abundance (up to 2 orders of magnitude) rather than a decrease, a result which confirms previous studies (Lepp and Dalgarno 1996). Moreover, they show that if, on the other hand, the gas is near-primordial, that is enriched only by large zero-metallicity stars, then it may well be under-abundant in nitrogen and hot cores formed by such gas will have a lower abundance of nitrogen-bearing compounds. Lintott and Viti (2006) found that their models were consistent with observations of the Cloverleaf quasar and J1409 if $\sim 10^7$ are present in the beam and yield column densities of $\sim 10^{17}$ and $\sim 10^{16}$ for the Cloverleaf and J1409 respectively, consistent with the observed HCN luminosities.

3 Discussion and conclusions

One of the limitations of using such chemical models is that the formation and properties of hot cores in high redshift galaxies have not as yet been explored. The physical parameters in gas at high redshift are likely to be very different from those in the Milky Way Galaxy.

Table 1 Calculated column densities for a variety of chemical species in cm^{-2} for *one* hot core assuming initial abundances as in CL02 and for different (reduced from solar) metallicities (Adapted from Table 2 of Lintott et al. 2005). The notation $a(b)$ stands for $a \times 10^b$

Species	1/100	1/500	1/1000
CO	6.5(15)	1.8(15)	1.1(15)
H2CO	2.9(14)	9.3(13)	4.2(13)
H2S	1.5(09)	2.3(09)	3.9(09)
CS	1.2(15)	5.7(14)	3.1(14)
H2CS	6.9(14)	1.1(14)	3.4(13)
SO	7.9(14)	7.8(12)	2.7(12)
SO2	7.1(14)	2.1(12)	3.7(11)
CH3OH	1.1(12)	3.1(10)	9.6(09)
HCN	7.6(10)	1.1(11)	3.9(11)

For example, the dust-to-gas ratio in the Milky Way is such that the optical extinction associated with a hot core is normally much larger than 100 visual magnitudes. In galaxies at high redshift, the metallicity and the dust abundances are likely to be much lower, implying that molecular abundances and visual extinctions may also be lower. Both of these parameters affect the initial collapse of the cloud, mainly through the thermal balance and ionization level, forming a star and its associated hot core. In high redshift galaxies, the cosmic microwave background temperature will be considerably higher than in the Milky Way, and may affect the formation of ices and desorption of weakly-bound molecules (such as CO) from ices.

High rates of star formation imply high ambient radiation fluxes. The high radiation fields may not affect molecules in a hot core if the dust shielding is adequate, but the enhanced cosmic ray flux will certainly drive chemistry more quickly (if that flux is not too great). Elemental abundances will certainly be lower than in the Milky Way Galaxy, but relative abundances are also likely to differ. If so, then the variety of molecules produced in the collapse phase will differ, and so may the physical and chemical nature of dust. This in turn may affect the dynamics of the collapse that leads to the stars as well as the timescales and hence the star formation rate.

In conclusion, the macroscopic process is crucially affected by the microscopic properties in the gas, including chemistry in the gas phase, on surfaces and within ices on interstellar dust grains, adsorption and desorption from ices, and local heating and cooling processes. The application of ideas about the formation of massive stars in the Galaxy to the very different conditions pertaining in high redshift galaxies requires a close collaboration between observational and theoretical astronomers, astrochemists, laboratory and theoretical chemists, and cosmologists.

Acknowledgements SV acknowledges financial support from an individual PPARC Advanced Fellowship.

References

Abel, T., Bryan, G.L., Norman, M.L.: Science **295**, 93 (2002)

Bertoldi, F., et al.: Astron. Astrophys. **409**, L47 (2003)

Beuther, H., Thorwirth, S., Zhang, Q., Hunter, T.R., Megeath, S.T., Walsh, A.J., Menten, K.M.: Astrophys. J. **627**, 834 (2005)

Brown, P.D., Charnley, S.B., Millar, T.J.: Mon. Not. R. Astron. Soc. **231**, 409 (1988)

Caselli, P., Walmsley, C.M., Terzieva, R., Herbst, E.: Astrophys. J. **499**, 234 (1998)

Chieffi, A., Limongi, M.: Astrophys. J. **577**, 281 (2002)

Gao, Y., Solomon, P.M.: Astrophys. J. **606**, 271 (2004)

García-Burillo, S., Martín-Pintado, J., Fuente, A., Usero, A., Neri, R.: Astrophys. J. Lett. **575**, 55 (2002)

Garrod, R.T., Herbst, E.: Astron. Astrophys. **457**, 927 (2006)

Hatchell, J., Thompson, M.A., Millar, T.J., MacDonald, G.H.: Astron. Astrophys. **338**, 713 (1998)

Heger, A., Woosley, S.E.: Astrophys. J. **567**, 532 (2002)

Lepp, S., Dalgarno, A.: Astron. Astrophys. **306**, L21 (1996)

Lintott, C.L., Viti, S., Williams, D.A., Rawlings, J.M.C., Ferreras, I.: Mon. Not. R. Astron. Soc. **360**, 1527 (2005)

Lintott, C.L., Viti, S.: Astrophys. J. **646**, L37 (2006)

Mao, R.Q., Henkel, C., Schulz, A., Zielinsky, M., Mauersberger, R., Störzer, H., Wilson, T.L., Gensheimer, P.: Astron. Astrophys. **358**, 433 (2000)

Martín, S., Martín-Pintado, J., Mauersberger, R.: Astron. Astrophys. **450**, L13 (2006)

Smail, I., Ivison, R.J., Blain, A.W., Kneib, J.-P.: Mon. Not. R. Astron. Soc. **331**, 495 (2002)

Umeda, H., Nomoto, K.: Astrophys. J. **565**, 385 (2002)

Viti, S., Collings, M.P., Dever, J.W., McCoustra, M.R.S., Williams, D.A.: Mon. Not. R. Astron. Soc. **354**, 1141 (2004)

Walter, F., et al.: Nature **424**, 406 (2003)

Weiß, A., Neininger, N., Henkel, C., Stutzki, J., Klein, U.: Astrophys. J. Lett. **554**, 143 (2001)

Wu, et al.: Astrophys. J. **635**, L173 (2005)

Dense molecular gas in a sample of LIRGs and ULIRGs: The low-redshift connection to the huge high-redshift starbursts and AGNs

Javier Graciá-Carpio · Santiago García-Burillo ·
Pere Planesas

Originally published in the journal Astrophysics and Space Science, Volume 313, Nos 1–3.
DOI: 10.1007/s10509-007-9629-y © Springer Science+Business Media B.V. 2007

Abstract The sample of nearby LIRGs and ULIRGs for which dense molecular gas tracers have been measured is building up, allowing for the study of the physical and chemical properties of the gas in the variety of objects in which the most intense star formation and/or AGN activity in the local universe is taking place. This characterisation is essential to understand the processes involved, discard others and help to interpret the powerful starbursts and AGNs at high redshift that are currently being discovered and that will routinely be mapped by ALMA. We have studied the properties of the dense molecular gas in a sample of 17 nearby LIRGs and ULIRGs through millimeter observations of several molecules (HCO^+, HCN, CN, HNC and CS) that trace different physical and chemical conditions of the dense gas in these extreme objects. In this paper we present the results of our HCO^+ and HCN observations. We conclude that the very large range of measured line luminosity ratios for these two molecules severely questions the use of a unique molecular tracer to derive the dense gas mass in these galaxies.

Keywords Galaxies: active · Galaxies: ISM · Galaxies: starburst · Infrared: galaxies · ISM: molecules · Radio lines: galaxies

J. Graciá-Carpio (✉) · S. García-Burillo · P. Planesas
Observatorio Astronómico Nacional, Alfonso XII, 3,
28014 Madrid, Spain
e-mail: j.gracia@oan.es

S. García-Burillo
e-mail: s.gburillo@oan.es

P. Planesas
e-mail: p.planesas@oan.es

1 Introduction

The origin (starburst and/or AGN) of the infrared luminosity in luminous and ultraluminous infrared galaxies (LIRGs: $10^{11} L_\odot \leq L_{ir} < 10^{12} L_\odot$ and ULIRGs: $L_{ir} \geq 10^{12} L_\odot$) has been the subject of intense debate since their discovery (Sanders et al. 1988; Genzel et al. 1998; Veilleux et al. 1999; Gao and Solomon 2004b). In order to derive the dominant contribution to their IR luminosities the molecular gas properties of these galaxies have been extensively analysed through millimeter observations (see the review of Sanders and Mirabel 1996). The higher star formation efficiency of the molecular gas (SFE $\propto L_{ir}/L_{CO}$) observed in LIRGs and ULIRGs compared to that found in spiral galaxies, led Sanders et al. (1991) to propose that a dust enshrouded AGN contribute significantly to L_{ir} in these galaxies. However, Solomon et al. (1992), and more recently Gao and Solomon (2004a), found that the L_{ir}/L_{HCN} luminosity ratio, considered as a measure of the SFE of the dense gas traced by the HCN(1–0) transition, is almost constant independently of L_{ir}. According to this result the dense molecular gas properties in LIRGs and ULIRGs are similar to those in normal spiral galaxies and, as a result, the contribution to L_{ir} from a dust enshrouded AGN in LIRGs and ULIRGs is not required. These conclusions strongly depend on the assumption that L_{HCN} is an unbiased tracer of dense molecular gas mass.

However, recent results have cast several doubts about the reliability of HCN as an unbiased tracer of the dense molecular gas mass in galaxies (Kohno et al. 2001; Usero et al. 2004; Kohno 2005), being LIRGs, ULIRGs and high-redshift galaxies a particular case (Graciá-Carpio et al. 2006; Imanishi et al. 2006; García-Burillo et al. 2006). The main concerns about the use of L_{HCN} as a quantitative probe of the dense gas mass come from the particular chemistry and excitation conditions of this molecule. HCN abundance can be significantly enhanced under the influence of X-ray chemistry driven by an embedded AGN (Lepp and Dalgarno 1996; Maloney et al. 1996) or in the molecular gas closely associated with high-mass star forming regions (Blake et al. 1987; Lahuis et al. 2006). In addition to that, the excitation of HCN lines might be affected by IR pumping through a 14 μm vibrational transition near strong mid-infrared sources (Aalto et al. 1995). All these effects will contribute to increase the total HCN(1–0) emission, breaking the claimed proportionality between L_{HCN} and the total dense molecular gas mass. In this context, the conclusions extracted by Gao and Solomon (2004b) about the starburst origin of the IR luminosity in LIRGs and ULIRGs can be questioned.

2 Sample selection and observations

In order to test if the HCN(1–0) emission is a fair tracer of the dense molecular gas mass we have conducted with the IRAM 30-meter telescope a dense molecular gas survey in a sample of 17 LIRGs and ULIRGs selected to cover homogeneously the L_{ir}[1] range between $10^{11.3} L_\odot$ and $10^{12.5} L_\odot$. All galaxies are located at distances larger than 50 Mpc to be confident that the total emission of the molecular gas can be measured in a single pointing (HCO$^+$(1–0) beam $\sim 28'' = 7$ kpc at 50 Mpc). Several molecules (HCO$^+$, HCN, CN, HNC and CS) and rotational transitions (J = 1–0, 3–2) were observed in 7 periods between November 2004 and November 2006. The results from our full observations will be discussed in a future paper. Here we will concentrate on our HCO$^+$ and HCN results.

3 Dense molecular gas in LIRGs and ULIRGs

In a recent article (Graciá-Carpio et al. 2006) we presented the results of our HCO$^+$ survey in LIRGs and ULIRGs and showed an intriguing trend between the HCN/HCO$^+$ J = 1–0 luminosity ratio and L_{ir}. In that paper we discussed that the observed trend could be the result of an anomalous excitation and/or chemistry of the HCN molecule (or, alternatively, of the HCO$^+$ molecule; see also Papadopoulos 2006). In Fig. 1a we show an updated version of the same plot including our recent HCN(1–0) reobservations of the Solomon et al. (1992) sample. With the addition of the new data the existence of a trend is confirmed. Independently of

the origin of this trend (see Graciá-Carpio et al. 2006 for a detailed discussion), it is evident that the properties of the dense molecular gas in LIRGs are different from those in ULIRGs. It is also clear that it is necessary to observe several dense gas tracers in order to characterise the molecular gas properties in these galaxies.

Figure 1b also shows that dense molecular gas properties seem to change with increasing L_{fir} from normal spiral galaxies to LIRGs, ULIRGs and high-redshift galaxies. We can see that the SFE of the dense gas traced by the HCN(1–0) line luminosity is significantly higher in LIRGs and ULIRGs than in normal spiral galaxies. We can interpret this result in three different ways. First, it may represent a real variation of the SFE of the molecular gas with L_{fir}; this is not surprising as the Initial Mass Function (IMF) and the Schmidt Law do not need to be the same in different extragalactic environments. Second, if we assume that the SFE of the dense gas is constant independently of L_{fir}, it may indicate that the SFE should be measured using a different tracer of the dense gas mass with a higher critical density than HCN(1–0). Third, Fig. 1b may indicate that there is an additional contribution to L_{fir} from a dust enshrouded AGN in the most IR luminous galaxies. This is probably true in the case of Palomar-Green QSOs and high-redshift galaxies (Rowan-Robinson 2000), but it is not clear in the case of LIRGs and ULIRGs. We should note that if L_{HCN} overestimates the dense molecular gas content at high IR luminosities, this would imply that the reported increase of the SFE of the dense gas as a function of L_{fir} would be even higher.

In an effort to constrain the relative abundances and excitation properties of HCN and HCO$^+$ we have studied their J = 3–2/1–0 luminosity ratios in Fig. 2a. The critical densities of HCN rotational transitions are higher than those of HCO$^+$ by a factor of ~ 6. That means that HCN and HCO$^+$ may trace different gas phases with different densities, and that if we want to model the rotational emission of these molecules we need to consider two molecular gas phases with different density and temperature. However, given the small number of line ratios available, we have taken a more simple approach to interpret our results and we have assumed that both molecules trace the same molecular gas phase (i.e. similar density and kinetic temperature). If we also assume an abundance ratio [HCN]/[HCO$^+$] ~ 1, and that collisional excitation dominates the rotational emission of these molecules, then a simple LVG analysis indicates that the HCO$^+$ molecule should be more excited (i.e. a higher J = 3–2/1–0 ratio) than HCN. In Fig. 2a we have highlighted in grey the region where HCO$^+$ (3–2)/HCO$^+$(1–0) > HCN(3–2)/HCN(1–0). About half of the galaxies fall into this region of 'expected excitation'. However, the other half fall into a region that was not expected by our simple LVG calculations. This 'unexpected excitation' can be explained if [HCN]/[HCO$^+$] ≥ 10 in

---

[1] All luminosities have been calculated assuming a flat Λ-dominated cosmology described by $H_0 = 71$ km s^{-1} Mpc^{-1} and $\Omega_m = 0.27$ (Spergel et al. 2003).

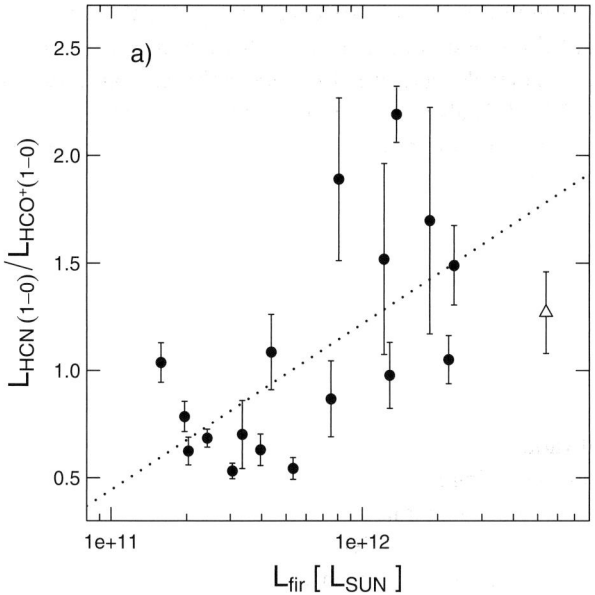

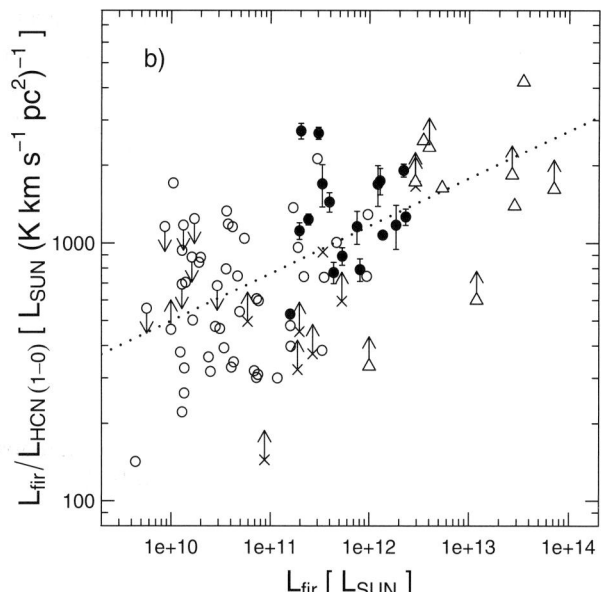

Fig. 1 a HCN(1–0)/HCO$^+$(1–0) luminosity ratio as a function of L_{fir} in our sample of LIRGs and ULIRGs (*black circles*) and the Cloverleaf quasar (*open triangle*; Solomon et al. 2003; Riechers et al. 2006). The observed trend indicates that the excitation and/or chemical properties of the dense molecular gas are different for LIRGs and ULIRGs (see Graciá-Carpio et al. 2006 for a more extended discussion). **b** L_{fir}/L_{HCN} luminosity ratio as a function of L_{fir} in a sample of normal spiral galaxies, LIRGs and ULIRGs (*open circles*; Gao and Solomon 2004a), our sample of LIRGs and ULIRGs (*black circles*), a sample of infrared-excess Palomar-Green QSOs (*crosses*; Evans et al. 2006) and a sample of high-redshift objects (*open trian-*

gles; Greve et al. 2006 and references therein). Contrary to previous results (Gao and Solomon 2004b), our new HCN(1–0) reobservations point to a significantly higher SFE of the dense gas for LIRGs and ULIRGs compared to normal less luminous spiral galaxies. Palomar-Green QSOs and high-redshift objects fall within the linear regression fit (*dotted line*) calculated for our sample of LIRGs and ULIRGs and Gao and Solomon (2004a) galaxies. The higher SFE traced by L_{HCN} in LIRGs and ULIRGs may be related to a real difference of the dense molecular gas properties in these galaxies or to an additional contribution to L_{fir} from a dust enshrouded AGN

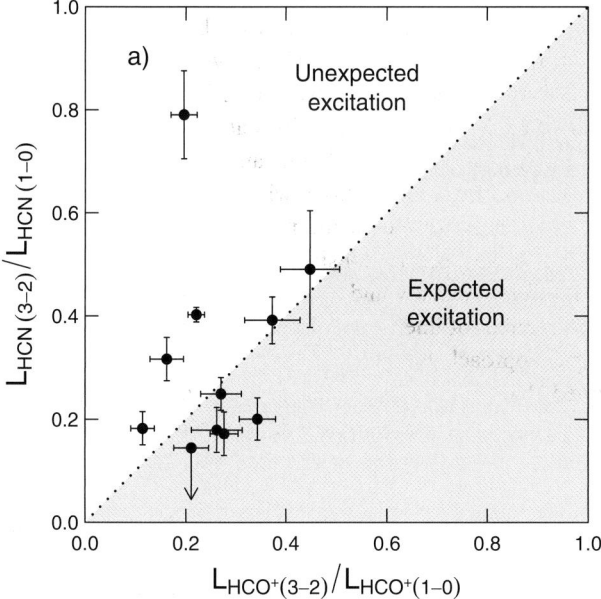

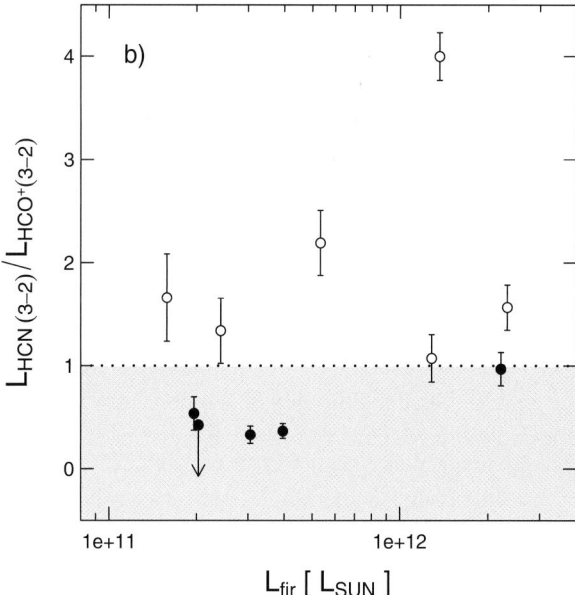

Fig. 2 a HCN(3–2)/HCN(1–0) luminosity ratio against the HCO$^+$(3–2)/HCO$^+$(1–0) luminosity ratio in our sample of LIRGs and ULIRGs. We can see that HCN and HCO$^+$ excitation is clearly subthermal for most galaxies. We have highlighted in *grey* the region of 'expected excitation' predicted by our simple LVG calculations assuming similar abundances for both molecules (see text). About half of the galaxies show an 'unexpected excitation' that can be explained if

[HCN]/[HCO$^+$] > 10. **b** HCN(3–2)/HCO$^+$(3–2) luminosity ratio as a function of L_{fir} in our sample of LIRGs and ULIRGs. We have highlighted in *grey* the region where HCN(3–2)/HCO$^+$(3–2) ≤ 1, as predicted by our LVG calculations. To explain the J = 3–2 line emission in those galaxies that do not satisfy the mentioned condition we need HCN to be overabundant with respect to HCO$^+$. Open circles represent those galaxies with 'unexpected excitation' in (**a**)

these galaxies, an abundance ratio much higher than those observed in Galactic star-forming regions (see for example Table 10 in Stäuber et al. 2006). This result does not depend on the density, kinetic temperature and column density of H_2 considered in the calculations. We note that in a two phase model the $[HCN]/[HCO^+]$ abundance problem would still appear in the densest phase.

We have adopted a similar approach in Fig. 2b where we have represented the $HCN(3–2)/HCO^+(3–2)$ luminosity ratio as a function of L_{fir} in our sample of LIRGs and ULIRGs. If we assume that both molecules trace the same molecular gas phase and have similar abundances, then the $HCN(3–2)/HCO^+(3–2)$ luminosity ratio should be <1. We have highlighted in grey the region where this condition is fulfilled. Again, half of the galaxies do not fall into the region predicted by our simple model. $HCN(3–2)/HCO^+$ $(3–2) \geq 1$ is predicted for $[HCN]/[HCO^+] \geq 7$, independently of the density, kinetic temperature and H_2 column density of the medium.

4 Dense molecular gas in high-redshift galaxies

Observations of dense molecular gas at high redshift are still rare and difficult. Only four high-redshift galaxies have been detected in HCN to date (Solomon et al. 2003; Vanden Bout et al. 2004; Carilli et al. 2005; Wagg et al. 2005), two in HCO^+ (Riechers et al. 2006; García-Burillo et al. 2006) and one in HNC and tentatively in CN (Guélin et al. 2007). In spite of the difficulties, it is possible to apply the same kind of analysis described above to high redshift objects. Wagg et al. (2005), García-Burillo et al. (2006) and Guélin et al. (2007) have used simple radiative transfer calculations to impose stringent constraints on $[HCN]/[CO]$, $[HCN]/[HCO^+]$, $[HCN]/[HNC]$ and $[HCN]/[CN]$ abundance ratios in the broad absorption line quasar APM 08279+5255 at $z = 3.9$. Their results point to HCN being overabundant with respect to HCO^+ and CN by a factor of ~ 10, while $[HCN]/[CO] \sim 10^{-2}–10^{-3}$ and $[HCN]/[HNC]$ ~ 1.6. Infrared pumping through higher vibrational transitions may also play a role in the excitation of some of these molecular lines and can equally explain the observed luminosity ratios. The infrared luminosity of APM 08279+5255 is dominated by the contribution of its AGN (Rowan-Robinson 2000), which makes this galaxy an ideal candidate to test the effects of the feedback of activity in the properties of the dense molecular gas. The fact that in this galaxy HCN seems to be overabundant with respect to HCO^+, as we found in some of the galaxies of our sample of LIRGs and ULIRGs, could indicate that similar processes dominate the dense molecular gas chemistry in these galaxies.

References

Aalto, S., Booth, R.S., Black, J.H., Johansson, L.E.B.: Molecular gas in starburst galaxies: line intensities and physical conditions. Astron. Astrophys. **300**, 369 (1995)

Blake, G.A., Sutton, E.C., Masson, C.R., Phillips, T.G.: Molecular abundances in OMC-1—The chemical composition of interstellar molecular clouds and the influence of massive star formation. Astrophys. J. **315**, 621 (1987)

Carilli, C.L., Solomon, P., Vanden Bout, P., Walter, F., Beelen, A., Cox, P., Bertoldi, F., Menten, K.M., Isaak, K.G., Chandler, C.J., Omont, A.: A search for dense molecular gas in high-redshift infrared-luminous galaxies. Astrophys. J. **618**, 586 (2005)

Evans, A.S., Solomon, P.M., Tacconi, L.J., Vavilkin, T., Downes, D.: Dense molecular gas and the role of star formation in the host galaxies of quasi-stellar objects. Astron. J. **132**, 2398 (2006)

Gao, Y., Solomon, P.M.: HCN survey of normal spiral, infrared-luminous, and ultraluminous galaxies. Astrophys. J. Suppl. Ser. **152**, 63 (2004a)

Gao, Y., Solomon, P.M.: The star formation rate and dense molecular gas in galaxies. Astrophys. J. **606**, 271 (2004b)

García-Burillo, S., Graciá-Carpio, J., Guélin, M., Neri, R., Cox, P., Planesas, P., Solomon, P.M., Tacconi, L.J., Vanden Bout, P.A.: A new probe of dense gas at high redshift: detection of HCO^+ (5–4) line emission in APM 08279+5255. Astrophys. J. **645**, L17 (2006)

Genzel, R., Lutz, D., Sturm, E., Egami, E., Kunze, D., Moorwood, A.F.M., Rigopoulou, D., Spoon, H.W.W., Sternberg, A., Tacconi-Garman, L.E., Tacconi, L., Thatte, N.: What powers ultraluminous IRAS galaxies? Astrophys. J. **498**, 579 (1998)

Graciá-Carpio, J., García-Burillo, S., Planesas, P., Colina, L.: Is HCN a true tracer of dense molecular gas in luminous and ultraluminous infrared galaxies? Astrophys. J. **640**, L135 (2006)

Greve, T.R., Hainline, L.J., Blain, A.W., Smail, I., Ivison, R.J., Papadopoulos, P.P.: A search for dense gas in luminous submillimeter galaxies with the 100 m green bank telescope. Astron. J. **132**, 1938 (2006)

Guélin, M., Salomé, P., Neri, R., García-Burillo, S., Graciá-Carpio, J., Cernicharo, J., Cox, P., Planesas, P., Solomon, P.M., Tacconi, L.J., Vanden Bout, P.: Detection of HNC and tentative detection of CN at $z = 3.9$. Astron. Astrophys. **462**, L45 (2007)

Imanishi, M., Nakanishi, K., Kohno, K.: Millimeter interferometric investigations of the energy sources of three ultraluminous infrared galaxies, UGC 5101, Markarian 273, and IRAS 17208-0014, based on HCN-to-HCO^+ ratios. Astron. J. **131**, 2888 (2006)

Kohno, K.: In: Hüttmeister, S., Manthey, E., Bomans, D., Weis, K. (eds.) AIP Conf. Proc. 783: The Evolution of Starbursts, p. 203 (2005)

Kohno, K., Matsushita, S., Vila-Vilaró, B., Okumura, S.K., Shibatsuka, T., Okiura, M., Ishizuki, S., Kawabe, R.: In: Knapen, J.H., Beckman, J.E., Shlosman, I., Mahoney, T.J. (eds.) ASP Conf. Ser. 249: The Central Kiloparsec of Starbursts and AGN: The La Palma Connection, p. 672 (2001)

Lahuis, F., Spoon, H.W.W., Tielens, A.G.G.M., Doty, S.D., Armus, L., Charmandaris, V., Houck, J.R., Stäuber, P., van Dishoeck, E.F.: Infrared molecular starburst fingerprints in deeply obscured (U)LIRG nuclei, astro-ph/0612748 (2006)

Lepp, S., Dalgarno, A.: X-ray-induced chemistry of interstellar clouds. Astron. Astrophys. **306**, L21 (1996)

Maloney, P.R., Hollenbach, D.J., Tielens, A.G.G.M.: X-ray-irradiated molecular gas. I. Physical processes and general results. Astrophys. J. **466**, 561 (1996)

Papadopoulos, P.P.: HCN versus HCO+ as dense molecular gas mass tracer in luminous infrared galaxies, astro-ph/0610477 (2006)

Riechers, D.A., Walter, F., Carilli, C.L., Weiss, A., Bertoldi, F., Menten, K.M., Knudsen, K.K., Cox, P.: First detection of HCO^+ emission at high redshift. Astrophys. J. **645**, L13 (2006)

Rowan-Robinson, M.: Hyperluminous infrared galaxies. Mon. Not. R. Astron. Soc. **316**, 885 (2000)

Sanders, D.B., Mirabel, I.F.: Luminous infrared galaxies. Annu. Rev. Astron. Astrophys. **34**, 749 (1996)

Sanders, D.B., Soifer, B.T., Elias, J.H., Madore, B.F., Matthews, K., Neugebauer, G., Scoville, N.Z.: Ultraluminous infrared galaxies and the origin of quasars. Astrophys. J. **325**, 74 (1988)

Sanders, D.B., Scoville, N.Z., Soifer, B.T.: Molecular gas in luminous infrared galaxies. Astrophys. J. **370**, 158 (1991)

Solomon, P.M., Downes, D., Radford, S.J.E.: Dense molecular gas and starbursts in ultraluminous galaxies. Astrophys. J. **387**, L55 (1992)

Solomon, P., Vanden Bout, P., Carilli, C., Guelin, M.: The essential signature of a massive starburst in a distant quasar. Nature **426**, 636 (2003)

Spergel, D.N., Verde, L., Peiris, H.V., Komatsu, E., Nolta, M.R., Bennett, C.L., Halpern, M., Hinshaw, G., Jarosik, N., Kogut, A., Limon, M., Meyer, S.S., Page, L., Tucker, G.S., Weiland, J.L., Wollack, E., Wright, E.L.: First-year Wilkinson microwave anisotropy probe (WMAP) observations: determination of cosmological parameters. Astrophys. J. Suppl. Ser. **148**, 175 (2003)

Stäuber, P., Benz, A.O., Jørgensen, J.K., van Dishoeck, E.F., Doty, S.D., van der Tak, F.F.S.: Tracing high energy radiation with molecular lines near deeply embedded protostars, astro-ph/0608393 (2006)

Usero, A., García-Burillo, S., Fuente, A., Martín-Pintado, J., Rodríguez-Fernández, N.J.: Molecular gas chemistry in AGN. I. The IRAM 30 m survey of NGC 1068. Astron. Astrophys. **419**, 897 (2004)

Vanden Bout, P.A., Solomon, P.M., Maddalena, R.J.: High-redshift HCN emission: dense star-forming molecular gas in IRAS F10214+4724. Astrophys. J. **614**, L97 (2004)

Veilleux, S., Kim, D.-C., Sanders, D.B.: Optical spectroscopy of the IRAS 1 Jy sample of ultraluminous infrared galaxies. Astrophys. J. **522**, 113 (1999)

Wagg, J., Wilner, D.J., Neri, R., Downes, D., Wiklind, T.: HCN $J = 5$–4 emission in APM 08279+5255 at $z = 3.91$. Astrophys. J. **634**, L13 (2005)